MACHINE TOOL DESIGN HANDBOOK

McGraw-Hill Offices:

New Delhi
New York
St Louis
San Francisco
Auckland
Bogotá
Guatemala
Hamburg
Lisbon
London
Madrid
Mexico
Montreal
Panama
Paris
San Juan
São Paulo
Singapore
Sydney
Tokyo
Toronto

MACHINE TOOL DESIGN HANDBOOK

CENTRAL MACHINE TOOL INSTITUTE
Bangalore

Tata McGraw-Hill Publishing Co. Ltd.
New Delhi

Central Machine Tool Institute
Bangalore

© 1982, Central Machine Tool Institute, Bangalore,
First Reprint 1983
Second Reprint 1985
Third Reprint 1986
Fourth Reprint 1987
Fifth Reprint 1988

This book can be exported from India only by
Tata McGraw-Hill Publishing Company Limited, New Delhi

Sponsoring Editor   : RAMAN KHANNA
Production Assistant   : ANSON BABU

This book has been jointly published by Tata McGraw-Hill
Publishing Company Limited, 4/12 Asaf Ali Road, New Delhi - 110002
and Central Machine Tool Institute, Bangalore and printed by **Mohan
Makhijani** at Rekha Printers Pvt. Ltd., A-102/1, Okhla Industrial **Estate,**
Phase II, New Delhi - 110020

FOREWORD

I am glad that the Central Machine Tool Institute is bringing out this Handbook on Machine Tool Design from the studies and compilations made during the short course of 12 years of its existence. The machine tool engineering industry has far too long relied on empiricism combined with high input of skill and tradition. It is only during the last two decades that systematic investigations and research efforts have been devoted to machine tools and metal cutting. Considerable information is now available on the behaviour and wear of cutting tools, dynamics and performance of machines, and on selection of machine tool elements and systems. This has become necessary because of the advancement in control systems, pressure of higher productivity and reliability for which a very much higher level of performance is demanded from today's machine tools.

The design of a machine tool is closely linked with its intended usage and this would vary from industry to industry, shop to shop and operator to operator. By and large, a machine tool therefore has to cater to large variables of environmental and working factors and these have to be considered during the design and manufacturing stages to result in a reliable and productive product.

The Government of India, realising the importance of the machine tool industry has given considerable encouragement to the promotion of the industry as well as the research and development efforts. There are now several leading faculties in machine tool engineering such as the IITs, research and development institutions like CMTI as well as R & D sections of machine tool industries contributing to the development in this field.

I am happy that the CMTI has taken upon itself the task of compiling various calculation methods, guidelines and performance aspects of machines which go in building a quality machine tool. I find the compilation exhaustive and should be useful as much to the practising professional engineer as to the student and the development engineer in this field. The Handbook covers various aspects of machine tools, parameters and working factors, materials and control systems. The tables and extracts from the national and international standards would also come in handy for the practising engineers. I understand that handbooks devoted exclusively for machine tools in the English language and in metric system are not many. Along with academic texts and monographs to cover various technical and scientific subjects, handbooks form a useful link for application of known technologies, theories and practices in industrial design and construction. These would also reflect the engineering and industrial experience of the country. It is commendable that CMTI has brought out this Handbook out of its rich collection of material and expertise acquired. This volume should therefore be very useful to our industry.

It is appropriate that the National Book Trust has supported this publication and I have no doubt that this Handbook will fulfil the long-standing demands of our engineers in this field, and form a useful addition to our technical publications.

DR. A. RAMACHANDRAN

PREFACE

The Machine Tool Design Handbook is a result of compilation for a number of years by the Central Machine Tool Institute, on design data, calculation methods and machine tool practices. It draws heavily on information collected as a result of the number of designs and machines tested during this period by the Institute. Even in the earlier stages it was observed that systems and practices in machine tool development were more of a proprietory nature and published information was difficult to get. Hence, systematic and intensive efforts were made to compile data for use by the design office of the Central Machine Tool Institute and for dissemination of this informatiom to the Indian machine tool industry.

In 1970, the Central Machine Tool Institute conducted the first course on machine tool design for practising designers from the Industry. To assist the participants, a six-volume course material was prepared, dealing with calculation data, design practices as well as theoretical notes on machine tool concepts and practices. These volumes produced in a mimeographic form have been received with great appreciations and several suggested that all this data be brought out in a concise handbook form. Greatly encouraged by this response, the publication of this Handbook was taken up in all earnestness. In this endeavour, the Institute also received a quick response from the National Book Trust of India for subsidising the publication of the Handbook. The book estimated originally at 740 pages has run into 980 pages, thanks to the enthusiasm with which the information has been collected and added particularly during the final stages of its compilation.

The Handbook is a comprehensive collection of useful design data and reference material needed by both the practising machine tool engineers and students of engineering. Subjects covered include the design of machine elements, design practices with practical data and guidelines on various aspects of machine tool design such as machine tool guideways, spindles, gear boxes, etc. Chapters have been exclusively devoted for hydraulics and electrical equipment for control of machine tools as well as those dealing with materials and heat treatment.

The Handbook is supplemented with excerpts from Indian as well as international standards on related topics. The metric system has been used throughout. This forms one of the major publications of the Central Machine Tool Institute and is planned to be a forerunner for a number of handbooks and monographs to be brought out by the Institute in the future. One on Numerical Controls and another on Production Engineering are currently under preparation.

It gives me a sense of great personal satisfaction having been associated with the establishment of the Institute from its inception and as the President of its Governing Council for several years, that this Handbook which is a result of the active work and studies at the Institute, is now made available to the Industry, the R & D engineers and students in the machine tool field. Comments and suggestions for improvement of the contents and materials would be exceedingly welcome and all of us in the Central Machine Tool Institute would feel fully rewarded if the book serves the needs of the Industry.

DR. S. M. PATIL

ACKNOWLEDGEMENTS

The publishers wish to express their appreciation and thanks to the following for permitting excerpts from relevant publications as indicated, in the handbook.

1 VERLAG HALLWAG, BERN UND STUTTGART; for Fig. 114 from the paper CALCULATION AND DESIGN OF RADIAL BEARINGS (BERECHNUNG UND KONSTRUKTION VON RADIALGLEITLAGERN) by Dipl. Ing. K Milowiz in BLAUE TR-REIHE, HEFT 3.

2 M/s. COOPER ENGINEERING LTD.; for Tables on MEEHANITE from the paper MATERIAL FOR MACHINE TOOLS by S. D. BHAGWAT, Manager, Meehanite metals division.

3 M/s. SKEFKO INDIA BEARING CO. LTD; for Tables 121 and 149 from SKF BEARINGS IN MACHINE TOOLS from their catalogue No. 2800E/GB600 of April 1970 and for relevant information on SKF bearings from SKF GENERAL CATALOGUE No. $\dfrac{\text{2580 E}}{\text{Reg. 872.2 65 1 69}}$

4 M/s. VEREINIGTE DREHBANK-FABRIKEN, for Fig. 82 from VDF NEWS 34, December 1968.

5 M/s. NEEDLE ROLLER BEARING CO LTD; THANA, for relevant information on NRB Needle roller bearings.

Data and information from the following INDIAN and DIN Standards have been reproduced with the permission of the respective National Standards Bodies.

Regarding the Indian Standards, references are invited to the respective standards, as listed below for further details. These are available from Indian Standards Institution, Bahadur Shah Zafar Marg, New Delhi and its regional and branch offices at Ahmedabad, Bangalore, Bombay, Calcutta, Chandigarh, Hyderabad, Kanpur, Madras and Patna.

INDIAN STANDARDS

SP : 5-1969 Guide to the use of International system (SI) units
IS : 28-1975 Phosphor bronze ingots and castings
IS : 210-1970 Grey iron castings

IS : 305-1961 Aluminium bronze ingots and castings

IS : 318-1962 Leaded tin bronze ingots and castings

IS : 319-1974 Free cutting brass bars, rods and sections

IS : 325-1970 Three phase induction motors

IS : 554-1975 Dimensions for pipe threads where pressure tight joints are required on the threads

IS : 617-1975 Aluminium and aluminium alloy ingots and castings for general engineering purposes

IS : 696-1972 Code of practice for general engineering drawings

IS : 733-1975 Wrought aluminium and aluminium alloy bars, rods and sections (for general engineering purposes).

IS : 736-1974 Wrought aluminium and aluminium alloys, plate (for general engineering purposes)

IS : 737-1974 Wrought aluminium and aluminium alloys, sheet and strip (for general engineering purposes)

IS : 919-1963 Recommendation for limits and fits for engineering

IS : 1030-1974 Carbon steel castings for general engineering purposes

IS : 1076-1967 Preferred numbers

IS : 1231-1974 Dimensions of 3 phase foot mounted induction motors

IS : 1356 (Part I)-1972 Electrical equipment of machines for general use

IS : 1363-1967 Black hexagon bolts, nuts and locknuts (dia. 6 to 39 *mm*) and black hexagon screws (dia. 6 to 24 *mm*)

IS : 1364-1967 Precision and semi-precision hexagon bolts, screws, nuts and locknuts (dia. 6 to 39 *mm*)

IS : 1365-1968 Slotted countersunk head and slotted raised countersunk head screws (dia. range 1.6 to 20 *mm*)

IS : 1366-1968 Slotted cheese head screws (dia. range 1.6 to 20 *mm*)

IS : 1367-1967 Technical supply conditions for threaded fasteners

IS : 1368-1967 Dimensions of ends of bolts and screws

IS : 1369-1975 Dimensions of screw thread runouts and undercuts

IS : 1570-1961 Schedules for wrought steels for general engineering purposes

IS : 1715-1973 Dimensions for self-holding tapers

IS : 1821-1967 Dimensions for clearance holes for metric bolts

IS : 1862-1975 Studs

IS : 1865-1974 Iron castings with spheroidal or nodular graphite

IS : 2013-1974 Dimensions for T-slots

IS : 2014-1962 T-bolts

IS : 2015-1962 T-nuts

IS : 2016-1967 Plain washers

IS : 2032-   Graphical symbols used in electrotechnology

IS : 2048-1975 Parallel keys and keyways

IS : 2062-1969 Structural steel (fusion welding quality)

IS : 2102-1969 Allowable deviations for dimensions without specified tolerances

IS : 2161-1962 Coolant pumps for machine tools

IS : 2182-1962 Recommendation for symbols to be given on indication plates of machine tools

IS : 2218-1962 Speeds for machine tools
IS : 2219-1962 Feeds for machine tools
IS : 2223-1971 Dimensions of flange mounted ac induction motors
IS : 2232-1967 Slotted and castle nuts
IS : 2243-1971 Drill chucks
IS : 2269-1967 Hexagon socket head cap screws
IS : 2294-1963 Woodruff keys and keyslots
IS : 2340-1972 Dimensions for self release 7/24 tapers for milling arbor and tool shanks
IS : 2388-1971 Slotted grub screws
IS : 2389-1968 Precision hexagon bolts, screws, nuts and locknuts (dia. 1.6 to 5 mm)
IS : 2393-1972 Cylindrical pins
IS : 2403-1964 Transmission steel roller chains and chain wheels
IS : 2473-1975 Dimensions for centre holes
IS : 2494-1974 V-belts for industrial purposes
IS : 2507-1975 Cold rolled steel strips for springs
IS : 2535-1969 Basic rack and modules of cylindrical gears for general engineering and heavy engineering
IS : 2540-1963 Dimensions for threaded centre holes
IS : 2582 (Part I)-1972 Dimensions for lathe spindle noses and face plates – spindle noses type A
IS : 2582 (Part II)-1972 Dimensions for lathe spindle noses and face plates – spindle noses camlock type
IS : 2582 (Part III)-1972 Dimensions for lathe spindle noses and face plates – spindle noses bayonet type
IS : 2585-1968 Blark square bolts and nuts (dia. range 6 to 39 mm) and black square screws (dia. range 6 to 24 mm)
IS : 2610-1964 Dimensions for straight sided splines for machine tools
IS : 2636-1972 Wing nuts
IS : 2642-1974 Sizes for machine tool tables
IS : 2643 (Part I)-1975 Pipe threads for fastening purposes – basic profile and dimensions
IS : 2687-1975 Cap nuts
IS : 2709-1964 Guide for the selection of fits
IS : 2710-1975 Parallel keys and keyways for machine tools
IS: 2769-1969 Sizes for squares and square holes for general engineering purposes
IS : 2990-1965 Dimensions for tenons
IS : 2996-1964 Mounting dimensions for grinding wheels
IS : 3063-1972 Single coil rectangular section spring washers for bolts, nuts and screws
IS : 3075-1965 Dimensions for circlips
IS : 3142-1965 V-grooved pulleys for V-belts – groove sections A, B, C, D and E
IS : 3230-1970 Recommendation for tapping drill sizes
IS : 3403-1966 Dimensions for knurling

IS : 3406 (Part I)-1975 Dimensions for countersinks and counterbores – Countersinks

IS : 3406 (Part II)-1975 Dimensions for countersinks and counterbores – Counterbores

IS : 3428-1966 Dimensions for relief grooves

IS : 3457-1966 Radii and chamfers for general engineering purposes

IS : 3458-1972 Tapers for general engineering purposes

IS : 3460-1972 Knurled nuts

IS : 3524-1966 Threaded taper pins

IS : 3624-1966 Bourdon tube pressure and vacuum gauges

IS : 3726-1972 Thumb screws

IS : 4009-1967 Grease nipples

IS : 4170-1967 Brass rods for general engineering purposes

IS : 4172-1967 Dimensions for radii under the head of bolts and screws

IS : 4215-1967 Specification for needle bearings

IS : 4218 (Part III)-1967 ISO metric screw threads – basic dimensions for design profiles

IS : 4398-1972 Carbon chromium steel for the manufacture of balls, rollers and bearing races

IS : 4454 (Part I)-1975 Steel wire for cold formed springs – patented and cold drawn steel wire, unalloyed

IS : 4499-1968 Dimensions for depth of holes for studs

IS : 4673-1968 Wick feed lubricators

IS : 4974-1968 Grease nipples, small

IS : 5107-1969 Recommendation on nominal pressures for oil-hydraulic system elements

IS : 5108-1969 Recommendation on nominal rates of flow for oil-hydraulic system elements

IS : 5109-1969 Recommendation on nominal bores for oil-hydraulic system elements

IS : 5129-1969 Rotary shaft oil seal units (related dimensions)

IS : 5308-1969 Slotted countersunk head and slotted raised countersunk head screws (small head series dia. 1.6 to 6 *mm*)

IS : 5370-1969 Plain washers with outside diameter ≈ 3×inside diameter

IS : 5519-1969 Deviations for untoleranced dimensions of grey iron castings

IS · 5559-1970 Oilers

IS : 5669-1970 General plan of boundary dimensions for radial rolling bearings

IS : 5692-1970 Tolerances for radial rolling bearings

IS : 5932-1970 Boundary dimensions for thrust ball bearings with flat seats

IS : 5933-1970 Tolerances for thrust ball bearings with flat seats

IS : 5934-1970 Chamfers and fillet radii for rolling bearings

IS : 5935-1970 Radial internal clearances in unloaded radial rolling bearings

IS : 6094-1971 Hexagon socket grub screws

IS : 6681-1972 Dimensions for self release 7/24 tapers for spindles for milling machines

IS : 6688-1972 Taper pins

IS :6689-1972 Hardened cylindrical pins
IS :6735-1972 Spring washers for screws with cylindrical head
IS :6761-1972 Countersunk head screws with hexagon socket
IS : 7008 (Part I)-1973 ISO Metric trapezoidal screw threads – basic and design profiles
IS : 7008 (Part II)-1973 ISO Metric trapezoidal screw threads – pitch diameter combinations
IS : 7008 (Part III)-1973 ISO Metric trapezoidal screw threads-basic dimensions for design profiles
IS : 7443-1974 Methods for load rating of worm gears
IS : 7461-1974 General plan of boundary dimensions for tapered roller bearings
IS : 7513-1974 Graphical symbols for fluid power systems
IS : 7790-1975 Domed cap nuts
IS : 7811-1975 Phosphor bronze rods and bars

Regarding the DIN Standards, references are invited to the standards as listed for further details. These standards are published by M/s. DEUTCHES INSTITUT FUER NORMUNG e.V., 1 BERLIN 30, POSTFACH 1107. DIN Standards may be ordered through ISI, New Delhi, who are the sole selling agent in India for these standards.

DIN STANDARDS

1. DIN   808-1972 Cardon joints; connecting dimensions, fixing, loading installation
2. DIN   910 (Bl. 1)-1973 Hexagon head pipe plugs
3. DIN   2089 (Sheet 1)-1963 Helical springs made from circular section wire and bar. Calculation and design of compression springs
4. DIN   2093-1967 Disc springs; dimensions and quality properties
5. DIN   2095-1973 Helical springs made of round wire, specifications for cold coiled compression springs
6. DIN   2353-1966 Non soldered taper bush type pipe unions, complete unions and survey
7. DIN   3852 (Bl. 2)-1964 Screwed plugs; tapped holes; with Whitworth pipe thread, general outlay of types
8. DIN   3870-1965 Non soldered and soldered nipple type fittings; union nuts
9. DIN   3901-1965 Non soldered taper bush type unions; double ended union
10. DIN   3902-1957 Straight pipe joints
11. DIN   3961-1953 Tolerances for spur gear tooth systems to DIN 867
12. DIN   3962 (Sheet 2)-1952 Tolerances for spur gear tooth systems to DIN 867. tolerances for individual errors. (Modules above 0.6 & upto 1.6)
13. DIN   3962 (Sheet 3)-1952 Tolerances for spur gear tooth system to DIN 867. tolerances for individual errors (Modules above 1.6 & upto 4)
14. DIN   3962 (Sheet 4)-1952 Tolerances for spur gear tooth system to DIN 867. tolerances for individual errors (Modules above 4 & upto 10)

15. DIN   3963-1953 Tolerances for spur gear tooth systems to DIN 867. Permissible tooth alignment error, permissible composite error, tooth thickness allowances.
16. DIN   3964 Gear tooth systems; centre distance allowance
17. DIN   3976-1963 Tolerances for spur gear tooth systems to DIN 867. Permissible tooth alignment error, permissible composite error, tooth distance allowance.
18. DIN   5508-1966 Bending radius for tubes. Design principle for rail vehicles
19. DIN   6327 (Sheet 1)-1972 Adjustable adapters for tools with Morse taper shanks, short types
20. DIN   6327 (Sheet 2)-1972 Adjustable adapters for tools with Morse taper shanks, long types
21. DIN   6327 (Sheet 3)-1972 Adjustable adapters for tools with Morse taper shanks, stepped type
22. DIN   6327 (Sheet 4)-1972 Adjustment nuts for adapters
23. DIN   6935-1969 Cold forming and cold bending of flat rolled sheets
24. DIN   55058-1973 Drill heads for adjustable adapters

CONTENTS

BASIC REFERENCE DATA (Contd.)

MACHINE ELEMENTS

MACHINE ELEMENTS (Contd.)

CONTENTS

MACHINE ELEMENTS (Contd.)

MACHINE TOOL DESIGN

MACHINING DATA

MATERIALS AND HEAT TREATMENT

ELECTRICAL EQUIPMENT FOR MACHINE TOOLS

OIL HYDRALICS IN MACHINE TOOLS

LIST OF INDIAN STANDARDS REVISED SINCE THE FIRST PUBLICATION OF THIS HANDBOOK

1. IS 210-1978   Grey iron castings
2. IS 325-1978   Three-phase induction motors
3. IS 1365-1978   Slotted countersunk head and slotted raised countersunk head screws (dia. range 1.6 to 20 mm)
4. IS 1368-1980   Dimensions of ends of bolts and screws
5. IS 2014-1977   T-bolts
6. IS 2015-1977   T-nuts
7. IS 2062-1980   Structural steel (fusion welding quality)
8. IS 2102 (Part I)—1980   Allowable deviations for dimensions
9. IS 2294-1980   Woodruff keys and keyslots
10. IS 2393-1980   Cylindrical pins
11. IS 2403-1975   Transmission steel roller chains and chain wheels
12. IS 2535-1978   Basic rack and modules of cylindrical gears for general engineering and heavy engineering
13. IS 3428-1980   Dimensions for relief grooves
14. IS 3624-1979   Bourdon tube pressure and vacuum gauges
15. IS 5129-1979   Rotary shaft oil seal units (related dimensions)
16. IS 5519-1979   Deviations for untoleranced dimensions of grey iron castings
17. IS 7461-1980   General plan of boundary dimensions for tapered roller bearings

LIST OF INDIAN STANDARDS AMENDED SINCE THE FIRST PUBLICATION OF THIS HANDBOOK—AMENDMENTS RECEIVED AFTER 1978.01.01

1. IS 554-1975   Amd. No. 1 Jan 1981
2. IS 617-1975   Amd. No. 1 Dec 1978
   No. 2 Jun 1980
3. IS 696-1972   Amd. No. 2 Sep 1979
   No. 3 Feb 1981
4. IS 736-1974   Amd. No. 1 Aug 1978
5. IS 919-1963   Amd. No. 1 Mar 1978
   No. 2 Feb 1980
6. IS 1369-1975   Amd. No.2 Apr 1979
7. IS 1862-1975   Amd. No. 1 Mar 1981
8. IS 2016-1967   Amd. No. 2 Mar 1979
   No. 3 Nov 1980
9. IS 2048-1975   Amd. No. 1 Feb 1978
10. IS 2161-1962   Amd. No. 1 Apr 1978
11. IS 2243-1971   Amd. No. 1 Dec 1977
12. IS 2340-1972   Amd. No. 2 Feb 1979
13. IS 2403-1975   Amd. No. 2 Nov 1980
14. IS 2494-1974   Amd. No. 2 Oct 1978
15. IS 2507-1975   Amd. No.1 Jun 1978
   No 2 Apr 1979

16. IS 2582 (Part II)—1972 Amd. No.1. Dec 1979
17. IS 2610-1964 Amd. No. 2 Feb 1981
18. IS 2710-1975 Amd. No.1 Jan 1979
19. IS 3093-1972 Amd. No. 1 Mar 1981
20. IS 3075-1965 Amd. No. 2 Nov 1979
21. IS 4398-1972 Amd. No. 2 Oct 1978
                         No. 3 Feb 1980
22. IS 4454 (Part I)—1975 Amd. No. 1 Jan 1978
23. IS 4673-1968 Amd. No.1 Apr 1981
24. IS 5370-1969 Amd. No.1 Jun 1979
25. IS 6094-1971 Amd. No.2 Apr 1981
26. IS 7008 (Part I) Amd. No.2 Mar 1978

BASIC REFERENCE DATA

MATHEMATICS

Numbers, Squares, Cubes, Roots,etc.   1 - 50

r	n²	n³	√n	³√n	1/n	π n	π n²/4	n
1	1	1	1.0000	1.0000	1.000000	3.142	0.8	1
2	4	8	1.4142	1.2599	0.500000	6.283	3.1	2
3	9	27	1.7321	1.4422	0.333333	9.425	7.1	3
4	16	64	2.0000	1.5874	0.250000	12.566	12.6	4
5	25	125	2.2361	1.7100	0.200000	15.708	19.6	5
6	36	216	2.4495	1.8171	0.166667	18.850	28.3	6
7	49	343	2.6458	1.9129	0.142857	21.991	38.5	7
8	64	512	2.8284	2.0000	0.125000	25.133	50.3	8
9	81	729	3.0000	2.0801	0.111111	28.274	63.6	9
10	100	1000	3.1623	2.1544	0.100000	31.416	78.5	10
11	121	1331	3.3166	2.2240	0.090909	34.558	95.0	11
12	144	1728	3.4641	2.2894	0.083333	37.699	113.1	12
13	169	2197	3.6056	2.3513	0.076923	40.841	132.7	13
14	196	2744	3.7417	2.4101	0.071429	43.982	153.9	14
15	225	3375	3.8730	2.4662	0.066667	47.124	176.7	15
16	256	4096	4.0000	2.5198	0.062500	50.265	201.1	16
17	289	4913	4.1231	2.5713	0.058824	53.407	227.0	17
18	324	5832	4.2426	2.6207	0.055556	56.549	254.5	18
19	361	6859	4.3589	2.6684	0.052632	59.690	283.5	19
20	400	8000	4.4721	2.7144	0.050000	62.832	314.2	20
21	441	9261	4.5826	2.7589	0.047619	65.973	346.4	21
22	484	10648	4.6904	2.8020	0.045455	69.115	380.1	22
23	529	12167	4.7958	2.8439	0.043478	72.257	415.5	23
24	576	13824	4.8990	2.8845	0.041667	75.398	452.4	24
25	625	15625	5.0000	2.9240	0.040000	78.540	490.9	25
26	676	17576	5.0990	2.9625	0.038462	81.681	530.9	26
27	729	19683	5.1962	3.0000	0.037037	84.823	572.6	27
28	784	21952	5.2915	3.0366	0.035714	87.965	615.8	28
29	841	24389	5.3852	3.0723	0.034483	91.106	660.5	29
30	900	27000	5.4772	3.1072	0.033333	94.248	706.9	30
31	961	29791	5.5678	3.1414	0.032258	97.389	754.8	31
32	1024	32768	5.6569	3.1748	0.031250	100.531	804.2	32
33	1089	35937	5.7446	3.2075	0.030303	103.673	855.3	33
34	1156	39304	5.8310	3.2396	0.029412	106.814	907.9	34
35	1225	42875	5.9161	3.2711	0.028571	109.956	962.1	35
36	1296	46656	6.0000	3.3019	0.027778	113.097	1017.9	36
37	1369	50653	6.0828	3.3322	0.027027	116.239	1075.2	37
38	1444	54872	6.1644	3.3620	0.026316	119.381	1134.1	38
39	1521	59319	6.2450	3.3912	0.025641	122.522	1194.6	39
40	1600	64000	6.3246	3.4200	0.025000	125.664	1256.6	40
41	1681	68921	6.4031	3.4482	0.024390	128.805	1320.3	41
42	1764	74088	6.4807	3.4760	0.023810	131.947	1385.4	42
43	1849	79507	6.5574	3.5034	0.023256	135.088	1452.2	43
44	1936	85184	6.6332	3.5303	0.022727	138.230	1520.5	44
45	2025	91125	6.7082	3.5569	0.022222	141.372	1590.4	45
46	2116	97336	6.7823	3.5830	0.021739	144.513	1661.9	46
47	2209	103823	6.8557	3.6088	0.021277	147.655	1734.9	47
48	2304	110592	6.9282	3.6342	0.020833	150.796	1809.6	48
49	2401	117649	7.0000	3.6593	0.020408	153.938	1885.7	49
50	2500	125000	7.0711	3.6840	0.020000	157.080	1963.5	50

Numbers, Squares, Cubes, Roots, etc. 50 - 100

n	n^2	n^3	$\sqrt{n}$	$\sqrt[3]{n}$	$1/n$	πn	$\pi n^2/4$	n
50	2500	125000	7.0711	3.6840	0.020000	157.080	1963.5	50
51	2601	132651	7.1414	3.7084	0.019608	160.221	2042.8	51
52	2704	140608	7.2111	3.7325	0.019231	163.363	2123.7	52
53	2809	148877	7.2801	3.7563	0.018868	166.504	2206.2	53
54	2916	157464	7.3485	3.7798	0.018519	169.646	2290.2	54
55	3025	166375	7.4162	3.8030	0.018182	172.788	2375.8	55
56	3136	175616	7.4833	3.8259	0.017857	175.929	2463.0	56
57	3249	185193	7.5498	3.8485	0.017544	179.071	2551.8	57
58	3364	195112	7.6158	3.8709	0.017241	182.212	2642.1	58
59	3481	205379	7.6811	3.8930	0.016949	185.354	2734.0	59
60	3600	216000	7.7460	3.9149	0.016667	188.496	2827.4	60
61	3721	226981	7.8102	3.9365	0.016393	191.637	2922.5	61
62	3844	238328	7.8740	3.9579	0.016129	194.779	3019.1	62
63	3969	250047	7.9373	3.9791	0.015873	197.920	3117.2	63
64	4096	262144	8.0000	4.0000	0.015625	201.062	3217.0	64
65	4225	274625	8.0623	4.0207	0.015385	204.204	3318.3	65
66	4356	287496	8.1240	4.0412	0.015152	207.345	3421.2	66
67	4489	300763	8.1854	4.0615	0.014925	210.487	3525.7	67
68	4624	314432	8.2462	4.0817	0.014706	213.628	3631.7	68
69	4761	328509	8.3066	4.1016	0.014493	216.770	3739.3	69
70	4900	343000	8.3666	4.1213	0.014286	219.911	3848.5	70
71	5041	357911	8.4261	4.1408	0.014085	223.053	3959.2	71
72	5184	373248	8.4853	4.1602	0.013889	226.195	4071.5	72
73	5329	389017	8.5440	4.1793	0.013699	229.336	4185.4	73
74	5476	405224	8.6023	4.1983	0.013514	232.478	4300.8	74
75	5625	421875	8.6603	4.2172	0.013333	235.619	4417.9	75
76	5776	438976	8.7178	4.2358	0.013158	238.761	4536.5	76
77	5929	456533	8.7750	4.2543	0.012987	241.903	4656.6	77
78	6084	474552	8.8318	4.2727	0.012821	245.044	4778.4	78
79	6241	493039	8.8882	4.2908	0.012658	248.186	4901.7	79
80	6400	512000	8.9443	4.3089	0.012500	251.327	5026.5	80
81	6561	531441	9.0000	4.3267	0.012346	254.469	5153.0	81
82	6724	551368	9.0554	4.3445	0.012195	257.611	5281.0	82
83	6889	571787	9.1104	4.3621	0.012048	260.752	5410.6	83
84	7056	592704	9.1652	4.3795	0.011905	263.894	5541.8	84
85	7225	614125	9.2195	4.3968	0.011765	267.035	5674.5	85
86	7396	636056	9.2736	4.4140	0.011628	270.177	5808.8	86
87	7569	658503	9.3274	4.4310	0.011494	273.319	5944.7	87
88	7744	681472	9.3808	4.4480	0.011364	276.460	6082.1	88
89	7921	704969	9.4340	4.4647	0.011236	279.602	6221.1	89
90	8100	729000	9.4868	4.4814	0.011111	282.743	6361.7	90
91	8281	753571	9.5394	4.4979	0.010989	285.885	6503.9	91
92	8464	778688	9.5917	4.5144	0.010870	289.027	6647.6	92
93	8649	804357	9.6437	4.5307	0.010753	292.168	6792.9	93
94	8836	830584	9.6954	4.5468	0.010638	295.310	6939.8	94
95	9025	857375	9.7468	4.5629	0.010526	298.451	7088.2	95
96	9216	884736	9.7980	4.5789	0.010417	301.593	7238.2	96
97	9409	912673	9.8489	4.5947	0.010309	304.734	7389.8	97
98	9604	941192	9.8995	4.6104	0.010204	307.876	7543.0	98
99	9801	970299	9.9499	4.6261	0.010101	311.018	7697.7	99
100	10000	1000000	10.0000	4.6416	0.010000	314.159	7854.0	100

Numbers, Squares, Cubes, Roots, etc.   100 - 150

n	n^2	n^3	$\sqrt{n}$	$\sqrt[3]{n}$	$1/n$	πn	$\pi n^2/4$	n
100	10000	1000000	10.0000	4.6416	0.010000	314.159	7854.0	100
101	10201	1030301	10.0499	4.6570	0.009901	317.301	8011.8	101
102	10404	1061208	10.0995	4.6723	0.009804	320.442	8171.3	102
103	10609	1092727	10.1489	4.6875	0.009709	323.584	8332.3	103
104	10816	1124864	10.1980	4.7027	0.009615	326.726	8494.9	104
105	11025	1157625	10.2470	4.7177	0.009524	329.867	8659.0	105
106	11236	1191016	10.2956	4.7326	0.009434	333.009	8824.7	106
107	11449	1225043	10.3441	4.7475	0.009346	336.150	8992.0	107
108	11664	1259712	10.3923	4.7622	0.009259	339.292	9160.9	108
109	11881	1295029	10.4403	4.7769	0.009174	342.434	9331.3	109
110	12100	1331000	10.4881	4.7914	0.009091	345.575	9503.3	110
111	12321	1367631	10.5357	4.8059	0.009009	348.717	9676.9	111
112	12544	1404928	10.5830	4.8203	0.008929	351.858	9852.0	112
113	12769	1442897	10.6301	4.8346	0.008850	355.000	10028.7	113
114	12996	1481544	10.6771	4.8488	0.008772	358.142	10207.0	114
115	13225	1520875	10.7238	4.8629	0.008696	361.283	10386.9	115
116	13456	1560896	10.7703	4.8770	0.008621	364.425	10568.3	116
117	13689	1601613	10.8157	4.8910	0.008547	367.566	10751.3	117
118	13924	1643032	10.8628	4.9049	0.008475	370.708	10935.9	118
119	14161	1685159	10.9087	4.9187	0.008403	373.850	11122.0	119
120	14400	1728000	10.9545	4.9324	0.008333	376.991	11309.7	120
121	14641	1771561	11.0000	4.9461	0.008264	380.133	11499.0	121
122	14884	1815848	11.0454	4.9597	0.008197	383.274	11689.9	122
123	15129	1860867	11.0905	4.9732	0.008130	386.416	11882.3	123
124	15376	1906624	11.1355	4.9866	0.008065	389.557	12076.3	124
125	15625	1953125	11.1803	5.0000	0.008000	392.699	12271.8	125
126	15876	2000376	11.2250	5.0133	0.007937	395.841	12469.0	126
127	16129	2048383	11.2694	5.0265	0.007874	398.982	12667.7	127
128	16384	2097152	11.3137	5.0397	0.007813	402.124	12868.0	128
129	16641	2146689	11.3578	5.0528	0.007752	405.265	13069.8	129
130	16900	2197000	11.4018	5.0658	0.007692	408.407	13273.2	130
131	17161	2248091	11.4455	5.0788	0.007634	411.549	13478.2	131
132	17424	2299968	11.4891	5.0916	0.007576	414.690	13684.8	132
133	17689	2352637	11.5326	5.1045	0.007519	417.832	13892.9	133
134	17956	2406104	11.5758	5.1172	0.007463	420.973	14102.6	134
135	18225	2460375	11.6190	5.1299	0.007407	424.115	14313.9	135
136	18496	2515456	11.6619	5.1426	0.007353	427.257	14526.7	136
137	18769	2571353	11.7047	5.1551	0.007299	430.398	14741.1	137
138	19044	2628072	11.7473	5.1676	0.007246	433.540	14957.1	138
139	19321	2685619	11.7898	5.1801	0.007194	436.681	15174.7	139
140	19600	2744000	11.8322	5.1925	0.007143	439.823	15393.8	140
141	19881	2803221	11.8743	5.2048	0.007092	442.965	15614.5	141
142	20164	2863288	11.9164	5.2171	0.007042	446.106	15836.8	142
143	20449	2924207	11.9583	5.2293	0.006993	449.248	16060.6	143
144	20736	2985984	12.0000	5.2415	0.006944	452.389	16286.0	144
145	21025	3048625	12.0416	5.2536	0.006897	455.531	16513.0	145
146	21316	3112136	12.0830	5.2656	0.006849	458.673	16741.5	146
147	21609	3176523	12.1244	5.2776	0.006803	461.814	16971.7	147
148	21904	3241792	12.1655	5.2896	0.006757	464.956	17203.4	148
149	22201	3307949	12.2066	5.3015	0.006711	468.097	17436.6	149
150	22500	3375000	12.2474	5.3133	0.006667	471.239	17671.5	150

MATHEMATICS

Numbers, Squares, Cubes, Roots, etc.   150 - 200

n	n^2	n^3	$\sqrt{n}$	$\sqrt[3]{n}$	$1/n$	πn	$\pi n^2/4$	n
150	22500	3375000	12.2474	5.3133	0.006667	471.239	17671.5	150
151	22801	3442951	12.2882	5.3251	0.006623	474.380	17907.9	151
152	23104	3511808	12.3288	5.3368	0.006579	477.522	18145.8	152
153	23409	3581577	12.3693	5.3485	0.006536	480.664	18385.4	153
154	23716	3652264	12.4097	5.3601	0.006494	483.805	18626.5	154
155	24025	3723875	12.4499	5.3717	0.006452	486.947	18869.2	155
156	24336	3796416	12.4900	5.3832	0.006410	490.088	19113.4	156
157	24649	3869893	12.5300	5.3947	0.006369	493.230	19359.3	157
158	24964	3944312	12.5698	5.4061	0.006329	496.372	19606.7	158
159	25281	4019679	12.6095	5.4175	0.006289	499.513	19855.7	159
160	25600	4096000	12.6491	5.4288	0.006250	502.655	20106.2	160
161	25921	4173281	12.6886	5.4401	0.006211	505.796	20358.3	161
162	26244	4251528	12.7279	5.4514	0.006173	508.938	20612.0	162
163	26569	4330747	12.7671	5.4626	0.006135	512.080	20867.2	163
164	26896	4410944	12.8062	5.4737	0.006098	515.221	21124.1	164
165	27225	4492125	12.8452	5.4848	0.006061	518.363	21382.5	165
166	27556	4574296	12.8841	5.4959	0.006024	521.504	21642.4	166
167	27889	4657463	12.9228	5.5069	0.005988	524.646	21904.0	167
168	28224	4741632	12.9615	5.5178	0.005952	527.788	22167.1	168
169	28561	4826809	13.0000	5.5288	0.005917	530.929	22431.8	169
170	28900	4913000	13.0384	5.5397	0.005882	534.071	22698.0	170
171	29241	5000211	13.0767	5.5505	0.005848	537.212	22965.8	171
172	29584	5088448	13.1149	5.5613	0.005814	540.354	23235.2	172
173	29929	5177717	13.1529	5.5721	0.005780	543.496	23506.2	173
174	30276	5268024	13.1909	5.5828	0.005747	546.637	23778.7	174
175	30625	5359375	13.2288	5.5934	0.005714	549.779	24052.8	175
176	30976	5451776	13.2665	5.6041	0.005682	552.920	24328.5	176
177	31329	5545233	13.3041	5.6147	0.005650	556.062	24605.7	177
178	31684	5639752	13.3417	5.6252	0.005618	559.203	24884.6	178
179	32041	5735339	13.3791	5.6357	0.005587	562.345	25164.9	179
180	32400	5832000	13.4164	5.6462	0.005556	565.487	25446.9	180
181	32761	5929741	13.4536	5.6567	0.005525	568.628	25730.4	181
182	33124	6028568	13.4907	5.6671	0.005495	571.770	26015.5	182
183	33489	6128487	13.5277	5.6774	0.005464	574.911	26302.2	183
184	33856	6229504	13.5647	5.6877	0.005435	578.053	26590.4	184
185	34225	6331625	13.6015	5.6980	0.005405	581.195	26880.3	185
186	34596	6434856	13.6382	5.7083	0.005376	584.336	27171.6	186
187	34969	6539203	13.6748	5.7185	0.005348	587.478	27464.6	187
188	35344	6644672	13.7113	5.7287	0.005319	590.619	27759.1	188
189	35721	6751269	13.7477	5.7388	0.005291	593.761	28055.2	189
190	36100	6859000	13.7840	5.7489	0.005263	596.903	28352.9	190
191	36481	6967871	13.8203	5.7590	0.005236	600.044	28652.1	191
192	36864	7077888	13.8564	5.7690	0.005208	603.186	28952.9	192
193	37249	7189057	13.8924	5.7790	0.005181	606.327	29255.3	193
194	37636	7301384	13.9284	5.7890	0.005155	609.469	29559.2	194
195	38025	7414875	13.9642	5.7989	0.005128	612.611	29864.8	195
196	38416	7529536	14.0000	5.8088	0.005102	615.752	30171.9	196
197	38809	7645373	14.0357	5.8186	0.005076	618.894	30480.5	197
198	39204	7762392	14.0712	5.8285	0.005051	622.035	30790.7	198
199	39601	7880599	14.1067	5.8383	0.005025	625.177	31102.6	199
200	40000	8000000	14.1421	5.8480	0.005000	628.319	31415.9	200

Numbers, Squares, Cubes, Roots, etc.   200 - 250

n	n^2	n^3	$\sqrt{n}$	$\sqrt[3]{n}$	$1/n$	πn	$\pi n^2/4$	n
200	40000	8000000	14.1421	5.8480	0.005000	628.319	31415.9	200
201	40401	8120601	14.1774	5.8578	0.004975	631.460	31730.9	201
202	40804	8242408	14.2127	5.8675	0.004950	634.602	32047.4	202
203	41209	8365427	14.2478	5.8771	0.004926	637.743	32365.5	203
204	41616	8489664	14.2829	5.8868	0.004902	640.885	32685.1	204
205	42025	8615125	14.3178	5.8964	0.004878	644.026	33006.4	205
206	42436	8741816	14.3527	5.9059	0.004854	647.168	33329.2	206
207	42849	8869743	14.3875	5.9155	0.004831	650.310	33653.5	207
208	43264	8998912	14.4222	5.9250	0.004808	653.451	33979.5	208
209	43681	9129329	14.4568	5.9345	0.004785	656.593	34307.0	209
210	44100	9261000	14.4914	5.9439	0.004762	659.734	34636.1	210
211	44521	9393931	14.5258	5.9533	0.004739	662.876	34966.7	211
212	44944	9528128	14.5602	5.9627	0.004717	666.018	35298.9	212
213	45369	9663597	14.5945	5.9721	0.004695	669.159	35632.7	213
214	45796	9800344	14.6287	5.9814	0.004673	672.301	35968.1	214
215	46225	9938375	14.6629	5.9907	0.004651	675.442	36305.0	215
216	46656	10077696	14.6969	6.0000	0.004630	678.584	36643.5	216
217	47089	10218313	14.7309	6.0092	0.004608	681.726	36983.6	217
218	47524	10360232	14.7648	6.0185	0.004587	684.867	37325.3	218
219	47961	10503459	14.7986	6.0277	0.004566	688.009	37668.5	219
220	48400	10648000	14.8324	6.0368	0.004545	691.150	38013.3	220
221	48841	10793861	14.8661	6.0459	0.004525	694.292	38359.6	221
222	49284	10941048	14.8997	6.0550	0.004505	697.434	38707.6	222
223	49729	11089567	14.9332	6.0641	0.004484	700.575	39057.1	223
224	50176	11239424	14.9666	6.0732	0.004464	703.717	39408.1	224
225	50625	11390625	15.0000	6.0822	0.004444	706.858	39760.8	225
226	51076	11543176	15.0333	6.0912	0.004425	710.000	40115.0	226
227	51529	11697083	15.0665	6.1002	0.004405	713.142	40470.8	227
228	51984	11852352	15.0997	6.1091	0.004386	716.283	40828.1	228
229	52441	12008989	15.1327	6.1180	0.004367	719.425	41187.1	229
230	52900	12167000	15.1658	6.1269	0.004348	722.566	41547.6	230
231	53361	12326391	15.1987	6.1358	0.004329	725.708	41909.6	231
232	53824	12487168	15.2315	6.1446	0.004310	728.849	42273.3	232
233	54289	12649337	15.2643	6.1534	0.004292	731.991	42638.5	233
234	54756	12812904	15.2971	6.1622	0.004274	735.133	43005.3	234
235	55225	12977875	15.3297	6.1710	0.004255	738.274	43373.6	235
236	55696	13144256	15.3623	6.1797	0.004237	741.416	43743.6	236
237	56169	13312053	15.3948	6.1885	0.004219	744.557	44115.0	237
238	56644	13481272	15.4272	6.1972	0.004202	747.699	44488.1	238
239	57121	13651919	15.4596	6.2058	0.004184	750.841	44862.7	239
240	57600	13824000	15.4919	6.2145	0.004167	753.982	45238.9	240
241	58081	13997521	15.5242	6.2231	0.004149	757.124	45616.7	241
242	58564	14172488	15.5563	6.2317	0.004132	760.265	45996.1	242
243	59049	14348907	15.5885	6.2403	0.004115	763.407	46377.0	243
244	59536	14526784	15.6205	6.2488	0.004098	766.549	46759.5	244
245	60025	14706125	15.6525	6.2573	0.004082	769.690	47143.5	245
246	60516	14886936	15.6844	6.2658	0.004065	772.832	47529.2	246
247	61009	15069223	15.7162	6.2743	0.004049	775.973	47916.4	247
248	61504	15252992	15.7480	6.2828	0.004032	779.115	48305.1	248
249	62001	15438249	15.7797	6.2912	0.004016	782.257	48695.5	249
250	62500	15625000	15.8114	6.2996	0.004000	785.398	49087.4	250

6

MATHEMATICS

Numbers, Squares, Cubes, Roots, etc.   250 - 300

n	n²	n³	√n	³√n	1/n	πn	πn²/4	n
250	62500	15625000	15.8114	6.2996	0.004000	785.398	49087.4	250
251	63001	15813251	15.8430	6.3080	0.003984	788.540	49480.9	251
252	63504	16003008	15.8745	6.3164	0.003968	791.681	49875.9	252
253	64009	16194277	15.9060	6.3247	0.003953	794.823	50272.6	253
254	64516	16387064	15.9374	6.3330	0.003937	797.965	50670.7	254
255	65025	16581375	15.9687	6.3413	0.003922	801.106	51070.5	255
256	65536	16777216	16.0000	6.3496	0.003906	804.248	51471.9	256
257	66049	16974593	16.0312	6.3579	0.003891	807.389	51874.8	257
258	66564	17173512	16.0624	6.3661	0.003876	810.531	52279.2	258
259	67081	17373979	16.0935	6.3743	0.003861	813.672	52685.3	259
260	67600	17576000	16.1245	6.3825	0.003846	816.814	53092.9	260
261	68121	17779581	16.1555	6.3907	0.003831	819.956	53502.1	261
262	68644	17984728	16.1864	6.3988	0.003817	823.097	53912.9	262
263	69169	18191447	16.2173	6.4070	0.003802	826.239	54325.2	263
264	69696	18399744	16.2481	6.4151	0.003788	829.380	54739.1	264
265	70225	18609625	16.2788	6.4232	0.003774	832.522	55154.6	265
266	70756	18821096	16.3095	6.4312	0.003759	835.664	55571.6	266
267	71289	19034163	16.3401	6.4393	0.003745	838.805	55990.2	267
268	71824	19248832	16.3707	6.4473	0.003731	841.947	56410.4	268
269	72361	19465109	16.4012	6.4553	0.003717	845.088	56832.2	269
270	72900	19683000	16.4317	6.4633	0.003704	848.230	57255.5	270
271	73441	19902511	16.4621	6.4713	0.003690	851.372	57680.4	271
272	73984	20123648	16.4924	6.4792	0.003676	854.513	58106.9	272
273	74529	20346417	16.5227	6.4872	0.003663	857.655	58534.9	273
274	75076	20570824	16.5529	6.4951	0.003650	860.796	58964.6	274
275	75625	20796875	16.5831	6.5030	0.003636	863.938	59395.7	275
276	76176	21024576	16.6132	6.5108	0.003623	867.080	59828.5	276
277	76729	21253933	16.6433	6.5187	0.003610	870.221	60262.8	277
278	77284	21484952	16.6733	6.5265	0.003597	873.363	60698.7	278
279	77841	21717639	16.7033	6.5343	0.003584	876.504	61136.2	279
280	78400	21952000	16.7332	6.5421	0.003571	879.646	61575.2	280
281	78961	22188041	16.7631	6.5499	0.003559	882.788	62015.8	281
282	79524	22425768	16.7929	6.5577	0.003546	885.929	62458.0	282
283	80089	22665187	16.8226	6.5654	0.003534	889.071	62901.8	283
284	80656	22906304	16.8523	6.5731	0.003521	892.212	63347.1	284
285	81225	23149125	16.8819	6.5808	0.003509	895.354	63794.0	285
286	81796	23393656	16.9115	6.5885	0.003497	898.495	64242.4	286
287	82369	23639903	16.9411	6.5962	0.003484	901.637	64692.5	287
288	82944	23887872	16.9706	6.6039	0.003472	904.779	65144.1	288
289	83521	24137569	17.0000	6.6115	0.003460	907.920	65597.2	289
290	84100	24389000	17.0294	6.6191	0.003448	911.062	66052.0	290
291	84681	24642171	17.0587	6.6267	0.003436	914.203	66508.3	291
292	85264	24897088	17.0880	6.6343	0.003425	917.345	66966.2	292
293	85849	25153757	17.1172	6.6419	0.003413	920.487	67425.6	293
294	86436	25412184	17.1464	6.6494	0.003401	923.628	67886.7	294
295	87025	25672375	17.1756	6.6569	0.003390	926.770	68349.3	295
296	87616	25934336	17.2047	6.6644	0.003378	929.911	68813.4	296
297	88209	26198073	17.2337	6.6719	0.003367	933.053	69279.2	297
298	88804	26463592	17.2627	6.6794	0.003356	936.195	69746.5	298
299	89401	26730899	17.2916	6.6869	0.003344	939.336	70215.4	299
300	90000	27000000	17.3205	6.6943	0.003333	942.478	70685.8	300

Numbers, Squares, Cubes, Roots, etc.   300 - 350

n	n^2	n^3	$\sqrt{n}$	$\sqrt[3]{n}$	$1/n$	πn	$\pi n^2/4$	n
300	90000	27000000	17.3205	6.6943	0.003333	942.478	70685.8	300
301	90601	27270901	17.3494	6.7018	0.003322	945.619	71157.9	301
302	91204	27543608	17.3781	6.7092	0.003311	948.761	71631.5	302
303	91809	27818127	17.4069	6.7166	0.003300	951.903	72106.6	303
304	92416	28094464	17.4356	6.7240	0.003289	955.044	72583.4	304
305	93025	28372625	17.4642	6.7313	0.003279	958.186	73061.7	305
306	93636	28652616	17.4929	6.7387	0.003268	961.327	73541.5	306
307	94249	28934443	17.5214	6.7460	0.003257	964.469	74023.0	307
308	94864	29218112	17.5499	6.7533	0.003247	967.611	74506.0	308
309	95481	29503629	17.5784	6.7606	0.003236	970.752	74990.6	309
310	96100	29791000	17.6068	6.7679	0.003226	973.894	75476.8	310
311	96721	30080231	17.6352	6.7752	0.003215	977.035	75964.5	311
312	97344	30371328	17.6635	6.7824	0.003205	980.177	76453.8	312
313	97969	30664297	17.6918	6.7897	0.003195	983.319	76944.7	313
314	98596	30959144	17.7200	6.7969	0.003185	986.460	77437.1	314
315	99225	31255875	17.7482	6.8041	0.003175	989.602	77931.1	315
316	99856	31554496	17.7764	6.8113	0.003165	992.743	78426.7	316
317	100489	31855013	17.8045	6.8185	0.003155	995.885	78923.9	317
318	101124	32157432	17.8326	6.8256	0.003145	999.026	79422.6	318
319	101761	32461759	17.8606	6.8328	0.003135	1002.168	79922.9	319
320	102400	32768000	17.8885	6.8399	0.003125	1005.310	80424.8	320
321	103041	33076161	17.9165	6.8470	0.003115	1008.451	80928.2	321
322	103684	33386248	17.9444	6.8541	0.003106	1011.593	81433.2	322
323	104329	33698267	17.9722	6.8612	0.003096	1014.734	81939.8	323
324	104976	34012224	18.0000	6.8683	0.003086	1017.876	82448.0	324
325	105625	34328125	18.0278	6.8753	0.003077	1021.018	82957.7	325
326	106276	34645976	18.0555	6.8824	0.003067	1024.159	83469.0	326
327	106929	34965783	18.0831	6.8894	0.003058	1027.301	83981.8	327
328	107584	35287552	18.1108	6.8964	0.003049	1030.442	84496.3	328
329	108241	35611289	18.1384	6.9034	0.003040	1033.584	85012.3	329
330	108900	35937000	18.1659	6.9104	0.003030	1036.726	85529.9	330
331	109561	36264691	18.1934	6.9174	0.003021	1039.867	86049.0	331
332	110224	36594368	18.2209	6.9244	0.003012	1043.009	86569.7	332
333	110889	36926037	18.2483	6.9313	0.003003	1046.150	87092.0	333
334	111556	37259704	18.2757	6.9382	0.002994	1049.292	87615.9	334
335	112225	37595375	18.3030	6.9451	0.002985	1052.434	88141.3	335
336	112896	37933056	18.3303	6.9521	0.002976	1055.575	88668.3	336
337	113569	38272753	18.3576	6.9589	0.002967	1058.717	89196.9	337
338	114244	38614472	18.3848	6.9658	0.002959	1061.858	89727.0	338
339	114921	38958219	18.4120	6.9727	0.002950	1065.000	90258.7	339
340	115600	39304000	18.4391	6.9795	0.002941	1068.142	90792.0	340
341	116281	39651821	18.4662	6.9864	0.002933	1071.283	91326.9	341
342	116964	40001688	18.4932	6.9932	0.002924	1074.425	91863.3	342
343	117649	40353607	18.5203	7.0000	0.002915	1077.566	92401.3	343
344	118336	40707584	18.5472	7.0068	0.002907	1080.708	92940.9	344
345	119025	41063625	18.5742	7.0136	0.002899	1083.849	93482.0	345
346	119716	41421736	18.6011	7.0203	0.002890	1086.991	94024.7	346
347	120409	41781923	18.6279	7.0271	0.002882	1090.133	94569.0	347
348	121104	42144192	18.6548	7.0338	0.002874	1093.274	95114.9	348
349	121801	42508549	18.6815	7.0406	0.002865	1096.416	95662.3	349
350	122500	42875000	18.7083	7.0473	0.002857	1099.557	96211.3	350

8　　　　　　　　　　　　　　**MATHEMATICS**

n	n^2	n^3	$\sqrt{n}$	$\sqrt[3]{n}$	$1/n$	πn	$\pi n^2/4$	n
350	122500	42875000	18.7083	7.0473	0.002857	1099.557	96211.3	350
351	123201	43243551	18.7350	7.0540	0.002849	1102.699	96761.8	351
352	123904	43614208	18.7617	7.0607	0.002841	1105.841	97314.0	352
353	124609	43986977	18.7883	7.0674	0.002833	1108.982	97867.7	353
354	125316	44361864	18.8149	7.0740	0.002825	1112.124	98423.0	354
355	126025	44738875	18.8414	7.0807	0.002817	1115.265	98979.8	355
356	126736	45118016	18.8680	7.0873	0.002809	1118.407	99538.2	356
357	127449	45499293	18.8944	7.0940	0.002801	1121.549	100098.2	357
358	128164	45882712	18.9209	7.1006	0.002793	1124.690	100659.8	358
359	128881	46268279	18.9473	7.1072	0.002786	1127.832	101222.9	359
360	129600	46656000	18.9737	7.1138	0.002778	1130.973	101787.6	360
361	130321	47045881	19.0000	7.1204	0.002770	1134.115	102353.9	361
362	131044	47437928	19.0263	7.1269	0.002762	1137.257	102921.7	362
363	131769	47832147	19.0526	7.1335	0.002755	1140.398	103491.1	363
364	132496	48228544	19.0788	7.1400	0.002747	1143.540	104062.1	364
365	133225	48627125	19.1050	7.1466	0.002740	1146.681	104634.7	365
366	133956	49027896	19.1311	7.1531	0.002732	1149.823	105208.8	366
367	134689	49430863	19.1572	7.1596	0.002725	1152.965	105784.5	367
368	135424	49836032	19.1833	7.1661	0.002717	1156.106	106361.8	368
369	136161	50243409	19.2094	7.1726	0.002710	1159.248	106940.6	369
370	136900	50653000	19.2354	7.1791	0.002703	1162.389	107521.0	370
371	137641	51064811	19.2614	7.1855	0.002695	1165.531	108103.0	371
372	138384	51478848	19.2873	7.1920	0.002688	1168.672	108686.5	372
373	139129	51895117	19.3132	7.1984	0.002681	1171.814	109271.7	373
374	139876	52313624	19.3391	7.2048	0.002674	1174.956	109858.4	374
375	140625	52734375	19.3649	7.2112	0.002667	1178.097	110446.6	375
376	141376	53157376	19.3907	7.2177	0.002660	1181.239	111036.5	376
377	142129	53582633	19.4165	7.2240	0.002653	1184.380	111627.9	377
378	142884	54010152	19.4422	7.2304	0.002646	1187.522	112220.8	378
379	143641	54439939	19.4679	7.2368	0.002639	1190.664	112815.4	379
380	144400	54872000	19.4936	7.2432	0.002632	1193.805	113411.5	380
381	145161	55306341	19.5192	7.2495	0.002625	1196.947	114009.2	381
382	145924	55742968	19.5448	7.2558	0.002618	1200.088	114608.4	382
383	146689	56181887	19.5704	7.2622	0.002611	1203.230	115209.3	383
384	147456	56623104	19.5959	7.2685	0.002604	1206.372	115811.7	384
385	148225	57066625	19.6214	7.2748	0.002597	1209.513	116415.6	385
386	148996	57512456	19.6469	7.2811	0.002591	1212.655	117021.2	386
387	149769	57960603	19.6723	7.2874	0.002584	1215.796	117628.3	387
388	150544	58411072	19.6977	7.2936	0.002577	1218.938	118237.0	388
389	151321	58863869	19.7231	7.2999	0.002571	1222.080	118847.2	389
390	152100	59319000	19.7484	7.3061	0.002564	1225.221	119459.1	390
391	152881	59776471	19.7737	7.3124	0.002558	1228.363	120072.5	391
392	153664	60236288	19.7990	7.3186	0.002551	1231.504	120687.4	392
393	154449	60698457	19.8242	7.3248	0.002545	1234.646	121304.0	393
394	155236	61162984	19.8494	7.3310	0.002538	1237.788	121922.1	394
395	156025	61629875	19.8746	7.3372	0.002532	1240.929	122541.7	395
396	156816	62099136	19.8997	7.3434	0.002525	1244.071	123163.0	396
397	157609	62570773	19.9249	7.3496	0.002519	1247.212	123785.8	397
398	158404	63044792	19.9499	7.3558	0.002513	1250.354	124410.2	398
399	159201	63521199	19.9750	7.3619	0.002506	1253.495	125036.2	399
400	160000	64000000	20.0000	7.3681	0.002500	1256.637	125663.7	400

Numbers, Squares, Cubes, Roots, etc.   400 - 450

n	n^2	n^3	$\sqrt{n}$	$\sqrt[3]{n}$	$1/n$	πn	$\pi n^2/4$	n
400	160000	64000000	20.0000	7.3681	0.002500	1256.637	125663.7	400
401	160801	64481201	20.0250	7.3742	0.002494	1259.779	126292.8	401
402	161604	64964808	20.0499	7.3803	0.002488	1262.920	126923.5	402
403	162409	65450827	20.0749	7.3864	0.002481	1266.062	127555.7	403
404	163216	65939264	20.0998	7.3925	0.002475	1269.203	128189.5	404
405	164025	66430125	20.1246	7.3986	0.002469	1272.345	128824.9	405
406	164836	66923416	20.1494	7.4047	0.002463	1275.487	129461.9	406
407	165649	67419143	20.1742	7.4108	0.002457	1278.628	130100.4	407
408	166464	67917312	20.1990	7.4169	0.002451	1281.770	130740.5	408
409	167281	68417929	20.2237	7.4229	0.002445	1284.911	131382.2	409
410	168100	68921000	20.2485	7.4290	0.002439	1288.053	132025.4	410
411	168921	69426531	20.2731	7.4350	0.002433	1291.195	132670.2	411
412	169744	69934528	20.2978	7.4410	0.002427	1294.336	133316.6	412
413	170569	70444997	20.3224	7.4470	0.002421	1297.478	133964.6	413
414	171396	70957944	20.3470	7.4530	0.002415	1300.619	134614.1	414
415	172225	71473375	20.3715	7.4590	0.002410	1303.761	135265.2	415
416	173056	71991296	20.3961	7.4650	0.002404	1306.903	135917.9	416
417	173889	72511713	20.4206	7.4710	0.002398	1310.044	136572.1	417
418	174724	73034632	20.4450	7.4770	0.002392	1313.186	137227.9	418
419	175561	73560059	20.4695	7.4829	0.002387	1316.327	137885.3	419
420	176400	74088000	20.4939	7.4889	0.002381	1319.469	138544.2	420
421	177241	74618461	20.5183	7.4948	0.002375	1322.611	139204.8	421
422	178084	75151448	20.5426	7.5007	0.002370	1325.752	139866.8	422
423	178929	75686967	20.5670	7.5067	0.002364	1328.894	140530.5	423
424	179776	76225024	20.5913	7.5126	0.002358	1332.035	141195.7	424
425	180625	76765625	20.6155	7.5185	0.002353	1335.177	141862.5	425
426	181476	77308776	20.6398	7.5244	0.002347	1338.318	142530.9	426
427	182329	77854483	20.6640	7.5302	0.002342	1341.460	143200.9	427
428	183184	78402752	20.6882	7.5361	0.002336	1344.602	143872.4	428
429	184041	78953589	20.7123	7.5420	0.002331	1347.743	144545.5	429
430	184900	79507000	20.7364	7.5478	0.002326	1350.885	145220.1	430
431	185761	80062991	20.7605	7.5537	0.002320	1354.026	145896.3	431
432	186624	80621568	20.7846	7.5595	0.002315	1357.168	146574.1	432
433	187489	81182737	20.8087	7.5654	0.002309	1360.310	147253.5	433
434	188356	81746504	20.8327	7.5712	0.002304	1363.451	147934.5	434
435	189225	82312875	20.8567	7.5770	0.002299	1366.593	148617.0	435
436	190096	82881856	20.8806	7.5828	0.002294	1369.734	149301.0	436
437	190969	83453453	20.9045	7.5886	0.002288	1372.876	149986.7	437
438	191844	84027672	20.9284	7.5944	0.002283	1376.018	150673.9	438
439	192721	84604519	20.9523	7.6001	0.002278	1379.159	151362.7	439
440	193600	85184000	20.9762	7.6059	0.002273	1382.301	152053.1	440
441	194481	85766121	21.0000	7.6117	0.002268	1385.442	152745.0	441
442	195364	86350888	21.0238	7.6174	0.002262	1388.584	153438.5	442
443	196249	86938307	21.0476	7.6232	0.002257	1391.726	154133.6	443
444	197136	87528384	21.0713	7.6289	0.002252	1394.867	154830.3	444
445	198025	88121125	21.0950	7.6346	0.002247	1398.009	155528.5	445
446	198916	88716536	21.1187	7.6403	0.002242	1401.150	156228.3	446
447	199809	89314623	21.1424	7.6460	0.002237	1404.292	156929.6	447
448	200704	89915392	21.1660	7.6517	0.002232	1407.434	157632.6	448
449	201601	90518849	21.1896	7.6574	0.002227	1410.575	158337.1	449
450	202500	91125000	21.2132	7.6631	0.002222	1413.717	159043.1	450

Numbers, Squares, Cubes, Roots,etc.   450 - 500

n	n^2	n^3	$\sqrt{n}$	$\sqrt[3]{n}$	$1/n$	πn	$\pi n^2/4$	n
450	202500	91125000	21.2132	7.6631	0.002222	1413.717	159043.1.	450
451	203401	91733851	21.2368	7.6688	0.002217	1416.858	159750.8	451
452	204304	92345408	21.2603	7.6744	0.002212	1420.000	160460.0	452
453	205209	92959677	21.2838	7.6801	0.002208	1423.141	161170.8	453
454	206116	93576664	21.3073	7.6857	0.002203	1426.283	161883.1	454
455	207025	94196375	21.3307	7.6914	0.002198	1429.425	162597.1	455
456	207936	94818816	21.3542	7.6970	0.002193	1432.566	163312.6	456
457	208849	95443993	21.3776	7.7026	0.002188	1435.708	164029.6	457
458	209764	96071912	21.4009	7.7082	0.002183	1438.849	164748.3	458
459	210681	96702579	21.4243	7.7138	0.002179	1441.991	165468.5	459
460	211600	97336000	21.4476	7.7194	0.002174	1445.133	166190.3	460
461	212521	97972181	21.4709	7.7250	0.002169	1448.274	166913.6	461
462	213444	98611128	21.4942	7.7306	0.002165	1451.416	167638.5	462
463	214369	99252847	21.5174	7.7362	0.002160	1454.557	168365.0	463
464	215296	99897344	21.5407	7.7418	0.002155	1457.699	169093.1	464
465	216225	100544625	21.5639	7.7473	0.002151	1460.841	169822.7	465
466	217156	101194696	21.5870	7.7529	0.002146	1463.982	170553.9	466
467	218089	101847563	21.6102	7.7584	0.002141	1467.124	171286.7	467
468	219024	102503232	21.6333	7.7639	0.002137	1470.265	172021.0	468
469	219961	103161709	21.6564	7.7695	0.002132	1473.407	172757.0	469
470	220900	103823000	21.6795	7.7750	0.002128	1476.549	173494.5	470
471	221841	104487111	21.7025	7.7805	0.002123	1479.690	174233.5	471
472	222784	105154048	21.7256	7.7860	0.002119	1482.832	174974.1	472
473	223729	105823817	21.7486	7.7915	0.002114	1485.973	175716.3	473
474	224676	106496424	21.7715	7.7970	0.002110	1489.115	176460.1	474
475	225625	107171875	21.7945	7.8025	0.002105	1492.257	177205.5	475
476	226576	107850176	21.8174	7.8079	0.002101	1495.398	177952.4	476
477	227529	108531333	21.8403	7.8134	0.002096	1498.540	178700.9	477
478	228484	109215352	21.8632	7.8188	0.002092	1501.681	179450.9	478
479	229441	109902239	21.8861	7.8243	0.002088	1504.823	180202.5	479
480	230400	110592000	21.9089	7.8297	0.002083	1507.964	180955.7	480
481	231361	111284641	21.9317	7.8352	0.002079	1511.106	181710.5	481
482	232324	111980168	21.9545	7.8406	0.002075	1514.248	182466.8	482
483	233289	112678587	21.9773	7.8460	0.002070	1517.389	183224.8	483
484	234256	113379904	22.0000	7.8514	0.002066	1520.531	183984.2	484
485	235225	114084125	22.0227	7.8568	0.002062	1523.672	184745.3	485
486	236196	114791256	22.0454	7.8622	0.002058	1526.814	185507.9	486
487	237169	115501303	22.0681	7.8676	0.002053	1529.956	186272.1	487
488	238144	116214272	22.0907	7.8730	0.002049	1533.097	187037.9	488
489	239121	116930169	22.1133	7.8784	0.002045	1536.239	187805.2	489
490	240100	117649000	22.1359	7.8837	0.002041	1539.380	188574.1	490
491	241081	118370771	22.1585	7.8891	0.002037	1542.522	189344.6	491
492	242064	119095488	22.1811	7.8944	0.002033	1545.664	190116.6	492
493	243049	119823157	22.2036	7.8998	0.002028	1548.805	190890.2	493
494	244036	120553784	22.2261	7.9051	0.002024	1551.947	191665.4	494
495	245025	121287375	22.2486	7.9105	0.002020	1555.088	192442.2	495
496	246016	122023936	22.2711	7.9158	0.002016	1558.230	193220.5	496
497	247009	122763473	22.2935	7.9211	0.002012	1561.372	194000.4	497
498	248004	123505992	22.3159	7.9264	0.002008	1564.513	194781.9	498
499	249001	124251499	22.3383	7.9317	0.002004	1567.655	195564.9	499
500	250000	125000000	22.3607	7.9370	0.002000	1570.796	196349.5	500

MATHEMATICS

11

Numbers, Squares, Cubes, Roots, etc.   500 - 550

n	n²	n³	√n	³√n	1/n	πn	πn²/4	n
500	250000	125000000	22.3607	7.9370	0.002000	1570.796	196349.5	500
501	251001	125751501	22.3830	7.9423	0.001996	1573.938	197135.7	501
502	252004	126506008	22.4054	7.9476	0.001992	1577.080	197923.5	502
503	253009	127263527	22.4277	7.9528	0.001988	1580.221	198712.8	503
504	254016	128024064	22.4499	7.9581	0.001984	1583.363	199503.7	504
505	255025	128787625	22.4722	7.9634	0.001980	1586.504	200296.2	505
506	256036	129554216	22.4944	7.9686	0.001976	1589.646	201090.2	506
507	257049	130323843	22.5167	7.9739	0.001972	1592.787	201885.8	507
508	258064	131096512	22.5389	7.9791	0.001969	1595.929	202683.0	508
509	259081	131872229	22.5610	7.9843	0.001965	1599.071	203481.7	509
510	260100	132651000	22.5832	7.9896	0.001961	1602.212	204282.1	510
511	261121	133432831	22.6053	7.9948	0.001957	1605.354	205084.0	511
512	262144	134217728	22.6274	8.0000	0.001953	1608.495	205887.4	512
513	263169	135005697	22.6495	8.0052	0.001949	1611.637	206692.4	513
514	264196	135796744	22.6716	8.0104	0.001946	1614.779	207499.1	514
515	265225	136590875	22.6936	8.0156	0.001942	1617.920	208307.2	515
516	266256	137388096	22.7156	8.0208	0.001938	1621.062	209117.0	516
517	267289	138188413	22.7376	8.0260	0.001934	1624.203	209928.3	517
518	268324	138991832	22.7596	8.0311	0.001931	1627.345	210741.2	518
519	269361	139798359	22.7816	8.0363	0.001927	1630.487	211555.6	519
520	270400	140608000	22.8035	8.0415	0.001923	1633.628	212371.7	520
521	271441	141420761	22.8254	8.0466	0.001919	1636.770	213189.3	521
522	272484	142236648	22.8473	8.0517	0.001916	1639.911	214008.4	522
523	273529	143055667	22.8692	8.0569	0.001912	1643.053	214829.2	523
524	274576	143877824	22.8910	8.0620	0.001908	1646.195	215651.5	524
525	275625	144703125	22.9129	8.0671	0.001905	1649.336	216475.4	525
526	276676	145531576	22.9347	8.0723	0.001901	1652.478	217300.8	526
527	277729	146363183	22.9565	8.0774	0.001898	1655.619	218127.8	527
528	278784	147197952	22.9783	8.0825	0.001894	1658.761	218956.4	528
529	279841	148035889	23.0000	8.0876	0.001890	1661.903	219786.6	529
530	280900	148877000	23.0217	8.0927	0.001887	1665.044	220618.3	530
531	281961	149721291	23.0434	8.0978	0.001883	1668.186	221451.7	531
532	283024	150568768	23.0651	8.1028	0.001880	1671.327	222286.5	532
533	284089	151419437	23.0868	8.1079	0.001876	1674.469	223123.0	533
534	285156	152273304	23.1084	8.1130	0.001873	1677.610	223961.0	534
535	286225	153130375	23.1301	8.1180	0.001869	1680.752	224800.6	535
536	287296	153990656	23.1517	8.1231	0.001866	1683.894	225641.8	536
537	288369	154854153	23.1733	8.1281	0.001862	1687.035	226484.5	537
538	289444	155720872	23.1948	8.1332	0.001859	1690.177	227328.8	538
539	290521	156590819	23.2164	8.1382	0.001855	1693.318	228174.7	539
540	291600	157464000	23.2379	8.1433	0.001852	1696.460	229022.1	540
541	292681	158340421	23.2594	8.1483	0.001848	1699.602	229871.1	541
542	293764	159220088	23.2809	8.1533	0.001845	1702.743	230721.7	542
543	294849	160103007	23.3024	8.1583	0.001842	1705.885	231573.9	543
544	295936	160989184	23.3238	8.1633	0.001838	1709.026	232427.6	544
545	297025	161878625	23.3452	8.1683	0.001835	1712.168	233282.9	545
546	298116	162771336	23.3666	8.1733	0.001832	1715.310	234139.8	546
547	299209	163667323	23.3880	8.1783	0.001828	1718.451	234998.2	547
548	300304	164566592	23.4094	8.1833	0.001825	1721.593	235858.2	548
549	301401	165469149	23.4307	8.1882	0.001821	1724.734	236719.8	549
550	302500	166375000	23.4521	8.1932	0.001818	1727.876	237582.9	550

Numbers, Squares, Cubes, Roots, etc.   550 - 600

n	n^2	n^3	$\sqrt{n}$	$\sqrt[3]{n}$	$1/n$	πn	$\pi n^2/4$	n
550	302500	166375000	23.4521	8.1932	0.001818	1727.876	237582.9	550
551	303601	167284151	23.4734	8.1982	0.001815	1731.018	238447.7	551
552	304704	168196608	23.4947	8.2031	0.001812	1734.159	239314.0	552
553	305809	169112377	23.5160	8.2081	0.001808	1737.301	240181.8	553
554	306916	170031464	23.5372	8.2130	0.001805	1740.442	241051.3	554
555	308025	170953875	23.5584	8.2180	0.001802	1743.584	241922.3	555
556	309136	171879616	23.5797	8.2229	0.001799	1746.726	242794.8	556
557	310249	172808693	23.6008	8.2278	0.001795	1749.867	243669.0	557
558	311364	173741112	23.6220	8.2327	0.001792	1753.009	244544.7	558
559	312481	174676879	23.6432	8.2377	0.001789	1756.150	245422.0	559
560	313600	175616000	23.6643	8.2426	0.001786	1759.292	246300.9	560
561	314721	176558481	23.6854	8.2475	0.001783	1762.433	247181.3	561
562	315844	177504328	23.7065	8.2524	0.001779	1765.575	248063.3	562
563	316969	178453547	23.7276	8.2573	0.001776	1768.717	248946.9	563
564	318096	179406144	23.7487	8.2621	0.001773	1771.858	249832.0	564
565	319225	180362125	23.7697	8.2670	0.001770	1775.000	250718.7	565
566	320356	181321496	23.7908	8.2719	0.001767	1778.141	251607.0	566
567	321489	182284263	23.8118	8.2768	0.001764	1781.283	252496.9	567
568	322624	183250432	23.8328	8.2816	0.001761	1784.425	253388.3	568
569	323761	184220009	23.8537	8.2865	0.001757	1787.566	254281.3	569
570	324900	185193000	23.8747	8.2913	0.001754	1790.708	255175.9	570
571	326041	186169411	23.8956	8.2962	0.001751	1793.849	256072.0	571
572	327184	187149248	23.9165	8.3010	0.001748	1796.991	256969.7	572
573	328329	188132517	23.9374	8.3059	0.001745	1800.133	257869.0	573
574	329476	189119224	23.9583	8.3107	0.001742	1803.274	258769.8	574
575	330625	190109375	23.9792	8.3155	0.001739	1806.416	259672.3	575
576	331776	191102976	24.0000	8.3203	0.001736	1809.557	260576.3	576
577	332929	192100033	24.0208	8.3251	0.001733	1812.699	261481.8	577
578	334084	193100552	24.0416	8.3300	0.001730	1815.841	262389.0	578
579	335241	194104539	24.0624	8.3348	0.001727	1818.982	263297.7	579
580	336400	195110000	24.0832	8.3396	0.001724	1822.124	264207.9	580
581	337561	196122941	24.1039	8.3443	0.001721	1825.265	265119.8	581
582	338724	197137368	24.1247	8.3491	0.001718	1828.407	266033.2	582
583	339889	198155287	24.1454	8.3539	0.001715	1831.549	266948.2	583
584	341056	199176704	24.1661	8.3587	0.001712	1834.690	267864.8	584
585	342225	200201625	24.1868	8.3634	0.001709	1837.832	268782.9	585
586	343396	201230056	24.2074	8.3682	0.001706	1840.973	269702.6	586
587	344569	202262003	24.2281	8.3730	0.001704	1844.115	270623.9	587
588	345744	203297472	24.2487	8.3777	0.001701	1847.256	271546.7	588
589	346921	204336469	24.2693	8.3825	0.001698	1850.398	272471.1	589
590	348100	205379000	24.2899	8.3872	0.001695	1853.540	273397.1	590
591	349281	206425071	24.3105	8.3919	0.001692	1856.681	274324.7	591
592	350464	207474688	24.3311	8.3967	0.001689	1859.823	275253.8	592
593	351649	208527857	24.3516	8.4014	0.001686	1862.964	276184.5	593
594	352836	209584584	24.3721	8.4061	0.001684	1866.106	277116.7	594
595	354025	210644875	24.3926	8.4108	0.001681	1869.248	278050.6	595
596	355216	211708736	24.4131	8.4155	0.001678	1872.389	278986.0	596
597	356409	212776173	24.4336	8.4202	0.001675	1875.531	279923.0	597
598	357604	213847192	24.4540	8.4249	0.001672	1878.672	280861.5	598
599	358801	214921799	24.4745	8.4296	0.001669	1881.814	281801.6	599
600	360000	216000000	24.4949	8.4343	0.001667	1884.956	282743.3	600

Numbers, Squares, Cubes, Roots, etc. 600 - 650

n	n^2	n^3	$\sqrt{n}$	$\sqrt[3]{n}$	$1/n$	πn	$\pi n^2/4$	n
600	360000	216000000	24.4949	8.4343	0.001667	1884.956	282743.3	600
601	361201	217081801	24.5153	8.4390	0.001664	1888.097	283686.6	601
602	362404	218167208	24.5357	8.4437	0.001661	1891.239	284631.4	602
603	363609	219256227	24.5561	8.4484	0.001658	1894.380	285577.8	603
604	364816	220348864	24.5764	8.4530	0.001656	1897.522	286525.8	604
605	366025	221445125	24.5967	8.4577	0.001653	1900.664	287475.4	605
606	367236	222545016	24.6171	8.4623	0.001650	1903.805	288426.5	606
607	368449	223648543	24.6374	8.4670	0.001647	1906.947	289379.2	607
608	369664	224755712	24.6577	8.4716	0.001645	1910.088	290333.4	608
609	370881	225866529	24.6779	8.4763	0.001642	1913.230	291289.3	609
610	372100	226981000	24.6982	8.4809	0.001639	1916.372	292246.7	610
611	373321	228099131	24.7184	8.4856	0.001637	1919.513	293205.6	611
612	374544	229220928	24.7386	8.4902	0.001634	1922.655	294166.2	612
613	375769	230346397	24.7588	8.4948	0.001631	1925.796	295128.3	613
614	376996	231475544	24.7790	8.4994	0.001629	1928.938	296092.0	614
615	378225	232608375	24.7992	8.5040	0.001626	1932.079	297057.2	615
616	379456	233744896	24.8193	8.5086	0.001623	1935.221	298024.0	616
617	380689	234885113	24.8395	8.5132	0.001621	1938.363	298992.4	617
618	381924	236029032	24.8596	8.5178	0.001618	1941.504	299962.4	618
619	383161	237176659	24.8797	8.5224	0.001616	1944.646	300933.9	619
620	384400	238328000	24.8998	8.5270	0.001613	1947.787	301907.1	620
621	385641	239483061	24.9199	8.5316	0.001610	1950.929	302881.7	621
622	386884	240641848	24.9399	8.5362	0.001608	1954.071	303858.0	622
623	388129	241804367	24.9600	8.5408	0.001605	1957.212	304835.8	623
624	389376	242970624	24.9800	8.5453	0.001603	1960.354	305815.2	624
625	390625	244140625	25.0000	8.5499	0.001600	1963.495	306796.2	625
626	391876	245314376	25.0200	8.5544	0.001597	1966.637	307778.7	626
627	393129	246491883	25.0400	8.5590	0.001595	1969.779	308762.8	627
628	394384	247673152	25.0599	8.5635	0.001592	1972.920	309748.5	628
629	395641	248858189	25.0799	8.5681	0.001590	1976.062	310735.7	629
630	396900	250047000	25.0998	8.5726	0.001587	1979.203	311724.5	630
631	398161	251239591	25.1197	8.5772	0.001585	1982.345	312714.9	631
632	399424	252435968	25.1396	8.5817	0.001582	1985.487	313706.9	632
633	400689	253636137	25.1595	8.5862	0.001580	1988.628	314700.4	633
634	401956	254840104	25.1794	8.5907	0.001577	1991.770	315695.5	634
635	403225	256047875	25.1992	8.5952	0.001575	1994.911	316692.2	635
636	404496	257259456	25.2190	8.5997	0.001572	1998.053	317690.4	636
637	405769	258474853	25.2389	8.6043	0.001570	2001.195	318690.2	637
638	407044	259694072	25.2587	8.6088	0.001567	2004.336	319691.6	638
639	408321	260917119	25.2784	8.6132	0.001565	2007.478	320694.6	639
640	409600	262144000	25.2982	8.6177	0.001563	2010.619	321699.1	640
641	410881	263374721	25.3180	8.6222	0.001560	2013.761	322705.2	641
642	412164	264609288	25.3377	8.6267	0.001558	2016.902	323712.8	642
643	413449	265847707	25.3574	8.6312	0.001555	2020.044	324722.1	643
644	414736	267089984	25.3772	8.6357	0.001553	2023.186	325732.9	644
645	416025	268336125	25.3969	8.6401	0.001550	2026.327	326745.3	645
646	417316	269586136	25.4165	8.6446	0.001548	2029.469	327753.2	646
647	418609	270840023	25.4362	8.6490	0.001546	2032.610	328774.7	647
648	419904	272097792	25.4558	8.6535	0.001543	2035.752	329791.8	648
649	421201	273359449	25.4755	8.6579	0.001541	2038.894	330810.5	649
650	422500	274625000	25.4951	8.6624	0.001538	2042.035	331830.7	650

Numbers, Squares, Cubes, Roots, etc.   650 - 700

n	n^2	n^3	$\sqrt{n}$	$\sqrt[3]{n}$	$1/n$	πn	$\pi n^2/4$	n
650	422500	274625000	25.4951	8.6624	0.001538	2042.035	331830.7	650
651	423801	275894451	25.5147	8.6668	0.001536	2045.177	332852.5	651
652	425104	277167808	25.5343	8.6713	0.001534	2048.318	333875.9	652
653	426409	278445077	25.5539	8.6757	0.001531	2051.460	334900.8	653
654	427716	279726264	25.5734	8.6801	0.001529	2054.602	335927.4	654
655	429025	281011375	25.5930	8.6845	0.001527	2057.743	336955.4	655
656	430336	282300416	25.6125	8.6890	0.001524	2060.885	337985.1	656
657	431649	283593393	25.6320	8.6934	0.001522	2064.026	339016.3	657
658	432964	284890312	25.6515	8.6978	0.001520	2067.168	340049.1	658
659	434281	286191179	25.6710	8.7022	0.001517	2070.310	341083.5	659
660	435600	287496000	25.6905	8.7066	0.001515	2073.451	342119.4	660
661	436921	288804781	25.7099	8.7110	0.001513	2076.593	343157.0	661
662	438244	290117528	25.7294	8.7154	0.001511	2079.734	344196.0	662
663	439569	291434247	25.7488	8.7198	0.001508	2082.876	345236.7	663
664	440896	292754944	25.7682	8.7241	0.001506	2086.018	346278.9	664
665	442225	294079625	25.7876	8.7285	0.001504	2089.159	347322.7	665
666	443556	295408296	25.8070	8.7329	0.001502	2092.301	348368.1	666
667	444889	296740963	25.8263	8.7373	0.001499	2095.442	349415.0	667
668	446224	298077632	25.8457	8.7416	0.001497	2098.584	350463.5	668
669	447561	299418309	25.8650	8.7460	0.001495	2101.725	351513.6	669
670	448900	300763000	25.8844	8.7503	0.001493	2104.867	352565.2	670
671	450241	302111711	25.9037	8.7547	0.001490	2108.009	353618.5	671
672	451584	303464448	25.9230	8.7590	0.001488	2111.150	354673.2	672
673	452929	304821217	25.9422	8.7634	0.001486	2114.292	355729.6	673
674	454276	306182024	25.9615	8.7677	0.001484	2117.433	356787.5	674
675	455625	307546875	25.9808	8.7721	0.001481	2120.575	357847.0	675
676	456976	308915776	26.0000	8.7764	0.001479	2123.717	358908.1	676
677	458329	310288733	26.0192	8.7807	0.001477	2126.858	359970.8	677
678	459684	311665752	26.0384	8.7850	0.001475	2130.000	361035.0	678
679	461041	313046839	26.0576	8.7893	0.001473	2133.141	362100.8	679
680	462400	314432000	26.0768	8.7937	0.001471	2136.283	363168.1	680
681	463761	315821241	26.0960	8.7980	0.001468	2139.425	364237.0	681
682	465124	317214568	26.1151	8.8023	0.001466	2142.566	365307.5	682
683	466489	318611987	26.1343	8.8066	0.001464	2145.708	366379.6	683
684	467856	320013504	26.1534	8.8109	0.001462	2148.849	367453.2	684
685	469225	321419125	26.1725	8.8152	0.001460	2151.991	368528.5	685
686	470596	322828856	26.1916	8.8194	0.001458	2155.133	369605.2	686
687	471969	324242703	26.2107	8.8237	0.001456	2158.274	370683.6	687
688	473344	325660672	26.2298	8.8280	0.001453	2161.416	371763.5	688
689	474721	327082769	26.2488	8.8323	0.001451	2164.557	372845.0	689
690	476100	328509000	26.2679	8.8366	0.001449	2167.699	373928.1	690
691	477481	329939371	26.2869	8.8408	0.001447	2170.841	375012.7	691
692	478864	331373888	26.3059	8.8451	0.001445	2173.982	376098.9	692
693	480249	332812557	26.3249	8.8493	0.001443	2177.124	377186.7	693
694	481636	334255384	26.3439	8.8536	0.001441	2180.265	378276.0	694
695	483025	335702375	26.3629	8.8578	0.001439	2183.407	379366.9	695
696	484416	337153536	26.3818	8.8621	0.001437	2186.548	380459.4	696
697	485809	338608873	26.4008	8.8663	0.001435	2189.690	381553.5	697
698	487204	340068392	26.4197	8.8706	0.001433	2192.832	382649.1	698
699	488601	341532099	26.4386	8.8748	0.001431	2195.973	383746.3	699
700	490000	343000000	26.4575	8.8790	0.001429	2199.115	384845.1	700

Numbers, Squares. Cubes, Roots,etc.   700 - 750

n	n^2	n^3	$\sqrt{n}$	$\sqrt[3]{n}$	$1/n$	πn	$\pi n^2/4$	n
700	490000	343000000	26.4575	8.8790	0.001429	2199.115	384845.1	700
701	491401	344472101	26.4764	8.8833	0.001427	2202.256	385945.4	701
702	492804	345948408	26.4953	8.8875	0.001425	2205.398	387047.4	702
703	494209	347428927	26.5141	8.8917	0.001422	2208.540	388150.8	703
704	495616	348913664	26.5330	8.8959	0.001420	2211.681	389255.9	704
705	497025	350402625	26.5518	8.9001	0.001418	2214.823	390362.5	705
706	498436	351895816	26.5707	8.9043	0.001416	2217.964	391470.7	706
707	499849	353393243	26.5895	8.9085	0.001414	2221.106	392580.5	707
708	501264	354894912	26.6083	8.9127	0.001412	2224.248	393691.8	708
709	502681	356400829	26.6271	8.9169	0.001410	2227.389	394804.7	709
710	504100	357911000	26.6458	8.9211	0.001408	2230.531	395919.2	710
711	505521	359425431	26.6646	8.9253	0.001406	2233.672	397035.3	711
712	506944	360944128	26.6833	8.9295	0.001404	2236.814	398152.9	712
713	508369	362467097	26.7021	8.9337	0.001403	2239.956	399272.1	713
714	509796	363994344	26.7208	8.9378	0.001401	2243.097	400392.8	714
715	511225	365525875	26.7395	8.9420	0.001399	2246.239	401515.2	715
716	512656	367061696	26.7582	8.9462	0.001397	2249.380	402639.1	716
717	514089	368601813	26.7769	8.9503	0.001395	2252.522	403764.6	717
718	515524	370146232	26.7955	8.9545	0.001393	2255.664	404891.6	718
719	516961	371694959	26.8142	8.9587	0.001391	2258.805	406020.2	719
720	518400	373248000	26.8328	8.9628	0.001389	2261.947	407150.4	720
721	519841	374805361	26.8514	8.9670	0.001387	2265.088	408282.2	721
722	521284	376367048	26.8701	8.9711	0.001385	2268.230	409415.5	722
723	522729	377933067	26.8887	8.9752	0.001383	2271.371	410550.4	723
724	524176	379503424	26.9072	8.9794	0.001381	2274.513	411686.9	724
725	525625	381078125	26.9258	8.9835	0.001379	2277.655	412824.9	725
726	527076	382657176	26.9444	8.9876	0.001377	2280.796	413964.5	726
727	528529	384240583	26.9629	8.9918	0.001376	2283.938	415105.7	727
728	529984	385828352	26.9815	8.9959	0.001374	2287.079	416248.5	728
729	531441	387420489	27.0000	9.0000	0.001372	2290.221	417392.8	729
730	532900	389017000	27.0185	9.0041	0.001370	2293.363	418538.7	730
731	534361	390617891	27.0370	9.0082	0.001368	2296.504	419686.1	731
732	535824	392223168	27.0555	9.0123	0.001366	2299.646	420835.2	732
733	537289	393832837	27.0740	9.0164	0.001364	2302.787	421985.8	733
734	538756	395446904	27.0924	9.0205	0.001362	2305.929	423138.0	734
735	540225	397065375	27.1109	9.0246	0.001361	2309.071	424291.7	735
736	541696	398688256	27.1293	9.0287	0.001359	2312.212	425447.0	736
737	543169	400315553	27.1477	9.0328	0.001357	2315.354	426603.9	737
738	544644	401947272	27.1662	9.0369	0.001355	2318.495	427762.4	738
739	546121	403583419	27.1846	9.0410	0.001353	2321.637	428922.4	739
740	547600	405224000	27.2029	9.0450	0.001351	2324.779	430084.0	740
741	549081	406869021	27.2213	9.0491	0.001350	2327.920	431247.2	741
742	550564	408518488	27.2397	9.0532	0.001348	2331.062	432412.0	742
743	552049	410172407	27.2580	9.0572	0.001346	2334.203	433578.3	743
744	553536	411830784	27.2764	9.0613	0.001344	2337.345	434746.2	744
745	555025	413493625	27.2947	9.0654	0.001342	2340.487	435915.6	745
746	556516	415160936	27.3130	9.0694	0.001340	2343.628	437086.6	746
747	558009	416832723	27.3313	9.0735	0.001339	2346.770	438259.2	747
748	559504	418508992	27.3496	9.0775	0.001337	2349.911	439433.4	748
749	561001	420189749	27.3679	9.0816	0.001335	2353.053	440609.2	749
750	562500	421875000	27.3861	9.0856	0.001333	2356.194	441786.5	750

Numbers, Squares, Cubes, Roots, etc: 750 - 800

n	n^2	n^3	$\sqrt{n}$	$\sqrt[3]{n}$	$1/n$	πn	$\pi n^2/4$	n
750	562500	421875000	27.3861	9.0056	0.001333	2356.194	441786.5	750
751	564001	423564751	27.4044	9.0896	0.001332	2359.336	442965.3	751
752	565504	425259008	27.4226	9.0937	0.001330	2362.478	444145.8	752
753	567009	426957777	27.4408	9.0977	0.001328	2365.619	445327.8	753
754	568516	428661064	27.4591	9.1017	0.001326	2368.761	446511.4	754
755	570025	430368875	27.4773	9.1057	0.001325	2371.902	447696.6	755
756	571536	432081216	27.4955	9.1098	0.001323	2375.044	448883.3	756
757	573049	433798093	27.5136	9.1138	0.001321	2378.186	450071.6	757
758	574564	435519512	27.5318	9.1178	0.001319	2381.327	451261.5	758
759	576081	437245479	27.5500	9.1218	0.001318	2384.469	452453.0	759
760	577600	438976000	27.5681	9.1258	0.001316	2387.610	453646.0	760
761	579121	440711081	27.5862	9.1298	0.001314	2390.752	454840.6	761
762	580644	442450728	27.6043	9.1338	0.001312	2393.894	456036.7	762
763	582169	444194947	27.6225	9.1378	0.001311	2397.035	457234.5	763
764	583696	445943744	27.6405	9.1418	0.001309	2400.177	458433.8	764
765	585225	447697125	27.6586	9.1458	0.001307	2403.318	459634.6	765
766	586756	449455096	27.6767	9.1498	0.001305	2406.460	460837.1	766
767	588289	451217663	27.6948	9.1537	0.001304	2409.602	462041.1	767
768	589824	452984832	27.7128	9.1577	0.001302	2412.743	463246.7	768
769	591361	454756609	27.7308	9.1617	0.001300	2415.885	464453.8	769
770	592900	456533000	27.7489	9.1657	0.001299	2419.026	465662.6	770
771	594441	458314011	27.7669	9.1696	0.001297	2422.168	466872.9	771
772	595984	460099648	27.7849	9.1736	0.001295	2425.310	468084.7	772
773	597529	461889917	27.8029	9.1775	0.001294	2428.451	469298.2	773
774	599076	463684824	27.8209	9.1815	0.001292	2431.593	470513.2	774
775	600625	465484375	27.8388	9.1855	0.001290	2434.734	471729.8	775
776	602176	467288576	27.8568	9.1894	0.001289	2437.876	472947.9	776
777	603729	469097433	27.8747	9.1933	0.001287	2441.017	474167.6	777
778	605284	470910952	27.8927	9.1973	0.001285	2444.159	475388.9	778
779	606841	472729139	27.9106	9.2012	0.001284	2447.301	476611.8	779
780	608400	474552000	27.9285	9.2052	0.001282	2450.442	477836.2	780
781	609961	476379541	27.9464	9.2091	0.001280	2453.584	479062.2	781
782	611524	478211768	27.9643	9.2130	0.001279	2456.725	480289.8	782
783	613089	480048687	27.9821	9.2170	0.001277	2459.867	481519.0	783
784	614656	481890304	28.0000	9.2209	0.001276	2463.009	482749.7	784
785	616225	483736625	28.0179	9.2248	0.001274	2466.150	483982.0	785
786	617796	485587656	28.0357	9.2287	0.001272	2469.292	485215.8	786
787	619369	487443403	28.0535	9.2326	0.001271	2472.433	486451.3	787
788	620944	489303872	28.0713	9.2365	0.001269	2475.575	487688.3	788
789	622521	491169069	28.0891	9.2404	0.001267	2478.717	488926.9	789
790	624100	493039000	28.1069	9.2443	0.001266	2481.858	490167.0	790
791	625681	494913671	28.1247	9.2482	0.001264	2485.000	491408.7	791
792	627264	496793088	28.1425	9.2521	0.001263	2488.141	492652.0	792
793	628849	498677257	28.1603	9.2560	0.001261	2491.283	493896.8	793
794	630436	500566184	28.1780	9.2599	0.001259	2494.425	495143.3	794
795	632025	502459875	28.1957	9.2638	0.001258	2497.566	496391.3	795
796	633616	504358336	28.2135	9.2677	0.001256	2500.708	497640.8	796
797	635209	506261573	28.2312	9.2716	0.001255	2503.849	498892.0	797
798	636804	508169592	28.2489	9.2754	0.001253	2506.991	500144.7	798
799	638401	510082399	28.2666	9.2793	0.001252	2510.133	501399.0	799
800	640000	512000000	28.2843	9.2832	0.001250	2513.274	502654.8	800

Numbers, Squares, Cubes, Roots, etc.   800 - 850

n	n^2	n^3	$\sqrt{n}$	$\sqrt[3]{n}$	$1/n$	πn	$\pi n^2/4$	n
800	640000	512000000	28.2843	9.2832	0.001250	2513.274	502654.8	800
801	641601	513922401	28.3019	9.2870	0.001248	2516.416	503912.2	801
802	643204	515849608	28.3196	9.2909	0.001247	2519.557	505171.2	802
803	644809	517781627	28.3373	9.2948	0.001245	2522.699	506431.8	803
804	646416	519718464	28.3549	9.2986	0.001244	2525.840	507693.9	804
805	648025	521660125	28.3725	9.3025	0.001242	2528.982	508957.6	805
806	649636	523606616	28.3901	9.3063	0.001241	2532.124	510222.9	806
807	651249	525557943	28.4077	9.3102	0.001239	2535.265	511489.8	807
808	652864	527514112	28.4253	9.3140	0.001238	2538.407	512758.2	808
809	654481	529475129	28.4429	9.3179	0.001236	2541.548	514028.2	809
810	656100	531441000	28.4605	9.3217	0.001235	2544.690	515299.7	810
811	657721	533411731	28.4781	9.3255	0.001233	2547.832	516572.9	811
812	659344	535387328	28.4956	9.3294	0.001232	2550.973	517847.6	812
813	660969	537367797	28.5132	9.3332	0.001230	2554.115	519123.8	813
814	662596	539353144	28.5307	9.3370	0.001229	2557.256	520401.7	814
815	664225	541343375	28.5482	9.3408	0.001227	2560.398	521681.1	815
816	665856	543338496	28.5657	9.3447	0.001225	2563.540	522962.1	816
817	667489	545338513	28.5832	9.3485	0.001224	2566.681	524244.6	817
818	669124	547343432	28.6007	9.3523	0.001222	2569.823	525528.8	818
819	670761	549353259	28.6182	9.3561	0.001221	2572.964	526814.5	819
820	672400	551368000	28.6356	9.3599	0.001220	2576.106	528101.7	820
821	674041	553387661	28.6531	9.3637	0.001218	2579.248	529390.6	821
822	675684	555412248	28.6705	9.3675	0.001217	2582.389	530681.0	822
823	677329	557441767	28.6880	9.3713	0.001215	2585.531	531973.0	823
824	678976	559476224	28.7054	9.3751	0.001214	2588.672	533266.5	824
825	680625	561515625	28.7228	9.3789	0.001212	2591.814	534561.6	825
826	682276	563559976	28.7402	9.3827	0.001211	2594.956	535858.3	826
827	683929	565609283	28.7576	9.3865	0.001209	2598.097	537156.6	827
828	685584	567663552	28.7750	9.3902	0.001208	2601.239	538456.4	828
829	687241	569722789	28.7924	9.3940	0.001206	2604.380	539757.8	829
830	688900	571787000	28.8097	9.3978	0.001205	2607.522	541060.8	830
831	690561	573856191	28.8271	9.4016	0.001203	2610.663	542365.3	831
832	692224	575930368	28.8444	9.4053	0.001202	2613.805	543671.5	832
833	693889	578009537	28.8617	9.4091	0.001200	2616.947	544979.1	833
834	695556	580093704	28.8791	9.4129	0.001199	2620.088	546288.4	834
835	697225	582182875	28.8964	9.4166	0.001198	2623.230	547599.2	835
836	698896	584277056	28.9137	9.4204	0.001196	2626.371	548911.6	836
837	700569	586376253	28.9310	9.4241	0.001195	2629.513	550225.6	837
838	702244	588480472	28.9482	9.4279	0.001193	2632.655	551541.1	838
839	703921	590589719	28.9655	9.4316	0.001192	2635.796	552858.3	839
840	705600	592704000	28.9828	9.4354	0.001190	2638.938	554176.9	840
841	707281	594823321	29.0000	9.4391	0.001189	2642.079	555497.2	841
842	708964	596947688	29.0172	9.4429	0.001188	2645.221	556819.0	842
843	710649	599077107	29.0345	9.4466	0.001186	2648.363	558142.4	843
844	712336	601211584	29.0517	9.4503	0.001185	2651.504	559467.4	844
845	714025	603351125	29.0689	9.4541	0.001183	2654.646	560793.9	845
846	715716	605495736	29.0861	9.4578	0.001182	2657.787	562122.0	846
847	717409	607645423	29.1033	9.4615	0.001181	2660.929	563451.7	847
848	719104	609800192	29.1204	9.4652	0.001179	2664.071	564783.0	848
849	720801	611960049	29.1376	9.4690	0.001178	2667.212	566115.8	849
850	722500	614125000	29.1548	9.4727	0.001176	2670.354	567450.2	850

MATHEMATICS

Numbers, Squares, Cubes, Roots, etc.   850 - 900

n	n^2	n^3	$\sqrt{n}$	$\sqrt[3]{n}$	$1/n$	πn	$\pi n^2/4$	n
850	722500	614125000	29.1548	9.4727	0.001176	2670.354	567450.2	850
851	724201	616295051	29.1719	9.4764	0.001175	2673.495	568786.1	851
852	725904	618470208	29.1890	9.4801	0.001174	2676.637	570123.7	852
853	727609	620650477	29.2062	9.4838	0.001172	2679.779	571462.8	853
854	729316	622835864	29.2233	9.4875	0.001171	2682.920	572803.4	854
855	731025	625026375	29.2404	9.4912	0.001170	2686.062	574145.7	855
856	732736	627222016	29.2575	9.4949	0.001168	2689.203	575489.5	856
857	734449	629422793	29.2746	9.4986	0.001167	2692.345	576834.9	857
858	736164	631628712	29.2916	9.5023	0.001166	2695.486	578181.9	858
859	737881	633839779	29.3087	9.5060	0.001164	2698.628	579530.4	859
860	739600	636056000	29.3258	9.5097	0.001163	2701.770	580880.5	860
861	741321	638277381	29.3428	9.5134	0.001161	2704.911	582232.2	861
862	743044	640503928	29.3598	9.5171	0.001160	2708.053	583585.4	862
863	744769	642735647	29.3769	9.5207	0.001159	2711.194	584940.2	863
864	746496	644972544	29.3939	9.5244	0.001157	2714.336	586296.6	864
865	748225	647214625	29.4109	9.5281	0.001156	2717.478	587654.5	865
866	749956	649461896	29.4279	9.5317	0.001155	2720.619	589014.1	866
867	751689	651714363	29.4449	9.5354	0.001153	2723.761	590375.2	867
868	753424	653972032	29.4618	9.5391	0.001152	2726.902	591737.8	868
869	755161	656234909	29.4788	9.5427	0.001151	2730.044	593102.1	869
870	756900	658503000	29.4958	9.5464	0.001149	2733.186	594467.9	870
871	758641	660776311	29.5127	9.5501	0.001148	2736.327	595835.2	871
872	760384	663054848	29.5296	9.5537	0.001147	2739.469	597204.2	872
873	762129	665338617	29.5466	9.5574	0.001145	2742.610	598574.7	873
874	763876	667627624	29.5635	9.5610	0.001144	2745.752	599946.8	874
875	765625	669921875	29.5804	9.5647	0.001143	2748.894	601320.5	875
876	767376	672221376	29.5973	9.5683	0.001142	2752.035	602695.7	876
877	769129	674526133	29.6142	9.5719	0.001140	2755.177	604072.5	877
878	770884	676836152	29.6311	9.5756	0.001139	2758.318	605450.9	878
879	772641	679151439	29.6479	9.5792	0.001138	2761.460	606830.8	879
880	774400	681472000	29.6648	9.5828	0.001136	2764.602	608212.3	880
881	776161	683797841	29.6816	9.5865	0.001135	2767.743	609595.4	881
882	777924	686128968	29.6985	9.5901	0.001134	2770.885	610980.1	882
883	779689	688465387	29.7153	9.5937	0.001133	2774.026	612366.3	883
884	781456	690807104	29.7321	9.5973	0.001131	2777.168	613754.1	884
885	783225	693154125	29.7489	9.6010	0.001130	2780.309	615143.5	885
886	784996	695506456	29.7658	9.6046	0.001129	2783.451	616534.4	886
887	786769	697864103	29.7825	9.6082	0.001127	2786.593	617926.9	887
888	788544	700227072	29.7993	9.6118	0.001126	2789.734	619321.0	888
889	790321	702595369	29.8161	9.6154	0.001125	2792.876	620716.7	889
890	792100	704969000	29.8329	9.6190	0.001124	2796.017	622113.9	890
891	793881	707347971	29.8496	9.6226	0.001122	2799.159	623512.7	891
892	795664	709732288	29.8664	9.6262	0.001121	2802.301	624913.0	892
893	797449	712121957	29.8831	9.6298	0.001120	2805.442	626315.0	893
894	799236	714516984	29.8998	9.6334	0.001119	2808.584	627718.5	894
895	801025	716917375	29.9166	9.6370	0.001117	2811.725	629123.6	895
896	802816	719323136	29.9333	9.6406	0.001116	2814.867	630530.2	896
897	804609	721734273	29.9500	9.6442	0.001115	2818.009	631938.4	897
898	806404	724150792	29.9666	9.6477	0.001114	2821.150	633348.2	898
899	808201	726572699	29.9833	9.6513	0.001112	2824.292	634759.6	899
900	810000	729000000	30.0000	9.6549	0.001111	2827.433	636172.5	900

Numbers, Squares, Cubes, Roots, etc.   900 - 950

n	n²	n³	√n	³√n	1/n	πn	πn²/4	n
900	810000	729000000	30.0000	9.6549	0.001111	2827.433	636172.5	900
901	811801	731432701	30.0167	9.6585	0.001110	2830.575	637587.0	901
902	813604	733870808	30.0333	9.6620	0.001109	2833.717	639003.1	902
903	815409	736314327	30.0500	9.6656	0.001107	2836.858	640420.7	903
904	817216	738763264	30.0666	9.6692	0.001106	2840.000	641839.9	904
905	819025	741217625	30.0832	9.6727	0.001105	2843.141	643260.7	905
906	820836	743677416	30.0998	9.6763	0.001104	2846.283	644683.1	906
907	822649	746142643	30.1164	9.6799	0.001103	2849.425	646107.0	907
908	824464	748613312	30.1330	9.6834	0.001101	2852.566	647532.5	908
909	826281	751089429	30.1496	9.6870	0.001100	2855.708	648959.6	909
910	828100	753571000	30.1662	9.6905	0.001099	2858.849	650388.2	910
911	829921	756058031	30.1828	9.6941	0.001098	2861.991	651818.4	911
912	831744	758550528	30.1993	9.6976	0.001096	2865.133	653250.2	912
913	833569	761048497	30.2159	9.7012	0.001095	2868.274	654683.6	913
914	835396	763551944	30.2324	9.7047	0.001094	2871.416	656118.5	914
915	837225	766060875	30.2490	9.7082	0.001093	2874.557	657555.0	915
916	839056	768575296	30.2655	9.7118	0.001092	2877.699	658993.0	916
917	840889	771095213	30.2820	9.7153	0.001091	2880.840	660432.7	917
918	842724	773620632	30.2985	9.7188	0.001089	2883.982	661873.9	918
919	844561	776151559	30.3150	9.7224	0.001088	2887.124	663316.7	919
920	846400	778688000	30.3315	9.7259	0.001087	2890.265	664761.0	920
921	848241	781229961	30.3480	9.7294	0.001086	2893.407	666206.9	921
922	850084	783777448	30.3645	9.7329	0.001085	2896.548	667654.4	922
923	851929	786330467	30.3809	9.7364	0.001083	2899.690	669103.5	923
924	853776	788889024	30.3974	9.7400	0.001082	2902.832	670554.1	924
925	855625	791453125	30.4138	9.7435	0.001081	2905.973	672006.3	925
926	857476	794022776	30.4302	9.7470	0.001080	2909.115	673460.1	926
927	859329	796597983	30.4467	9.7505	0.001079	2912.256	674915.4	927
928	861184	799178752	30.4631	9.7540	0.001078	2915.398	676372.3	928
929	863041	801765089	30.4795	9.7575	0.001076	2918.540	677830.8	929
930	864900	804357000	30.4959	9.7610	0.001075	2921.681	679290.9	930
931	866761	806954491	30.5123	9.7645	0.001074	2924.823	680752.5	931
932	868624	809557568	30.5287	9.7680	0.001073	2927.964	682215.7	932
933	870489	812166237	30.5450	9.7715	0.001072	2931.106	683680.5	933
934	872356	814780504	30.5614	9.7750	0.001071	2934.248	685146.8	934
935	874225	817400375	30.5778	9.7785	0.001070	2937.389	686614.7	935
936	876096	820025856	30.5941	9.7819	0.001068	2940.531	688084.2	936
937	877969	822656953	30.6105	9.7854	0.001067	2943.672	689555.2	937
938	879844	825293672	30.6268	9.7889	0.001066	2946.814	691027.9	938
939	881721	827936019	30.6431	9.7924	0.001065	2949.956	692502.1	939
940	883600	830584000	30.6594	9.7959	0.001064	2953.097	693977.8	940
941	885481	833237621	30.6757	9.7993	0.001063	2956.239	695455.2	941
942	887364	835896888	30.6920	9.8028	0.001062	2959.380	696934.1	942
943	889249	838561807	30.7083	9.8063	0.001060	2962.522	698414.5	943
944	891136	841232384	30.7246	9.8097	0.001059	2965.663	699896.6	944
945	893025	843908625	30.7409	9.8132	0.001058	2968.805	701380.2	945
946	894916	846590536	30.7571	9.8167	0.001057	2971.947	702865.4	946
947	896809	849278123	30.7734	9.8201	0.001056	2975.088	704352.1	947
948	898704	851971392	30.7896	9.8236	0.001055	2978.230	705840.5	948
949	900601	854670349	30.8058	9.8270	0.001054	2981.371	707330.4	949
950	902500	857375000	30.8221	9.8305	0.001053	2984.513	708821.8	950

MATHEMATICS

Numbers, Squares, Cubes, Roots, etc.   950 - 1000

n	n^2	n^3	$\sqrt{n}$	$\sqrt[3]{n}$	$1/n$	πn	$\pi n^2/4$	n
950	902500	857375000	30.8221	9.8305	0.001053	2984.513	708821.8	950
951	904401	860085351	30.8383	9.8339	0.001052	2987.655	710314.9	951
952	906304	862801408	30.8545	9.8374	0.001050	2990.796	711809.5	952
953	908209	865523177	30.8707	9.8408	0.001049	2993.938	713305.7	953
954	910116	868250664	30.8869	9.8443	0.001048	2997.079	714803.4	954
955	912025	870983875	30.9031	9.8477	0.001047	3000.221	716302.8	955
956	913936	873722816	30.9192	9.8511	0.001046	3003.363	717803.7	956
957	915849	876467493	30.9354	9.8546	0.001045	3006.504	719306.1	957
958	917764	879217912	30.9516	9.8580	0.001044	3009.646	720810.2	958
959	919681	881974079	30.9677	9.8614	0.001043	3012.787	722315.8	959
960	921600	884736000	30.9839	9.8648	0.001042	3015.929	723822.9	960
961	923521	887503681	31.0000	9.8683	0.001041	3019.071	725331.7	961
962	925444	890277128	31.0161	9.8717	0.001040	3022.212	726842.0	962
963	927369	893056347	31.0322	9.8751	0.001038	3025.354	728353.9	963
964	929296	895841344	31.0483	9.8785	0.001037	3028.495	729867.4	964
965	931225	898632125	31.0644	9.8819	0.001036	3031.637	731382.4	965
966	933156	901428696	31.0805	9.8854	0.001035	3034.779	732899.0	966
967	935089	904231063	31.0966	9.8888	0.001034	3037.920	734417.2	967
968	937024	907039232	31.1127	9.8922	0.001033	3041.062	735936.9	968
969	938961	909853209	31.1288	9.8956	0.001032	3044.203	737458.2	969
970	940900	912673000	31.1448	9.8990	0.001031	3047.345	738981.1	970
971	942841	915498611	31.1609	9.9024	0.001030	3050.486	740505.6	971
972	944784	918330048	31.1769	9.9058	0.001029	3053.628	742031.6	972
973	946729	921167317	31.1929	9.9092	0.001028	3056.770	743559.2	973
974	948676	924010424	31.2090	9.9126	0.001027	3059.911	745088.4	974
975	950625	926859375	31.2250	9.9160	0.001026	3063.053	746619.1	975
976	952576	929714176	31.2410	9.9194	0.001025	3066.194	748151.4	976
977	954529	932574833	31.2570	9.9227	0.001024	3069.336	749685.3	977
978	956484	935441352	31.2730	9.9261	0.001022	3072.478	751220.8	978
979	958441	938313739	31.2890	9.9295	0.001021	3075.619	752757.8	979
980	960400	941192000	31.3050	9.9329	0.001020	3078.761	754296.4	980
981	962361	944076141	31.3209	9.9363	0.001019	3081.902	755836.6	981
982	964324	946966168	31.3369	9.9396	0.001018	3085.044	757378.3	982
983	966289	949862087	31.3528	9.9430	0.001017	3088.186	758921.6	983
984	968256	952763904	31.3688	9.9464	0.001016	3091.327	760466.5	984
985	970225	955671625	31.3847	9.9497	0.001015	3094.469	762012.9	985
986	972196	958585256	31.4006	9.9531	0.001014	3097.610	763561.0	986
987	974169	961504803	31.4166	9.9565	0.001013	3100.752	765110.5	987
988	976144	964430272	31.4325	9.9598	0.001012	3103.894	766661.7	988
989	978121	967361669	31.4484	9.9632	0.001011	3107.035	768214.4	989
990	980100	970299000	31.4643	9.9666	0.001010	3110.177	769768.7	990
991	982081	973242271	31.4802	9.9699	0.001009	3113.318	771324.6	991
992	984064	976191488	31.4960	9.9733	0.001008	3116.460	772882.1	992
993	986049	979146657	31.5119	9.9766	0.001007	3119.602	774441.1	993
994	988036	982107784	31.5278	9.9800	0.001006	3122.743	776001.7	994
995	990025	985074875	31.5436	9.9833	0.001005	3125.885	777563.8	995
996	992016	988047936	31.5595	9.9866	0.001004	3129.026	779127.5	996
997	994009	991026973	31.5753	9.9900	0.001003	3132.168	780692.8	997
998	996004	994011992	31.5911	9.9933	0.001002	3135.309	782259.7	998
999	998001	997002999	31.6070	9.9967	0.001001	3138.451	783828.2	999
1000	1000000	1000000000	31.6228	10.0000	0.001000	3141.593	785398.2	1000

Table of Logarithms 100 - 150

N.	L.	0	1	2	3	4	5	6	7	8	9
100	00 000	043	087	130	173	217	260	303	346	389
101	432	475	518	561	604	647	689	732	775	817
102	860	903	945	988	*030	*072	*115	*157	*199	*242
103	01 284	326	368	410	452	494	536	578	620	662
104	703	745	787	828	870	912	953	995	*036	*078
105	02 119	160	202	243	284	325	366	407	449	490
106	531	572	612	653	694	735	776	816	857	898
107	938	979	*019	*060	*100	*141	*181	*222	*262	*302
108	03 342	383	423	463	503	543	583	623	663	703
109	743	782	822	862	902	941	981	*021	*060	*100
110	04 139	179	218	258	297	336	376	415	454	493
111	532	571	610	650	689	727	766	805	844	883
112	922	961	999	*038	*077	*115	*154	*192	*231	*269
113	05 308	346	385	423	461	500	538	576	614	652
114	690	729	767	805	843	881	918	956	994	*032
115	06 070	108	145	183	221	258	296	333	371	408
116	446	483	521	558	595	633	670	707	744	781
117	819	856	893	930	967	*004	*041	*078	*115	*151
118	07 188	225	262	298	335	372	408	445	482	518
119	555	591	628	664	700	737	773	809	846	882
120	918	954	990	*027	*063	*099	*135	*171	*207	*243
121	08 279	314	350	386	422	458	493	529	565	600
122	636	672	707	743	778	814	849	884	920	955
123	991	*026	*061	*096	*132	*167	*202	*237	*272	*307
124	09 342	377	412	447	482	517	552	587	621	656
125	691	726	760	795	830	864	899	934	968	*003
126	10 037	072	106	140	175	209	243	278	312	346
127	380	415	449	483	517	551	585	619	653	687
128	721	755	789	823	857	890	924	958	992	*025
129	11 059	093	126	160	193	227	261	294	327	361
130	394	428	461	494	528	561	594	628	661	694
131	727	760	793	826	860	893	926	959	992	*024
132	12 057	090	123	156	189	222	254	287	320	352
133	385	418	450	483	516	548	581	613	646	678
134	710	743	775	808	840	872	905	937	969	*001
135	13 033	066	098	130	162	194	226	258	290	322
136	354	386	418	450	481	513	545	577	609	640
137	672	704	735	767	799	830	862	893	925	956
138	988	*019	*051	*082	*114	*145	*176	*208	*239	*270
139	14 301	333	364	395	426	457	489	520	551	582
140	613	644	675	706	737	768	799	829	860	891
141	922	953	983	*014	*045	*076	*106	*137	*168	*198
142	15 229	259	290	320	351	381	412	442	473	503
143	534	564	594	625	655	685	715	746	776	806
144	836	866	897	927	957	987	*017	*047	*077	*107
145	16 137	167	197	227	256	286	316	346	376	406
146	435	465	495	524	554	584	613	643	673	702
147	732	761	791	820	850	879	909	938	967	997
148	17 026	056	085	114	143	173	202	231	260	289
149	319	348	377	406	435	464	493	522	551	580
150	609	638	667	696	725	754	782	811	840	869

P. P.

	44	43	42
1	4.4	4.3	4.2
2	8.8	8.6	8.4
3	13.2	12.9	12.6
4	17.6	17.2	16.8
5	22.0	21.5	21.0
6	26.4	25.8	25.2
7	30.8	30.1	29.4
8	35.2	34.4	33.6
9	39.6	38.7	37.8

	41	40	39
1	4.1	4.0	3.9
2	8.2	8.0	7.8
3	12.3	12.0	11.7
4	16.4	16.0	15.6
5	20.5	20.0	19.5
6	24.6	24.0	23.4
7	28.7	28.0	27.3
8	32.8	32.0	31.2
9	36.9	36.0	35.1

	38	37	36
1	3.8	3.7	3.6
2	7.6	7.4	7.2
3	11.4	11.1	10.8
4	15.2	14.8	14.4
5	19.0	18.5	18.0
6	22.8	22.2	21.6
7	26.6	25.9	25.2
8	30.4	29.6	28.8
9	34.2	33.3	32.4

	35	34	33
1	3.5	3.4	3.3
2	7.0	6.8	6.6
3	10.5	10.2	9.9
4	14.0	13.6	13.2
5	17.5	17.0	16.5
6	21.0	20.4	19.8
7	24.5	23.8	23.1
8	28.0	27.2	26.4
9	31.5	30.6	29.7

	32	31	30
1	3.2	3.1	3.0
2	6.4	6.2	6.0
3	9.6	9.3	9.0
4	12.8	12.4	12.0
5	16.0	15.5	15.0
6	19.2	18.6	18.0
7	22.4	21.7	21.0
8	25.6	24.8	24.0
9	28.8	27.9	27.0

Table of Logarithms 150 - 200

N.	L. 0	1	2	3	4	5	6	7	8	9
150	17 609	638	667	696	725	754	782	811	840	869
151	898	926	955	984	*013	*041	*070	*099	*127	*156
152	18 184	213	241	270	298	327	355	384	412	441
153	469	498	526	554	583	611	639	667	696	724
154	752	780	808	837	865	893	921	949	977	*005
155	19 033	061	089	117	145	173	201	229	257	285
156	312	340	368	396	424	451	479	507	535	562
157	590	618	645	673	700	728	756	783	811	838
158	866	893	921	948	976	*003	*030	*058	*085	*112
159	20 140	167	194	222	249	276	303	330	358	385
160	412	439	466	493	520	548	575	602	629	656
161	683	710	737	763	790	817	844	871	898	925
162	952	978	*005	*032	*059	*085	*112	*139	*165	*192
163	21 219	245	272	299	325	352	378	405	431	458
164	484	511	537	564	590	617	643	669	696	722
165	748	775	801	827	854	880	906	932	958	985
166	22 011	037	063	089	115	141	167	194	220	246
167	272	298	324	350	376	401	427	453	479	505
168	531	557	583	608	634	660	686	712	737	763
169	789	814	840	866	891	917	943	968	994	*019
170	23 045	070	096	121	147	172	198	223	249	274
171	300	325	350	376	401	426	452	477	502	528
172	553	578	603	629	654	679	704	729	754	779
173	805	830	855	880	905	930	955	980	*005	*030
174	24 055	080	105	130	155	180	204	229	254	279
175	304	329	353	378	403	428	452	477	502	527
176	551	576	601	625	650	674	699	724	748	773
177	797	822	846	871	895	920	944	969	993	*018
178	25 042	066	091	115	139	164	188	212	237	261
179	285	310	334	358	382	406	431	455	479	503
180	527	551	575	600	624	648	672	696	720	744
181	768	792	816	840	864	888	912	935	959	983
182	26 007	031	055	079	102	126	150	174	198	221
183	245	269	293	316	340	364	387	411	435	458
184	482	505	529	553	576	600	623	647	670	694
185	717	741	764	788	811	834	858	881	905	928
186	951	975	998	*021	*045	*068	*091	*114	*138	*161
187	27 184	207	231	254	277	300	323	346	370	393
188	416	439	462	485	508	531	554	577	600	623
189	646	669	692	715	738	761	784	807	830	852
190	875	898	921	944	967	989	*012	*035	*058	*081
191	28 103	126	149	171	194	217	240	262	285	307
192	330	353	375	398	421	443	466	488	511	533
193	556	578	601	623	646	668	691	713	735	758
194	780	803	825	847	870	892	914	937	959	981
195	29 003	026	048	070	092	115	137	159	181	203
196	226	248	270	292	314	336	358	380	403	425
197	447	469	491	513	535	557	579	601	623	645
198	667	688	710	732	754	776	798	820	842	863
199	885	907	929	951	973	994	*016	*038	*060	*081
200	30 103	125	146	168	190	211	233	255	276	298

P. P.

	29	28
1	2.9	2.8
2	5.8	5.6
3	8.7	8.4
4	11.6	11.2
5	14.5	14.0
6	17.4	16.8
7	20.3	19.6
8	23.2	22.4
9	26.1	25.2

	27	26
1	2.7	2.6
2	5.4	5.2
3	8.1	7.8
4	10.8	10.4
5	13.5	13.0
6	16.2	15.6
7	18.9	18.2
8	21.6	20.8
9	24.3	23.4

	25
1	2.5
2	5.0
3	7.5
4	10.0
5	12.5
6	15.0
7	17.5
8	20.0
9	22.5

	24	23
1	2.4	2.3
2	4.8	4.6
3	7.2	6.9
4	9.6	9.2
5	12.0	11.5
6	14.4	13.8
7	16.8	16.1
8	19.2	18.4
9	21.6	20.7

	22	21
1	2.2	2.1
2	4.4	4.2
3	6.6	6.3
4	8.8	8.4
5	11.0	10.5
6	13.2	12.6
7	15.4	14.7
8	17.6	16.8
9	19.8	18.9

Table of Logarithms 200 - 250

N.	L.	0	1	2	3	4	5	6	7	8	9
200	30	103	125	146	168	190	211	233	255	276	298
201		320	341	363	384	406	428	449	471	492	514
202		535	557	578	600	621	643	664	685	707	728
203		750	771	792	814	835	856	878	899	920	942
204		963	984	*006	*027	*048	*069	*091	*112	*133	*154
205	31	175	197	218	239	260	281	302	323	345	366
206		387	408	429	450	471	492	513	534	555	576
207		597	618	639	660	681	702	723	744	765	785
208		806	827	848	869	890	911	931	952	973	994
209	32	015	035	056	077	098	118	139	160	181	201
210		222	243	263	284	305	325	346	366	387	408
211		428	449	469	490	510	531	552	572	593	613
212		634	654	675	695	715	736	756	777	797	818
213		838	858	879	899	919	940	960	980	*001	*021
214	33	041	062	082	102	122	143	163	183	203	224
215		244	264	284	304	325	345	365	385	405	425
216		445	465	486	506	526	546	566	586	606	626
217		646	666	686	706	726	746	766	786	806	826
218		846	866	885	905	925	945	965	985	*005	*025
219	34	044	064	084	104	124	143	163	183	203	223
220		242	262	282	301	321	341	361	380	400	420
221		439	459	479	498	518	537	557	577	596	616
222		635	655	674	694	713	733	753	772	792	811
223		830	850	869	889	908	928	947	967	986	*005
224	35	025	044	064	083	102	122	141	160	180	199
225		218	238	257	276	295	315	334	353	372	392
226		411	430	449	468	488	507	526	545	564	583
227		603	622	641	660	679	698	717	736	755	774
228		793	813	832	851	870	889	908	927	946	965
229		984	*003	*021	*040	*059	*078	*097	*116	*135	*154
230	36	173	192	211	229	248	267	286	305	324	342
231		361	380	399	418	436	455	474	493	511	530
232		549	568	586	605	624	642	661	680	698	717
233		736	754	773	791	810	829	847	866	884	903
234		922	940	959	977	996	*014	*033	*051	*070	*088
235	37	107	125	144	162	181	199	218	236	254	273
236		291	310	328	346	365	383	401	420	438	457
237		475	493	511	530	548	566	585	603	621	639
238		658	676	694	712	731	749	767	785	803	822
239		840	858	876	894	912	931	949	967	985	*003
240	38	021	039	057	075	093	112	130	148	166	184
241		202	220	238	256	274	292	310	328	346	364
242		382	399	417	435	453	471	489	507	525	543
243		561	578	596	614	632	650	668	686	703	721
244		739	757	775	792	810	828	846	863	881	899
245		917	934	952	970	987	*005	*023	*041	*058	*076
246	39	094	111	129	146	164	182	199	217	235	252
247		270	287	305	322	340	358	375	393	410	428
248		445	463	480	498	515	533	550	568	585	602
249		620	637	655	672	690	707	724	742	759	777
250		794	811	829	846	863	881	898	915	933	950

P. P.

	22	21
1	2.2	2.1
2	4.4	4.2
3	6.6	6.3
4	8.8	8.4
5	11.0	10.5
6	13.2	12.6
7	15.4	14.7
8	17.6	16.8
9	19.8	18.9

	20
1	2.0
2	4.0
3	6.0
4	8.0
5	10.0
6	12.0
7	14.0
8	16.0
9	18.0

	19
1	1.9
2	3.8
3	5.7
4	7.6
5	9.5
6	11.4
7	13.3
8	15.2
9	17.1

	18
1	1.8
2	3.6
3	5.4
4	7.2
5	9.0
6	10.8
7	12.6
8	14.4
9	16.2

	17
1	1.7
2	3.4
3	5.1
4	6.8
5	8.5
6	10.2
7	11.9
8	13.6
9	15.3

Table of Logarithms 250 - 300

N.	L. 0	1	2	3	4	5	6	7	8	9
250	39 794	811	829	846	863	881	898	915	933	950
251	967	985	*002	*019	*037	*054	*071	*088	*106	*123
252	40 140	157	175	192	209	226	243	261	278	295
253	312	329	346	364	381	398	415	432	449	466
254	483	500	518	535	552	569	586	603	620	637
255	654	671	688	705	722	739	756	773	790	807
256	824	841	858	875	892	909	926	943	960	976
257	993	*010	*027	*044	*061	*078	*095	*111	*128	*145
258	41 162	179	196	212	229	246	263	280	296	313
259	330	347	363	380	397	414	430	447	464	481
260	497	514	531	547	564	581	597	614	631	647
261	664	681	697	714	731	747	764	780	797	814
262	830	847	863	880	896	913	929	946	963	979
263	996	*012	*029	*045	*062	*078	*095	*111	*127	*144
264	42 160	177	193	210	226	243	259	275	292	308
265	325	341	357	374	390	406	423	439	455	472
266	488	504	521	537	553	570	586	602	619	635
267	651	667	684	700	716	732	749	765	781	797
268	813	830	846	862	878	894	911	927	943	959
269	975	991	*008	*024	*040	*056	*072	*088	*104	*120
270	43 136	152	169	185	201	217	233	249	265	281
271	297	313	329	345	361	377	393	409	425	441
272	457	473	489	505	521	537	553	569	584	600
273	616	632	648	664	680	696	712	727	743	759
274	775	791	807	823	838	854	870	886	902	917
275	933	949	965	981	996	*012	*028	*044	*059	*075
276	44 091	107	122	138	154	170	185	201	217	232
277	248	264	279	295	311	326	342	358	373	389
278	404	420	436	451	467	483	498	514	529	545
279	560	576	592	607	623	638	654	669	685	700
280	716	731	747	762	778	793	809	824	840	855
281	871	836	902	917	932	948	963	979	994	*010
282	45 025	040	056	071	086	102	117	133	148	163
283	179	194	209	225	240	255	271	286	301	317
284	332	347	362	378	393	408	423	439	454	469
285	484	500	515	530	545	561	576	591	606	621
286	637	652	667	682	697	712	728	743	758	773
287	788	803	818	834	849	864	879	894	909	924
288	939	954	969	984	*000	*015	*030	*045	*060	*075
289	46 090	105	120	135	150	165	180	195	210	225
290	240	255	270	285	300	315	330	345	359	374
291	389	404	419	434	449	464	479	494	509	523
292	538	553	568	583	598	613	627	642	657	672
293	687	702	716	731	746	761	776	790	805	820
294	835	850	864	879	894	909	923	938	953	967
295	982	997	*012	*026	*041	*056	*070	*085	*100	*114
296	47 129	144	159	173	188	202	217	232	246	261
297	276	290	305	319	334	349	363	378	392	407
298	422	436	451	465	480	494	509	524	538	553
299	567	582	596	611	625	640	654	669	683	698
300	712	727	741	756	770	784	799	813	828	842

P. P.

18		17		16		15		14	
1	1.8	1	1.7	1	1.6	1	1.5	1	1.4
2	3.6	2	3.4	2	3.2	2	3.0	2	2.8
3	5.4	3	.5.1	3	4.8	3	4.5	3	4.2
4	7.2	4	6.8	4	6.4	4	6.0	4	5.6
5	9.0	5	8.5	5	8.0	5	7.5	5	7.0
6	10.8	6	10.2	6	9.6	6	9.0	6	8.4
7	12.6	7	11.9	7	11.2	7	10.5	7	9.8
8	14.4	8	13.6	8	12.8	8	12.0	8	11.2
9	16.2	9	15.3	9	14.4	9	13.5	9	12.6

Table of Logarithms 300 - 350

N.	L. 0	1	2	3	4	5	6	7	8	9
300	47 712	727	741	756	770	784	799	813	828	842
301	857	871	885	900	914	929	943	958	972	986
302	48 001	015	029	044	058	073	087	101	116	130
303	144	159	173	187	202	216	230	244	259	273
304	287	302	316	330	344	359	373	387	401	416
305	430	444	458	473	487	501	515	530	544	558
306	572	586	601	615	629	643	657	671	686	700
307	714	728	742	756	770	785	799	813	827	841
308	855	869	883	897	911	926	940	954	968	982
309	996	*010	*024	*038	*052	*066	*080	*094	*108	*122
310	49 136	150	164	178	192	206	220	234	248	262
311	276	290	304	318	332	346	360	374	388	402
312	415	429	443	457	471	485	499	513	527	541
313	554	568	582	596	610	624	638	651	665	679
314	693	707	721	734	748	762	776	790	803	817
315	831	845	859	872	886	900	914	927	941	955
316	969	982	996	*010	*024	*037	*051	*065	*079	*092
317	50 106	120	133	147	161	174	188	202	215	229
318	243	256	270	284	297	311	325	338	352	365
319	379	393	406	420	433	447	461	474	488	501
320	515	529	542	556	569	583	596	610	623	637
321	651	664	678	691	705	718	732	745	759	772
322	786	799	813	826	840	853	866	880	893	907
323	920	934	947	961	974	987	*001	*014	*028	*041
324	51 055	068	081	095	108	121	135	148	162	175
325	188	202	215	228	242	255	268	282	295	308
326	322	335	348	362	375	388	402	415	428	441
327	455	468	481	495	508	521	534	548	561	574
328	587	601	614	627	640	654	667	680	693	706
329	720	733	746	759	772	786	799	812	825	838
330	851	865	878	891	904	917	930	943	957	970
331	983	996	*009	*022	*035	*048	*061	*075	*088	*101
332	52 114	127	140	153	166	179	192	205	218	231
333	244	257	270	284	297	310	323	336	349	362
334	375	388	401	414	427	440	453	466	479	492
335	504	517	530	543	556	569	582	595	608	621
336	634	647	660	673	686	699	711	724	737	750
337	763	776	789	802	815	827	840	853	866	879
338	892	905	917	930	943	956	969	982	994	*007
339	53 020	033	046	058	071	084	097	110	122	135
340	148	161	173	186	199	212	224	237	250	263
341	275	288	301	314	326	339	352	364	377	390
342	403	415	428	441	453	466	479	491	504	517
343	529	542	555	567	580	593	605	618	631	643
344	656	668	681	694	706	719	732	744	757	769
345	782	794	807	820	832	845	857	870	882	895
346	908	920	933	945	958	970	983	995	*008	*020
347	54 033	045	058	070	083	095	108	120	133	145
348	158	170	183	195	208	220	233	245	258	270
349	283	295	307	320	332	345	357	370	382	394
350	407	419	432	444	456	469	481	494	506	518

P. P.

15
1	1.5
2	3.0
3	4.5
4	6.0
5	7.5
6	9.0
7	10.5
8	12.0
9	13.5

14
1	1.4
2	2.8
3	4.2
4	5.6
5	7.0
6	8.4
7	9.8
8	11.2
9	12.6

13
1	1.3
2	2.6
3	3.9
4	5.2
5	6.5
6	7.8
7	9.1
8	10.4
9	11.7

12
1	1.2
2	2.4
3	3.6
4	4.8
5	6.0
6	7.2
7	8.4
8	9.6
9	10.8

Table of Logarithms 350 - 400

N.	L. 0	1	2	3	4	5	6	7	8	9
350	54 407	419	432	444	456	469	481	494	506	518
351	531	543	555	568	580	593	605	617	630	642
352	654	667	679	691	704	716	728	741	753	765
353	777	790	802	814	827	839	851	864	876	888
354	900	913	925	937	949	962	974	986	998	*011
355	55 023	035	047	060	072	084	096	108	121	133
356	145	157	169	182	194	206	218	230	242	255
357	267	279	291	303	315	328	340	352	364	376
358	388	400	413	425	437	449	461	473	485	497
359	509	522	534	546	558	570	582	594	606	618
360	630	642	654	666	678	691	703	715	727	739
361	751	763	775	787	799	811	823	835	847	859
362	871	883	895	907	919	931	943	955	967	979
363	991	*003	*015	*027	*038	*050	*062	*074	*086	*098
364	56 110	122	134	146	158	170	182	194	205	217
365	229	241	253	265	277	289	301	312	324	336
366	348	360	372	384	396	407	419	431	443	455
367	467	478	490	502	514	526	538	549	561	573
368	585	597	608	620	632	644	656	667	679	691
369	703	714	726	738	750	761	773	785	797	808
370	820	832	844	855	867	879	891	902	914	926
371	937	949	961	972	984	996	*008	*019	*031	*043
372	57 054	066	078	089	101	113	124	136	148	159
373	171	183	194	206	217	229	241	252	264	276
374	287	299	310	322	334	345	357	368	380	392
375	403	415	426	438	449	461	473	484	496	507
376	519	530	542	553	565	576	588	600	611	623
377	634	646	657	669	680	692	703	715	726	738
378	749	761	772	784	795	807	818	830	841	852
379	864	875	887	898	910	921	933	944	955	967
380	978	990	*001	*013	*024	*035	*047	*058	*070	*081
381	58 092	104	115	127	138	149	161	172	184	195
382	206	218	229	240	252	263	274	286	297	309
383	320	331	343	354	365	377	388	399	410	422
384	433	444	456	467	478	490	501	512	524	535
385	546	557	569	580	591	602	614	625	636	647
386	659	670	681	692	704	715	726	737	749	760
387	771	782	794	805	816	827	838	850	861	872
388	883	894	906	917	928	939	950	961	973	984
389	995	*006	*017	*028	*040	*051	*062	*073	*084	*095
390	59 106	118	129	140	151	162	173	184	195	207
391	218	229	240	251	262	273	284	295	306	318
392	329	340	351	362	373	384	395	406	417	428
393	439	450	461	472	483	494	506	517	528	539
394	550	561	572	583	594	605	616	627	638	649
395	660	671	682	693	704	715	726	737	748	759
396	770	780	791	802	813	824	835	846	857	868
397	879	890	901	912	923	934	945	956	966	977
398	988	999	*010	*021	*032	*043	*054	*065	*076	*086
399	60 097	108	119	130	141	152	163	173	184	195
400	206	217	228	239	249	260	271	282	293	304

P. P.

13

1	1.3
2	2.6
3	3.9
4	5.2
5	6.5
6	7.8
7	9.1
8	10.4
9	11.7

12

1	1.2
2	2.4
3	3.6
4	4.8
5	6.0
6	7.2
7	8.4
8	9.6
9	10.8

11

1	1.1
2	2.2
3	3.3
4	4.4
5	5.5
6	6.6
7	7.7
8	8.8
9	9.9

10

1	1.0
2	2.0
3	3.0
4	4.0
5	5.0
6	6.0
7	7.0
8	8.0
9	9.0

Table of Logarithms 400 - 450

N.	L. 0	1	2	3	4	5	6	7	8	9
400	60 206	217	228	239	249	260	271	282	293	304
401	314	325	336	347	358	369	379	390	401	412
402	423	433	444	455	466	477	487	498	509	520
403	531	541	552	563	574	584	595	606	617	627
404	638	649	660	670	681	692	703	713	724	735
405	746	756	767	778	788	799	810	821	831	842
406	853	863	874	885	895	906	917	927	938	949
407	959	970	981	991	*002	*013	*023	*034	*045	*055
408	61 066	077	087	098	109	119	130	140	151	162
409	172	183	194	204	215	225	236	247	257	268
410	278	289	300	310	321	331	342	352	363	374
411	384	395	405	416	426	437	448	458	469	479
412	490	500	511	521	532	542	553	563	574	584
413	595	606	616	627	637	648	658	669	679	690
414	700	711	721	731	742	752	763	773	784	794
415	805	815	826	836	847	857	868	878	888	899
416	909	920	930	941	951	962	972	982	993	*003
417	62 014	024	034	045	055	066	076	086	097	107
418	118	128	138	149	159	170	180	190	201	211
419	221	232	242	252	263	273	284	294	304	315
420	325	335	346	356	366	377	387	397	408	418
421	428	439	449	459	469	480	490	500	511	521
422	531	542	552	562	572	583	593	603	613	624
423	634	644	655	665	675	685	696	706	716	726
424	737	747	757	767	778	788	798	808	818	829
425	839	849	859	870	880	890	900	910	921	931
426	941	951	961	972	982	992	*002	*012	*022	*033
427	63 043	053	063	073	083	094	104	114	124	134
428	144	155	165	175	185	195	205	215	225	236
429	246	256	266	276	286	296	306	317	327	337
430	347	357	367	377	387	397	407	417	428	438
431	448	458	468	478	488	498	508	518	528	538
432	548	558	568	579	589	599	609	619	629	639
433	649	659	669	679	689	699	709	719	729	739
434	749	759	769	779	789	799	809	819	829	839
435	849	859	869	879	889	899	909	919	929	939
436	949	959	969	979	988	998	*008	*018	*028	*038
437	64 048	058	068	078	088	098	108	118	128	137
438	147	157	167	177	187	197	207	217	227	237
439	246	256	266	276	286	296	306	316	326	335
440	345	355	365	375	385	395	404	414	424	434
441	444	454	464	473	483	493	503	513	523	532
442	542	552	562	572	582	591	601	611	621	631
443	640	650	660	670	680	689	699	709	719	729
444	738	748	758	768	777	787	797	807	816	826
445	836	846	856	865	875	885	895	904	914	924
446	933	943	953	963	972	982	992	*002	*011	*021
447	65 031	040	050	060	070	079	089	099	108	118
448	128	137	147	157	167	176	186	196	205	215
449	225	234	244	254	263	273	283	292	302	312
450	321	331	341	350	360	369	379	389	398	408

P. P.

11
1 | 1.1
2 | 2.2
3 | 3.3
4 | 4.4
5 | 5.5
6 | 6.6
7 | 7.7
8 | 8.8
9 | 9.9

10
1 | 1.0
2 | 2.0
3 | 3.0
4 | 4.0
5 | 5.0
6 | 6.0
7 | 7.0
8 | 8.0
9 | 9.0

9
1 | 0.9
2 | 1.8
3 | 2.7
4 | 3.6
5 | 4.5
6 | 5.4
7 | 6.3
8 | 7.2
9 | 8.1

MATHEMATICS

Table of Logarithms   450 - 500

N.	L.	0	1	2	3	4	5	6	7	8	9	P. P.
450	65	321	331	341	350	360	369	379	389	398	408	
451		418	427	437	447	456	466	475	485	495	504	
452		514	523	533	543	552	562	571	581	591	600	
453		610	619	629	639	648	658	667	677	686	696	
454		706	715	725	734	744	753	763	772	782	792	
455		801	811	820	830	839	849	858	868	877	887	
456		896	906	916	925	935	944	954	963	973	982	**10**
457		992	*001	*011	*020	*030	*039	*049	*058	*068	*077	1|1.0
458	66	087	096	106	115	124	134	143	153	162	172	2|2.0
459		181	191	200	210	219	229	238	247	257	266	3|3.0
460		276	285	295	304	314	323	332	342	351	361	4|4.0
461		370	380	389	398	408	417	427	436	445	455	5|5.0
462		464	474	483	492	502	511	521	530	539	549	6|6.0
463		558	567	577	586	596	605	614	624	633	642	7|7.0
464		652	661	671	680	689	699	708	717	727	736	8|8.0
465		745	755	764	773	783	792	801	811	820	829	9|9.0
466		839	848	857	867	876	885	894	904	913	922	
467		932	941	950	960	969	978	987	997	*006	*015	
468	67	025	034	043	052	062	071	080	089	099	108	
469		117	127	136	145	154	164	173	182	191	201	
470		210	219	228	237	247	256	265	274	284	293	
471		302	311	321	330	339	348	357	367	376	385	**9**
472		394	403	413	422	431	440	449	459	468	477	1|0.9
473		486	495	504	514	523	532	541	550	560	569	2|1.8
474		578	587	596	605	614	624	633	642	651	660	3|2.7
475		669	679	688	697	706	715	724	733	742	752	4|3.6
476		761	770	779	788	797	806	815	825	834	843	5|4.5
477		852	861	870	879	888	897	906	916	925	934	6|5.4
478		943	952	961	970	979	988	997	*006	*015	*024	7|6.3
479	68	034	043	052	061	070	079	088	097	106	115	8|7.2
480		124	133	142	151	160	169	178	187	196	205	9|8.1
481		215	224	233	242	251	260	269	278	287	296	
482		305	314	323	332	341	350	359	368	377	386	
483		395	404	413	422	431	440	449	458	467	476	
484		485	494	502	511	520	529	538	547	556	565	
485		574	583	592	601	610	619	628	637	646	655	**8**
486		664	673	681	690	699	708	717	726	735	744	1|0.8
487		753	762	771	780	789	797	806	815	824	833	2|1.6
488		842	851	860	869	878	886	895	904	913	922	3|2.4
489		931	940	949	958	966	975	984	993	*002	*011	4|3.2
490	69	020	028	037	046	055	064	073	082	090	099	5|4.0
491		108	117	126	135	144	152	161	170	179	188	6|4.8
492		197	205	214	223	232	241	249	258	267	276	7|5.6
493		285	294	302	311	320	329	338	346	355	364	8|6.4
494		373	381	390	399	408	417	425	434	443	452	9|7.2
495		461	469	478	487	496	504	513	522	531	539	
496		548	557	566	574	583	592	601	609	618	627	
497		636	644	653	662	671	679	688	697	705	714	
498		723	732	740	749	758	767	775	784	793	801	
499		810	819	827	836	845	854	862	871	880	888	
500		897	906	914	923	932	940	949	958	966	975	

Table of Logarithms 500 - 550

N.	L. 0	1	2	3	4	5	6	7	8	9
500	69 897	906	914	923	932	940	949	958	966	975
501	984	992	*001	*010	*018	*027	*036	*044	*053	*062
502	70 070	079	088	096	105	114	122	131	140	148
503	157	165	174	183	191	200	209	217	226	234
504	243	252	260	269	278	286	295	303	312	321
505	329	338	346	355	364	372	381	389	398	406
506	415	424	432	441	449	458	467	475	484	492
507	501	509	518	526	535	544	552	561	569	578
508	586	595	603	612	621	629	638	646	655	663
509	672	680	689	697	706	714	723	731	740	749
510	757	766	774	783	791	800	808	817	825	834
511	842	851	859	868	876	885	893	902	910	919
512	927	935	944	952	961	969	978	986	995	*003
513	71 012	020	029	037	046	054	063	071	079	088
514	096	105	113	122	130	139	147	155	164	172
515	181	189	198	206	214	223	231	240	248	257
516	265	273	282	290	299	307	315	324	332	341
517	349	357	366	374	383	391	399	408	416	425
518	433	441	450	458	466	475	483	492	500	508
519	517	525	533	542	550	559	567	575	584	592
520	600	609	617	625	634	642	650	659	667	675
521	684	692	700	709	717	725	734	742	750	759
522	767	775	784	792	800	809	817	825	834	842
523	850	858	867	875	883	892	900	908	917	925
524	933	941	950	958	966	975	983	991	999	*008
525	72 016	024	032	041	049	057	066	074	082	090
526	099	107	115	123	132	140	148	156	165	173
527	181	189	198	206	214	222	230	239	247	255
528	263	272	280	288	296	304	313	321	329	337
529	346	354	362	370	378	387	395	403	411	419
530	428	436	444	452	460	469	477	485	493	501
531	509	518	526	534	542	550	558	567	575	583
532	591	599	607	616	624	632	640	648	656	665
533	673	681	689	697	705	713	722	730	738	746
534	754	762	770	779	787	795	803	811	819	827
535	835	843	852	860	868	876	884	892	900	908
536	916	925	933	941	949	957	965	973	981	989
537	997	*006	*014	*022	*030	*038	*046	*054	*062	*070
538	73 078	086	094	102	111	119	127	135	143	151
539	159	167	175	183	191	199	207	215	223	231
540	239	247	255	263	272	280	288	296	304	312
541	320	328	336	344	352	360	368	376	384	392
542	400	408	416	424	432	440	448	456	464	472
543	480	488	496	504	512	520	528	536	544	552
544	560	568	576	584	592	600	608	616	624	632
545	640	648	656	664	672	679	687	695	703	711
546	719	727	735	743	751	759	767	775	783	791
547	799	807	815	823	830	838	846	854	862	870
548	878	886	894	902	910	918	926	933	941	949
549	957	965	973	981	989	997	*005	*013	*020	*028
550	74 036	044	052	060	068	076	084	092	099	107

P. P.

9
1	0.9
2	1.8
3	2.7
4	3.6
5	4.5
6	5.4
7	6.3
8	7.2
9	8.1

8
1	0.8
2	1.6
3	2.4
4	3.2
5	4.0
6	4.8
7	5.6
8	6.4
9	7.2

7
1	0.7
2	1.4
3	2.1
4	2.8
5	3.5
6	4.2
7	4.9
8	5.6
9	6.3

Table of Logarithms 550 - 600

N.	L.	0	1	2	3	4	5	6	7	8	9
550	74	036	044	052	060	068	076	084	092	099	107
551		115	123	131	139	147	155	162	170	178	186
552		194	202	210	218	225	233	241	249	257	265
553		273	280	288	296	304	312	320	327	335	343
554		351	359	367	374	382	390	398	406	414	421
555		429	437	445	453	461	468	476	484	492	500
556		507	515	523	531	539	547	554	562	570	578
557		586	593	601	609	617	624	632	640	648	656
558		663	671	679	687	695	702	710	718	726	733
559		741	749	757	764	772	780	788	796	803	811
560		819	827	834	842	850	858	865	873	881	889
561		896	904	912	920	927	935	943	950	958	966
562		974	981	989	997	*005	*012	*020	*028	*035	*043
563	75	051	059	066	074	082	089	097	105	113	120
564		128	136	143	151	159	166	174	182	189	197
565		205	213	220	228	236	243	251	259	266	274
566		282	289	297	305	312	320	328	335	343	351
567		358	366	374	381	389	397	404	412	420	427
568		435	442	450	458	465	473	481	488	496	504
569		511	519	526	534	542	549	557	565	572	580
570		587	595	603	610	618	626	633	641	648	656
571		664	671	679	686	694	702	709	717	724	732
572		740	747	755	762	770	778	785	793	800	808
573		815	823	831	838	846	853	861	868	876	884
574		891	899	906	914	921	929	937	944	952	959
575		967	974	982	989	997	*005	*012	*020	*027	*035
576	76	042	050	057	065	072	080	087	095	103	110
577		118	125	133	140	148	155	163	170	178	185
578		193	200	208	215	223	230	238	245	253	260
579		268	275	283	290	298	305	313	320	328	335
580		343	350	358	365	373	380	388	395	403	410
581		418	425	433	440	448	455	462	470	477	485
582		492	500	507	515	522	530	537	545	552	559
583		567	574	582	589	597	604	612	619	626	634
584		641	649	656	664	671	678	686	693	701	708
585		716	723	730	738	745	753	760	768	775	782
586		790	797	805	812	819	827	834	842	849	856
587		864	871	879	886	893	901	908	916	923	930
588		938	945	953	960	967	975	982	989	997	*004
589	77	012	019	026	034	041	048	056	063	070	078
590		085	093	100	107	115	122	129	137	144	151
591		159	166	173	181	188	195	203	210	217	225
592		232	240	247	254	262	269	276	283	291	298
593		305	313	320	327	335	342	349	357	364	371
594		379	386	393	401	408	415	422	430	437	444
595		452	459	466	474	481	488	495	503	510	517
596		525	532	539	546	554	561	568	576	583	590
597		597	605	612	619	627	634	641	648	656	663
598		670	677	685	692	699	706	714	721	728	735
599		743	750	757	764	772	779	786	793	801	808
600		815	822	830	837	844	851	859	866	873	880

P. P.

8

1	0.8
2	1.6
3	2.4
4	3.2
5	4.0
6	4.8
7	5.6
8	6.4
9	7.2

7

1	0.7
2	1.4
3	2.1
4	2.8
5	3.5
6	4.2
7	4.9
8	5.6
9	6.3

Table of Logarithms 600 - 650

N.	L. 0	1	2	3	4	5	6	7	8	9
600	77 815	822	830	837	844	˙851	859	866	873	880
601	887	895	902	909	916	924	931	938	945	952
602	960	967	974	981	988	996	*003	*010	*017	*025
603	78 032	039	046	053	061	068	075	082	089	097
604	104	111	118	125	132	140	147	154	161	168
605	176	183	190	197	204	211	219	226	233	240
606	247	254	262	269	276	283	290	297	305	312
607	319	326	333	340	347	355	362	369	376	383
608	390	398	405	412	419	426	433	440	447	455
609	462	469	476	483	490	497	504	512	519	526
610	533	540	547	554	561	569	576	583	590	597
611	604	611	618	625	633	640	647	654	661	668
612	675	682	689	696	704	711	718	725	732	739
613	746	753	760	767	774	781	789	796	803	810
614	817	824	831	838	845	852	859	866	873	880
615	888	895	902	909	916	923	930	937	944	951
616	958	965	972	979	986	993	*000	*007	*014	*021
617	79 029	036	043	050	057	064	071	078	085	092
618	099	106	113	120	127	134	141	148	155	162
619	169	176	183	190	197	204	211	218	225	232
620	239	246	253	260	267	274	281	288	295	302
621	309	316	323	330	337	344	351	358	365	372
622	379	386	393	400	407	414	421	428	435	442
623	449	456	463	470	477	484	491	498	505	511
624	518	525	532	539	546	553	560	567	574	581
625	588	595	602	609	616	623	630	637	644	650
626	657	664	671	678	685	692	699	706	713	720
627	727	734	741	748	754	761	768	775	782	789
628	796	803	810	817	824	831	837	844	851	858
629	865	872	879	886	893	900	906	913	920	927
630	934	941	948	955	962	969	975	982	989	996
631	80 003	010	017	024	030	037	044	051	058	065
632	072	079	085	092	099	106	113	120	127	134
633	140	147	154	161	168	175	182	188	195	202
634	209	216	223	229	236	243	250	257	264	271
635	277	284	291	298	305	312	318	325	332	339
636	346	353	359	366	373	380	387	393	400	407
637	414	421	428	434	441	448	455	462	468	475
638	482	489	496	502	509	516	523	530	536	543
639	550	557	564	570	577	584	591	598	604	611
640	618	625	632	638	645	652	659	665	672	679
641	686	693	699	706	713	720	726	733	740	747
642	754	760	767	774	781	787	794	801	808	814
643	821	828	835	841	848	855	862	868	875	882
644	889	895	902	909	916	922	929	936	943	949
645	956	963	969	976	983	990	996	*003	*010	*017
646	81 023	030	037	043	050	057	064	070	077	084
647	090	097	104	111	117	124	131	137	144	151
648	158	164	171	178	184	191	198	204	211	218
649	224	231	238	245	251	258	265	271	278	285
650	291	298	305	311	318	325	331	338	345	351

P. P.

8

1	0.8
2	1.6
3	2.4
4	3.2
5	4.0
6	4.8
7	5.6
8	6.4
9	7.2

7

1	0.7
2	1.4
3	2.1
4	2.8
5	3.5
6	4.2
7	4.9
8	5.6
9	6.3

6

1	0.6
2	1.2
3	1.8
4	2.4
5	3.0
6	3.6
7	4.2
8	4.8
9	5.4

MATHEMATICS

Table of Logarithms 650 - 700

N.	L. 0	1	2	3	4	5	6	7	8	9	P. P.
650	81 291	298	305	311	318	325	331	338	345	351	
651	358	365	371	378	385	391	398	405	411	418	
652	425	431	438	445	451	458	465	471	478	485	
653	491	498	505	511	518	525	531	538	544	551	
654	558	564	571	578	584	591	598	604	611	617	
655	624	631	637	644	651	657	664	671	677	684	
656	690	697	704	710	717	723	730	737	743	750	
657	757	763	770	776	783	790	796	803	809	816	
658	823	829	836	842	849	856	862	869	875	882	
659	889	895	902	908	915	921	928	935	941	948	
660	954	961	968	974	981	987	994	*000	*007	*014	7
661	82 020	027	033	040	046	053	060	066	073	079	1 | 0.7
662	086	092	099	105	112	119	125	132	138	145	2 | 1.4
663	151	158	164	171	178	184	191	197	204	210	3 | 2.1
664	217	223	230	236	243	249	256	263	269	276	4 | 2.8
665	282	289	295	302	308	315	321	328	334	341	5 | 3.5
666	347	354	360	367	373	380	387	393	400	406	6 | 4.2
667	413	419	426	432	439	445	452	458	465	471	7 | 4.9
668	478	484	491	497	504	510	517	523	530	536	8 | 5.6
669	543	549	556	562	569	575	582	588	595	601	9 | 6.3
670	607	614	620	627	633	640	646	653	659	666	
671	672	679	685	692	698	705	711	718	724	730	
672	737	743	750	756	763	769	776	782	789	795	
673	802	808	814	821	827	834	840	847	853	860	
674	866	872	879	885	892	898	905	911	918	924	
675	930	937	943	950	956	963	969	975	982	988	
676	995	*001	*008	*014	*020	*027	*033	*040	*046	*052	
677	83 059	065	072	078	085	091	097	104	110	117	
678	123	129	136	142	149	155	161	168	174	181	
679	187	193	200	206	213	219	225	232	238	245	
680	251	257	264	270	276	283	289	296	302	308	6
681	315	321	327	334	340	347	353	359	366	372	1 | 0.6
682	378	385	391	398	404	410	417	423	429	436	2 | 1.2
683	442	448	455	461	467	474	480	487	493	499	3 | 1.8
684	506	512	518	525	531	537	544	550	556	563	4 | 2.4
685	569	575	582	588	594	601	607	613	620	626	5 | 3.0
686	632	639	645	651	658	664	670	677	683	689	6 | 3.6
687	696	702	708	715	721	727	734	740	746	753	7 | 4.2
688	759	765	771	778	784	790	797	803	809	816	8 | 4.8
689	822	828	835	841	847	853	860	866	872	879	9 | 5.4
690	885	891	897	904	910	916	923	929	935	942	
691	948	954	960	967	973	979	985	992	998	*004	
692	84 011	017	023	029	036	042	048	055	061	067	
693	073	080	086	092	098	105	111	117	123	130	
694	136	142	148	155	161	167	173	180	186	192	
695	198	205	211	217	223	230	236	242	248	255	
696	261	267	273	280	286	292	298	305	311	317	
697	323	330	336	342	348	354	361	367	373	379	
698	386	392	398	404	410	417	423	429	435	442	
699	448	454	460	466	473	479	485	491	497	504	
700	510	516	522	528	535	541	547	553	559	566	

Tables of Logarithms 700 - 750

N.	L. 0	1	2	3	4	5	6	7	8	9
700	84 510	516	522	528	535	541	547	553	559	566
701	572	578	584	590	597	603	609	615	621	628
702	634	640	646	652	658	665	671	677	683	689
703	696	702	708	714	720	726	733	739	745	751
704	757	763	770	776	782	788	794	800	807	813
705	819	825	831	837	844	850	856	862	868	874
706	880	887	893	899	905	911	917	924	930	936
707	942	948	954	960	967	973	979	985	991	997
708	85 003	009	016	022	028	034	040	046	052	058
709	065	071	077	083	089	095	101	107	114	120
710	126	132	138	144	150	156	163	169	175	181
711	187	193	199	205	211	217	224	230	236	242
712	248	254	260	266	272	278	285	291	297	303
713	309	315	321	327	333	339	345	352	358	364
714	370	376	382	388	394	400	406	412	418	425
715	431	437	443	449	455	461	467	473	479	485
716	491	497	503	509	516	522	528	534	540	546
717	552	558	564	570	576	582	588	594	600	606
718	612	618	625	631	637	643	649	655	661	667
719	673	679	685	691	697	703	709	715	721	727
720	733	739	745	751	757	763	769	775	781	788
721	794	800	806	812	818	824	830	836	842	848
722	854	860	866	872	878	884	890	896	902	908
723	914	920	926	932	938	944	950	956	962	968
724	974	980	986	992	998	*004	*010	*016	*022	*028
725	86 034	040	046	052	058	064	070	076	082	088
726	094	100	106	112	118	124	130	136	141	147
727	153	159	165	171	177	183	189	195	201	207
728	213	219	225	231	237	243	249	255	261	267
729	273	279	285	291	297	303	308	314	320	326
730	332	338	344	350	356	362	368	374	380	386
731	392	398	404	410	415	421	427	433	439	445
732	451	457	463	469	475	481	487	493	499	504
733	510	516	522	528	534	540	546	552	558	564
734	570	576	581	587	593	599	605	611	617	623
735	629	635	641	646	652	658	664	670	676	682
736	688	694	700	705	711	717	723	729	735	741
737	747	753	759	764	770	776	782	788	794	800
738	806	812	817	823	829	835	841	847	853	859
739	864	870	876	882	888	894	900	906	911	917
740	923	929	935	941	947	953	958	964	970	976
741	982	988	994	999	*005	*011	*017	*023	*029	*035
742	87 040	046	052	058	064	070	075	081	087	093
743	099	105	111	116	122	128	134	140	146	151
744	157	163	169	175	181	186	192	198	204	210
745	216	221	227	233	239	245	251	256	262	268
746	274	280	286	291	297	303	309	315	320	326
747	332	338	344	349	355	361	367	373	379	384
748	390	396	402	408	413	419	425	431	437	442
749	448	454	460	466	471	477	483	489	495	500
750	506	512	518	523	529	535	541	547	552	558

P. P.

7
1	0.7
2	1.4
3	2.1
4	2.8
5	3.5
6	4.2
7	4.9
8	5.6
9	6.3

6
1	0.6
2	1.2
3	1.8
4	2.4
5	3.0
6	3.6
7	4.2
8	4.8
9	5.4

5
1	0.5
2	1.0
3	1.5
4	2.0
5	2.5
6	3.0
7	3.5
8	4.0
9	4.5

MATHEMATICS

Tables of Logarithms   750 - 800

N.	L. 0	1	2	3	4	5	6	7	8	9
750	87 506	512	518	523	529	535	541	547	552	558
751	564	570	576	581	587	593	599	604	610	616
752	622	628	633	639	645	651	656	662	668	674
753	679	685	691	697	703	708	714	720	726	731
754	737	743	749	754	760	766	772	777	783	789
755	795	800	806	812	818	823	829	835	841	846
756	852	858	864	869	875	881	887	892	898	904
757	910	915	921	927	933	938	944	950	955	961
758	967	973	978	984	990	996	*001	*007	*013	*018
759	88 024	030	036	041	047	053	058	064	070	076
760	081	087	093	098	104	110	116	121	127	133
761	138	144	150	156	161	167	173	178	184	190
762	195	201	207	213	218	224	230	235	241	247
763	252	258	264	270	275	281	287	292	298	304
764	309	315	321	326	332	338	343	349	355	360
765	366	372	377	383	389	395	400	406	412	417
766	423	429	434	440	446	451	457	463	468	474
767	480	485	491	497	502	508	513	519	525	530
768	536	542	547	553	559	564	570	576	581	587
769	593	598	604	610	615	621	627	632	638	643
770	649	655	660	666	672	677	683	689	694	700
771	705	711	717	722	728	734	739	745	750	756
772	762	767	773	779	784	790	795	801	807	812
773	818	824	829	835	840	846	852	857	863	868
774	874	880	885	891	897	902	908	913	919	925
775	930	936	941	947	953	958	964	969	975	981
776	986	992	997	*003	*009	*014	*020	*025	*031	*037
777	89 042	048	053	059	064	070	076	081	087	092
778	098	104	109	115	120	126	131	137	143	148
779	154	159	165	170	176	182	187	193	198	204
780	209	215	221	226	232	237	243	248	254	260
781	265	271	276	282	287	293	298	304	310	315
782	321	326	332	337	343	348	354	360	365	371
783	376	382	387	393	398	404	409	415	421	426
784	432	437	443	448	454	459	465	470	476	481
785	487	492	498	504	509	515	520	526	531	537
786	542	548	553	559	564	570	575	581	586	592
787	597	603	609	614	620	625	631	636	642	647
788	653	658	664	669	675	680	686	691	697	702
789	708	713	719	724	730	735	741	746	752	757
790	763	768	774	779	785	790	796	801	807	812
791	818	823	829	834	840	845	851	856	862	867
792	873	878	883	889	894	900	905	911	916	922
793	927	933	938	944	949	955	960	966	971	977
794	982	988	993	998	*004	*009	*015	*020	*026	*031
795	90 037	042	048	053	059	064	069	075	080	086
796	091	097	102	108	113	119	124	129	135	140
797	146	151	157	162	168	173	179	184	189	195
798	200	206	211	217	222	227	233	238	244	249
799	255	260	266	271	276	282	287	293	298	304
800	309	314	320	325	331	336	342	347	352	358

P. P.

6	
1	0.6
2	1.2
3	1.8
4	2.4
5	3.0
6	3.6
7	4.2
8	4.8
9	5.4

5	
1	0.5
2	1.0
3	1.5
4	2.0
5	2.5
6	3.0
7	3.5
8	4.0
9	4.5

Tables of Logarithms 800 - 850

N.	L.	0	1	2	3	4	5	6	7	8	9	P. P.
800	90	309	314	320	325	331	336	342	347	352	358	
801		363	369	374	380	385	390	396	401	407	412	
802		417	423	428	434	439	445	450	455	461	466	
803		472	477	482	488	493	499	504	509	515	520	
804		526	531	536	542	547	553	558	563	569	574	
805		580	585	590	596	601	607	612	617	623	628	
806		634	639	644	650	655	660	666	671	677	682	
807		687	693	698	703	709	714	720	725	730	736	
808		741	747	752	757	763	768	773	779	784	789	
809		795	800	806	811	816	822	827	832	838	843	
810		849	854	859	865	870	875	881	886	891	897	6
811		902	907	913	918	924	929	934	940	945	950	1|0.6
812		956	961	966	972	977	982	988	993	998	*004	2|1.2
813	91	009	014	020	025	030	036	041	046	052	057	3|1.8
814		062	068	073	078	084	089	094	100	105	110	4|2.4
815		116	121	126	132	137	142	148	153	158	164	5|3.0
816		169	174	180	185	190	196	201	206	212	217	6|3.6
817		222	228	233	238	243	249	254	259	265	270	7|4.2
818		275	281	286	291	297	302	307	312	318	323	8|4.8
819		328	334	339	344	350	355	360	365	371	376	9|5.4
820		381	387	392	397	403	408	413	418	424	429	
821		434	440	445	450	455	461	466	471	477	482	
822		487	492	498	503	508	514	519	524	529	535	
823		540	545	551	556	561	566	572	577	582	587	
824		593	598	603	609	614	619	624	630	635	640	
825		645	651	656	661	666	672	677	682	687	693	
826		698	703	709	714	719	724	730	735	740	745	
827		751	756	761	766	772	777	782	787	793	798	
828		803	808	814	819	824	829	834	840	845	850	
829		855	861	866	871	876	882	887	892	897	903	
830		908	913	918	924	929	934	939	944	950	955	5
831		960	965	971	976	981	986	991	997	*002	*007	1|0.5
832	92	012	018	023	028	033	038	044	049	054	059	2|1.0
833		065	070	075	080	085	091	096	101	106	111	3|1.5
834		117	122	127	132	137	143	148	153	158	163	4|2.0
835		169	174	179	184	189	195	200	205	210	215	5|2.5
836		221	226	231	236	241	247	252	257	262	267	6|3.0
837		273	278	283	288	293	298	304	309	314	319	7|3.5
838		324	330	335	340	345	350	355	361	366	371	8|4.0
839		376	381	387	392	397	402	407	412	418	423	9|4.5
840		428	433	438	443	449	454	459	464	469	474	
841		480	485	490	495	500	505	511	516	521	526	
842		531	536	542	547	552	557	562	567	572	578	
843		583	588	593	598	603	609	614	619	624	629	
844		634	639	645	650	655	660	665	670	675	681	
845		686	691	696	701	706	711	716	722	727	732	
846		737	742	747	752	758	763	768	773	778	783	
847		788	793	799	804	809	814	819	824	829	834	
848		840	845	850	855	860	865	870	875	881	886	
849		891	896	901	906	911	916	921	927	932	937	
850		942	947	952	957	962	967	973	978	983	988	

Tables of Logarithms　850 - 900

N.	L. 0	1	2	3	4	5	6	7	8	9	P. P.
850	92 942	947	952	957	962	967	973	978	983	988	
851	993	998	*003	*008	*013	*018	*024	*029	*034	*039	
852	93 044	049	054	059	064	069	075	080	085	090	
853	095	100	105	110	115	120	125	131	136	141	
854	146	151	156	161	166	171	176	181	186	192	
855	197	202	207	212	217	222	227	232	237	242	**6**
856	247	252	258	263	268	273	278	283	288	293	
857	298	303	308	313	318	323	328	334	339	344	1 0.6
858	349	354	359	364	369	374	379	384	389	394	2 1.2
859	399	404	409	414	420	425	430	435	440	445	3 1.8
860	450	455	460	465	470	475	480	485	490	495	4 2.4
861	500	505	510	515	520	526	531	536	541	546	5 3.0
862	551	556	561	566	571	576	581	586	591	596	6 3.6
863	601	606	611	616	621	626	631	636	641	646	7 4.2
864	651	656	661	666	671	676	682	687	692	697	8 4.8
865	702	707	712	717	722	727	732	737	742	747	9 5.4
866	752	757	762	767	772	777	782	787	792	797	
867	802	807	812	817	822	827	832	837	842	847	
868	852	857	862	867	872	877	882	887	892	897	
869	902	907	912	917	922	927	932	937	942	947	
870	952	957	962	967	972	977	982	987	992	997	
871	94 002	007	012	017	022	027	032	037	042	047	**5**
872	052	057	062	067	072	077	082	086	091	096	1 0.5
873	101	106	111	116	121	126	131	136	141	146	2 1.0
874	151	156	161	166	171	176	181	186	191	196	3 1.5
875	201	206	211	216	221	226	231	236	240	245	4 2.0
876	250	255	260	265	270	275	280	285	290	295	5 2.5
877	300	305	310	315	320	325	330	335	340	345	6 3.0
878	349	354	359	364	369	374	379	384	389	394	7 3.5
879	399	404	409	414	419	424	429	433	438	443	8 4.0
880	448	453	458	463	468	473	478	483	488	493	9 4.5
881	498	503	507	512	517	522	527	532	537	542	
882	547	552	557	562	567	571	576	581	586	591	
883	596	601	606	611	616	621	626	630	635	640	
884	645	650	655	660	665	670	675	680	685	689	
885	694	699	704	709	714	719	724	729	734	738	
886	743	748	753	758	763	768	773	778	783	787	**4**
887	792	797	802	807	812	817	822	827	832	836	1 0.4
888	841	846	851	856	861	866	871	876	880	885	2 0.8
889	890	895	900	905	910	915	919	924	929	934	3 1.2
890	939	944	949	954	959	963	968	973	978	983	4 1.6
891	988	993	998	*002	*007	*012	*017	*022	*027	*032	5 2.0
892	95 036	041	046	051	056	061	066	071	075	080	6 2.4
893	085	090	095	100	105	109	114	119	124	129	7 2.8
894	134	139	143	148	153	158	163	168	173	177	8 3.2
895	182	187	192	197	202	207	211	216	221	226	9 3.6
896	231	236	240	245	250	255	260	265	270	274	
897	279	284	289	294	299	303	308	313	318	323	
898	328	332	337	342	347	352	357	361	366	371	
899	376	381	386	390	395	400	405	410	415	419	
900	424	429	434	439	444	448	453	458	463	468	

Tables of Logarithms 900 - 950

N.	L. 0	1	2	3	4	5	6	7	8	9	P. P.
900	95 424	429	434	439	444	448	453	458	463	468	
901	472	477	482	487	492	497	501	506	511	516	
902	521	525	530	535	540	545	550	554	559	564	
903	569	574	578	583	588	593	598	602	607	612	
904	617	622	626	631	636	641	646	650	655	660	
905	665	670	674	679	684	689	694	698	703	708	
906	713	718	722	727	732	737	742	746	751	756	
907	761	766	770	775	780	785	789	794	799	804	
908	809	813	818	823	828	832	837	842	847	852	
909	856	861	866	871	875	880	885	890	895	899	
910	904	909	914	918	923	928	933	938	942	947	
911	952	957	961	966	971	976	980	985	990	995	**5**
912	999	*004	*009	*014	*019	*023	*028	*033	*038	*042	1 | 0.5
913	96 047	052	057	061	066	071	076	080	085	090	2 | 1.0
914	095	099	104	109	114	118	123	128	133	137	3 | 1.5
915	142	147	152	156	161	166	171	175	180	185	4 | 2.0
916	190	194	199	204	209	213	218	223	227	232	5 | 2.5
917	237	242	246	251	256	261	265	270	275	280	6 | 3.0
918	284	289	294	298	303	308	313	317	322	327	7 | 3.5
919	332	336	341	346	350	355	360	365	369	374	8 | 4.0
920	379	384	388	393	398	402	407	412	417	421	9 | 4.5
921	426	431	435	440	445	450	454	459	464	468	
922	473	478	483	487	492	497	501	506	511	515	
923	520	525	530	534	539	544	548	553	558	562	
924	567	572	577	581	586	591	595	600	605	609	
925	614	619	624	628	633	638	642	647	652	656	
926	661	666	670	675	680	685	689	694	699	703	
927	708	713	717	722	727	731	736	741	745	750	
928	755	759	764	769	774	778	783	788	792	797	
929	802	806	811	816	820	825	830	834	839	844	
930	848	853	858	862	867	872	876	881	886	890	
931	895	900	904	909	914	918	923	928	932	937	
932	942	946	951	956	960	965	970	974	979	984	**4**
933	988	993	997	*002	*007	*011	*016	*021	*025	*030	1 | 0.4
934	97 035	039	044	049	053	058	063	067	072	077	2 | 0.8
935	081	086	090	095	100	104	109	114	118	123	3 | 1.2
936	128	132	137	142	146	151	155	160	165	169	4 | 1.6
937	174	179	183	188	192	197	202	206	211	216	5 | 2.0
938	220	225	230	234	239	243	248	253	257	262	6 | 2.4
939	267	271	276	280	285	290	294	299	304	308	7 | 2.8
940	313	317	322	327	331	336	340	345	350	354	8 | 3.2
941	359	364	368	373	377	382	387	391	396	400	9 | 3.6
942	405	410	414	419	424	428	433	437	442	447	
943	451	456	460	465	470	474	479	483	488	493	
944	497	502	506	511	516	520	525	529	534	539	
945	543	548	552	557	562	566	571	575	580	585	
946	589	594	598	603	607	612	617	621	626	630	
947	635	640	644	649	653	658	663	667	672	676	
948	681	685	690	695	699	704	708	713	717	722	
949	727	731	736	740	745	749	754	759	763	768	
950	772	777	782	786	791	795	800	804	809	813	

MATHEMATICS

Tables of Logarithms 950 - 1000

N.	L.	0	1	2	3	4	5	6	7	8	9	P. P.
950	97	772	777	782	786	791	795	800	804	809	813	
951		818	823	827	832	836	841	845	850	855	859	
952		864	868	873	877	882	886	891	896	900	905	
953		909	914	918	923	928	932	937	941	946	950	
954		955	959	964	968	973	978	982	987	991	996	
955	98	000	005	009	014	019	023	028	032	037	041	
956		046	050	055	059	064	068	073	078	082	087	
957		091	096	100	105	109	114	118	123	127	132	
958		137	141	146	150	155	159	164	168	173	177	
959		182	186	191	195	200	204	209	214	218	223	
960		227	232	236	241	245	250	254	259	263	268	**5**
961		272	277	281	286	290	295	299	304	308	313	1 0.5
962		318	322	327	331	336	340	345	349	354	358	2 1.0
963		363	367	372	376	381	385	390	394	399	403	3 1.5
964		408	412	417	421	426	430	435	439	444	448	4 2.0
965		453	457	462	466	471	475	480	484	489	493	5 2.5
966		498	502	507	511	516	520	525	529	534	538	6 3.0
967		543	547	552	556	561	565	570	574	579	583	7 3.5
968		588	592	597	601	605	610	614	619	623	628	8 4.0
969		632	637	641	646	650	655	659	664	668	673	9 4.5
970		677	682	686	691	695	700	704	709	713	717	
971		722	726	731	735	740	744	749	753	758	762	
972		767	771	776	780	784	789	793	798	802	807	
973		811	816	820	825	829	834	838	843	847	851	
974		856	860	865	869	874	878	883	887	892	896	
975		900	905	909	914	918	923	927	932	936	941	
976		945	949	954	958	963	967	972	976	981	985	
977		989	994	998	*003	*007	*012	*016	*021	*025	*029	
978	99	034	038	043	047	052	056	061	065	069	074	
979		078	083	087	092	096	100	105	109	114	118	
980		123	127	131	136	140	145	149	154	158	162	**4**
981		167	171	176	180	185	189	193	198	202	207	1 0.4
982		211	216	220	224	229	233	238	242	247	251	2 0.8
983		255	260	264	269	273	277	282	286	291	295	3 1.2
984		300	304	308	313	317	322	326	330	335	339	4 1.6
985		344	348	352	357	361	366	370	374	379	383	5 2.0
986		388	392	396	401	405	410	414	419	423	427	6 2.4
987		432	436	441	445	449	454	458	463	467	471	7 2.8
988		476	480	484	489	493	498	502	506	511	515	8 3.2
989		520	524	528	533	537	542	546	550	555	559	9 3.6
990		564	568	572	577	581	585	590	594	599	603	
991		607	612	616	621	625	629	634	638	642	647	
992		651	656	660	664	669	673	677	682	686	691	
993		695	699	704	708	712	717	721	726	730	734	
994		739	743	747	752	756	760	765	769	774	778	
995		782	787	791	795	800	804	808	813	817	822	
996		826	830	835	839	843	848	852	856	861	865	
997		870	874	878	883	887	891	896	900	904	909	
998		913	917	922	926	930	935	939	944	948	952	
999		957	961	965	970	974	978	983	987	991	996	
1000	00	000	004	009	013	017	022	026	030	035	039	

Trigonometric Tables   Sine—Cosine

Sin ↓→	0′	10′	20′	30′	40′	50′	60′	
0	0,00000	0,00291	0,00582	0.00873	0,01164	0,01454	0,01745	89
1	0,01745	0,02036	0,02327	0,02618	0,02908	0,03199	0,03490	88
2	0,03490	0,03781	0,04071	0,04362	0,04653	0,04943	0,05234	87
3	0,05234	0,05524	0,05814	0,06105	0,06395	0,06685	0,06976	86
4	0,06976	0,07266	0,07556	0,07846	0,08136	0,08426	0,08716	85
5	0,08716	0,09005	0,09295	0,09585	0,09874	0,10164	0,10453	84
6	0,10453	0,10742	0,11031	0,11320	0,11609	0,11898	0,12187	83
7	0,12187	0,12476	0,12764	0,13053	0,13341	0,13629	0,13917	82
8	0,13917	0,14205	0,14493	0,14781	0,15069	0,15356	0,15643	81
9	0,15643	0,15931	0,16218	0,16505	0,16792	0,17078	0,17365	80
10	0,17365	0,17651	0,17937	0,18224	0,18509	0,18795	0,19081	79
11	0,19081	0,19366	0,19652	0,19937	0,20222	0,20507	0,20791	78
12	0,20791	0,21076	0,21360	0,21644	0,21928	0,22212	0,22495	77
13	0,22495	0,22778	0,23062	0,23345	0,23627	0,23910	0,24192	76
14	0,24192	0,24474	0,24756	0,25038	0,25320	0,25601	0,25882	75
15	0,25882	0,26163	0,26443	0,26724	0,27004	0,27284	0,27564	74
16	0,27564	0,27843	0,28123	0,28402	0,28680	0,28959	0,29237	73
17	0,29237	0,29515	0,29793	0,30071	0,30348	0,30625	0,30902	72
18	0,30902	0,31178	0,31454	0,31730	0,32006	0,32282	0,32557	71
19	0,32557	0,32832	0,33106	0,33381	0,33655	0,33929	0,34202	70
20	0,34202	0,34475	0,34748	0,35021	0,35293	0,35565	0,35837	69
21	0,35837	0,36108	0,36379	0,36650	0,36921	0,37191	0,37461	68
22	0,37461	0,37730	0,37999	0,38268	0,38537	0,38805	0,39073	67
23	0,39073	0,39341	0,39608	0,39875	0,40141	0,40408	0,40674	66
24	0,40674	0,40939	0,41204	0,41469	0,41734	0,41998	0,42262	65
25	0,42262	0,42525	0,42788	0,43051	0,43313	0,43575	0,43837	64
26	0,43837	0,44098	0,44359	0,44620	0,44880	0,45140	0,45399	63
27	0,45399	0,45658	0,45917	0,46175	0,46433	0,46690	0,46947	62
28	0,46947	0,47204	0,47460	0,47716	0,47971	0,48226	0,48481	61
29	0,48481	0,48735	0,48989	0,49242	0,49495	0,49748	0,50000	60
30	0,50000	0,50252	0,50503	0,50754	0,51004	0,51254	0,51504	59
31	0,51504	0,51753	0,52002	0,52250	0,52498	0,52745	0,52992	58
32	0,52992	0,53238	0,53484	0,53730	0,53975	0,54220	0,54464	57
33	0,54464	0,54708	0,54951	0,55194	0,55436	0,55678	0,55919	56
34	0,55919	0,56160	0,56401	0,56641	0,56880	0,57119	0,57358	55
35	0,57358	0,57596	0,57833	0,58070	0,58307	0,58543	0,58779	54
36	0,58779	0,59014	0,59248	0,59482	0,59716	0,59949	0,60182	53
37	0,60182	0,60414	0,60645	0,60876	0,61107	0,61337	0,61566	52
38	0,61566	0,61795	0,62024	0,62251	0,62479	0,62706	0,62932	51
39	0,62932	0,63158	0,63383	0,63608	0,63832	0,64056	0,64279	50
40	0,64279	0,64501	0,64723	0,64945	0,65166	0,65386	0,65606	49
41	0,65606	0,65825	0,66044	0,66262	0,66480	0,66697	0,66913	48
42	0,66913	0,67129	0,67344	0,67559	0,67773	0,67987	0,68200	47
43	0,68200	0,68412	0,68624	0,68835	0,69046	0,69256	0,69466	46
44	0,69466	0,69675	0,69883	0,70091	0,70298	0,70505	0,70711	45
	60′	50′	40′	30′	20′	10′	0	

←↑
Cos

Trigonometric Tables   Cosine — Sine

Cos ↓→	0′	10′	20′	30′	40′	50′	60′	
0	1,00000	1,00000	0,99998	0,99996	0,99993	0,99989	0,99985	89
1	0,99979	0,99979	0,99973	0,99966	0,99958	0,99949	0,99939	88
2	0,99939	0,99929	0,99917	0,99905	0,99892	0,99878	0,99863	87
3	0,99863	0,99847	0,99831	0,99813	0,99795	0,99776	0,99756	86
4	0,99756	0,99736	0,99714	0,99692	0,99668	0,99644	0,99619	85
5	0,99619	0,99594	0,99567	0,99540	0,99511	0,99482	0,99452	84
6	0,99452	0,99421	0,99390	0,99357	0,99324	0,99290	0,99255	83
7	0,99255	0,99219	0,99182	0,99144	0,99106	0,99067	0,99027	82
8	0,99027	0,98986	0,98944	0,98902	0,98858	0,98814	0,98769	81
9	0,98769	0,98723	0,98676	0,98629	0,98580	0,98531	0,98481	80
10	0,98481	0,98430	0,98378	0,98325	0,98272	0,98218	0,98163	79
11	0,98163	0,98107	0,98050	0,97992	0,97934	0,97875	0,97815	78
12	0,97815	0,97754	0,97692	0,97630	0,97566	0,97502	0,97437	77
13	0,97437	0,97371	0,97304	0,97237	0,97169	0,97100	0,97030	76
14	0,97030	0,96959	0,96887	0,96815	0,96742	0,96667	0,96593	75
15	0,96593	0,96517	0,96440	0,96363	0,96285	0,96206	0,96126	74
16	0,96126	0,96046	0,95964	0,95882	0,95799	0,95715	0,95630	73
17	0,95630	0,95545	0,95459	0,95372	0,95284	0,95195	0,95106	72
18	0,95106	0,95015	0,94924	0,94832	0,94740	0,94646	0,94552	71
19	0,94552	0,94457	0,94361	0,94264	0,94167	0,94068	0,93969	70
20	0,93969	0,93869	0,93769	0,93667	0,93565	0,93462	0,93358	69
21	0,93358	0,93253	0,93148	0,93042	0,92935	0,92827	0,92718	68
22	0,92718	0,92609	0,92499	0,92388	0,92276	0,92164	0,92050	67
23	0,92050	0,91936	0,91822	0,91706	0,91590	0,91472	0,91355	66
24	0,91355	0,91236	0,91116	0,90996	0,90875	0,90753	0,90631	65
25	0,90631	0,90507	0,90383	0,90259	0,90133	0,90007	0,89879	64
26	0,89879	0,89752	0,89623	0,89493	0,89363	0,89232	0,89101	63
27	0,89101	0,88968	0,88835	0,88701	0,88566	0,88431	0,88295	62
28	0,88295	0,88158	0,88020	0,87882	0,87743	0,87603	0,87462	61
29	0,87462	0,87321	0,87178	0,87036	0,86892	0,86748	0,86603	60
30	0,86603	0,86457	0,86310	0,86163	0,86015	0,85866	0,85717	59
31	0,85717	0,85567	0,85416	0,85264	0,85112	0,84959	0,84805	58
32	0,84805	0,84650	0,84495	0,84339	0,84182	0,84025	0,83867	57
33	0,83867	0,83708	0,83549	0,83389	0,83228	0,83066	0,82904	56
34	0,82904	0,82741	0,82577	0,82413	0,82248	0,82082	0,81915	55
35	0,81915	0,81748	0,81580	0,81412	0,81242	0,81072	0,80902	54
36	0,80902	0,80730	0,80558	0,80386	0,80212	0,80038	0,79864	53
37	0,79864	0,79688	0,79512	0,79335	0,79158	0,78980	0,78801	52
38	0,78801	0,78622	0,78442	0,78261	0,78079	0,77897	0,77715	51
39	0,77715	0,77531	0,77347	0,77162	0,76977	0,76791	0,76604	50
40	0,76604	0,76417	0,76229	0,76041	0,75851	0,75661	0,75471	49
41	0,75471	0,75280	0,75088	0,74896	0,74703	0,74509	0,74314	48
42	0,74314	0,74120	0,73924	0,73728	0,73531	0,73333	0,73135	47
43	0,73135	0,72937	0,72737	0,72537	0,72337	0,72136	0,71934	46
44	0,71934	0,71732	0,71529	0,71325	0,71121	0,70916	0,70711	45
	60′	50′	40′	30′	20′	10′	0′	

←↑
Sin

Trigonometric Tables — Tangent—Cotangent

Tan ↓→	0′	10′	20′	30′	40′	50′	60′	
0	0,00000	0,00291	0,00582	0,00873	0,01164	0,01455	0,01746	89
1	0,01746	0,02036	0,02328	0,02619	0,02910	0,03201	0,03492	88
2	0,03492	0,03783	0,04075	0,04366	0,04658	0,04949	0,05241	87
3	0,05241	0,05533	0,05824	0,06116	0,06408	0,06700	0,06993	86
4	0,06993	0,07285	0,07578	0,07870	0,08163	0,08456	0,08749	85
5	0,08749	0,09042	0,09335	0,09629	0,09923	0,10216	0,10510	84
6	0,10510	0,10805	0,11099	0,11394	0,11688	0,11983	0,12278	83
7	0,12278	0,12574	0,12869	0,13165	0,13461	0,13758	0,14054	82
8	0,14054	0,14351	0,14648	0,14945	0,15243	0,15540	0,15838	81
9	0,15838	0,16137	0,16435	0,16734	0,17033	0,17333	0,17633	80
10	0,17633	0,17933	0,18233	0,18534	0,18835	0,19136	0,19438	79
11	0,19438	0,19740	0,20042	0,20345	0,20648	0,20952	0,21256	78
12	0,21256	0,21560	0,21864	0,22169	0,22475	0,22781	0,23087	77
13	0,23087	0,23393	0,23700	0,24008	0,24316	0,24624	0,24933	76
14	0,24933	0,25242	0,25552	0,25862	0,26172	0,26483	0,26795	75
15	0,26795	0,27107	0,27419	0,27732	0,28046	0,28360	0,28675	74
16	0,28675	0,28990	0,29305	0,29621	0,29938	0,30255	0,30573	73
17	0,30573	0,30891	0,31210	0,31530	0,31850	0,32171	0,32492	72
18	0,32492	0,32814	0,33136	0,33460	0,33783	0,34108	0,34433	71
19	0,34433	0,34758	0,35085	0,35412	0,35740	0,36068	0,36397	70
20	0,36397	0,36727	0,37057	0,37388	0,37720	0,38053	0,38386	69
21	0,38386	0,38721	0,39055	0,39391	0,39727	0,40065	0,40403	68
22	0,40403	0,40741	0,41081	0,41421	0,41763	0,42105	0,42447	67
23	0,42447	0,42791	0,43136	0,43481	0,43828	0,44175	0,44523	66
24	0,44523	0,44872	0,45222	0,45573	0,45924	0,46277	0,46631	65
25	0,46631	0,46985	0,47341	0,47698	0,48055	0,48414	0,48773	64
26	0,48773	0,49134	0,49495	0,49858	0,50222	0,50587	0,50953	63
27	0,50953	0,51319	0,51688	0,52057	0,52427	0,52798	0,53171	62
28	0,53171	0,53545	0,53920	0,54296	0,54673	0,55051	0,55431	61
29	0,55431	0,55812	0,56194	0,56577	0,56962	0,57348	0,57735	60
30	0,57735	0,58124	0,58513	0,58905	0,59297	0,59691	0,60086	59
31	0,60086	0,60483	0,60881	0,61280	0,61681	0,62083	0,62487	58
32	0,62487	0,62892	0,63299	0,63707	0,64117	0,64528	0,64941	57
33	0,64941	0,65355	0,65771	0,66189	0,66608	0,67028	0,67451	56
34	0,67451	0,67875	0,68301	0,68728	0,69157	0,69588	0,70021	55
35	0,70021	0,70455	0,70891	0,71329	0,71769	0,72211	0,72654	54
36	0,72654	0,73100	0,73547	0,73996	0,74447	0,74900	0,75355	53
37	0,75355	0,75812	0,76272	0,76733	0,77196	0,77661	0,78129	52
38	0,78129	0,78598	0,79070	0,79544	0,80020	0,80498	0,80978	51
39	0,80978	0,81461	0,81946	0,82434	0,82923	0,83415	0,83910	50
40	0,83910	0,84407	0,84906	0,85408	0,85912	0,86419	0,86929	49
41	0,86929	0,87441	0,87955	0,88473	0,88992	0,89515	0,90040	48
42	0,90040	0,90569	0,91090	0,91633	0,92170	0,92709	0,93252	47
43	0,93252	0,93797	0,94345	0,94896	0,95451	0,96008	0,96569	46
44	0,96569	0,97133	0,97700	0,98270	0,98843	0,99420	1,00000	45
	60′	50′	40′	30′	20′	10′	0′	

←↑
Cot

Trigonometric Tables   Cotangent— Tangent

Cot ↓→	0′	10′	20′	30′	40′	50′	60′	
0	∞	343,77371	171,88540	114,58865	85,93979	68,75009	57,28996	89
1	57,28996	49,10388	42,96408	38,18846	34,36777	31,24158	28,63625	88
2	28,63625	26,43160	24,54176	22,90377	21,47040	20,20555	19,08114	87
3	19,08114	18,07498	17,16934	16,34986	15,60478	14,92442	14,30067	86
4	14,30067	13,72674	13,19688	12,70621	12,25051	11,82617	11,43005	85
5	11,43005	11,05943	10,71191	10,38540	10,07803	9,78817	9,51436	84
6	9,51436	9,25530	9,00983	8,77689	8,55555	8,34496	8,14435	83
7	8,14435	7,95302	7,77035	7,59575	7,42871	7,26873	7,11537	82
8	7,11537	6,96823	6,82694	6,69116	6,56055	6,43484	6,31375	81
9	6,31375	6,19703	6,08444	5,97576	5,87080	5,76937	5,67128	80
10	5,67128	5,57638	5,48451	5,39552	5,30928	5,22566	5,14455	79
11	5,14455	5,06584	4,98940	4,91516	4,84300	4,77286	4,70463	78
12	4,70463	4,63825	4,57363	4,51071	4,44942	4,38969	4,33148	77
13	4,33148	4,27471	4,21933	4,16530	4,11256	4,06107	4,01078	76
14	4,01078	3,96165	3,91364	3,86671	3,82083	3,77595	3,73205	75
15	3,73205	3,68909	3,64705	3,60588	3,56557	3,52609	3,48741	74
16	3,48741	3,44951	3,41236	3,37594	3,34023	3,30521	3,27085	73
17	3,27085	3,23714	3,20406	3,17159	3,13972	3,10842	3,07768	72
18	3,07768	3,04749	3,01783	2,98869	2,96004	2,93189	2,90421	71
19	2,90421	2,87700	2,85023	2,82391	2,79802	2,77254	2,74748	70
20	2,74748	2,72281	2,69853	2,67462	2,65109	2,62791	2,60509	69
21	2,60509	2,58261	2,56046	2,53865	2,51715	2,49597	2,47509	68
22	2,47509	2,45451	2,43422	2,41421	2,39449	2,37504	2,35585	67
23	2,35585	2,33693	2,31826	2,29984	2,28167	2,26374	2,24604	66
24	2,24604	2,22857	2,21132	2,19430	2,17749	2,16090	2,14451	65
25	2,14451	2,12832	2,11233	2,09654	2,08094	2,06553	2,05030	64
26	2,05030	2,03526	2,02039	2,00569	1,99116	1,97680	1,96261	63
27	1,96261	1,94858	1,93470	1,92098	1,90741	1,89400	1,88073	62
28	1,88073	1,86760	1,85462	1,84177	1,82906	1,81649	1,80405	61
29	1,80405	1,79174	1,77955	1,76749	1,75556	1,74375	1,73205	60
30	1,73205	1,72047	1,70901	1,69766	1,68643	1,67530	1,66428	59
31	1,66428	1,65337	1,64256	1,63185	1,62125	1,61074	1,60033	58
32	1,60033	1,59002	1,57981	1,56969	1,55966	1,54972	1,53987	57
33	1,53987	1,53010	1,52043	1,51084	1,50133	1,49190	1,48256	56
34	1,48256	1,47330	1,46411	1,45501	1,44598	1,43703	1,42815	55
35	1,42815	1,41934	1,41061	1,40195	1,39336	1,38484	1,37638	54
36	1,37638	1,36800	1,35968	1,35142	1,34323	1,33511	1,32704	53
37	1,32704	1,31904	1,31110	1,30323	1,29541	1,28764	1,27994	52
38	1,27994	1,27230	1,26471	1,25717	1,24969	1,24227	1,23490	51
39	1,23490	1,22758	1,22031	1,21310	1,20593	1,19882	1,19175	50
40	1,19175	1,18474	1,17777	1,17085	1,16398	1,15715	1,15037	49
41	1,15037	1,14363	1,13694	1,13029	1,12369	1,11713	1,11061	48
42	1,11061	1,10414	1,09770	1,09131	1,08496	1,07864	1,07237	47
43	1,07237	1,06613	1,05994	1,05378	1,04766	1,04158	1,03553	46
44	1,03553	1,02952	1,02355	1,01761	1,01170	1,00583	1,00000	45
	60′	50′	40′	30′	20′	10′	0′	

←↑
Tan

Involute Function 10° - 15°

Min.	10°	11°	12°	13°	14°	15°	Min.
0	0.0017941	0.0023941	0.0031171	0.0039754	0.0049819	0.0061498	0
1	0.0018031	0.0024051	0.0031302	0.0039909	0.0050000	0.0061707	1
2	0.0018122	0.0024161	0.0031434	0.0040065	0.0050182	0.0061917	2
3	0.0018213	0.0024272	0.0031566	0.0040221	0.0050364	0.0062127	3
4	0.0018305	0.0024383	0.0031699	0.0040377	0.0050546	0.0062337	4
5	0.0018397	0.0024495	0.0031832	0.0040534	0.0050729	0.0062548	5
6	0.0018489	0.0024607	0.0031966	0.0040692	0.0050912	0.0062760	6
7	0.0018581	0.0024719	0.0032100	0.0040849	0.0051096	0.0062972	7
8	0.0018674	0.0024831	0.0032234	0.0041007	0.0051280	0.0063184	8
9	0.0018767	0.0024944	0.0032369	0.0041166	0.0051465	0.0063397	9
10	0.0018860	0.0025057	0.0032504	0.0041325	0.0051650	0.0063611	10
11	0.0018954	0.0025171	0.0032639	0.0041484	0.0051835	0.0063825	11
12	0.0019048	0.0025285	0.0032775	0.0041644	0.0052021	0.0064039	12
13	0.0019142	0.0025399	0.0032911	0.0041804	0.0052208	0.0064254	13
14	0.0019237	0.0025513	0.0033048	0.0041965	0.0052395	0.0064470	14
15	0.0019332	0.0025628	0.0033185	0.0042126	0.0052582	0.0064686	15
16	0.0019427	0.0025744	0.0033322	0.0042288	0.0052770	0.0064902	16
17	0.0019523	0.0025859	0.0033460	0.0042450	0.0052958	0.0065119	17
18	0.0019619	0.0025975	0.0033598	0.0042612	0.0053147	0.0065337	18
19	0.0019715	0.0026091	0.0033736	0.0042775	0.0053336	0.0065555	19
20	0.0019812	0.0026208	0.0033875	0.0042938	0.0053526	0.0065773	20
21	0.0019909	0.0026325	0.0034014	0.0043101	0.0053716	0.0065992	21
22	0.0020006	0.0026443	0.0034154	0.0043266	0.0053907	0.0066211	22
23	0.0020103	0.0026560	0.0034294	0.0043430	0.0054098	0.0066431	23
24	0.0020201	0.0026678	0.0034434	0.0043595	0.0054289	0.0066652	24
25	0.0020299	0.0026797	0.0034575	0.0043760	0.0054481	0.0066873	25
26	0.0020398	0.0026916	0.0034716	0.0043926	0.0054674	0.0067094	26
27	0.0020496	0.0027035	0.0034858	0.0044092	0.0054867	0.0067316	27
28	0.0020596	0.0027154	0.0035000	0.0044259	0.0055060	0.0067539	28
29	0.0020695	0.0027274	0.0035142	0.0044426	0.0055254	0.0067762	29
30	0.0020795	0.0027394	0.0035285	0.0044593	0.0055448	0.0067985	30
31	0.0020895	0.0027515	0.0035428	0.0044761	0.0055643	0.0068209	31
32	0.0020995	0.0027636	0.0035572	0.0044929	0.0055838	0.0068434	32
33	0.0021096	0.0027757	0.0035716	0.0045098	0.0056034	0.0068659	33
34	0.0021197	0.0027879	0.0035860	0.0045267	0.0056230	0.0068884	34
35	0.0021298	0.0028001	0.0036005	0.0045437	0.0056427	0.0069110	35
36	0.0021400	0.0028123	0.0036150	0.0045607	0.0056624	0.0069337	36
37	0.0021502	0.0028246	0.0036296	0.0045777	0.0056822	0.0069564	37
38	0.0021605	0.0028369	0.0036441	0.0045948	0.0057020	0.0069791	38
39	0.0021707	0.0028493	0.0036588	0.0046120	0.0057218	0.0070019	39
40	0.0021810	0.0028616	0.0036735	0.0046291	0.0057417	0.0070248	40
41	0.0021914	0.0028741	0.0036882	0.0046464	0.0057617	0.0070477	41
42	0.0022017	0.0028865	0.0037029	0.0046636	0.0057817	0.0070706	42
43	0.0022121	0.0028990	0.0037177	0.0046809	0.0058017	0.0070936	43
44	0.0022226	0.0029115	0.0037325	0.0046983	0.0058218	0.0071167	44
45	0.0022330	0.0029241	0.0037474	0.0047157	0.0058420	0.0071398	45
46	0.0022435	0.0029367	0.0037623	0.0047331	0.0058622	0.0071630	46
47	0.0022541	0.0029494	0.0037773	0.0047506	0.0058824	0.0071862	47
48	0.0022646	0.0029620	0.0037923	0.0047681	0.0059027	0.0072095	48
49	0.0022752	0.0029747	0.0038073	0.0047857	0.0059230	0.0072328	49
50	0.0022859	0.0029875	0.0038224	0.0048033	0.0059434	0.0072561	50
51	0.0022965	0.0030003	0.0038375	0.0048210	0.0059638	0.0072796	51
52	0.0023073	0.0030131	0.0038527	0.0048387	0.0059843	0.0073030	52
53	0.0023180	0.0030260	0.0038679	0.0048564	0.0060048	0.0073266	53
54	0.0023288	0.0030389	0.0038831	0.0048742	0.0060254	0.0073501	54
55	0.0023396	0.0030518	0.0038984	0.0048921	0.0060460	0.0073738	55
56	0.0023504	0.0030648	0.0039137	0.0049099	0.0060667	0.0073975	56
57	0.0023613	0.0030778	0.0039291	0.0049279	0.0060874	0.0074212	57
58	0.0023722	0.0030908	0.0039445	0.0049458	0.0061081	0.0074450	58
59	0.0023831	0.0031039	0.0039599	0.0049638	0.0061289	0.0074688	59

Involute Functions 16°-21°

Min.	16°	17°	18°	19°	20°	21°	Min.
0	0.0074927	0.0090247	0.0107604	0.0127151	0.0149044	0.0173449	0
1	0.0075166	0.0090519	0.0107912	0.0127496	0.0149430	0.0173878	1
2	0.0075406	0.0090792	0.0108220	0.0127842	0.0149816	0.0174308	2
3	0.0075647	0.0091065	0.0108528	0.0128188	0.0150203	0.0174738	3
4	0.0075888	0.0091339	0.0108838	0.0128535	0.0150591	0.0175169	4
5	0.0076130	0.0091614	0.0109147	0.0128883	0.0150979	0.0175601	5
6	0.0076372	0.0091889	0.0109458	0.0129232	0.0151369	0.0176034	6
7	0.0076614	0.0092164	0.0109769	0.0129581	0.0151758	0.0176468	7
8	0.0076857	0.0092440	0.0110081	0.0129931	0.0152149	0.0176902	8
9	0.0077101	0.0092717	0.0110393	0.0130281	0.0152540	0.0177337	9
10	0.0077345	0.0092994	0.0110706	0.0130632	0.0152932	0.0177773	10
11	0.0077590	0.0093272	0.0111019	0.0130984	0.0153325	0.0178209	11
12	0.0077835	0.0093551	0.0111333	0.0131336	0.0153719	0.0178646	12
13	0.0078081	0.0093830	0.0111648	0.0131689	0.0154113	0.0179084	13
14	0.0078327	0.0094109	0.0111964	0.0132043	0.0154507	0.0179523	14
15	0.0078574	0.0094390	0.0112280	0.0132398	0.0154903	0.0179963	15
16	0.0078822	0.0094670	0.0112596	0.0132753	0.0155299	0.0180403	16
17	0.0079069	0.0094952	0.0112913	0.0133108	0.0155696	0.0180844	17
18	0.0079318	0.0095234	0.0113231	0.0133465	0.0156094	0.0181286	18
19	0.0079567	0.0095516	0.0113550	0.0133822	0.0156492	0.0181728	19
20	0.0079817	0.0095799	0.0113869	0.0134180	0.0156891	0.0182172	20
21	0.0080067	0.0096083	0.0114189	0.0134538	0.0157291	0.0182616	21
22	0.0080317	0.0096367	0.0114509	0.0134897	0.0157692	0.0183061	22
23	0.0080568	0.0096652	0.0114830	0.0135257	0.0158093	0.0183506	23
24	0.0080820	0.0096937	0.0115151	0.0135617	0.0158495	0.0183953	24
25	0.0081072	0.0097223	0.0115474	0.0135978	0.0158898	0.0184400	25
26	0.0081325	0.0097510	0.0115796	0.0136340	0.0159301	0.0184848	26
27	0.0081578	0.0097797	0.0116120	0.0136702	0.0159705	0.0185296	27
28	0.0081832	0.0098085	0.0116444	0.0137065	0.0160110	0.0185746	28
29	0.0082087	0.0098373	0.0116769	0.0137429	0.0160516	0.0186196	29
30	0.0082342	0.0098662	0.0117094	0.0137794	0.0160922	0.0186647	30
31	0.0082597	0.0098951	0.0117420	0.0138159	0.0161329	0.0187099	31
32	0.0082853	0.0099241	0.0117747	0.0138525	0.0161737	0.0187551	32
33	0.0083110	0.0099532	0.0118074	0.0138891	0.0162145	0.0188004	33
34	0.0083367	0.0099823	0.0118402	0.0139258	0.0162554	0.0188458	34
35	0.0083625	0.0100115	0.0118730	0.0139626	0.0162964	0.0188913	35
36	0.0083883	0.0100407	0.0119059	0.0139994	0.0163375	0.0189369	36
37	0.0084142	0.0100700	0.0119389	0.0140364	0.0163786	0.0189825	37
38	0.0084401	0.0100994	0.0119720	0.0140734	0.0164198	0.0190282	38
39	0.0084661	0.0101288	0.0120051	0.0141104	0.0164611	0.0190740	39
40	0.0084921	0.0101583	0.0120382	0.0141475	0.0165024	0.0191199	40
41	0.0085182	0.0101878	0.0120715	0.0141847	0.0165439	0.0191659	41
42	0.0085444	0.0102174	0.0121048	0.0142220	0.0165854	0.0192119	42
43	0.0085706	0.0102471	0.0121381	0.0142593	0.0166269	0.0192580	43
44	0.0085969	0.0102768	0.0121715	0.0142967	0.0166686	0.0193042	44
45	0.0086232	0.0103066	0.0122050	0.0143342	0.0167103	0.0193504	45
46	0.0086496	0.0103364	0.0122386	0.0143717	0.0167521	0.0193968	46
47	0.0086760	0.0103663	0.0122722	0.0144093	0.0167939	0.0194432	47
48	0.0087025	0.0103963	0.0123059	0.0144470	0.0168359	0.0194897	48
49	0.0087290	0.0104263	0.0123396	0.0144847	0.0168779	0.0195363	49
50	0.0087556	0.0104564	0.0123734	0.0145225	0.0169200	0.0195829	50
51	0.0087823	0.0104865	0.0124073	0.0145604	0.0169621	0.0196296	51
52	0.0088090	0.0105167	0.0124412	0.0145983	0.0170044	0.0196765	52
53	0.0088358	0.0105469	0.0124752	0.0146363	0.0170467	0.0197233	53
54	0.0088626	0.0105773	0.0125093	0.0146744	0.0170891	0.0197703	54
55	0.0088895	0.0106076	0.0125434	0.0147126	0.0171315	0.0198174	55
56	0.0089164	0.0106381	0.0125776	0.0147508	0.0171740	0.0198645	56
57	0.0089434	0.0106686	0.0126119	0.0147891	0.0172166	0.0199117	57
58	0.0089704	0.0106991	0.0126462	0.0148275	0.0172593	0.0199590	58
59	0.0089975	0.0107298	0.0126806	0.0148659	0.0173021	0.0200063	59

Involute Functions 22° - 27°

Min.	22°	23°	24°	25°	26°	27°	Min.
0	0.0200538	0.0230491	0.0263497	0.0299753	0.0339470	0.0382866	0
1	0.0201013	0.0231015	0.0264074	0.0300386	0.0340162	0.0383621	1
2	0.0201489	0.0231541	0.0264652	0.0301020	0.0340856	0.0384378	2
3	0.0201966	0.0232067	0.0265231	0.0301655	0.0341550	0.0385136	3
4	0.0202444	0.0232594	0.0265810	0.0302291	0.0342246	0.0385895	4
5	0.0202922	0.0233122	0.0266391	0.0302928	0.0342942	0.0386655	5
6	0.0203401	0.0233651	0.0266973	0.0303566	0.0343640	0.0387416	6
7	0.0203881	0.0234181	0.0267555	0.0304205	0.0344339	0.0388179	7
8	0.0204362	0.0234711	0.0268139	0.0304844	0.0345038	0.0388942	8
9	0.0204844	0.0235242	0.0268723	0.0305485	0.0345739	0.0389706	9
10	0.0205326	0.0235775	0.0269308	0.0306127	0.0346441	0.0390472	10
11	0.0205809	0.0236308	0.0269894	0.0306769	0.0347144	0.0391239	11
12	0.0206293	0.0236842	0.0270481	0.0307413	0.0347847	0.0392006	12
13	0.0206778	0.0237376	0.0271069	0.0308058	0.0348552	0.0392775	13
14	0.0207264	0.0237912	0.0271658	0.0308703	0.0349258	0.0393545	14
15	0.0207750	0.0238449	0.0272248	0.0309350	0.0349965	0.0394316	15
16	0.0208238	0.0238986	0.0272839	0.0309997	0.0350673	0.0395088	16
17	0.0208726	0.0239524	0.0273430	0.0310646	0.0351382	0.0395862	17
18	0.0209215	0.0240063	0.0274023	0.0311295	0.0352092	0.0396636	18
19	0.0209704	0.0240603	0.0274617	0.0311946	0.0352803	0.0397411	19
20	0.0210195	0.0241144	0.0275211	0.0312597	0.0353515	0.0398188	20
21	0.0210686	0.0241686	0.0275806	0.0313250	0.0354228	0.0398966	21
22	0.0211178	0.0242228	0.0276403	0.0313903	0.0354942	0.0399745	22
23	0.0211671	0.0242772	0.0277000	0.0314557	0.0355658	0.0400524	23
24	0.0212165	0.0243316	0.0277598	0.0315213	0.0356374	0.0401306	24
25	0.0212660	0.0243861	0.0278197	0.0315869	0.0357091	0.0402088	25
26	0.0213155	0.0244407	0.0278797	0.0316527	0.0357810	0.0402871	26
27	0.0213651	0.0244954	0.0279398	0.0317185	0.0358529	0.0403655	27
28	0.0214148	0.0245502	0.0279999	0.0317844	0.0359249	0.0404441	28
29	0.0214646	0.0246050	0.0280602	0.0318504	0.0359971	0.0405227	29
30	0.0215145	0.0246600	0.0281206	0.0319166	0.0360694	0.0406015	30
31	0.0215644	0.0247150	0.0281810	0.0319828	0.0361417	0.0406804	31
32	0.0216145	0.0247702	0.0282416	0.0320491	0.0362142	0.0407594	32
33	0.0216646	0.0248254	0.0283022	0.0321156	0.0362868	0.0408385	33
34	0.0217148	0.0248807	0.0283630	0.0321821	0.0363594	0.0409177	34
35	0.0217651	0.0249361	0.0284238	0.0322487	0.0364322	0.0409970	35
36	0.0218154	0.0249916	0.0284848	0.0323154	0.0365051	0.0410765	36
37	0.0218659	0.0250471	0.0285458	0.0323823	0.0365781	0.0411561	37
38	0.0219164	0.0251028	0.0286069	0.0324492	0.0366512	0.0412357	38
39	0.0219670	0.0251585	0.0286681	0.0325162	0.0367244	0.0413155	39
40	0.0220177	0.0252143	0.0287294	0.0325833	0.0367977	0.0413954	40
41	0.0220685	0.0252703	0.0287908	0.0326506	0.0368712	0.0414754	41
42	0.0221193	0.0253263	0.0288523	0.0327179	0.0369447	0.0415555	42
43	0.0221703	0.0253824	0.0289139	0.0327853	0.0370183	0.0416358	43
44	0.0222213	0.0254386	0.0289756	0.0328528	0.0370921	0.0417161	44
45	0.0222724	0.0254948	0.0290373	0.0329205	0.0371659	0.0417966	45
46	0.0223236	0.0255512	0.0290992	0.0329882	0.0372399	0.0418772	46
47	0.0223749	0.0256076	0.0291612	0.0330560	0.0373139	0.0419579	47
48	0.0224262	0.0256642	0.0292232	0.0331239	0.0373881	0.0420387	48
49	0.0224777	0.0257208	0.0292854	0.0331920	0.0374624	0.0421196	49
50	0.0225292	0.0257775	0.0293476	0.0332601	0.0375368	0.0422006	50
51	0.0225808	0.0258343	0.0294100	0.0333283	0.0376113	0.0422818	51
52	0.0226325	0.0258912	0.0294724	0.0333967	0.0376859	0.0423630	52
53	0.0226843	0.0259482	0.0295349	0.0334651	0.0377606	0.0424444	53
54	0.0227361	0.0260053	0.0295976	0.0335336	0.0378354	0.0425259	54
55	0.0227881	0.0260625	0.0296603	0.0336023	0.0379103	0.0426075	55
56	0.0228401	0.0261197	0.0297231	0.0336710	0.0379853	0.0426892	56
57	0.0228922	0.0261771	0.0297860	0.0337399	0.0380605	0.0427710	57
58	0.0229444	0.0262345	0.0298490	0.0338088	0.0381357	0.0428530	58
59	0.0229967	0.0262920	0.0299121	0.0338778	0.0382111	0.0429351	59

MATHEMATICS

Involute Functions 28° - 33°

Min.	28°	29°	30°	31°	32°	33°	Min.
0	0.0430172	0.0481636	0.0537515	0.0598086	0.0663640	0.0734489	0
1	0.0430995	0.0482530	0.0538485	0.0599136	0.0664776	0.0735717	1
2	0.0431819	0.0483426	0.0539457	0.0600189	0.0665914	0.0736946	2
3	0.0432645	0.0484323	0.0540430	0.0601242	0.0667054	0.0738177	3
4	0.0433471	0.0485221	0.0541404	0.0602297	0.0668195	0.0739409	4
5	0.0434299	0.0486120	0.0542379	0.0603354	0.0669337	0.0740643	5
6	0.0435128	0.0487020	0.0543356	0.0604412	0.0670481	0.0741878	6
7	0.0435957	0.0487922	0.0544334	0.0605471	0.0671627	0.0743115	7
8	0.0436789	0.0488825	0.0545314	0.0606532	0.0672774	0.0744354	8
9	0.0437621	0.0489730	0.0546295	0.0607594	0.0673922	0.0745594	9
10	0.0438454	0.0490635	0.0547277	0.0608657	0.0675072	0.0746835	10
11	0.0439289	0.0491542	0.0548260	0.0609722	0.0676223	0.0748079	11
12	0.0440124	0.0492450	0.0549245	0.0610788	0.0677376	0.0749324	12
13	0.0440961	0.0493359	0.0550231	0.0611856	0.0678530	0.0750570	13
14	0.0441799	0.0494269	0.0551218	0.0612925	0.0679686	0.0751818	14
15	0.0442639	0.0495181	0.0552207	0.0613995	0.0680843	0.0753068	15
16	0.0443479	0.0496094	0.0553197	0.0615067	0.0682002	0.0754319	16
17	0.0444321	0.0497008	0.0554188	0.0616140	0.0683162	0.0755571	17
18	0.0445163	0.0497924	0.0555181	0.0617215	0.0684324	0.0756826	18
19	0.0446007	0.0498840	0.0556175	0.0618291	0.0685487	0.0758082	19
20	0.0446853	0.0499758	0.0557170	0.0619368	0.0686652	0.0759339	20
21	0.0447699	0.0500677	0.0558166	0.0620447	0.0687818	0.0760598	21
22	0.0448546	0.0501598	0.0559164	0.0621527	0.0688986	0.0761859	22
23	0.0449395	0.0502519	0.0560164	0.0622609	0.0690155	0.0763121	23
24	0.0450245	0.0503442	0.0561164	0.0623692	0.0691326	0.0764385	24
25	0.0451096	0.0504367	0.0562166	0.0624777	0.0692498	0.0765651	25
26	0.0451948	0.0505292	0.0563169	0.0625863	0.0693672	0.0766918	26
27	0.0452801	0.0506219	0.0564174	0.0626950	0.0694848	0.0768187	27
28	0.0453656	0.0507147	0.0565180	0.0628039	0.0696024	0.0769457	28
29	0.0454512	0.0508076	0.0566187	0.0629129	0.0697203	0.0770729	29
30	0.0455369	0.0509006	0.0567196	0.0630221	0.0698383	0.0772003	30
31	0.0456227	0.0509938	0.0568206	0.0631314	0.0699564	0.0773278	31
32	0.0457086	0.0510871	0.0569217	0.0632408	0.0700747	0.0774555	32
33	0.0457947	0.0511806	0.0570230	0.0633504	0.0701931	0.0775833	33
34	0.0458808	0.0512741	0.0571244	0.0634602	0.0703117	0.0777113	34
35	0.0459671	0.0513678	0.0572259	0.0635700	0.0704304	0.0778395	35
36	0.0460535	0.0514616	0.0573276	0.0636801	0.0705493	0.0779678	36
37	0.0461401	0.0515555	0.0574294	0.0637902	0.0706684	0.0780963	37
38	0.0462267	0.0516496	0.0575313	0.0639005	0.0707876	0.0782249	38
39	0.0463135	0.0517438	0.0576334	0.0640110	0.0709069	0.0783537	39
40	0.0464004	0.0518381	0.0577356	0.0641216	0.0710265	0.0784827	40
41	0.0464874	0.0519326	0.0578380	0.0642323	0.0711461	0.0786118	41
42	0.0465745	0.0520271	0.0579405	0.0643432	0.0712659	0.0787411	42
43	0.0466618	0.0521218	0.0580431	0.0644542	0.0713859	0.0788706	43
44	0.0467491	0.0522167	0.0581458	0.0645654	0.0715060	0.0790002	44
45	0.0468366	0.0523116	0.0582487	0.0646767	0.0716263	0.0791300	45
46	0.0469242	0.0524067	0.0583518	0.0647882	0.0717467	0.0792600	46
47	0.0470120	0.0525019	0.0584549	0.0648998	0.0718673	0.0793901	47
48	0.0470998	0.0525973	0.0585582	0.0650116	0.0719880	0.0795204	48
49	0.0471878	0.0526928	0.0586617	0.0651235	0.0721089	0.0796508	49
50	0.0472759	0.0527884	0.0587652	0.0652355	0.0722300	0.0797814	50
51	0.0473641	0.0528841	0.0588690	0.0653477	0.0723512	0.0799122	51
52	0.0474525	0.0529799	0.0589728	0.0654600	0.0724725	0.0800431	52
53	0.0475409	0.0530759	0.0590768	0.0655725	0.0725940	0.0801742	53
54	0.0476295	0.0531721	0.0591809	0.0656851	0.0727157	0.0803055	54
55	0.0477182	0.0532683	0.0592852	0.0657979	0.0728375	0.0804369	55
56	0.0478070	0.0533647	0.0593896	0.0659108	0.0729595	0.0805685	56
57	0.0478960	0.0534612	0.0594941	0.0660239	0.0730816	0.0807003	57
58	0.0479851	0.0535578	0.0595988	0.0661371	0.0732039	0.0808322	58
59	0.0480743	0.0536546	0.0597036	0.0662505	0.0733263	0.0809643	59

Involute Functions 34° - 39°

Min.	34°	35°	36°	37°	38°	39°	Min.
0	0.0810966	0.0893423	0.0982240	0.1077822	0.1180605	0.1291056	0
1	0.0812290	0.0894850	0.0983776	0.1079475	0.1182382	0.1292965	1
2	0.0813616	0.0896279	0.0985315	0.1081130	0.1184161	0.1294876	2
3	0.0814943	0.0897710	0.0986855	0.1082787	0.1185942	0.1296789	3
4	0.0816273	0.0899142	0.0988397	0.1084445	0.1187725	0.1298704	4
5	0.0817604	0.0900576	0.0989941	0.1086106	0.1189510	0.1300622	5
6	0.0818936	0.0902012	0.0991487	0.1087769	0.1191297	0.1302542	6
7	0.0820271	0.0903450	0.0993035	0.1089434	0.1193087	0.1304464	7
8	0.0821606	0.0904889	0.0994584	0.1091101	0.1194878	0.1306389	8
9	0.0822944	0.0906331	0.0996136	0.1092770	0.1196672	0.1308316	9
10	0.0824283	0.0907774	0.0997689	0.1094440	0.1198468	0.1310245	10
11	0.0825624	0.0909218	0.0999244	0.1096113	0.1200266	0.1312177	11
12	0.0826967	0.0910665	0.1000802	0.1097788	0.1202066	0.1314110	12
13	0.0828311	0.0912113	0.1002361	0.1099465	0.1203869	0.1316046	13
14	0.0829657	0.0913564	0.1003922	0.1101144	0.1205673	0.1317985	14
15	0.0831005	0.0915016	0.1005485	0.1102825	0.1207480	0.1319925	15
16	0.0832354	0.0916469	0.1007050	0.1104508	0.1209289	0.1321868	16
17	0.0833705	0.0917925	0.1008616	0.1106193	0.1211100	0.1323814	17
18	0.0835058	0.0919382	0.1010185	0.1107880	0.1212913	0.1325761	18
19	0.0836413	0.0920842	0.1011756	0.1109570	0.1214728	0.1327711	19
20	0.0837769	0.0922303	0.1013328	0.1111261	0.1216546	0.1329663	20
21	0.0839127	0.0923765	0.1014903	0.1112954	0.1218366	0.1331618	21
22	0.0840486	0.0925230	0.1016479	0.1114649	0.1220188	0.1333575	22
23	0.0841847	0.0926696	0.1018057	0.1116347	0.1222012	0.1335534	23
24	0.0843210	0.0928165	0.1019637	0.1118046	0.1223838	0.1337495	24
25	0.0844575	0.0929635	0.1021219	0.1119747	0.1225666	0.1339459	25
26	0.0845941	0.0931106	0.1022804	0.1121451	0.1227497	0.1341425	26
27	0.0847309	0.0932580	0.1024389	0.1123156	0.1229330	0.1343394	27
28	0.0848679	0.0934055	0.1025977	0.1124864	0.1231165	0.1345365	28
29	0.0850050	0.0935533	0.1027567	0.1126573	0.1233002	0.1347338	29
30	0.0851424	0.0937012	0.1029159	0.1128285	0.1234842	0.1349313	30
31	0.0852799	0.0938493	0.1030753	0.1129999	0.1236683	0.1351291	31
32	0.0854175	0.0939975	0.1032348	0.1131715	0.1238527	0.1353271	32
33	0.0855553	0.0941460	0.1033946	0.1133433	0.1240373	0.1355254	33
34	0.0856933	0.0942946	0.1035545	0.1135153	0.1242221	0.1357239	34
35	0.0858315	0.0944435	0.1037147	0.1136875	0.1244072	0.1359226	35
36	0.0859699	0.0945925	0.1038750	0.1138599	0.1245924	0.1361216	36
37	0.0861084	0.0947417	0.1040356	0.1140325	0.1247779	0.1363208	37
38	0.0862471	0.0948910	0.1041963	0.1142053	0.1249636	0.1365202	38
39	0.0863859	0.0950406	0.1043572	0.1143784	0.1251495	0.1367199	39
40	0.0865250	0.0951903	0.1045184	0.1145516	0.1253357	0.1369198	40
41	0.0866642	0.0953402	0.1046797	0.1147250	0.1255221	0.1371199	41
42	0.0868036	0.0954904	0.1048412	0.1148987	0.1257087	0.1373203	42
43	0.0869431	0.0956406	0.1050029	0.1150726	0.1258955	0.1375209	43
44	0.0870829	0.0957911	0.1051648	0.1152466	0.1260825	0.1377218	44
45	0.0872228	0.0959418	0.1053269	0.1154209	0.1262698	0.1379228	45
46	0.0873628	0.0960926	0.1054892	0.1155954	0.1264573	0.1381242	46
47	0.0875031	0.0962437	0.1056517	0.1157701	0.1266450	0.1383257	47
48	0.0876435	0.0963949	0.1058144	0.1159451	0.1268329	0.1385275	48
49	0.0877841	0.0965463	0.1059773	0.1161202	0.1270210	0.1387296	49
50	0.0879249	0.0966979	0.1061404	0.1162955	0.1272094	0.1389319	50
51	0.0880659	0.0968496	0.1063037	0.1164711	0.1273980	0.1391344	51
52	0.0882070	0.0970016	0.1064672	0.1166468	0.1275869	0.1393372	52
53	0.0883483	0.0971537	0.1066309	0.1168228	0.1277759	0.1395402	53
54	0.0884898	0.0973061	0.1067947	0.1169990	0.1279652	0.1397434	54
55	0.0886314	0.0974586	0.1069588	0.1171754	0.1281547	0.1399469	55
56	0.0887732	0.0976113	0.1071231	0.1173520	0.1283444	0.1401506	56
57	0.0889152	0.0977642	0.1072876	0.1175288	0.1285344	0.1403546	57
58	0.0890574	0.0979173	0.1074523	0.1177058	0.1287246	0.1405588	58
59	0.0891998	0.0980705	0.1076171	0.1178831	0.1289150	0.1407632	59

MATHEMATICS

Involute Functions 40° - 45°

Min.	40°	41°	42°	43°	44°	45°	Min.
0	0.1409679	0.1537017	0.1673658	0.1820235	0.1977439	0.2146018	0
1	0.1411729	0.1539217	0.1676017	0.1822766	0.1980153	0.2148929	1
2	0.1413780	0.1541419	0.1678380	0.1825300	0.1982871	0.2151843	2
3	0.1415835	0.1543623	0.1680745	0.1827837	0.1985591	0.2154760	3
4	0.1417891	0.1545831	0.1683113	0.1830377	0.1988315	0.2157681	4
5	0.1419950	0.1548040	0.1685484	0.1832920	0.1991042	0.2160605	5
6	0.1422012	0.1550253	0.1687857	0.1835465	0.1993772	0.2163533	6
7	0.1424076	0.1552468	0.1690234	0.1838014	0.1996505	0.2166464	7
8	0.1426142	0.1554685	0.1692613	0.1840566	0.1999242	0.2169398	8
9	0.1428211	0.1556906	0.1694994	0.1843121	0.2001982	0.2172336	9
10	0.1430282	0.1559128	0.1697379	0.1845678	0.2004724	0.2175277	10
11	0.1432355	0.1561354	0.1699767	0.1848239	0.2007471	0.2178222	11
12	0.1434432	0.1563582	0.1702157	0.1850803	0.2010220	0.2181170	12
13	0.1436510	0.1565812	0.1704550	0.1853369	0.2012972	0.2184121	13
14	0.1438591	0.1568046	0.1706946	0.1855939	0.2015728	0.2187076	14
15	0.1440675	0.1570281	0.1709344	0.1858512	0.2018487	0.2190035	15
16	0.1442761	0.1572520	0.1711746	0.1861087	0.2021249	0.2192996	16
17	0.1444849	0.1574761	0.1714150	0.1863666	0.2024014	0.2195962	17
18	0.1446940	0.1577005	0.1716557	0.1866248	0.2026783	0.2198930	18
19	0.1449033	0.1579251	0.1718967	0.1868832	0.2029554	0.2201903	19
20	0.1451129	0.1581500	0.1721380	0.1871420	0.2032329	0.2204878	20
21	0.1453227	0.1583752	0.1723795	0.1874011	0.2035108	0.2207857	21
22	0.1455328	0.1586006	0.1726214	0.1876604	0.2037889	0.2210840	22
23	0.1457431	0.1588263	0.1728635	0.1879201	0.2040674	0.2213826	23
24	0.1459537	0.1590523	0.1731059	0.1881801	0.2043462	0.2216815	24
25	0.1461645	0.1592785	0.1733486	0.1884404	0.2046253	0.2219808	25
26	0.1463756	0.1595050	0.1735915	0.1887010	0.2049047	0.2222805	26
27	0.1465869	0.1597318	0.1738348	0.1889619	0.2051845	0.2225805	27
28	0.1467985	0.1599588	0.1740783	0.1892230	0.2054646	0.2228808	28
29	0.1470103	0.1601861	0.1743221	0.1894845	0.2057450	0.2231815	29
30	0.1472223	0.1604136	0.1745662	0.1897463	0.2060257	0.2234826	30
31	0.1474347	0.1606414	0.1748106	0.1900084	0.2063068	0.2237840	31
32	0.1476472	0.1608695	0.1750553	0.1902709	0.2065882	0.2240857	32
33	0.1478600	0.1610979	0.1753003	0.1905336	0.2068699	0.2243878	33
34	0.1480731	0.1613265	0.1755455	0.1907966	0.2071520	0.2246903	34
35	0.1482864	0.1615554	0.1757911	0.1910599	0.2074344	0.2249931	35
36	0.1485000	0.1617846	0.1760369	0.1913236	0.2077171	0.2252962	36
37	0.1487138	0.1620140	0.1762830	0.1915875	0.2080001	0.2255997	37
38	0.1489279	0.1622437	0.1765294	0.1918518	0.2082835	0.2259036	38
39	0.1491422	0.1624737	0.1767761	0.1921163	0.2085672	0.2262078	39
40	0.1493568	0.1627039	0.1770230	0.1923812	0.2088512	0.2265124	40
41	0.1495716	0.1629344	0.1772703	0.1926464	0.2091356	0.2268173	41
42	0.1497867	0.1631652	0.1775179	0.1929119	0.2094203	0.2271226	42
43	0.1500020	0.1633963	0.1777657	0.1931777	0.2097053	0.2274282	43
44	0.1502176	0.1636276	0.1780138	0.1934438	0.2099907	0.2277342	44
45	0.1504335	0.1638592	0.1782622	0.1937102	0.2102764	0.2280406	45
46	0.1506496	0.1640910	0.1785109	0.1939769	0.2105624	0.2283473	46
47	0.1508659	0.1643232	0.1787599	0.1942440	0.2108487	0.2286543	47
48	0.1510825	0.1645556	0.1790092	0.1945113	0.2111354	0.2289618	48
49	0.1512994	0.1647882	0.1792588	0.1947790	0.2114225	0.2292695	49
50	0.1515165	0.1650212	0.1795087	0.1950469	0.2117098	0.2295777	50
51	0.1517339	0.1652544	0.1797589	0.1953152	0.2119975	0.2298862	51
52	0.1519515	0.1654879	0.1800093	0.1955838	0.2122855	0.2301950	52
53	0.1521694	0.1657217	0.1802601	0.1958527	0.2125739	0.2305042	53
54	0.1523875	0.1659557	0.1805111	0.1961220	0.2128626	0.2308138	54
55	0.1526059	0.1661900	0.1807624	0.1963915	0.2131516	0.2311238	55
56	0.1528246	0.1664246	0.1810141	0.1966613	0.2134410	0.2314341	56
57	0.1530435	0.1666595	0.1812660	0.1969315	0.2137307	0.2317447	57
58	0.1532626	0.1668946	0.1815182	0.1972020	0.2140207	0.2320557	58
59	0.1534821	0.1671301	0.1817707	0.1974728	0.2143111	0.2323671	59

Involute Functions 46° - 51°

Min.	46°	47°	48°	49°	50°	51°	Min.
0	0.2326789	0.2520640	0.2728545	0.2951571	0.3190890	0.3447792	0
1	0.2329910	0.2523987	0.2732135	0.2955422	0.3195024	0.3452231	1
2	0.2333034	0.2527338	0.2735729	0.2959279	0.3199162	0.3456675	2
3	0.2336163	0.2530693	0.2739328	0.2963140	0.3203306	0.3461124	3
4	0.2339295	0.2534051	0.2742930	0.2967005	0.3207454	0.3465579	4
5	0.2342430	0.2537414	0.2746537	0.2970875	0.3211608	0.3470038	5
6	0.2345570	0.2540781	0.2750148	0.2974749	0.3215766	0.3474503	6
7	0.2348713	0.2544151	0.2753764	0.2978628	0.3219930	0.3478974	7
8	0.2351859	0.2547526	0.2757383	0.2982512	0.3224098	0.3483450	8
9	0.2355010	0.2550904	0.2761007	0.2986400	0.3228271	0.3487931	9
10	0.2358163	0.2554287	0.2764635	0.2990292	0.3232449	0.3492417	10
11	0.2361321	0.2557673	0.2768268	0.2994190	0.3236632	0.3496909	11
12	0.2364482	0.2561064	0.2771904	0.2998092	0.3240820	0.3501406	12
13	0.2367647	0.2564458	0.2775545	0.3001998	0.3245013	0.3505908	13
14	0.2370816	0.2567856	0.2779190	0.3005909	0.3249211	0.3510416	14
15	0.2373988	0.2571258	0.2782840	0.3009825	0.3253414	0.3514929	15
16	0.2377165	0.2574665	0.2786493	0.3013745	0.3257621	0.3519448	16
17	0.2380344	0.2578075	0.2790151	0.3017670	0.3261834	0.3523972	17
18	0.2383528	0.2581489	0.2793814	0.3021599	0.3266052	0.3528501	18
19	0.2386715	0.2584907	0.2797480	0.3025533	0.3270275	0.3533036	19
20	0.2389906	0.2588329	0.2801151	0.3029472	0.3274503	0.3537576	20
21	0.2393101	0.2591755	0.2804826	0.3033416	0.3278736	0.3542122	21
22	0.2396299	0.2595185	0.2808506	0.3037364	0.3282973	0.3546673	22
23	0.2399501	0.2598619	0.2812189	0.3041316	0.3287216	0.3551229	23
24	0.2402707	0.2602058	0.2815877	0.3045274	0.3291464	0.3555791	24
25	0.2405916	0.2605500	0.2819570	0.3049236	0.3295717	0.3560359	25
26	0.2409130	0.2608946	0.2823267	0.3053202	0.3299975	0.3564931	26
27	0.2412347	0.2612396	0.2826968	0.3057174	0.3304238	0.3569510	27
28	0.2415567	0.2615850	0.2830673	0.3061150	0.3308506	0.3574093	28
29	0.2418792	0.2619309	0.2834383	0.3065130	0.3312779	0.3578683	29
30	0.2422020	0.2622771	0.2838097	0.3069116	0.3317057	0.3583277	30
31	0.2425252	0.2626237	0.2841815	0.3073106	0.3321341	0.3587878	31
32	0.2428488	0.2629708	0.2845538	0.3077101	0.3325629	0.3592483	32
33	0.2431728	0.2633182	0.2849265	0.3081100	0.3329922	0.3597094	33
34	0.2434971	0.2636661	0.2852997	0.3085105	0.3334221	0.3601711	34
35	0.2438218	0.2640143	0.2856733	0.3089113	0.3338524	0.3606333	35
36	0.2441469	0.2643630	0.2860473	0.3093127	0.3342833	0.3610961	36
37	0.2444724	0.2647121	0.2864218	0.3097146	0.3347147	0.3615594	37
38	0.2447982	0.2650616	0.2867967	0.3101169	0.3351466	0.3620233	38
39	0.2451245	0.2654115	0.2871721	0.3105197	0.3355790	0.3624878	39
40	0.2454511	0.2657618	0.2875479	0.3109229	0.3360119	0.3629527	40
41	0.2457781	0.2661125	0.2879241	0.3113267	0.3364454	0.3634183	41
42	0.2461055	0.2664636	0.2883008	0.3117309	0.3368793	0.3638844	42
43	0.2464332	0.2668151	0.2886779	0.3121356	0.3373138	0.3643511	43
44	0.2467614	0.2671671	0.2890554	0.3125408	0.3377488	0.3648183	44
45	0.2470899	0.2675194	0.2894334	0.3129464	0.3381843	0.3652861	45
46	0.2474188	0.2678722	0.2898119	0.3133525	0.3386203	0.3657544	46
47	0.2477481	0.2682254	0.2901908	0.3137591	0.3390568	0.3662233	47
48	0.2480778	0.2685790	0.2905701	0.3141662	0.3394939	0.3666928	48
49	0.2484078	0.2689330	0.2909499	0.3145738	0.3399315	0.3671628	49
50	0.2487383	0.2692874	0.2913301	0.3149819	0.3403695	0.3676334	50
51	0.2490691	0.2696422	0.2917108	0.3153904	0.3408082	0.3681045	51
52	0.2494003	0.2699975	0.2920919	0.3157994	0.3412473	0.3685763	52
53	0.2497319	0.2703531	0.2924735	0.3162089	0.3416870	0.3690485	53
54	0.2500639	0.2707092	0.2928555	0.3166189	0.3421271	0.3695214	54
55	0.2503963	0.2710657	0.2932380	0.3170293	0.3425678	0.3699948	55
56	0.2507290	0.2714226	0.2936209	0.3174403	0.3430091	0.3704688	56
57	0.2510622	0.2717800	0.2940043	0.3178517	0.3434508	0.3709433	57
58	0.2513957	0.2721377	0.2943881	0.3182637	0.3438931	0.3714185	58
59	0.2517296	0.2724959	0.2947724	0.3186761	0.3443359	0.3718942	59

Involute Functions 52° - 57°

Min.	52°	53°	54°	55°	56°	57°	Min.
0	0.3723704	0.4020203	0.4339041	0.4682169	0.5051766	0.5450273	0
1	0.3728473	0.4025329	0.4344555	0.4688106	0.5058164	0.5457175	1
2	0.3733247	0.4030461	0.4350076	0.4694050	0.5064569	0.5464085	2
3	0.3738026	0.4035599	0.4355604	0.4700001	0.5070983	0.5471005	3
4	0.3742812	0.4040744	0.4361138	0.4705960	0.5077405	0.5477933	4
5	0.3747603	0.4045894	0.4366679	0.4711926	0.5083835	0.5484870	5
6	0.3752400	0.4051051	0.4372227	0.4717900	0.5090273	0.5491816	6
7	0.3757203	0.4056214	0.4377782	0.4723881	0.5096719	0.5498771	7
8	0.3762012	0.4061384	0.4383343	0.4729869	0.5103173	0.5505735	8
9	0.3766826	0.4066559	0.4388911	0.4735865	0.5109635	0.5512708	9
10	0.3771646	0.4071741	0.4394487	0.4741868	0.5116106	0.5519689	10
11	0.3776472	0.4076930	0.4400069	0.4747879	0.5122585	0.5526680	11
12	0.3781304	0.4082124	0.4405657	0.4753897	0.5129071	0.5533679	12
13	0.3786141	0.4087325	0.4411253	0.4759923	0.5135566	0.5540688	13
14	0.3790984	0.4092552	0.4416856	0.4765956	0.5142069	0.5547705	14
15	0.3795834	0.4097746	0.4422465	0.4771996	0.5148581	0.5554731	15
16	0.3800689	0.4102966	0.4428081	0.4778044	0.5155100	0.5561767	16
17	0.3805549	0.4108192	0.4433705	0.4784100	0.5161628	0.5568811	17
18	0.3810416	0.4113424	0.4439335	0.4790163	0.5168164	0.5575864	18
19	0.3815289	0.4118663	0.4444972	0.4796234	0.5174708	0.5582927	19
20	0.3820167	0.4123908	0.4450616	0.4802312	0.5181260	0.5589998	20
21	0.3825051	0.4129160	0.4456267	0.4808398	0.5187821	0.5597078	21
22	0.3829941	0.4134418	0.4461924	0.4814492	0.5194390	0.5604168	22
23	0.3834837	0.4139682	0.4467589	0.4820593	0.5200967	0.5611267	23
24	0.3839739	0.4144953	0.4473261	0.4826701	0.5207553	0.5618374	24
25	0.3844647	0.4150230	0.4478940	0.4832817	0.5214147	0.5625491	25
26	0.3849561	0.4155514	0.4484626	0.4838941	0.5220749	0.5632617	26
27	0.3854481	0.4160804	0.4490318	0.4845073	0.5227360	0.5639752	27
28	0.3859406	0.4166101	0.4496018	0.4851212	0.5233979	0.5646896	28
29	0.3864338	0.4171403	0.4501725	0.4857359	0.5240606	0.5654050	29
30	0.3869275	0.4176713	0.4507439	0.4863513	0.5247242	0.5661213	30
31	0.3874219	0.4182029	0.4513159	0.4869675	0.5253886	0.5668384	31
32	0.3879168	0.4187351	0.4518887	0.4875845	0.5260538	0.5675565	32
33	0.3884123	0.4192680	0.4524622	0.4882022	0.5267199	0.5682756	33
34	0.3889085	0.4198015	0.4530364	0.4888207	0.5273868	0.5689955	34
35	0.3894052	0.4203357	0.4536113	0.4894400	0.5280546	0.5697164	35
36	0.3899025	0.4208705	0.4541869	0.4900601	0.5287232	0.5704382	36
37	0.3904004	0.4214060	0.4547632	0.4906809	0.5293927	0.5711609	37
38	0.3908990	0.4219421	0.4553403	0.4913026	0.5300630	0.5718846	38
39	0.3913981	0.4224789	0.4559180	0.4919249	0.5307342	0.5726092	39
40	0.3918978	0.4230164	0.4564965	0.4925481	0.5314062	0.5733347	40
41	0.3923982	0.4235545	0.4570757	0.4931721	0.5320791	0.5740612	41
42	0.3928991	0.4240932	0.4576555	0.4937968	0.5327528	0.5747886	42
43	0.3934007	0.4246326	0.4582361	0.4944223	0.5334274	0.5755169	43
44	0.3939028	0.4251727	0.4588175	0.4950486	0.5341028	0.5762462	44
45	0.3944056	0.4257134	0.4593995	0.4956757	0.5347791	0.5769764	45
46	0.3949089	0.4262548	0.4599823	0.4963035	0.5354563	0.5777076	46
47	0.3954129	0.4267969	0.4605657	0.4969322	0.5361343	0.5784397	47
48	0.3959175	0.4273396	0.4611499	0.4975616	0.5368132	0.5791727	48
49	0.3964227	0.4278830	0.4617349	0.4981918	0.5374929	0.5799067	49
50	0.3969285	0.4284270	0.4623205	0.4988228	0.5381735	0.5806417	50
51	0.3974349	0.4289717	0.4629069	0.4994546	0.5388550	0.5813776	51
52	0.3979419	0.4295171	0.4634940	0.5000872	0.5395373	0.5821144	52
53	0.3984496	0.4300631	0.4640818	0.5007206	0.5402205	0.5828522	53
54	0.3989578	0.4306098	0.4646703	0.5013548	0.5409046	0.5835910	54
55	0.3994667	0.4311572	0.4652596	0.5019897	0.5415895	0.5843307	55
56	0.3999762	0.4317052	0.4658496	0.5026255	0.5422753	0.5850713	56
57	0.4004863	0.4322540	0.4664403	0.5032621	0.5429620	0.5858129	57
58	0.4009970	0.4328033	0.4670318	0.5038995	0.5436495	0.5865555	58
59	0.4015084	0.4333534	0.4676240	0.5045376	0.5443380	0.5872991	59

Areas and Volumes

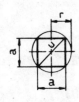

SQUARE

Circumference $s = 4a$

Area $A = a^2 = \dfrac{u^2}{2}$

Diagonal $u = a\sqrt{2} \approx 1.414\,a$

RECTANGLE

Circumference $s = 2\,(a + b)$

Area $A = ab$

Diagonal $u = \sqrt{a^2 + b^2}$

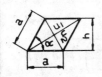

RHOMBUS

Circumference $s = 4a$

Area $A = ah = a^2 \sin\alpha = \dfrac{u_1 u_2}{2}$

Height $h = a \sin\alpha$

Diagonal $u_1 = a\sqrt{2\,(1 + \cos\alpha)}$

$u_2 = a\sqrt{2\,(1 - \cos\alpha)}$

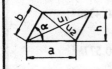

PARALLELOGRAM

Circumference $s = 2\,(a + b)$

Area $A = ah = ab \sin\alpha$

Height $h = b \sin\alpha$

Diagonal $u_1 = \sqrt{a^2 + b^2 + 2ab\,\cos\alpha}$

$u_2 = \sqrt{a^2 + b^2 - 2ab\,\cos\alpha}$

Areas and Volumes

(contd.)

TRIANGLE

Circumference $s = a + b + c$

Area $A = \dfrac{ch}{2} = \frac{1}{2}\, bc \sin a$

Centre of gravity $OT = t/3$

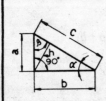

RIGHTANGLED TRIANGLE

Area $A = \dfrac{ab}{2} = \dfrac{ch}{2}$ $c^2 = a^2 + b^2$

$\sin a = \dfrac{a}{c};\quad \cos a = \dfrac{b}{c};$

$\tan a = \dfrac{a}{b};\quad \cot a = \dfrac{b}{a}$

Height $h = b\,\sin a = a\,\sin \beta$

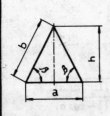

ISOSCELES TRIANGLE

Area $A = \dfrac{ah}{2} = \dfrac{ab\,\sin \beta}{2}$

Height $h = \sqrt{b^2 - \left(\dfrac{a}{2}\right)^2} = b\,\sin \beta$

Side $b = \sqrt{h^2 + \left(\dfrac{a}{2}\right)^2} = \dfrac{a}{2\cos\beta}$

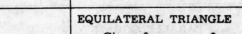

EQUILATERAL TRIANGLE

Circumference $s = 3a$

Area $A = \dfrac{a^2}{4}\sqrt{3} \approx 0.433a^2$

$\quad\quad = \dfrac{h^2}{3}\sqrt{3} \approx 0.577h^2$

Height $h = \dfrac{a}{2}\sqrt{3} \approx 0.866a$

Areas and Volumes

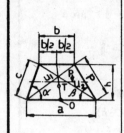

TRAPEZIUM

Circumference $s = a + b + c + d$

Area $A = Ph = Pc \sin a$

Mean width $P = \dfrac{a + b}{2}$

Height $h = c \sin a = d \sin \beta$

Diagonal $u_1 = \sqrt{a^2 + d^2 - 2ad \cos \beta}$

$u_2 = \sqrt{a^2 + c^2 - 2ac \cos a}$

Centre of gravity $OT = \tfrac{1}{3} h \, \dfrac{2a + b}{a + b}$

REGULAR POLYGON

n = number of sides

Central angle $a = 360°/n$

Circumference $s = na$

Area $A = \tfrac{1}{2} nar_2$

CIRCLE

Circumference $s = \pi d$

Area $A = \dfrac{\pi d^2}{4}$

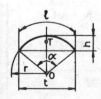

ARC OF A CIRCLE

Length of arc $l = \dfrac{\pi}{180} ra = r \, \text{arc} \, a$

$= 0.01745 \, r a$

Height of arc $h = r - \tfrac{1}{2} \sqrt{4 r^2 - t^2}$

$= r \left(1 - \cos \dfrac{a}{2} \right)$

Length of chord $t = 2\sqrt{h(2r - h)} = 2r \sin \dfrac{a}{2}$

Radius $r = \dfrac{t^2}{8h} + \dfrac{h}{2}$

Centre of gravity $OT = r \cdot \dfrac{t}{l}$

Areas and Volumes

SECTOR OF A CIRCLE

$$\text{Area } A = \frac{lr}{2} = \frac{\pi r^2 \cdot a}{360} = \tfrac{1}{2}\, r^2 \text{ arc } a$$

Centre of gravity $OT = \tfrac{2}{3}\, r \cdot \dfrac{t}{l}$

SEGMENT OF A CIRCLE

$$\text{Area } A = \tfrac{1}{2}\,[rl - t\,(r - h)]$$
$$= \tfrac{1}{2}r^2\,(\text{arc } a - \sin a)$$

Centre of gravity $OT = \dfrac{t^3}{12\,A}$

$(A = \text{Surface Area})$

ANNULUS

$$\text{Area } A = \frac{\pi D^2}{4} - \frac{\pi d^2}{4} = \pi d_s a$$

CUBE

Surface area $A = 6a^2$
Volume $V = a^3$
Space diagonal $u = a\sqrt{3} \approx 1.732a$
Face diagonal $u_s = a\sqrt{2} \approx 1.414a$

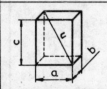

RECTANGULAR PRISM

Surface area $A = 2(ab + bc + ca)$
Volume $V = abc$
Space diagonal $u = \sqrt{a^2 + b^2 + c^2}$

RIGHT POLYGONAL PRISM

$S = \text{Area of base}$
$s = \text{Circumference of base}$
Surface Area $A = 2S + sh$
Volume $V = Sh$

Areas and Volumes

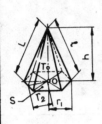

REGULAR PYRAMID

S = Area of base
s = Circumference of base
r_1 = Radius of circumscribed circle
r_2 = Radius of inscribed circle

Surface Area $A = S + \dfrac{sL}{2} = \dfrac{s}{2}(r_2 + L)$

Volume $V = \dfrac{Sh}{3}$

Height of slant face $L = \sqrt{h^2 + r_2^2}$
Height of slant edge $l = \sqrt{h^2 + r_1^2}$
Centre of gravity $OT = h/4$

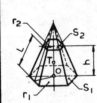

FRUSTUM OF A REGULAR PYRAMID

S_1 = Area of bottom base
S_2 = Area of top face
s_1 = Circumference of base
s_2 = Circumference of top face

Surface area $A = S_1 + S_2 + \dfrac{L}{2}(s_1 + s_2)$

Volume $V = \dfrac{h}{3}(S_1 + S_2 + \sqrt{S_1 S_2})$

Height of slant face $L = \sqrt{(r_1 - r_2)^2 + h^2}$

Centre of gravity OT

$$= \frac{h}{4} \frac{S_1 + 2\sqrt{S_1 S_2} + 3 S_2}{S_1 + \sqrt{S_1 S_2} + S_2}$$

CYLINDER

Surface area $A = \dfrac{\pi d^2}{2} + \pi dh$

Volume $V = \dfrac{\pi d^2}{4} h$

Areas and Volumes

(contd.)

CONE

Surface area $A = \dfrac{\pi d^2}{4} + \dfrac{\pi d L}{2}$

Volume $V = \dfrac{\pi d^2}{4} \cdot \dfrac{h}{3}$

Length of generator $L = \sqrt{r^2 + h^2}$

Centre of gravity $OT = \dfrac{h}{4}$

FRUSTUM OF A CONE

Surface area $S = \dfrac{\pi D^2}{4} + \dfrac{\pi d^2}{4} + \dfrac{\pi}{2}(D+d)L$

Volume $V = \dfrac{\pi h}{12}(D^2 + Dd + d^2)$

Length of generator $L = \sqrt{\left(\dfrac{D-d}{2}\right)^2 + h^2}$

Centre of gravity $OT = \dfrac{h}{4} \cdot \dfrac{R^2 + 2Rr + 3r^2}{R^2 + Rr + r^2}$

SPHERE

Surface area $A = \pi d^2$

Volume $V = \dfrac{\pi d^3}{6}$

SEGMENT OF A SPHERE

Surface area $A = \dfrac{\pi t^2}{4} + 2\pi r h$

Volume $V = \dfrac{\pi t^2}{4} \cdot \dfrac{h}{2} + \dfrac{\pi h^3}{6}$

Radius $r = \dfrac{t^2}{8h} + \dfrac{h}{2}$

Centre of gravity $OT = \dfrac{3(2r - h^2)}{4(3r - h)}$

Areas and Volumes

SECTION OF A SPHERE

Surface area $A = \dfrac{\pi t_1^2}{4} + \dfrac{\pi t_2^2}{4} + 2\pi rh$

Volume $V = \dfrac{h}{2}\left(\dfrac{\pi t_1^2}{4} + \dfrac{\pi t_2^2}{4} + \dfrac{\pi h^2}{3}\right)$

SECTOR OF A SPHERE

Surface area $A = 2\pi rh + \frac{1}{2}\pi rt$

Volume $V = \frac{2}{3}\pi r^2 h$

Diameter $t = 2\sqrt{h(2r-h)}$

Centre of gravity $OT = \frac{3}{4}\left(r - \dfrac{h}{2}\right)$

RING WITH CIRCULAR SECTION

Surface area $A = \pi^2 dD$

Volume $V = \dfrac{\pi^2}{4}d^2 D$

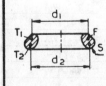

ROTARY RING WITH ARBITRARY SECTION

$T_1 =$ Centre of gravity of contour line of section

$T_2 =$ Centre of gravity of surface of section

$s =$ Circumference of section

$F =$ Area of section

Surface Area $A = s\pi d_1$

Volume $V = F\pi d_2$

MECHANICS

SI Units

BASE UNITS

QUANTITY	UNIT	SYMBOL
Length	Metre	m
Mass	Kilogram	kg
Time	Second	s
Electric current	Ampere	A
Thermodynamic temp	Kelvin	K
Luminous intensity	Candela	cd
Plane angle *	Radian	rad
Solid angle *	Steradian	sr

* Supplementary Units

MULTIPLES OF SI UNITS

MULITIPLYING FACTOR	PREFIX	
	UNIT	SYMBOL
10^{12}	Tera	T
10^9	Giga	G
10^6	Mega	M
10^3	Kilo	k
10^2	Hecto	h
10	Deca	da
10^{-1}	Deci	d
10^{-2}	Centi	c
10^{-3}	Milli	m
10^{-6}	Micro	μ
10^{-9}	Nano	n
10^{-12}	Pico	p
10^{-15}	Femto	f
10^{-18}	Atto	a

DERIVED UNITS

QUANTITY	UNIT	SYMBOL	IN TERMS OF BASE OR DERIVED SI UNITS
Frequency	Hertz	Hz	$1\ Hz = 1\ c/s$
Force	Newton	N	$1\ N = 1\ kgm/s^2$
Pressure and stress	Pascal	Pa	$1\ Pa = 1\ N/m^2$
Work, energy	Joule	J	$1\ J = 1\ Nm$
Power	Watt	W	$1\ W = 1\ J/s$
Quantity of Electricity	Coulomb	C	$1\ C = 1\ As$
Electric potential	Volt	V	$1\ V = 1\ W/A$
Electric Capacitance	Farad	F	$1\ F = 1\ As/V$
Electric resistance	Ohm	Ω	$1\ \Omega = 1\ V/A$
Electric conductance	Siemens	S	$1\ S = 1\ \Omega^{-1}$
Magnetic flux	Weber	Wb	$1\ Wb = 1\ Vs$
Magnetic flux density	Tesla	T	$1\ T = 1\ Wb/m^2$
Inductance	Henry	H	$1\ H = 1\ Vs/A$
Luminous flux	Lumen	lm	$1\ lm = 1\ cd.sr$
Illuminance	Lux	lx	$1\ lx = 1\ lm/m^2$

EXAMPLES OF MULTIPLES AND SUBMULTIPLES OF SI UNITS				
Quantity	SI unit	Preferred multiple of SI unit	Other Units which may be used	Remarks
Plane angle	rad (Radian)	m rad μ rad	$^\circ$ (Degree) (Minute) $''$ (Second)	—
Solid angle	sr (Steradian)	—	—	—
Length	m (Metre)	km $\quad$ cm mm μm nm	$^\circ$A for wave length	$1\,^\circ$A $= 10^{-10}m$
Area	m^2	km^2 $\quad$ dm^2 cm^2 mm^2	are (hectare)	1 are $= 100 m^2$ 1 hectare $= 100$ ares $= 10^4 m^2$
Volume	m^3	dm^3 cm^3 mm^3	l (Litre)	$1\ l = 1000\ cm^3$ $1\ ml = 1\ cm^3$
Time	s (Second)	ks $\quad$ ms μs ns	d (Day) h (Hour) $min.$ (Minute)	—
Angular Velocity	rad/s	—	—	—
Velocity	m/s	—	—	$1\ km/h = \dfrac{1}{3.6}\ m/s$
Acceleration	m/s^2	—	—	$g = 9.80665\ m/s^2$
Frequency	Hz	—	—	$1\ Hz = 1\ c/s$
Rotational frequency	Hz	—	rpm	$1\ rpm = \dfrac{1}{60}\ Hz$
Wave length	m	—	$^\circ$A	$1\,^\circ$A $= 10^{-10}m$
Mass	kg (Kilogram)	Mg $\quad$ g mg μg	t (Tonne)	1 tonne $= 1000\ kg$
Mass density	kg/m^3	kg/dm^3 mg/m^3 g/cm^3	$tonne/m^3$ kg/l	—
Momentum	$kg.\,m/s$	—	—	—
Moment of momentum Angular momentum	$kg.m^2/s$	—	—	—

EXAMPLES OF MULTIPLES AND SUBMULTIPLES OF SI UNITS (contd.)

Quantity	SI Unit	Preferred multiple of SI unit	Other units which may be used	Remarks
Moment of inertia	$kg.m^2$	—	$kg\,f.m.s^2$	$1\ kgf.m.s^2$ $= 9.80665\ kg.m^2$
Force	N (Newton)	MN mN kN μN	kgf	$1\ kgf = 9.80665\ N$
Moment of force	$N.m$	$MN.m$ $mN.m$ $kN.m$ $\mu N.m$	$kgf.m$	$1\ kgf.m$ $= 9.80665\ N.m$
Pressure	Pa (Pascal)	$G\,Pa$ mPa $M\,Pa$ μPa $k\,Pa$	hbar bar	$1\ bar = 100\ kPa$ $= 10^5 N/m$ $1\ atm \approx 0.1\ N/mm^2$
Stress	Pa N/m^2	$G\,Pa$ $M\,Pa$ $m\,Pa$ N/mm^2 $\mu\,Pa$ $k\,Pa$	—	$1\ Pa = 1\ N/m^2$ $1\ M\,Pa = 1\ N/mm^2$
Dynamic viscosity	$Pa.s$	$m\,Pa.s$	—	$1\ cP† = 1\ m\,Pa.s$
Kinematic viscosity	m^2/s	mm^2/s	—	$1\ cSt† = 1\ mm^2/s$
Surface tension	N/m	$m\ N/m$	—	—
Work, energy	J (Joule)	TJ mJ GJ MJ kJ	$Wh, kWh,$ MWh, GWh and TWh in electrical and $kgf.m$ in mechanical fields	$1\ J = 1\ Nm$ $1\ kWh = 3.6\ MJ$ $1\ kgf.m = 9.80665 J$
Power	W (Watt)	GW mW MW μW kW	$kgf.m/s$	$1\ W = 1\ J/s$ $= 1\ Nm/s$ $1\ kgf.m/s = 9.08665 W$
Thermodynamic temperature	K(Kelvin)	—	—	—
Celsius temperature	°C	Note: The Celsius temperature t is equal to the difference $t = T - To$ between two thermodynamic temperatures T and To, where $To = 273.15K$.		
Temp. interval	K or °C	—	—	—
Linear expansion coefficient	K^{-1}	—	—	—
Quantity of heat	J	TJ mJ GJ MJ kJ	—	—

† Poise and stokes belong to *CGS* units. They should not be used with SI units.

EXAMPLES OF MULTIPLES AND SUBMULTIPLES OF SI UNITS (concld.)				
Quantity	*SI unit*	*Preferred multiple of SI unit*	*Other Units which may be used*	*Remarks*
Heat flow rate	*W*	*kW*	—	1 *W* = 1 *J/s*
Electric current	*A* (Ampere)	*kA* *mA* *μA* *nA* *pA*	—	—
Electric charge	*C* (Coulomb)	*kC* *μC* *nC* *pC*	—	—
Electric potential or electromotive force	*V* (Volt)	*MV* *mV* *kV* *μV*	—	—
Capacitance	*F* (Farad)	*mF* *μF* *nF* *pF*	—	—
Magnetic flux density	*T* (Tesla)	*mT* *μT* *nT*	—	—
Magnetic flux	*Wb* (Weber)	*mWb*	—	—
Self inductance, mutual inductance	*H* (Henry)	*mH* *μH* *nH* *pH*	—	—
Electric resistance, impedance	Ω (Ohm)	*G* Ω *m* Ω *M* Ω *μ* Ω *k* Ω	—	—
Conductance Admittance Susceptance	*S* (Siemens)	*kS* *mS* *μS*	—	1 *S* = 1 Ω^{-1}
Reluctance	H^{-1}	—	—	—
Permeance	*H*	—	—	—
Active power	*W*	*TW* *mW* *GW* *μW* *MW* *nW* *kW*	—	—
Apparant power	*VA*	*MVA* *kVA*	—	—

EQUIVALENTS OF COMMONLY USED MEASURES AND UNITS

LINEAR MEASURE

1 μin	=	0.0254 μm
1 in	=	25.4 mm
1 μm	=	39.37 μin
1 mm	=	0.03937 in
1 m	=	3.28084 ft
1 ft	=	0.3048 m

SQUARE MEASURE

1 in^2	=	645.16 mm^2
	=	6.4516 cm^2
1 ft^2	=	0.092903 m^2
1 mm^2	=	0.00155 in^2
1 cm^2	=	0.155 in^2
1 m^2	=	10.7639 ft^2

CUBIC MEASURE

1 in^3	=	16.387064 cm^3
1 ft^3	=	0.0283168 m^3
	=	28.3168 dm^3 (litr.)
1 $Imp.\ gal.$	=	4.54609 dm^3 (litr.)
1 $U.S.\ gal.$	=	3.78541 dm^3 (litr.)
1 cm^3	=	0.06102 in^3
1 dm^3 (litr.)	=	0.0353147 ft^3
	=	0.219969 $Imp.\ gal.$
	=	0.264172 $U.S.\ gal.$
1 m^3	=	35.3147 ft^3

MOMENT OF INERTIA

1 in^4	=	41.62314 cm^4
1 cm^4	=	0.024 in^4

MASS

1 pound	=	0.4535924 kg
1 ton	=	1.01605 tonnes
1 kg	=	2.20462 pounds
1 tonne	=	0.98421 ton

WEIGHT, FORCE

1 poundal	=	0.031081 lbf
	=	0.0140981 kgf
	=	0.138255 N
1 lbf	=	32.174 poundals (*pdl*)
	=	4.44822 N
	=	0.4535924 kgf
1 kgf	=	9.80665 N
	=	2.20462 lbf
	=	70.932 poundals (*pdl*)
1 N	=	0.101972 kgf
	=	0.224809 lbf
	=	7.2330 poundals (*pdl*)

WORK, ENERGY

1 $lbf.\ in$	=	1.152125·$kgf.\ cm$
	=	0.112985 J
1 $lbf.\ ft$	=	0.138255 $kgf.\ m$
	=	1.35582 J
1 $kgf.\ m$	=	7.2330 $lbf.\ ft$
	=	86.796 1$lbf.\ in$
	=	9.80666 J
1 J	=	0.101972 $kgf.\ m$
	=	0.73756 $lbf.\ ft$
	=	8.85072 $lbf.\ in$

WEIGHT PER UNIT LENGTH

1 lbf per ft run	=	1.48816 kgf per m run
1 kgf per m run	=	0.67197 lbf per ft run

PRESSURE, STRESS

1 lbf/in^2	=	0.070307 kgf/cm^2
	=	6894.76 N/m^2
1 lbf/ft^2	=	4.88243 kgf/m^2
	=	47.8803 N/m^2
1 kgf/cm^2	=	98066.5 Pa
	=	14.2233 lbf/in^2
1 kgf/m^2	=	9.80665 Pa
	=	0.20482 lbf/ft^2
1 N/m^2	=	0.101976 kgf/m^2
	=	0.0001450375 lbf/in^2

ATMOSPHERIC PRESSURE

1 in of Hg	=	0.0345259 kgf/cm^2
	=	3386.39 N/m^2
	=	0.0338639 bar
	=	0.0334211 std. atm
	=	0.49116 lbf/in^2
1 mm of Hg	=	0.0013595 kgf/cm^2
	=	133.322 Pa
	=	0.00133322 bar
	=	0.00131579 std. atm
	=	0.019337 lbf/in^2
1 std. atm	=	14.69598 lbf/in^2
	=	101325 Pa
	=	1.03322 kgf/cm^2
	=	1.01325 bars
	=	29.9213 in of Hg
	=	760 mm of Hg

MASS DENSITY

1 lb/in^3	=	0.0276799 kg/cm^3
1 lb/ft^3	=	16.0185 kg/m^3
1 kg/cm^3	=	36.1273 lbs/in^3
1 kg/m^3	=	0.062428 lbs/ft^3

EQUIVALENTS OF COMMONLY USED MEASURES AND UNITS

(concld.)

SPEED

1 ft/min	=	0.00508 m/sec
1 m/sec	=	196.85040 ft/min
1 $radian/sec$	=	9.549 rpm
1 rpm	=	0.1047 $radian/sec$

POWER

1 $lbf. ft/sec$	=	1.35582 W
1 $H.P.$	=	550 lbf. ft/sec
	=	76.04025 $kgf. m/sec$
	=	745.70 W
	=	1.01387 metric $H.P.$
1 metric $H.P.$	=	75 $kgf. m/sec$
	=	735.50 W
	=	0.98632 $H.P.$

ANGULAR MEASURE

1 radian	=	57.296°(57°17' 45")
1 degree	=	0.0174533 radian

TEMPERATURE

°C	=	$(°F - 32)5/9$
	=	$°K - 273.15$
°F	=	$9/5 × °K - 459.67$
	=	$9/5 × °C + 32$
°K	=	$°C + 273.15$
	=	$5/9(°F + 459.67)$

VISCOSITY

1 poise	=	0.1 $N sec/m^2$
	=	100 centipoises
1 stoke	=	$10^{-4} m^2/sec$
	=	100 centistokes

HEAT

1 $B Th U$	=	1055.06 J
	=	778.2 $lbf.ft$
	=	0.251996 $kcal$
	=	5/9 CHU
1 CHU	=	1.8 $B Th U$
	=	1400.4 $lbf. ft$
	=	0.4535928 $kcal$
	=	1899.108 J
1 $kcal$	=	3.96832 $B Th U$
	=	2.20462 CHU
	=	4186.8 J
	=	3087.351 $lbf. ft$

CALORIFIC VALUE

1 $B Th U/lb$	=	5/9 $kcal/kg$
1 CHU/lb	=	$kcal/kg$
1 $kcal/kg$	=	1 CHU/lb
	=	1.8 $B Th U/lb$

STANDARD GRAVITY

g	=	32.174 ft per sec per sec
	=	980.665 cm per sec per sec

CONVERSION CHART
(PRESSURES & STRESSES)

$$lbf/in^2 \rightleftharpoons Tonf/in^2 \rightleftharpoons kgf/mm^2 \rightleftharpoons N/mm^2$$

lbf/in^2	$Tonf/in^2$	kgf/mm^2	N/mm^2 $\approx 10atm$	lbf/in^2	$Tonf/in^2$	kgf/mm^2	N/mm^2 $\approx 10atm$
10	0.0045	0.007	0.069	**100**	0.0446	0.0703	0.689
11	0.0049	0.0077	0.076	101.5	0.0453	0.0714	**0.7**
11.2	**0.005**	0.0079	0.077	110	0.0491	0.077	0.758
12	0.0054	0.0084	0.083	112	**0.05**	0.079	0.772
13	0.0058	0.0091	0.09	113.8	0.0508	**0.08**	0.784
14	0.0062	0.0098	0.096	116	0.0519	0.0816	**0.8**
14.22	0.0063	**0.01**	0.098	120	0.0536	0.0840	0.827
14.5	0.0065	0.0102	**0.1**	125	0.0558	0.0877	0.861
15	0.0067	0.0105	0.103	128	0.0571	**0.09**	0.882
16	0.0071	0.0112	0.110	130	0.0580	0.0913	0.896
17	0.0076	0.0119	0.117	130.5	0.0583	0.0918	**0.9**
18	0.008	0.0127	0.124	134.4	**0.06**	0.0945	0.926
19	0.0085	0.0134	0.131	140	0.0625	0.0984	0.965
20	0.0089	0.0141	0.138	142.2	0.0635	**0.1**	0.980
22.4	**0.01**	0.0157	0.154	145	0.0647	0.102	**1**
28.45	0.0127	**0.02**	0.196	**150**	0.0670	0.105	1.03
29	0.0129	0.0204	**0.2**	156.8	**0.07**	0.110	1.08
30	0.0134	0.0211	0.207	160	0.0714	0.112	1.10
40	0.0179	0.0281	0.276	170	0.0759	0.119	1.17
42.67	0.019	**0.03**	0.294	179.2	**0.08**	0.126	1.23
43.51	0.0194	0.0306	**0.3**	180	0.0804	0.127	1.24
44.8	**0.02**	0.0315	0.309	190	0.0848	0.134	1.31
50	0.0223	0.0352	0.345	**200**	0.0893	0.141	1.38
56.89	0.0254	**0.04**	0.392	201.6	**0.09**	0.142	1.39
58	0.0259	0.0408	**0.4**	224	**0.1**	0.157	1.54
60	0.0268	0.0422	0.413	**250**	0.112	0.176	1.72
67.2	**0.03**	0.0472	0.463	284.5	0.127	**0.2**	1.96
70	0.0313	0.0492	0.482	290	0.129	0.204	**2**
71.12	0.0318	**0.05**	0.490	**300**	0.134	0.211	2.07
72.52	0.0324	0.0510	**0.5**	**350**	0.156	0.246	2.41
80	0.0357	0.0562	0.551	**400**	0.179	0.281	2.76
85.4	0.0381	**0.06**	0.588	426.7	0.190	**0.3**	2.94
87	0.0388	0.0612	**0.6**	435.1	0.194	0.306	**3**
89.6	**0.04**	0.0630	0.617	448	**0.2**	0.315	3.09
90	0.0402	0.0633	0.620	**450**	0.201	0.316	3.10
99.56	0.0444	**0.07**	0.686	**500**	0.223	0.352	3.44

CONVERSION CHART
(PRESSURES & STRESSES)

$$lbf/in^2 \rightleftharpoons Tonf/in^2 \rightleftharpoons kgf/mm^2 \rightleftharpoons N/mm^2$$

(contd.)

lbf/in^2	$Tonf/in^2$	kgf/mm^2	N/mm^2 $\approx 10atm$	lbf/in^2	$Tonf/in^2$	kgf/mm^2	N/mm^2 $\approx 10atm$
550	0.246	0.385	3.79	1792	0.8	1.26	12.35
568.9	0.254	0.4	3.92	1800	0.804	1.27	12.40
580.1	0.259	0.408	4	1900	0.848	1.34	13.09
600	0.268	0.422	4.13	2000	0.893	1.41	13.78
650	0.290	0.457	4.48	2016	0.9	1.42	13.89
672	0.3	0.472	4.63	2240	1	1.57	15.43
700	0.313	0.492	4.82	2500	1.12	1.76	17.22
711.2	0.318	0.5	4.90	2845	1.27	2	19.60
725.2	0.324	0.510	5	2901	1.29	2.04	20
750	0.335	0.527	5.17	3000	1.34	2.11	20.67
800	0.357	0.562	5.51	3500	1.56	2.46	24.11
850	0.379	0.598	5.82	4000	1.79	2.81	27.56
853.4	0.381	0.6	5.88	4267	1.90	3	29.40
870.2	0.388	0.612	6	4351	1.94	3.06	30
896	0.4	0.630	6.17	4480	2	3.15	30.87
900	0.402	0.633	6.20	4500	2.01	3.18	31.05
950	0.424	0.668	6.54	5000	2.23	3.52	34.45
995.5	0.444	0.7	6.86	5500	2.46	3.85	37.89
1000	0.446	0.703	6.89	5689	2.54	4	39.20
1015	0.453	0.714	7	5801	2.59	4.08	40
1100	0.491	0.770	7.58	6000	2.68	4.22	41.34
1120	0.5	0.790	7.72	6500	2.90	4.57	44.78
1138	0.508	0.8	7.84	6720	3	4.72	46.30
1160	0.519	0.816	8	7000	3.13	4.92	48.23
1200	0.536	0.840	8.27	7112	3.18	5	49.00
1280	0.571	0.9	8.82	7252	3.24	5.10	50
1300	0.580	0.913	8.96	7500	3.35	5.27	51.67
1305	0.583	0.918	9	8000	3.57	5.62	55.12
1344	0.6	0.945	9.26	8500	3.79	5.98	58.56
1400	0.625	0.984	9.65	8534	3.81	6	58.80
1422	0.635	1	9.79	8702	3.88	6.12	60
1450	0.647	1.02	10	8960	4	6.30	61.73
1500	0.670	1.05	10.33	9000	4.02	6.33	62.01
1568	0.7	1.10	10.80	9500	4.24	6.68	65.45
1600	0.714	1.12	11.02	9956	4.44	7	68.60
1700	0.759	1.19	11.71	10000	4.46	7.03	68.90

CONVERSION CHART

(PRESSURES & STRESSES)

$lbf/in^2 \rightleftharpoons Tonf/in^2 \rightleftharpoons kgf/mm^2 \rightleftharpoons N/mm^2$

(contd.)

lbf/in^2	$Tonf/in^2$	kgf/mm^2	N/mm^2 $\approx 10atm$	lbf/in^2	$Tonf/in^2$	kgf/mm^2	N/mm^2 $\approx 10atm$
10150	4.53	7.14	**70**	23000	10.27	16.17	158.5
11000	4.91	7.70	75.79	23210	10.36	16.31	**160**
11200	**5**	7.90	77.17	24000	10.71	16.87	165.4
11380	5.03	**8**	78.40	24180	10.79	**17**	166.6
11600	5.19	8.16	**80**	24640	**11**	17.32	169.5
12000	5.36	8.40	82.68	24660	11.01	17.33	**170**
12800	5.71	**9**	88.20	**25000**	11.16	17.58	172.2
13000	5.80	9.13	89.57	25600	11.46	**18**	176.4
13050	5.83	9.18	**90**	26000	11.61	18.28	179.1
13440	**6**	9.45	92.60	26110	11.65	18.35	**180**
14000	6.25	9.84	96.46	26880	**12**	18.90	185.2
14220	6.35	**10**	98.00	27000	12.05	18.98	186.0
14500	6.47	10.20	**100**	27020	12.06	**19**	186.2
15000	6.70	10.50	103.3	27600	12.32	19.40	**190**
15650	6.98	**11**	107.8	28000	12.50	19.68	192.9
15680	**7**	11.02	108.0	28450	12.70	**20**	196.0
15950	7.12	11.22	**110**	29000	12.946	20.39	199.8
16000	7.14	11.25	110.2	29010	12.951	20.40	**200**
17000	7.59	11.90	117.1	29120	**13**	20.47	200.6
17070	7.62	**12**	117.6	29870	13.33	**21**	205.8
17400	7.77	12.24	**120**	**30000**	13.39	21.10	206.7
17920	**8**	12.60	123.5	30460	13 60	21.41	**210**
18000	8.04	12.70	124.0	31000	13.84	21.79	213.6
18490	8.25	**13**	127.4	31290	13.97	**22**	215.6
18850	8.42	13.26	**130**	31360	**14**	22.05	216.1
19000	8.48	13.40	130.9	31910	14.24	22.43	**220**
19910	8.89	**14**	137.2	32000	14.29	22.50	220.5
20000	8.93	14.10	137.8	32710	14.60	**23**	225.4
20160	**9**	14.17	138.9	33000	14.73	23.20	227.4
20300	9.06	14.27	**140**	33360	14.89	23.45	**230**
21000	9.38	14.76	144.7	33600	**15**	23.62	231.5
21330	9.52	**15**	147.0	34000	15.18	23.90	234.3
21760	9.71	15.29	**150**	34140	15.24	**24**	235.2
22000	9.82	15.47	151.6	34810	15.54	24.47	**240**
22400	**10**	15.70	154.3	**35000**	15.63	24.61	241.1
22760	10.16	**16**	156.8	35560	15.87	**25**	245.0

CONVERSION CHART
(PRESSURES & STRESSES)
$lbf/in^2 \rightleftharpoons Tonf/in^2 \rightleftharpoons kgf/mm^2 \rightleftharpoons N/mm^2$

(concld.)

lbf/in^2	$Tonf/in^2$	kgf/mm^2	N/mm^2 $\approx 10atm$	lbf/in^2	$Tonf/in^2$	kgf/mm^2	N/mm^2 $\approx 10atm$
35840	16	25.20	246.9	85340	38.10	60	588.0
36000	16.07	25.31	248.0	87020	38.85	61.20	600
36260	16.19	25.49	250	89600	40	63.00	617.3
36980	16.51	26	254.8	90000	40.18	63.30	620.1
37000	16.52	26.01	254.9	99560	44.45	70	686.0
37710	16.83	26.61	260	100000	44.64	70.30	689.0
38000	16.96	26.72	261.8	101530	45.32	71.40	700
38080	17	26.77	262.4	110000	49.11	77.00	757.9
38400	17.14	27	264.6	112000	50	79.00	771.7
39000	17.41	27.42	268.7	113790	50.80	80	784.0
39160	17.48	27.53	270	116030	51.80	81.60	800
39820	17.78	28	274.4	120000	53.57	84.00	862.8
40000	17.86	28.10	275.6	128010	57.15	90	882.0
40320	18	28.35	277.8	130000	58.04	91.30	895.7
40610	18.13	28.55	280	130530	58.27	91.80	900
41000	18.30	28.82	282.5	134400	60	94.50	926.0
41250	18.41	29	284.2	140000	62.50	98.40	964.6
42000	18.75	29.53	289.4	142230	63.50	100	980.0
42060	18.78	29.57	290	145040	64.75	102.0	1000
42560	19	29.92	293.2	150000	66.96	105.0	1033
42670	19.05	30	294.0	156460	69.85	110	1078
43000	19.20	30.23	296.3	156800	70	110.2	1080
43510	19.42	30.60	300	159540	71.22	112.2	1100
44000	19.64	30.93	303.2	160000	71.43	112.5	1102
44800	20	31.50	308.7	170000	75.89	119.0	1171
45000	20.09	31.63	310.5	170680	76.20	120	1176
50000	22.32	35.20	344.5	174050	77.70	122.4	1200
56000	25	39.37	385.8	179200	80	126.0	1235
56890	25.40	40	391.2	180000	80.36	127.0	1240
58010	25.90	40.80	400	184900	82.55	130	1274
60000	26.79	42.20	413.4	188550	84.17	132.6	1300
67200	30	47.20	463.0	190000	84.82	134.0	1309
70000	31.25	49.20	482.3	199130	88.89	140	1372
71120	31.75	50	490.0	200000	89.29	141.0	1378
72520	32.37	51.00	500	201600	90	141.7	1389
80000	35.71	56.20	551.2	203050	90.65	142.8	1400

CENTRE OF GRAVITY, MOMENT OF INERTIA AND SECTION MODULUS

Figure	Centre of Gravity	Moment of Inertia	Section Modulus in Bending
	$\dfrac{h}{2}$	$\dfrac{bh^3}{12}$	$\dfrac{bh^2}{6}$
	$\dfrac{h}{2}$	$\dfrac{h^4}{12}$	$\dfrac{h^3}{6}$
	$\dfrac{h}{2}\sqrt{2}$	$\dfrac{h^4}{12}$	$\dfrac{h^3}{12}\sqrt{2}=0.1179h^3$
	$\dfrac{H}{2}$	$\dfrac{b}{12}(H^3-h^3)$	$\dfrac{b}{6H}(H^3-h^3)$
	$\dfrac{a}{2}$	$\dfrac{1}{12}\left(a^4-\dfrac{3\pi}{16}d^4\right)$	$\dfrac{1}{6a}\left(a^4-\dfrac{3\pi}{16}d^4\right)$
	$\dfrac{A}{2}$	$\dfrac{A^4-a^4}{12}$	$\dfrac{1}{6}\times\dfrac{A^4-a^4}{A}$
	$\dfrac{A}{2}\sqrt{2}$	$\dfrac{A^4-a^4}{12}$	$\dfrac{A^4-a^4}{12A}\sqrt{2}$ $=0.1179\times\dfrac{A^4-a^4}{A}$
	$\dfrac{2}{3}h$	$\dfrac{bh^3}{36}$	$\dfrac{bh^2}{24}$

CENTRE OF GRAVITY, MOMENT OF INERTIA AND SECTION MODULUS

(contd)

Figure	Centre of Gravity	Moment of Inertia	Section Modulus in Bending
	$\dfrac{3b+2b_1}{3(2b+b_1)}h$	$\dfrac{6b^2+6bb_1+b_1^2}{36(2b+b_1)}h^3$	$\dfrac{6b^2+6bb_1+b_1^2}{12(3b+2b_1)}h^2$
	$e_1 = 0.866r$ $e_2 = r$	$I_1 = I_2 = \dfrac{5\sqrt{3}}{16}r^4$ $= 0.5413r^4$	$Z_1 = 0.625\,r^3$ $Z_2 = 0.5413r^3$
	$e_1 = 0.924r$ $e_2 = r$	$I_1 = I_2 = 0.6381r^4$	$Z_1 = 0.6906r^3$ $Z_2 = 0.6381r^3$
	$\dfrac{d}{2}$	$I = \dfrac{\pi d^4}{64} = 0.0491d^4$ Polar moment of intertia $Ip = \dfrac{\pi d^4}{32} = 0.1d^4$	$Z = \dfrac{\pi d^3}{32} = 0.0982d^3$ Section modulus in torsion $Zp = \dfrac{\pi d^3}{16} = 0.2d^3$
	$\dfrac{D}{2}$	$I = \dfrac{\pi}{64}(D^4 - d^4)$ $\approx 0.05(D^4 - d^4)$ Polar moment of intertia $Ip = \dfrac{\pi}{32}(D^4 - d^4)$	$Z = \dfrac{\pi}{32} \times \dfrac{D^4 - d^4}{D}$ Section modulus in torsion $Zp = \dfrac{\pi}{16} \times \dfrac{D^4 - d^4}{D}$
	$\dfrac{h}{2}$	$\dfrac{1}{12}\left[\dfrac{3\pi}{16}(d_1^4 - d^4)\right.$ $+ b(h^3 - d_1^3)$ $\left. + b^3(h - d_1)\right]$	$\dfrac{1}{6h}\left[\dfrac{3\pi}{16}(d_1^4 - d^4)\right.$ $+ b(h^3 - d_1^3)$ $\left. + b^3(h - d_1)\right]$

CENTRE OF GRAVITY, MOMENT OF
INERTIA AND SECTION MODULUS

(concld.)

Figure			
Centre of Gravity	$e = \dfrac{H}{2}$	$e = \dfrac{H}{2}$	$e_1 = \dfrac{aH^2 + bd^2}{2(aH + bd)}$ $e_2 = H - e_1$
Moment of Inertia	$I = \dfrac{BH^3 - bh^3}{12}$	$I = \dfrac{BH^3 + bh^3}{12}$	$I = \dfrac{Be_1^3 - bh^3 + ae_2^3}{3}$
Section Modulus in Bending	$Z = \dfrac{BH^3 - bh^3}{6H}$	$Z = \dfrac{BH^3 + bh^3}{6H}$	$Z_1 = \dfrac{I}{e_1} \quad Z_2 = \dfrac{I}{e_2}$

Bending Moment, Slope and Deflection

TYPE OF LOADING	BENDING MOMENT AND SLOPE	DEFLECTION
	$M = -Wx$ $(M_{max})_B = -Wl$ $\theta_A = \dfrac{Wl^2}{2EI}$	$Y = \dfrac{W}{6EI}(x^3 - 3l^2x + 2l^3)$ $(Y_{max})_A = \dfrac{Wl^3}{3EI}$
	$M_{A\,to\,B} = 0$ $M_{B\,to\,C} = -W(x-b)$ $(M_{max})_C = -Wa$ $\theta_{A\,to\,B} = \dfrac{Wa^2}{2EI}$	$Y_{A\,to\,B} = \dfrac{W}{6EI}(-a^3 + 3a^2l - 3a^2x)$ $Y_{B\,to\,C} = \dfrac{W}{6EI}\left[(x-b)^3 - 3a^2(x-b) + 2a^3\right]$ $(Y_{max})_A = \dfrac{W}{6EI}(3a^2l - a^3)$
	$M = -\dfrac{Wx^2}{2l}$ $(M_{max})_B = -\dfrac{Wl}{2}$ $\theta_A = \dfrac{Wl^2}{6EI}$	$Y = \dfrac{W}{24EIl}(x^4 - 4l^3x + 3l^4)$ $(Y_{max})_A = \dfrac{Wl^3}{8EI}$

Bending Moment, Slope and Deflection

(contd.)

TYPE OF LOADING	BENDING MOMENT AND SLOPE	DEFLECTION
	$M = M_o$ $(M_{max})_{A\,to\,B} = M_o$ $\theta_A = -\dfrac{M_o l}{EI}$	$Y = -\dfrac{M_o}{2EI}(l^2 - 2lx + x^2)$ $(Y_{max})_A = -\dfrac{M_o l^2}{2EI}$
	$M_{A\,to\,B} = 0$ $(M_{max})_{B\,to\,C} = M_o$ $\theta_{A\,to\,B} = -\dfrac{M_o a}{EI}$	$Y_{A\,to\,B} = -\dfrac{M_o a}{EI}\left(l - \dfrac{a}{2} - x\right)$ $Y_{B\,to\,C} = -\dfrac{M_o}{2EI}\left[(x-l+a)^2 - 2a(x-l+a) + a^2\right]$ $(Y_{max})_A = -\dfrac{M_o a}{EI}\left(l - \dfrac{a}{2}\right)$
	$M_{A\,to\,B} = \dfrac{W}{2}\,x$ $M_{B\,to\,C} = \dfrac{W}{2}(l-x)$ $(M_{max})_B = \dfrac{Wl}{4}$ $\theta_C = -\theta_A = \dfrac{Wl^2}{16EI}$	$Y_{A\,to\,B} = \dfrac{W}{48EI}(3l^2 x - 4x^3)$ $(Y_{max})_B = \dfrac{Wl^3}{48EI}$

(contd.)

Bending Moment, Slope and Deflection

TYPE OF LOADING	BENDING MOMENT AND SLOPE	DEFLECTION
	$M_{A\,to\,B} = \dfrac{Wb}{l}\,x$ $M_{B\,to\,C} = \dfrac{Wa}{l}(l-x)$ $(M_{max})_B = \dfrac{Wab}{l}$ $\theta_A = -\dfrac{W}{6EI}\left(bl - \dfrac{b^3}{l}\right)$ $\theta_C = \dfrac{W}{6EI}\left(2bl + \dfrac{b^3}{l} - 3b^2\right)$	$Y_{A\,to\,B} = \dfrac{W\,bx}{6EI\,l}\,[2l(l-x)-b^2-(l-x)^2]$ $Y_{B\,to\,C} = \dfrac{Wa\,(l-x)}{6EI\,l}\,[2\,lb-b^2-(l-x)^2]$ $Y_{max} = \dfrac{Wab}{27EI\,l}(a+2b)\sqrt{3a(a+2b)}$ at $x = \sqrt{\tfrac{1}{3}a\,(a+2b)}$ when $a>b$
	$M = \dfrac{W}{2}\left(x - \dfrac{x^2}{l}\right)$ $(M_{max})_{x=\frac{l}{2}} = \dfrac{Wl}{8}$ $\theta_B = -\theta_A = \dfrac{Wl^2}{24EI}$	$Y = \dfrac{Wx}{24EI\,l}\,(l^3-2lx^2+x^3)$ $(Y_{max})_{x=\frac{l}{2}} = \dfrac{5}{384}\left(\dfrac{Wl^3}{EI}\right)$

Bending Moment, Slope and Deflection

(contd.)

TYPE OF LOADING	BENDING MOMENT AND SLOPE	DEFLECTION
	$M = M_o\left(1 - \dfrac{x}{l}\right)$ $(M_{max})_A = M_o$ $\theta_A = -\dfrac{M_o l}{3EI}$ $\theta_B = \dfrac{M_o l}{6EI}$	$Y = -\dfrac{M_o}{6EI}\left(3x^2 - \dfrac{x^3}{l} - 2lx\right)$ $Y_{max} = 0.0642\,\dfrac{M_o l^2}{EI}$ at $x = 0.422\,l$
	$M_{A\,to\,B} = -\dfrac{M_o}{l}x$ $M_{B\,to\,C} = M_o\left(1 - \dfrac{x}{l}\right)$ $(M_{max})_{\text{left of B}} = -\dfrac{M_o a}{l}$ $(M_{max})_{\text{right of B}} = M_o\left(1 - \dfrac{a}{l}\right)$ $\theta_A = \dfrac{M_o}{6EI}\left(2l - 6a + \dfrac{3a^2}{l}\right)$ $\theta_B = \dfrac{M_o}{EI}\left(a - \dfrac{a^2}{l} - \dfrac{l}{3}\right)$ $\theta_C = \dfrac{M_o}{6EI}\left(1 - \dfrac{3a^2}{l}\right)$	$Y_{A\,to\,B} = -\dfrac{M_o}{6EI}\left[\left(6a - \dfrac{3a^2}{l} - 2l\right)x - \dfrac{x^3}{l}\right]$ $Y_{B\,to\,C} = -\dfrac{M_o}{6EI}\left[3a^2 + 3x^2 - \dfrac{x^3}{l} - \left(2l + \dfrac{3a^2}{l}\right)x\right]$ $Y_B = \dfrac{M_o a}{3EIl}\left(3al - 2a^2 - l^2\right)$

Bending Moment, Slope and Deflection

(contd.)

TYPE OF LOADING	BENDING MOMENT AND SLOPE	DEFLECTION
	$M_{A\,to\,B} = Wz$ $M_{B\,to\,C} = \dfrac{Wbx}{a}$ $(M_{max})_B = Wb$ $\theta_A = \dfrac{Wb}{6EI}(2l+b)$ $\theta_B = \dfrac{Wab}{3EI} \qquad \theta_C = \dfrac{Wab}{6EI}$	$Y_{B\,to\,C} = -\ \dfrac{Wbx\,(x^2-a^2)}{6EI\,a}$ $Y_{A\,to\,B} = \dfrac{W}{6EI}\,[z^3-bz\,(2l-b)+2b^2]$ $(Y_{max})_A = \dfrac{Wb^2l}{3EI}$
	$M_{A\,to\,B} = -Wx$ $M_{B\,to\,C} = Wa$ $\theta_B = -\ \theta_C = \dfrac{Wal}{2EI}$	$Y_{A\,to\,B} = \dfrac{W(a-x)}{6\,EI}\,[3a(l+a-x)-(a-x^2)]$ $Y_{B\,to\,C} = -\ \dfrac{Waz(l-z)}{2\,EI}$ $Y_{x=\frac{l}{2}} = -\ \dfrac{Wal^2}{8\,EI}$ $Y_A = Y_D = \dfrac{Wa^2}{6\,EI}\,(2a+3l)$

Bending Moment, Slope and Deflection

(contd.)

TYPE OF LOADING	BENDING MOMENT AND SLOPE	DEFLECTION
	$M_{A\ to\ B} = -\dfrac{wx^2}{2}$ $M_{B\ to\ C} = \dfrac{wx^2}{2} - \dfrac{W}{2}(x-a)$ $\theta_B = -\theta_C = \dfrac{Wl}{24\,EIL}(6a^2-l^2)$	$Y_{A\ to\ B} = \dfrac{W(a-x)}{24\,EIL}\cdot G$ where $G = [6a^2(l+a-x)-(a-x)^2(3a+x)-l^3]$ $Y_{B\ to\ C} = \dfrac{Wz(l-z)}{24EIL}\cdot F$ where $F = [z\,(l-z)+l^2-6a^2]$ $Y_A = Y_D = \dfrac{Wa}{24\,EIL}[3a^2(a+2l)-l^3]$
	$M_{A\ to\ B} = -\dfrac{Wab^2}{l^2} + \dfrac{Wb^2}{l^3}(3a+b)\,x$ $M_{B\ to\ C} = -\dfrac{Wab^2}{l^2} + \dfrac{Wb^2}{l^3}(3a+b)\,x - W(x-a)$ $(M_{max})_B = \dfrac{Wab^2}{l^2} + \dfrac{Wb^2}{l^3}(3a+b)\,a$ $(M_{max})_A = -M_1 = -\dfrac{Wab^2}{l^2}$; $a<b$ $(M_{max})_C = -M_2 = -\dfrac{Wa^2b}{l^2}$; $a>b$	$Y_{A\ to\ B} = \dfrac{Wb^2x^2}{6\,EI\,l^3}(3ax+bx-3al)$ $Y_{B\ to\ C} = \dfrac{Wa^2(l-x)^2}{6\,EI\,l^3}[(3b+a)(l-x)-3bl]$ $Y_{max} = \dfrac{2W}{3EI}\dfrac{a^3b^2}{(3a+b)^2}$ at $x=\dfrac{2al}{3a+b}$ if $a>b$ $Y_{max} = \dfrac{2W}{3EI}\cdot\dfrac{a^2b^3}{(3b+a)^2}$ at $x=\left(l-\dfrac{2bl}{3b+a}\right)$ if $a<b$.

Bending Moment, Slope and Deflection (contd.)

TYPE OF LOADING	BENDING MOMENT AND SLOPE	DEFLECTION
	$M_{A \text{ to } B} = \frac{W}{8}(4x - l)$ $M_{B \text{ to } C} = \frac{W}{8}(3l - 4x)$ $(M_{max})_B = \frac{Wl}{8}$ $(M_{max})_{A,C} = -\frac{Wl}{8}$	$Y_{A \text{ to } B} = \frac{W}{48EI}(3lx^2 - 4x^3)$ $(Y_{max})_B = \frac{Wl^3}{192EI}$
	$M = \frac{W}{2}\left(x - \frac{x^2}{l} - \frac{l}{6}\right)$ $(M_{max})_{x=\frac{l}{2}} = \frac{Wl}{24}$ $(M_{max})_{A,B} = -\frac{Wl}{12}$	$Y = \frac{Wx^2}{24EIl}(2lx - l^2 - x^2)$ $(Y_{max})_{x=\frac{l}{2}} = \frac{Wl^3}{384EI}$
	$M_{A \text{ to } B} = -M_1 + R_1 x$ $M_{B \text{ to } C} = -M_1 + R_1 x + M_o$ $(M_{max})_{\text{right of B}}$ $= M_o\left(\frac{4a}{l} - \frac{9a^2}{l^2} + \frac{6a^3}{l^3}\right)$	$Y_{A \text{ to } B} = -\frac{1}{6EI}(3M_1 x^2 - R_1 x^3)$ $Y_{B \text{ to } C} = -\frac{1}{6EI}\left[(M_o - M_1)(3x^2 - 6lx + 3l^2)\right.$ $\left. - R_1(3l^2 x - x^3 - 2l^3)\right]$

contd. in p 78

Bending Moment, Slope and Deflection   (contd.)

TYPE OF LOADING	BENDING MOMENT AND SLOPE	DEFLECTION
	(M_{max}) left of B $$= -M_o \left(\frac{4a}{l} - \frac{9a^2}{l^2} + \frac{6a^3}{l^3} - 1 \right)$$ $$M_1 = -\frac{M_o}{l^2}(4la - 3a^2 - l^2)$$ $$M_2 = \frac{M_o}{l^2}(2la - 3a^2)$$ $$R_1 = -\frac{6M_o}{l^3}(al - a^2)$$ $$R_2 = \frac{6M_o}{l^3}(al - a^2)$$	$$+Y_{max} \text{ at } x = \left[l - \frac{2M_2}{R_2} \right] \text{ if } a < \frac{2l}{3}$$ $$-Y_{max} \text{ at } x = \left[\frac{2M_1}{R_1} \right] \text{ if } a > \frac{l}{3}$$
	$$M_{A\,to\,B} = \frac{5}{16} Wx$$ $$M_{B\,to\,C} = W\left(\frac{l}{2} - \frac{11}{16}x \right)$$ $$(M_{max})_B = \frac{5}{32} Wl$$ $$(M_{max})_C = \frac{3}{16} Wl \quad \theta_A = -\frac{Wl^2}{32EI}$$	$$Y_{A\,to\,B} = \frac{W}{96EI}(5x^3 - 3l^2x)$$ $$Y_{B\,to\,C} = \frac{W}{96EI}\left[5x^3 - 16(x - \frac{l}{2})^3 - 3l^2x \right]$$ $$Y_{max} = 0.00932 \frac{Wl^3}{EI} \quad \text{at } x = 0.447l$$

(contd.)

Bending Moment, Slope and Deflection

TYPE OF LOADING	BENDING MOMENT AND SLOPE	DEFLECTION
	$M = W\left(\dfrac{3x}{8} - \dfrac{x^2}{2l}\right)$ $(M_{max})_{x=\frac{3l}{8}} = \dfrac{9}{128}\,Wl$ $(M_{max})_B = -\dfrac{Wl}{8} \qquad \theta_A = -\dfrac{Wl^2}{48EI}$	$Y = \dfrac{W}{48EIl}\,(3lx^3 - 2x^4 - l^3x)$ $Y_{max} = 0{\cdot}0054\,\dfrac{Wl^3}{EI}$ at $x = 0{\cdot}4215l$
	$M_{A\,to\,B} = R_1 x$ $M_{B\,to\,C} = R_1 x - W\,(x - l + a)$ $(M_{max})_B = R_1(l - a)$ $(M_{max})_C = -M_2$ $\qquad = \dfrac{W}{2l^2}\,(a^3 + 2al^2 - 3a^2l)$ $\theta_A = \dfrac{W}{4EI}\left(\dfrac{a^3}{l} - a^2\right)$ $R_1 = \dfrac{W}{2}\left(\dfrac{3a^2l - a^3}{l^3}\right) \quad R_2 = W - R_1$	$Y_{A\,to\,B} = \dfrac{1}{6EI}\left[R_1\,(x^3 - 3l^2x) + 3\,W\,a^2x\right]$ $Y_{B\,to\,C} = \dfrac{1}{6EI}\left[R_1\,(x^3 - 3l^2x) \right.$ $\left. \qquad\qquad + W[3a^2x - (x-b)^3]\right]$ Y_{max} is at $x = l\sqrt{1 - \dfrac{2l}{3l-a}}\;$ if $a < 0{\cdot}586l$ Y_{max} is at $x = \dfrac{l(l^2 + b^2)}{3l^2 - b^2}\;$ if $a > 0{\cdot}586l$ $(Y_{max})_B = 0{\cdot}0098\,\dfrac{Wl^3}{EI}\;$ if $a = 0{\cdot}586l$

Bending Moment, Slope and Deflection

TYPE OF LOADING	BENDING MOMENT AND SLOPE	DEFLECTION
	$M = \dfrac{M_o}{2}\left(2 - \dfrac{3x}{l}\right)$ $(M_{max})_A = M_o$ $(M_{max})_B = -\dfrac{M_o}{2}$ $\theta_A = -\dfrac{M_o l}{4EI}$	$Y = \dfrac{M_o}{4EI}\left(2x^2 - \dfrac{x^3}{l} - lx\right)$ $(Y_{max})_{x=\frac{l}{3}} = \dfrac{M_o l^2}{27EI}$
	$M_{A\,to\,B} = \dfrac{3M_o}{2l}\dfrac{(l^2-a^2)}{l^2}x$ $M_{B\,to\,C} = -\dfrac{3M_o}{2l}\dfrac{(l^2-a^2)x}{l^2} + M_o$ $(M_{max})\text{ right of B} = M_o\left[1 - \dfrac{3a(l^2-a^2)}{2l^3}\right]$ $(M_{max})_C = -\dfrac{M_o}{2}\left(1 - \dfrac{3a^2}{l^2}\right)$ when $a < 0.275\,l$ $(M_{max})\text{ left of B} = -\dfrac{3}{2}\cdot\dfrac{M_o a}{l}\left(\dfrac{l^2-a^2}{l^2}\right)$ when $a > 0.275\,l$ $\theta_A = \dfrac{M_o}{EI}\left(a - \dfrac{l}{4} - \dfrac{3a^2}{4l}\right)$	$Y_{A\,to\,B} = \dfrac{M_o}{EI}\left[\dfrac{l^2-a^2}{4l^3}(3l^2x - x^3) - (l-a)x\right]$ $Y_{B\,to\,C} = \dfrac{M_o}{EI}\left[\dfrac{l^2-a^2}{4l^3}(3l^2x - x^3)\right.$ $\left. - lx + \dfrac{1}{2}(x^2 + a^2)\right]$

Bending Moment, Slope and Deflection (contd.)

TYPE OF LOADING	BENDING MOMENT AND SLOPE	DEFLECTION
	$M_{A,D} = \dfrac{w}{8}(3\,lx - 4x^2)$ $(M_{max})_{x=\frac{3l}{8}} = \dfrac{9\,w\,l^2}{128}$ $(M_{max})_B = -\dfrac{wl^2}{8}$ $\theta_C = -\theta_A = \dfrac{wl^3}{48\,\mathrm{EI}}$	$Y = \dfrac{w}{48\,\mathrm{EI}}(l^3x - 3lx^3 + 2x^4)$ $Y_{max} = 0.00541\,\dfrac{wl^4}{EI}$ at $x = 0.4215\,l$
	$(M_{max})_{B,D} = \dfrac{Wab^2}{2l^3}(2l+a)$ $(M_{max})_C = -\dfrac{Wab}{2l^2}(2l-b)$ $R_A = R_E = \dfrac{Wb^2}{2l^3}(2l+a)$ $R_C = \dfrac{Wa}{l^3}(3l^2 - a^2)$	$Y_{A\,to\,B} = \dfrac{x}{6\,EI}\left[(l^2-x^2)R_A - \dfrac{Wb^3}{l}\right]$ $Y_{B\,to\,C} = \dfrac{l-x}{6\,EIl}\left[Wab\,(l+a) \right.$ $\left. - Wa\,(l-x)^2 - \dfrac{Wab}{2l^2}(l+a)(l+x)x \right]$

(concld.)

Bending Moment, Slope and Deflection

THEOREM OF THREE MOMENTS

$$\frac{M_1 l_1}{I_1} + 2M_2\left(\frac{l_1}{I_1} + \frac{l_2}{I_2}\right) + \frac{M_3 l_2}{I_2} = -\frac{6A_1 a_1}{I_1 l_1} - \frac{6A_2 b_2}{I_2 l_2}$$

$$\theta = -\frac{M_2 l_1}{3 EI_1} + \frac{M_1 l_1}{6 EI_1} + \frac{A_1 a_1}{EI_1 l_1}$$

$$\theta' = +\frac{A_2 b_2}{EI_2 l_2} + \frac{M_2 l_2}{3 EI_2} + \frac{M_3 l_2}{6 EI_2}$$

A_1 and A_2 are the areas of BM diagram for spans l_1 and l_2 respectively

DRAWING STANDARDS

DRAWING SHEET SIZES (As per IS 696 - 1972)

(1) PRINCIPLES OF BASIC PAPER SIZES

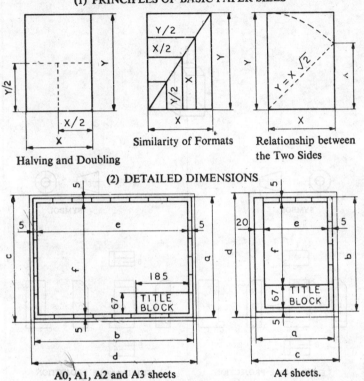

Halving and Doubling

Similarity of Formats

Relationship between
the Two Sides

(2) DETAILED DIMENSIONS

A0, A1, A2 and A3 sheets

A4 sheets.

Designation of Sheet	Trimmed Sheet Size (Cut Print Size)		Untrimmed Sheet Size (Original Sheet Size)		No. of Zones		Surface Area of Trimmed print sq. m
	a	b	c	d	e	f	
A0	841	1189	880	1230	16	12	1
A1	594	841	625	880	12	8	0.5
A2	420	594	450	625	8	6	0.25
A3	297	420	330	450	8	6	0.125
A4	210	297	240	330	4	4	0.0625

PRINCIPLES OF ORTHOGRAPHIC PROJECTIONS

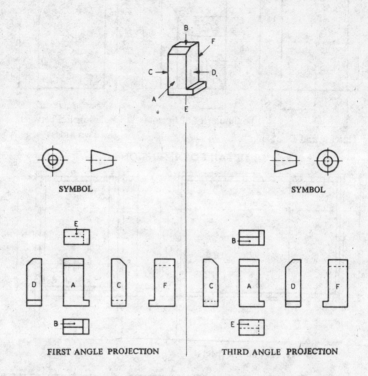

SYMBOL   SYMBOL

FIRST ANGLE PROJECTION   THIRD ANGLE PROJECTION

DESIGNATION OF VIEWS

View in direction A = View from front or front view
View in direction B = View from above or top view
View in direction C = View from Left or side view
View in direction D = View from right or side view
View in direction E = View from below or bottom view
View in direction F = View from rear or rear view

	CONVENTIONAL REPRESENTATION (As per IS 696 - 1972)				
Title	Actual Projection	Convention	Title	Actual Projection	Convention
Screw Threads			Flats and Square ends		
Small Intersections			Radial Ribs		Spoke A omitted · Spoke B revolved
			Section of Ribs		
Coil Springs		ϕ	End View of Splines		
		$\square$	End View of Serrations		
			Spiral Springs		
			Bearings		
Holes on Linear Pitch			Straight Knurling		
Holes on Circular Pitch			Diamond Knurling		
Repeated Parts			Disc Springs		

CONVENTIONAL REPRESENTATION (As per IS 696-1972)

(concld.)

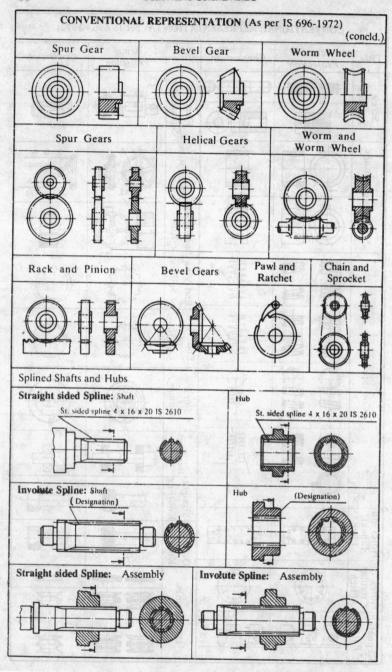

Spur Gear	Bevel Gear	Worm Wheel

Spur Gears	Helical Gears	Worm and Worm Wheel

Rack and Pinion	Bevel Gears	Pawl and Ratchet	Chain and Sprocket

Splined Shafts and Hubs

Straight sided Spline: Shaft

St. sided spline 4 x 16 x 20 IS 2610

Hub

St. sided spline 4 x 16 x 20 IS 2610

Involute Spline: Shaft
(Designation)

Hub   (Designation)

Straight sided Spline: Assembly

Involute Spline: Assembly

SYMBOLS FOR MACHINE TOOL INDICATION PLATES
(As per IS 2182 - 1962)

Meaning	Symbol	Remarks
Direction of rectilinear motion		—
Rectilinear motion in two directions		—
Interrupted rectilinear motion (JOG)		—
Limited rectilinear motion		—
Limited rectilinear motion and return		—
Oscillating rectilinear motion (continuous)		—
Direction of continuous rotation		—
Rotation in two directions		—
Direction of interrupted rotation (JOG)		—
Limited rotation		—
Limited rotation & return		—
Oscillating rotation		—
Direction of spindle rotation		—
Number of revolutions per minute (spindle speed)	x ⭕ /min	Number of revolutions x to be given according to the case, either with the symbol itself or in the numerical table accompanying it.

SYMBOLS FOR MACHINE TOOL INDICATION PLATES
(As per IS 2182 - 1962)
(contd.)

Meaning		Symbol	Remarks
Feed per revolution		$\text{MM} \times mm / \circlearrowleft$	Value of the feed x to be given according to the case, either with the symbol itself or in the numerical table accompanying it.
Feed per minute		$\text{MM} \times mm / min$	Value of the feed x to be given according to the case, either with the symbol itself or in the numerical table accompanying it.
Normal feed		$\text{MM} \quad ^1/_1$	—
Reduced feed		$\text{MM} \quad ^1/_x$	$1/x$ being the ratio between reduced feed and normal feed
Rapid feed		$\text{MM} \quad ^x/_1$	$x/1$ being the ratio between rapid feed and normal feed.
Longitudinal feed			—
Transverse feed			—
Vertical feed			—
Rapid Traverse	Longitudinal		—
	Transverse		—
	Vertical		—

SYMBOLS FOR MACHINE TOOL INDICATION PLATES

(As per IS 2182 - 1962) (contd.)

Meaning		Symbol	Remarks
Threading	Right hand	▢〰	—
	Left hand	▢〰	—
Increase of value (speed for instance)		+	—
Decrease of value (speed for instance)		—	—
Speed of planing cut		x m / min	Speed x, in metres per minute, to be given according to the case, either with the symbol itself or in the accompanying numerical table.
Speed of turning cut		x m / min	x being the speed in metres per minute.
Speed of drilling cut		x m / min	x being the speed in metres per minute.
Speed of grinding cut		x m / min	x being the speed in metres per minute.
Speed of milling cut		x m / min	x being the speed in metres per minute.
Conventional milling			—
Climb milling (down milling)			—

SYMBOLS FOR MACHINE TOOL INDICATION PLATES
(As per IS 2182 - 1962)
(contd.)

Meaning	Symbol	Remarks
Electric motor		
Rectangular table		—
Circular table		—
Turning spindle		The sketch may be replaced by another which represents more truly the actual element of the machine.
Drilling spindle		The sketch may be replaced by another which represents more truly the actual element of the machine.
Milling spindle		The sketch may be replaced by another which represents more truly the actual element of the machine.
Grinding spindle		The sketch may be replaced by another which represents more truly the actual element of the machine.
Pump (General symbol)		—
Coolant pump		—
Lubricating pump		—

SYMBOLS FOR MACHINE TOOL INDICATION PLATES
(As per IS 2182 - 1962)   (contd.)

Meaning	Symbol	Remarks
Hydraulic system pump		—
Hydraulic motor		—
Tracer		—
Stepless regulation		—
Adjustable		To be used on another symbol representing the element to be adjusted.
Lock or tighten		—
Unlock, unclamp (chuck open)		—
Brake on		—
Brake off		—
Automatic cycle		—

SYMBOLS FOR MACHINE TOOL INDICATION PLATES

(As per IS 2182 - 1962)   (contd.)

Meaning	Symbol	Remarks
Hand control		—
Start, on		The short bar should be indicated in green colour on the control button.
Stop, off		Ring shaped symbol in red on the control button.
Start and stop with same control		Symbol on the control button
In action as long as button is operated		Symbol on the control button
Emergency stop, master stop		Big button of red colour
Clutch engaged		—
Clutch disengaged		—
Half nut engaged		—
Half nut disengaged		—

SYMBOLS FOR MACHINE TOOL INDICATION PLATES
(As per IS 2182 - 1962)   (contd.)

Meaning	Symbol	Remarks
Engage tracer		—
Disengage tracer		—
Do not change speed while in motion		—
Change speed while in motion only		—
Shear pin construction		—
Attention (high voltage) danger		Arrow in red colour
Caution		In yellow colour
Main switch		Arrow in red colour
Coolant		—

SYMBOLS FOR MACHINE TOOL INDICATION PLATES
(As per IS 2182 - 1962)

(concld.)

Meaning	Symbol	Remarks
Machine lighting		—
Weights		x being the value of the weight, followed by the symbol of the unit used.
Refilling		—
Full level		—
Drain		—
Oil, lubricant		—
Blowing unit		—
Suction unit		—

GRAPHICAL SYMBOLS FOR KINEMATIC SCHEMES OF MACHINE TOOLS

Symbol	Nomenclature	Symbol	Nomenclature	Symbol	Nomenclature
General				**Operating Levers**	
	Axis or Centreline		Rotation of shaft a) Clockwise b) Counter clockwise c) Both		Shaft end for detachable handle
	Shaft, spindle, connecting rod, etc.		Fixed attachment or support of shaft, etc.		Hand crank
	Nature and direction of motion		Fixed support for rod in reciprocating motion a) Sliding support b) Rolling support		Hand wheel
or	Rectilinear motion in one direction		Support for lever, etc. a) Rigidly supported b) Freely supported		Control lever
	Reciprocating motion		Ball and socket joint		Eccentric
or	Rotary motion in one direction		Flywheel on shaft		Adjustable stops
	Rotary motion, reversible				
	Rocking motion				
	Shifting or change over (gear, belt, etc.)				

GRAPHICAL SYMBOLS FOR KINEMATIC SCHEMES OF MACHINE TOOLS (contd.)

Symbol	Nomenclature
Springs	
	Compression springs
	Tension springs
	Spiral springs
	Conical springs
	Leaf springs
	Disc springs (Belleville washers)
Electric Motors	
	General symbol
	Foot mounted
	Flange mounted
	Built in

Symbol	Nomenclature
Bearings General (Type not specified)	
	Radial
	Radial thrust Single sided
	Double sided
	Thrust Single sided
	Double sided
Sliding Bearings	
	Radial
	Radial thrust Single sided
	Double sided
	Thrust Single sided
	Double sided

Symbol	Nomenclature
Antifriction Bearings	
	Radial General symbol
	Radial roller bearing
	Radial thrust Single sided
	Double sided
	Radial Self aligning
	Radial thrust roller bearing Single sided
	Thrust bearing Single sided
	Double sided

GRAPHICAL SYMBOLS FOR KINEMATIC SCHEMES OF MACHINE TOOLS (contd.)

Shaft Mountings

Symbol	Nomenclature
	Freely mounted
	Rigidly fixed
	Sliding

Couplings

Symbol	Nomenclature
	Flange coupling
	Flange coupling with protection against overload
	Flexible coupling
	Universal coupling
	Telescopic coupling
	Muff coupling
	Claw clutch or toothed coupling

Clutches

Symbol	Nomenclature	Symbol	Nomenclature
	Multiplate clutch General symbol without indicating the type		Overrunning clutch (or unidirectional clutch)
	Single sided, General symbol		Cone clutch Single sided
	Single sided, Electromagnetic		Double sided
	Single sided, Hydraulic or pneumatic		Disc or single plate clutch Single sided
	Double sided General symbol		Double sided
	Double sided Electromagnetic		Disc clutch with friction jaws
	Double sided Hydraulic or Pneumatic		Disc clutch with expanding rings

GRAPHICAL SYMBOLS FOR KINEMATIC SCHEMES OF MACHINE TOOLS (contd.)

Symbol	Nomenclature	Symbol	Nomenclature	Symbol	Nomenclature
	Clutches		**Cams**		**Link Mechanisms**
	Centrifugal clutch		Disc cams		Connecting rod with adjustable block variable stroke
	Overload clutch (or safety clutch)		Single edged		Crank mechanism with rocker arm
	Ball and spring type		Double edged		a) With constant stroke
	Brakes		Followers for cams		b) With variable stroke
	Disc brake		a) Pointed b) Flat faced c) Roller type		**Ratchet and Geneva Mechanism**
	General symbol		Drum cams		Pawl and ratchet
	Electromagnetic		Cylindrical		
	Hydraulic or Pneumatic		Conical		
			Flat plate cam (Cam with reciprocating motion)		

GRAPHICAL SYMBOLS FOR KINEMATIC SCHEMES OF MACHINE TOOLS (contd.)

Symbol	Nomenclature	Symbol	Nomenclature	Symbol	Nomenclature
Ratchet and Geneva Mechanism		**Belt and Chain Transmission**		**Belt and Chain Transmission**	
	Geneva mechanism *a)* With external engagement *b)* With internal engagement		Flat belt transmission *a)* Open drive *b)* With idler pulley		Belt shifter Chain transmission General symbol
	Belt pulley on shaft *a)* Fixed (or fast) *b)* Free (or loose) V-belt transmission		Infinitely variable V-belt transmission Stepped V-pulley		**Gear Transmission**
					Cylindrical gears General symbol without indicating the type *a)* External gearing *b)* Internal gearing *c)* Planetary gearing

GRAPHICAL SYMBOLS FOR KINEMATIC SCHEMES OF MACHINE TOOLS (concld.)

Gear Transmission

Symbol	Nomenclature
	Bevel gears. General Symbol without indicating the type
	Rack and pinion. General Symbol without indicating the type of teeth (Rack and worm-same symbol to be used)
	Cluster gears
	Worm and worm wheel

Symbol	Nomenclature
	Spiral gearing or Crossed helical gears
	Gear shifter

Power Screws

Symbol	Nomenclature
	Power screw LH (Lead Screw)
	Nut on lead screw solid (non-split)
	Nut on lead screw solid (non-split) with recirculating balls
	Split or half nut on lead screw

Spindle Noses of Machine Tools

Symbol	Nomenclature
	Turning
	Chucking
	Collet work
	Drilling
	Boring with face plate
	Milling
	Grinding

GEOMETRICAL TOLERANCES
(SYMBOLS AND INTERPRETATIONS)

Features	Tol. Type	Symbol	Figure	Interpretation
For single features	Form	—		**STRAIGHTNESS:** The axis of a rotating part or the considered line on a flat surface is contained within the tolerance zone.
		◯		**CIRCULARITY (ROUNDNESS):** The Circumference of any circular cross-section is contained between two Coplanar concentric circles with the tolerance zone corresponding to the difference between the two radii.
		⌒		**PROFILE OF ANY LINE:** The considered profile is contained within two lines which envelop circles of diameter equal to tolerance zone and having their centres on a line having the correct profile.
		▱		**FLATNESS:** The surface is contained between two parallel planes apart by a distance equal to the tolerance zone.
		⌀		**CYLINDRICITY:** The considered cylindrical surface is contained within two coaxial cylinders, the tolerance zone corresponding to the difference between the two radii.
		⌓	SPHERE ø t	**PROFILE OF ANY SURFACE:** The considered surface is contained within two surfaces enveloping spheres of diameter equal to the tolerance zone and having their centres on a surface having the correct profile.
For related features	Attitude	//		**PARALLELISM:** The considered line or surface is contained within two parallel lines or planes which are theoretically parallel to a datum line or surface and the distance between the two defining the tolerance zone.

GEOMETRICAL TOLERANCES
(SYMBOLS AND INTERPRETATIONS)

(concld.)

Feature	Tol. Type	Symbol	Figure	Interpretation
For related features	Attitude	⊥		**PERPENDICULARITY (SQUARENESS):** The considered line or surface is contained within two parallel lines or planes which are theoretically perpendicular to a datum line or surface and the distance between the two defining the tolerance zone.
		∠		**ANGULARITY:** The considered line or surface is contained within two parallel lines or planes which are theoretically inclined at the specified angle to a datum line or surface and the distance between the two defining the tolerance zone.
	Composite	↗		**RUNOUT:** It is the maximum permissible variation of position of any considered feature with respect to a fixed point during one complete revolution about the datum axis.
	Location	⊕		**POSITION:** The considered point, line or a surface is contained within a tolerance zone defined by a geometric configuration whose principle plane or axis passes through the true specified position.
		◎		**CONCENTRICITY AND COAXIALITY:** The axis of a considered feature is contained within a tolerance zone, the axis of which coincides with the datum axis.
		⚌		**SYMMETRY:** The considered plane is contained within two parallel planes which are symmetrically disposed about the datum plane and the distance between the planes defining the tolerance zone.

SCREW THREADS AND THREADED FASTENERS

ISO METRIC SCREW THREADS (As per IS 4218 - 1967)

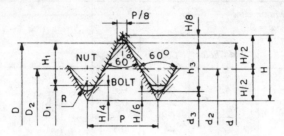

NUT DIMENSIONS

$D_1 = D - 1.08253\,P$

$D_2 = D - 0.64952\,P$

$H = 0.86603\,P$

$H_1 = 0.54127\,P = \dfrac{5}{8}\,H$

$R = \dfrac{H}{6} = 0.14434\,P$

BOLT DIMENSIONS

$d_2 = d - 0.64952\,P$

$d_3 = d - 1.22687\,P$

$h_3 = 0.61343\,P$

Major Dia. $d=D$		Pitch P		Stress Area mm^2	Tap Drill Size
Preferred	2nd Choice	Coarse	Fine		
1	—	0.25	—	0.46	0.75
1	—	—	0.2	0.53	0.8
—	1.1	0.25	—	0.59	0.85
—	1.1	—	0.2	0.67	0.9
1.2	—	0.25	—	0.73	0.95
1.2	—	—	0.2	0.82	1
—	1.4	0.3	—	0.98	1.1
—	1.4	—	0.2	1.17	1.2
1.6	—	0.35	—	1.27	1.25
1.6	—	—	0.2	1.59	1.4
—	1.8	0.35	—	1.7	1.45
—	1.8	—	0.2	2.04	1.6
2	—	0.4	—	2.07	1.6
2	—	—	0.25	2.45	1.75
—	2.2	0.45	—	2.48	1.75
—	2.2	—	0.25	3.03	1.95
2.5	—	0.45	—	3.39	2.05
2.5	—	—	0.35	3.7	2.15
3	—	0.5	—	5.03	2.5
3	—	—	0.35	5.61	2.7
—	3.5	0.6	—	6.78	2.9
—	3 5	—	0.35	7.9	3.2
4	—	0.7	—	8.78	3.3

ISO METRIC SCREW THREADS (As per IS 4218 - 1967) (contd.)

Major Dia. d = D		Pitch P		Stress Area mm²	Tap Drill Size
Preferred	2nd Choice	Coarse	Fine		
4	—	—	0.5	9.79	3.5
—	4.5	0.75	—	11.3	3.8
—	4.5	—	0.5	12.8	4
5	—	0.8	—	14.2	4.2
5	—	—	0.5	16.1	4.5
6	—	1	—	20.1	5
6	—	—	0.75	22	5.2
—	7	1	—	28.9	6
—	7	—	0.75	31.3	6.2
8	—	1.25	—	36.6	6.8
8	—	—	0.75	41.8	7.2
8	—	—	1	39.2	7
10	—	1.5	—	58	8.5
10	—	—	0.75	67.9	9.2
10	—	—	1	64.5	9
10	—	—	1.25	61.2	8.8
12	—	1.75	—	84.3	10.25
12	—	—	1	96.1	11
12	—	—	1.25	92.1	10.8
12	—	—	1.5	88.1	10.5
—	14	2	—	115	12
—	14	—	1	134	13
—	14	—	1.5	125	12.5
16	—	2	—	157	14
16	—	—	1	178	15
16	—	—	1.5	167	14.5
—	18	2.5	—	192	15.5
—	18	—	1	229	17
—	18	—	1.5	216	16.5
—	18	—	2	204	16·
20	—	2.5	—	245	17.5
20	—	—	1	285	—
20	—	—	1.5	272	18.5
20	—	—	2	258	18
—	22	2.5	—	303	19.5
—	22	—	1	348	—
—	22	—	1.5	333	20.5
—	22	—	2	318	20
24	—	3	—	353	21
24	—	—	1	418	—
24	—	—	1.5	401	22.5

ISO METRIC SCREW THREADS (As per IS 4218 - 1967) (concld.)

Major Dia. d= D		Pitch P		Stress Area mm²	Tap Drill Size
Preferred	2nd Choice	Coarse	Fine		
24	—	—	2	384	22
—	27	3	—	459	24
—	27	—	1	533	—
—	27	—	1.5	514	25.5
—	27	—	2	496	25
30	—	3.5	—	561	26.5
30	—	—	1	663	—
30	—	—	1.5	642	28.5
30	—	—	2	621	28
—	33	3.5	—	694	29.5
—	33	—	1.5	785	31.5
—	33	—	2	761	31
—	35	—	1.5	886	33.5
36	—	4	—	817	32
36	—	—	1.5	940	34.5
36	—	—	2	915	34
36	—	—	3	865	33
—	39	4	—	976	35
—	39	—	1.5	1110	37.5
—	39	—	2	1080	37
—	39	—	3	1030	36
42	—	4.5	—	1120	37.5
42	—	—	1.5	1290	40.5
42	—	—	2	1260	40
42	—	—	3	1210	39
42	—	—	4	1150	38
—	45	4.5	—	1300	40.5
—	45	—	1.5	1490	43.5
—	45	—	2	1460	43
—	45	—	3	1400	42
—	45	—	4	1340	41
48	—	5	—	1470	43
48	—	—	1.5	1710	46.5
48	—	—	2	1670	46
48	—	—	3	1600	45
48	—	—	4	1540	44
—	52	5	—	1760	47
—	52	—	1.5	2010	50.5
—	52	—	2	1970	50
—	52	—	3	1900	49
—	52	—	4	1830	48

ISO METRIC TRAPEZOIDAL SCREW THREADS (As per IS 7008-1973)

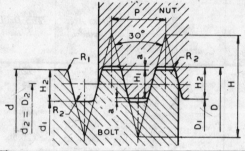

$$H_1 = 0.5P$$
$$H_2 = H_1 + a$$
$$H = 1.866P$$
$$d_2 = D_2 = d - 0.5P$$
$$R_1 max = 0.5a$$
$$R_2 max = a$$

Nominal dia. d (Major dia. of bolt)		Pitch P	Major dia. of nut D	Minor dia		Pitch dia. $d_2 = D_2$	Crest allowance a	Core area cm^2
Preferred	2nd choice			Bolt d_1	Nut D_1			
8	—	1.5	8.3	6.2	6.5	7.25	0.15	0.3
—	9	1.5	9.3	7.2	7.5	8.25	0.15	0.41
—	9	2	9.5	6.5	7	8	0.25	0.33
10	—	1.5	10.3	8.2	8.5	9.25	0.15	0.53
10	—	2	10.5	7.5	8	9		0.44
—	11	2	11.5	8.5	9	10		0.57
—	11	3	11.5	7.5	8	9.5		0.44
12	—	2	12.5	9.5	10	11		0.71
12	—	3	12.5	8.5	9	10.5		0.57
—	14	2	14.5	11.5	12	13	0.25	1.04
—	14	3	14.5	10.5	11	12.5		0.87
16	—	2	16.5	13.5	14	15		1.43
16	—	4	16.5	11.5	12	14		1.04
—	18	2	18.5	15.5	16	17		1.9
—	18	4	18.5	13.5	14	16		1.43
20	—	2	20.5	17.5	18	19		2.4
20	—	4	20.5	15.5	16	18		1.9
—	22	3	22.5	18.5	19	20.5		2.69
—	22	5	22.5	16.5	17	19.5		2.14
—	22	8	23	13	14	18	0.5	1.33
24	—	3	24.5	20.5	21	22.5	0.25	3.3
24	—	5	24.5	18.5	19	21.5	0.25	2.69
24	—	8	25	15	16	20	0.5	1.77
—	26	3	26.5	22.5	23	24.5	0.25	3.98
—	26	5	26.5	20.5	21	23.5	0.25	3.3
—	26	8	27	17	18	22	0.5	2.27
28	—	3	28.5	24.5	25	26.5	0.25	4.71
28	—	5	28.5	22.5	23	25.5	0.25	3.98
28	—	8	29	19	20	24	0.5	2.84
—	30	3	30.5	26.5	27	28.5	0.25	5.51
—	30	6	31	23	24	27	0.5	4.15
—	30	10	31	19	20	25	0.5	2.84

ISO METRIC TRAPEZOIDAL SCREW THREADS (As per IS 7008-1973)

Nominal dia. d (Major dia. of bolt)		Pitch P	Major dia. of nut D	Minor dia		Pitch dia. $d_2 = D_2$	Crest allowance a	Core area cm^2
Preferred	2nd choice			Bolt d_1	Nut D_1			
32	—	3	32.5	28.5	29	30.5	0.25	6.38
32	—	6	33	25	26	29	0.5	4.91
32	—	10	33	21	22	27	0.5	3.46
—	34	3	34.5	30.5	31	32.5	0.25	7.3
—	34	6	35	27	28	31	0.5	5.73
—	34	10	35	23	24	29	0.5	4.12
36	—	3	36.5	32.5	33	34.5	0.25	8.3
36	—	6	37	29	30	33	0.5	6.61
36	—	10	37	25	26	31	0.5	4.91
—	38	3	38.5	34.5	35	36.5	0.25	9.35
—	38	7	39	30	31	34.5	0.5	7.1
—	38	10	39	27	28	33	0.5	5.73
40	—	3	40.5	36.5	37	38.5	0.25	10.46
40	—	7	41	32	33	36.5	0.5	8.04
40	—	10	41	29	30	35	0.5	6.61
—	42	3	42.5	38.5	39	40.5	0.25	11.64
—	42	7	43	34	35	38.5	0.5	9.1
—	42	10	43	31	32	37	0.5	7.55
44	—	3	44.5	40.5	41	42.5	0.25	12.38
44	—	7	45	36	37	40.5	0.5	10.18
—	46	3	46.5	42.5	43	44.5	0.25	14.19
—	46	8	47	37	38	42	0.5	10.75
48	—	3	48.5	44.5	45	46.5	0.25	15.55
48	—	8	49	39	40	44	0.5	11.95
—	50	3	50.5	46.5	47	48.5	0.25	16.98
—	50	8	51	41	42	46	0.5	13.2
52	—	3	52.5	48.5	49	50.5	0.25	18.47
52	—	8	53	43	44	48	0.5	14.52
—	55	3	55.5	51.5	52	53.5	0.25	20.83
—	55	9	56	45	46	50.5	0.5	15.9
60	—	3	60.5	56.5	57	58.5	0.25	25.07
60	—	9	61	50	51	55.5	0.5	19.63
—	65	4	65.5	60.5	61	63	0.25	28.75
—	65	10	66	54	55	60	0.5	22.9
70	—	4	70.5	65.5	66	68	0.25	33.7
70	—	10	71	59	60	65	0.5	27.34
—	75	4	75.5	70.5	71	73	0.25	39.04
—	75	10	76	64	65	70	0.5	32.17
80	—	4	80.5	75.5	76	78	0.25	44.77
80	—	10	81	69	70	75	0.5	37.3
—	85	4	85.5	80.5	81	83	0.25	50.9
90	—	4	90.5	85.5	86	88	0.25	57.41
—	95	4	95.5	90.5	91	93	0.25	64.33
100	—	4	100.5	95.5	96	98	0.25	71.63

Size d (inch)	Size d (mm)	Pitch P	TPI Z	Stress. Area mm²	Tap Drill Size
1/2	12.7	2.117	12	78.5	10.3
9/16	14.288	2.117	12	104.5	11.8
5/8	15.875	2.309	11	131	13.25
3/4	19.05	2.54	10	196	16.25
7/8	22.225	2.822	9	272	19
1	25.4	3.175	8	357	21.75
1 1/8	28.575	3.629	7	448	24.25
1 1/4	31.75	3.629	7	576	27.25
1 3/8	34.925	4.233	6	685	30
1 1/2	38.1	4.233	6	836	33
1 5/8	41.275	5.08	5	948	35
1 3/4	44.450	5.08	5	1125	38.5
1 7/8	47.625	5.644	4½	1275	41
2	50.8	5.644	4½	1490	44
2 1/4	57.15	6.35	4	1890	—
2 1/2	63.5	6.35	4	2400	—
2 3/4	69.85	7.257	3½	2880	—
3	76.2	7.257	3½	3520	—
3 1/4	82.55	7.815	3¼	4130	—
3 1/2	88.9	7.815	3¼	4880	—
3 3/4	95.25	8.467	3	5600	—
4	101.6	8.467	3	6480	—
4 1/2	114.3	8.835	2⅞	8320	—
5	127	9.236	2⅝	10400	—
5 1/2	139.7	9.676	2⅝	12650	—
6	152.4	10.16	2¼	15200	—

WHITWORTH THREADS

$$P = \frac{25.4}{Z} \qquad r = 0.13733P \qquad H = 0.96049P \qquad H_1 = 0.64033P$$

Size d (inch)	Size d (mm)	Pitch P	TPI Z	Stress Area mm²	Tap Drill Size
1/16	1.588	0.423	60	—	1.15
3/32	2.381	0.529	48	—	1.85
1/8	3.175	0.635	40	—	2.5
5/32	3.969	0.794	32	—	3.2
3/16	4.763	1.058	24	—	3.7
7/32	5.556	1.058	24	—	4.5
1/4	6.35	1.27	20	17.5	4.9
5/16	7.938	1.411	18	29.5	6.4
3/8	9.525	1.588	16	44	7.7
7/16	11.112	1.814	14	60.8	9.1

FASTENING PIPE THREADS
(As per IS 2643 - 1975)

$H = 0.960491\,P$
$h = 0.640327\,P$
$r = 0.137329\,P$

Size Designation	Major diameter $d=D$	Pitch P	Threads per inch	Tap drill size
G 1/8	9.728	0.907	28	8.8
G 1/4	13.157	1.337	19	11.75
G 3/8	16.662	1.337	19	15.25
G 1/2	20.955	1.814	14	19
G 3/4	26.441	1.814	14	24.25
G 1	33.249			30.75
G 1 1/4	41.91		11	39
G 1 1/2	47.803			45
G 2	59.614	2.309		57
G 2 1/4	65.71			—
G 2 1/2	75.184			—
G 3	87.884			—
G 3 1/2	100.33			—
G 4	113.03			—
G 5	138.43			—
G 6	163.83			—

TAPER 1:16 ON DIA
AXIS OF THREAD
55°   P   NUT   BOLT
d_2 d_1

55°   P   H   NUT   BOLT   r
$d_2 = D_2$ $d_1 = D_1$ $d = D$

TAPER PIPE THREADS
(As per IS 554 - 1975)

(All dimensions in mm)

Size Designation	Pitch P	Threads per inch	Gauge plane dia. D	Nom. distance of gauge plane, l	Nom. useful length of threads, L
R 1/8	0.907	28	9.728	4	6.5
R 1/4	1.337	19	13.157	6	9.7
R 3/8	1.337	19	16.662	6.4	10.1
R 1/2	1.814	14	20.955	8.2	13.2
R 3/4	1.814	14	26.441	9.5	14.5
R 1			33.249	10.4	16.8
R 1 1/4			41.91	12.7	19.1
R 1 1/2		11	47.803	12.7	19.1
R 2	2.309		59.614	15.9	23.4
R 2 1/2			75.184	17.5	26.7
R 3			87.884	20.6	29.8
R 3 1/2			100.33	22.2	31.4
R 4			113.03	25.4	35.8
R 5			138.43		
R 6			163.83	28.6	40.1

*An external taper pipe thread is designated R........
An internal taper pipe thread is designated R_C........

BLACK HEXAGON BOLTS AND SCREWS (As per IS 1363 - 1967)

(All dimensions in mm)

Nom. Size	d Max.	d Min.	s	e Min.	k js16	a* Max.	b** Min. b₁	b** Min. b₂	b** Min. b₃	Lt Bolt From	Lt Bolt Upto	Lt Screw From	Lt Screw Upto
M 6	6.48	5.7	10	10.89	4	4	18	—	—	25	100	10	40
M 8	8.9	7.64	13	14.2	5.5	4.5	22	28	—	30	120	12	50
M 10	10.9	9.64	17	18.72	7	5	26	32	—	35	150	16	60
M 12	13.1	11.57	19	20.88	8	6	30	36	49	40	300	20	80
M 16	17.1	15.57	24	26.17	10	7.5	38	44	57	50	300	25	80
M 20	21.3	19.48	30	32.95	13	9	46	52	65	60	400	45	80
M 24	25.3	23.48	36	39.55	15	11	54	60	73	70	400	55	80
M 30	31.3	29.48	46	50.85	19	—	66	72	85	90	400	—	—
M 36	37.6	35.38	55	60.79	23	—	78	84	97	110	400	—	—

(s column tolerance: 19, h 15 / 19, h 14 ; < VI)

Notes:

* Dimension *a* is applicable to screws only.

** Dimension *b* is applicable to bolts only. b_1 for $L \leq 130$, b_2 for $130 < L \leq 200$ and b_3 for $L > 200$.

† Lengths are: in steps of 2 between 10 and 16; in steps of 5 between 20 and 90; in steps of 10 between 100 and 200 and in steps of 20 between 200 and 400.

PRECISION AND SEMI - PRECISION HEXAGON BOLTS AND SCREWS
(As per IS 1364 - 1967; 2389 - 1968)

(All dimensions in mm)

Size@@ Coarse	Size@@ Fine	d h13	s@	e Min. Precision	e Min. Semi Precision	k†	dw Min.	a* max.	b Min.** b1	b2	b3	L†† Bolt From	Bolt Upto	Screw From	Screw Upto
M 1.6	—	1.6	3.2	3.48	—	1.1	—	0.8	9	—	—	12	16	3	12
M 2	—	2	4	4.38	—	1.4	—	0.9	10	—	—	12	16	3	16
M 2.5	—	2.5	5	5.51	—	1.7	—	1.2	11	—	—	14	25	3	25
M 3	—	3	5.5	6.08	—	2	—	1.2	12	—	—	16	25	4	25
M 4	—	4	7	7.74	—	2.8	6.3	1.5	14	—	—	20	70	5	70
M 5	—	5	8	8.87	—	3.5	7.2	1.8	16	—	—	25	75	8	75
M 6	—	6	10	11.05	10.89	4	9	2.5	18	24	—	25	80	10	80
M 8	M 8×1	8	13	14.38	14.2	5.5	11.7	3	22	28	—	30	90	12	90
M 10	M 10×1.25	10	17	18.9	18.72	7	15.3	4	26	32	—	35	120	16	100
M 12	M 12×1.25	12	19	21.1	20.88	8	17.1	5	30	36	49	40	120	20	100
M 16	M 16×1.5	16	24	26.75	26.17	10	21.6	6.5	38	44	57	50	200	25	120
M 20	M 20×1.5	20	30	33.53	32.95	13	27	7.5	46	52	65	60	300	45	150
M 24	M 24×2	24	36	39.98	39.55	15	32.4		54	60	73	70	300	55	150
M 30	M 30×2	30	46	51.28	50.85	19	41.4	13	66	72	85	90	300	60	150
M 36	M 36×3	36	55	61.31	60.79	23	49.5	14	78	84	97	110	300	60	150

@@Semi precision bolts and screws available in sizes M6 & above only.

@Tolerance on s

For precision class— h12 for s ≤8

h13 for 8 < s ≤32; h14 for s >32

For semiprecision class -

h14 for s≤19; h15 for s >19

*Dimension a for screws only.

**Dimension b for bolts only.

b1 for L≤130, b2 for 130<L≤200, b3 for L>200

†Tolerance on k-precision class-js14 semiprecision class -js15.

††Lengths are: in steps of 1 between 3 and 6; in steps of 2 between 8 and 16; in steps of 5 between 20 and 90; in steps of 10 between 100 and 200; in steps of 20 between 200 and 300.

HEXAGON SOCKET HEAD CAP SCREWS (As Per IS 2269 - 1967)

(All dimensions in mm)

Size		d h13	D h13	s D12	k h13	t Max.	t Min.	b* b_1	b* b_2	L** js15 From	L** js15 Upto
Coarse	Fine										
M3	—	3	5.5	2.5	3	1.7	1.3	12	—	4	40
M4	—	4	7	3	4	2.4	2	14	—	6	50
M5	—	5	8.5	4	5	3.1	2.7	16	—	10	60
M6	—	6	10	5	6	3.78	3.3	18	—	12	60
M8	M8×1	8	13	6	8	4.78	4.3	22	—	14	100
M10	M10×1.25	10	16	8	10	6.25	5.5	26	—	14	120
M12	M12×1.25	12	18	10	12	7.5	6.6	30	—	20	120
M16	M16×1.5	16	24	14	16	9.7	8.8	38	44	30	150
M20	M20×1.5	20	30	17	20	11.8	10.7	46	52	40	180
M24	M24×2	24	36	19	24	14	12.9	54	60	60	200
M30	M30×2	30	45	22	30	18.2	17.1	66	72	80	200
M36	M36×3	36	54	27	36	22.1	20.8	78	84	100	200

Notes:

*b_1 is applicable for $L \leq 125$ mm and b_2 for $L > 125$ mm.

**Where $L \leq b$, the screws are considered as fully threaded.

Lengths are: in steps of 1 between 4 and 6; in steps of 2 between 8 and 16; in steps of 5 between 20 and 90; in steps of 10 between 90 and 200.

SLOTTED COUNTERSUNK HEAD AND SLOTTED RAISED COUNTERSUNK HEAD SCREWS - SMALL HEAD SERIES

(As per IS 5308 - 1969)

*Dimension *a* is applicable for Type A only, which are fully threaded.

@Dimension *b* is applicable for Type B only, which are partially threaded.

@@Lengths are : in steps of 1 between 3 and 10; in steps of 2 between 12 and 16; in steps of 5 between 20 and 80; in steps of 10 between 90 & 110.

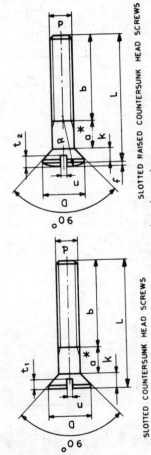

SLOTTED COUNTERSUNK HEAD SCREWS

SLOTTED RAISED COUNTERSUNK HEAD SCREWS

(All dimensions in *mm*)

Size	d Max.	d Min.	D Max.	D Min.	k Max.	k Min.	f ≈	R ≈	n Max.	n Min.	t_1 Max.	t_1 Min.	t_2 Max.	t_2 Min.	a^* Max.	$b@$ Min.	$L@@$ Type A From	Type A Upto	Type B From	Type B Upto
M1.6	1.6	1.46	3	2.86	0.7	0.62	0.4	3	0.5	0.36	0.51	0.3	0.83	0.64	0.7	15	3	16	—	—
M2	2	1.86	3.5	3.32	0.75	0.65	0.5	4	0.6	0.46	0.64	0.4	1.04	0.8	0.8	16	4	20	20 only	
M2.5	2.5	2.36	4.5	4.32	1	0.87	0.6	5	0.8	0.66	0.8	0.5	1.3	1	0.9	18	5	25	25 only	
M3	3	2.86	5	4.82	1	0.87	0.75	6	1	0.86	0.96	0.6	1.56	1.2	1	19	6	30	25	30
M4	4	3.82	7	6.78	1.5	1.38	1	8	1.2	1.06	1.28	0.8	2.08	1.6	1.4	22	8	40	30	40
M5	5	4.82	9	8.78	2	1.78	1.25	10	1.51	1.26	1.6	1	2.6	2	1.6	25	10	50	30	50
M6	6	5.82	10	9.78	2	1.78	1.5	12	1.91	1.66	1.92	1.2	3.12	2.4	2	28	10	80	35	80

SLOTTED COUNTERSUNK HEAD AND SLOTTED RAISED COUNTERSUNK HEAD SCREWS (concld.)

(As per IS 1365 - 1968)

(All dimensions in mm)

Size	d Max.	d Min.	D Max	D Min.	k Max.	k Min.	f	R	n Max.	n Min.	t_1 Max.	t_1 Min.	t_2 Max.	t_2 Min.	a^* Max.	$b@$ Min.	L @@ Type A From	Type A Upto	Type B From	Type B Upto
M1.6	1.6	1.46	3.2	2.72	0.8	0.72	0.4	≈ 3	0.6	0.46	0.51	0.32	0.83	0.64	0.7	15	3	16	—	—
M2	2	1.86	4	3.4	1	0.9	0.5	≈ 4	0.7	0.56	0.64	0.4	1.04	0.8	0.8	16	4	20	20 only	
M2.5	2.5	2.36	5	4.25	1.25	1.12	0.6	≈ 5	0.8	0.66	0.8	0.5	1.3	1	0.9	18	5	25	25 only	
M3	3	2.86	6	5.1	1.5	1.35	0.75	≈ 6	1	0.86	0.96	0.6	1.56	1.2	1	19	6	30	25	30
M4	4	3.82	8	6.8	2	1.8	1	≈ 8	1.2	1.06	1.28	0.8	2.08	1.6	1.4	22	8	40	30	40
M5	5	4.82	10	8.5	2.5	2.25	1.25	≈ 10	1.51	1.26	1.6	1	2.6	2	1.6	25	10	50	30	50
M6	6	5.82	12	10.2	3	2.7	1.5	≈ 12	1.91	1.66	1.92	1.2	3.12	2.4	2	28	10	80	35	80
M8 / M8×1	8	7.78	16	14	4	3.6	2	≈ 16	2.31	2.06	2.56	1.6	4.16	3.2	2.5	34	12	90	45	90
M10 / M10×1.25	10	9.78	20	17.5	5	4.5	2.5	20	2.81	2.56	3.2	2	5.2	4	3	40	16	100	50	100
M12 / M12×1.25	12	11.73	24	21	6	5.4	3	25	3.31	3.06	3.84	2.4	6.24	4.8	3.5	46	20	100	60	100
M16 / M16×1.5	16	15.73	32	28	8	7.2	4	32	4.37	4.07	5.12	3.2	8.32	6.4	4	58	25	110	75	110
M20 / M20×1.5	20	19.67	40	35	10	9	5	40	5.37	5.07	6.4	4	10.4	8	5	70	45	110	90	110

(Refer page 113 for figure)

SLOTTED CHEESE HEAD SCREWS (As per IS 1366 - 1968)

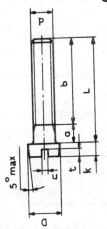

Notes:
1. *Dimension *a* is applicable for Type A only, which are fully threaded.
2. @Dimension *b* is applicable for Type B only, which are partially threaded.
3. @@Lengths are: in steps of 1 between 3 and 6; in steps of 2 between 8 and 16; in steps of 5 between 20 and 90; in steps of 10 between 90 and 110.

(All dimensions in mm)

Size (Coarse)	Size (Fine)	d Max.	d Min.	D Max.	D Min.	k Max.	k Min.	n Max.	n Min.	t Max.	t Min.	a* Max.	b@ Min.	L@@ Type A From	Type A Upto	Type B From	Type B Upto
M 1.6	—	1.6	1.46	3	2.6	1	0.86	0.6	0.46	0.65	0.45	0.7	15	3	16	—	—
M 2	—	2	1.86	3.8	3.32	1.3	1.05	0.7	0.56	0.85	0.6	0.8	16	4	20	18	20
M 2.5	—	2.5	2.36	4.5	4.02	1.6	1.35	0.8	0.66	1	0.7	0.9	18	5	25	20	25
M 3	—	3	2.86	5.5	5.02	2	1.75	1	0.86	1.3	0.9	1	19	6	30	25	30
M 4	—	4	3.82	7	6.42	2.6	2.35	1.2	1.06	1.6	1.2	1.4	22	8	40	25	40
M 5	—	5	4.82	8.5	7.92	3.3	3	1.51	1.26	2	1.5	1.6	25	10	50	30	50
M 6	—	6	5.82	10	9.42	3.9	3.6	1.91	1.66	2.3	1.8	2	28	10	80	35	80
M 8	M 8×1	8	7.78	13	12.3	5	4.7	2.31	2.06	2.8	2.3	2.5	34	12	90	40	90
M 10	M 10×1.25	10	9.78	16	15.3	6	5.7	2.81	2.56	3.2	2.7	3	40	16	100	45	100
M 12	M 12×1.25	12	11.73	18	17.3	7	6.64	3.31	3.06	3.8	3.2	3.5	46	20	100	55	100
M 16	M 16×1.5	16	15.73	24	23.16	9	8.64	4.37	4.07	4.6	4	4	58	25	110	65	110
M 20	M 20×1.5	20	19.67	30	29.16	11	10.57	5.37	5.07	5.6	5	5	70	50	110	80	110

COUNTERSUNK HEAD SCREWS WITH HEXAGON SOCKET
(As per IS 6761 - 1972)

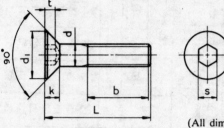

(All dimensions in *mm)

d h13	b(+2.5P)*		d₁ h14	k	s D12	t IT14	L**			
	b_1	b_2					Fully threaded		Partially threaded	
							From	Upto	From	Upto
M 3	12	—	6	1.7	2	1.2	8	40	25	40
M 4	14	—	8	2.3	2.5	1.8	8	50	30	50
M 5	16	—	10	2.8	3	2.3	8	50	35	50
M 6	18	24	12	3.3	4	2.5	8	55	40	55
M 8 M 8×1	22	28	16	4.4	5	3.5	12	60	45	60
M 10 M 10×1.25	26	32	20	5.5	6	4.4	15	100	45	100
M 12 M 12×1.5	30	36	24	6.5	8	4.6	18	140	55	140
M 16 M 16×1.5	38	44	30	7.5	10	5.3	30	150	55	150
M 20 M 20×2	46	52	36	8.5	12	5.9	35	150	80	150
M 24 M 24×2	54	60	39	14	14	10.3	45	150	100	150

*P is the pitch of screw thread.   **Lengths are: in steps of 2 between 8 and 20;
b_1 for $L < 130$ and   in steps of 5 between 20 and 60;
b_2 for $130 < L \leqslant 200$.   in steps of 10 between 60 and 150.

SLOTTED GRUB SCREWS (As per IS 2388 - 1971)

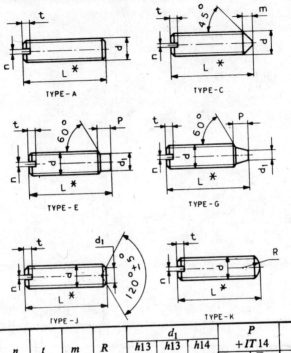

Size d	n H14	t js15	m	R	d_1 h13 Type E	d_1 h13 Type G	d_1 h14 Type J	P +IT 14 Type E	P +IT 14 Tpye G	L* js15 From	L* js15 Upto
M 1	0.25	0.6	0.5	—	—	—	—	—	—	2	4
M 1.2	0.25	0.6	0.6	—	—	—	—	—	—	2	4
M 1.6	0.3	0.8	0.8	—	—	—	—	—	—	2	6
M 2	0.3	0.8	1	—	1.3	—	—	1.5	—	3	8
M 2.5	0.4	1	1.3	—	1.6	—	—	2	—	4	10
M 3	0.5	1.2	1.5	2.25	2	—	1.4	2.5	—	4	12
M 4	0.6	1.4	2	3	2.5	—	2	3	—	5	14
M 5	0.8	1.8	2.5	3.75	3.5	1.6	2.5	3	4	5	16
M 6	1	2	2.5	4.5	4	2.4	3	3.5	4.5	6	20
M 8	1.2	2.5	3	6	6	3.7	5	5	5	8	25
M 10	1.6	3	4	7.5	7	4.4	6	5.5	6	10	35
M 12	2	4	5	9	9	5.1	8	7	8	12	45
M 16	2.5	4.5	6	12	12	7.4	10	9	10	16	50
M 20	3	5	7	15	15	9.7	14	9	12	20	65
M 24	4	6	8	18	—	—	16	—	—	25	70

*Lengths are : in steps of 1 between 2 and 6;   in steps of 2 between 6 and 16;
in steps of 5 between 20 and 70.   (All dimensions in *mm*)

HEXAGON SOCKET GRUB SCREW (As per IS 6094 - 1971):

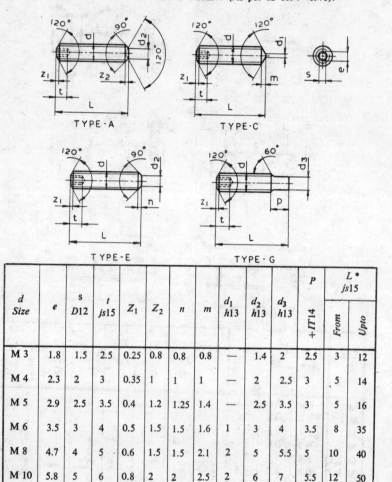

TYPE-A   TYPE-C

TYPE-E   TYPE-G

d Size	e	s D12	t js15	Z_1	Z_2	n	m	d_1 h13	d_2 h13	d_3 h13	P +IT14	L * js15 From	Upto
M 3	1.8	1.5	2.5	0.25	0.8	0.8	0.8	—	1.4	2	2.5	3	12
M 4	2.3	2	3	0.35	1	1	1	—	2	2.5	3	5	14
M 5	2.9	2.5	3.5	0.4	1.2	1.25	1.4	—	2.5	3.5	3	5	16
M 6	3.5	3	4	0.5	1.5	1.5	1.6	1	3	4	3.5	8	35
M 8	4.7	4	5	0.6	1.5	1.5	2.1	2	5	5.5	5	10	40
M 10	5.8	5	6	0.8	2	2	2.5	2	6	7	5.5	12	50
M 12	7	6	8	1	2	2	2.8	2	8	8.5	7	16	60
M 16	9.4	8	10	1.1	3	3	3.5	4	10	12	9	25	70
M 20	11.7	10	12	1.2	3	3	4	6	14	15	9	30	90
M 24	14	12	15	1.5	4	4	5	8	16	18	11	40	100

* Lengths are: in steps of 1 between 3 and 6   (All dimensions in *mm*)
              in steps of 2 between 6 and 16;
              in steps of 5 between 20 and 80;
              in steps of 10 between 80 and 100.

THUMB SCREWS (As per IS 3726 - 1972)

TYPE A TYPE B TYPE C TYPE D TYPE E

d h13	a h14	D js15	b Min.	k**	m js15	n Max.	n Min.	t	e	L* Type A & C From	L* Type A & C Upto	L* Type B & D From	L* Type B & D Upto	L* Type E @ From	L* Type E @ Upto	L* Type E @@ From	L* Type E @@ Upto
M1.6	3.8	7.5	5	2	5	0.6	0.5	1	1	8	—	2	6	3	12	6	12
M2	4.5	9	6	2	5.3	0.6	0.5	1.1	1.5	10	—	3	8	4	14	8	14
M2.5	5	11	8	2.5	6.5	0.8	0.6	1.5	2	12	—	3	10	5	16	10	16
M3	6	12	9	2.5	7.5	1	0.8	1.8	2	14	16	4	12	6	20	12	20
M4	8	16	12	3.5	9.5	1.2	1	2.2	3	—	—	5	16	8	25	20	25
M5	10	20	15	4	11.5	1.4	1.2	2.8	3	20	—	6	16	10	25	20	25
M6	12	24	18	5	15	1.8	1.6	3.5	4	25	—	8	20	12	30	20	30
M8	16	30	24	6	18	2.2	2	4.5	5	30	—	12	25	16	30	25	30
M10	20	36	30	8	23	2.8	2.5	6	6	40	—	20	30	20	40	30	40

Type A: Thumb screw partially threaded.
Type B: Thumb screw fully threaded
Type C: Slotted thumb screw partially threaded.
Type D: Slotted thumb screw fully threaded.
Type E: Flat thumb screw.

**Tolerance on k is js14 applicable for type E only
@ Fully threaded.
@@ Partially threaded.
*Lengths are: in steps of 1 between 2 and 6; in steps of 2 between 6 and 16; in steps of 5 between 20 and 30; in steps of 10 between 30 and 40.

(All dimensions in mm)

STUDS (As per IS 1862—1975)

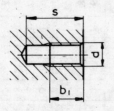

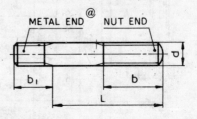

HOLE FOR STUD@@

(All dimensions in *mm*)

d h13	b min. L ≤ 125	b min. 125 < L ≤ 200	b min. L > 200	b₁ js16 Type A	b₁ js16 Type B	b₁ js16 Type C	s Type A	s Type B	s Type C	L* Types A & B From	L* Types A & B upto	L* Type C From	L* Type C Upto
M 3	12	—	—	3	4.5	—	6	8	—	14	25	—	—
M 4	14	—	—	4	6	8	8	10	12	16	70	25	60
M 5	16	—	—	5	7.5	—	9	12	—	20	80	—	—
M 6	18	—	—	6	9	—	11	14	—	25	85	—	—
M 8	22	—	—	8	12	16	13	17	21	30	110	40	100
M 8×1													
M 10	26	32	—	10	15	—	16	21	—	30	150		
M 10×1.25							15	20					
M 12	30	36	—	12	18	24	18	24	30	35	170	80	140
M 12×1.25							17	23	29				
M 16	38	44	57	16	24	32	23	31	39	45	300	100	160
M 16×1.5							22	30	38				
M 20	46	52	65	20	30	40	28	38	48	55	300	120	200
M 20×1.5							26	36	46				
M 24	54	60	73	24	36	—	33	45	—	65	300	—	—
M 24×2							31	43					
M 30	66	72	85	30	45	—	40	55	—	75	300	—	—
M 30×2							37	52					
M 36	78	84	97	36	54	—	47	65	—	90	300	—	—
M 36×3							45	63					

Type A - for use in steel
Type B - for use in CI
Type C - for use in Al alloys

@Interference threads
as per IS 2186 - 1967.
@@As per IS 4499 - 1968

* Normally available lengths : For
types A & B; 14, 16, 20, 25 to
90 in steps of 5, 90 to 200 in
steps of 10 and 200 to 300 in
steps of 25.
For Type C: 25, 30, 40, 60 to
160 in steps of 20 and 200.

HEXAGON NUTS, LOCK NUTS - BLACK, SEMI-PRECISION AND PRECISION GRADES (As per IS 1363 - 1967, 1364 - 1967 & 2389 - 1968)

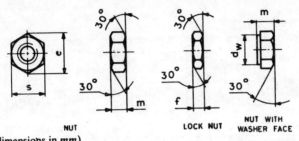

NUT   LOCK NUT   NUT WITH WASHER FACE

(All dimensions in *mm*)

Nominal Size d**		Tol. on d* Black Grade	m@	f@	s@@	dw	e Min	
Coarse	Fine						Precision	Black & Semi-Precision
M 1.6	—	—	1.3	1	3.2	—	3.48	—
M 2	—	—	1.6	1.2	4	—	4.38	—
M 2.5	—	—	2	1.6	5	—	5.51	—
M 3	—	—	2.4	1.6	5.5	—	6.08	—
M 4	—	—	3.2	2	7	—	7.74	—
M 5	—	—	4	2.5	8	—	8.87	—
M 6	—	+0.48 −0.3	5	3	10	9	11.05	10.89
M 8	M 8×1	+0.9 −0.36	6.5	4	13	11.7	14.38	14.2
M 10	M 10×1.25	+0.9 −0.36	8	5	17	15.3	18.9	18.72
M 12	M 12×1.25	+1.1 −0.43	10	7	19	17.1	21.1	20.88
M 16	M 16×1.5	+1.1 −0.43	13	8	24	21.6	26.75	26.17

HEXAGON NUTS, LOCK NUTS - BLACK, SEMI-PRECISION AND PRECISION GRADES (As per IS 1363 - 1967, 1364 - 1967 & 2389 - 1968)

(All dimensions in *mm*)

(concld.)

Nominal Size d**		Tol. on d* Black Grade	m@	f@	s @@	dw	e Min.	
Coarse	Fine						Precision	Black & Semi-Precision
M 20	M 20×1.5	+1.3 —0.52	16	9	30	27	33.53	32.95
M 24	M 24×2	+1.3 —0.52	19	10	36	32.4	39.98	39.55
M 30	M 30×2	+1.3 —0.52	24	12	46	41.4	51.28	50.85
M 36	M 36×3	+1.6 —0.62	29	14	55	49.5	61.31	60.79

@Tolerance on *m* & *f*: Precision grade - *h*14
   Semi-precision grade - *h*16 & black *js* 16
@@Tolerance on *s* : Precision grade
        M 1.6 to M 5 - *h*12
        M 6 to M 36, for *s* ≤ 32 - *h*13
               for *s* > 32 - *h*14
  Black & semi-precision grade
        M 6 to M 36, for *s* ≤ 19 - *h*14
               for *s* > 19 - *h*15
*Tolerance on *d*: Precision and semi-precision grades - *h*13
**Black and semi-precision grades: M 6 and above only.
Nuts with washer face are available only for precision grade and for sizes M 6 and above.
Nuts with single chamfer are also available in black and semi-precision grades.

CAP NUTS (As per IS 2687 - 1975)

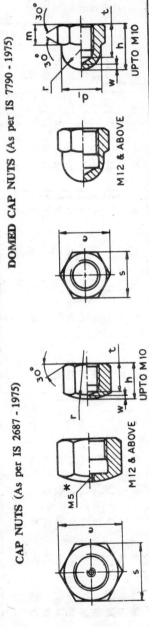

(All dimensions in *mm*)

Size	W Min.	t js15	r	s h13	e Min.	h h14
M 6	1.5	7	12	10	11.05	9
M 8	1.5	9.5	15	13	14.38	12
M 10	2	11	20	17	18.9	14
M 12	2	13.5	25	19	21.1	16
M 16	2	17	30	24	26.75	20
M 20	2.5	21	35	30	33.53	25
M 24	3	24	40	36	39.98	30
M 30	3	28	60	46	51.28	34
M 36	4	36	70	55	61.31	44

* For Sizes M 30 and M 36 only, and for sealing Sl.Hd.Grub screw 'Å M 5×6 to be used.

DOMED CAP NUTS (As per IS 7790 - 1975)

Size	W Min.	h h13	t js15	m js15	d_1	r	s@	e Min.
M 6	2	12	8	5	9.5	4.75	10	11.05
M 8	2	15	11	6.5	12.5	6.25	13	14.38
M 10	2	18	13	8	16	8	17	18.9
M 12	3	22	16	10	18	9	19	21.1
M 16	3	28	21	13	23	11.5	24	26.75
M 20	3	34	26	16	28	14	30	33.53
M 24	3	42	31	19	34	17	36	39.98

@ Tolerance on s: s ≤ 32-h13
                   s > 32-h14

CAP NUTS AND DOMED CAP NUTS ARE AVAILABLE IN PRECISION GRADE ONLY

WING NUTS (As per IS 2636 - 1972)

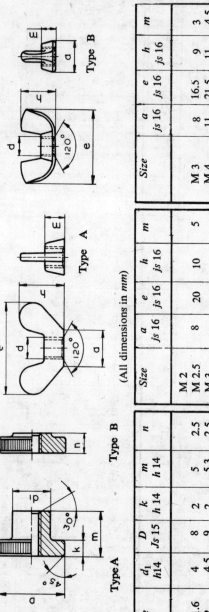

Type B

Size	a js 16	e js 16	h js 16	m
M 3	8	16.5	9	3
M 4	11	21.5	11	4.5
M 5	11	21.5	11	4.5
M 6	12	27	13	5
M 8	14	31	16	6
M 10	17	36	18	7
M 12	22	47.5	23	9.5
M 16	29.5	68	35	13.5
M 20	29.5	68	35	13.5

Type A

Size	a js 16	e js 16	h js 16	m
M 2	8	20	10	5
M 2.5				
M 3				
M 4				
M 5	10	25	12	6
M 6	12	32	16	8
M 8	16	40	20	10
M 10	20	50	25	12
M 12	23	64	32	14
M 16	28	72	36	16
M 20	36	90	45	20
M 24	45	112	56	24

(All dimensions in *mm*)

KNURLED NUTS (As per IS 3460 - 1972)

Type A   Type B

Size	d₁ h14	D Js 15	k h14	m h14	n
M 1.6	4	8	2	5	2.5
M 2	4.5	9	2	5.3	2.5
M 2.5	5	11	2.5	7	3
M 3	6	12	2.5	7.5	3
M 4	8	16	3.5	9.5	4
M5	10	20	4	11.5	5
M 6	12	24	5	15	6
M 8	16	30	6	18	8
M 10	20	36	8	23	10

THREAD RUNOUT AND UNDERCUT

(As per IS 1369 - 1975)

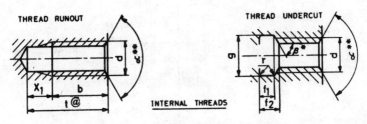

THREAD RUNOUT   THREAD UNDERCUT

INTERNAL THREADS

(All dimensions in *mm*)

Pitch of the thread P	Thread diameter d for coarse pitch	Thread runout		Thread undercut				
		X_1 Min		f_1 Max		f_2 Max.	r $\approx 0.5P$	g $H13$
		Long	Normal	Long $\approx 6P$	Normal $\approx 4P$	Normal		
0.35	M 1.6	—	1.7	—	—	1.9	0.2	
0.4	M 2	2.6	1.8	—	—	2.2	0.2	$d+0.2$
0.45	M 2.5	3.4	2.5	—	—	2.4	0.2	
0.5	M 3	3.8	2.8	3	2	2.7	0.3	
0.7	M 4	4.8	3.4	4.2	2.8	3.8	0.4	
0.8	M 5	5.2	3.6	4.8	3.2	4.2	0.4	
1	M 6	6.5	4.5	6	4	5.2	0.5	
1.25	M 8	7.5	5	7.5	5	6.7	0.6	$d+0.5$
1.5	M 10	8.5	5.5	9	6	7.8	0.8	
1.75	M 12	9.5	6	10.5	7	9.1	1	
2	M 16	10.5	6.5	12	8	10.3	1	
2.5	M 20	12.5	7.5	15	10	13	1.2	
3	M 24	14.5	8.5	18	12	15.2	1.6	
3.5	M 30	17	10	21	14	17.7	1.6	
4	M 36	19	11	24	16	20	2	

* β should be 60° for thread machining and 30° for thread forming.
** α is normally 120°; $\alpha = 60°$ is recommended when a stud with thread
  runout is used. A counterbore is recommended for light metal studs.
@ $t \geqslant b + X_1$ or $b + f_2$ where b is the useful length of thread.
  f_2 max. is based on a minimum transition angle of 30°. 'Normal'
  values of f_1 and X_1 is recommended in general. 'Long' values of
  f_1 and X_1 are recommended for blind holes where threads are cut
  with taps; and for product grade B.

THREAD RUNOUT AND UNDERCUT

(As per IS 1369 - 1975)

(contd.)

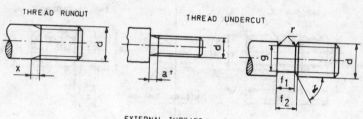

EXTERNAL THREADS

(All dimensions in *mm*)

Pitch of the thread P	Thread diameter d for coarse pitch	Thread runout X Max Normal $\approx 2.5\,P$	Thread undercut						Runout distance, $a^\dagger$ Max	
			f_1 Min		f_2 Max		r	g	Long	Normal
			Normal	Short	Normal $\approx 3P$	Short $\approx 2P$	$\approx 0.5P$	$h\,13$	$\approx 4P$	$\approx 3P$
0.35	M 1.6	0.8	0.7	0.4	1.1	0.7	0.2	$d-0.6$	—	1.05
0.4	M 2	1	0.8	0.5	1.2	0.8	0.2	$d-0.7$	—	1.2
0.45	M 2.5	1	1	0.5	1.4	0.9	0.2	$d-0.7$	—	1.35
0.5	M 3	1.2	1.1	0.5	1.5	1	0.3	$d-0.8$	—	1.5
0.7	M 4	1.6	1.5	0.8	2.1	1.4	0.4	$d-1.1$	—	2.1
0.8	M 5	2	1.7	0.9	2.4	1.6	0.4	$d-1.3$	3.2	2.4
1	M 6	2.5	2.1	1.1	3	2	0.5	$d-1.6$	4	3
1.25	M 8	3	2.7	1.5	3.8	2.5	0.6	$d-2$	5	4
1.5	M 10	3.5	3.2	1.8	4.5	3	0.8	$d-2.3$	6	4.5
1.75	M 12	4	3.9	2.1	5.3	3.5	1	$d-2.6$	7	5.3
2	M 16	5	4.5	2.5	6	4	1	$d-3$	8	6
2.5	M 20	6	5.6	3.2	7.5	5	1.2	$d-3.6$	10	7.5
3	M 24	7	6.7	3.7	9	6	1.6	$d-4.4$	12	9
3.5	M 30	8	7.7	4.7	10.5	7	1.6	$d-5$	14	10.5
4	M 36	10	9	5	12	8	2	$d-5.7$	16	12

† Dimension a applicable when threads are cut upto the head.

* $\propto$ is 60° for 'plunge cut' undercut and 30° for 'profile cut' undercut. Further, for turned parts it is 60° and for copy turned parts 30°.

1. 'Normal' values of X and a to be used for all types of fasteners in product grades P, S & B

2. 'Long' values of a to be used for all fasteners in product grade B.

3. f_1 min. based on a minimum transition angle of 30°

THREAD RUNOUT AND UNDERCUT

(As per IS 1369 - 1975)   (concld.)

Width of runout for various angles of throat or taper lead for ISO Metric Screw Threads

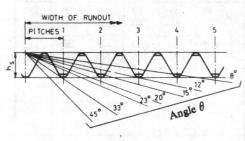

h_s— Theoretical depth of thread

Tool	Angle θ	Width of runout in terms of pitch
Dies	45°	0.61
	33°	0.95
	20°	1.69
	15°	2.3
Taps	23°	1.45
	12°	2.89
	8°	4.36
	4°	8.77

(All dimensions in *mm*)

CLEARANCE HOLES FOR METRIC BOLTS

(As per IS 1821 - 1967)

Bolt dia *d*	Clearance hole dia. D			Chamfer *Z*
	Fine *H*12	Medium *H*13	Coarse *H*14	
1.6	1.7	1.8	2	0.1
2	2.2	2.4	2.6	0.2
2.5	2.7	2.9	3.1	0.2
3	3.2	3.4	3.6	0.2
4	4.3	4.5	4.8	0.3
5	5.3	5.5	5.8	0.3
6	6.4	6.6	7	0.4
8	8.4	9	10	0.6
10	10.5	11	12	0.6
12	13	14	15	1.2
16	17	18	19	1.2
20	21	22	24	1.2
24	25	26	28	1.2
30	31	33	35	2
36	37	39	42	2

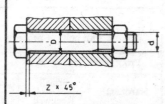

MECHANICAL PROPERTIES OF THREADED FASTENERS

(As per IS 1367 - 1967)

BOLTS (EXTERNALLY THREADED COMPONENTS):

Property			Property class*					
			4.6	4.8	6.6	8.8	10.9	12.9
Tensile strength kgf/mm^2	R_m	Min.	40	40	60	80	100	120
		Max.	55	55	80	100	120	140
Hardness — Brinell	HB	Min.	110	110	170	225	280	330
		Max.	170	170	245	300	365	425
Hardness — Rockwell	HRB	Min.	62	62	88	—	—	—
		Max.	88	88	102	—	—	—
	HRC	Min.	—	—	—	18	27	34
		Max.	—	—	—	3ł	38	44
Yield stress kgf/mm^2	R_e	Min.	24	32	36	—	—	—
Stress at permanent limit kgf/mm^2	$R_{0.2}$	Min.	—	—		64	90	108
Proof stress S_p	kgf/mm^2		22.6	29.1	33.9	58.2	79.2	95
	S_p/R_e		0.94	0.91	0.94	0.94	0.88	0.83
Elongation %	Min.		25	14	16	12	9	8
Impact strength $kgf.m/cm^2$	Min.		—	—	—	6	4	3

NUTS (INTERNALLY THREADED COMPONENTS):

Property			Property class**			
			4	6	8	12
For use with bolts of property class. @			4.6 4.8	6.6	8.8	10.9 12.9
Proof load stress kgf/mm^2			40	60	80	120
Hardness	Brinell	HB Max.	302	302	302	353
	Rockwell	HRC Max.	30	30	30	36

* The first digit of the symbol indicates 1/10th of the min. tensile strength, while the second digit indicates 1/10th the ratio between min. yield stress and the min. tensile strength expressed as a percentage.

** The symbol corresponds to 1/10th of the proof load stress.

@ Nuts of a higher strength property class may be substituted for nuts of a lower strength property class and not vice-versa.

PERMISSIBLE LOAD CARRYING CAPACITY FOR THREADED FASTENERS

SYMBOL Size	8.8			6.6			4.8			4.6		
	Max. Tightening torque * kgf.m	Static tensile load @ kgf.	Bolt subjected to varying load @ @ ≈ kgf.	Max. Tightening torque * kgf.m	Static tensile load @ kgf.	Bolt subjected to varying load @ @ ≈ kgf.	Max. Tightening torque * kgf.m	Static tensile load @ kgf.	Bolt subjected to varying load @ @ ≈ kgf.	Max. Tightening torque * kgf.m	Static tensile load @ kgf.	Bolt subjected to varying load @ @ ≈ kgf.
M 3	0.06	100	67	0.04	63	42	0.03	45	30	—	—	—
M 4	0.14	176	118	0.09	110	74	0.06	79	53	—	—	—
M 5	0.28	284	190	0.17	177	120	0.13	128	86	—	—	—
M 6	0.47	402	270	0.29	250	168	0.21	181	121	0.17	141	95
M 8	1.15	732	490	0.71	456	306	0.51	329	220	0.4	256	172
M 8×1	1.2	784	525	0.76	490	328	0.54	353	236	0.42	274	184
M 10	2.25	1160	780	1.4	725	485	1.02	522	350	0.8	406	272
M 10×1.25	2.36	1224	820	1.48	765	513	1.06	551	370	0.83	428	287
M 12	3.94	1686	1130	2.5	1050	705	1.78	759	510	1.38	590	396
M 12×1.25	4.2	1842	1230	2.64	1150	770	1.9	829	555	1.48	645	432
M 14	6.25	2300	1540	3.92	1440	965	2.8	1035	695	2.2	805	540
M 14×1.5	6.7	2500	1670	4.2	1560	1046	3.04	1125	755	2.36	875	585
M 16	9.7	3140	2100	6.1	1960	1310	4.35	1413	950	3.4	1099	735
M 16×1.5	10.2	3340	2240	6.4	2080	1400	4.6	1503	1000	3.6	1169	785
M 18	13.3	3840	2570	8.35	2400	1610	6	1728	1160	4.7	1344	900
M 18×1.5	14.8	4320	2900	9.3	2700	1810	6.65	1944	1300	5.2	1512	1010
M 20	18.9	4900	3280	11.8	3060	2040	8.5	2205	1480	6.6	1715	1150
M 20×1.5	20.7	5440	3640	13	3400	2280	9.3	2448	1640	7.25	1904	127.0

PERMISSIBLE LOAD CARRYING CAPACITY FOR THREADED FASTENERS (concld.)

SYMBOL	8.8			6.6			4.8			4.6		
Size	Max. Tightening torque* kgf.m	Static tensile load @ kgf.	Bolt subjected to varying load @@ ≈ kgf	Max. Tightening torque* kgf.m	Static tensile load @ kgf	Bolt subjected to varying load @@ ≈ kgf	Max. Tightening torque* kgf.m	Static tensile load @ kgf	Bolt subjected to varying load @@ ≈ kgf.	Max. Tightening torque* kgf.m	Static tensile load @ kgf	Bolt subjected to varying load @@ ≈ kgf
M 22	25.6	6060	4050	—	—	—	11.5	2727	1830	8.95	2121	1420
M 22×1.5	27.8	6660	4450	—	—	—	12.5	2997	2000	9.7	2331	1560
M 24	32.8	7060	4750	—	—	—	14.7	3177	2140	11.5	2471	1660
M 24×2	35	7680	5150	—	—	—	15.8	3456	2320	12.3	2688	1800
M 27	47.7	9180	6150	—	—	—	21.5	4131	2770	16.7	3213	2150
M 27×2	50.8	9920	6650	—	—	—	22.8	4464	3000	17.8	3472	2320
M 30	65	11220	7500	—	—	—	29.2	5049	3380	22.8	3927	2630
M 30×2	70.5	12420	8350	—	—	—	31.8	5589	3740	24.7	4347	2900
M 33	88	13880	9300	—	—	—	39.5	6246	4200	30.6	4858	3260
M 33×2	95	15220	10200	—	—	—	43	6849	4600	33.2	5327	3570
M 36	113	16340	10900	—	—	—	51	7353	4920	39.5	5719	3840
M 36×3	119	17300	11600	—	—	—	53.4	7785	5200	42	6055	4050

Notes: 1. * Static tensile loads induced by applied torque assumed equal to permissible static load carrying capacity of fasteners
2. @@Permissible loads in bolts subjected to variable loading assumed to be approx. 2/3 the static tensile load.
3. Rigidity in compression is assumed to be greater than that in tension.
4. Values applicable to bolts subjected to tension only.
5. @Permissible stresses assumed for calculation: 2000 kgf/cm² for 8.8; 900 kgf/cm² for 4.8; 1250 kgf/cm² for 6.6;
700 kgf/cm² for 4.6.

PLAIN WASHERS
(As per IS 5370 - 1969)

d	D	s	Suitable for bolt size
2.8	8	0.8	M 2.5
3.2	9	0.8	M 3
4.3	12	1	M 4
5.3	15	1.5	M 5
6.4	18	1.5	M 6
8.4	25	2	M 8
10.5	30	2.5	M 10
13	40	3	M 12
17	50	3	M 16
21	60	4	M 20

For Type A ∇∇ (unmachined)
For Type B 1.6 ∇

PUNCHED WASHERS
(As per IS 2016 - 1967)

Type A: For Hexagon bolts and screws.

Type B: For round and cheese head screws

(All dimensions in *mm*)

d	D Type A	D Type B	s	Suitable for bolt or screw size
1.8	4	3.5	0.4	M 1.6
2.4	5	4.5	0.4	M 2
2.9	6.5	5	0.5	M 2.5
3.4	7	6	0.5	M 3
4.5	9	8	0.8	M 4
5.5	10	9.5	1	M 5
6.6	12.5	11	1.6	M 6
9	17	14	1.6	M 8
11	21	18	2	M 10
14	24	20	2.5	M 12
18	30	27	3.15	M 16
22	37	33	3.15	M 20
26	44	—	4	M 24
33	56	—	4	M 30
39	66	—	5	M 36

MACHINED WASHERS
(As per IS 2016 - 1967)

d H12	D Basic	D Tol.	s Basic	s Tol.	Suitable for Bolt or Screw size
1.7	4	0 / −0.3	0.3	±0.1	M 1.6
2.2	5		0.3		M 2
2.7	6.5		0.5		M 2.5
3.2	7		0.5		M 3
4.3	9		0.8		M 4
5.3	10		1		M 5
6.4	12.5		1.6	+0.2	M 6
8.4	17	0 / −0.4	1.6		M 8
10.5	21	0 / −0.5	2		M 10
13	24		2.5	+0.3	M 12
17	30		3		M 16
21	37	0 / −0.8	3		M 20
25	44		4		M 24
31	56	0 / −1	4	±0.6	M 30
37	66		5		M 36

SPRING WASHERS FOR SCREWS WITH CYLINDRICAL HEAD (As per IS 6735 - 1972)

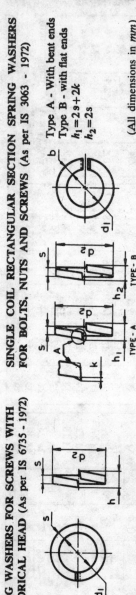

Nom. Size	d1 Basic	d1 Tol.+	d2 Max.	h Min.	h Max.	s	Screw Size
2	2.1	0.3	3.4	1.2	1.4	0.9	M 2
2.5	2.6	0.3	4.4	1.6	1.8	1	M 2.5
3	3.1	0.3	5.6	2	2.36	1	M 3
4	4.1	0.3	7	2.4	2.83	1.2	M 4
5	5.1	0.3	8.8	3.2	3.78	1.6	M 5
6	6.1	0.4	9.9	3.2	3.78	1.6	M 6
8	8.1	0.4	12.7	4	4.72	2	M 8
10	10.2	0.5	16	5	5.9	2.5	M 10
12	12.2	0.5	18	5	5.9	2.5	M 12
16	16.2	0.8	24.4	7	8.25	3.5	M 16
20	20.2	1	30.6	9	10.6	4.5	M 20
24	24.5	1	35.9	10	11.8	5	M 24
30	30.5	1.2	44.2	12	14.2	6	M 30
36	36.5	1.2	52.3	14	16.5	7	M 36

SINGLE COIL RECTANGULAR SECTION SPRING WASHERS FOR BOLTS, NUTS AND SCREWS (As per IS 3063 - 1972)

Type A - With bent ends
Type B - with flat ends
$h_1 = 2s + 2k$
$h_2 = 2s$

(All dimensions in mm)

Nom. Size	d1 Basic	d1 Tol.+	d2 Max.	b Basic	b Tol.+−	s Basic	s Tol.+−	k	Bolt, Nut or Screw size
2	2.1	0.3	4.4	0.9	0.1	0.5	0.1	—	M 2
2.5	2.6	0.3	5.1	1	0.1	0.6	0.1	—	M 2.5
3	3.1	0.3	6.2	1.3	0.1	0.8	0.1	—	M 3
4	4.1	0.3	7.6	1.5	0.1	0.9	0.1	0.15	M 4
5	5.1	0.3	9.2	1.8	0.1	1.2	0.1	0.15	M 5
6	6.1	0.4	11.8	2.5	0.15	1.6	0.1	0.2	M 6
8	8.2	0.4	14.8	3	0.15	2	0.1	0.3	M 8
10	10.2	0.6	18.1	3.5	0.2	2.2	0.15	0.3	M 10
12	12.2	0.8	21.1	4	0.2	2.5	0.15	0.4	M 12
16	16.2	0.8	27.4	5	0.2	3.5	0.2	0.4	M 16
20	20.2	1	33.6	6	0.2	4	0.2	0.4	M 20
24	24.5	1	40	7	0.25	5	0.2	0.5	M 24
30	30.5	1.2	48.2	8.	0.25	6	0.2	0.8	M 30
36	36.5	1.2	58.2	10	0.25	6	0.2	0.8	M 36

GENERAL ENGINEERING TABLES

DIMENSIONS FOR CENTRE HOLES (As per IS 2473 - 1975)

(All dimensions in mm)

d	TYPE A			TYPE B			TYPE R				
	D	t ref.	a**	D	t ref.	a**	D	t ref.	r Max.	r Min.	a**
1	2.12	1.9	3	3.15	2.2	3.5	2.12	1.9	3.15	2.5	3
(1.25)	2.65	2.3	4	4	2.7	4.5	2.65	2.3	4	3.15	4
1.6	3.35	2.9	5	5	3.4	5.6	3.35	2.9	5	4	5
2	4.25	3.7	6	6.3	4.3	6.6	4.25	3.7	6.3	5	6
2.5	5.3	4.6	7	8	5.4	8.3	5.3	4.6	8	6.3	7
3.15	6.7	5.9	9	10	6.8	10	6.7	5.8	10	8	9
4	8.5	7.4	11	12.5	8.6	12.7	8.5	7.4	12.5	10	11
(5)	10.6	9.2	14	16	10.8	15.6	10.6	9.2	16	12.5	14
6.3	13.2	11.5	18	18	12.9	20	13.2	11.4	20	16	18
(8)	17	14.8	22	22.4	16.4	25	17	14.7	25	20	22
10	21.2	18.4	28	28	20.4	31	21.2	18.3	31.5	25	28

Sizes within brackets are non-preferred and should be avoided where possible.

* Dimension *l* depends on the corresponding length of the respective centre drills. It shall not, even in case of drilling with resharpened centre drills, be less than the value. *t*. This is necessary to ensure that the point of the centre does not touch the bottom of the centre hole.

** Parting off dimension, when the centre hole is required to be removed from the finished workpiece.

TYPE A

TYPE B

TYPE R

CENTRE HOLES

(All dimensions in *mm*)   (concld.)

LIMITS OF WORKPIECE DIAMETERS FOR DIFFERENT SIZES OF CENTRE HOLES				REPRESENTATION OF CENTRE HOLES ON DRAWINGS	
Centre hole dia d	Diameter of workpiece		Weight of manufactured workpiece kg. Max.		Centre hole **must** remain in the finished part
	over	upto			
1	6	10	—		Centre hole **may** remain in the finished part
1.6	10	16	—		
2.5	16	40	—		Centre hole **must not** remain in the finished part
4	40	80	—		
6.3	—	—	1600		
10	—	—	4000		
Values applicable for all types of centre holes (types A, B & R) VALUES ARE FOR INFORMATION ONLY.					Ground centre holes

THREADED CENTRE HOLES (As per IS 2540 - 1963)

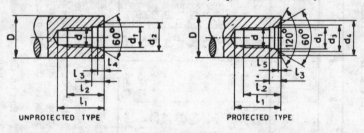

UNPROTECTED TYPE   PROTECTED TYPE

d	d_1	d_2	d_3	d_4	l_1 Min.	l_2	l_3	l_3 @ Tol.	l_4	l_5	D Min.
M 4	4.3	6.5	—	—	13	9	3.5	+0.5	2	—	—
M 5	5.3	8	7.4	9	16	12	4	+0.5	2.5	0.5	10
M 6	6.4	10	9.4	11	20	16	4.5	+0.5	2.5	0.5	14
M 8	8.4	12	10.8	14.5	25	20	5.5	+1	3.2	1	17
M 10	10.5	15	13.8	17.5	30	24	7	+1	4	1	22
M 12	13	18	16.8	20.5	36	28	8	+1	4.5	1	30
M 16	17	24	21.7	28.5	40	32	10	+1	6	2	38
M 20	21	29	26.7	33.5	50	40	12	+1	7	2	50
M 24	25	35	32.7	39.5	63	50	14	+1	9	2	85
M 30	31	44	41.7	48.5	80	65	18	+1	11	2	130

@ Applicable to unprotected type only.

COUNTERSINKS [As per IS 3406 (Part 1) - 1975]

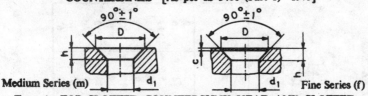

Medium Series (m) Fine Series (f)

Type A: FOR SLOTTED COUNTERSUNK HEAD AND SLOTTED RAISED COUNTERSUNK HEAD SCREWS
(As per IS 1365 - 1968) (All dimensions in *mm*)

Nominal Size	Suitable for Screw	Medium Series (m)		Fine Series (f)		$C^{+0.1}_{0}$	h ≈
		d_1 H13	D H13	d_1 H12	D H12		
1.6	M 1.6	1.8	3.7	1.7	3.4	0.2	1
2	M 2	2.4	4.6	2.2	4.3	0.2	1.2
2.5	M 2.5	2.9	5	2.7	5.3	0.2	1.5
3	M 3	3.4	6.6	3.2	6.3	0.2	1.7
4	M 4	4.5	9	4.3	8.3	0.3	2.3
5	M 5	5.5	11	5.3	10.4	0.3	2.8
6	M 6	6.6	13	6.4	12.4	0.3	3.3
8	M 8 M 8×1	9	17.2	8.4	16.5	0.4	4.4
10	M 10 M 10×1.25	11	21.5	10.5	20.5	0.5	5.5
12	M 12 M 12×1.25	14	26	13	25	0.5	6.5
16	M 16 M 16×1.5	18	34	17	33	0.5	8.5
20	M 20 M 20×1.5	22	42	21	41	0.5	10.5

Type B: FOR SLOTTED COUNTERSUNK HEAD AND SLOTTED RAISED COUNTERSUNK HEAD SCREWS, SMALL HEAD SERIES (As per IS 5308 - 1969)

Nominal Size	Suitable for Screws	Medium Series (m)		Fine Series (f)		$C^{+0.1}_{0}$	h ≈
		d_1 H13	D H13	d_1 H12	D H12		
1.6	M 1.6	1.8	3.5	1.7	3.2	0.2	0.9
2	M 2	2.4	4	2.2	3.7	0.2	1
2.5	M 2.5	2.9	5.1	2.7	4.8	0.2	1.2
3	M 3	3.4	5.6	3.2	5.3	0.2	1.2
4	M 4	4.5	7.8	4.3	7.3	0.3	1.8
5	M 5	5.5	10	5.3	9.4	0.3	2.3
6	M 6	6.6	11	6.4	10.4	0.3	2.3

SPOT FACING

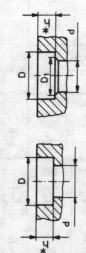

Nom. size of screw	d	D
M 3	3.4	9
M 4	4.5	10
M 5	5.5	11
M 6	6.6	13
M 8	9	18
M 10	11	22
M 12	14	26
M 16	18	33
M 20	22	40
M 24	26	48
M 30	33	61

COUNTER BORES (As per IS 3406 Part II - 1975)

* (1) For slotted cheese head screws as per IS 1366 - 1968
* (2) For socket head cap screws as per IS 2269 - 1967
** 90° counterbore or rounded. Under 12 mm thread diameter, it is only deburred.

(All dimensions in *mm*)

Nominal size	Suitable for screw	Medium series(m) d H13	Medium series(m) D H13	Fine series (f) d H12	Fine series (f) D H13	D_1**	h* 1)	h* 2)	Tol. on h
1.6	M 1.6	—	—	1.7	3.2	—	1.2	—	+0.2 / 0
2	M 2	—	—	2.2	4	—	1.6	—	
2.5	M 2.5	—	—	2.7	4.8	—	2	—	
3	M 3	3.4	6.3	3.2	5.9	—	2.4	3.4	
4	M 4	4.5	8	4.3	7.4	—	3.2	4.6	
5	M 5	5.5	9.5	5.3	8.9	—	4	5.7	
6	M 6	6.6	11	6.4	10.4	—	4.7	6.8	+0.4 / 0
8	M 8 × 1	9	14.5	8.4	13.5	—	6	9	
10	M 10 × 1.25	11	17.5	10.5	16.5	16	7	11	
12	M 12 × 1.25	14	20	13	19	20	8	13	
16	M 16 × 1.5	18	26	17	25	24	10.5	17.5	
20	M 20 × 1.5	22	33	21	31	28	12.5	21.5	
24	M 24 × 2	26	39	25	37	36	—	25.5	
30	M 30 × 2	33	48	31	46	42	—	32	+0.6 / 0
36	M 36 × 3	39	57	37	55	—	—	38	

TAPERS FOR GENERAL ENGINEERING PURPOSES (As per IS 3458 - 1966)

Nominal taper 1:x	Cone angle α	α/2 (Setting angle)	Setting value for α/2 for sinebar of 100 mm length	Remarks and examples of application
1 : 0.289	**120°**	60°	86.603	Protective countersink for centre holes.
1 : 0.5	**90°**	45°	70.711	Valve cones, tip of lathe centres, countersunk screws.
1 : 0.596	**80°**	40°	64.279	Self-tapping screws.
1 : 0.866	**60°**	30°	50	Vee-slots, centre holes, tip of lathe centres, countersunk screws.
1 : 1.207	**45°**	22°30'	38.268	Countersunk rivets, countersunk round head rivets.
1 : 1.374	**40°**	20°	34.202	Collet chucks.
(1 : 1.5)	36°52'12''	18°26'6''	31.623	Conical seals.
1 : 1.866	**30°**	15°	25.882	Rough conical countersunk screws, centering cones on cutter supports.
(1 : 2.255)	**25°**	12°30'	21.644	Taper bushes for pipe joints.
(1 : 2.352)	**24°**	12°	20.791	Pipe fittings.
3.5 : 12	16°35'40''	8°17'50''	14.431	Steep angle tapers, milling spindle noses and milling cutters.
(1 : 3.429)	14°15'	7°7'30''	12.403	Spindle noses and clamping device flanges on machine tools.
1 : 5	11°25'16''	5°42'38''	9.95	Axially stressed machine parts in torsion, pivot journals, friction clutches, pulley bores, locking devices for abrasive discs, hose connections for pneumatic tools etc.

TAPER ON CONE.

Bracketed values are non-preferred and are meant for replacements only.

Values in bold letters are basic values

TAPER ON CONE FRUSTUM

TAPERS FOR GENERAL ENGINEERING PURPOSES

(concld.)

Nominal taper 1 : x	Cone angle ∝	∝/2 (Setting angle)	Setting value for ∝/2 for sinebar of 100 mm length	Remarks and examples of application.
(1 : 6)	9°31′38″	4°45′49″	8.305	Conical seals for cocks, die-sinking mills.
1 : 10	5°43′30″	2°51′45″	4.994	Machine parts stressed across or along the axis or in torsion, taper shaft ends, adjustable bearing bushes.
1 : 12	4°46′18″	2°23′9″	4.163	Antifriction bearings.
(1 : 16)	3°34′48″	1°47′24″	3.123	Fitting unions with whitworth pipe threads, metric taper threads, taper pipe threads.
1 : 20	2°51′52″	1°25′56″	2.499	Metric tapers, tool tapers, tool shanks and taper nose spindles for machine tools.
1 : 19.212	2°58′54″	1°29′27″	2.602	Morse Taper No. 0
1 : 20.047	2°51′26″	1°25′43″	2.493	Morse Taper No. 1
1 : 20.02	2°51′40″	1°25′50″	2.497	Morse Taper No. 2 — Tool tapers, tool shanks and taper-nose spindles for machine tools.
1 : 19.922	2°52′32″	1°26′16″	2.509	Morse Taper No. 3
1 : 19.254	2°58′30″	1°29′15″	2.596	Morse Taper No. 4
1 : 19.002	3°0′52″	1°30′26″	2.63	Morse Taper No. 5
1 : 19.18	2°59′12″	1°29′36″	2.606	Morse Taper No. 6
1 : 30	1°54′34″	57′17″	1.666	Bores in shell reamers and shell drills.
1 : 50	1°8′46″	34′23″	1	Taper pins, conical pipe threads.

RELIEF GROOVES

(All dimensions in *mm*)

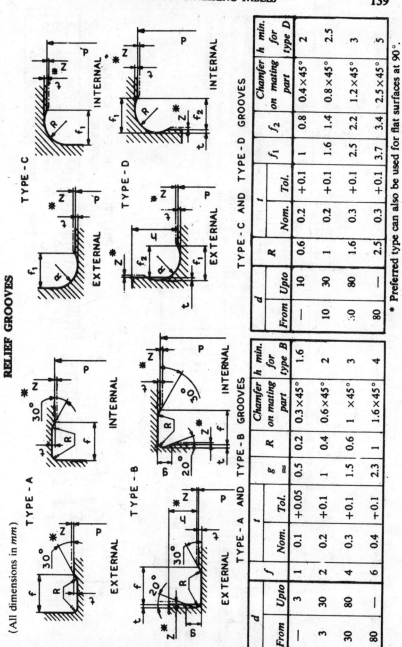

TYPE-A AND TYPE-B GROOVES

d		f	t		g ≈	R	Chamfer on mating part	h min. for type B
From	Upto		Nom.	Tol.				
—	3	1	0.1	+0.05	0.5	0.2	0.3×45°	1.6
3	30	2	0.2	+0.1	1	0.4	0.6×45°	2
30	80	4	0.3	+0.1	1.5	0.6	1 ×45°	3
80	—	6	0.4	+0.1	2.3	1	1.6×45°	4

TYPE-C AND TYPE-D GROOVES

d		R	t		f_1	f_2	Chamfer on mating part	h min. for type D
From	Upto		Nom.	Tol.				
—	10	0.6	0.2	+0.1	1	0.8	0.4×45°	2
10	30	1	0.2	+0.1	1.6	1.4	0.8×45°	2.5
30	80	1.6	0.3	+0.1	2.5	2.2	1.2×45°	3
80	—	2.5	0.3	+0.1	3.7	3.4	2.5×45°	5

* Preferred type can also be used for flat surfaces at 90°.

RELIEF GROOVES

(All dimensions in *mm*)

(concld.)

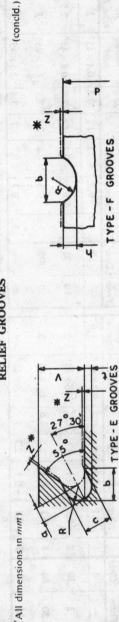

TYPE - E GROOVES

V		C	t	a	b	R
From	Upto		≈	≈		
6	16	2	0.5	1.2	2.5	1
20	32	3.2	0.8	2	4	1.6
40	50	5	1.4	2.8	6	2.5

Note: Type E for 55° dovetail guides.

TYPE - F GROOVES

Note: Type F recommended for components with the same diameters but containing different tolerances or surface roughness. Also used on stepped cylindrical surfaces if the step height is less that the minimum recommended for type D.

d		b	h		R
From	Upto		Nom.	Tol.	
—	10	0.9	0.2		0.6
10	30	1.4	0.2	+0.1	1
30	80	2	0.3	0	1.6
80	—	2.8	0.4		2.5

RADII AND CHAMFERS (As per IS 3457 - 1966)

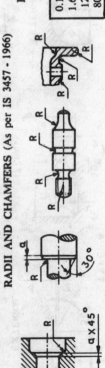

PREFERRED SIZES OF RADII AND CHAMFERS

0.1	0.2	0.25	0.3	0.4	0.6	0.8	1	1.2
1.6	2	2.5	3	4	5	6	8	10
12	16	20	25	32	40	50	63	70
80	100	125	160	200	250	320	—	—

KNURLING (As per IS 3403 - 1966)

(All dimensions in mm)

STRAIGHT KNURLING

CROSS KNURLING

DIAMOND KNURLING

EFFECT OF KNURLING

CHAMFER APPROX. P

UP TO ¼ P

Turned Diameter D		Straight Knurling — For all Materials				Cross Knurling — For Vulcanite, etc.			Diamond Knurling — For Light Alloys, Brass, Fibre, etc.			Diamond Knurling — For Steel		
		Pitch P for Width b.												
Over	Upto	-/2	2/6	6/16	16/32	-/6	6/16	16/32	-/6	6/16	16/32	-/6	6/16	16/32
–	8	0.5	0.5	0.5	0.5	0.5	0.5	0.5	0.5	0.5	0.5	0.5	0.5	0.5
8	16	0.5	0.5	0.5	0.5	0.5	0.5	0.5	0.5	0.5	0.5	0.8	0.8	0.8
16	32	0.5	0.5	0.8	0.8	0.5	0.8	0.8	0.5	0.8	0.8	0.8	1	1
32	63	0.5	0.5	0.8	1	0.5	0.8	1	0.5	0.8	1	0.8	1	1.2
63	100	0.8	0.8	0.8	1	0.8	1	1	0.8	0.8	1	0.8	1.2	1.6
100		0.8	1	1	1	0.8	1	1.2	0.8	1	1.2	1	1.2	2

PREFERRED NUMBERS (As per IS 1076 - 1967)

PREFERRED NUMBERS OF BASIC SERIES

R 5	R 10	R 20	R 40	R 5	R 10	R 20	R 40
1	1	1	1				3.35
			1.06			3.55	3.55
		1.12	1.12				3.75
			1.18	4	4	4	4
	1.25	1.25	1.25				4.25
			1.32			4.5	4.5
		1.4	1.4				4.75
			1.5		5	5	5
1.6	1.6	1.6	1.6				5.3
			1.7			5.6	5.6
		1.8	1.8				6
			1.9	6.3	6.3	6.3	6.3
	2	2	2				6.7
			2.12			7.1	7.1
		2.24	2.24				7.5
			2.36		8	8	8
2.5	2.5	2.5	2.5				8.5
			2.65			9	9
		2.8	2.8				9.5
			3	10	10	10	10
	3.15	3.15	3.15				

SPEEDS AND FEEDS FOR MACHINE TOOLS

Speeds for Machine Tools (As per IS 2218 - 1962)

Basic value of feed R 20	Allowable limits of speeds			
	Total tolerance (Mechanical + Electrical)		Mechanical tolerance only	
	Min.($\approx -2\%$)	Max.($\approx +6\%$)	Min.($\approx -2\%$)	Max.($\approx +3\%$)
100	98	106	98	103
112	110	119	110	116
125	123	133	123	130
140	138	150	138	145
160	155	168	155	163
180	174	188	174	183
200	196	212	196	206
224	219	237	219	231
250	246	266	246	259
280	276	299	276	290
315	310	335	310	326
355	348	376	348	365
400	390	422	390	410
450	438	473	438	460
500	491	531	491	516
560	551	596	551	579
630	618	669	618	650
710	694	750	694	729
800	778	842	778	818
900	873	945	873	918
1000	980	1060	980	1030

Speeds for machine tools in number of revolutions or strokes per minute.

(1) The table of basic values may be extended by multiplying or dividing the values by 10,100, etc.

(2) The basic values are to be selected such that they correspond to
  (i) the whole range of values given in R 20 series or
  (ii) the values obtained by anyone of the following derived series R 20/2 (...710...); R 20/3 (...710......); R 20/4 (...355......) R 20/4 (...710......); R 20/6 (...710......).
The values of these derived series are obtained by selecting every 2nd, 3rd, 4th or 6th term in both directions of the values specified within the brackets. Series R 20/2, R 20/3 and R 20/6 contain the values 355, 710, 1400 and 2800, while series R 20/4 contains either 355 and 1400 or 710 and 2800. These values nearly correspond to the full load speeds of asynchronous motors.

(3) BASIC VALUE is the speed under full load, that is indicated on the name plate of the machine.

(4) ACTUAL VALUE of spindle speed under load is equal to the spindle speed at no load $\times N_c/N_v$ where N_c = Speed of the motor under load indicated in the name plate of the motor. N_v = speed of the motor measured when the machine is running under no load.

SPEEDS AND FEEDS FOR MACHINE TOOLS

Feeds for Machine Tools (As per IS 2219 - 1962)   (contd.)

Basic value of feed R 20	Feeds for machine tools in millimetres			
	Allowable limits of feeds			
	Per minute		Per revolution or stroke	
	Total tolerance (Mechanical + Electrical)		Mechanical tolerance only	
	Min.($\approx -2\%$)	Max.($\approx +6\%$)	Min.($\approx -2\%$)	Max.($\approx +3\%$)
1	0.98	1.06	0.98	1.03
1.12	1.1	1.19	1.1	1.16
1.25	1.23	1.33	1.23	1.3
1.4	1.38	1.5	1.38	1.45
1.6	1.55	1.68	1.55	1.65
1.8	1.74	1.88	1.74	1.83
2	1.96	2.12	1.96	2.06
2.24	2.19	2.37	2.19	2.31
2.5	2.46	2.66	2.46	2.59
2.8	2.76	2.99	2.76	2.9
3.15	3.1	3.35	3.1	3.26
3.55	3.48	3.76	3.48	3.65
4	3.9	4.22	3.9	4.1
4.5	4.38	4.73	4.38	4.6
5	4.91	5.31	4.91	5.16
5.6	5.51	5.96	5.51	5.79
6.3	6.18	6.69	6.18	6.5
7.1	6.94	7.5	6.94	7.29
8	7.78	8.42	7.78	8.18
9	8.73	9.45	8.73	9.18
10	9.8	10.6	9.8	10.3

(1) The table of basic values may be extended by multiplying or dividing the values by 10,100 etc.

(2) The basic values of feeds are to be selected such that they either correspond to (a) the whole range of values given in R 20 ; R 10 or R 5 basic series preferred numbers, or (b) the values obtained by anyone of the following derived series.

(i) R 20/3 (......1......) and (ii) R 10/3 (......1......).

The values of these derived series are obtained by selecting every third term in both directions of the value specified within the brackets.

(3) BASIC VALUE is the value of feed under load, indicated on the nameplate of the machine.

(4) ACTUAL VALUE of the feed under load = feed on no load $\times$ N_c/N_v, where N_c = Speed of the motor under load, indicated on the nameplate of the motor and N_v = Speed of the motor measured when the machine is running under no load.

SPEEDS AND FEEDS FOR MACHINE TOOLS
(contd.)

Preferred Basic Values for Speeds and Feeds of Machine Tools at Full Load, according to ISO Recommendations

SPEEDS IN REVOLUTIONS OR STROKES PER MINUTE

R20	R20/2 (...710...)	R20/3 (...710...)	R20/4 (...355...)	R20/4 (...710...)	R20/6 (...710...)	R 20	R20/2 (...710...)	R20/3 (...710...)	R20/4 (...355...)	R20/4 (...710...)	R20/6 (...710...)
1.12*	1.26	1.41	1.58	1.58	2	1.12	1.26	1.41	1.58	1.58	2
10						355	355	355	355		355
11.2	11.2	11.2		11.2	11.2	400					
12.5						450	450			450	
14	14		14			500		500			
16		16				560	560		560		
18	18			18		630					
20						710	710	710		710	710
22.4	22.4	22.4	22.4		22.4	800					
25						900	900		900		
28	28			28		1000		1000			
31.5		31.5				1120	1120			1120	
35.5	35.5		35.5			1250					
40						1400	1400	1400	1400		1400
45	45	45		45	45	1600					
50						1800	1800			1800	
56	56		56			2000		2000			
63		63				2240	2240		2240		
71	71			71		2500					
80						2800	2800	2800		2800	2800
90	90	90	90		90	3150					
100						3550	3550		3550		
112	112			112		4000		4000			
125		125				4500	4500			4500	
140	140		140			5000					
160						5600	5600	5600	5600		5600
180	180	180		180	180	6300					
200						7100	7100			7100	
224	224		224			8000		8000			
250		250				9000	9000		9000		
280	280			280		10000					
315											

*This row represents the ratio of progression of the respective series.

SPEEDS AND FEEDS FOR MACHINE TOOLS

(concld.)

Preferred Basic Values for Speeds and Feeds of Machine Tools at Full Load, according to ISO Recommendations

FEEDS FOR MACHINE TOOLS IN mm/REVOLUTION OR STROKE

R5	R10	R10/3 (...1...)	R20	R20/3 (...1...)	R5	R10	R10/3 (...1...)	R20	R20/3 (...1...)
1.6*	1.26	2	1.12	1.41	1.6	1.26	2	1.12	1.41
0.01	0.01		0.01					0.355	0.355
			0.011	0.011	0.4	0.4		0.4	
	0.012		0.012					0.45	
			0.014			0.5	0.5	0.5	0.5
0.016	0.016	0.016	0.016	0.016				0.56	
			0.018		0.63	0.63		0.63	
	0.02		0.02					0.71	0.71
			0.022	0.022		0.8		0.8	
0.025	0.025		0.025	0.025				0.9	
			0.028		1	1	1	1	1
	0.031	0.031	0.031	0.031				1.12	
			0.035			1.25		1.25	
0.04	0.04		0.04					1.4	1.4
			0.045	0.045	1.6	1.6		1.6	
	0.05		0.05					1.8	
			0.056			2	2	2	2
0.063	0.063	0.063	0.063	0.063				2.24	
			0.071		2.5	2.5		2.5	
	0.08		0.08					2.8	2.8
			0.09	0.09		3.15		3.15	
0.1	0.1		0.1					3.55	
			0.112		4	4	4	4	4
	0.125	0.125	0.125	0.125				4.5	
			0.14			5		5	
0.16	0.16		0.16					5.6	5.6
			0.18	0.18	6.3	6.3		6.3	
	0.2		0.2					7.1	
			0.224			8	8	8	8
0.25	0.25	0.25	0.25	0.25				9	
			0.28		10	10		10	
	0.315		0.315						

* This row represents the ratio of progression of the respective series.

T SLOTS
(As per IS 2013 - 1974)

(All dimensions in mm)

Nom. Size a	b_1 Min.	b_1 Max.	c Min.	c Max.	h_1 Min.	h_1 Max.	n_1†† Max.	n_2†† Max.	n_3†† Max.	Z_1
10	16	18	7	8	17	21	1	0.6	1	0.5
12	19	21	8	9	20	25				
14	23	25	9	11	23	28			1.6	
18	30	32	12	14	30	36	1.6			
22	37	40	16	18	38	45		1	2.5	
28	46	50	20	22	48	56				
36	56	60	25	28	61	71	2.5			
42	68	72	32	35	74	85		1.6	4	1

† Tolerance on a

  Guiding slots-H8

  Fixing slots-H12

†† Height of chamfer at 45°

   or radius of rounding

T BOLTS
(As per IS 2014 - 1962)

T NUTS
(As per IS 2015 - 1962)

(All dimensions in mm)

Nom. Size a (T-NUT)	T BOLT d	Tol.t on a for T-nut	U*	b₂** Min.	n₄	L@ L₁ From	L@ L₁ Upto	L@ L₂ From	L@ L₂ Upto	S	k	Tol. on S & K	h₂	n₅	Z₂
10	M 8	−0.3 −0.5	12	25	1.6	25	—	32	80	15	6	0 −0.5	12	1.6	0.4
12	M 10	−0.3 −0.5	14	30	1.6	25	32	40	125	18	7	0 −0.5	14	1.6	0.4
14	M 12	−0.3 −0.6	16	35	2.5	32	40	50	160	22	8	0 −0.5	16	2.5	0.4
18	M 16	−0.3 −0.6	20	50	2.5	40	50	65	200	28	10	0 −0.5	20	2.5	0.4
22	M 20	−0.3 −0.6	25	60	2.5	50	65	80	260	34	14	0 −0.5	28	2.5	0.4
28	M 24	−0.4 −0.7	36	75	4	65	80	100	320	43	18	0 −1	36	4	0.4
36	M 30	−0.4 −0.7	45	90	4	100	—	125	400	53	23	0 −1	44	4	0.4
42	M 36	−0.4 −0.7	60	110	4	125	—	160	500	64	28	0 −1	52	6	0.6

* For bolts with length L_1
** For bolts with length L_2

@ Available lengths: 25, 32, 40, 50, 65, 80, 100, 125, 160, 200, 260, 320, 400 & 500.

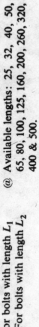

T BOLT

T NUT

TENONS (As per IS 2990-1965)

$\frac{6.3}{\bigvee}\left(\frac{1.6}{\bigvee}\ \frac{0.8}{\bigvee}\right)$

(All dimensions in *mm*)

Nom. size. *a*	Figure and dimensions	Size of slot in fixture
10 and 12		Fixing screw *CH.HD.SCR.* *M6×l@*
14 and 18		Fixing screw *HEX.SOC.HD.* *SCR.M6×l@*
22, 28. 36 and 42.		Fixing screw *HEX.SOC.HD.* *SCR.M8×l@*

@Length *l* as required.

CIRCLIPS (As per IS 3075 - 1965)

External Circlips, Type A - Light Series   (All dimensions in *mm*)

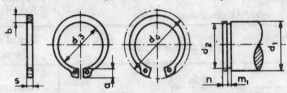

Shaft dia. d_1	Circlip details						Shaft details				Axial force kgf
	s h11	a Max.	b ≈	d_3	Tol. on d_3	d_4 Expanded	d_2	Tol. on d_2	m_1 H13	n Min.	
10		3.3	1.8	9.3	+0.15 −0.30	17.6	9.6			0.6	153
11		3.3	1.8	10.2		18.6	10.5			0.75	210
12		3.3	1.8	11		19.6	11.5		1.1		230
13	1	3.4	2	11.9		20.8	12.4	h11		0.9	300
14		3.5	2.1	12.9	+0.18 −0.36	22	13.4				325
15		3.6	2.2	13.8		23.2	14.3			1.1	400
16		3.7	2.2	14.7		24.4	15.2			1.2	490
17		3.8	2.3	15.7		25.6	16.2				520
18		3.9	2.4	16.5		26.8	17				690
19		3.9	2.5	17.5		27.8	18				725
20		4	2.6	18.5		29	19			1.5	770
21		4.1	2.7	19.5		30.2	20		1.3		805
22	1.2	4.2	2.8	20.5		31.4	21				845
24		4.4	3	22.2		33.8	22.9				1010
25		4.4	3	23.2	+0.21 −0.42	34.8	23.9			1.7	1060
26		4.5	3.1	24.2		36	24.9				1100
28		4.7	3.2	25.9		38.4	26.6				1500
29		4.8	3.4	26.9		39.6	27.6	h12		2.1	1560
30	1.5	5	3.5	27.9		41	28.6		1.6		1620
32		5.2	3.6	29.6		43.4	30.3			2.6	2100
34		5 4	3.8	31.5		45.8	32.3				2220
35		5.6	3.9	32.2	+0.25 −0.50	47.2	33				2670
36		5.6	4	33.2		48.2	34			3	2760
38		5.8	4.2	35.2		50.6	36				2910
40	1.75	6	4.4	36.5		53	37.5		1.85		3810
42		6.5	4.5	38.5	+0.39 −0.78	56	39.5			3.8	4000
45		6.7	4.7	41.5		59.4	42.5				4300
48		6.9	5	44.5		62.8	45.5				4600
50		6.9	5.1	45.8		64.8	47				5700
52	2	7	5.2	47.8		67	49		2.15	4.5	5950
55		7.2	5.4	50.8	+0.46 −0.92	70.4	52				6300

CIRCLIPS (As per IS 3075 - 1965)

(contd.)

External Circlips, Type A - Light Series (All dimensions in mm)

Shaft dia. d_1	Circlip details						Shaft details				Axial force kgf
	s $h11$	a Max.	b $\approx$	d_3	Tol. on d_3	d_4 Expanded	d_2	Tol. on d_2	m_1 $H13$	n Min.	
56		7.3	5.5	51.8		71.6	53				6400
58		7.3	5.6	53.8		73.6	55				6650
60	2	7.4	5.8	55.8		75.8	57		2.15		6900
62		7.5	6	57.8		78	59				7100
63		7.6	6.2	58.8		79.2	60				7250
65		7.8	6.3	60.8		81.6	62			4.5	7500
68		8	6.5	63.5	+0.46	85	65				7840
70		8.1	6.6	65.5	−0.92	87.2	67	$h12$			8050
72		8.2	6.8	67.5		89.4	69				8300
75	2.5	8.4	7	70.5		92.8	72		2.65		8600
78		8.6	7.3	73.5		96.2	75				9000
80		8.6	7.4	74.5		98.2	76.5				10700
82		8.7	7.6	76.5		101	78.5				11000
85		8.7	7.8	79.5		104	81.5				11400
88		8.8	8	82.5		107	84.5			5.3	11900
90	3	8.8	8.2	84.5		109	86.5		3.15		12100
95		9.4	8.6	89.5		115	91.5				12800
100		9.6	9	94.5		121	96.5				13500
105		9.9	9.3	98	+0.54	126	101				16200
110		10.1	9.6	103	−1.08	132	106				17000
115		10.6	9.8	108		138	111				17800
120		11	10.2	113		143	116				18500
125		11.4	10.4	118		149	121			6	19300
130		11.6	10.7	123		155	126				20100
135		11.8	11	128		160	131				20900
140	4	12	11.2	133		165	136	$h13$			21700
145		12.2	11.5	138		171	141		4.15		22500
150		13	11.8	142		177	145				28900
155		13	12	146	+0.63	182	150				30000
160		13.3	12.2	151	−1.26	188	155				31000
165		13.5	12.5	155.5		193	160				32000
170		–	12.9	160.5		197	165			7.5	32900
175			Max	165.5		202	170				33800
180		–	13.5	170.5		208	175				34800
185			Max	175.5		213	180				

CIRCLIPS (As per IS 3075 - 1965)

Internal Circlips, Type B — Light Series   (All dimensions in *mm*)

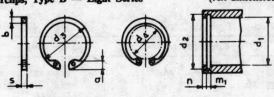

| Bore dia. d_1 | Circlip details | | | | | | Bore details | | | | Axial force kgf |
	$h11$	a Max.	b $\approx$	d_3	Tol. on d_3	d_4 Compressed	d_2	Tol. on d_2	m_1 H13	n Min.	
10		3.2	1.4	10.8		3.1	10.4				160
11		3.3	1.5	11.8		3.9	11.4			0.6	176
12		3.4	1.7	13		4.7	12.5			0.75	240
13		3.6	1.8	14.1	+0.36	5.3	13.6			0.9	314
14		3.7	1.9	15.1	−0.18	6	14.6				336
15		3.7	2	16.2		7	15.7	*H*11		1.1	422
16	1	3.8	2	17.3		7.7	16.8		1.1	1.2	515
17		3.9	2.1	18.3		8.4	17.8				547
18		4.1	2.2	19.5		8.9	19				725
19		4.1	2.2	20.5		9.8	20				764
20		4.2	2.3	21.5		10.6	21			1.5	805
21		4.2	2.4	22.5	+0.42	11.6	22				845
22		4.2	2.5	23.5	−0.21	12.6	23				882
24		4.4	2.6	25.9		14.2	25.2				1160
25		4.5	2.7	26.9		15	26.2			1.8	1200
26	1.2	4.7	2.8	27.9		15.6	27.2		1.3		1250
28		4.8	2.9	30.1		17.4	29.4			2.1	1580
30		4.8	3	32.1		19.4	31.4				1690
32		5.4	3.2	34.4		20.2	33.7	*H*12		2.6	2200
34		5.4	3.3	36.5	+0.50	22.2	35.7				2320
35		5.4	3.4	37.8	−0.25	23.2	37				2820
36	1.5	5.4	3.5	38.8		24.2	38		1.6	3	2900
37		5.5	3.6	39.8		25	39				2980
38		5.5	3.7	40.8		26	40				3070
40		5.8	3.9	43.5	+0.78	27.4	42.5				4050
42	1.75	5.9	4.1	45.5	−0.39	29.2	44.5				4250
45		6.2	4.3	48.5		31.6	47.5		1.85	3.8	4520
47		6.4	4.4	50.5		33.2	49.5				4720
48		6.4	4.5	51.5		34.6	50.5				4820
50		6.5	4.6	54.2	+0.92	36	53				6070
52	2	6.7	4.7	56.2	−0.46	37.6	55		2.15	4.5	6300
55		6.8	5	59.2		40.4	58				6650
56		6.8	5.1	60.2		41.4	59				6750

CIRCLIPS (As per IS 3075 - 1965)

(contd.)

Internal Circlips, Type B Light Series (All dimensions in mm)

| Bore dia. d_1 | Circlip details | | | | | | d_2 | Bore details | | | Axial force kgf |
	s h11	a Max.	b ≈	d_3	Tol. on d_3	d_4 Compressed		Tol. on d_2	m_1 H13	n Min.	
58	2	6.9	5.2	62.2		43.2	61				7000
60		7.3	5.4	64.2		44.4	63		2.15		7250
62		7.3	5.5	66.2		46.4	65				7480
63		7.3	5.6	67.2	+0.92	47.4	66			4.5	7580
65		7.6	5.8	69.2	−0.46	48.8	68				7820
68		7.8	6.1	72.5		51.4	71				8170
70		7.8	6.2	74.5		53.4	73				8420
72	2.5	7.8	6.4	76.5		55.4	75	H12			8650
75		7.8	6.6	79.5		58.4	78		2.65		9000
78		8.5	6.8	82.5		60	81				9350
80		8.5	7	85.5		62	83.5				11200
82		8.5	7	87.5		64	85.5				11500
85		8.6	7.2	90.5		66.8	88.5				11900
88		8.6	7.4	93.5		69.8	91.5				12300
90		8.6	7.6	95.5		71.8	93.5			5.3	12600
92	3	8.7	7.8	97.5	+1.08	73.6	95.5		3.15		12900
95		8.8	8.1	100.5	−0.54	76.4	98.5				13300
98		9	8.3	103.5		79	101.5				13700
100		9	8.4	105.5		81	103.5				14000
102		9.2	8.5	108		82.6	106				16300
105		9.2	8.7	112		85.6	109				16800
108		9.5	8.9	115		88	112				17300
110		10.4	9	117		88.2	114				17600
112		10.5	9.1	119		90	116				17900
115		10.5	9.3	122		93	119				18400
120		11	9.7	127		97	124			6	19200
125		11	10	132		102	129				19900
130		11	10.2	137		107	134				20700
135		11.2	10.5	142		112	139				21500
140	4	11.2	10.7	147	+1.26	117	144	H13	4.15		22300
145		11.4	10.9	152	−0.63	122	149				23100
150		12	11.2	158		125	155				30000
155		12	11.4	164		130	160				30900
160		13	11.6	169		133	165				31900
165		13	11.8	174.5		138	170				32900
170		—	12.2 Max	179.5		145	175			7.5	33900
175			12.7 Max	184.5	+1.44 −0.72	149	180				34800

CIRCLIPS (As per IS 3075 - 1965)

(concld.)

Internal Circlips, Type B — Light Series

(All dimensions in mm)

Bore dia. d_1	Circlip details						Bore details				
	s h11	a Max.	b ≈	d_3	Tol. on d_3	d_4 Compressed	d_2	Tol. on d_2	m H13	n Min.	Axial force kgf
180	4	–	13.2 Max.	189.5	+1.44 −0.72	153	185	H13	4.15	7.5	34500
185			13.7 Max.	194.5		157	190				35000

External Circlips, Type C

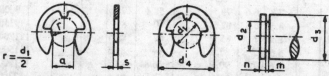

$$r = \frac{d_1}{2}$$

Nom. size d_1	Circlip details				Shaft details					
	d_4 Expanded	a H10	s	Tol. on s	d_3 From	d_3 Upto	d_2 h11	m	Tol. on m	n Min.
0.8	2	0.58	0.2		1	1.4	0.8	0.24	+0.02 0	0.4
1.2	3	1.01	0.3		1.4	2	1.2	0.34		0.6
1.5	4	1.28	0.4		2	2.5	1.5	0.44		0.8
1.9	4.5	1.61	0.5		2.5	3	1.9	0.54		1
2.3	6	1.94	0.6	±0.02	3	4	2.3	0.64		1
3.2	7	2.7	0.6		4	5	3.2	0.64		1
4	9	3.34	0.7		5	7	4	0.74	+0.03 0	1.2
5	11	4.11	0.7		6	8	5	0.74		1.2
6	12	5.26	0.7		7	9	6	0.74		1.2
7	14	5.84	0.9		8	11	7	0.94		1.5
8	16	6.52	1		9	12	8	1.05		1.8
9	18.5	7.63	1.1		10	14	9	1.15		2
10	20	8.32	1.2	±0.03	11	15	10	1.25	+0.06 0	2
12	23	10.45	1.3		13	18	12	1.35		2.5
15	29	12.61	1.5		16	24	15	1.55		3
19	37	15.92	1.75		20	31	19	1.8		3.5
24	44	21.88	2		25	38	24	2.05		4

For diameter of shaft 8mm and above, preferably Type A Circlip to be used.

CYLINDERICAL AND TAPER PINS

(All dimensions in *mm*)

d	L_1	D Min.	V Min.	Cylindrical pin From	Cylindrical pin Upto	Taper pin From	Taper pin Upto
1.5	0.5	4	5	4	16	8	25
2	0.6	6	6.5	6	25	10	35
2.5	0.7	6	6.5	6	25	10	35
3	0.8	8	7	8	30	12	45
4	1	11	8.5	8	45	14	55
5	1.2	17	10.5	10	50	20	60
6	1.5	23	13	12	60	25	90
8	1.8	30	15	14	80	25	130
10	2	45	17.5	20	100	30	160
12	2.5	75	23	25	150	35	180
16	3	110	26	30	180	40	200
20	4	160	—	40	200	45	200
25	—	—	—	50	200	50	200

Length L * js15

* Lengths are: in steps of 1 between 4 and 6;
in steps of 2 between 6 and 16;
in steps of 5 between 20 and 80;
in steps of 10 between 80 and 200;

CYLINDRICAL PINS (IS 2393 - 1972)

TAPER PINS (IS 6688 - 1972)

TYPE - A

TYPE - B

C = 0.2L

SPLIT TAPER PIN

HARDENED CYLINDRICAL PINS (IS 6689 - 1972)

THREADED TAPER PINS (As per IS 3524 - 1966)

EXTERNAL THREADED TAPER PIN

INTERNAL THREADED TAPER PIN

(All dimensions in mm)

EXTERNAL THREADED TAPER PIN

Nom. Size	d_1 Max.	d_1 Min.	b	C	d_2	d_3	L@ From	L@ Upto
6	6	5.95	15	20	M 6	4.5	40	50
8	8	7.94	18	24	M 8	6	50	60
10	10	9.94	20	27	M 10	7	60	80
12	12	11.93	22	30	M 12	9	80	110
16	16	15.93	28	38	M 16	12	100	140
20	20	19.92	28	38	M 16	12	120	180
25	25	24.92	32	43	M 20	15	140	225
30	30	29.92	38	51	M 24	18	160	250
40	40	39.9	45	63	M 30	23	180	350

INTERNAL THREADED TAPER PIN

d_2	d_3	t_1	t_2	t_3	L@ From	L@ Upto
M 4	4.3	6	10	1	25	100
M 5	5.3	8	13	1.2	25	120
M 6	6.4	10	16	1.2	30	160
M 8	8.4	12	18	1.2	35	160
M 10	10.5	16	23	1.6	40	200
M 12	13	20	27	1.6	50	225
M 16	17	24	33	2	55	225
M 20	21	32	41	2	60	250
M 24	25	36	46	2.5	70	250

Note: @Lengths are: in steps of 5 between 25 and 60; in steps of 10 between 60 and 120; in steps of 20 between 120 and 200; in steps of 25 between 200 and 250; in steps of 50 between 250 and 350.

SPACE REQUIRED FOR SPANNERS

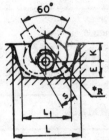

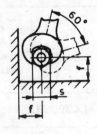

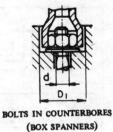

BOLTS IN COUNTERBORES
(BOX SPANNERS)

BOLTS IN POCKETS OR CORNERS
*R FOR SEMICIRCULAR POCKET

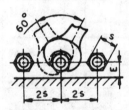

BOLTS ON LINEAR PITCH

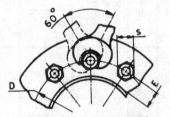

BOLTS ON CIRCULAR PITCH

(All dimensions in *mm*)

Bolt Nom. Size d	Width Across Flats s	Open Jaw Spanners										Box Spanners
		Bolts in pockets or corners					Bolts on circular pitch					Hexagonal nuts or screws in counter bores
							Number of bolts					
		E = K	f	L	L₁	R	4	6	8	10	12	
							Min. pitch circle dia. D					D₁ H12
M 3	5.5	5	8.5	20	15	10	12	18	25	31	38	—
M 4	7	6	10	26	20	13	15	23	31	39	46	—
M 5	8	7	10	32	25	16	17	26	36	45	54	—
M 6	10	9	13	36	26	18	21	32	45	56	68	20
M 8	13	11	16	50	39	25	28	43	55	70	85	25
M 10	17	14	20	60	45	30	35	55	76	95	115	30
M 12	19	16	22	68	50	34	39	61	84	110	135	32
M 16	24	18	28	80	60	40	49	77	110	135	165	40
M 20	30	22	34	100	75	50	62	96	135	170	205	48
M 24	36	25	40	120	95	60	74	120	160	200	245	56
M 30	46	32	50	150	115	75	94	150	205	255	310	70
M 36	55	40	60	180	140	90	115	180	245	305	375	80

COLD FORMING AND COLD BENDING OF FLAT ROLLED SHEETS AND STRIPS (As per DIN 6935 - 1969)

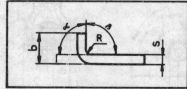

α : Bending angle
β : Opening angle
Bending angle α may have any value between 0 and 180°.
Thickness S will be reduced upto 20% (≈) in the bent portion.

MINIMUM ALLOWABLE BENDING RADIUS R
(All dimensions in mm)

Material	Tensile strength kgf/mm²	From	—	0.5	0.8	1	1.5	2.5	3	4	5	6	7	8
		Upto	0.5	0.8	1	1.5	2.5	3	4	5	6	7	8	10
Steel	≤ 40		0.6	1	1.6	1.6	2.5	3	5	6	8	10	12	16
									6	8	10	12	16	20
	>40 ≤65		1	1.6	2	3	5	6	7	9	10	12	16	20
									8	10	12	16	20	25
	>65		1.2	2	2.5	4	6	8	10	12	—	—	—	—
Copper	≤25		0.3	0.4	0.5	0.8	1.2	1.6	2	3	—	—	—	—
	>25		0.4	0.6	0.8	1.2	2	2.5	3	4	—	—	—	—
Brass	≤40		0.6	1	1.2	2	3	4	5	6.5	—	—	—	—
	>40		0.8	1.2	1.6	2.5	4	4.5	6	7.5	—	—	—	—
Aluminium			0.6	1	1	1.6	2.5	4	6	10	—	—	—	—

Values are applicable for bending angle α ≤ 120°. For α > 120°, the value of radius for the next higher range should be used.

Values of radii indicated are applicable to forming and bending at right angles to the direction of rolling. For Steels upto 65 kgf/mm² and thickness S ≥ 3, the higher values indicated are applicable when forming and bending are parallel to the direction of rolling. Bending at right angles to the direction of rolling is preferred.

Actual values of radii should be selected from the preferred values of bending radii.

COLD FORMING AND COLD BENDING OF FLAT ROLLED SHEETS AND STRIPS (As per DIN 6935 - 1969)　　(concld.)

PREFERRED BENDING RADII　　　　(All dimensions in *mm*)

1	1.6	2.5	4	6	10	16	20	25	32	40	50	63	80	100

Minimum Leg Length and Permissible Deviations on Opening Angle and Radius
For press brake forming of sheet and plate, the minimum leg length *b* can be taken as equal to 4R.

PERMISSIBLE DEVIATION ON OPENING ANGLE β:

Length of smaller leg, *b*	From	—	10	30	80	180
	Upto	10	30	80	180	—
Permissible deviation.		$\pm 2°30'$	$\pm 2°$	$\pm 1°45'$	$\pm 1°30'$	$\pm 1°$

Values indicated are for $R/S = 4$. When $R/S > 4$, larger variation must be allowed owing to spring back.

PERMISSIBLE DEVIATION OF BENDING RADIUS R

Bending radius R	From	—	1	3	6	10	18	30	50
	Upto	1	3	6	10	18	30	50	100
Permissible deviation		+0.5 0	+0.8 0	+1 0	+1.5 0	+2 0	+2.5 0	+3 0	+4 0

CALCULATION OF LENGTH BEFORE BENDING:

 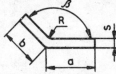

Length before bending,　$L = a + b + V$.

V is a compensating value varying with the bending angle.

V is +ve when β is between 0° and 65° and —ve when $\beta > 65°$.

For $\beta \leqslant 90°$,　$V = \pi \left[\dfrac{180 - \beta}{180} \right] \left[R + \dfrac{S}{2}\,k \right] - 2(R + S)$

For $\beta > 90° \leqslant 165°$,　$V = \pi \left[\dfrac{180 - \beta}{180} \right] \left[R + \dfrac{S}{2}\,k \right] - 2(R + S)\tan\left[\dfrac{180 - \beta}{2} \right]$

For $\beta > 165°$ and upto 180°,　$V = 0$.

The values of the correction factor k may be taken from the table.

CORRECTION FACTOR k.

R/S	From	0.65	1	1.5	2.4	3.8
	Upto	1	1.5	2.4	3.8	—
k		0.6	0.7	0.8	0.9	1

MACHINE ELEMENTS

SHAFTS, KEYS AND SPLINES

Nomenclature:

A	Area of M/I diagram, kgf/mm^2
a, b	Distances, mm
d	Outside diameter of shaft, mm
d_1	Inside diameter of shaft, mm
E	Young's modulus for material of shaft, kgf/mm^2
I	Moment of inertia of the shaft, mm^4
K	Factor for end condition
l	Length of shaft between bearings, mm
M	Bending moment, $kgf.mm$
n	Speed of shaft, rpm
n_c	Critical speed of shaft neglecting self weight, rpm
n_1	Critical speed of shaft due to self weight only, rpm
P	Power transmitted, kW
P_{cr}	Critical load in buckling that would result in failure, kgf
r	Least radius of gyration of the cross-section: $d/4$ for solid circular cross-section; $\sqrt{(d^2+d_1^2)}/4$ for hollow circular cross-section, mm
S	Slenderness ratio $= l/r$
T	Twisting moment, $kgf.mm = 975000P/n$
W	Total load acting on the shaft, kgf
y	Static deflection under load, mm
Z	Section modulus, mm^3
Δ	Sum of the moments of M/I diagram about a point on the elastic curve, kgf/mm
θ	Angular deflection of shaft, degrees
σ_b	Bending strength, kgf/mm^2
σ_{ba}	Allowable bending stress, kgf/mm^2
σ_c	Ultimate compressive strength, kgf/mm^2
σ_{sa}	Allowable shear strength, kgf/mm^2
σ_u	Ultimate tensile strength, kgf/mm^2
σ_y	Yield strength, kgf/mm^2

Shafts are rotating members mounted in suitable bearings and usually carry power transmitting elements like gears, pulleys, nuts, flywheels, etc., and are subjected to bending, torsion, buckling or shear. They should be checked for strength, stiffness, torsional stiffness, buckling and critical speed depending upon the particular requirement.

Table 1 Allowable bending stress σ_{ba} kgf/mm^2 for steel shafts

Material of shaft and heat treatment		With keyway			Without keyway			Stepped shaft (a)* (b)*		
		Diameter of shaft d, mm								
		30	50	100	30	50	100	30	50	100
C14	Normalised $\sigma_u = 37-45$	5.5	5.2	4.8	7.2	6.8	6	8.8 (6.4)	7.6 (5.6)	6.8 (4.8)
C14	Case hardened $\sigma_u = 50$	7	6.5	6	9	8.5	7.5	11 (8)	9.5 (7)	8.5 (6)
C45	Normalised $\sigma_u = 63-71$	7.5	7	6.5	10	9.5	8.5	11.5 (9)	10 (8)	9 (7)
C45	Hardened & tempered $\sigma_u = 70-85$	8.5	8	7	11.5	10.5	10	13.5 (10.5)	11.5 (9)	10 (8)
17Mn1Cr95 13Ni3Cr80 Case hardened $\sigma_u = 80-85$		9	8.5	8	12	11	10	14 (11)	12 (9.5)	10.5 (8.5)
15Ni2Cr1Mo15 Case hardened $\sigma_u = 110$ 40Ni2Cr1Mo28 Hardened & tempered $\sigma_u = 100-155$		9.5	9	8.5	13	12	11	15 (11.5)	13 (10)	11 (9)

Notes: 1) The allowable bending stress values given in the table correspond to smooth, steady running of shafts. If loads on the shaft are intensely variable or if peak loads occur, then the allowable values should be reduced.

  *2) For stepped shafts two separate values are indicated. While the first one gives the allowable bending stress for stepped shafts with radiused transition as in (a) the bracketed values are applicable for stepped shafts with a relief at the transition as in (b). These allowable values are accurate for $l/d \leqslant 1.2$ and in case of (a), it is applicable only for $r_a/d \geqslant 0.05$. For other cases these are only approximate. For a very accurate and detailed calculation when the loads are fluctuating, repeating or reversing, it is necessary to take into account factors like stress concentration, size correction, surface roughness, etc., for the endurance limit of the material of the shaft.

In machine tools normally, during design of the shafts, requirements from stiffness considerations are more stringent than strength considerations. Invariably this leads to a much larger size of shaft than would be necessary strictly from strength considerations. Therefore to make the calculations simple, dynamic character of load is not taken into account and stress concentration factors are not included in the equations, though some account of this has been made in the values of allowable stresses recommended.

Calculation procedure for strength:

i) If the shaft is subjected to a combined bending and twisting moment, the required diameter of shaft is calculated from

$$\sigma_{ba} = \frac{\sqrt{M^2 + 0.45\,T^2}}{Z}$$

where $\sqrt{M^2 + 0.45\,T^2}$ is the equivalent bending moment.

$$M = \sqrt{M_x^2 + M_y^2}$$ resultant of bending moment in the X and Y planes which are at right angles.

Values of σ_{ba} are listed in Table 1.

ii) If the shaft is subjected to a pure twisting moment, diameter of the shaft from the strength point of view is approximately calculated, for steels with ultimate tensile strength $\sigma_u = 50 - 80\,kgf/mm^2$, from Table 2.

Table 2

Sl. No.	d	Load and Bending Moment	$\sigma_{sa}, kgf/mm^2$
1	$100\sqrt[3]{\dfrac{P}{n}}$	Steady load with small bending moment	5
2	$110\sqrt[3]{\dfrac{P}{n}}$	Fluctuating load with small bending moment or steady load with considerable bending moment	3.7
3	$120\sqrt[3]{\dfrac{P}{n}}$	Fluctuating load with considerable bending moment or steady load with large bending moment (long length of shaft)	2.85

For other values of σ_{sa}, say σ_{sn}, the diameter obtained from Table 2 should be multiplied by the factor $(\sigma_{sa} / \sigma_{sn})^{1/3}$

Calculation procedure for stiffness:

Heavy loads will deform the shafts excessively, impairing the normal working of gears and bearings. This results in non-uniform load distribution between the gears leading to concentration of pressure at the edges. Calculation for stiffness necessitates calculation of shaft deflection and slope due to bending. Formulae for slope and deflection of the commonly encountered cases are given in pages 71 - 82

If two or more loads are acting, then the principle of superposition may be used to get the resultant slope and deflection.

For stepped shafts, the deflection may be calculated by any of the following methods:

   a) Macauley's method
   b) Area-moment method
   c) Conjugate beam method
   d) Castigliano's method
   e) Graphical integration method

An example of calculation of deflection of a stepped shaft using the area-moment method is given below.

The area-moment method of determining the deflection of a shaft is based on the principle that the vertical distance of any point P on the elastic curve from the tangent at any other point Q on the elastic curve is equal to the moment of the area of the M/I diagram between the points P and Q about an ordinate through the point P.

A simply supported stepped shaft is loaded by a concentrated load of 200 kgf as shown in Fig. 1. It is required to calculate the deflection at the free end C.

By summing up the moments of areas above line. $A'C'$ with reference to the ordinate at C, Δ_1 is calculated to be 5660 kgf/mm .

By summing up the moments of areas above line $A'B'$ with reference to ordinate at B', Δ_2 is calculated to be 4109 kgf/mm , Δ_3 is determined by proportion $A'C'\Delta_2/A'B' = 5115\ kgf/mm$, deflection at $C' = \Delta_1 - \Delta_3 /E = 0.026\ mm$ or 26 μm

Note: If any point of the M/I diagram is negative then the moment of that part of M/I diagram should be taken as negative.

TORSIONAL STIFFNESS:

Shafts are checked for torsional stiffness using the following formulae:

For solid shafts.

For $\theta \leqslant 0.25°$/metre length, $d \geqslant 128 \times ^4\sqrt{P/n}$

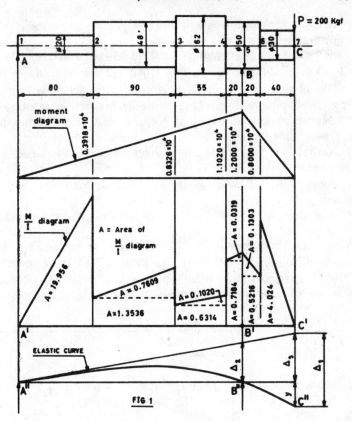

FIG 1

Section	Length	Diameter	(I) × 10⁻⁵
1-2	80	20	0.0785
2-3	90	48	2.6057
3-4	55	62	7.2533
4-5	20	50	3.0679
5-6	20	50	3.0679
6-7	40	30	0.3976

Section	(Moment × 10⁻⁴	$\dfrac{M}{I} \times 10^2$
1	0	0
2	0.392	49.89
		1.504
3	0.8326	3.195
		1.148
4	1.102	1.519
		3.592
5	1.2	3.911
		—
6	0.8	2.608
		20.12
7	0	0

For $\theta \leqslant 0.35^{\circ}$/metre length, $d \geqslant 118 \times {}^4\sqrt{P/n}$

For hollow shafts:

For $\theta \leqslant 0.25^{\circ}$/metre length, $(d^4 - d_1{}^4) \geqslant 1.959 \times 10^8 \, (P/n)$

For $\theta \leqslant 0.35^{\circ}$/meter length, $(d^4 - d_1{}^4) \geqslant 2.742 \times 10^8 \, (P/n)$

Permissible values of deflection, slope and angle of twist which are generally adopted in machine tool design practice are as follows:

1) Maximum deflection of shaft not to exceed 2×10^{-4} times the span between bearings.

2) Deflection of shaft in microns under the gear not to exceed 10 times the module of gear.

3) Slope of the shaft where gear is mounted, not to exceed 1×10^{-3} radian.

4) The angle of twist should not exceed 0.25° to 0.35° per metre length of shaft.

5) Maximum slope of shaft at bearings is limited by the permissible misalignment of bearings. Permissible misalignment of rolling bearings is dependent upon type of bearing, internal clearance of bearing during operation, bearing size, its internal

Table 3

Sl. No.	Type	Normal limit of misalignment (radians)	Remarks
1	Single row ball bearings	1×10^{-3}	Can be exceeded under occasional peak conditions
2	Angular contact and duplex ball bearings	0.3×10^{-3}	Greater misalignment can be critical under heavy axial load
3	Double row self alignment ball bearings	40×10^{-3}	Ultimately limited by ball/outer raceway contact position
4	Double row radial ball bearings and thrust bearings	0.2×10^{-3}	Should be closely controlled if normal reliability is anticipated
5	Spherical roller bearings	9×10^{-3}	Ultimately limited by roller/outer raceway contact position
6	Cylindrical and tapered roller bearings	0.3×10^{-3}	Greater misalignment can be critical under heavy load
7	Gears	1×10^{-3}	—

design together with forces and moments acting on it. Due to complex relationship between these factors, exact values of permissible misalignment cannot be obtained but Table 3 gives recommended permissible misalignment values under normal service conditions. If the actual value of misalignment is greater than that permitted in Table 3, in a particular design, then one has to increase the shaft diameter suitably.

Critical speed:

The critical speed formulae given in Table 4 apply to shafts of uniform cross-section carrying static load or carrying uniformly distributed loads, covering different conditions of bearing arrangements. If the bearings are self aligning or very short, shaft is considered to be simply supported at the ends.

If the bearings are long and stiff, the shaft is considered to be fixed. The formulae for static load and distributed load apply to vertical shafts as well as horizontal shafts, the critical speeds being the same in both cases. These formulae hold good for steel shafts, with a modulus of elasticity $E = 2.1 \times 10^4$ kgf/mm^2 and specific weight $= 7.85 \times 10^{-6}$ kgf/mm^3.

If the shaft carries a distributed load or a number of point loads, then the shaft will have infinite number of critical speeds, the first critical speed is usually the important one in engineering work. The actual speed should be at least $\pm$ 20% away from the critical speed.

Rayleigh method for critical speed:

This is the most commonly used method for obtaining the critical speed of a shaft. The weight of the shaft is usually replaced by a number of concentrated loads, which are added to the concentrated loads already acting on the shaft. Thus, the system is reduced to that of a weightless shaft carrying a number of concentrated loads. Referring to Fig. 2, W_1 and W_2 are concentrated loads, y_1 and y_2 are deflections at A and B respectively, when W_1 and W_2 act simultaneously. The method can be extended to include many concentrated loads.

$$n_c = 946.48 \sqrt{\frac{\Sigma W y}{\Sigma W y^2}} \quad rpm$$

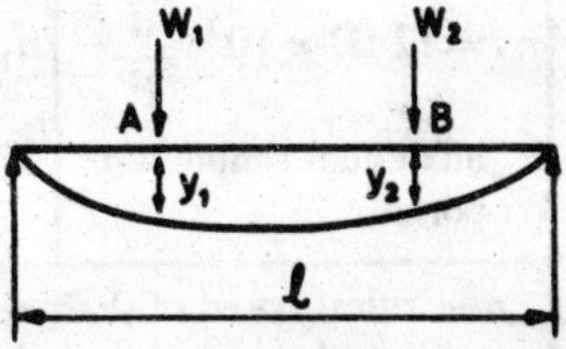

Fig. 2

Table 4

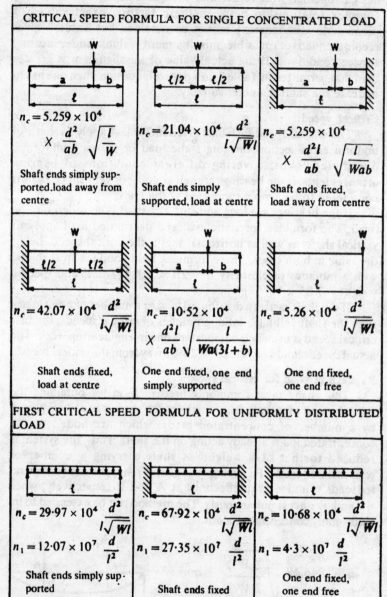

CRITICAL SPEED FORMULA FOR SINGLE CONCENTRATED LOAD		
$n_c = 5.259 \times 10^4 \times \dfrac{d^2}{ab}\sqrt{\dfrac{l}{W}}$ Shaft ends simply supported, load away from centre	$n_c = 21.04 \times 10^4 \dfrac{d^2}{l\sqrt{Wl}}$ Shaft ends simply supported, load at centre	$n_c = 5.259 \times 10^4 \times \dfrac{d^2 l}{ab}\sqrt{\dfrac{l}{Wab}}$ Shaft ends fixed, load away from centre
$n_c = 42.07 \times 10^4 \dfrac{d^2}{l\sqrt{Wl}}$ Shaft ends fixed, load at centre	$n_c = 10.52 \times 10^4 \times \dfrac{d^2 l}{ab}\sqrt{\dfrac{l}{Wa(3l+b)}}$ One end fixed, one end simply supported	$n_c = 5.26 \times 10^4 \dfrac{d^2}{l\sqrt{Wl}}$ One end fixed, one end free

FIRST CRITICAL SPEED FORMULA FOR UNIFORMLY DISTRIBUTED LOAD		
$n_c = 29.97 \times 10^4 \dfrac{d^2}{l\sqrt{Wl}}$ $n_1 = 12.07 \times 10^7 \dfrac{d}{l^2}$ Shaft ends simply supported	$n_c = 67.92 \times 10^4 \dfrac{d^2}{l\sqrt{Wl}}$ $n_1 = 27.35 \times 10^7 \dfrac{d}{l^2}$ Shaft ends fixed	$n_c = 10.68 \times 10^4 \dfrac{d^2}{l\sqrt{Wl}}$ $n_1 = 4.3 \times 10^7 \dfrac{d}{l^2}$ One end fixed, one end free

n_c—Critical speed of shaft neglecting self weight, *rpm*
n_1—Critical speed of shaft due to self weight, *rpm*

For this method it is necessary to draw the static deflection curve.

For a stepped shaft, the critical speed can be calculated by the Rayleigh method after calculating the static deflections at each and every change of cross-section.

Calculation for stability in buckling:

Long and slender shafts, axially loaded, should also be checked for stability in buckling.

RANKINE'S FORMULA FOR BUCKLING:

$$P_{cr} = \frac{\pi d^2 \sigma_c / 4}{\left[l + K \left(\frac{l}{r} \right)^2 \right]}$$

where K depends upon the material and end condition.
For steel columns $K = 1/25000$ for both ends of column fixed; $1/12500$ for one end fixed, one end hinged; $1/6250$ for both ends hinged.
Rankine's formula should be used if the slenderness ratio S is in the range of 20 - 100.

EULER'S FORMULA FOR BUCKLING

$$P_{cr} = \frac{C \pi^2 E I}{l^2}$$

where factor C depends on different end conditions, the values of which are given in Table 5. Euler's formula should be used only if S exceeds the values given in Table 6.

Table 5 Value of factor C for Euler's and Johnson's formula.

Both ends fixed	One end fixed, one end hinged	Both ends hinged	One end fixed, one end free
4	2	1	0.25

Table 6 Limiting values of S for Euler's formula

Material	Flat ends	Hinged ends	Round ends
Structrual steel	195	155	120
C.I.	120	100	75

J.B.JOHNSON'S FORMULA FOR BUCKLING:

$$P_{cr} = \frac{\pi d^2 \sigma_y}{4}\left(1 - \frac{Q}{4r^2}\right)$$

where $Q = \sigma_y l^2 / C\pi^2 E$; values of C are given in Table 5
If $Q/r^2 > 2$ then Euler's formula should be used
If $Q/r^2 < 2$ then J. B. Johnson's formula should be used.

FACTOR OF SAFETY FOR BUCKLING STABILITY:

If the load conditions and physical qualities of material are accurately known, a factor of safety as low as 1.25 may be used if it is important to keep the weight of the shaft to be minimum. For normal work and for steady loads, a factor of safety of 2-2.5 may be used.

Checking calculations for Keys:

NOMENCLATURE:

A	Bearing area of the key, mm^2
b	Width of key, mm
d	Shaft diameter, mm
F	Tangential force, kgf
l	Total length of key, mm
t	Height of key, mm
t_1	Depth of keyway in shaft, mm
t_2	Depth of keyway in hub, mm
p	Bearing pressure, kgf/mm^2
T	Maximum torque, $kgf.mm$

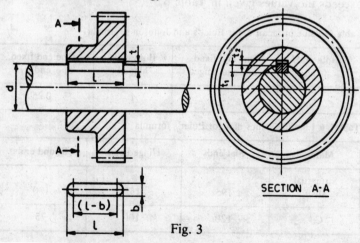

SECTION A-A

Fig. 3

Keys are to be checked for bearing pressure on the critical bearing area. Parallel keys for machine tools have more depth in shaft and less in the hub. For such keys, the critical bearing area will be the contact area between the key and the hub.

In such cases,

Bearing pressure $p = F/A$

where, $F = 2T/d$

and $A = t_2 (l-b)$, for round ended keys

           $= t_2 l$, for square ended keys

Parellel keys for general engineering purposes have almost equal depth in shaft and hub. For such keys the critical area is the contact area between the key and shaft/hub.

In such cases,

Bearing pressure, $p = F/A$

where, $F = 2T/d$

and $A = t (l-b)/2$ for round ended keys

     $= tl/2$ for square ended keys.

ALLOWABLE BEARING PRESSURE:

For soft keys made out of steel having a minimum tensile strength of $58\ kgf/mm^2$, the maximum bearing pressure allowed is 8 to $10\ kgf/mm^2$ (lower value for medium working conditions and higher value for good working conditions).

For hardened keys, the maximum bearing pressure can be allowed upto 10 to $12\ kgf/mm^2$.

Note: It is not necessary to check the keys in shear since the shear area will be more than double that of the crushing area.

Checking calculations for splines:

NOMENCLATURE:

c	Chamfer of edges in hub and spline, *mm*
D	Major diameter, *mm*
d	Minor diameter, *mm*
F	Tangential force, *kgf*
h	Total height of the effective contact surfaces between shaft and hub, $mm = n' h'$
h'	Height of one contact surface, *mm*
l	Length of hub, *mm*
n	Number of splines
n'	Number of effective contact surfaces $\approx$ (2/3 to 3/4).n
T	Maximum torque, *kgf.mm*

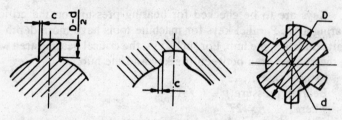

Fig. 4

Splines have to be checked for bearing pressure on the contact surfaces between the shaft and the hub.

Bearing pressure $p = F/h.l$

where $\quad F = 4T/(D+d)$

$$h' = \frac{(D-d)}{2} - 2c$$

ALLOWABLE BEARING PRESSURE:

The maximum bearing pressure on the contacting surfaces of splines should be within the specified limits of the values given in Table 7.

Table 7 Permissible values of bearing pressure in *kgf/mm²* in splines of spline shafts

Type of connection	Working conditions	Flanks of splines in shaft & hub unhardened	Flanks of splines in shaft & hub hardened
Sliding, without load	Heavy (shocks in both directions, vibrations, insufficient lubrication, less accurate manufacture)	1.5-2	2-3.5
	Medium	2-3	3-6
	Good	2.5-4	4-7
Sliding, with load	Heavy	—	0.3-1
	Medium	—	0.5-1.5
	Good	—	1-2
Stationary	Heavy	3.5-5	4-7
	Medium	6-10	10-14
	Good	8-12	12-20

PARALLEL KEYS AND KEYWAYS

Non sliding keys

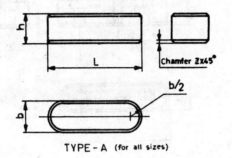

TYPE - A (for all sizes)

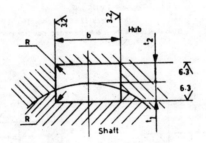

Kèyway for non-sliding key

Sliding keys

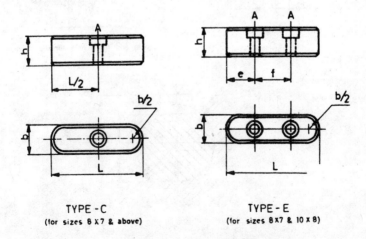

TYPE - C
(for sizes 8 X 7 & above)

TYPE - E
(for sizes 8X7 & 10 X 8)

Sliding keys   (concld.)

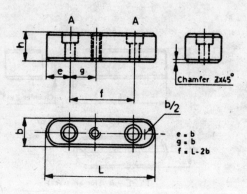

e = b
g = b
f = L-2b

Chamfer 2x45°

TYPE-E
(for sizes 12X8 & above)
(also available without the threaded hole for jacking screw)

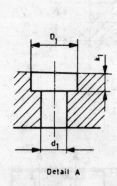

Detail A

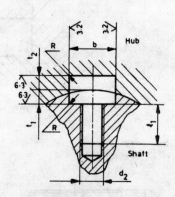

Keyway and fixing hole for sliding keys

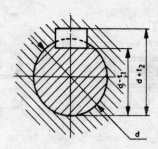

Table 8 Dimensions for parallel keys and keyways (as per IS 2048 - 1975 and IS 2710 - 1975)

Key b (h9)	Key h [1]	Z Min.	Z Max.	Keyway t₁ IS 2710	Keyway t₁ as per [2] IS 2048	Keyway t₂ IS 2710	Keyway t₂ as per [2] IS 2048	Tol. on t₁ & t₂	R Min.	R Max.	b [3]	Sliding key d₂	Sliding key l₁	Hole in key d₁H12	Hole in key D₁H13	Hole in key k₁	Size of Ch. hd. scr.
4	4	0.16	0.25	3	2.5	1.1	1.8	+0.1	0.08	0.16	4	–	–	–	–	–	–
5	5	0.25	0.4	3.8	3	1.3	2.3	0	0.16	0.25	5	–	–	–	–	–	–
6	6	0.25	0.4	4.4	3.5	1.7	2.8		0.16	0.25	6	–	–	–	–	–	–
8	7	0.25	0.4	5.4	4	1.7	3.3		0.16	0.25	8	M3	4	3.2	5.9	2.4	M3x8
10	8	0.4	0.6	6	5	2.1	3.3		0.25	0.4	10	M3	5	3.2	5.9	2.4	M3x10
12	8	0.4	0.6	6	5	2.1	3.3		0.25	0.4	12	M4	6	4.3	7.4	3.2	M4x10
14	9	0.4	0.6	6.5	5.5	2.6	3.8		0.25	0.4	14	M5	6	5.3	8.9	4	M5x10
16	10	0.4	0.6	7.5	6	2.6	4.3	+0.2	0.25	0.4	16	M5	6	5.3	8.9	4	M5x10
18	11	0.4	0.6	8	7	3.1	4.4	0	0.25	0.4	18	M6	6	6.4	10.4	4.7	M6x12
20	12	0.6	0.8	8	7.5	4.1	4.9		0.4	0.6	20	M6	6	6.4	10.4	4.7	M6x12
22	14	0.6	0.8	10	9	4.1	5.4		0.4	0.6	22	M6	8	6.4	10.4	4.7	M6x16
25	14	0.6	0.8	10	9	4.1	5.4		0.4	0.6	25	M8	9	8.4	13.5	6	M8x16
28	16	0.6	0.8	11	10	5.1	6.4		0.4	0.6	28	M10	9	10.5	16.5	7	M10x16
32	18	0.6	0.8	13	11	5.2	7.4		0.4	0.6	32	M10	10	10.5	16.5	7	M10x20

* Refer page 178

3) Tolerance on width b of keyway:

Type of fit	in shaft	in hub
Running	H9	D10
Light drive	N9	Js9
Force	P9	P9

Notes:

1) Tolerance on h: for square keys - h9 for rectangular keys - h11

2) The depths t₁ and t₂ as per IS 2710 provide an approximate bearing area of 70% and 30% in the shaft and hub respectively and the depths t₁ and t₂ as per IS 2048 provide an approximate bearing area of 50% each in the shaft and hub

4) Sliding keys are available in sizes 8x7 and above only.

Table 8   Dimensions for parallel keys and keyways (As per IS 2048 - 1975 and IS 2710 - 1975) (concld.)

Key		h 1)	Length of Key, L, (As per IS 2710) 6) 7)						Length of key, L, (As per IS 2048) 6) 7)						Suitable for shaft dia. 8) d	
b h9			Type A		Type C		Type E		Type A		Type C		Type E		Above	Upto
			Min.	Max.	Min.	Max.	Min.	Max.	Min.	Max.	Min.	Max.	Min.	Max.		
4		4	10	45	—	—	—	—	8	45	—	—	—	—	10	12
5		5	12	56	—	—	—	—	10	56	—	—	—	—	12	17
6		6	16	70	—	—	—	—	14	70	—	—	—	—	17	22
8		7	20	90	20	36	40	90	18	90	18	36	40	90	22	30
10		8	25	110	25	45	50	110	22	110	22	45	50	110	30	38
12		8	32	140	32	50	56	140	28	140	28	50	56	140	38	44
14		9	40	160	40	56	63	160	36	160	36	56	63	160	44	50
16		10	45	180	45	63	70	180	45	180	45	63	70	180	50	58
18		11	50	200	50	70	80	200	50	200	50	70	80	200	58	65
20		12	56	220	56	80	90	220	56	220	56	80	90	220	65	75
22		14	63	250	63	90	100	250	63	250	63	90	100	250	75	85
25		14	70	250	70	90	100	250	70	280	70	90	100	280	85	95
28		16	80	250	80	100	110	250	80	320	80	100	110	320	95	110
32		18	90	250	90	100	110	250	90	360	90	100	110	360	110	130

5) Tolerance on dimensions e, f and L/2 which are applicable for both keys and keyways is as follows:

    Keys with screws M3, M4 & M5 : ± 0.05; Keys with screws M6, M8 & M10 : ± 0.1

6) The preferred lengths of Keys are 8, 10, 12, 14, 16, 18, 20, 22, 25, 28, 32, 36, 40, 45, 50, 56, 63, 70, 80, 90, 100, 110, 125, 140, 160, 180, 200, 220, 250, 280, 320 and 360.

7) Tolerance on length of keys and keyways (+ ve for keyways and -ve for keys): 10 to 28 : 0.2; 32 to 80 : 0.3; 90 to 360 : 0.5.

8) Smaller section keys may be used if adequate for torque transmission.

9) Sliding keys are used when the hub is required to slide over the shaft, e.g., clutches

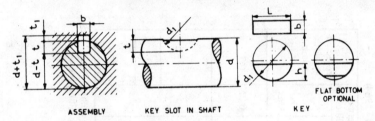

ASSEMBLY   KEY SLOT IN SHAFT   KEY   FLAT BOTTOM OPTIONAL

Table 9 Woodruff keys (As per IS 2294 - 1963)

Key section		Shaft dia. d		d_1		L	t		t_1	Radius r^*	
b $h9$	h $h12$	From	Upto	Nom.	Tol.	$\approx$	Nom.	Tol.	$+0.1$ 0	Nom.	Tol. **
1	1.4	3	4	4		3.82	1		0.6		
1.5	2.6	4	6	7		6.76	2		0.8		
2		6	8				1.8		1		
	3.7	8	10	10		9.66	2.9			0.2	0.1
3							2.8				
	5			13	-0.1	12.65	4.1		1.1		
	5	10	12	13		12.65	4.1				
4	6.5			16		15.72	5.6	$+0.1$			
	6.5	12	17	16		15.72	5.4				
5	7.5			19		18.57	6.4		1.3		
	7.5			19		18.57	6				
6@	9	17	22	22		21.63	7.5		1.7	0.4	0.2
	10@			25	-0.2	24.49	8.5				
	9	22	30	22	-0.1	21.63	7.5				
8	11			28		27.35	9.5	$+0.2$			
	11			28	-0.2	27.35	9.1				
10	13	30	38	32		31.43	11.1		2.1		

Note: 1. * For key, either chamfer or radius; for keyway in shaft & hub, radius only.
2. ** Tolerance on r : for key, +ve; for keyway,-ve.
3. @ Key section 6 x 10 is non-preferred.
4. Key sizes given are for torque transmission. When keys are used for location only, a smaller key maybe used for a given shaft diameter.

Table 10 Straight sided splines for machine tools—4 splined (As per IS 2610 - 1964)

Designation N x d x D	B	External splines – Type A – d₁ Min	g Max @@ @	e Max	f Min	Type B – h	Type B – r₁ Max	Type M – m	Type M – n	Type M – r₂	Internal splines – k Max	Internal splines – r₃ Max	Projected tip width of hob
4 x 16 x 20	6	15	0.3	1.78	2.87	6.7	0.15	5.64	1.7	0.3	0.3	0.25	0.7
4 x 18 x 22		16.9		1.84	4.35	7.7							
4 x 21 x 25	8	20.1		1.64	5	8.9		7.52					
4 x 24 x 28		23		1.7	7.3	10.4			1.63				
4 x 28 x 32	10	26.8		2.19	7.39	12.1		9.4		0.6			1
4 x 32 x 38		30.3		2.7	9.56	14.2			2.55				
4 x 36 x 42	12	34.5	0.5	2.52	11.03	15.9	0.25	11.28			0.5	0.4	
4 x 42 x 48		40.2		2.7	15.41	19							
4 x 46 x 52	14	44.4		2.56	16.79	20.7		13.16	3.4	1			
4 x 52 x 60		49.5		2.5	21.63	23.7							
4 x 56 x 65	16	56.2		3.05	23.26	26.4		15.04	2.98				1.3
4 x 62 x 70		59.5		4.39	23.61	28.3			3.4				1.6
4 x 68 x 78		64.4		4.84	27.57	31.2			4.25				

Notes: Type A is obtained only by hobbing and type B by straddle milling. If the flanks are to be ground, type M undercut is produced by milling the roots of the splines as shown.

@ Values are based on the generating process.

@@ Rounding in place of chamfer is permissible

BASIC SPLINE PROFILE

INTERNAL

EXTERNAL

TYPE-B

TYPE-A

TYPE-M

INTERNAL SPLINE

k x 45°

Table 11 Straight sided splines for machine tools — 6 splined (As per IS 2610 - 1964)

Designation $N \times d \times D$	B	Type A d_1* Min.	Type A g** Max.	Type A e Max.	Type A f Min.	Type B h	Type B r_1 Max.	Type M m	Type M n	Type M r_2	Internal k Max.	Internal r_3 Max.	Projected tip width of hob
6 x 21 x 25	5	19.5	0.3	1.98	1.95	9.7	0.15	4.7	1.7		0.3	0.2	0.7
6 x 23 x 28	6	21.3		2.3	1.34	11			2.13				
6 x 26 x 32	6	23.4	0.4	2.94	1.65	11.8		5.64		0.6	0.4	0.3	1
6 x 28 x 34	7	25.9			1.7	12.9							
6 x 32 x 38	8	29.9	0.5	2.92	2.83	14.8		6.58	2.55				
6 x 36 x 42	8	33.7			4.95	16.5		7.52					
6 x 42 x 48	10	39.94		2.94	6.02	19.3	0.25	9.4					
6 x 46 x 52	12	44.16		3.08	5.81	21.1		11.28	3.4	1			
6 x 52 x 60	14	49.5		3.56	5.89	23.9		13.16	2.98		0.5	0.4	1.3
6 x 58 x 65	14	55.74		3.98	8.29	26.7			3.4				
6 x 62 x 70		59.5		4.13	8.03	28.6			4.25				1.6
6 x 68 x 78	16	64.4		4.86	9.73	31.4		15.04					
6 x 72 x 82		68.3		4.45	12.67	33.4			5.1	1.6			2
6 x 78 x 90		73		5.83	13.07	36.2			5.53				
6 x 82 x 95		76.6		6.44	13.96	38			5.1				
6 x 88 x 100		82.9		6.07	17.84	41.3							

Notes: Type A is obtained only by hobbing. Type B is obtained by straddle milling if the flanks are to be ground. Type M undercut is produced by milling the roots of the splines as shown.

* d_1. These values are based on the generating process.

** g-Rounding in place of chamfer is permissible.

Tolerances for straight sided splines (4 & 6 splines)

Tolerance for		d^*	D	B
Shaft	Sliding fit	$g6$	$a11$	$h9$
	push fit	$j6$		
Hub		$H7$	$H13$	$D9$

* For requirements of high accuracies, or in case of small length of bore, a tolerance of $H6$ instead of $H7$ and $h6$ instead of $j6$ may be specified.

For long bores and large splined shafts, a tolerance of $f7$ instead of $g6$ is recommended.

Maximum permissible alignment error

Nominal dia of shaft, *mm*		Max. permissible alignment error, *mm/m*
Over	Upto	
—	32	0.35
32	65	0.5
65	105	0.65
105	145	0.85

Rotary shaft seals

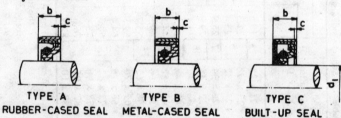

TYPE A
RUBBER-CASED SEAL

TYPE B
METAL-CASED SEAL

TYPE C
BUILT-UP SEAL

Table 12 Rotary shaft seals

Shaft dia. d_1	Nominal bore dia. of housing	Types A & B $b \pm 0.2$	Min.	Shaft dia. d_1	Nominal bore dia. of housing	Types A & B $b \pm 0.2$	Min.
6	16	7	0.3	8	22	7	0.3
6	22			8	24		
7	16			9	22		
7	22			9	24		
8	16			9	26		

Table 12 Rotary shaft seals (contd.)

Shaft dia. d_1	Nominal bore dia of housing	Types A & B $b \pm 0.2$	c Min.
10	19		
	22		
	24		
	26		
11	22		
	26		
12	22		
	24		
	28		
	30		
14	24		
	28		
	30		
	35		
15	24		
	26		
	30	7	0.3
	32		
	35		
16	28		
	30		
	32		
	35		
17	28		
	30		
	32		
	35		
	40		
18	30		
	32		
	35		
	40		
20	30		
	32		
	35		
	40		
	47		

Shaft dia. d_1	Nominal bore dia of housing	$b \pm 0.2$ Types A & B	$b \pm 0.2$ Type C	c Min.
22	32		—	
	35		—	
	40		9	0.3
	47		9	
24	35		—	
	37		—	
	40		9	
	47		9	
25	35		—	
	40		—	
	42		—	
	47		9	
	52		9	
26	37		—	
	42		9	
	47		9	
28	40	7	—	
	47		—	
	52		9	
30	40		—	
	42		—	
	47		—	
	52		9	0.4
	62		9	
32	45		—	
	47		9	
	52		9	
35	47		—	
	50		—	
	52		9	
	62		9	
36	47		—	
	50		—	
	52		9	
	62		9	
38	52		—	
	55		9	

Table 12 Rotary shaft seals (concld.)

Shaft dia. d_1	Nominal bore dia. of housing	$b\pm0.2$ Types A & B	Type C	c Min.		Shaft dia. d_1	Nominal bore dia. of housing	$b\pm0.2$ Types A & B	Type C	c Min.
38	62		9				75		—	
40	52	7	—			60	80	8		0.4
	55						85		10	
	62						90			
	72		9			62	85			
42	55		—				90			
	62		10			63	85			
	72						90			
45	60		—			65	85			
	62						90			
	65		10				100			
	72					68	90			
48	62		—				100			
	72		10			70	90	10	12	0.5
50	65		—	0.4			100			
	68					72	95			
	72	8					100			
	80		10			75	95			
52	68		—				100			
	72		10			78	100			
55	70		—			80	100			
	72						110			
	80		10			85	110			
	85						120			
56	70		—			90	110			
	72						120			
	80		10			95	120	12	15	0.8
	85						125			
58	72		—			100	120			
	80		10				125			
							130			

Mounting details

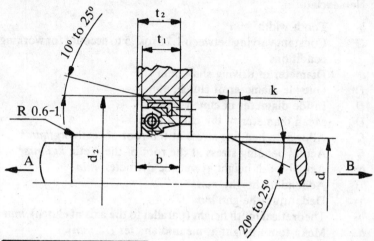

Shaft dia. d_1	Chamfer width k
upto 30	3
30-60	4
60-120	6
120-250	8
above 250	10

Surface speed m/sec	Surface roughness Ra
upto 2	0.8
2-4	0.4
4-12	0.2-0.4

Shaft details:

1. Tolerance on shaft dia. d_1 is $h8$ to $h11$.
2. Shafts for speeds between 4 and 12 $m/sec.$ should be hardened to HRC 45 to 60 and ground
3. Shafts for speeas between 8 and 12 $m/sec.$should be hard chrome plated to a thickness of 0.02 to 0.03 mm and polished.

Housing details:

1. Tolerance on housing dia. d_2 is $H8$
2. t_1 min = 0.85 b, $t_2 = b+0.3$
   In case of special design, $t = b + 1.5$ to 2
3. Surface roughness should be $Ra = 1.6$ to $6.3 \mu m$

For mounting shaft in the direction A, it is recommended that the edge of shaft be rounded off; for mounting shaft in the direction B, it is recommended that the edge of shaft be chamfered.

TOOTHED CLUTCHES

Nomenclature:

b	Tooth width, *mm*
C	Constant varying between 1.25 to 2.5 to account for working conditions
d	Diameter of driving shaft, *mm*
D	Outside diameter of clutch, *mm*
D_i	Inside diameter of clutch, *mm*
D_m	Mean diameter of the clutch, *mm*
f_a	Allowable bending stress at the root of teeth, *kgf/mm²*
f_b	Actual bending stress at the root of the teeth, *kgf/mm²*
h	Actual tooth height at outside diameter, *mm*
h_1	Addendum height, *mm*
h_2	Dedendum height, *mm*
h_a	Theoretical tooth height (parallel to the axis of clutch), *mm*
h_m	Mean tooth height at mean diameter D_m, *mm*
h_n	Effective tooth height, *mm*
h_s	Theoretical tooth height perpendicular to root face, *mm*
K	Constant varying between 2 and 3 to take into account the manufacturing errors.
n	Speed of driving shaft, *rpm*
p	Boss length of the driving part of clutch, *mm*
p_a	Allowable bearing pressure between teeth, *kgf/mm²*
p_b	Actual bearing pressure between teeth, *kgf/mm²*
P	Power transmitted, *kW*
q	Boss length of the driven part of clutch, *mm*
r	Radius of rounding of tooth, *mm*
S	Tooth thickness, *mm*
S_a	Width of tooth tip surfaces after chamfering, *mm*
S_m	Tooth thickness at root at diameter D_m, *mm*
S_u	Tooth gap thickness, *mm*
t	Tooth pitch on outside diameter D, *mm*
t_p	Chord length between two teeth at outside diameter D for measurement of teeth, *mm*
t_s	Tooth pitch on diameter $D/\cos\alpha$ *mm*
t_{sec}	Clutch engagement time, *seconds*
T	Design torque, *kgf. mm*
T_r	Transmitted torque, *kgf.mm*
V	Clearance between tip and root of teeth of driving and driven parts, *mm*

W	Section modulus at the root of the tooth, mm^3
X	Height of tooth chamfer, mm
X_1	Height of rounding at tooth root, mm
z	No. of teeth in the clutch
$\propto$	Angle of tip face or root face of tooth, *degrees*
β	Tip angle of tooth, *degrees*
β_1	Tooth flank angle, *degrees*
γ	Angular pitch of tooth, *degrees*
Δn	Difference in speed between driving and driven parts of clutch, *rpm*
μ_1	Friction coefficient between clutch body and shaft as well as key.
μ_2	Friction coefficient between clutch teeth
ϕ_2	$\tan^{-1}\mu_2$

Toothed clutches are used where quick engagement and disengagement of driving and driven shafts are desired.

Clutch engagement during running is generally possible when relative velocity is less than $0.7\,m/sec$. If surface speed exceeds this value, then the clutch can be engaged only at rest. Following factors determine the suitability of engagement during running.

a) shape, number of teeth, and the ratio of tooth width to tooth gap width;
b) inertia loads on driving and driven parts;
c) speed of engagement and value of engaging force;
d) synchronization of speeds;
e) ease of operation; etc.

Fine toothed clutches with large number of teeth are used for quick engagement. These types of clutches carry relatively small loads. Clutches with trapezoidal tooth profile and without side clearance are used for engagement during small relative speeds or in stationary conditions. However they can also be used for medium speed applications 0.3 to 0.5 m/sec. by providing suitable tooth chamfer (refer Table 13, type 5).

Normally a clutch is held in engaged position with the help of a shifting lever provided with a spring loaded detent. In cases where β_1 is greater than 5^0, it is necessary to provide an axial force to keep them engaged. Disengaged position of the lever should also be ensured and in this position, a minimum clearance of $1mm$ is provided between the driving and driven parts of the clutch.

The bearing length of the sliding part of the clutch should be about 2 to 2.5 times the seating diameter to avoid rapid wear as well as to ensure a reliable operation.

Case hardening or direct hardening steels can be selected depending on the function of the clutch. The material of the clutch need not be heat treated for clutches which are lightly loaded or which are rarely operated, that too while at rest.

Depending on the function, most suitable tooth profile is selected from Table 13. Guidelines for values of tooth proportions are given in Tables 14 to 20.

Table 13 Survey of tooth profiles used in toothed clutches

Sl. No.	Type		Tooth Shape	Properties
1	Fine tooth clutches.	Bi-directional $\beta = 60^\circ, 90^\circ$	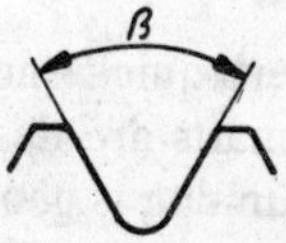	Quick engagement, fine adjustment. 90° angle preferred for safety clutch applications. Commonly used number of teeth are from 11 to 165. Tip and root surfaces should be tapered towards the axis for proper engagement (see Fig. 5) Holding force is required to keep the clutch engaged.
2	Fine tooth clutches.	Uni-directional $\beta = 45^\circ, 60^\circ$	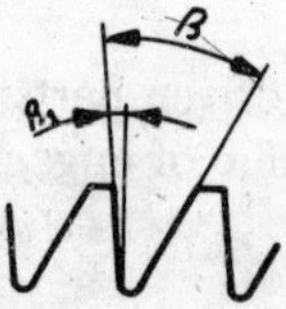	Transmits power in one direction only. Quick engagement. Tip and root surfaces are tapered towards the axis. Holding force not required to keep the clutch engaged. Driving tooth flank angle β_1 (0° to 5°) is given for easy disengagement.
3	Fine tooth clutches.	Uni-directional with reduced teeth $\beta = 45^\circ$	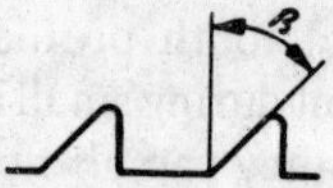	Clutch has four teeth, remaining portion is milled out. Can be engaged or disengaged quickly. Load carrying capacity is small.
4	Trapezoidal teeth without side clearance.		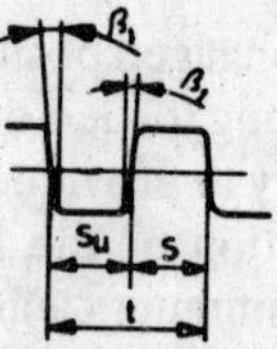	High precision clutches with odd number of teeth 5, 7 or 9 and tooth gaps are of equal width. Clutch is self holding at $\beta_1 = 5^\circ$. Transmits large torque. When used as a safety clutch, $\beta_1 = 30^\circ$ or 45°.

Table 13 Survey of tooth profiles used in toothed clutches (contd.)

Sl. No.	Type	Tooth Shape	Properties
5	Trapezoidal teeth chamfered on face	β_1 120° S_u s t	Same properties as those of type 4. Tooth face has 120° chamfer and tooth gap is wider than tooth thickness to allow easy engagement during running.
6	Safety clutch with 3 teeth $\beta_1 = 45°$	45°	Clutch has three teeth. Used only as a safety clutch. Considerable axial force is required to keep the clutch in engagement.
7	Flat teeth with side clearance		Manufacture and inspection of flat teeth with $z = 2, 3, 6$ is simple and does not require special milling cutters or inspection gauges. Clearance on tooth side is essential.
8	Uni-directional with top face relieved	β_1 β_2	Clutch transmits considerable load. Transmits power only in one direction. Easy engagement. Costly to manufacture because of the additional operation of relieving the top surface.

Calculation of tooth parameters:

A. FINE TOOTH CLUTCH BI-DIRECTIONAL:

Values of D, D_i, z and β (60° or 90°) to be selected based on design considerations. Other parameters to be calculated with the help of Tables 14 and 15.

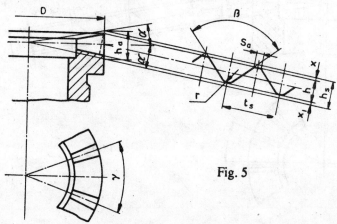

Fig. 5

Table 14

Symbol	Formula	
t	$t = \pi D / z$	
γ	$\gamma = 360 / z$ $\gamma/4 = 90 / z$	
	$\beta = 60^\circ$	$\beta = 90^\circ$
*$\propto$	$\sin\alpha = \dfrac{\tan \gamma/4}{\tan \beta/2}$	$\sin\alpha = \tan \gamma/4$
* t_s	$t_s = t/\cos\alpha$	
h_a	$h_a = D \tan\alpha$	
* h_s	$h_s = D \sin\alpha = D \dfrac{\tan \gamma/4}{\tan \beta/2}$	
* h	$h = h_s - (X + X_1)$	

Table 15

$r \pm 0.03$	S_a	$\beta = 60^\circ$		$\beta = 90^\circ$	
		X	X_1	X	X_1
0.1	0.2	0.17	0.1	0.1	0.04
0.25	0.5	0.43	0.25	0.25	0.1
0.5	1	0.86	0.5	0.5	0.2
1	2	1.73	1	1	0.4

* These values have to be specified on the drawings.

B. FINE TOOTH CLUTCH—UNI-DIRECTIONAL:

Value of D, z and β (45° or 60°) to be selected on design considerations. Values of r, S_a, X and X_1 to be selected from Table 16. All other parameters to be calculated from Table 17.

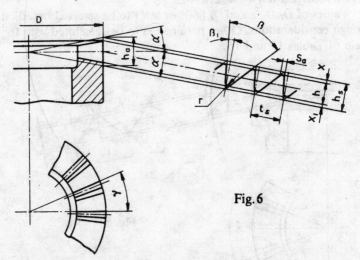

Fig. 6

Table 16

Symbol	Formula
* t	$t = \dfrac{D}{z}$
γ	$\gamma = \dfrac{360^\circ}{z}$
* α	$\sin\alpha = \dfrac{\tan\gamma/2}{\tan(\beta - \beta_1) + \tan\beta_1}$
* t_s	$t_s = \dfrac{t}{\cos\alpha}$
h_a	$h_a = D\,\tan\alpha$
* h_s	$h_s = D\,\sin\alpha$
* h	$h = h_s - (X + X_1)$

Table 17

$r \pm 0.03$	S_a		$\beta = 45^\circ$		$\beta = 60^\circ$	
	β		X	X_1	X	X_1
	45°	60°				
0.1	0.3	0.35	0.3	0.14	0.2	0.09
0.25	0.8	0.9	0.8	0.35	0.52	0.18
0.5	1.5	1.8	1.5	0.7	1.04	0.36

* These values have to be specified on the drawing.
   The tolerance on angle α is $\pm 5'$.

C. MULTI TOOTH CLUTCHES WITH TRAPEZOIDAL TEETH—WITHOUT SIDE CLEARANCE

Values of D, z, β, (10°, 16°, 30°, 45°, 60°, 90°) to be selected based on design consideration. Other parameters to be calculated from Table 18 and Table 19.

Table 18

Symbol	Formula
* t	$t = \dfrac{\pi D}{z}$ $\dfrac{t}{2} = S_u = S$
h_n	from Table—19
* h	from Table—19
* h_1	$h_1 = \dfrac{h_n}{2}$
h_2	$h_2 = h - h_1$
* γ	$\gamma = \dfrac{360^\circ}{z}$
* t_p	$t_p = D\,\sin\dfrac{\gamma}{4}$

Table 19

h_n	3	4	5	6	8
h	3.5	4.5	5.5	6.5	8.5
r	0.25				
S_a	0.5 x 45°				
h_n	10	12	16	20	25
h	11	13	17	21	26.5
r	0.5				
S_a	1 x 45°				

* These values have to be specified on the drawing.

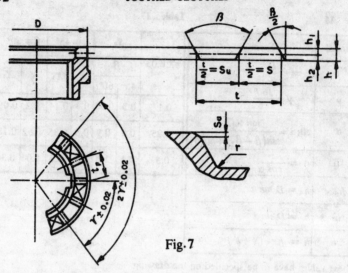

Fig. 7

This type of clutch has normally odd number of teeth 5, 7 or 9 to enable two tooth sides to be simultaneously machined in one pass of the cutter as shown in Fig. 8. This however requires accurate positioning of cutter with respect to clutch axis and accurate indexing. Position of cutter should be checked right after milling first tooth gap. Method of inspection is shown in Fig. 9. Inspection value of Mx is taken from Table 20.

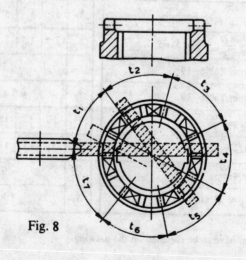

Fig. 8

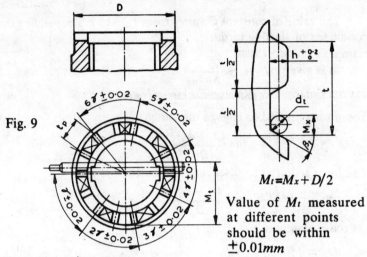

Fig. 9

$$M_t = M_x + D/2$$

Value of M_t measured at different points should be within $\pm 0.01mm$

Table 20   Values of M_x

h	d_t	β_1					
		5^0	8^0	15^0	$22^0\ 30'$	30^0	45^0
3.5	4	4.008	4.02	4.071	4.165	4.309	4.828
4.5	5	5.01	5.025	5.088	5.206	5.387	6.036
5.5	6	6.011	6.029	6.106	6.247	6.464	7.243
6.5	7	7.013	7.034	7.123	7.288	7.541	8.45
8.5	9	9.017	9.044	9.159	9.371	9.696	10.864
11	12	12.023	12.059	12.212	12.494	12.928	14.485
13	14	14.027	14.069	14.247	14.577	15.083	16.899
17	18	18.034	18.088	18.317	18.742	19.392	21.728
21	22	22.042	22.108	22.388	22.906	23.702	26.556
26.5	28	28.053	28.138	28.494	29.153	30.166	33.799

Strength Calculations for clutch teeth:

DESIGN TORQUE:

Transmitted torque $T = 975,000\ P/n$

Design torque $T_d = C.\ T$

The value of constant C varies from 1.25 to 2.5 depending on conditions of starting torque.

CLUTCH ENGAGEMENT TIME:

t_{sec} is given by $\quad z \geqslant \dfrac{60}{\Delta n \, t_{sec}}$

CLUTCH ENGAGEMENT AND DISENGAGEMENT FORCE:

The force required to engage the clutch.

$$F_{eng} = 2 \, T_d \left[\frac{\mu_1}{d} + tan \frac{(\beta_1 + \phi_2)}{D_m} \right]$$

The force required to disengage the clutch.

$$F_{dis} = 2 \, T_d \left[\frac{\mu_1}{d} - tan \frac{(\beta_1 - \phi_2)}{D_m} \right]$$

CLUTCH PARAMETERS:

$$
\begin{aligned}
D &= (1.8 \text{ to } 2.5) \, d \\
b &= (0.125 \text{ to } 0.2) \, D \\
h &= (0.6 \text{ to } 1) \, b \\
D_m &= (D + D_i)/2 \\
p &= (1.3 \text{ to } 1.7) d \\
q &= (2 \text{ to } 2.5) d
\end{aligned}
$$

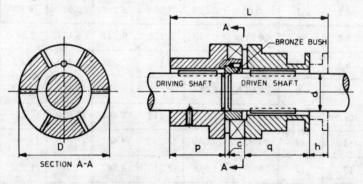

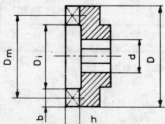

Fig. 10

Flank angle of tooth- β_1

For self locking of the clutch

$$\tan \beta_1 \leq \frac{\mu_2 \left(1 + \dfrac{\mu_1}{\mu_2} \dfrac{D_m}{d} \right)}{1 - \mu_1 \mu_2 \dfrac{D_m}{d}}$$

However since the product $\mu_1 \cdot \mu_2$ being very small, the denominator is taken as 1 and flank angle can be calculated from the equation

$$\tan \beta_1 \leq \mu_2 \left(1 + \frac{\mu_1}{\mu_2} \frac{D_m}{d} \right)$$

BEARING PRESSURE BETWEEN TEETH:

Bearing Pressure $\quad p_b = \dfrac{2 T_d}{b \, D_m \, h_m \, Z}$

where $\quad p_a = 8$ to $12 \ kgf/mm^2$, for clutches engaged at rest.

$\qquad = 2$ to $3 \ kgf/mm^2$ for clutches engaged during running.

BENDING STRESS IN THE TEETH:

The maximum bending stress in the teeth

$$f_b = \frac{2 T_d \cdot h_m \cdot K}{D_m \cdot Z \cdot W} \leq f_a$$

UNIDIRECTIONAL CLUTCHES

Unidirectional clutches are generally classified into two types i.e.,toothed clutches and friction clutches. The friction type clutches in general and the ones using rollers or cams in particular are popular owing to the possibility of operation at high speeds, minimum lost motion, extremely quiet operation with smaller impact loads and comparatively longer life. Toothed clutches are used for low speeds. They are noisy in overrunning and engagement is accompanied by shock loads.

The use of roller type unidirectional clutches (Fig. 16) is wide spread in machine building owing to their simplicity in design and manufacture though they are larger in size when compared to cam (sprag) type clutches (Fig. 11) for the same torque transmission.

Fig.11

Roller type unidirectional clutches:

NOMENCLATURE

a	Minimum distance from the centre of the inner ring to the contacting surface, mm
b	Width of the outer ring, mm
d_r	Roller diameter, mm
d_1	Diameter of the spring loaded pin, mm
D	Nominal diameter of the clutch, mm
D_o	Outside diameter of the outer ring, mm
E	Modulus of elasticity, kgf/mm^2
f_{all}	Allowable stress, kgf/mm^2
f_b	Bending stress, kgf/mm^2
f_e	Combined stress, kgf/mm^2
f_t	Tensile stress, kgf/mm^2
F	Resultant force, kgf
F_n	Normal force, kgf
F_t	Tangential force, kgf
$\left.\begin{array}{c}F_{t1}\\F_{t2}\end{array}\right\}$	Circumferential forces at points 1 and 2, kgf

l Length of the roller, mm

M_1 & M_2 Bending moments at points 1 and 2, $kgf.\ mm$

P_o Pressure on the projected area of the roller, kgf/mm^2

R Nominal radius of the clutch, mm

R_m Mean radius of the outer ring, mm

t Thickness of the outer ring, mm

T Transmitted torque, $kgf.\ mm$

z Number of rollers

α Wedging angle for the roller, *degrees*

β Friction angle, *degrees*

2γ Angular disposition of the rollers, *degrees*.

MATERIAL.

The material of the clutch parts should be such that it is possible to achieve a hardness of HRC 63-64 at surfaces contacting with rollers. The recommended material is direct hardening ball bearing steel, such as $105Cr1Mn60$

WEDGING ANGLE α:

The choice of the wedging angle (Fig. 12) depends upon the coefficient of friction at the contacting surfaces and lubrication conditions. The coefficient of friction may vary from 0.03 (for polished steel surfaces) to 0.1. For hardened steel surfaces with surface roughness Ra = 0.4, the value of coefficient of friction is around 0.07 which corresponds to a friction angle $\beta = 4^0$ Calculations given here are based upon this value of β and since the limiting condition for slip-free torque transmission being:

$$tan\ \alpha/2\ < tan\beta_{min}$$ the value of wedging angle $\alpha \leqslant 8^0$

For clutches used as friction ratchets with a high frequency of engagement, angle $\alpha\ =\ 6^0$ to 7^0 is recommended.

THE SHAPE OF THE CONTACTING SURFACE:

If the wedging angle has to remain constant from no load contact point to full load contact point the shape of the curve BC (Fig. 12) has to be a logarithmic spiral. This can be for all practical purposes replaced by an arc of a circle with radius R' and having the centre at O which is obtained by the intersection of the normal to the contact surface at B and the perpendicular to BO at O. But in practice for the type of construction shown in Fig. 12 a straight line perpendicular to O B at B is selected as the contact line for the inner ring owing to its simplicity. But this leads to a larger wedging angle as the wear takes place. Sometimes hardened inserts are used for the contact surface on the inner ring (Fig. 13).

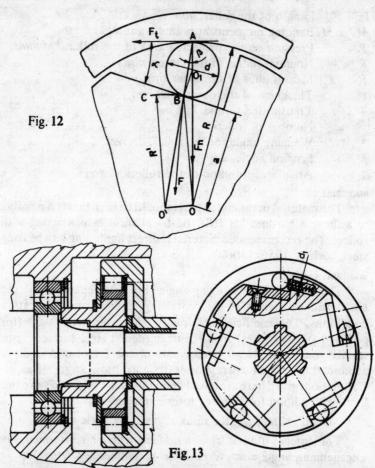

Fig. 12

Fig. 13

CALCULATION OF THE ROLLER SIZE:

This is based upon the resultant force F (Fig. 12) created due to wedging of the roller between contacting surfaces.

$$T = F_t . z . R \quad ....(1)$$

From Fig. 12

$$F_t = F \sin \beta \quad ....(2)$$

$$F_n = F_t / tan \beta \quad ......(3)$$

Since $\alpha/2 \leq \beta$ is the limiting condition for wedging to take place.

$$F = \frac{T}{z R \sin \beta} = \frac{1}{z \sin \alpha/2} \cdot \frac{T}{R} \quad ......(4)$$

The distance a is given by

$$a = \left(R - \frac{dr}{2}\right) \cos \alpha - \frac{dr}{2} \quad \ldots\ldots\ldots(5)$$

The tolerance on the values a, R and D will decide the deviation from the intended initial wedging angle α.

To simplify the calculation of roller dimensions and permissible load transmitted by one roller the following relationship is introduced.

Resultant force $F = p_o l \, d_r \ldots\ldots(6)$

d_r should be 1/3 to 1/5 R

$P_o = 5 kgf/mm^2$, for hardened steel (HRC 63−64)

Therefore $d_r l = 2 \times 10^{-1} F \ldots\ldots(7)$

To suit this product a roller size is chosen from the Table 21.

STRESS IN THE OUTER RING:

The outer ring is stressed in combined tension and bending. The maximum stresses occur at the outer fibre at point 1 (Fig. 14), and at inner fibre at point 2, however the magnitudes of these stresses are different.

The thickness of the ring should be such that the combined stresses do not exceed the permissible value $f_{all} = 25 \, kgf/mm^2$.

To simplify the procedure the thickness is calculated considering the maximum bending moment at point 1 with the allowable stress in bending for the given materials $f_b = 20 \, kgf/mm^2$

$$t = \frac{KF_n}{2b} + \sqrt{\frac{KF_n D}{b}} \quad \ldots\ldots\ldots\ldots(8)$$

where K the constant which is given in Table 22 corresponding to different numbers of rollers and calculated for $f_b = 20 \, kgf/mm^2$.

Table 21   All dimensions in mm   **Table 22**

d_r	l	d_r	l
3	5	12	18
4	6	13	20
4	8	14	20
5	8	15	22
5	10	16	24
5.5	8	17	24
6	8	18	26
6	12	19	28
6.5	9	20	30
7	10	22	34
7	14	23	34
7.5	11	24	36
8	12	25	36
9	14	26	40
10	14	28	44
11	15	30	48

z	K
3	2.85×10^{-4}
4	2.05×10^{-4}
5	1.61×10^{-4}
6	1.34×10^{-4}
8	0.99×10^{-4}
10	0.79×10^{-4}
12	0.66×10^{-4}

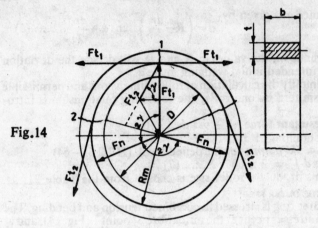

Fig.14

Then the following equations could be used to find the stress at point 1.

$$F_{t1} = 0.5 \, F_n \, \tan\gamma \qquad \dots\dots\dots(9)$$

$$f_t = F_{t1} \quad bt \qquad \dots\dots\dots(10)$$

$$M_1 = R_m \left(\frac{z}{2\pi} F_n - F_{t1} \right) \qquad \dots\dots(11)$$

and $f_b = 6M_1/bt^2$.(12)

Combined stress at point 1, $f_e = f_t + f_b \leq f_{all}$
also at point 2,

$$F_{t2} = 0.5 \, F_n / \sin\gamma \qquad \dots\dots(13)$$

$$f_t = F_{t2}/bt \dots\dots(14)$$

$$M_2 = R_m \left(\frac{z}{2\pi} F_n - F_{t2} \right) \qquad \dots\dots(15)$$

and $f_b = 6M_2/bt^2$ ......(16)

combined stress at point 2, $f_e = f_t + f_b \leq f_{all}$
In the above equations
$$R_m = \frac{1}{2} (D + t)$$

and $\gamma = 180/z$

OTHER CONSIDERATIONS:

Clutch life in terms of total number of engagements.
If the contact stress calculated from

$$P_{max} = 0.59 \, \sqrt{F_n \, E / l d_r} \qquad \dots\dots(17)$$

does not exceed 150 kgf/mm^2 and if the recommended materials are used with the corresponding values of surface hardness, a life of approximately 5×10^6 engagement cycles can easily be obtained. For precision applications the contact stress to be $\not> 120$ kgf/mm^2.

MAXIMUM RECOMMENDED OVER-RUNNING SPEED:
While overrunning, the rollers and the rings are subjected to wear due to the slipping of the rollers over contact surfaces. The recommended maximum sliding speed is $4 m/sec.$ at the working surface of the outer ring. For higher speeds, designs which are contactless while over-running are to be used (Fig. 15).

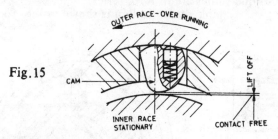

Fig. 15

FREQUENCY OF ENGAGEMENT:
The dynamic wedging action of the rollers increases the normal force at very high frequency of engagement cycles. Table 23 gives the recommended maximum frequency of operation and the values of maximum average slip during engagement.

Table 23

Nominal dia. D, mm	32	40	50	65	80	100	125	160	200
Max. no. of engagements per min.	250	200	160	125	100	80	65	50	40
Max. slip*	3^0	3^0	$2^0 30'$	2^0	$1^0 30'$	1^0	$45'$	$45'$	$30'$

*Based on a value of $\alpha = 6°$

SPRINGS:
Individual springs for each roller are suggested so as to eliminate non-uniform bearing resulting from dimensional deviations of the clutch elements. The spring force should be slightly more than the one required to overcome frictional resistance, weight of the roller and the effect of its centrifugal force. In clutches used for indexing, larger spring forces should be used to minimise the lost motion.

The torque transmission to the inner hub is usually through a key and that from the outer ring is preferably by facial slots. The fits should be tight fit to press fit. For lubrication, acid and water free oil with viscosity between 20 to 37 cSt (20 to 37 mm^2/sec) at 50° C should be used. Upto $2m/sec$ sliding velocity at the inner ring ball bearing grease could be used.

The dimensions of the clutch for a given transmitting torque T and a chosen outside diameter D_o can be found from Table 24. These dimensions are safe from material strength point of view and ensure a reliable functioning of the clutch and hence can be used whenever possible.

The torque values are in R10, series.

A typical example of application is shown in Fig. 16.

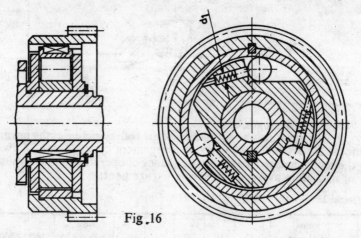

Fig. 16

Table 24 Dimensions of clutches

Maximum torque T kgf. mm $\times 10^2$	Nominal diameter D $H7$ mm	Outside diameter D_o, mm (minimum size)	Rollers Diameter d_r mm	Length l mm	Number z	Diameter of pin d_1, mm	0 a -0.05
6.3	40	55	5	8	3	3	14.85
12.5	40	55	6.5	9	3	3	13.34
25	50	65	8	12	3	4	16.8
40	50	65	8	12	5	4	16.8
63	56	70	9	14	5	4	18.77
100	70	85	11	15	5	5	23.7
160	70	85	12	18	6	6	22.7
250	90	110	14	20	6	6	30.6
400	100	120	16	24	6	8	33.6
630	120	140	16	24	8	8	43.5
1000	140	165	19	28	8	8	50.4
1600	160	190	22	34	8	8	57.3

SAFETY CLUTCHES

Safety clutches are used to disengage the driven part from the driving part when the transmitted load or torque exceeds a permissible limit. The precision with which the safety clutches limit the transmitted torque is the index of their accuracy.

Safety clutches are commonly classified as:

Release type: which disengage by snap-action when over-loaded and require manual re-engaging when the overload is released. The associated axial motion during snap action disengagement can be made use of to operate a limit switch for disconnecting the power supply to the electric motor and/or for actuating a visual or audible alarm signal.

Slipping type: which slip when overloaded and re-engage automatically the moment overload is released.

Slipping type clutches are subjected to heavy wear and heat generation depending upon the duration and frequency of overload. The following types are commonly used.

(a) toothed type, (b) ball type, (c) friction disc type (refer chapter friction clutches)

Toothed type safety clutches:

The basic design formulae to be used for this type of clutch are given in section *toothed clutches*. Certain specific points to be observed when these are used as safety clutches are given below.

Nomenclature:

d Shaft diameter, *mm*

D_m Mean diameter of the teeth, *mm*

$\mu_1 = tan\phi_1$, Coefficient of friction at the clutch sliding surface on the shaft

$\mu_2 = tan\phi_2$, Coefficient of friction at the teeth

F_{dis} Spring force while disengaging, *kgf*

F_1 Spring force required in the absence of friction, *kgf*

L Length of the sliding member, *mm*

T_{dis} Limiting torque at which the clutch commences to disengage (nominal rating of the clutch), *kgf. mm*

β_1 Tooth flank angle, *degrees*

$$T_{dis} = F_{dis} \cdot \frac{D_m}{2} \left[tan(\beta_1 - \phi_2) - \frac{D_m \mu_1}{d} \right]$$

Actual disengaging torque may vary considerably from the nominal value (i.e., it may increase or decrease) depending upon

    i) the condition of lubrication

    ii) pitch accuracy of the teeth

iii) error of co-axiality of the driving and the driven shafts

iv) L/d ratio of the sliding member and its fit on the shaft

A provision is usually made for variation of the spring force for setting a wide range of limiting torque.

Tooth flank angle β_1 for maximum efficiency is given by

$$\eta_{max} = \frac{F_{dis}}{F_1} = \frac{tan(\beta_1 - \phi_2) - (D_m \mu_1 / d)}{tan\beta_1}$$

$\mu_1, \mu_2 = 0.1$ for lubricated condition.

0.15 to 0.3 for poorly lubricated condition.

In practice the value of β_1 is selected between 45^0 and 60^0 (larger values for poor lubrication conditions), 45^0 being very common as it requires about half the spring force necessary for a clutch with $\beta_1 = 60^0$ and gives a satisfactory performance. Angles below 30^0 may lead to self locking even at very low spring force.

The length of the sliding member is chosen to provide an L/d ratio of 1.5 to 2 to eliminate self locking. A fit of $H7/f7$ is recommended between the sliding member and the shaft. The minimum number of teeth recommended is 3. Higher number of teeth require increased precision in manufacture. Figure 17 shows an example of application.

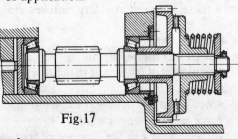

Fig.17

Ball type clutches:

Ball type clutches are used with advantage where frequency of overloading is less and loads are relatively small. When used for larger loads or frequent operations, a premature failure of the clutch may occur owing to considerably high contact stresses.

Ball type clutches are easier to manufacture. The friction conditions are more favourable (combined rolling and sliding) than in a toothed type clutch and the disengagement torque is almost independent of the lubricating conditions.

An insert pad is recommended to be interposed between the ball and the spring to prevent the end coils of the spring moving into the clearance around the ball causing a seizure and rendering the clutch inoperative.

Nomenclature:

μ Coefficient of friction (assumed same at all sliding surfaces)
F_t Tangential force on the ball, *kgf*
P Spring force in assembled condition, *kgf*
R Pitch circle radius of the balls, *mm*
T Maximum transmitted torque, *kgf. mm*
z Number of springs
α Half cone angle of the ball seating, *degrees*

Design calculations:

A force analysis of the ball in Fig. 18 gives

$$P = F_t \left[\frac{\sin\alpha - \mu\cos\alpha}{\mu\sin\alpha + \cos\alpha} - \mu \right]$$

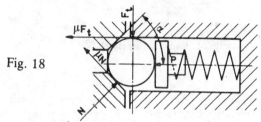

Fig. 18

The above equation can be written as

$$P = F_t \, C$$

where C is a constant for the clutch. If $\mu = 0.1$ we get the following values for C for different values of α.

Table 25

α	30°	37° 30'	45°	60°
C	0.35	0.52	0.72	1.3

A value of $\alpha = 45°$ is recommended.

F_t is calculated as $F_t = T / Rz$

The three variants of the ball type clutch are shown in Figs.19, 20 and 21.

For the type shown in Fig.20, the spring force P $CF_t/2$ as the torque is transmitted by two rows of balls.

Basic parameters for ball clutches are tabulated in Table 26 The values correspond to Figs. 22, 23 and 24.

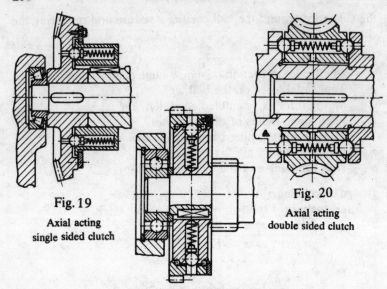

Fig. 19

Axial acting
single sided clutch

Fig. 20

Axial acting
double sided clutch

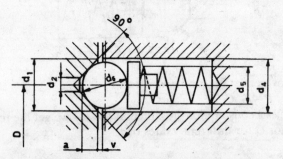

Fig. 21. Radial acting clutch

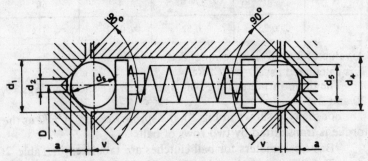

Fig. 22. Axial acting single sided clutch

Fig. 23. Axial acting double sided clutch

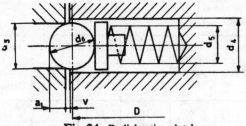

Fig 24 Radial acting clutch

Table 26

Size	Pitch dia. $D\ mm$	Spring force $Pkgf$	Ball dia d_6 inch	No. of balls		Nominal torque $kgf.\ mm$	
				Min.	Max.	Min.	Max.
1	40	2	3/8	2	8	100	400
		4				200	800
		8	5/8		4	400	800
		14.5				800	1600
2	50	2	3/8	2	10	125	630
		4				250	1250
		8	5/8		6	500	1600
		14.5				1000	3150
3	63	2	3/8	2	12	160	1000
		4				315	2000
		8	5/8		8	630	2500
		14.5				1250	5000
4	80	2	3/8	2	16	200	1600
		4				400	3150
		8	5/8		10	800	4000
		14.5				1600	8000

Table 26 (contd.)

Size	Pitch dia. $D\ mm$	Spring force $Pkgf$	Ball dia. d_6 inch	No. of balls Min.	Max.	Nominal torque kgf. mm Min.	Max
5	100	2	3/8	2	20	250	2500
		4				500	5000
		8	5/8		12	1000	6300
		14.5				2000	12500
6	125	2	3/8	2	24	315	4000
		4				630	8000
		8	5/8		16	1250	10000
		14.5				2500	20000
7	160	2	3/8	2	32	400	6300
		4				800	12500
		8	5/8		20	1600	16000
		14.5				3150	31500
8	200	2	3/8	2	40	500	10000
		4				1000	20000
		8	5/8		28	2000	28000
		14.5				4000	56000
9	250	2	3/8	2	—	630	—
		4				1250	—
		8	5/8		36	2500	45000
		14.5				5000	90000

Table 26 (concld.)

Size	Pitch dia. D, mm	Spring force $P\,kgf$	Ball dia. d_6 inch	No. of balls Min.	No. of balls Max.	Nominal torque kgf. mm Min.	Nominal torque kgf. mm Max.
10	315	2	3/8	2	—	800	—
		4	3/8			1600	—
		8	5/8		45	3150	71000
		14.5	5/8			6300	140000

Table 27 Dimensions of ball, detent, etc.

Ball diameter, d inch.	Ball diameter, d mm	d_1	d_2	d_3	d_4	a	a_1	V	d_5	P	F_t
3/8	9.525	8	5	6.75	10	2	1.5	0.5	6.3	2 4	2.5 5
5/8	15.875	13	8	11.25	16.5	3.2	2.4	0.8	12.6	8 14.5	10 18

Suitable length of spring is selected as required. Refer *Helical compression springs*

FRICTION CLUTCHES

Nomenclature:

a	Arm length of operating lever, mm
b	Active width of friction surfaces, mm
b_2	Width of cross section of operating lever, mm
b_2'	Width of hub surface in common with pin, mm
c	Thickness of the friction disc, mm
c_1	Thickness of end pressure plate, mm
C_1	Driving system coefficient (**Table** 28)
C_2	Driven system coefficient (**Table** 29)
C_3	Wear coefficient (**Table** 30)
C_4	Frequency coefficient (**Table** 31)
C	Sum total of coefficients $C_1 + C_2 + C_3 + C_4$
d	Diameter of shaft, mm
d_1	Diameter of the lever pin, mm
D_1	Diameter at which torque is transmitted in hub slot, mm
e	Bearing length between lug of disc and slot in hub, mm
E	Modulus of elasticity, kgf/mm^2
f_{ba}	Allowable bending stress, kgf/mm^2
f_{ca}	Allowable bearing stress between pin and lever, kgf/mm^2
f_{ca}'	Allowable bearing stress between pin and hub surface, kgf/mm^2
f_{ca}''	Allowable bearing stress between lug of disc and hub slots, kgf/mm^2
f_{sa}	Allowable shear stress for the pin, kgf/mm^2
f_{ta}	Allowable torsional stress, kgf/mm^2
F_a	Axial force between friction surfaces, kgf
F_n	Normal force between the surfaces in friction contact, kgf
F_p	Force on the lever pin, kgf
F_z	Force required to shift the lever during engagement, kgf
h_2	Height of cross section of operating lever, mm
i	Number of friction surfaces
i_l	Number of operating levers
i_t	Number of lugs in friction discs
K	Speed factor
K_1	Engagement factor
l	Arm length of operating lever, mm
m_1	Number of driving discs
m_2	Number of driven discs
n	Speed, rpm

p Pressure between friction discs, kgf/mm^2
p_a Allowable pressure between friction discs, kgf/mm^2
p_b Basic allowable pressure, kgf/mm^2 (Table 32)
P Power to be transmitted, kW
r_o Outer radius of friction disc, mm
r_i Inner radius of friction disc, mm
$r_m = \left(r_o + r_i\right)\big/ 2$ Mean radius, mm
T_d Design torque, $kgf.mm$
T_r Rated torque, $kgf.mm$
α Half cone angle, *degrees*
μ Coefficient of friction between friction surfaces (**Table 32**)
ψ Ratio b_2/h_2

In friction clutches the torque is transmitted due to axial pressure between the friction discs. A friction clutch connects a rotating shaft of a machine to a stationary shaft so that it may be made to rotate at the speed of the rotating shaft without shock and transmit torque. The torque which can be transmitted depends upon the coefficient of friction of the materials in contact, the normal pressure on these materials, the mean radius of the tangential force and the ability of the clutch elements and the housing to dissipate the heat generated during slipping period. The two common types of friction clutches discussed here are

    i) Disc clutches and
    ii) Cone clutches.

These friction clutches may also be used as brakes.

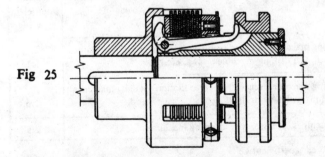

Fig 25

Disc clutches: (Fig.25)

Disc clutches are widely used in industry. They present a large friction area in a small space and establish a uniform pressure distribution for effective torque transmission and are not subjected to centrifugal effects.

The multiplate disc clutch consists of a series of friction discs alternately connected to the driving and the driven shafts. One set is directly connected to the driven shaft by means of splines while the mating set is keyed to the inner surface of the hub which is in turn rigidly connected to the driving shaft. The contact surfaces between the discs provide for necessary friction surfaces.

The axial force necessary, to establish tangential friction between the friction discs for the transmission of torque, is applied through an end pressure plate by means of springs or toggles in the assembly.

DESIGN TORQUE.

$$T_d = C \, T_r \quad \ldots\ldots(1)$$
$$\text{where } T_r = 974000 \, P/n \quad \ldots(2)$$

Table 28 Driving system coefficient C_1

Type of driving system	Coefficient C_1	
	For direct coupling or when a gear box is used	When a Vee or Flat belt transmission is used
Electric motor directly connected to mains	0.5	0.33
Machine with slow starting torque characteristic such as ring wound electric motor and line transmissions	0.33	

Table 29 Driven System Coefficient C_2

Type of driven system	Coefficient C_2
Metal cutting machine tools and wood working machine tools (lathes, milling machines, drilling machines, saws, etc.)	1.25
Metal cutting machine tools with reverse motion (planing machines), heavy machinery, heavy drills	1.6
Forging machines, presses, wire drawing machines.	2.5

Table 30 Wear coefficient C_3

Speed rpm	100	160	240	400	620	1000	1400	1800
C_3	0.1	0.13	0.16	0.2	0.25	0.32	0.38	0.43

Table 31 Frequency coefficient for sliding of friction surfaces C_4

Number of engagements during a normal 8 hr daily run.	1	8	16	32	48	96	240	480
C_4	0	0.2	0.55	0.75	0.9	1.2	1.8	2

Table 32 Allowable pressure and coefficient of friction

Friction pair	Running condition	Allowable bearing pressure p_b kgf/mm²	Coefficient of friction
Hardened steel/ Hardened steel	Wet Dry	0.06 to 0.08 0.025 to 0.03	0.05 to 0.08 0.18
Steel/Sintered Bronze	Wet Dry	0.05 to 0.2 0.05 to 0.08	0.06 to 0.08 0.12 to 0.18
Steel/Ferrodo	Dry	0.02 to 0.025	0.25 to 0.45
Steel/Fibre	Dry	0.035 to 0.04	0.2
Cast Iron/ Cast Iron	Dry Wet	0.025 to 0.03 0.06 to 0.08	0.15 —

BASIC PARAMETERS: (Fig 26)

The number of friction surfaces required in the clutch to transmit torque T_d are given by

$$i \; = \; T_d \; / \; 2\pi.p.b\mu \; r_m^2, \qquad \text{where } b = r_o - r_i \quad ....(3)$$

and the axial force required for gripping the discs is given by

$$F_a \; = \; p\pi(r_o^2 - r_i^2) \qquad\qquad ....(4)$$

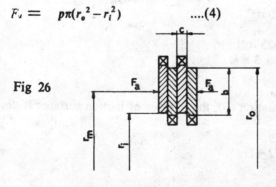

Fig 26

Table 33 Speed factor

Peripheral speed at mean radius r_m m/sec.	Speed factor K
upto 2.5	1
2.5 to 5	0.8
5 to 10	0.65
10 to 15	0.55

If the frequency of clutch engagement is high, about 50 to 100 times per hour, the allowable pressure p_a is further reduced by about 25% of the value obtained by equation (5).

PRESSURE BETWEEN DISCS:

The pressure between discs depends on the material of sliding surfaces, the peripheral speed of clutch at the mean radius r_m and also on the condition whether it is running wet or dry.

$$p_a = K.p_b \qquad ....(5)$$

Values of p_b and K are taken from Tables 32 and 33 respectively.

FRICTION COEFFICIENT FOR DISC SURFACES:

Friction characteristics depend on material pairing of the friction surfaces, design of friction faces, surface finish, sliding speed, temperature existing on the disc surface and lubrication conditions. Values of friction coefficient for different materials in surface contact (friction pair) running wet or dry is given in Table 32.

DESIGN OF CLUTCH PARTS:

Shaft diameter:

$$d = \left[\frac{495 \times 10^4 \times P \times C}{n \times f_{ta}} \right]^{1/3} \qquad ......(6)$$

Disc Dimensions:

$r_i \approx 2d$

$r_o \approx (1.25 \text{ to } 1.8) r_i$

$c \approx 1 \text{ to } 3 \text{ mm}$ ....(7)

$c_1 \approx 3c$

Number of discs:

From equation (3), the number of friction surfaces is determined, then

$m_1 = i/2$

$m_2 = i/2 + 1$ ....(8)

For smoother engagement of discs and small axial force F_a for proper gripping, the number of discs should be large. However, too many discs may result in rubbing in the disengaged condition and lead to heat generation. A decision about the number of discs therefore depends upon other considerations such as space, cost, etc.

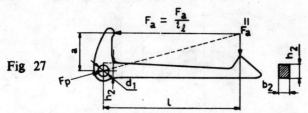

Fig 27

Dimensions of operating lever for manual operation (Fig. 27):

Dimensions a and l are fixed during design after studying the space limitations.

$$h_2 = \left[\frac{60 F_a \cdot a}{f_{ba} \cdot \psi i_l}\right]^{1/2} \quad \ldots\ldots(9)$$

$$b_2 = \psi \, h_2 \qquad \ldots\ldots(10)$$

values of ψ may be taken to be between 0.8 and 1

With increasing wear in discs, the total thickness of the plates gradually gets reduced and hence the gripping force F_a in the engaged condition also gets reduced and may fall to such a value that the clutch starts slipping. This calls for a readjustment of plates. To avoid a rapid decrease in F_a it is recommended to design the lever such that in the engaged condition the deflection at the end of its longer arm is in the range of 0.5 to 1.5 mm.

$$y = \frac{6 F_a \, a l^2}{E \, b_2 h_2^3 i_l} = 0.5 \text{ to } 1.5 \text{ mm} \qquad \ldots\ldots(11)$$

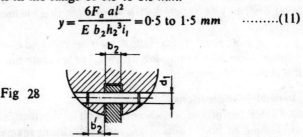

Fig 28

Design of lever pin: (Fig. 28)

The diameter of the lever pin in double shear is given by

$$d_1 = \left[\frac{2 F_p}{\pi f_{sd}}\right]^{\frac{1}{2}} \quad \ldots\ldots(12), \qquad \text{where } F_p = \frac{F_a}{i_l \cdot l} \sqrt{a^2 + l^2} \quad \ldots\ldots(13)$$

The pin diameter for bearing against lever

$$d_1 = F_p \,/\, b_2 f_{ca} \qquad \qquad \dots\dots\dots(14)$$

and for bearing against hub surface

$$d_1 = F_p \,/\, 2b_2' f_{ca}' \qquad \qquad \dots\dots\dots(15)$$

The larger of the values of d_1 thus obtained from equations 12, 14 and 15 is chosen as the design diameter of the pin.

Fig 29

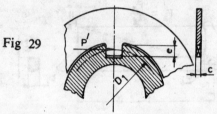

Bearing pressure between the lugs of discs and hub slots (Fig. 29)

$$f''_{ca} = \frac{P'}{ce} = \frac{2T_d}{m_1 i_i' ce D_1} \leqslant 2-2 \cdot 5 \qquad \dots\dots(16)$$

Force required on the shifter for operating the lever during clutch engagement:

Resistance to the clutch engagement is offered by the friction force between the levers and shifter ring. Hence, the force required to operate the lever during engagements is given by

$$F_z = F_a \, a \, \mu \,/\, l \qquad \qquad \dots\dots\dots(17)$$

To get a gradual increase of the force on the lever, the shifter ring is provided with a cam surface at its actuating point. Alternatively the lever end may be tapered with the shifter ring surface remaining cylindrical.

For large sized clutches the actuating lever end is provided with a roller. Normally springs are provided to separate the plates when the clutch is disengaged.

Thermal considerations:

Clutch performance is also influenced by the heat generated due to rubbing of plates during engagement, disengagement and idle running conditions. In critical cases, clutch design has to be checked for these effects.

Hydraulic, Pneumatic and Electromagnetic clutches:

The basic calculations for hydraulic and pneumatic clutches are the same as that for mechanically operated clutches with the difference that the engagement and disengagement operations are performed by hydraulic or pneumatic means, instead of mechanically operated levers.

A typical drawing of a multidisc friction clutch with hydraulically operated multidisc friction brake is shown in Fig. 30

Multidisc friction clutches operated by energizing and de-energizing electromagnets are known as electro-magnetic clutches and are also widely used in machine tools.

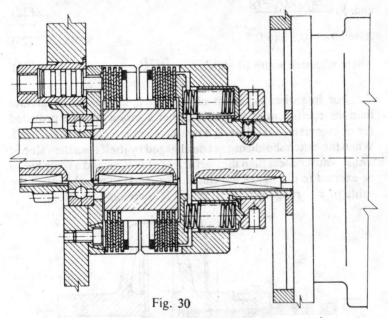

Fig. 30

Cone Clutches:

In cone clutches (Fig. 31) the wedging action of the parts increases the normal force on the lining which results in increased tangential friction force and the torque.

Assuming that the normal wear is proportional to the product of normal pressure and the radius

$$T_r = \mu \frac{(r_o + r_i)}{2K_1 \sin\alpha} F_a \qquad \qquad ....(18)$$

where K_1 is the engagement factor, the value of which varies from 1.3 to 1.5 for metal cutting machine tools.

If the clutch is engaged when both the driving and the driven parts are stationary,

$$F_a = F_n \cdot \sin\alpha . \qquad \qquad ....(19)$$

However if one of the parts is rotating with the other stationary

$$F_a = F_n(\sin\alpha + \mu\cos\alpha) \qquad \dots(20)$$

Experiments have shown that the term $\mu\cos\alpha$ is only partially effective, therefore

$$F_a = F_n\left[\sin\alpha + \frac{\mu\cos\alpha}{4}\right] \qquad \dots(21)$$

and $F_n = \dfrac{2\pi p_a\, r_i(r_o - r_i)}{\sin\alpha} \qquad \dots(22)$

Also $b/r_m = 0.3$ to 0.5 $\qquad\qquad \dots(23)$

where effective width of clutch $b = \dfrac{r_o - r_i}{\sin\alpha} \qquad \dots(24)$

For free disengagement of the clutch, $\tan\alpha$ must be greater than the coefficient of friction. Further a spring must be provided for engagement, for exerting force F_a when the clutch is engaged. When the clutch should not get disengaged by itself then the value of $\tan\alpha$ must be less than the coefficient of friction and a force has to be exerted to disengage the clutch. For metal to metal contact the value of α ranging from 8° to 10° is commonly adopted.

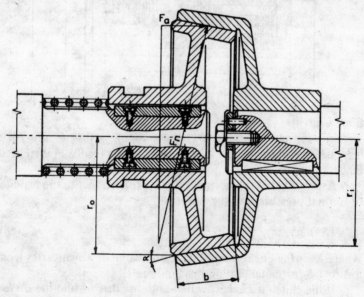

Fig. 31

BRAKES

The function of brakes is to bring to rest quickly the spindle and other transmission devices, for the purpose of shifting gears, to engage speed, to change workpieces, to change tools or any similar functions. Brakes may not be necessary where spindle speeds are low and the transmission devices use only sleeve bearings or when the cutting time is sufficiently large compared to the time lost during the spindle coasting to rest. Modern high speed machine spindles provided with antifriction bearings have a considerably reduced internal resistance. Further it is very common to mount heavy masses on the spindles resulting in an increased kinetic energy. These considerations call for the provision of brakes a must in modern machine tools.

The equation of motion during braking is

$$(M_b + M_r)\, dt = I_r d\omega$$

where M_b is the braking moment, M_r is the braking moment due to internal resistance of the transmission system, both reckoned at the axis of the braking shaft. I_r is the moment of inertia of the moving system also reckoned to the axis of the braking shaft.

$$I_r = \Sigma_1^n \frac{I_1 \omega_1^2}{\omega_b}$$

where $I_1, I_2, ... I_n$ are the moments of inertia of the rotating masses to be braked and $\omega_1, \omega_2, ... \omega_n$ are the respective angular velocities and ω_b is the angular velocity of the shaft to be braked.

The time required for braking is given by

$$T_b = \frac{\Sigma_1^n I_1 \omega_1^2}{M_b \omega_b \left[1 + \dfrac{M_r}{M_b} \right]}$$

From this it is evident that the braking time is smaller in the following cases:

  i) smaller the kinetic energy of the mass to be braked
 ii) greater the braking moment
iii) greater the angular velocity of the shaft to be braked
iv) greater the internal resistance of the system.

It is therefore advantageous to position the brake on the shaft rotating at the highest speed. Normally, the brakes are situated near the driving motor.

Various types of brakes are used in machine tools such as shoe, cone, disc, band, etc. They are operated either manually, pneumatically, hydraulically or electrically.

Figs. 32 and 33 show a shoe type and a band type brake respectively.

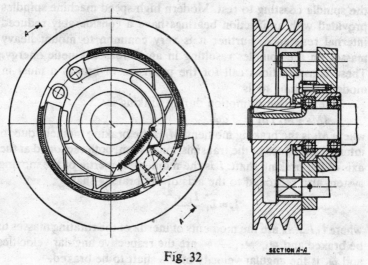

Fig. 32

SECTION A-A

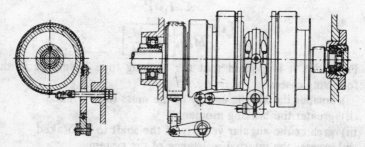

Fig. 33

UNIVERSAL JOINTS

Nomenclature:

f	Shock factor
f_α	Correction factor for torque transmitted
h	Life, *hours*
M	Torque to be transmitted, *kgf. m*
M_c	Design torque, *kgf. m*
n	Speed, *rpm*
P	Power to be transmitted, *kW*
P_c	Design Power, *kW*
α, β	Angle of inclination of shafts, *degrees*
η	Correction factor for power transmitted

Universal joint (also known as Hooke's joint):

This is a type of coupling used for connecting two shafts which are offset and parallel, and intersecting. Commercially many forms of this coupling are available. One well known form is shown in figure. A single universal joint is not used where absolute uniformity of motion is essential for the driven shaft. Lack of uniformity in the speed of driven shaft resulting from the use of a single universal joint may be avoided if two shafts are connected with an intermediate shaft and using two universal joints arranged properly.

Universal joints are normally manufactured either with plain bush bearings or with needle roller bearings. Table 34 shows an example of universal joint with leading dimensions for shaft diameters 6 to 50 *mm*. Material of the fork pieces of the joint should conform to steel with minimum tensile strength of 60 *kgf/mm²*.

Table 35 indicates the dimensions and strength requirements for taper pins required to fasten the shaft to the joint.

LOAD CARRYING CAPACITY:

Figure 34 shows torque and power transmission capacity of single universal joint with plain bearings in continuous duty for an angle of inclination $\alpha = 10°$ with respect to speed. For higher values of angle of inclination α, the power to be transmitted P is to be divided by correction factor η (Fig. 35) to arrive at design power P_c. For angle of inclination α between $0°$ and $5°$, $\eta = 1.25$; and for angle of inclination α between $5°$ and $10°$ correction factor is to be obtained by linear interpolation.

Figure 36 shows torque transmission capacity of single universal joint with needle bearings in continuous duty for an angle of inclination α upto $10°$ with respect to product of speed and the life of neeale bearing. In this case the torque to be transmitted M has to be multiplied by shock factor f (whose value varies from 1 to 3 depending upon the nature of load and location of the joint) and corection factor f_α (Fig. 37) for values of angle of inclination greater than $10°$ to arrive at design torque M

Double joints should be loaded only upto 90% of the values obtained for single joints.

Example - 1 (Refer Figs. 34 and 35):
　　Power to be transmitted $P = 1.5\,kW$
　　Speed $n = 250\,rpm$
　　Angle of inclination　$\alpha = 22° \, 30'$
　　Type of joint: Joint with plain bush bearings
　Procedure: Correction factor from Fig. 35　$\eta = 0.45$,
Design power $P_c = \dfrac{P}{\eta} = \dfrac{1.5}{0.45} = 3.3kW$

Figure 34 gives for $P = 3.3\,kW$ and $n = 250\,rpm$, single universal joint of size 32 x 63 or 40 x 63 which has a permissible torque of 12.8 *kgf.m*

Example - 2　(Refer Figs. 36 and 37)
　　Torque to be transmitted $M = 7\,kgf.m$
　　Speed $n = 1400\,rpm$
　　Life $h = 500\,hours$
　　Angle of inclination $\alpha = 20°$
　　Type of joint: Joint with needle bearings
　　Shock factor $f_z = 1.5$
　Procedure: Correction factor $f_\alpha = 1.1$ (from Fig. 37)
Design torque $M_c = M \cdot f_z \, f_\alpha$
　$7 \times 1.5 \times 1.1 = 11.6\,kgf.m$
$h \times n = 500 \times 1400 = 70 \times 10^4$

Fig. 36 gives for $M_c = 11.6\,kgf.m$ and $hn = 70 \times 10^4$ single universal joint of size　32 x 63 or 40 x 63.

Guidelines for use:

The application of a single universal joint leads to non-uniform torque transmission, the non-uniformity increases with an increase in the angle of inclination α.

If this is not permissible, two single joints or a double joint has to be used. In case of two single joints it is to be ensured that the axes of the forks lie in the same plane in the intermediate shaft as in the case of double joint (Fig. 38a). Otherwise non-uniform torque transmission will occur.

For the same reason the angles of inclinations at both ends of

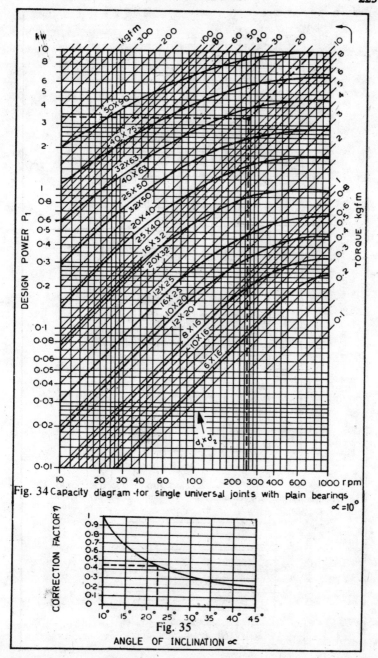

Fig. 34 Capacity diagram for single universal joints with plain bearings
$\alpha = 10°$

Fig. 35
ANGLE OF INCLINATION α

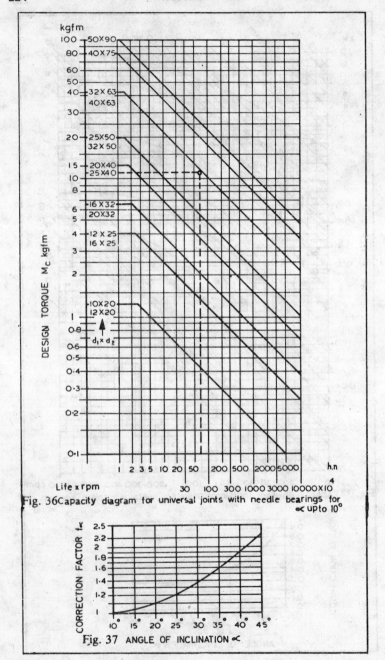

Fig. 36 Capacity diagram for universal joints with needle bearings for ∝ upto 10°

Fig. 37 ANGLE OF INCLINATION ∝

the intermediate shaft with the shafts to be coupled should be same
and should lie in the same plane (Figs. 39a, 39b and 39c). In case
where there is a likelyhood of movement in the positions of the
driving and driven shafts the movement should be such that the
shaft axes remain parallel to each other (Fig. 40)

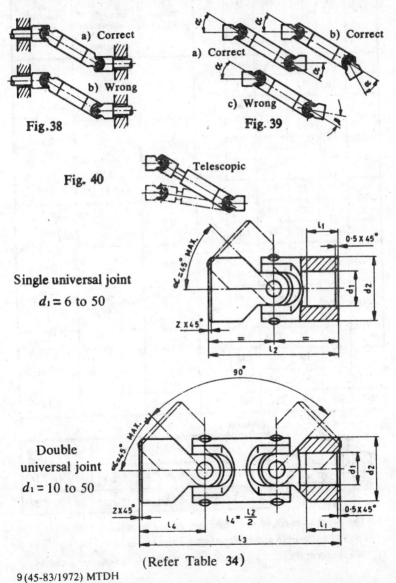

a) Correct

b) Wrong

Fig. 38

a) Correct

b) Correct

c) Wrong

Fig. 39

Fig. 40

Telescopic

Single universal joint
$d_1 = 6$ to 50

Double
universal joint
$d_1 = 10$ to 50

(Refer Table 34)

Table 34

d_1 H7	d_2 k11	l_1	l_2 +1	l_3 ±1	Z	Maximum allowable rotational play		Tolerance on coaxiality of the two bores
						Test torque kgf.m	Angular rotational play at an angle of inclination of 0° in minutes	
6		9	34	—	0.5	0.02	45	0.06
8	16	11	40	—				
10	20	15	52	74				
		13	48	—		0.04	40	
12	25	18	62	88				
		15	56	86				
16	32	22	74	104		0.1	32	0.09
		19	68			0.17	28	
20	40	25	86	124	1			
		23	82	128		0.34	25	
25	50	32	108	156				
		29	105	160		0.54	20	
32	63	40	132	188				
		36	130	198		1.5	18	0.12
40	75	50	166	238				
		44	160	245		2.2	16	
50	90	54	190	290		2.8	14	0.15

Table 35

d_1	6	8	10	12	16	20	25	32	40	50
d_3	2	3	4	5	6	8	10	12	14	16
W	4.5	5	6	7.5	9	11	15	18	22	27

1) The shear strength of the taper pin should be between 16 to 25 kgf/mm^2 in order to transmit the given torque.

2) The length of the taper pin should conform to dia. d_2 indicated in Table 34

BELTS

Three types of belts are normally used in power transmission.

1. Flat belts
2. V belts
3. Timing belts

Selection of the proper type of belt depends upon speed, load capacity, and maintenance requirements.

FLAT BELTS

The main difference between flat belts and other belts is that flat belts can provide a greater reduction ratio (20:1 and above) and operate under high speeds (about 50 m/sec.). Most efficient transmission is obtained at a speed range of 17.5 to 22.5 m/sec. Beyond a speed of 30 m/sec., there will be an excessive belt slip. Flat belts are used for power transmission between two or more parallel or skewed shafts, where a slight variation in transmission ratio is tolerable. Flat belts can also be used in crossed belt drives. Wherever possible pulley diameters larger than the recommended minimum should be used.

Leather, rubber, fabric and nylon cored chrome leather are some of the materials normally used for the manufacture of belts. Nylon cored chrome leather belts (also known as sandwiched belts) have a higher power transmission capacity due to their high coefficient of friction. The belts also possess excellent resistance to climatic conditions. For power ratings and other design parameters manufacturers catalogues may be referred.

V BELTS

Nomenclature:

C Centre distance, *mm*

d Pitch diameter of smaller pulley, *mm*

D Pitch diameter of larger pulley, *mm*

F_a Correction factor according to service, Table 44

F_c Length correction factor, Table 38

F_d Correction factor for arc of contact, Table 43

i Speed ratio

kW Rating of V belts, kW taking into account additional power per belt for given speed ratio, Tables 39 to 42A

L Pitch length of the belt, *mm*

n Rotational speed of low speed shaft, *rpm*

N Rotational speed of high speed shaft, *rpm*
N_b Number of belts
P_t Power to be transmitted, *kW*
v Velocity of the belt, *m/sec.*
Θ Arc of contact, degrees

V belts transmit a higher torque with lesser width and tension than normal flat belts due to the wedging action of the belt in the groove. This action enables V belts to be used at comparatively smaller arcs of contact than flat belts, permitting short centre distances and higher reduction ratios. V belts operate satisfactorily at speeds between 5 and 30 *m/sec.*, but maximum efficiency is obtained at speeds between 20 and 25 *m/sec.* Due to increased bending and windage losses combined with the loss due to wedging action, the efficiency of the drive is about 3 per cent lower than that for flat drives. Table 36 gives the formulae for calculation of various parameters in a V belt drive. Selection of the most favourable V belt section is facilitated by the use of Fig. 41. Nominal dimensions of standard V belts are given in Table 37. Belt sections smaller than A are also available and for details manufacturer's catalogues may be referred. Nominal lengths are indicated in Table 38. Power ratings are specified in Tables 39 to 42. Values of power correction factor F_d and service factor F_a are indicated in Tables 43 and 44 respectively.

The dimensions of the pulley groove is given in Tables 45 and 46. The pitch diameter of the pulley is measured over two rollers placed in the groove. Roller dimensions and the measurement on rollers are indicated in Table 48. The commonly used materials for pulley are cast iron, mild steel and aluminium.

Static balancing of pulleys is adequate for speeds below 10 *m/sec.* or when the ratio of face width to outside diameter is less than the value in relation to the peripheral speed as given in Table 47, otherwise dynamic balancing is necessary.

When the driver or the driven unit is not adjustable for belt tension a grooved idler is used on the inside of the drive. Such an idler should have a diameter larger than the smaller pulley on the drive. If the design does not permit a grooved idler, a flat back bent idler may be used but this reduces the efficiency and life of the belt.

Belts are marked with a symbol such as C 2032-50 indicating the belt cross section, the nominal inside length and a grading number. A grading number of 50 means that the measured pitch length of the belt equals the nominal pitch length. A deviation of

2.5 *mm* from nominal pitch length is represented by one unit and the grading number will increase or decrease, as the length is more or less, for example *C* 2032-51 and *C* 2031-49.

Belts running in multigrooved pulleys should be in matched sets to ensure even distribution of load. The grading number of the belt mark is referred to and the maximum variation in the length of the belt in a matched set should be as specified in Table 38.

Table 36

Sl.No.	Description	Equation
1	speed ratio, i	N/n
2	Pitch diameter of larger pulley, D	$i \times d$
3	Belt speed, v	$\dfrac{3.14 \times d \times N}{60,000}$
4	Pitch length of the belt, L	$2C + 1.57\,(D+d) + \dfrac{(D-d)^2}{4C}$
5	Centre distance, C	$A + \sqrt{(A^2 - B)}$ where $A = L/4 - 0.3925\,(D+d)$ $B = \dfrac{(D-d)^2}{8}$
6	Arc of contact, Θ	$2\cos^{-1}\left[\dfrac{(D-d)}{2C}\right]$
7	Number of belts, N_b	$\dfrac{P_t \times F_a}{kW \times F_c \times F_d}$
8	Installation and take up allowance for centre distance C Lower limit Upper Limit	 $-0.015L$ $+0.03L$

Table 37 Nominal dimensions of standard V belts   (All dimensions in *mm*)

	Belt section	Top width, W	Thickness, T
	A	13	8
	B	17	11
	C	22	14
	D	·32	19

Table 38 Nominal lengths and length correction factor: A, B, C & D section belts

Nom. inside length	Nom. lengths and length correction factors								Pitch length limits	Max. length variation in a matched set
	A		B		C		D			
mm	mm	Fc	mm	Fc	mm	Fc	mm	Fc	mm	mm
610	645	0.8							+11.4	
660	696	0.81							− 6.4	
711	747	0.82								
787	823	0.84								
813	848	0.85							+12.5	
889	925	0.87	932	0.81					− 7.5	
914	950	0.87	958	0.82						2.5
965	1001	0.88	1008	0.83						
991	1026	0.88	—	—					+14	
1016	1051	0.89	1059	9.84					− 8.9	
1067	1102	0.9	1110	0.85						
1092	1128	0.9	1136	0.86						
1168	1204	0.92	1211	0.87						
1219	1255	0.93	1262	0.88	1275	0.79				
1295	1331	0.94	1339	0.89	1351	0.8				
1372	—	—	1415	0.9	—	—			+16	
1397	1433	0.96	1440	0.9	1453	—			− 9	
1422	1458	0.96	1466	0.9	—	—				
1473	1509	0.97	—	—	—	—				5
1524	1560	0.98	1567	0.92	1580	0.82				
1600	1636	0.99	—	—	—	—				
1626	1661	0.99	1669	0.93	1681	0.83				
1651	1687	1	1694	0.94	—	—			+17.8	
1727	1763	1	1770	0.95	1783	0.85			−12.5	
1778	1814	1.01	1821	0.95	1834	—				
1905	1941	1.02	1948	0.97	1961	0.87				
1981	2017	1.03	2024	0.98	—	—				
2032	2068	1.04	2075	0.98	2088	0.88				
2057	2093	1.04	2101	0.98	2113	0.89			+ 30	7.5
2159	2195	1.05	2202	0.99	2215	0.9			−16	
2286	2322	1.06	2329	1	2342	0.91				
2438	2474	1.08	2482	1.01	2494	0.92				
2464	—	—	2507	1.02	—	—				
2540	—	—	2583	1.03	—	—			+33	
2667	2703	1.1	2710	1.04	2723	0.94			−18	
2845	2880	1.11	2888	1.05	2901	0.95				10
3048	3084	1.13	3091	1.07	3104	0.97	3127	0.86		
3150	—	—	—	—	3205	0.97	—	—		
3251	3287	1.14	3294	1.08	3307	0.98	3330	0.87	+38	
3404	—	—	—	—	3459	0.99	—	—	− 21	
3658	3693	1.14	3701	1.11	3713	1	3736	0.9		
4013	—	—	4056	1.13	4069	1.02	4092	0.92	+43−24	

Table 39 Power ratings, in kilowatts, for 'A' section V belts, (13mm wide)with 180° arc of contact on smaller pulley

Speed of faster shaft rpm	Smaller pulley diameter, mm								
	75	80	85	90	100	106	112	118	125
720	0.44	0.51	0.57	0.65	0.79	0.86	0.93	1.04	1.15
960	0.54	0.63	0.73	0.82	1	1.09	1.18	1.31	1.44
1440	0.73	0.86	0.99	1.12	1.38	1.5	1.63	1.81	2
2880	1.09	1.32	1.55	1.77	2.19	2.39	2.6	2.89	3.16
200	0.16	0.19	0.21	0.23	0.28	0.3	0.32	0.35	0.38
400	0.28	0.32	0.37	0.41	0.45	0.53	0.57	0.63	0.69
600	0.39	0.45	0.51	0.57	0.69	0.75	0.81	0.89	0.98
800	0.48	0.56	0.63	0.71	0.86	0.94	1.02	1 13	1.25
1000	0.56	0.66	0.75	0.85	1.04	1.13	1.22	1.36	1.49
1200	0.64	0.75	0.87	0.98	1.19	1.3	1.41	1.57	1.73
1400	0.72	0.84	0.97	1.1	1.35	1.47	1.6	1.77	1.95
1600	0.78	0.92	1.07	1.21	1.49	1.63	1.77	1.97	2.16
1800	0.84	1	1.16	1.32	1.63	1.77	1.92	2.15	2.35
2000	0.9	1.07	1.25	1.42	1.75	1.92	2.07	2.31	2.54
2200	0.95	1.14	1.33	1.51	1.86	2.04	2.21	2.47	2.71
2400	1	1.2	1.4	1.6	1.98	2.16	2.34	2.61	2.86
2600	1.04	1.26	1.47	1.67	2.07	2.27	2.45	2.73	3
2800	1.08	1.3	1.53	1.74	2.16	2.36	2.56	2.85	3.12
3000	1.11	1.35	1.58	1.8	2.24	2.45	2.65	2.94	3.21
3200	1.14	1.39	1.63	1.86	2.3	2.52	2.73	3.02	3.3
3400	1.16	1.42	1.67	1.91	2.36	2.58	2.79	3.09	3.36
3600	1.19	1.45	1.7	1.95	2.41	2.62	2.83	3.13	3.4
3800	1.19	1.46	1.72	1.98	2.45	2.66	2.87	3.16	3.42
4000	1.2	1.47	1.74	1.99	2.47	2.68	2.89	3.17	3.42
4200	1.2	1.48	1.75	2.01	2.48	2.69	2.89	3.15	3.39
4400	1.19	1.47	1.75	2.01	2.47	2.68	2.87	3.12	3.33
4600	1.18	1.46	1.74	2	2.45	2.65	2.83	3.07	3.26*
4800	1.16	1.45	1.72	1.98	2.42	2.61	2.78	2.99	3.15*
5000	1.14	1.42	1.69	1.95	2.38	2.56	2.71	2.89*	
5200	1.1	1.39	1.66	1.9	2.31	2.48	2.62		
5400	1.07	1.35	1.61	1.85	2.24	2.39	2.51*		
5600	1.02	1.3	1.56	1.78	2.14	2.27*	2.37*		
5800	0.97	1.25	1.49	1.7	2.03*	2.14*			
6000	0.9	1.17	1.41	1.61	1.9*				
6200	0.84	1.1	1.32	1.51	1.76*				
6400	0.76	1.01	1.22	1.39*					
6600	0.67	0.91	1.11	1.26*					
6800	0.58	0.81	0.98*	1.11*					
7000	0.48	0.69	0.84*						
7200	0.37	0.56*	0.69*						
7400	0.25	0.42*							
7600	0.11	0.27*							

Note: Consideration should be given to the loads that can be imposed on bearings and shafts etc., while using these ratings. *These values correspond to belt speeds above 30m/sec.

Table 39A Power ratings, in kilowatts, for 'A' section V belts,
(13*mm* wide) with 180° arc of contact on smaller pulley

Speed of faster shaft *rpm*	Additional power per belt for speed ratio								
	1.02 to 1.04	1.05 to 1.08	1.09 to 1.12	1.13 to 1.18	1.19 to 1.24	1.25 to 1.34	1.35 to 1.51	1.52 to 1.99	2 and above
720	0.008	0.015	0.022	0.637	0.045	0.06	0.067	0.082	0.089
960	0.016	0.03	0.045	0.052	0.067	0.082	0.089	0.104	0.119
1440	0.022	0.037	0.06	0.075	0.096	0.119	0.134	0.156	0.179
2880	0.037	0.075	0.119	0.156	0.2	0.24	0.28	0.32	0.36
200	0	0.008	0.008	0.008	0.015	0.015	0.022	0.022	0.022
400	0.008	0.008	0.015	0.002	0.03	0.03	0.037	0.045	0.052
600	0.008	0.015	0.022	0.03	0.045	0.052	0.06	0.067	0.075
800	0.008	0.022	0.03	0.045	0.052	0.067	0.075	0.089	0.096
1000	0.015	0.03	0.045	0.052	0.067	0.082	0.096	0.12	0.13
1200	0.02	0.03	0.05	0.07	0.08	0.1	0.12	0.13	0.15
1400	0.02	0.04	0.06	0.08	0.1	0.12	0.13	0.16	0.18
1600	0.02	0.04	0.07	0.09	0.12	0.13	0.16	0.18	0.2
1800	0.02	0.05	0.08	0.1	0.13	0.15	0.18	0.2	0.22
2000	0.03	0.05	0.08	0.12	0.14	0.16	0.19	0.22	0.26
2200	0.03	0.06	0.09	0.12	0.16	0.19	0.22	0.25	0.28
2400	0.03	0.07	0.1	0.13	0.16	0.2	0.23	0.27	0.3
2600	0.04	0.08	0.12	0.14	0.18	0.22	0.23	0.29	0.33
2800	0.04	0.08	0.12	0.16	0.19	0.23	0.28	0.31	0.35
3000	0.04	0.08	0.13	0.16	0.21	0.25	0.29	0.34	0.37
3200	0.04	0.09	0.13	0.18	0.22	0.27	0.31	0.36	0.4
3400	0.04	0.1	0.14	0.19	0.24	0.28	0.33	0.38	0.43
3600	0.05	0.1	0.16	0.2	0.25	0.3	0.35	0.4	0.45
3800	0.05	0.11	0.16	0.21	0.27	0.32	0.37	0.43	0.48
4000	0.05	0.12	0.16	0.22	0.28	0.34	0.39	0.45	0.5
4200	0.06	0.12	0.18	0.23	0.29	0.35	0.41	0.47	0.53
4400	0.06	0.12	0.19	0.25	0.31	0.37	0.43	0.49	0.55
4600	0.07	0.13	0.19	0.25	0.32	0.39	0.45	0.51	0.57
4800	0.07	0.13	0.2	0.27	0.34	0.4	0.47	0.54	0.6
5000	0.07	0.14	0.21	0.28	0.35	0.42	0.48	0.56	0.63
5200	0.08	0.14	0.22	0.29	0.37	0.43	0.51	0.58	0.66
5400	0.08	0.15	0.22	0.3	0.37	0.45	0.53	0.6	0.68
5600	0.08	0.16	0.23	0.31	0.39	0.47	0.54	0.63	0.7
5800	0.08	0.16	0.25	0.32	0.4	0.48	0.57	0.65	0.73
6000	0.08	0.16	0.25	0.34	0.42	0.5	0.59	0.67	0.75
6200	0.09	0.17	0.26	0.34	0.43	0.52	0.6	0.69	0.78
6400	0.09	0.18	0.27	0.36	0.45	0.54	0.63	0.72	0.81
6600	0.09	0.19	0.28	0.37	0.46	0.55	0.64	0.74	0.83
6800	0.1	0.19	0.28	0.38	0.48	0.57	0.66	0.76	0.85
7000	0.1	0.19	0.29	0.39	0.48	0.59	0.69	0.78	0.88
7200	0.1	0.2	0.3	0.4	0.5	0.6	0.7	0.81	0.9
7400	0.1	0.21	0.31	0.41	0.51	0.62	0.72	0.83	0.93
7600	0.1	0.21	0.32	0.43	0.53	0.63	0.74	0.85	0.95

Table 40

Power ratings, in kilowatts, for 'B' section V belts,
(17 mm wide) with 180° arc of contact on smaller pulley

Speed of faster shaft rpm	Smaller pulley pitch diameter, mm								
	125	132	140	150	160	170	180	190	200
720	1.36	1.48	1.66	1.96	2.13	2.36	2.6	2.83	3.1
960	1.69	1.85	2.08	2.45	2.68	2.98	3.27	3.56	3.91
1440	2.24	2.46	2.77	3.3	3.6	4	4.39	4.77	5.23
2880	3.03	3.34	3.79	4.48	4.85	5.32	5.72	6.08	6.43
200	0.5	0.54	0.6	0.69	0.75	0.82	0.89	0.97	1.05
400	0.87	0.94	1.04	1.22	1.33	1.46	1.6	1.74	1.91
600	1.19	1.29	1.45	1.7	1.85	2.05	2.24	2.44	2.98
800	1.48	1.61	1.81	2.13	2.33	2.57	2.83	3.08	3.39
1000	1.74	1.91	2.15	2.54	2.77	3.07	3.37	3.67	4.03
1200	1.99	2.18	2.45	2.9	3.17	3.52	3.87	4.18	4.62
1400	2.21	2.42	2.73	3.24	3.53	3.92	4.31	4.69	5.15
1600	2.4	2.63	2.98	3.53	3.86	4.29	4.71	5.11	5.6
1800	2.57	2.82	3.2	3.8	4.15	4.6	5.04	5.47	5.97
2000	2.71	2.98	3.39	4.02	4.38	4.86	5.32	5.76	6.27
2200	2.83	3.12	3.53	4.2	4.59	5.07	5.53	5.97	6.48
2400	2.93	3.22	3.65	4.34	4.73	5.22	5.68	6.11	6.59
2600	2.99	3.3	3.74	4.44	4.82	5.31	5.76	6.16	6.61
2800	3.03	3.34	3.79	4.48	4.86	5.33	5.76	6.13	6.51
3000	3.03	3.34	3.8	4.47	4.84	5.29	5.67	6	6.31*
3200	3	3.31	3.75	4.41	4.77	5.17	5.5	5.76*	
3400	2.94	3.24	3.68	4.3	4.62	4.97	5.24*	—	—
3600	2.84	3.14	3.54	4.12	4.4	4.69*	—	—	—
3800	2.7	2.98	3.36	3.89	4.11*	—	—	—	—
4000	2.53	2.79	3.14	3.58*	—	—	—	—	—
4200	2.31	2.55	2.86*	—					
4400	2.05	2.27*	2.86*	—					
4600	1.74*	1.93*	—	—					
4800	1.4*	—	—	—					

Note:

Consideration should be given to the loads that can be imposed on bearings and shafts etc., while using these ratings.

* These values correspond to belt speeds above 30 m/sec.

Table 40A

Power ratings, in kilowatts, for 'B' section V belts,
(17*mm* wide) with 180° arc of contact on smaller pulley

Speed of faster shaft rpm	Additional power per belt for speed ratio								
	1.02 to 1.04	1.05 to 1.08	1.09 to 1.12	1.13 to 1.18	1.19 to 1.24	1.25 to 1.34	1.35 to 1.51	1.52 to 1.99	2 and above
720	0.03	0.05	0.08	0.1	0.13	0.16	0.19	0.21	0.23
960	0.04	0.08	0.1	0.14	0.18	0.21	0.25	0.28	0.31
1440	0.05	0.1	0.11	0.21	0.26	0.31	0.37	0.43	0.48
2880	0.1	0.2	0.3	0.43	0.53	0.63	0.74	0.84	0.95
200	0.01	0.01	0.02	0.03	0.04	0.04	0.05	0.06	0.07
400	0.01	0.03	0.04	0.06	0.08	0.09	0.1	0.12	0.13
600	0.02	0.04	0.07	0.09	0.11	0.13	0.16	0.18	0.19
800	0.03	0.06	0.09	0.12	0.15	0.18	0.2	0.23	0.26
1000	0.04	0.08	0.11	0.15	0.19	0.22	0.25	0.29	0.33
1200	0.04	0.09	0.13	0.18	0.22	0.26	0.31	0.35	0.4
1400	0.05	0.1	0.16	0.2	0.25	0.31	0.36	0.41	0.46
1600	0.06	0.12	0.18	0.23	0.29	0.35	0.41	0.47	0.53
1800	0.07	0.13	0.19	0.26	0.33	0.39	0.46	0.53	0.59
2000	0.08	0.15	0.22	0.29	0.37	0.44	0.51	0.58	0.66
2200	0.08	0.16	0.24	0.32	0.4	0.48	0.56	0.64	0.72
2400	0.09	0.18	0.26	0.35	0.44	0.53	0.61	0.7	0.79
2600	0.1	0.2	0.28	0.38	0.48	0.57	0.66	0.76	0.86
2800	0.1	0.2	0.31	0.41	0.51	0.61	0.72	0.82	0.92
3000	0.11	0.22	0.33	0.44	0.55	0.66	0.77	0.88	0.98
3200	0.12	0.23	0.35	0.47	0.58	0.7	0.82	0.93	1.05
3400	0.13	0.25	0.37	0.5	0.62	0.75	0.87	0.99	1.12
3600	0.13	0.26	0.4	0.53	0.66	0.79	0.92	1.05	1.19
3800	0.14	0.28	0.42	0.56	0.69	0.84	0.97	1.11	1.25
4000	0.15	0.29	0.44	0.58	0.73	0.88	1.02	1.17	1.31
4200	0.16	0.31	0.46	0.61	0.77	0.92	1.07	1.23	1.38
4400	0.16	0.32	0.48	0.64	0.81	0.96	1.13	1.29	1.45
4600	0.17	0.34	0.51	0.67	0.84	1.01	1.18	1.34	1.51
4800	0.18	0.35	0.53	0.7	0.88	1.05	1.23	1.4	1.58

Table 41

Power ratings, in kilowatts, for 'C' section V belts,
(22 *mm* wide) with 180° arc of contact on smaller pulley

Speed of faster shaft rpm	Smaller pulley pitch diameter, *mm*								
	200	212	224	236	250	265	280	300	315
720	3.97	4.38	4.89	5.3	6.01	6.5	7.21	7.98	8.28
960	4.85	5.38	6.01	6.52	7.39	8.05	8.8	9.77	10.2
1440	6.14	6.81	7.68	8.28	9.4	10.1	11.1	12.1	12.5
100	0.82	0.89	0.98	1.05	1.18	1.27	1.39	1.53	1.6
200	1.45	1.59	1.75	1.89	2.12	2.28	2.51	2.77	2.9
300	2.01	2.21	2.45	2.64	2.98	3.21	3.54	3.91	4.1
400	2.53	2.78	3.09	3.34	3.77	4.08	4.5	4.98	5.22
500	3.01	3.32	3.7	4	4.52	4.88	5.4	5.98	6.26
600	3.46	3.82	4.26	4.61	5.22	5.65	6.25	6.92	7.23
700	3.89	4.29	4.79	5.19	5.88	6.37	7.04	7.83	8.2
800	4.28	4.74	5.29	5.74	6.5	7.04	7.83	8.58	9.02
900	4.65	5.14	5.76	6.24	7.08	7.68	8.43	9.4	9.84
1000	4.99	5.53	6.19	6.7	7.61	8.2	9.1	10.07	10.5
1100	5.29	5.88	6.58	7.14	8.13	8.72	9.69	10.7	11.1
1200	5.58	6.2	6.94	7.53	8.5	9.25	10.2	11.2	11.6
1300	5.84	6.48	7.26	7.9	8.95	9.62	10.6	11.6	12.1
1400	6.06	6.73	7.53	8.2	9.25	9.99	11	12	12.5
1500	6.26	6.94	7.83	8.43	9.54	10.3	11.3	12.2	12.7
1600	6.41	7.12	7.98	8.65	9.77	10.4	11.5	12.5	12.9
1700	6.54	7.26	8.2	8.8	9.92	10.6	11.6	12.5	13
1800	6.63	7.35	8.28	8.95	9.99	10.7	11.6	12.5	12.9
1900	6.69	7.42	8.35	8.95	9.99	10.7	11.6	12.5	12.7*
2000	6.7	7.43	8.35	8.95	9.99	10.6	11.3	12.1*	
2100	6.69	7.42	8.28	8.87	9.84	10.4	11.1*		
2200	6.62	7.35	8.2	8.72	9.69	10.2*			
2300	6.52	7.21	7.98	8.5	9.7*	9.84*	—	—	—
2400	6.38	7.04	7.76	8.28	9.02*	—	—	—	—
2500	6.18	6.82	7.46	7.98*	—				
2600	5.94	6.53	7.17	7.53*					
2700	5.65	6.2	6.76*						
2800	5.32	5.51*							
2900	4.92*	5.35*							
3000	4.48*								
3100	5 99*								

Note:

Consideration should be given to the loads that can be imposed on bearings and shafts etc., while using these ratings.

* These values correspond to belt speeds above 30 *m/sec*.

Table 41A

Power ratings, in kilowatts, for 'C section V belts,
(22 mm wide) with 180° arc of contact on smaller pulley

Speed of faster shaft rpm	Additional power per belt for speed ratio								
	1.02 to 1.04	1.05 to 1.08	1.09 to 1.12	1.13 to 1.18	1.19 to 1.24	1.25 to 1.34	1.35 to 1.51	1.52 to 1.99	2 and above
720	0.07	0.14	0.22	0.29	0.37	0.44	0.51	0.58	0.66
960	0.1	0.19	0.29	0.38	0.48	0.59	0.68	0.78	0.88
1440	0.14	0.29	0.44	0.59	0.73	0.88	1.03	1.17	1.32
100	0.01	0.02	0.03	0.04	0.05	0.06	0.07	0.08	0.09
200	0.02	0.04	0.06	0.08	0.1	0.12	0.14	0.16	0.19
300	0.03	0.06	0.09	0.12	0.15	0.19	0.22	0.25	0.28
400	0.04	0.08	0.12	0.16	0.2	0.25	0.28	0.33	0.37
500	0.05	0.1	0.15	0.2	0.25	0.31	0.36	0.41	0.45
600	0.06	0.12	0.19	0.25	0.31	0.37	0.43	0.48	0.55
700	0.07	0.14	0.22	0.28	0.36	0.43	0.5	0.57	0.64
800	0.08	0.16	0.25	0.33	0.41	0.48	0.57	0.65	0.73
900	0.09	0.19	0.28	0.37	0.45	0.55	0.64	0.73	0.82
1000	0.1	0.2	0.31	0.41	0.51	0.61	0.71	0.81	0.92
1100	0.11	0.22	0.33	0.45	0.56	0.67	0.78	0.89	1.01
1200	0.12	0.25	0.37	0.48	0.61	0.73	0.85	0.98	1.1
1300	0.13	0.26	0.4	0.53	0.66	0.79	0.92	1.06	1.19
1400	0.14	0.28	0.42	0.56	0.7	0.84	1	1.14	1.28
1500	0.15	0.31	0.45	0.61	0.76	0.92	1.07	1.22	1.37
1600	0.16	0.33	0.48	0.65	0.81	0.98	1.14	1.3	1.46
1700	0.17	0.34	0.51	0.69	0.87	1.04	1.21	1.38	1.66
1800	0.19	0.37	0.55	0.73	0.92	1.1	1.28	1.46	1.65
1900	0.19	0.39	0.58	0.78	0.97	1.16	1.35	1.54	1.74
2000	0.2	0.41	0.61	0.81	1.01	1.22	1.42	1.63	1.83
2100	0.22	0.43	0.64	0.86	1.07	1.28	1.49	1.71	1.92
2200	0.22	0.45	0.67	0.89	1.12	1.34	1.57	1.79	2.01
2300	0.22	0.47	0.7	0.93	1.17	1.4	1.63	1.87	2.1
2400	0.25	0.49	0.73	0.98	1.22	1.46	1.71	1.95	2.19
2500	0.25	0.51	0.76	1.01	1.27	1.52	1.78	2.04	2.29
2600	0.26	0.53	0.79	1.06	1.32	1.59	1.85	2.12	2.38
2700	0.28	0.55	0.82	1.1	1.37	1.65	1.92	2.19	2.47
2800	0.28	0.57	0.85	1.14	1.42	1.71	1.99	2.27	2.57
2900	0.3	0.59	0.89	1.18	1.48	1.77	2.07	2.36	2.65
3000	0.31	0.61	0.92	1.22	1.53	1.83	2.13	2.44	2.74
3100	0.31	0.63	0.95	1.26	1.57	1.89	2.21	2.52	2.83

Table 42   Power ratings, in kilowatts, for 'D' section V belts, (32 *mm* wide) with 180° arc of contact on smaller pulley

Speed of faster shaft rpm	Smaller pulley diameter, *mm*									
	350	375	400	425	450	475	500·	525	560	600
585	11.8	13.3	15	16.6	18.2	19.8	21.3	22.7	24.7	26.9
720	13.4	15.3	17.2	18.9	20.7	22.4	24.1	25.7	27.8	29.9
960	15.4	17.7	19.8	21.8	23.6	25.4	27	28.6	30.3	32.2*
1440	15.7	17.5	19.3*	20.6*	—	—	—	—	—	—
50	1.65	1.84	2.04	2.22	2.42	2.6	2.79	2.98	3.24	3.54
100	2.93	3.3	3.65	4.01	4.37	4.73	5.08	5.43	5.91	6.47
150	4.09	4.61	5.13	5.64	6.16	6.67	7.18	7.68	8.35	9.17
200	5.16	5.84	6.51	7.18	7.83	8.5	9.17	9.84	10.7	11.7
250	6.17	6.99	7.83	8.65	9.47	10.2	11	11.9	12.9	14.2
300	7.12	8.13	9.02	9.99	11	11.9	12.8	13.7	15.1	16.5
350	8.05	9.17	10.2	11.3	12.4	13.5	14.5	15.6	16.4	18.6
400	8.87	10.1	11.3	12.6	13.8	14.9	16.1	17.3	18.9	20.7
450	9.69	11.1	12.4	13.7	15.1	16.3	17.6	18.9	20.6	22.5
500	10.5	11.9	13.4	14.8	16.3	17.7	19.1	20.4	22.2	24,3
550	11.2	12.8	14.4	16	17.4	18.9	20.4	21.8	23.7	25.9
600	11.9	13.6	15.3	16.9	18.5	20.1	21.6	23.1	25.1	27.4
650	12.5	14.3	16.1	17.8	19.5	21.2	22.7	24.3	26.4	28.6
700	13.1	15.1	16.9	18.6	20.4	22.1	23.7	25.4	27.4	29.8
750	13.7	15.7	17.6	19.5	21.3	23	24.7	26.2	28.4	30.6
800	14.2	16.3	18.2	20.1	22	23.7	25.4	27.1	29.2	31.3
850	14.6	16.8	18.8	20.7	22.6	24.4	26.1	27.7	29.7	31.8
900	15.1	17.2	19.3	21.3	23.1	24.9	26.6	28.2	30.1	32.1
950	15.4	17.6	19.7	21.7	23.6	25.4	27	28.5	30.3	32.2
1000	15.7	17.9	20.1	22.1	23.9	25.7	27.2	28.7	30.4	32.1*
1050	16	18.2	20.3	22.3	24.2	25.8	27.3	28.7	30.3*	
1100	16.1	18.4	20.5	22.4	24.2	25.9	27.3	28.5*	29.9*	
1150	16.3	18.5	20.6	22.5	24.2	25.7	27.1*			
1200	16.3	18.6	20.6	22.4	24.1	25.6	26.7*			
1250	16.3	18.5	20.5	22.3	23.8	25.1*				
1300	16.2	18.4	20.4	22	23.4*	24.6*				
1350	16.1	18.2	20.1	21.6*	22.9*					
1400	15.9	17.9	19.7	21.1*						
1450	15.6	17.5	19.2*	20.5*						
1500	15.2	17.1	18.6*							
1550	14.8	16.6*	17.9*							
1600	14.3	15.9*								
1650	13.7*	15.1*								
1700	13*									
1750	12.3*									

Note: Consideration should be given to the loads that can be imposed on bearings and shafts etc., while using these ratings.

*   These values correspond to belt speeds above 30*m/sec.*

Table 42A Power ratings, in kilowatts, for 'D' section V belts, (32 *mm* wide) 180° arc of contact on smaller pulley

Speed of faster shaft rpm	Additional power per belt for speed ratio								
	1.02 to 1.04	1.05 to 1.08	1.09 to 1.12	1.13 to 1.18	1.19 to 1.24	1.25 to 1.34	1.35 to 1.51	1.52 to 1.99	2 and above
585	0.22	0.43	0.63	0.84	1.06	1.27	1.48	1.69	1.89
720	0.26	0.52	0.78	1.04	1.3	1.56	1.82	2.08	2.34
960	0.34	0.69	1.04	1.39	1.73	2.08	2.42	2.77	3.12
1440	0.51	1.04	1.56	2.07	2.6	3.12	3.63	4.14	4.68
50	0.01	0.04	0.05	0.07	0.09	0.11	0.13	0.14	0.16
100	0.04	0.07	0.1	0.14	0.18	0.22	0.25	0.29	0.33
150	0.05	0.11	0.16	0.22	0.27	0.33	0.38	0.43	0.48
200	0.07	0.14	0.22	0.29	0.36	0.43	0.51	0.57	0.65
250	0.09	0.18	0.27	0.36	0.45	0.54	0.63	0.72	0.81
300	0.11	0.22	0.32	0.43	0.54	0.65	0.76	0.87	0.98
350	0.13	0.25	0.38	0.51	0.63	0.76	0.88	1.01	1.13
400	0.14	0.29	0.43	0.57	0.72	0.87	1.01	1.16	1.3
450	0.16	0.33	0.48	0.65	0.81	0.98	1.13	1.3	1.46
500	0.18	0.36	0.54	0.72	0.9	1.08	1.26	1.44	1.63
550	0.2	0.4	0.6	0.79	0.99	1.19	1.39	1.59	1.78
600	0.22	0.43	0.65	0.87	1.08	1.3	1.51	1.73	1.95
650	0.23	0.47	0.7	0.94	1.17	1.41	1.64	1.87	2.11
700	0.25	0.51	0.75	1.01	1.26	1.51	1.77	2.02	2.27
750	0.27	0.54	0.81	1.08	1.35	1.63	1.89	2.16	2.43
800	0.29	0.57	0.87	1.16	1.44	1.73	2.02	2.3	2.6
850	0.31	0.61	0.92	1.22	1.54	1.84	2.15	2.45	2.76
900	0.33	0.65	0.97	1.3	1.63	1.95	2.27	2.6	2.92
950	0.34	0.69	1.03	1.37	1.72	2.06	2.39	2.74	3.08
1000	0.36	0.72	1.08	1.44	1.8	2.16	2.52	2.89	3.24
1050	0.38	0.76	1.13	1.51	1.89	2.27	2.65	3.03	3.41
1100	0.4	0.8	1.19	1.59	1.98	2.38	2.77	3.17	3.57
1150	0.42	0.83	1.25	1.66	2.07	2.49	2.9	3.32	3.73
1200	0.43	0.87	1.3	1.73	2.16	2.6	3.03	3.46	3.89
1250	0.45	0.9	1.35	1.8	2.25	2.71	3.15	3.61	4.06
1300	0.47	0.94	1.4	1.87	2.34	2.81	3.28	3.75	4.22
1350	0.48	0.98	1.46	1.95	2.43	2.92	3.41	3.89	4.38
1400	0.51	1.01	1.51	2.02	2.53	3.03	3.53	4.04	4.54
1450	0.52	1.04	1.57	2.09	2.62	3.14	3.66	4.18	4.71
1500	0.54	1.08	1.62	2.16	2.71	3.24	3.79	4.33	4.87
1550	0.56	1.12	1.68	2.24	2.8	3.36	3.91	4.47	5.03
1600	0.57	1.16	1.73	2.31	2.89	3.46	4.03	4.62	5.19
1650	0.6	1.19	1.78	2.38	2.98	3.57	4.16	4.76	5.35
1700	0.61	1.23	1.83	2.45	3.06	3.68	4.29	4.91	5.52
1750	0.63	1.26	1.89	2.53	3.15	3.79	4.41	5.05	5.68

Table 43 Power correction factor F_d based on arc of contact angle on smaller pulley.

Arc of contact	factor F_d	Arc of contact	factor F_d
180°	1	133°	0.87
177°	0.99	130°	0.86
174°	0.98	127°	0.85
171°	0.98	123°	0.83
169°	0.97	120°	0.82
166°	0.97	117°	0.81
163°	0.96	113°	0.8
160°	0.95	110°	0.78
157°	0.94	106°	0.77
154°	0.93	103°	0.75
151°	0.93	99°	0.73
148°	0.92	95°	0.72
145°	0.91	91°	0.7
142°	0.9	87°	0.68
139°	0.89	83°	0.65
136°	0.88	—	—

For intermediate values of arc of contact, F_d may be taken proportionately. For angles below 120° manufacturer to be consulted.

Table 44 Service factor F_a

Type of service	Type of driven machine	Normal torque, squirrel cage, synchronous and split phase ac motors and shunt wound dc motors			High torque single phase series wound and slip ring ac motors series and compound wound dc motors, shafts, clutches, brakes and direct on line starting.		
		upto 10 hrs./day	10 to 16 hrs./day	over 16 hrs./day	Upto 10 hrs./day	10 to 16 hrs./day	over 16 hrs./day
Light duty	Light to medium machine tools, centrifugal pumps and compressors with a rating upto 7.5 kW	1	1.1	1.2	1.1	1.2	1.3
Medium duty	Medium to heavy machine tools, positive displacement rotary pumps. punches and shears	1.1	1.2	1.3	1.2	1.3	1.4

If an idler pulley is used in the drive, the values given in the adjacent table should be added to the F_a values of the above table.

Idler pulley correction

Position of idler	Slack side	Tight side
Inside	0	0.1
Outside	0.1	0.2

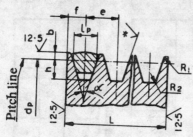

$$L=(x-1)e+2f$$
where
$x=$ Number of grooves

* Surface roughness $Ra=1.6\mu m$ for peripheral speeds upto 10 *m/sec.* and $Ra=0.8\mu m$ for peripheral speeds over 10 *m/sec.*

Table 45 Dimensions for standard V grooved pulleys

Groove section	l_p	b Min.	h Min.	e Nom.	e Tol.[†]	f Nom.	f Tol.	R_1 Min.	R_2 Min.	Tol. on α
A (13 x 8)	11	3.3	8.7	15	±0.3	10		1	1	±1°
B (17 x 11)	14	4.2	10.8	19	±0.4	12.5	+2 −1	1	1.6	±1°
C (22 x 14)	19	5.7	14.3	25.5	±0.5	17		1.6	2	±30′
D (32 x 19)	27	8.1	19.9	37	+0.6	24	+3 −1	2	3	±30′

† Tolerance applies to the distance between any two grooves, consecutive or not. Tolerance for side wobble and for runout is
0.001 *mm* per *mm* of pulley diameter for diameter upto 500
0.0015 *mm* per *mm* of pulley diameter for diameter over 500

Table 46

Groove angle α	Range of pitch diameters for groove section			
	A	B	C	D
38°	over 125 (200)	over 200 (200)	over 300 (355)	over 500
36°	—	—	Up to 300	Up to 500
34°	Up to 125	Up to 200	—	—

In order to reduce the wear on the belt used on smaller pulleys with a groove angle of 38° it is recommended that the pulley pitch diameter should not be less than the minimum values given within brackets for medium and heavy loads.

Table 47 Pulleys requiring dynamic balancing

OD of pulley	Ratio of face width to OD at and above which dynamic balancing is necessary at $v=20$	rpm at 20m/sec.	OD of pulley	Ratio of face width to OD at and above which dynamic balancing is necessary at $v=20$	rpm at 20m/sec.
75	1.4	5020	250	0.6	1505
100	1.26	3750	300	0.53	1255
125	1	3010	375	0.49	1000
150	0.9	2510	450	0.47	835
200	0.75	1875	—	—	—

1. For pulley diameters not included, the ratio may be taken proportionally
2. For speeds other than 20 m/sec., the ratio is inversely proportional to the square of the speed

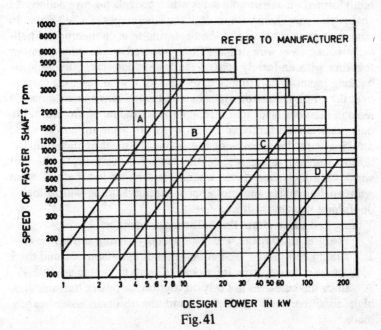

Fig. 41

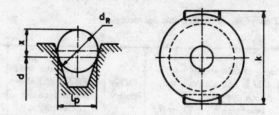

Pitch diameter of the groove $d = k - 2x$

Table 48 Inspection of pitch diameter by cylindrical rollers

Groove section	Diameter of roller d_R, mm	Correction $2x$, mm
A	11.6 h9	15
B	14.7 h11	19
C	20 h11	26
D	28.5 h11	37

TIMING BELT:

Timing belt is a moulded endless belt with uniformly spaced teeth formed on one or both sides which provide positive action of a chain drive combined with most of the advantages of a belt drive. In construction it consists of a load carrying tension member of helically wound steel wire or glass fibre, encased in a backing member together with uniformly spaced teeth moulded integrally with the backing member and finally covered with a wear resisting surface over the entire underside of the belt and teeth. Since an inextensible tension member is located on the dedendum line of the tooth, the belt pitch is sufficiently constant to ensure accurate and positive meshing with the grooves of the pulley. In most applications the tensile strength rating will be the basis for determining the belt width. In a high ratio drive, where less than six teeth are in mesh with smaller pulley the tooth shear strength rating will probably determine the belt width.

Timing belt drive offers the following advantages:

1. There is no slippage, creep or speed variation during drive.

2. Timing belt can be used at belt speeds upto $80 m/sec.$ and for a wide range of power from fractional kilowatt to as high as $400 kW.$

3. Since the belt does not rely on friction for power transmission, high initial tension is unnecessary and this results in lesser bearing loads.

4. Compact design is possible due to greater flexibility of the belt allowing for short centre distance, smaller pulleys and also due to a high power to weight ratio of the belt.

5. The tension member ensures no belt stretch and hence installation on fixed centre without any provision for belt adjustment is possible.

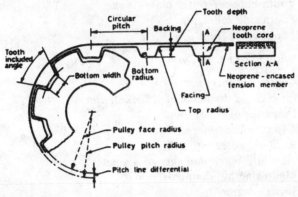

Timing belts are used to control and regulate drives that require a synchronised drive over several shafts such as in business machines; in high speed drives of the order of 20,000 *rpm* and more, such as in grinding machines and centrifuges; and in power drives such as in main and auxiliary drives of machine tools.

The design of a timing belt drive consists in selecting a suitable belt width and is easily accomplished from the data available from manufacturer's catalogue.

ROLLER CHAINS

Nomenclature:

b_1 Width between inner plates, *mm*
b_2 Width over inner link, *mm*
b_3 Width between outer plates, *mm*
ba Tooth side relief, *mm*
bf_1 Tooth width, *mm*
$\left.\begin{array}{c} bf_2 \\ bf_3 \end{array}\right\}$ Widths over teeth, *mm*
C Centre distance, *mm*
d Pitch circle diameter of sprocket, *mm*
d_1 Diameter of roller, *mm*
da Tip diameter, *mm*
df Root diameter, *mm*
dg Absolute maximum shroud diameter, *mm*
dR Measuring pin diameter, *mm*
Fa Service factor
ha Height of tooth above pitch polygon, *mm*
h_2 Maximum inner plate depth, *mm*
i Speed ratio
L Length of chain in multiples of pitch
L_1 Length of chain, *mm*
MR_e Measurement over pin for even number of teeth sprocket, *mm*
MR_o Measurement over pin for odd number of teeth sprocket, *mm*
n Speed of low speed shaft, *rpm*
N Speed of high speed shaft, *rpm*
p Pitch of the chain or Chordal pitch of the sprocket, *mm*
p_t Transverse pitch, *mm*
P Power transmitted, *kW*-from manufacturer's catalogue
P_d Design power, *kW*
P_t Power to be transmitted, *kW*
ra Shroud radius, *mm*
ra_{act} Actual shroud radius provided, *mm*
re_1 Minimum tooth flank radius, *mm*
re_2 Maximum tooth flank radius, *mm*
ri_1 Minimum roller seating radius, *mm*
ri_2 Maximum roller seating radius, *mm*
rx Minimum tooth side radius, *mm*

v Velocity of the chain, $m/sec.$
z Number of teeth on the sprocket under consideration
z_1 Number of teeth on smaller sprocket
z_2 Number of teeth on larger sprocket
α_1 Minimum roller seating angle, $degrees$
α_2 Maximum roller seating angle, $degrees$

Roller chain provides a positive drive at short centre distances. The flexibility of the chain enables it to absorb relatively high shock loads as compared to a gear drive. Roller chain drive is selected in general for low and medium speed applications.

Even though high grade roller chains can be operated at speeds as high as $20m/sec.$, it is recommended not to exceed a speed of $12m/sec.$ unless the chain is fully enclosed and well lubricated. Speed ratios greater than 7:1 are not generally recommended and it is desirable to compound two or more drives for higher ratios to obtain maximum service life. While using higher ratios the angle of wrap around smaller sprocket should not be less than 120° . The centre distance between shafts should preferably be 30 to 50 times the chain pitch; 40 times the chain pitch is about normal and 80 times the pitch is the recommended maximum. In high pulsating loads it is recommended to limit the centre distance to be about 20 to 30 times the pitch. Smaller pitch chains permit a greater number of teeth for an allowable sprocket diameter resulting in a smooth and noise-free operation due to reduced chordal action. The minimum centre distance that can be used is 1.5 times diameter of the large sprocket being used. The dimensions of standard chains are given in Table 49. The formulae for a chain drive design is given in Table 50. Length of the chain L to be preferably rounded off to an even number in which case all the links in the chain will be of normal type. However when the length is rounded off to an odd number, it will be necessary to use one cranked link.

Sprocket selection:

The sprocket profile dimensions can be calculated from Table 51. The actual tooth gap form will lie within maximum and minimum flank radii and should blend smoothly with the roller seating curve subtending the respective angles (Figs.45and46) The minimum number of teeth on smaller sprocket is generally recommended to be 17, however in practice, it is advisable to limit it to 21. Further,for very low rotational speeds of the order of 10 to 100 rpm the minimum can be as low as 9 teeth. The maximum number of teeth in the larger sprocket should be kept below 125, otherwise, a small amount of pitch elongation will cause the chain to ride the sprocket long before it is actually worn out. It is a good practice to use odd number of teeth on smaller sprocket.

For speeds upto 3 *m/sec.* and *z* upto 25, sprockets are generally made of mild steel. However, (i) when the speed exceeds 600 *rpm* and *z* upto 30, (ii) when the speed ratio exceeds 4:1 or (iii) the operating conditions are abrasive and corrosive, heat treatable carbon steel hardened and tempered to *HRC* 40-45 is used for small sprocket while the large sprockets are normally made of cast iron. When a sprocket is selected from manufacturer's ready stock, the bore should be checked for correct accomodation of both the shaft and key, with the bore generally not more than 0.7 times the hub diameter.

The design should generally provide for an adjustable centre distance to take up any slack due to elongation from wear and the actual range of adjustment should be atleast one to one and a half pitch. For fixed centre drives an idler sprocket with a minimum of 17 teeth is used on the slack side such that atleast 3 teeth are in full engagement with the chain.

The sprockets should lie in a vertical plane. Most satisfactory results are obtained when the slack side of the chain is at the bottom. When the slack side is on the top or for drives on steep inclination some means are to be provided to take up the slack.

Lubrication:

Roller chain drives operate successfully only with adequate lubrication. Oil cups are to be located such that the oil drops at the middle of the lower span. Lubrication is most effective when the oil is fed continuously to the driven side into the gap between the plates of roller chain. Table 53 lists some recommended methods of lubrication.

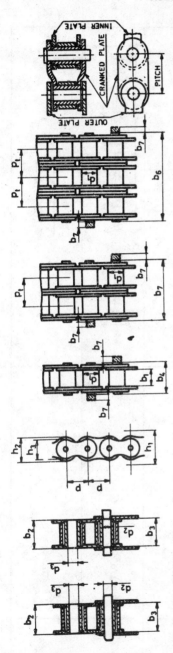

Table 49 Chain dimensions and breaking loads of base chains

ISO Chain No.	p	d_1 Max.	b_1 Min.	d_2 Max.	h_2 Max.	p_t	b_2 Max.	b_3 Min.	b_4 Max.	b_5 Max.	b_6 Max.	Min. breaking load, kgf.		
												Simple	Duplex	Triplex
05 B	8	5	3	2.31	7.11	5.64	4.77	4.99	8.6	14.3	19.9	460	800	1140
06 B	9.525	6.35	5.72	3.28	8.26	10.24	8.53	8.66	13.5	23.8	34	910	1730	2540
08 B	12.7	8.51	7.75	4.45	11.81	13.92	11.3	11.43	17	31	44.9	1820	3180	4540
10 B	15.875	10.16	9.65	5.08	14.73	16.59	13.28	13.41	19.6	36.2	52.8	2270	4540	6810
12 B	19.05	12.07	11.63	5.72	16.13	19.46	15.62	15.75	22.7	42.2	61.7	2950	5900	8850
16 B	25.4	15.88	17.02	8.28	21.08	31.88	25.45	25.58	36.1	68	99.9	4310	8620	12930
20 B	31.75	19.05	19.56	10.19	26.42	36.45	29.01	29.14	43.2	79.7	116.1	6580	13160	19740
24 B	38.1	25.4	25.4	14.63	33.4	48.36	37.92	38.05	53.4	101.8	150.2	9980	19960	29940
28 B	44.45	27.94	30.99	15.9	37.08	59.56	46.58	46.71	65.1	124.7	184.3	13160	26320	39480

Table 50

Sl. No.	Description	Equation
1.	Design power, P_d	$P_t \times F_a$
2.	Selection of suitable P and z_1	$P_d < P$
3.	Speed ratio, i	N/n
4.	Number of teeth on larger sprocket, z_2	$i \times z_1$
5.	Length of the chain, L	$\dfrac{2C}{p} + \dfrac{z_1 + z_2}{2} + \left(\dfrac{z_2 - z_1}{2\pi}\right)^2 \dfrac{p}{C}$
6.	L_1	$L \times p$
7.	Centre distance, C	$p/4\left[L - \dfrac{z_1 + z_2}{2} + \sqrt{\left(L - \dfrac{z_1 + z_2}{2}\right)^2 - 8\left(\dfrac{z_2 - z_1}{2\pi}\right)^2}\,\right]$
8.	Chain velocity, v	$\dfrac{\pi N p}{60{,}000 \sin\frac{180}{z_1}}$

Table 51

Sl. No.	Description	Equation	Tolerance
1	Pitch circle diameter, d	$\dfrac{p}{\sin\dfrac{180}{z}}$	$\begin{matrix}0\\-0.25\end{matrix}$ when $df \leqq 127$
2	Root diameter, df	$d - d_1$	$\begin{matrix}0\\-0.30\end{matrix}$ $127 < df \leqq 250$ $h\,11$ $\quad df > 250$
3	Tip diameter, da Max. Min.	$d + 1.25p - d_1$ $d + p\left(1 - \dfrac{1.6}{z}\right) - d_1$	
4	Tooth height above pitch polygon, ha Max. Min.	$0.625p - 0.5d_1 + \dfrac{0.8p}{z}$ $0.5(p - d_1)$	—
5	Roller sitting radius, ri_1 ri_2	$0.505d_1$ $(0.505d_1 + 0.069\sqrt[3]{d_1})$	Suffixes 1 and 2 refer to Min. and Max. tooth gap forms respectively. Fig. 45 and 46
6	Roller sitting angle, α_1 α_2	$140° - \dfrac{90°}{z}$ $120° - \dfrac{90°}{z}$	

Table 51 (contd.)

Sl. No.	Description	Equation	Tolerance
7.	Tooth flank radius, re_1 re_2	$0.12d_1(z+2)$ $0.008d_1(z^2+180)$	Suffixes 1 and 2 refer to Min. and Max. tooth gap forms respectively. Fig. 45 and 46
8.	Tooth width, bf_1 Simplex Duplex Triplex		$p \leqslant 12.7$ | $p > 12.7$ $0.93b_1$ | $0.95b_1$ $0.91b_1$ | $0.93b_1$ $0.88b_1$ | $0.93b_1$ b_1 from Table 49	h 14
9.	Width over teeth, bf_2 or bf_3	(No. of strands -1)$(p_t + bf_1)$ p_t from Table 49	—
10.	Minimum tooth radius	$r_x = p$	—
11.	Absolute maximum shroud diameter, dg	$p.\cot\dfrac{180°}{z} - 1.05h_2 - 1 - 2ra_{act}$	—
12.	Measuring pin diameter, dR	$dR = d_1$	$+\,0.1$ 0

Table 51 (contd.)

Sl. No.	Description	Equation	Tolerance
13.	Measurement over pin, Fig. 43. MR_e MR_o	$$\dfrac{d+dR}{d\cos\dfrac{90°}{z}}+dR$$	—
14.	Axial runout.	—	Total indicator reading $< 0.0009\,df + 0.08$ subject to 1.14 mm Max.
15.	Concentricity between bore and root diameter.	—	Total indicator reading $< 0.0009\,df + 0.08$ subject to 0.76 mm Max.

Tooth gap from (Fig. 43 and 44)   Wheel rim profile (Fig. 47)

Table 52 Service factor for roller chain drive, Fa

Type of load	Conditions of Service	
	8 hrs. day	24 hrs. day
Uniform load	1	1.2
Moderate shock	1.2	1.4
Heavy shock	1.4	1.7

Table 53 Lubrication of chain drive

Chain speed, m/sec.	Method of lubrication
Upto 4	Light dropwise lubrication, 4 to 14 drops/min.
4 to 7	Rapid drop (20 drops/min) or continuous with shallow bath
above 7	Forced feed lubrication

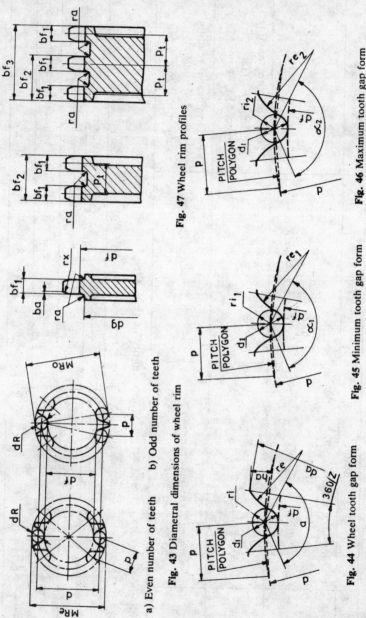

Fig. 47 Wheel rim profiles

Fig. 46 Maximum tooth gap form

Fig. 45 Minimum tooth gap form

Fig. 43 Diametral dimensions of wheel rim

a) Even number of teeth   b) Odd number of teeth

Fig. 44 Wheel tooth gap form

Nomenclature for Toothed Gearing (Spur, Helical, Crossed Helical, Bevel and Worm Gears)

Symbol	Description
a	Centre distance, mm
a_c	Centre distance between internal gear and its cutter mm
A	A factor, kgf/mm
b	Face width, mm
B	Factor of safety, mm
c	Tip clearance, mm
C	Strength factor, kgf/mm^2
d	Diameter of reference or pitch circle, mm
d_a	Tip (addendum) circle diameter, mm
	Throat circle dia. (in case of wormwheel), mm
d_e	External diameter, mm
d_f	Root (dedendum) circle diameter, mm
d_R	Diameter of the ball or roller used for measurement, mm
d_s	Diameter of the ball or roller touching the pitch circle (pitch line for rack), mm
f	Allowable tangential load, kgf/mm
F	Load (force, resistance), kgf
f_a	Tolerance on centre distance, μm
F_d	Dynamic load, kgf
f_d	Dynamic load, kgf/mm
f_{da}	Tolerance on addendum diameter, μm
f_e	Base pitch error, μm
f_f	Involute profile error, μm
f_g	Tolerance on the distance from tip (crown) to mounting surface, μm
f_i''	Tooth to tooth error - double flank, μm
F_i''	Total composite error - double flank, μm
f_l	Tolerance on the distance from cone apex to mounting face, μm
f_r	Radial runout, μm
f_s	Tolerance on chordal tooth thickness, μm
f_t	Adjacent pitch error, μm
f_x	Run out of mounting face, μm
f_{xa}	Run out of face and back cone, μm
F_n	Normal force (resultant), kgf

F_s	Separating (radial) force (at the mean dia. in case of bevel gears), *kgf*
F_t	Tangential force (at mean dia. in case of bevel gear), *kgf*
	(also cumulative pitch error where relevant), *μm*
f_w	Tolerance on base tangent length, *μm*
F_w	Limiting wear load normal to the tooth surface, *kgf*
F_x	Axial force (at mean dia. in case of bevel gears), *kgf*
f_β	Tooth alignment error
$f_{\delta a}$	Tolerance on tip angle, *mins.*
f_σ	Tolerance on back cone angle, *mins.*
g	Distance from tip to mounting surface, *mm*
G	Distance from apex to tip, *mm*
h	Tooth depth, *mm*
H	Height of pitch line of rack from the datum, *mm*
h_a	Addendum, *mm*
h_a	Chordal tooth height, *mm*
h_f	Dedendum, *mm*
i	Transmission ratio

$$\text{for a gear pair, } i = \frac{\text{no. of teeth on gear}}{\text{no. of teeth on pinion}}$$

$$\text{for a system, } i = \frac{rpm \text{ of input shaft}}{rpm \text{ of output shaft}}$$

j_{ni}	Gearing with increased backlash, *μm*
j_{nr}	Gearing with reduced backlash, *μm*
j_{ns}	Gearing with standard backlash, *μm*
j_{nz}	Gearing with zero backlash, *μm*
k	Addendum reduction coefficient
K	Load stress factor, *kgf/mm²*
l	Distance (distance from cone apex to mounting surface in case of bevel gears), *mm*
l_r	Length of root of worm wheel teeth, *mm*
L	Logarithmic coefficient
ΣL	Sum of logarithmic coefficients
m	Module, *mm*
M	Moment, *kgf.mm*
m_n	Normal module, *mm*

M_R	Dimension over rollers or balls, *mm*
$\triangle M_R$	Tolerance on M_R, *μm*
m_t	Transverse module, *mm*
m_x	Axial module, *mm*
n	Speed, *rpm*
p	Pitch, *mm*
P	Power to be transmitted, *kW*
p_n	Normal pitch, *mm*
p_t	Transverse pitch, *mm*
p_x	Axial pitch, *mm*
p_z	Lead of helix, *mm*
q	Diametral quotient
r	Radius *mm*
R	Cone distance, *mm*
R_a	Cosine of helix angle (cos β)
R_m	Mean cone distance, *mm*
s	Thickness of tooth, *mm*
$\overline{s}$	Chordal tooth thickness, *mm*
s_a	Thickness of tooth at the tip, *mm*
S_b	Allowable stress in continuous bending, *kgf/mm²*
s_f	Thickenss of tooth at the root, *mm*
$\overline{s}_f$	Thickness of tooth at the root per unit module, *mm*
S_s	Bending stress factor
S_w	Surface stress factor
s_x	Tooth thickness in axial plane, *mm*
T	Torque, *kgf.mm*
v	Pitch line velocity, *m/sec*
v_t	Tangential sliding velocity, *m/sec*
v_s	Relative sliding velocity, *m/sec*
W_k	Base tangent length over k no. of teeth, *mm*
$W_k{}^*$	Base tangent length over k no. of teeth for uncorrected gears of unit module, *mm*
x	Addendum modification coefficient
X_s	Speed factor for strength
X_w	Speed factor for wear
y	Centre distance modification coefficient
Y	Form factor
Y_z	Zone factor
z	No. of teeth or starts

α	Pressure angle, *deg*
α_n	Normal pressure angle, *deg*
α_R	Pressure angle at the centre of the roller, *deg*
α_{Rn}	Normal pressure angle at the centre of the ball or roller, *deg*
α_{Rt}	Transverse pressure angle at the centre of the ball or roller, *deg*
α_t	Transverse pressure angle, *deg*
α_x	Axial pressure angle, *deg*
β	Helix angle (mean tooth spiral angle in case of spiral bevel gears), *deg*
γ	Lead angle, *deg*
δ	Cone angle, *deg*
δ_a	Tip angle, *deg*
δ_f	Root angle, *deg*
$\triangle s$	Minimum thinning of teeth, *μm*
ϵ_∞	Transverse contact ratio
ϵ_β	Overlap ratio
ϵ_γ	Overall contact ratio
η	Efficiency
η_F	Efficiency of planetary drive with carrier fixed
θ	$\cos^{-1}\left(\dfrac{d_b}{d_a}\right)$, *deg*
θ_a	Addendum angle, *deg*
θ_f	Dedendum angle, *deg*
μ	Coefficient of friction
π	3.1416
ρ	Radius of curvature, *mm*
σ	Back cone angle, *deg*
Σ	Shaft angle, *deg*
ϕ	$\tan^{-1}\dfrac{\mu}{\cos \alpha_n}$, *deg*
ψ	Semitooth angle, *rad*

SUFFIXES AND INDICES

Suffixes:

1	Pinion (worm)
2	Gear
b	Base circle (except in S_b)
c	Cutter
$c1$	Cutter for cutting pinion
$c2$	Cutter for cutting internal gear
m	Mean (average) no. or value
v	Virtual number or value
A	Bearing A
B	Bearing B

Indices

Index' refers to working values

(eg. a' = working centre distance)

S_o correction - Corrected gears with $x_1 + x_2 = 0$

S correction - Corrected gears with $x_1 + x_2 \neq 0$

BASIC RACK PROFILE FOR GEARS WITH INVOLUTE TOOTH PROFILE

The rack tooth profile is used as basis for defining the standard tooth dimensions of gears with involute tooth profile. The basic rack is an imaginery rack having the standard basic rack tooth profile in the normal section.

The profile dimensions of gear teeth are based on module. Figure 48 shows the basic rack profile and graphical details of some of the parameters used in gearing. The basic rack profile is valid for involute spur, helical, bevel, cylindrical worm gears and racks in the module range of $0.6 \leqslant m \leqslant 50$.

Preferred series of modules normally used for gears are given in Table 54

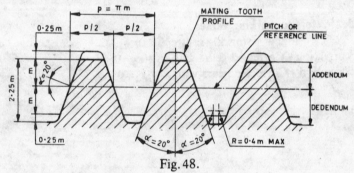

Fig. 48.

Table 54   Modules, m in the order of preference

I	II	I	II	I	II
0.6		2.5		8	
	0.7		2.75		9
0.8		3		10	
1			3.5	12	
1.25		4			14
1.5			4.5	16	
	1.75	5			18
2		6		20	
	2.25		7		

1) The profile of standard basic rack refers to cylindrical gears with involute teeth.

2) For profile with relief, relief is applied in principle to the tips of teeth.

3) In helical gears, the modules specified in the table refer to normal modules.

4) For axial module and normal module worms, the modules specified in the table refer to axial module and normal module respectively.

5) For bevel gears, the modules specified in the table are at the maximum pitch circle diameter. Root clearance is $0.2m$ max. and root radius is $0.35m$ max.

SPUR AND HELICAL GEARS

Gears are toothed wheels used for transmitting motion or power. If the teeth are cut parallel to the axis on a cylindrical surface, then they are called spur gears and if the teeth are inclined to the axis with a definite inclination they are called helical gears. These are used for transmitting motion with their rotational axes parallel. The tooth form based on involute curve to a base circle is commonly used in machine tools. The tooth form (profile) can be modified to get the required properties. Then the gears are said to be corrected. Changing the centre distance slightly will only give rise to a different working pressure angle without affecting the uniformity of motion transmission. In a helical gear pair the hands of helix are always opposite.

The teeth of spur and helical gears may fail due to a bad design, weak material, deviations from normal service or improper heat treatment. The typical failures are breaking off, pitting, seizure, surface abrasion, etc. Each gear pair is to be checked for safety in bending and wear load acting on them. In addition proper reduction in unit sliding velocities and compressive stresses reduce the surface abrasion and wear. Proper lubrication and correct selection of lubricants will prevent seizure of gears.

Spur gears

STANDARD SPUR GEARING

General formulae for calculation of dimensions of standard spur gearing are shown in Table 55. These equations are applicable when the axis of basic rack profile (i.e., the line on which tooth thickness is equal to tooth gap thickness) is tangential to the pitch circle of the gear as shown in Fig.49.

CORRECTED SPUR GEARING

In many cases, it is advantageous to correct the gear (see Fig.50) by shifting the profile of the cutter by an amount $m.x$. The gear tooth profile remains a part of the involute curve as in the case of standard uncorrected gearing, but a different portion of the involute curve of the same base circle is used as active profile. The correction is said to be positive or negative depending upon whether the cutter profile is shifted away from or towards the axis of the gear. This type of correction can be achieved by use of normal machinery and tools and does not call for any special equipment.

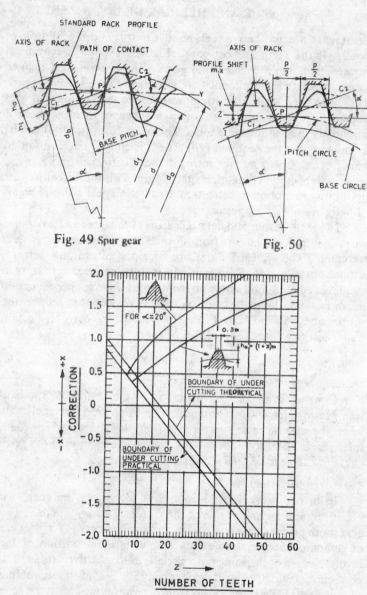

Fig. 49 Spur gear   Fig. 50

Fig. 51

Table 55 Spur gears without correction

Sl. No.	Description	Symbol	Equation
1	Number of teeth on pinion	z_1	To be selected as per requirement
2	Number of teeth on gear	z_2	
3	Module	m	
4	Pressure angle	α	$\alpha = 20°$
5	Addendum	h_a	$h_a = 1m$
6	Dedendum	h_f	$h_f = 1.25m$
7	Tip clearance	c	$c = 0.25\,m$
8	Total tooth height	h	$h = h_a + h_f = 2.25\ m$
9	Pitch diameter	d	$d = mz$
10	Base circle diameter	d_b	$d_b = d\cos\alpha$
11	Outside diameter	d_a	$d_a = d + 2h_a = m(z+2)$
12	Root diameter	d_f	$d_f = d - 2h_f = m(z-2.5)$
13	Centre distance	a	$a = (d_1 + d_2)/2 = m(z_1 + z_2)/2$
14	Thickness of tooth per module at the root	$\bar{s}_f$	From Fig. 58
15	Thickness of the tooth at the root	s_f	$m\bar{s}_f$
16	Sum of logarithmic coefficients for determining the allowable tangential tooth load.	ΣL	

The profile shift correction (also called *addendum modification*) can be used to avoid undercutting of gears having small number of teeth (i.e., below 17 teeth for 20° pressure angle). Correction is also resorted to for obtaining a given gear ratio with a prescribed centre distance. Positive correction is particularly advantageous with respect to surface pressure and bending stress in case of gears with small number of teeth. However a positive correction also reduces the tip thickness of tooth, and too large a positive correction can lead to pointed teeth. Fig.51 shows limiting values of profile correction for different number of teeth. In drives, where both the gears are given large positive correction, the values of contact ratio (i.e., the path of contact/base pitch) can reduce to a considerable extent and it is therefore necessary to check for contact ratio atleast in critical cases.

Thus, the correction of gears can have different effect on different parameters of gear engagement. Therefore it is necessary to strike a balance while selecting correction values to obtain a desired property. Fig. 52. gives values of sum of unit corrections on the given pair for achieving various running properties. Fig. 53 gives distribution of correction factors for speed reducing gear pairs. Similiarly, Fig. 54 gives distribution of correction factors for speed increasing gear pairs. Normally distribution of correction factors are to be made as per the above figures. In case the gear design becomes critical then selection of correction factor on the gear under consideration is left to the designers' choice.

Distribution of correction when
    1) centre distance is given
    2) centre distance can be chosen

In the first case, (y/z_m) is calculated and the corresponding values of $(x_1+x_2)/z_m$ is found from Table 72 or from equations (1) and (2) of Table 72. Then (x_1+x_2) is calculated. The value $(x_1+x_2)/2$ is located at the point z_m in Fig.53 or 54 depending upon the case i.e., speed decreasing or speed increasing gear pair. A pairing line is drawn through this point and the values of x_1 and x_2 corresponding to z_1 and z_2 are found out.

In the second case, the correction (x_1+x_2) is chosen depending upon the property desired of the gear pair from Fig. 52. and distributed in the same manner as in case (1) depending upon whether it is a speed increasing or decreasing gear pair.

When z_2 is greater than 150, z_m is taken as equal to $(z_1+150)/2$ to fix up the pairing line and x_1 corresponding to z_1 is found out. Then x_2 is determined from equation $x_2 = (x_1+x_2) - x_1$.

In case of helical gears number of teeth z_v to be used instead of z.

General formulae for calculation of dimensions of corrected gearing are given in Table 56.

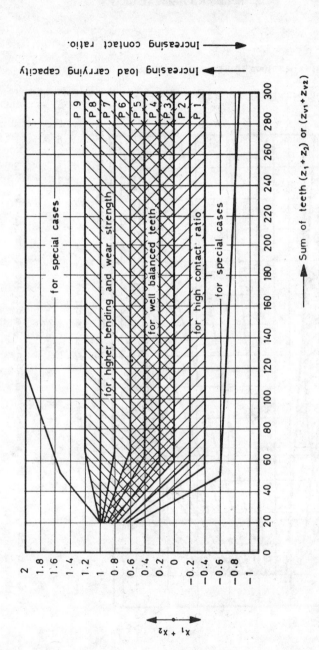

Fig. 52 Total correction factor

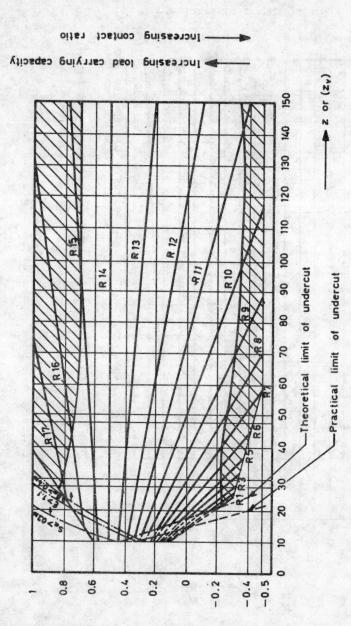

If x_2 or x_1 and x_3 lie in the shaded region the gears have to be checked for tooth interference.

Fig. 53 Distribution of correction factor for reduction drives

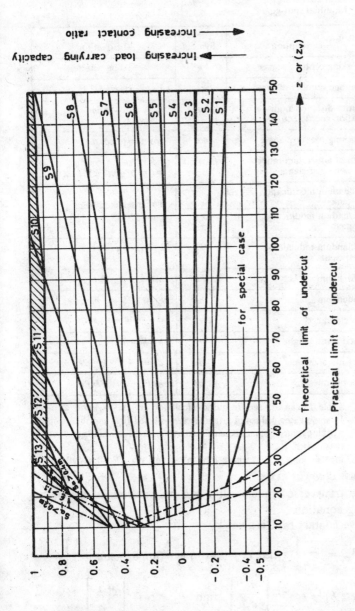

Increasing contact ratio ⟶

◀— Increasing load carrying capacity

z or (z_v)

S 8
S 7
S 6
S 5
S 4
S 3
S 2
S 1

S 9

S 10

S 11

S 12

S 13

$S_a > 0.18$
$i_e > 1.1$

$S_a > 0.2a$
$i_e < 1.1$

Theoretical limit of undercut

Practical limit of undercut

for special case

If x_2 or x_1 and x_2 lie in the shaded region the gears have to be checked for tooth interference.

Fig. 54 Distribution of correction factor for step up drives

Table 56   Modified spur gears

Sl. No.	Description	Symbol	Equation
1	Working centre distance	a'	To be selected
2	Average number of teeth	z_m	$z_m = (z_1 + z_2)/2$
3	Centre distance modification coefficient	y	$y = (a' - a)/m$
4	Working pressure angle	α'	$\cos \alpha' = (a.\cos \alpha)/a'$
5	Sum of addendum modification coefficients	$(x_1 + x_2)$	$(x_1 + x_2)/z_m$ $= (inv\, \alpha' - inv\, \alpha)/\tan \alpha$
6	Addendum modification for pinion	x_1	To be selected from Figs. 52, 53, 54
7	Addendum modification for gear	x_2	
8	Addendum reduction coefficient	k	$k = (x_1 + x_2 - y)$
9	Addendum	h_a	$h_a = (1 + x - k)\, m$
10	Dedendum	h_f	$h_f = (1.25 - x)\, m$
11	Tip clearance	c	$c = 0.25\, m$
12	Total tooth height	h	$h = h_a + h_f$
13	Outside diameter	d_a	$d_a = (z + 2 + 2x - 2k)\, m$
14	Root diameter	d_f	$d_f = (z - 2.5 + 2x)\, m$

Notes: 1   Table 72 can be used for quick calculation of $(x_1 + x_2)$ and α'

      2.   In some cases, values of x_1 and x_2 are selected first and then suitable working centre distance is calculated. Above formulae hold good even in these cases.

      3. $inv\, \alpha = \tan \alpha - \alpha$, where α is in radians; inv $20° = 0.014904$.

TRANSVERSE CONTACT RATIO

The transverse contact ratio can be determined using the following equation

Transerve contact ratio is given by

$$\epsilon_\alpha = \frac{1}{2\pi} \left[z_1(\tan \theta_1 - \tan \alpha') + z_2(\tan \theta_2 - \tan \alpha') \right]$$

where $\theta_1 = \cos^{-1} \left[\dfrac{d_{b1}}{d_{a1}} \right]$ and $\theta_2 = \cos^{-1} \left[\dfrac{d_{b2}}{d_{a2}} \right]$

For smooth transmission of movement, value of contact ratio should preferably be more than 1.2

The transverse contact ratio can also be determined graphically.

TOOTH TIP THICKNESS

Positive correction of gears reduces tooth tip thickness and care should be taken to see that the tip thickness is not reduced below $0.3\,m$ for power transmitting gear drives. Fig. 51 shows two curves for limiting values of positive correction-one for pointed teeth and the other for the tip thickness of $0.3\,m$.

Tip thickness is given by

$$s_a = \frac{d_a}{2}\left[\frac{(\pi + 4\,x\,tan\,\alpha)}{z} - 2(inv\,\theta - inv\,\alpha)\right]$$

where θ values are the same as in equation for transverse contact ratio.

Module m is generally chosen from standard series (Refer Table 54)

PRESSURE ANGLE α

Various values of pressure angles ($14\frac{1}{2}°$, $15°$, $20°$ etc.) are in use. But $20°$ pressure angle is most commonly used and tables in this handbook apply for gears of $20°$ pressure angle. The term pressure angle denotes the pressure angle of basic rack profile but the working pressure angle α' can be different if gears are corrected to work on a modified centre distance.

MODIFIED SPUR GEARS WITHOUT ALTERING THE CENTRE DISTANCE (S_o correction)

In this case $x_1 = -x_2$, with $x_1 + x_2 = 0$, $y = 0$, and $k = 0$. Also $a' = a$ and $\alpha' = \alpha$. Remaining values can be calculated from the set of formulae in Table 56. This is frequently carried out to prevent root undercutting when a pinion of considerably small number of teeth is used.

BASE TANGENT LENGTH

Spur gear tooth thickness is generally determined by measuring the base tangent length W_k across a certain number of teeth k determined by using Fig. 55. depending on the number of teeth and unit correction factor. For spur gears, it is enough to use only the lower part of this figure. General formula for calculation of base tangent length W_k is

$$W_k = m\,cos\alpha[(k-0.5)\pi + z\,inv\alpha] + 2mx\,sin\alpha = W_k^*m + 2mx\,sin\alpha$$

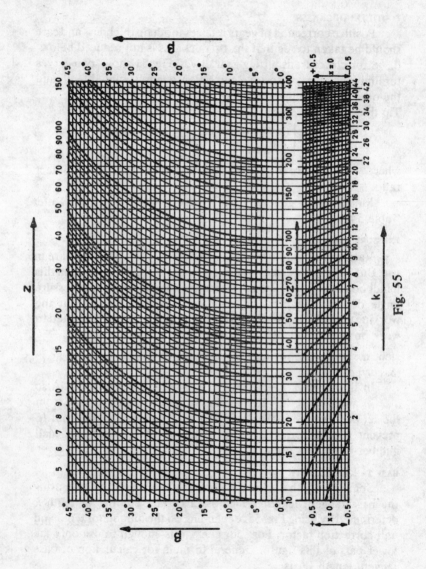

Fig. 55

Where W_k^* is the value of base tangent length for uncorrected gears of unit module. Values of W_k^* can be read from Tables 60 to 64. Values for other modules can be obtained by multiplying the table value by module. Values for corrected gears can be obtained by adding the value of $2.x.m.\sin \alpha$

CHORDAL TOOTH THICKNESS AND CHORDAL HEIGHT

Whenever it is not possible to measure the base tangent length over a certain number of teeth, for spur and helical gears, the inspection of the gear can be done by measuring the chordal tooth thickness over a single tooth at a pre-determined depth from the addendum circle.

The following equations are used (measured in normal plane)

$$\overline{s} = m_n z_v \sin\psi$$

$$\overline{h}_a = m_n \left[1 + x - k + \frac{z_v}{2}(1 - \cos\psi) \right]$$

$$\text{where } \psi = \frac{\pi}{2z_v} + \frac{2x \tan\alpha_n}{z_v}$$

MEASUREMENT OVER ROLLERS

Alternatively, spur and helical gears can also be inspected for tooth thickness by measuring over rollers. For values of measurement, chapter on *measurement over rollers* may be referred.

TOLERANCES FOR SPUR GEARS

The DIN standard specifications 3962 to 3967 have specified tolerances for 12 classes of accuracy of gears. Tables 65 to 67 cover the tolerance values for classes 5 to 10. Table below shows a guide to the selection of class of accuracy depending upon the peripheral speed of the gears

Peripheral speed, v m/s	Upto 3	3 to 6	6 to 12	12 to 20
Class of accuracy	10	8	6	5

Tolerances on following values have been listed.

Description	Symbol	Refer Table no.
1. Tolerance on base tangent length	f_w	65
2. Radial runout	f_r	
3. Involute profile error	f_f	66
4. Adjacent pitch error	f_t	

Description	Symbol	Refer Table no.
5. Base pitch error	f_e ⎫	
6. Cumulative pitch error	F_t ⎬	66
7. Total composite error, double flank	F_i'' ⎭	
8. Tooth to tooth error, double flank	f_i'' ⎫	
9. Tooth alignment error	f_β ⎬	65
10. Tolerance on centre distance	f_a	67
11. Tolerance on chordal tooth thickness	f_s	65

It is a normal practice to inspect only the base tangent length, radial runout on the gear and centre distance between the shafts. Other values are checked only where higher precision gears are involved.

CHOICE OF FITS FOR GEAR TEETH

Running clearance between the teeth of a mating gear pair is determined by the combined effect of tolerance on centre distance and the tolerance on the base tangent length. Tolerance on the centre distance is determined from Table 67 depending on the class of accuracy and the value of centre distance. The amount of clearance in the fit can be controlled by selecting from various tolerance zones prescribed for base tangent length (a to h) for external gears. Normally selected fit for base tangent length or chordal tooth thickness is a combination of zones c and d. Table 65 lists the values of tolerance, the upper limit being derived by the top value of fit d and the lower limit by bottom value of fit c.

TOLERANCE FOR OUTSIDE DIAMETER

Normally, a tolerance of $h9$ is specified for outside diameter of gears.

STRENGTH CALCULATION

A modified form of Buckingham method has been adopted for the calculation of strength of gear teeth. The values given are valid for gears having 20° pressure angle.

The basic equations used are the following:
Total dynamic load/mm face width is

$$f_d = \frac{F_t}{b} + \frac{\dfrac{F_t}{b} + A}{1 + \dfrac{0.76\sqrt{\dfrac{F_t}{b} + A}}{v}}$$

A is a factor which depends upon the elastic modulus of the materials of the gear pair and the effective error during action for a given pressure angle.

The permissible total load in wear and bending are governed by the following equations:

Permissible wear load, kgf/mm facewidth

$$= 2Km_n \; z_1 z_2 / (z_1 + z_2)$$

Permissible load in bending, kgf/mm face width

$$= \pi S_b m_n Y$$

Y is a form factor of the gear under consideration which is dependent upon number of teeth of gears under mesh.

For corrected gears the permissible load in bending is obtained by multiplying the above equation by a factor $(S_f \text{ corrected} / \bar{S}_f)^2$

For convenience of calculation the above equations have been converted to logarithmic form in Figures 56 & 57 and Tables 69 & 71 so that multiplication and division are replaced by addition and subtraction.

Calculation of tooth strength in bending

Allowable stresses in continuous reverse bending for different materials are given in Table 68. Table 69 gives values of logarithmic coefficient L for different values of S_b, m, and the number of teeth of gears in mesh. These three independent values of L when added together will give ΣL in bending. Corresponding to this value of ΣL, on the scale for f_d in Figs. 56 or 57, the maximum dynamic load per mm face width is obtained. If the factor of safety assumed is B, then the allowable dynamic load is found by dividing the maximum dynamic load by B. If the ordinate corresponding to this value is drawn to meet the corresponding pitch line velocity of the gear, the value of the ordinate at this point f will give the allowable tangential tooth load per mm width.

The face width is obtained by:

$$b = F_t / f$$

It is recommended to use a minimum factor of safety of 1.25 in all cases and for shock loads such as in milling machines, the value of factor of safety is generally 2.

Effect of tooth correction on bending

Depending on whether the correction is positive or negative the value of ΣL as calculated has to be respectively increased or decreased. The value of logarithmic coefficient L to be added or

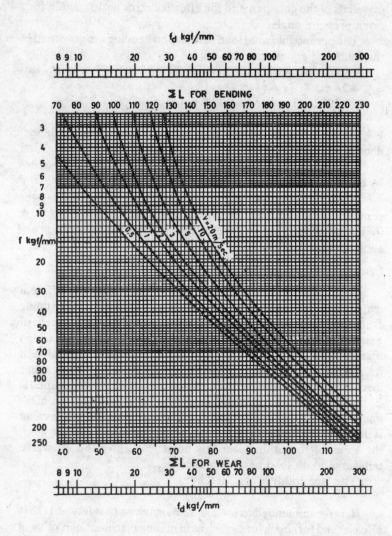

Fig. 56 Graphical determination of tangential tooth load for gears of accuracy corresponding to DIN 8 or 9 class for $\alpha = 20^\circ$

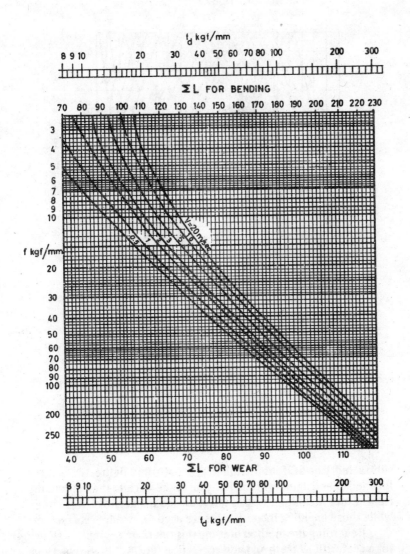

Fig. 57 Graphical determination of tangential tooth load for gears of accuracy corresponding to DIN 6 class for $\alpha = 20°$

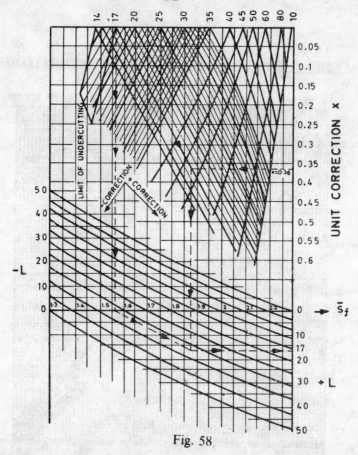

Fig. 58

subtracted can be found from Fig. 58. In this figure two sets of inclined lines originate at the top, one to the left and the other to the right. Lines inclining towards left are used for negative correction, while those inclining towards right are used for positive correction.

Regarding the method of using this figure an example is shown for a gear with 17 teeth and unit correction for $x = +0.36$. The value of L for this gear is obtained by drawing two vertical lines one from $z = 17$ till it meets $\bar{s}_f$ line and the other from the point of intersection of the horizontal from $x = 0.36$ and the right side inclined line for $z = 17$.

From the point of intersection of the first vertical line and the $\bar{s}_f$ line, a line parallel to the coefficient line is drawn till it meets the second vertical line. The horizontal from the point of intersection of these two lines will give the value of L as $+17$.

Calculation of tooth strength from wear considerations:

The values of load stress factor K depend on the materials of the gears in mesh and their heat treatment and is tabulated in Table 70.

The logarithmic coefficients corresponding to K, m and the number of teeth of gears under mesh are given in Table 71. The three independent values of logarithmic coefficients are added to get ΣL in wear. The total width of gear b required from wear considerations is then calculated in the same manner as in bending. A total width b corresponding to the larger value of the two as calculated for bending and wear, is finally chosen as the design width.

Effect of gear tooth correction on wear

It is known that positive correction improves the wear resistance properties of the gear. Maximum wear resistance is obtained when the gear pair is corrected for equalising specific sliding values at roots of their teeth. However, there is not sufficient information to quantitatively estimate the effects of correction on wear resistance of tooth. It is therefore not considered here.

Helical gears

Profile dimensions of helical gear teeth are based on normal module m_n. Coefficients x, y, and profile dimensions (addendum, dedendum) are calculated in the normal plane of the teeth. Centre distance a, and other properties of gearing like contact ratio are calculated in the transverse plane using transverse module m_t. Reference to Fig. 59 may be made for a graphical detail of some of the parameters used under helical gearing.

BASE TANGENT LENGTH

Number of teeth k over which the base tangent length is measured is determined by using Fig. 55 depending on the number of teeth, helix angle, and the unit correction factor. General formula for claculation of W_k

$$W_k = m_n \cos\alpha_n [(k - 0.5)\pi + z\ inv\alpha_t] + 2m_n x \sin\alpha_n$$

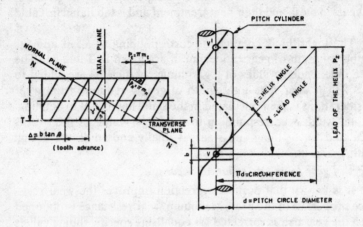

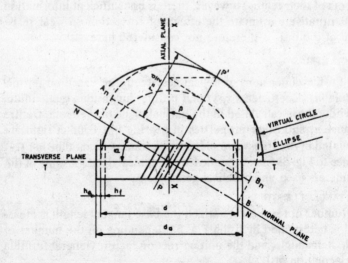

Fig. 59

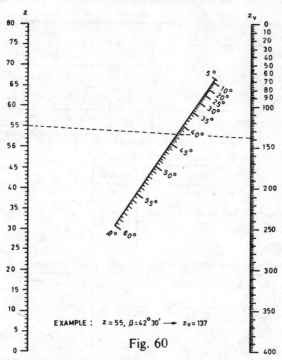

EXAMPLE : $z = 55$, $\beta = 42°30'$ ⟶ $z_v = 137$

Fig. 60

Table 57 General formulae for helical gears

Sl. No.	Description	Symbol	Equation
1	Number of teeth on pinion	z_1	
2	Number of teeth on gear	z_2	To be selected as per
3	Helix angle	β	requirement
4	Normal module	m_n	
5	Normal pressure angle	α_n	$\alpha_n = 20°$
6	Transverse module	m_t	$m_t = m_n . \sec \beta$
7	Pressure angle in transverse plane	α_t	$\tan \alpha_t = \tan \alpha_n . \sec \beta$
8	Pitch diameter	d	$d = z.m_t = z.m_n.\sec \beta$
9	Base circle diameter	d_b	$d_b = d \cos \alpha_t$
10	Lead	p_z	$p_z = \pi d.\cot \beta$
11*	Virtual number of teeth	z_v	$z_v = z / (\cos^3 \beta)$

* Virtual number of teeth can be determined from nomogram shown in Fig. 60.

Table 58 Helical gears without correction

Sl. No.	Description	Symbol	Equation
1	Addendum	ha	$ha = m_n$
2	Dedendum	hf	$hf = 1.25\, m_n$
3	Outside diameter	da	$da = d + 2\, m_n$
4	Root diameter	df	$df = d - 2.5\, m_n$
5	Centre distance	a	$a = (d_1 + d_2)/2$ $= (z_1 + z_2)\, m_n / 2\cos \beta$

Table 59 Helical gears with correction

Sl. No.	Description	Symbol	Equation
1	Working centre distance	a'	To be selected
2	Average virtual number of teeth	z_{vm}	$z_{vm} = (z_{v1} + z_{v2})/2$
3	Centre distance modification coefficient	y	$y = (a' - a)/m_n$
4*	Working normal pressure angle	α'_n	$y/z_{vm} = (\cos \alpha_n / \cos \alpha'_n) - 1$
5*	Sum of addendum modification coefficient	$(x_1 + x_2)$	$\dfrac{(x_1 + x_2)}{z_{vm}} = \dfrac{(inv\, \alpha'_n - inv\, \alpha_n)}{\tan \alpha_n}$
6	Addendum modification coefficient for pinion	x_1	To be selected from Figs. 52, 53, 54.
7	Addendum modification coefficient for gear	x_2	
8	Addendum reduction coefficient	k	$k = (x_1 + x_2 - y)$
9	Addendum	ha	$ha = (1 + x - k)\, m_n$
10	Dedendum	hf	$hf = (1.25 - x) m_n$
11	Outside diameter	da	$da = d + 2ha$
12	Root diameter	df	$df = d - 2hf$

* These equations give sufficiently accurate values for all helical gearing with usual backlash and allows for quick calculations using Table 59 for interpolation purposes. However, if a minimum backlash is essential, then the calculations are to be made in the transverse section using following formulae:

$$\frac{x_1 + x_2}{z_m} = \frac{inv\, \alpha'_t - inv\, \alpha_t}{\tan \alpha_n} \; ; \quad \frac{y}{z_m} = \frac{1}{\cos \beta} \left(\frac{\cos \alpha_t}{\cos \alpha'_t} - 1 \right)$$

$$\text{and} \quad \cos \alpha'_t = \frac{d_{b1} + d_{b2}}{2a'} \quad \text{where} \quad z_m = \frac{z_1 + z_2}{2}$$

CONTACT RATIO

Transverse contact ratio ϵ_α is given by formula

$$\epsilon_\alpha = \frac{1}{2\pi} \left[z_1(tan\,\theta_1 - tan\,\alpha'_t) + z_2(tan\,\theta_2 - tan\,\alpha'_t) \right]$$

θ_1 is given by $cos\theta_1 = \left[\dfrac{d_{b1}}{d_{a1}} \right]$

θ_2 is given by $cos\theta_2 = \left[\dfrac{d_{b2}}{d_{a2}} \right]$

The working transverse pressure angle is given by

$$cos\alpha'_t = \frac{d_{b1} + d_{b2}}{2a'}$$

Overlap ratio ϵ_β is given by formula

$$\epsilon_\beta = \frac{b\,sin\beta}{\pi\,m_n}$$

Overlap ratios greater than one are recommended.

Overall contact ratio $\epsilon_\gamma = \epsilon_\alpha + \epsilon_\beta$

TOLERANCES FOR HELICAL GEARS

In case of helical gears, the tolerances for radial runout f_r, double flank total composite error F_i'' and double flank tooth to tooth error f_i'' are related to pressure angle in transverse plane. They can be considered as straight spur gears with pressure angle and allowable errors have to be multiplied by a factor $R_a = cos\,\beta$

The other individual errors f_β, f_i, f_e, f_f and F_t remain unaltered.

The base tangent length tolerances f_w should also be multiplied by the factor R_a.

STRENGTH CALCULATION OF HELICAL GEARS

The same tables and figures used for strength calculation of spur gears are also used for calculation of strength of helical gears. Only normal module m_n and virtual number of teeth z_{v1} and z_{v2} are used (instead of m, z_1 and z_2) for determining the logarithmic coefficients, whereas pitch line velocities are considered in transverse plane.

Table 60

Base tangent length W_k^* for spur gears, 7 to 46 teeth measured over k teeth of uncorrected gear, pressure angle = 20°, m = 1							
No. of teeth z	BASE TANGENT LENGTH OVER k TEETH						
	1 Tooth	2 Teeth	3 Teeth	4 Teeth	5 Teeth	6 Teeth	7 Teeth
7	1.57410	4.52624	7.47837				
8	1.58811	4.54024	7.49237				
9	1.60212	4.55425	7.50638				
10	1.61612	4.56825	7.52038				
11	1.63013	4.58226	7.53439				
12	1.64413	4.59626	7.54840				
13	1.65814	4.61027	7.56240				
14	1.67214	4.62427	7.57641				
15	1.68615	4.63828	7.59041				
16	1.70015	4.65229	7.60442				
17	1.71416	4.66629	7.61842				
18		4.68030	7.63243	10.58456			
19		4.69430	7.64643	10.59857			
20		4.70831	7.66044	10.61257			
21		4.72231	7.67444	10.62658			
22		4.73632	7.68845	10.64058			
23		4.75032	7.70246	10.65459			
24		4.76433	7.71646	10.66859			
25		4.77834	7.73047	10.68260			
26		4.79234	7.74447	10.69660			
27			7.75848	10.71061	13.66274		
28			7.77248	10.72462	13.67675		
29			7.78649	10.73862	13.69075		
30			7.80049	10.75263	13.70476		
31			7.81450	10.76663	13.71876		
32			7.82851	10.78064	13.73277		
33			7.84251	10.79464	13.74677		
34			7.85652	10.80865	13.76078		
35			7.87052	10.82265	13.77479		
36				10.83666	13.78879	16.74092	
37				10.85066	13.80280	16.75493	
38				10.86467	13.81680	16.76893	
39				10.87868	13.83081	16.78294	
40				10.89268	13.84481	16.79694	
41				10.90669	13.85882	16.81095	
42				10.92069	13.87282	16.82496	
43				10.93470	13.88683	16.83896	
44				10.94870	13.90084	16.85297	
45					13.91484	16.86697	19.81910
46					13.92885	16.88098	19.83311

Table 61

Base tangent length W_k* for spur gears, 47 to 85 teeth measured over k teeth of uncorrected gear, pressure angle = 20°, m = 1						
No. of teeth z	BASE TANGENT LENGTH OVER k TEETH					
	5 Teeth	6 Teeth	7 Teeth	8 Teeth	9 Teeth	10 Teeth	11 Teeth
47	13.94285	16.89498	19.84711				
48	13.95686	16.90899	19.86112				
49	13.97086	16.92299	19.87513				
50	13.98487	16.93700	19.88913				
51	13.99887	16.95101	19.90314				
52	14.01288	16.96501	19.91714				
53	14.02689	16.97902	19.93115				
54		16.99302	19.94515	22.89728			
55		17.00703	19.95916	22.91129			
56		17.02103	19.97316	22.92530			
57		17.03504	19.98717	22.93930			
58		17.04904	20.00118	22.95331			
59		17.06305	20.01518	22.96731			
60		17.07706	20.02919	22.98132			
61		17.09106	20.04319	22.99532			
62		17.10507	20.05720	23.00933			
63			20.07120	23.02333	25.97547		
64			20.08521	23.03734	25.98947		
65			20.09921	23.05135	26.00348		
66			20.11322	23.06535	26.01748		
67			20.12723	23.07936	26.03149		
68			20.14123	23.09336	26.04549		
69			20.15524	23.10737	26.05950		
70			20.16924	23.12137	26.07350		
71			20.18325	23.13538	26.08751		
72				23.14938	26.10152	29.05365	
73				23.16339	26.11552	29.06765	
74				23.17740	26.12953	29.08166	
75				23.19140	26.14353	29.09566	
76				23.20541	26.15754	29.10967	
77				23.21941	26.17154	29.12368	
78				23.23342	26.18555	29.13768	
79				23.24742	26.19955	29.15169	
80				23.26143	26.21356	29.16569	
81					26.22757	29.17970	32.13183
82					26.24157	29.19370	32.14583
83					26.25558	29.20771	32.15984
84					26.26958	29.22171	32.17385
85					26.28359	29.23572	32.18785

Table 62

Base tangent length W_k* for spur gears, 86 to 125 teeth measured over k teeth of uncorrected gear, pressure angle = 20°, m = 1

No. of teeth z	BASE TANGENT LENGTH OVER k TEETH						
	9 Teeth	10 Teeth	11 Teeth	12 Teeth	13 Teeth	14 Teeth	15 Teeth
86	26.29759	29.24973	32.20186				
87	26.31160	29.26373	32.21586				
88	26.32560	29.27774	32.22987				
89	26.33961	29.29174	32.24387				
90		29.30575	32.25788	35.21001			
91		29.31975	32.27188	35.22402			
92		29.33376	32.28589	35.23802			
93		29.34776	32.29990	35.25203			
94		29.36177	32.31390	35.26603			
95		29.37577	32.32791	35.28004			
96		29.38978	32.34191	35.29404			
97		29.40379	32.35592	35.30805			
98		29.41779	32.36992	35.32205			
99			32.38393	35.33606	38.28819		
100			32.39793	35.35007	38.30220		
101			32.41194	35.36407	38.31620		
102			32.42595	35.37808	38.33021		
103			32.43995	35.39208	38.34421		
104			32.45396	35.40609	38.35822		
105			32.46796	35.42009	38.37222		
106			32.48197	35.43410	38.38623		
107			32.49597	35.44810	38.40024		
108				35.46211	38.41424	41.36637	
109				35.47612	38.42825	41.38038	
110				35.49012	38.44225	41.39438	
111				35.50413	38.45626	41.40839	
112				35.51813	38.47026	41.42239	
113				35.53214	38.48427	41.43640	
114				35.54614	38.49827	41.45041	
115				35.56015	38.51228	41.46441	
116				35.57415	38.52629	41.47842	
117					38.54029	41.49242	44.44455
118					38.55430	41.50643	44.45856
119					38.56830	41.52043	44.47257
120					38.58231	41.53444	44.48657
121					38.59631	41.54844	44.50058
122					38.61032	41.56245	44.51458
123					38.62432	41.57646	44.52859
124					38.63833	41.59046	44.54259
125					38.65234	41.60447	44.55660

Table 63

Base tangent length W_k^* for spur gears, 126 to 165 teeth measured over k teeth of uncorrected gear, pressure angle = $20°$, m = 1

No. of teeth z	BASE TANGENT LENGTH OVER k TEETH						
	14 Teeth	15 Teeth	16 Teeth	17 Teeth	18 Teeth	19 Teeth	20 Teeth
126	41.61847	44.57060	47.52274				
127	41.63248	44.58461	47.53674				
128	41.64648	44.59861	47.55075				
129	41.66049	44.61262	47.56475				
130	41.67449	44.62663	47.57876				
131	41.68850	44.64063	47.59276				
132	41.70251	44.65464	47.60677				
133	41.71651	44.66864	47.62077				
134	41.73052	44.68265	47.63478				
135		44.69665	47.64879	50.60092			
136		44.71066	47.66279	50.61492			
137		44.72466	47.67680	50.62893			
138		44.73867	47.69080	50.64293			
139		44.75268	47.70481	50.65594			
140		44.76668	47.71881	50.67094			
141		44.78069	47.73282	50.68495			
142		44.79469	47.74682	50.69896			
143		44.80870	47.76083	50.71296			
144			47.77483	50.72697	53.67910		
145			47.78884	50.74097	53.69310		
146			47.80285	50.75498	53.70711		
147			47.81685	50.76898	53.72111		
148			47.83086	50.78299	53.73512		
149			47.84486	50.79699	53.74913		
150			47.85887	50.81100	53.76313		
151			47.87287	50.82501	53.77714		
152			47.88688	50.83901	53.79114		
153				50.85302	53.80515	56.75728	
154				50.86702	53.81915	56.77128	
155				50.88103	53.83316	56.78529	
156				50.89503	53.84716	56.79930	
157				50.90904	53.86117	56.81330	
158				50.92304	53.87518	56.82731	
159				50.93705	53.88918	56.84131	
160				50.95105	53.90319	56.85532	
161				50.96506	53.91719	56.86932	59.83546
162					53.93120	56.88333	59.84947
163					53.94520	56.89733	59.86347
164					53.95921	56.91134	59.86347
165					53.97321	56.92535	59.87748

Table 64

No. of teeth z	Base tangent length W_k^* for spur gears, 166 to 204 teeth measured over k teeth of uncorrected gear, pressure angle = 20°, m = 1						
	BASE TANGENT LENGTH OVER k TEETH						
	18 Teeth	19 Teeth	20 Teeth	21 Teeth	22 Teeth	23 Teeth	24 Teeth
166	53.98722	56.93935	59.89148				
167	54.00123	56.95336	59.90549				
168	54.01523	56.96736	59.91949				
169	54.02924	56.98137	59.93350				
170	54.04324	56.99537	59.94750				
171		57.00938	59.96151	62.91364			
172		57.02338	59.97552	62.92765			
173		57.03739	59.98952	62.94165			
174		57.05140	60.00353	62.95566			
175		57.06540	60.01753	62.96966			
176		57.07941	60.03154	62.98367			
177		57.09341	60.04554	62.99767			
178		57.10742	60.05955	63.01168			
179		57.12142	60.07355	63.02569			
180			60.08756	63.03969	65.99182		
181			60.10157	63.05370	66.00583		
182			60.11557	63.06770	66.01983		
183			60.12958	63.08171	66.03384		
184			60.14358	63.09571	66.04785		
185			60.15759	63.10972	66.06185		
186			60.17159	63.12372	66.07586		
187			60.18560	63.13773	66.08986		
188			60.19960	63.15174	66.10387		
189				63.16574	66.11787	69.07000	
190				63.17975	66.13188	69.08401	
191				63.19375	66.14588	69.09802	
192				63.20776	66.15989	69.11202	
193				63.22176	66.17389	69.12603	
194				63.23577	66.18790	69.14003	
195				63.24977	66.20191	69.15404	
196				63.26378	66.21591	69.16804	
197				63.27779	66.22992	69.18205	
198					66.2432	69.19605	
199					66.25793	69.21006	72.14819
200					66.27193	69.22407	72.16219
201					66.28594	69.23807	72.17620
202					66.29994	69.25208	72.19020
203					66.31395	69.26608	72.20421
204					66.32796	69.28009	72.21821
							72.23222

Table 65   Tolerances for spur and helical gears

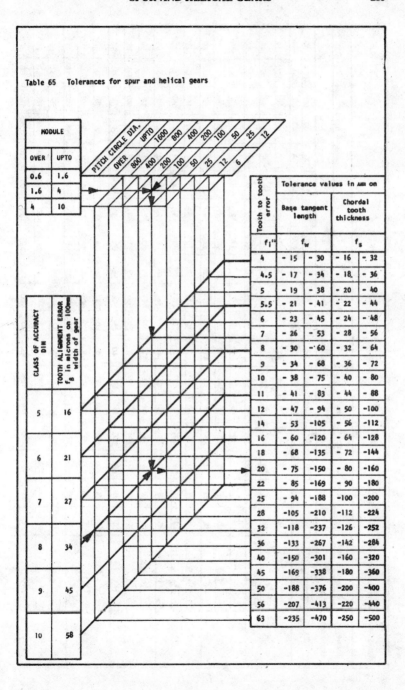

Tooth to tooth error	Tolerance values in μm on			
	Base tangent length		Chordal tooth thickness	
f_i''	f_w		f_s	
4	- 15	- 30	- 16	- 32
4.5	- 17	- 34	- 18	- 36
5	- 19	- 38	- 20	- 40
5.5	- 21	- 41	- 22	- 44
6	- 23	- 45	- 24	- 48
7	- 26	- 53	- 28	- 56
8	- 30	- 60	- 32	- 64
9	- 34	- 68	- 36	- 72
10	- 38	- 75	- 40	- 80
11	- 41	- 83	- 44	- 88
12	- 47	- 94	- 50	-100
14	- 53	-105	- 56	-112
16	- 60	-120	- 64	-128
18	- 68	-135	- 72	-144
20	- 75	-150	- 80	-160
22	- 85	-169	- 90	-180
25	- 94	-188	-100	-200
28	-105	-210	-112	-224
32	-118	-237	-126	-252
36	-133	-267	-142	-284
40	-150	-301	-160	-320
45	-169	-338	-180	-360
50	-188	-376	-200	-400
56	-207	-413	-220	-440
63	-235	-470	-250	-500

MODULE	
OVER	UPTO
0.6	1.6
1.6	4
4	10

CLASS OF ACCURACY DIN	TOOTH ALIGNMENT ERROR f_B in microns on 100mm width of gear
5	16
6	21
7	27
8	34
9	45
10	58

Table 66 Tolerances for spur and helical gears

Class of Accuracy DIN	Kind of Error	over 0.6 upto 1.6 mm						over 1.6 upto 4 mm						over 4 upto 10 mm					
	PITCH CIRCLE DIA. 'd' (mm) OVER	12	25	50	100	200	400	12	25	50	100	200	400	25	50	100	200	400	800
	UPTO	25	50	100	200	400	800	25	50	100	200	400	800	50	100	200	400	800	1600
5	f_t, f_e, f_f	4	4.5	5	5.5	7	8	4.5	5	5.5	6	7	9	6	7	8	9	10	12
	P_t	14	16	18	20	25	28	16	18	20	22	25	32	18	22	25	32	36	45
	f_r, F_i''	12	14	16	18	20	22	14	16	18	20	22	25	18	20	22	25	28	32
6	f_t, f_e, f_f	6	6	7	8	9	11	6	7	8	9	10	12	9	10	11	12	14	18
	F_t	20	22	25	28	36	40	22	25	28	32	36	45	28	32	36	40	50	63
	f_r, F_i''	18	20	22	25	28	32	20	22	25	28	32	36	25	28	32	36	40	45
7	f_t, f_e, f_f	8	9	10	11	12	16	9	10	11	12	14	18	12	14	16	18	20	25
	P_t^m	28	32	36	40	50	63	32	36	40	45	50	63	40	45	50	56	71	90
	f_r, P_i	25	28	32	36	40	45	28	32	36	40	45	50	36	40	45	50	56	63
8	f_t, f_e, f_f	11	12	14	16	18	22	12	14	16	18	20	25	18	20	22	25	28	36
	P_t^m	40	45	50	56	71	80	45	50	56	63	71	90	56	63	71	80	100	125
	f_r, P_i	36	40	45	50	56	63	40	45	50	56	63	71	50	56	63	71	80	90
9	f_t, f_e, f_f	16	18	20	22	25	32	18	20	22	25	28	36	25	28	32	36	40	50
	P_t^m	56	63	71	80	100	125	63	71	80	90	100	126	80	90	100	125	140	180
	f_r, P_i	50	56	63	71	80	90	56	63	71	80	90	100	71	80	90	100	110	125
10	f_t, f_e, f_f	25	28	32	36	40	50	28	32	36	40	45	56	40	45	50	56	63	80
	P_t^m	90	100	110	125	160	200	100	110	125	140	160	200	125	140	160	180	220	280
	f_r, P_i	71	80	90	100	110	125	80	90	100	110	125	140	100	110	125	140	160	180

Module "m"

Table 67 Tolerances for centre distance

Class of Accuracy DIN		Center distance					
	over	6.3	16	40	100	250	630
	upto	16	40	100	250	630	1600
	Fit			J			
5	Centre distance tolerance values ($\pm f_a$) in μm	9	11	14	18	22	28
6		12	16	20	25	32	40
7		18	22	28	36	45	56
8		25	32	40	50	63	80
9		36	45	56	71	90	110
10		50	63	80	100	125	160

Table 68 Values of allowable stress S_b in continuous reversed bending, kgf/mm^2

Material	I.S.Designation	Hardness HB core	UTS kgf/mm^2 core	Allowable stress in continuous reversed bending, kgf/mm^2
Cast Iron	Grade–20	190	20	9
	Grade–25	210	25	13
Direct hardening steel	C40	175	63	30
		200	67	34
		230	77	38
	C45	190	65	32
		200	67	34
		230	77	38
	30Ni4Cr1	445	155	32
	40Ni2Cr1Mo28	255	87	29
		315	107	32
		335	117	35
		370	127	38
		445	155	47
Case hardening steel	C14	140	50	25
	13Ni3Cr80	255	85	37
	15Ni2Cr1Mo15	330	110	50
	17Mn1Cr95	240	80	35

Table 69   Logarithmic coefficient for bending strength $\alpha = 20°$

(Rotated label at left of mating-gear columns: *No. of teeth on the gear under consideration*)

L	Sb kgf/mm²	m	\multicolumn{9}{No. of teeth on the mating gear}									m	Sb kgf/mm²	L
			17	20	30	40	50	60	100	150	150			
0	5	1												51
1	5.12		17										16.2	51
2	5.25		18	17									16.5	52
3	5.37		19	18									17	53
4	5.5		20	19	17							3.5	17.4	54
5	5.62		21	20	18	17							17.8	55
6	5.75		22	21	19	18	17	17					18.2	56
7	5.87		23	22	20	19	18	18					18.5	57
8	6		24	23	21	20	19	19	17	17			19	58
9	6.15		25	24	22	21	20	20	18	18			19.4	59
10	6.3	1.25	26	25	23	22	21	21	19	19	17	4	19.8	60
11	6.45		27	26	24	23	22	22	20	-	18		20.3	61
12	6.6		28	27	25	24	23	23	21	20	19		20.8	62
13	6.75		29	28	26	25	24	24	22	21	20		21.3	63
14	6.9		30	29	27	26	25	25	23	22	21		21.8	64
15	7.05		31	30	28	27	26	26	24	23	22	4.5	22.3	65
16	7.2		33	32	30	28	27	27	26	25	23		22.8	66
17	7.35		34	33	31	29	28	28	27	26	24		23.3	67
18	7.5	1.5	36	35	32	30	30	29	28	27	25		23.9	68
19	7.7		38	37	34	32	31	30	29	28	26		24.5	69
20	7.9		40	39	35	33	32	32	30	29	27	5	25.1	70
21	8.1		45	42	38	35	34	33	31	30	28		25.7	71
22	8.3		49	46	40	37	36	35	32	32	29		26.3	72
23	8.5		55	51	44	40	38	37	34	33	30		26.9	73
24	8.7	1.75	62	58	48	43	44	39	36	34	32	5.5	27.5	74
25	8.9		73	66	53	46	44	42	38	36	33		28.1	75
26	9.1		87	79	60	53	48	46	41	39	35		28.7	76
27	9.3		106	92	68	59	54	51	45	42	37		29.4	77
28	9.5		146	115	81	69	61	57	50	46	40	6	30.1	78
29	9.75		-	154	100	80	70	65	56	51	43		30.8	79
30	10	2	-	-	134	98	85	77	62	58	47		31.5	80
31	10.25					134	105	91	72	66	53	6.5	32.3	81
32	10.5						150	120	87	78	59		33.1	82
33	10.75							170	110	95	68		33.8	83
34	11								160	130	80		34.6	84
35	11.25	2.25									100	7	35.4	85
36	11.5										143		36.2	86
37	11.75												37	87
38	12												37.9	88
39	12.3												38.8	89
40	12.6	2.5										8	39.7	90
41	12.9												40.6	91
42	13.2												41.5	92
43	13.5												42.5	93
44	13.8	2.75											43.5	94
45	14.1											9	44.5	95
46	14.4												45.6	96
47	14.7												46.7	97
48	15	3											47.8	98
49	15.4												48.9	99
50	15.8											10	50	100

Table 70 Values of load stress factor K

Material of Gear pair		HB (Surface)	Heat treatment	Load stress factor K kgf/mm^2 for $\alpha = 20°$
C I and C I	Grade 20 Grade 25	190 210		0.05 0.063
	Steel and Steel	200 and 200 250 and 200 300 and 200 250 and 250 300 and 250 350 and 250 300 and 300 350 and 300 400 and 300 350 and 350 400 and 350 400 and 400	Normalised condition	0.055 0.072 0.091 0.091 0.113 0.137 0.137 0.163 0.177 0.192 0.222 0.256

Steel and Steel of equal hardness:

	Material	HB (Surface)	Heat treatment	Load stress factor
Direct hardening steel	$C40$	545 590	Flame or Induction hardened	0.47 0.54
	$C45$	560 625		0.5 0.6
	$30Ni4Cr1$	445		0.32
	$40Ni2Cr1Mo\underline{28}$	545 615		0.47 0.58
Case hardening steel	$C14$ $13Ni3Cr\underline{80}$ $15Ni2Cr1Mo\underline{15}$ $17Mn1Cr\underline{95}$	590 650	Case hardened (case depth $\approx 0.2m$ + grinding allowance for gears with $m \leqslant 8$)	0.54 0.64

Table 71 Logarithemic coefficient L for surface (stress) $\alpha = 20°$

L	K kgf/mm²	m		No. of teeth on the mating gear												
				17	18	19	20	22	25	30	35	40	50	70	100	150
0	0.09	1														
1																
2	0.1															
3				17	17											
4	0.11			19	18	17										
5		1.25		21	19	18	17									
6	0.12			23	21	20	19	17								
7				27	23	22	20	19	17							
8	0.13			30	27	25	22	20	19	17						
9		1.5		35	30	28	25	22	20	18	17					
10	0.14			40	35	31	28	25	22	19	18	17				
11	0.15			45	40	35	31	28	23	20	19	18	17			
12	0.16	1.75		55	45	40	35	31	26	22	20	19	18	17		
13				70	55	50	40	35	29	24	22	20	19	18		
14	0.17			90	70	65	50	40	32	27	24	22	20	19	17	
15	0.18	2		120	90	80	65	47	36	30	26	24	21	20	18	17
16	0.19				120	100	80	55	40	33	28	26	22	21	19	18
17	0.2					130	100	67	47	36	30	28	24	22	20	19
18	0.21	2.25					130	80	55	40	32	30	26	23	21	20
19	0.22							100	67	46	36	32	28	24	22	21
20	0.23	2.5						135	80	54	40	34	30	25	23	22
21	0.24								100	63	45	37	32	26	24	23
22	0.25	2.75							122	72	50	41	34	28	25	24
23	0.26								160	82	56	46	37	30	26	25
24	0.27	3								100	64	51	41	32	28	26
25	0.28									120	74	57	46	34	30	27
26	0.29										85	65	51	37	32	28
27	0.3	3.5									100	75	57	40	34	30
28	0.32										120	87	65	43	37	32
29	0.34											100	74	46	40	34
30	0.36	4										120	85	50	43	36
31	0.37												97	55	46	38
32	0.39												112	60	50	40
33	0.41	4.5											135	66	55	43
34	0.43													74	60	46
35	0.45	5												83	65	49
36	0.47													92	70	52
37	0.49	5.5												102	77	56
38	0.52													115	85	60
39	0.54	6												130	95	64
40	0.57														108	69
41	0.6	6.5													125	75
42	0.62	7														80
43	0.65															
44	0.68															
45	0.71	8														
46	0.75															
47	0.78															
48	0.82	9														
49	0.86															
50	0.9	10														

No. of teeth on the gear under consideration

Table 72 Modification values for involute gears (pressure angle $\alpha = 20°$)

Basic equations: $\dfrac{x_1 + x_2}{z_m} = \dfrac{\mathrm{inv}\,\alpha' - \mathrm{inv}\,\alpha}{\tan\alpha}$ (1); $\quad \dfrac{y}{z_m} = \dfrac{\cos\alpha}{\cos\alpha'} - 1$ (2)

Diff	$\dfrac{x_1+x_2}{z_m}$	α'	$\dfrac{y}{z_m}$	Diff	Diff	$\dfrac{x_1+x_2}{z_m}$	α'	$\dfrac{y}{z_m}$	Diff
66	-0.02036	16° 00'	-0.02244	82	145	0.02238	23° 00'	0.02084	127
68	-0.0197	10'	-0.02162	83	147	0.02383	10'	0.02211	128
69	-0.01902	20'	-0.02079	84	150	0.0253	20'	0.02339	129
71	-0.01833	30'	-0.01995	85	153	0.0268	30'	0.02468	130
73	-0.01762	40'	-0.0191	86	154	0.02833	40'	0.02598	132
74	-0.01689	50'	-0.01824	87	158	0.02987	50'	0.0273	132
75	-0.01615	17° 00'	-0.01737	88	159	0.03145	24° 00'	0.02862	134
77	-0.0154	10'	-0.01649	89	162	0.03304	10'	0.02996	135
79	-0.01463	20'	-0.01560	89	165	0.03466	20'	0.03131	136
80	-0.01384	30'	-0.01471	91	167	0.03631	30'	0.03267	138
82	-0.01304	40'	-0.01380	92	170	0.03798	40'	0.03405	139
84	-0.01222	50'	-0.01288	93	173	0.03968	50'	0.03544	140
85	-0.01138	18° 00'	-0.01195	94	175	0.04141	25° 00'	0.03684	141
87	-0.01053	10'	-0.01101	95	178	0.04316	10'	0.03825	142
88	-0.00966	20'	-0.01006	96	180	0.04494	20'	0.03967	144
91	-0.00878	30'	-0.00910	97	183	0.04674	30'	0.04111	145
92	-0.00787	40'	-0.00813	98	186	0.04857	40'	0.04256	147
94	-0.00695	50'	-0.00715	99	189	0.05043	50'	0.04403	147
95	-0.00601	19° 00'	-0.00616	100	191	0.05232	26° 00'	0.04550	150
98	-0.00506	10'	-0.00516	101	195	0.05423	10'	0.04700	150
99	-0.00408	20'	0.00415	102	197	0.05618	20'	0.04850	151
101	-0.00309	30'	-0.00313	103	200	0.05815	30'	0.05001	153
103	-0.00208	40'	-0.00210	105	203	0.06015	40'	0.05154	155
105	-0.00105	50'	-0.00105	105	206	0.06218	50'	0.05309	155
107	0.00000	20° 00'	0.00000	106	209	0.06424	27° 00'	0.05464	157
109	0.00107	10'	0.00106	108	212	0.06633	10'	0.05621	159
110	0.00216	20'	0.00214	109	215	0.06845	20'	0.05780	159
113	0.00326	30'	0.00323	109	218	0.07060	30'	0.05939	161
115	0.00439	40'	0.00432	111	221	0.07278	40'	0.06100	163
116	0.00554	50'	0.00543	112	225	0.07499	50'	0.06263	164
119	0.00670	21° 00'	0.00655	113	227	0.07724	28° 00'	0.06427	165
121	0.00789	10'	0.00768	114	231	0.07951	10'	0.06592	167
123	0.00910	20'	0.00882	115	234	0.08182	20'	0.06759	168
125	0.01033	30'	0.00997	116	238	0.08816	30'	0.06927	170
127	0.01158	40'	0.01113	117	240	0.08654	40'	0.07097	171
130	0.01285	50'	0.01230	119	244	0.08894	50'	0.07268	172
131	0.01415	22° 00'	0.01349	120	247	0.09138	29° 00'	0.07440	174
134	0.01546	10'	0.01469	120	251	0.09385	10'	0.07614	176
136	0.01680	20'	0.01589	122	253	0.09636	20'	0.07790	176
138	0.01816	30'	0.01711	124	258	0.09889	30'	0.07966	179
140	0.01954	40'	0.01835	124	261	0.10147	40'	0.08145	180
144	0.02094	50'	0.01959	125	265	0.10408	50'	0.08325	181
	0.02238	23° 00'	0.02084			0.10673	30° 00'	0.08506	

INTERNAL GEARS

An internal gear is one with teeth cut into the bore of a component. The teeth may be straight or helical. An internal gear pair consists of an internal gear and an external gear (pinion).

Internal gears possess several advantages when properly applied and hence are frequently used for reduction gears and planetary gear combinations.

Compared to a conventional external gear drive, internal gear drive

    i.   is more compact
   ii.   provides smoother drive
  iii.   has stronger tooth form

Internal gear is also used as a gear coupling with 1:1 ratio drive. In this case, since the gear and pinion are coaxial, modification is not necessary. In an internal gear pair drive, the pinion and gear rotate in the same direction. When an internal gear pair is used as a reduction drive employing a small pinion and a large internal gear, the question of design as regards the tooth shape and action does not require special consideration. When the number of teeth in the pinion closely approaches the number of teeth in the mating internal gear, then it is necessary to check the design of tooth shape and action for eliminating tooth interference as the teeth come out of engagement.

Normally internal gears are manufactured by gear shaping method. However, internal gears with helical teeth having helix angles above 30° should be avoided due to the limitations of kinematics of the gear shaper mechanism. They can also be manufactured by other processes like broaching, skiving, etc. Medium and large sized gears can be ground.

Internal gearing is designed either as normal uncorrected gearing or as S_o-corrected gearing. S-correction is used only in exceptional cases.

Determination of addendum, tip circle diameters and choice of number of teeth

The addendum circle diamters of pinion and gear depend on the method of manufacture, shape of the cutter, difference in number of teeth of gear and pinion, or of gear and shaping cutter. In addition to the usual undercutting at the root of the pinion teeth caused by the cutter tip as in external gears, there are other problems that may occur in internal gears such as (i) involute undercutting, (ii) root undercutting and (iii) tip interference.

Care should therefore be taken to avoid these problems while designing internal gears and the addendum, the number of teeth on both the pinion and the gear cutter should be chosen accordingly. (See under *Limiting values for number of teeth*)

Correction of internal gears

The correction on internal gear is positive when it is applied radially inwards, and negative when applied radially outwards. If the sum of the addendum modification coefficients x_1 and x_2 is zero (S_o-correction), the gears work at the theoretical pressure angle and pitch diameters of generation, in the same manner as external gears. It is necessary to remember that when the internal gear is given a negative correction (radially outwards), its profile is improved due to the increased radius of curvature of the tooth profile but the circular tooth thickness at the pitch circle reduces. If an internal gear pair is to work at other than the theoretical centre distance a, the working centre distance a' should preferably be greater than a.

S_o Correction:

To improve the contact conditions, sometimes the internal gearing is designed as a S_o-corrected gearing with positive correction on the pinion and equal negative correction on the gear.

S Correction:

This correction is to be used only in exceptional cases, and that too preferably for increasing the centre distance. In any case, the internal gear has to be corrected radially outwards, i.e. with negative corrction.

Limiting values for number of teeth (Internal spur gear pair)

The problems that occur in cutting of internal gears (using a disc type shaper cutter), assembly and working of an internal gear pair can be eliminated in case of $20°$ full-depth involute system if the following conditions are satisfied.

$$Z_2 - Z_{c2} > 12$$
$$Z_2 - Z_1 > 12$$
$$Z_{c2} > 16$$
$$Z_1 > 14$$

In certain cases it is possible to go beyond these limits. But it will necessitate the use of pinion and cutters with profile modifications. Further, the pinion and the internal gear tip diamters should satisfy the following conditions:

$$d_{a1} \leqslant \sqrt{d_{b1}^2 + \left[2(a_c - a)\sin\alpha + \sqrt{d_{ac2}^2 - d_{bc2}^2}\right]^2}$$

when $z_{e2} < z_1$

$$d_{a2} \geqslant \sqrt{d_{b2}^2 + (d_2 \sin\alpha - 2h_{ac1}\,\mathrm{cosec}\,\alpha)^2}$$

(when the pinion is cut by a rack type cutter or hob)

$$\geqslant \sqrt{d_{b2}^2 + \left[(d_2 + d_{c1})\sin\alpha - \sqrt{d_{ac1}^2 - d_{bc1}^2}\right]^2}$$

(when the pinion is cut by a disc type shaper cutter)

$$\geqslant \sqrt{d_{b2}^2 + \left[2a\sin\alpha + \sqrt{d_{t1}^2 - d_{b1}^2}\right]^2}$$

(when the internal gear is cut by a disc type shaper cutter)

where

$$d_{t1} = \sqrt{\left[d_1 - (d_{f1} + 2c)\right]^2 \cot^2\alpha + (d_{f1} + 2c)^2}$$

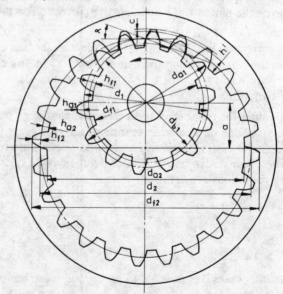

Fig. 61   Nomenclature for internal gears

Fig. 61 shows the important parameters for both the gear and the pinion. Table 73 gives the calculation of geometrical parameters.

Table 73 Calculation of geometrical parameters of internal gear pair.

Sl. No.	Nomenclature	Pinion	Internal gear	Cutter for cutting internal gear
1	Pressure angle		$\alpha_n = 20°$	
2	Pitch circle diameter	$d_1 = mz_1$	$d_2 = mz_2$	$d_{c2} = m \cdot z_{c2}$
3	Base circle diameter	$d_{b1} = d_1 \cos\alpha_n$	$d_{b2} = d_2 \cdot \cos\alpha_n$	$d_{bc2} = d_{c2} \cdot \cos\alpha_n$
4	Tip circle diameter	$d_{a1} = d_1 + 2h_a$	$d_{a2} = d_2 - 2h_a$	$d_{ac2} = d_{c2} + 2h_{ac2}$
5	Root circle diameter	$d_{f1} = d_1 - 2h_f$	$d_{f2} = d_2 + 2h_f$	$d_{fc2} = d_{c2} - 2h_{fc2}$
6	Theoritical centre distance	$a = 0.5\,m(z_2 - z_1)$		$a_c = 0.5\ m\ (z_2 - z_{c2})$
7	Working centre distance	—	a'	
8	Transmission ratio		$i = \dfrac{z_2}{z_1}$	—
9	Working pressure angle		$\cos\alpha'_n = \dfrac{a\,\cos\alpha_n}{a'}$	—
10	Sum of unit corrections		$(x_1 + x_2) = \dfrac{-(z_2 - z_1)\,(inv\alpha'_n - inv\alpha_n)}{2\,tan\alpha_n}$	—
11	Theoritical tooth depth	$h_1 = 0.5\,(d_{a1} - d_{f1})$	$h_2 = 0.5(d_{f2} - d_{a2})$	$h_{c2} = 2.5m$
12	Clearance		$0.25\ m$	

MEASUREMENT OVER ROLLERS FOR GEARS

I Spur and helical gears

Table 74 Spur and helical gears (with 2 measuring rollers or balls in diametrically opposite tooth spaces)

	Spur Gears	Helical Gears
Even number of teeth Figs. 62 & 63	$M_R = \dfrac{m_n z \cos\alpha_n}{\cos\alpha_{Rn}} \pm d_R$ $\Delta M_R = \dfrac{f_w}{\sin\alpha_{Rn}}$	$M_R = \dfrac{m_t . z . \cos\alpha_t}{\cos\alpha_{Rt}} \pm d_R$ $\Delta M_R = \dfrac{f_w}{(\sin\alpha_{Rt}\,\cos\beta_b)}$
Odd number of teeth Figs. 64 & 65	$M_R = \dfrac{m_n z \cos\alpha_n \cos\left(\frac{90}{z}\right)^\circ}{\cos\alpha_{Rn}} \pm d_R$ $\Delta M_R = \dfrac{f_w . \cos\left(\frac{90}{z}\right)^\circ}{\sin\alpha_{Rn}}$	$M_R = \dfrac{m_t . z \cos\alpha_t \cos\left(\frac{90}{z}\right)^\circ}{\cos\alpha_{Rt}} \pm d_R$ $\Delta M_R = \dfrac{f_w . \cos\left(\frac{90}{z}\right)^\circ}{(\sin\alpha_{Rt} . \cos\beta_b)}$

Note: For external gears $+d_R$, For internal gears $-d_R$

Table 75 Calculation of angles α_{Rn} and α_{Rt} (Refer Fig. 66)

	External gears	Internal gears
Spur gears	$Inv\alpha_{Rn} = Inv\alpha_n + \dfrac{d_R}{m_n z \cos\alpha_n}$ $- \dfrac{\pi}{2z} \pm \dfrac{2x\tan\alpha_n}{z}$	$Inv\alpha_{Rn} = Inv\alpha_n - \dfrac{d_R}{m_n . z . \cos\alpha_n}$ $+ \dfrac{\pi}{2z} \pm \dfrac{2x\tan\alpha_n}{z}$
Helical gears	$Inv\alpha_{Rt} = Inv\alpha_t + \dfrac{d_R}{m_n . z . \cos\alpha_n}$ $- \dfrac{\pi}{2z} \pm \dfrac{2x\tan\alpha_n}{z}$	$Inv\alpha_{Rt} = Inv\alpha_t - \dfrac{d_R}{m_n . z . \cos\alpha_n}$ $+ \dfrac{\pi}{2z} \pm \dfrac{2x\tan\alpha_n}{z}$

1. Absolute value of x should be used in the above equations
2. + sign when correction is applied radially outwards
3. − sign when correction is applied radially inwards
4. Both limits of tolerance $\triangle M_R$ will have − sign for external gears
5. Both limits of tolerance $\triangle M_R$ will have + sign for internal gears
6. Recommended values of diameter of roller d_R are 1.728 m_n for external gears and 1.44 m_n for internal gears.

   (without addendum modification)

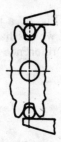

VEN NO. OF TEETH-EXTERNAL

Fig. 62.

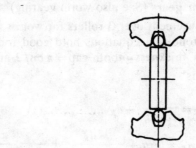

EVEN NO. OF TEETH-INTERNAL

Fig. 63.

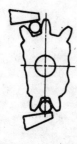

ODD NO. OF TEETH-EXTERNAL

Fig. 64.

ODD NO. OF TEETH-INTERNAL

Fig. 65.

d R
PITCH CIRCLE
BASE CIRCLE
INV α
α R
INV α R

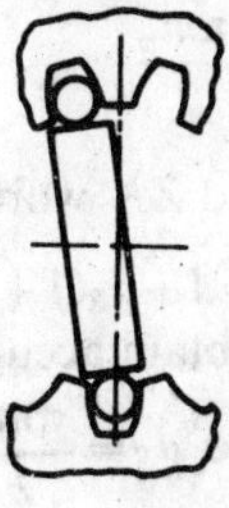

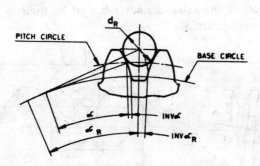

Fig. 66.

II Worm gears (See also worm gearing)

Measurement over 3 rollers for worms (Fig. 67 a)
The following equations hold good for worms having axial
tooth thickness = tooth gap = $\pi\, m_x/2$ at reference cylinder

ZI worms

$$M_R = \frac{d_b}{\cos\alpha_{R_t}} + d_R \quad \text{...............(1)}$$

$$Inv\alpha_{R_t} = inv\alpha_t + \frac{d_R}{m_x z \cos\gamma_b} + \frac{s}{d} - \frac{\pi}{z} \quad \text{............(2)}$$

where $\tan\alpha_t = \tan\alpha_n \cdot \operatorname{cosec}\gamma$

$$\tan\gamma_b = \frac{m_x \cdot z}{d \cdot \cos\alpha_t}$$

$$s = s_x \cot\gamma$$

$$s_x = \frac{\pi m_x}{2}$$

ZN, ZA and ZK worms:

$$M_R = d + d_R(1 + \sin\alpha_n) \quad \text{.........(3)}$$
To obtain accurate results $d_R = d_S$

where $d_S = \frac{\pi m_x'}{2} \cos\gamma \sec\alpha_n$

very small difference $(d_R - d_S)$ can be accounted for by adding to the
RHS of the above equation (3) an amount equal to

$$(d_R - d_S)\cot\alpha_n \cos\alpha_n$$

If the selected rollers do not project beyond the tip of the worm
teeth, a ball of the same diameter has to be used as shown in
Fig. 67(b) to measure $M_R/2$

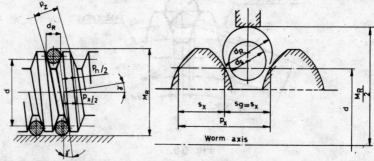

Fig. 67(a) Measurement over
three rollers for worms

Fig. 67(b) Ball gauging of worm

III Spur and helical racks: *Measurement over rollers.* (Fig. 68)

Equations

$$d_s = \frac{\pi m_n' . \sec\alpha_n}{2} \quad (d_R \text{ should be very close to } ds)$$

$$M_R = H + 0.5\left[d_R(1 + cosec\alpha_n) - \frac{\pi m_n}{2} cot\alpha_n \right]$$

Tolerance on $M_R(\triangle M_R)$

1. Class of accuracy of rack is same as class of accuracy of mating pinion.

2. Find the equivalent dia. of rack $d = \dfrac{l}{\pi}$ where l is the design length of rack.

3. $\triangle M_R = \dfrac{f_s}{2 \, tan\alpha_n}$ where f_s is chordal tooth thickness tolerance for gear of diameter d.

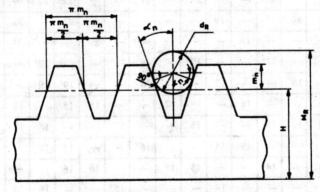

Fig. 68 Measurement over rollers for racks

GEAR RATIOS FOR GEAR BOXES

Table 76 gives the number of teeth of one of the gears constituting the gear pair for different values of sum of number of teeth of the gear pair to obtain different gear ratios in the preferred R40 series.

Table 76 No. of teeth of one of the gears constituting the gear pair

Ratio	Total number of teeth on the gear pair $z = (z_1 + z_2)$												
	50	51	52	53	54	55	56	57	58	59	60	61	62
1	25*	—	26*	—	27*	—	28*	—	29*	—	30*	—	31*
1.06	—	25	25	—	26	—	27	—	28	—	29	—	30
1.12	—	24	—	25*	—	26	—	27	—	28	—	29	29
1.18	23	—	24	—	25	25	—	26	—	27	—	28	—
1.25	22	—	23	—	24	—	25	—	26	26	—	27	—
1.32	—	22	—	23	—	—	24	—	25*	—	26	26	27
1.4	21	—	—	22	—	23	—	24	24	—	25*	—	26
1.5	20*	—	21	21	—	22*	—	23	23	—	24*	—	25
1.6	19	—	20*	—	21	21	—	22	—	—	23	—	24
1.7	—	19	—	—	20*	—	21	21	—	22	22	—	23
1.8	18	18	—	19	—	—	20*	—	—	21	—	22	22
1.9	—	—	18	—	—	19	—	—	20*	—	—	21	—
2	—	17*	—	—	18*	—	—	19*	—	—	20*	—	—
2.12	16	—	—	17	—	—	18	—	—	19	19	—	20
2.24	—	—	16	—	—	17	—	—	18	18	—	19	19
2.36	15	15	—	—	16	—	—	17	—	—	18	18	—
2.5	—	—	15	15	—	—	16*	—	—	17	17	—	—
2.65	—	14	—	—	15	15	—	—	16	16	—	—	17
2.8	—	—	—	14	—	—	—	15*	—	—	16	16	—
3	—	—	—	—	—	—	14*	—	—	—	15*	—	—
3.15	—	—	—	—	—	—	—	—	14	14	—	—	15
3.35	—	—	—	—	—	—	—	—	—	—	14	14	—

Notes: Only values marked* are exact ratios.

Other values listed have a maximum error of $\pm 2\%$

Table 76 No. of teeth of one of the gears constituting the gear pair (contd.)

Ratio	Total number of teeth on the gear pair $z = (z_1 + z_2)$											
	63	64	65	66	67	68	69	70	71	72	73	74
1	—	32*	—	33*	—	34*	—	35*	—	36*	—	37*
1.06	—	31	—	32	—	33	—	34	—	35	—	36
1.12	30	30	—	31	—	32	—	33	—	34	—	35
1.18	29	—	30	30	31	31	—	32	—	33	—	34
1.25	28*	—	29	—	30	30	31	31	—	32*	—	33
1.32	27	—	28	—	29	29	30	30	—	31	—	32
1.4	26	—	27	—	28	—	29	29	—	30*	—	31
1.5	25	—	26*	—	27	27	—	28*	—	29	29	—
1.6	24	—	25*	—	26	26	—	27	27	28	28	—
1.7	—	24	24	—	25	25	—	26	26	27	27	—
1.8	—	23	23	—	24	24	—	25*	—	26	26	—
1.9	22	22	—	23	23	—	24	24	—	25	25	—
2	21*	—	—	22*	—	—	23*	—	—	24*	—	25
2.12	20	—	21	21	—	22	22	—	23	23	—	24
2.24	—	20	20	—	—	21	—	—	22	22	—	23
2.36	19	19	—	—	20	20	—	21	21	—	22	22
2.5	18*	—	—	19	19	—	20	20*	20	—	21	21
2.65	—	—	18	18	—	—	19	19	—	20	20*	20
2.8	—	17	17	—	—	18	18	—	—	19	19	—
3	—	16*	—	—	17	17*	17	—	18	18*	18	—
3.15	15	—	—	16	16	—	—	17	17	—	—	18
3.35	—	—	—	15	—	—	16	16	—	—	17	17
3.55	14	14	15	—	—	15	15	—	—	16	16	—
3.75	—	—	—	14	14	—	—	—	15	15	—	—
4	—	—	—	—	—	—	14	14*	14	—	—	15

Notes: Only values marked* are exact ratios.

Other values listed have a maximum error of ±2%

Table 76 No. of teeth of one of the gears constituting the gear pair (contd.)

Ratio	Total number of teeth on gear pair $z = (z_1 + z_2)$												
	75	76	77	78	79	80	81	82	83	84	85	86	87
1	—	38*	—	39*	—	40*	—	41*	—	42*	—	43*	—
1.06	—	37	37	38	38	39	39	40	40	41	41	42	42
1.12	—	36	36	37	37	38	38	39	39	40	40	—	41
1.18	—	35	35	36	36	37	37	38	38	—	39	—	40
1.25	33	34	34	35	35	—	36*	—	37	37	38	38	39
1.32	32	33	33	34	34	—	35	35	36	36	37	37	—
1.4	31	32	32	—	33	33	34	34	—	35*	—	36	36
1.5	30*	—	31	31	—	32*	—	33	33	34	34*	34	35
1.6	29	29	—	30*	—	31	31	—	32	32	33	33	—
1.7	28	28	—	29	29	30	30*	30	31	31	—	32	32
1.8	27	27	—	28	28	—	29	29	30	30*	30	31	31
1.9	26	26	—	27	27	—	28	28	29	29	29	30	30*
2	25*	25	26	26*	26	27	27*	27	28	28*	28	29	29*
2.12	24	—	25	25*	25	—	26	26	—	27	27	—	28
2.24	23	—	24	24	—	25	25*	25	—	26	26	—	27
2.36	—	—	23	23	—	24	24	—	25	25*	25	—	26
2.5	—	22	22*	22	—	23	23	—	24	24*	24	—	25
2.65	—	21	21	—	—	22	22	—	23	23	23	—	24
2.8	20	20*	20	—	21	21	—	—	22	22	—	—	23
3	19	19*	19	—	20	20*	20	—	21	21*	21	—	22
3.15	18	—	—	19	19	19	—	20	20*	20	—	21	21
3.35	17	—	—	18	18	—	—	19	19	—	—	20	20*
3.55	—	—	17	17	—	—	18	18	18	—	—	19	19
3.75	16	16*	16	—	—	17	17	17	—	—	18	18	—
4	15*	15	—	—	16	16*	16	—	—	17	17*	17	—

Notes: Only values marked* are exact ratios.

Other values listed have a maximum error of $\pm 2\%$

Table 76 No. of teeth of one of the gears constituting the gear pair (contd.)

Ratio	Total number of teeth on the gear pair $z = (z_1 + z_2)$												
	88	89	90	91	92	93	94	95	96	97	98	99	100
1	44*	—	45*	—	46*	—	47*	—	48*	—	49*	49	50*
1.06	43	43	44	44	45	45	46	46	47	47	48	48	49
1.12	—	42	—	43	43	44	44	45	45	46	46	47	47
1.18	40	41	41	42	42	43	43	44	44	—	45	45	46
1.25	39	40	40*	40	41	41	42	42	43	43	44	44*	44
1.32	38	38	39	39	40	40	—	41	41	42	42	43	43
1.4	37	37	—	38	38	39	39	40	40*	40	41	41	42
1.5	35	36	36*	36	37	37	38	38*	38	39	39	40	40*
1.6	34	34	35	35*	35	36	36	—	37	37	38	38	38
1.7	33	33	33	34	34	—	35	35	36	36	36	37	37
1.8	—	32	32	—	33	33	34	34	34	35	35*	35	36
1.9	30	31	31	31	32	32	32	33	33	—	34	34	—
2	29	30	30*	30	31	31*	31	32	32*	32	33	33*	33
2.12	28	—	29	29	—	30	30	—	31	31	31	32	32
2.24	27	—	28	28	—	29	29	29	30	30	30	—	31
2.36	26	—	27	27	—	28	28	28	—	29	29	—	30
2.5	25	—	26	26*	26	—	27	27	—	28	28*	28	—
2.65	24	—	25	25	25	—	26	26	26	—	27	27	—
2.8	23	—	24	24	24	—	25	25*	25	—	26	26	26
3	22*	22	—	23	23*	23	—	24	24*	24	—	25	25*
3.15	21	—	22	22	22	—	—	23	23	—	—	24	24
3.35	20	—	21	21	21	—	—	22	22	22	—	23	23
3.55	—	—	20	20*	20	—	—	21	21	21	—	22	22
3.75	—	19	19	19	—	—	20	20*	20	—	—	21	21
4	—	18	18*	18	—	—	19	19*	19	—	—	20	20*

Notes: Only values marked* are exact ratios.

Other values listed have a maximum error of $\pm 2\%$

Table 76 No. of teeth of one of the gears constituting the gear pair (contd.)

Ratio	Total number of teeth of gear pair $z = (z_1 + z_2)$									
	101	102	103	104	105	106	107	108	109	110
1	50	51*	51	52*	52	53*	53	54*	54	55*
1.06	49	50	50*	50	51	51	52	52	53	53
1.12	48	48	49	49	50	50*	50	51	51	52
1.18	46	47	47	48	48	49	49	50	50*	50
1.25	45	45	46	46	47	47	48	48*	48	49
1.32	44	44	44	45	45	46	46	47	47	47
1.4	42	43	43	43	44	44	45	45*	45	46
1.5	40	41	41	42	42*	42	43	43	44	44*
1.6	39	39	40	40*	40	41	41	42	42	42
1.7	37	38	38	39	39	39	40	40*	40	41
1.8	36	36	37	37	—	38	38	39	39	39
1.9	35	35	—	36	36	37	37	37	38	38
2	34	34*	34	35	35*	35	36	36*	36	37
2.12	32	33	33	33	34	34	34	35	35	35
2.24	31	—	32	32	32	33	33	33	34	34
2.36	30	30	31	31	31	—	32	32	32	33
2.5	29	29	—	30	30*	30	31	31	31	31
2.65	28	28	28	—	29	29	29	30	30	30
2.8	—	27	27	27	28	28	28	—	29	29
3	25	—	26	26*	26	—	27	27*	27	—
3.15	24	—	25	25	25	—	26	26	26	—
3.35	23	—	24	24	24	24	—	25	25	25
3.55	22	—	—	23	23	23	—	24	24	24
3.75	21	—	22	22	22	22	—	23	23	23
4	20	—	—	21	21*	21	—	—	22	22*

Notes: Only values marked* are exact ratios.

Other values listed have a maximum error of $\pm 2\%$

Table 76 No. of teeth of one of the gears constituting the gear pair (concld.)

Ratio	Total number of teeth on the gear pair $z = (z_1 + z_2)$									
	111	112	113	114	115	116	117	118	119	120
1	55	56*	56	57*	57	58*	58	59*	59	60*
1.06	54	54	55	55	56	56	57	57	58	58
1.12	52	53	53	54	54	55	55	56	56	57
1.18	51	51	52	52	53	53	54	54	55	55
1.25	49	50	50	51	51	52	52*	52	53	53
1.32	48	48	49	49	50	50*	50	51	51	52
1.4	46	47	47	48	48	48	49	49	50	50*
1.5	44	45	45	46	46*	46	47	47	48	48*
1.6	43	43	43	44	44	45	45*	45	46	46
1.7	41	41	42	42	43	43	43	44	44	44
1.8	40	40*	40	41	41	41	42	42	43	43
1.9	38	39	39	39	40	40*	40	41	41	41
2	37*	37	38	38*	38	39	39*	39	40	40*
2.12	36	36	36	37	37	37	38	38	38	38
2.24	34	35	35	35	—	36	36	36	37	37
2.36	33	33	34	34	34	35	35	35	35	36
2.5	32	32*	32	33	33	33	33	34	34*	34
2.65	30	31	31	31	—	32	32	32	33	33
2.8	29	—	30	30*	30	—	31	31	31	32
3	28	28*	28	—	29	29*	29	—	30	30*
3.15	27	27	27	—	28	28	28	—	29	29
3.35	—	26	26	26	—	27	27	27	27	28
3.55	—	25	25	25	25	—	26	26	26	26
3.75	—	—	24	24*	24	—	25	25	25	25
4	22	—	—	23	23*	23	—	—	24	24*

Notes: Only values marked* are exact ratios.

Other values listed have a maximum error of $\pm 2\%$

CROSSED HELICAL GEARS

When two helical gears, mounted on non-parallel, non-intersecting shafts mesh with each other, they form a pair of crossed helical gearing.

The direction of rotation of the driven gear depends upon the position, hand of helix and the direction of rotation of the driver. The helix angles may be equal or different. When crossed helical gears have helix angles of the same hand, then the one with greater helix angle should always be the driver. In most cases the helix angles on a pair of crossed helical gears are of the same hand. When the angle between the shafts is small, however, the helix angles may be of opposite hand.

Since the theoretical contact between a pair of crossed helical gears is a point, the useful face width of the gears depends upon the distance the point of contact travels across the faces of the gears. Additional face width does not increase the load carrying capacity of crossed helical gears as it does with spur and helical gears.

Table 77 gives the values of coefficient of friction and Table 78 gives the tooth parameters and equations.

Table 77 Coefficient of friction for crossed helical gears

vs m/sec.	μ	vs m/sec.	μ	vs m/sec.	μ
0	0.2	0.875	0.0383	8.75	0.0545
0.05	0.1209	1	0.0365	10	0.0582
0.1	0.0993	1.25	0.0358	12.5	0.065
0.15	0.0859	1.5	0.033	15	0.0712
0.2	0.0764	2	0.0327	20	0.0822
0.25	0.0693	2.5	0.0335	25	0.0919
0.3	0.0637	3	0.0349	30	0.1007
0.35	0.0591	3.5	0.0366	35	0.1088
0.4	0.0553	4	0.0384	40	0.1163
0.45	0.0522	4.5	0.0402	45	0.1233
0.5	0.0495	5	0.042	50	0.13
0.625	0.0444	6.25	0.0465	—	—
0.75	0.0408	7.5	0.0506	—	—

Table 78 Tooth parameters and equations for crossed helical gears.

Sl. No.	Nomenclature	Symbol	Formulae
1	Number of teeth on pinion	z_1	—
2	Number of teeth on gear	z_2	—
3	Normal module	m_n	—
4	Normal pressure angle	α_n	—
5	Transverse module	m_t	$m_n/cos\beta$
6	Transverse pressure angle	α_t	$tan^{-1}\dfrac{tan\alpha_n}{cos\beta}$
7	Shaft angle	Σ	$\beta_1 + \beta_2$ for same hand of helix. $\beta_1 \sim \beta_2$ for opposite hands of helix.
8	Pitch circle diameter	d	$z.m_t$
9	Centre distance	a	$0.5(d_1+d_2)$
10	Minimum working face width	b'	$\dfrac{2m_n\,sin\beta}{tan\alpha_n}$ or $\leqslant 3m_n$
11	Tangential or sliding component of velocity	v_t	$v.sin\beta$, v is the peripheral velocity
12	Relative sliding velocity of gear pair	v_s	$v_{t1}+v_{t2}=\dfrac{v_1\,sin\Sigma}{cos\beta_2}=\dfrac{v_2\,sin\Sigma}{cos\beta_1}$
13	Tangential force	$F_{t1}*$	$\dfrac{975000}{n_1}\dfrac{2}{d_1}\cdot P=F_{x2}$
14	Axial force	$F_{x1}*$	$F_{t1}\,tan(\beta_1-\phi)=F_{t2}$ where $\phi=tan^{-1}\dfrac{\mu}{cos\alpha_n}$
15	Separating force	$F_{s1}*$	$F_{x1}\dfrac{tan\alpha_n}{sin\beta_1}=F_{s2}$
16	Resultant force normal to the tooth surface	F_n*	$\sqrt{F_{t1}^2+F_{x1}^2+F_{s1}^2}$
17	Efficiency of the drive	η	$\dfrac{tan(\beta_1-\phi)}{tan\beta_1}$
18	Self locking in forward drive	—	$\beta_1=\phi$
19	Condition for maximum efficiency	—	$\beta_1=\dfrac{\Sigma+\phi}{2}$; $\beta_2=\dfrac{\Sigma-\phi}{2}$
20	Maximum efficiency	η_{max}	$\dfrac{1-0.5\,tan\,\phi\,tan\,\Sigma/2}{1+0.5\,tan\,\phi\,tan\,\Sigma/2}$

* For shaft angle $\Sigma=90°$

WORM GEARING

Worm gearing is used to transmit motion between two non-intersecting, non parallel shafts. Most commonly used angle between the shafts is 90°. Generally, the worm is cylindrical and worm wheel is globoidal. Such a pair is called *Single enveloping worm gearing*. Description and calculations given further are applicable only for single enveloping worm gearing with 90° shaft angle. Single enveloping worm gearing has line contact between mating teeth. A worm and worm wheel have the same hand of helix.

Types of cylindrical worms

Depending upon the form of thread profile, worms are classified into four types, i.e. *ZA, ZN, ZI* and *ZK*. The selection of appropriate profile is usually determined by economics of manufacture. The working properties are not so much governed by the thread profile, since the shape of worm wheel hob must match the worm profile perfectly. A standard axial module is usually chosen. A standard normal module is chosen only when the method of manufacturing both the worm and worm wheel favours it.

TYPE ZA

Worm thread is straight sided in the axial section (Fig. 69). Thread profile in transverse plane is an Archimedian spiral, therefore these worms are also called *Spiral worms*. The standard pressure angle is taken in the axial plane. The tooth flank profile is

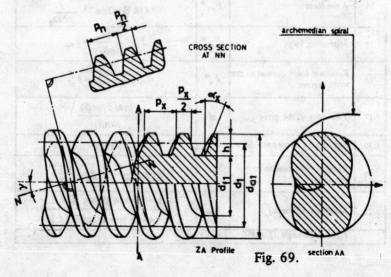

Fig. 69.

obtained when a trapezoidal turning tool is applied with its root cutting edge parallel to the worm axis and the plane containing the cutting edges includes worm axis.

TYPE ZN

Worm thread is straight sided in the normal section (Fig. 70). The standard pressure angle is taken in the normal plane. The tooth profile is obtained by tilting a turning tool of trapezoidal profile by the mean lead angle about its line of symmetry such that this line of symmetry passes through the worm axis and is perpendicular to it. For other modes of manufacture, refer Fig. 70.

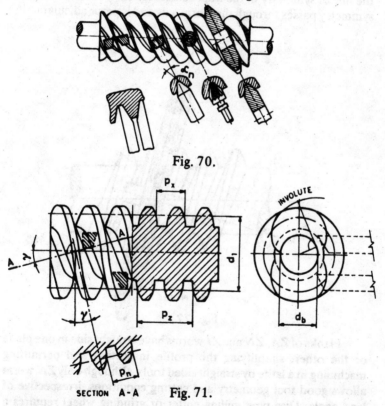

Fig. 70.

Fig. 71.

SECTION A-A

TYPE ZI

Worm thread is an involute curve in the transverse section (Fig. 71). These worms are also known as *Involute helicoid worms*. The worm is in effect a helical gear with high helix angle and a low number of teeth. The straight generator is tangential to the involute base cylinder about the worm axis. The standard pressure angle is taken in the normal plane. The tooth profile is obtained when a trapezoidal turning tool is applied with its cutting edges lying in a plane parallel to the worm axis, either above or below the worm axis. ISI has standardised on this profile.

TYPE ZK

Worm thread is produced by disc type rotary tool having straight sided profile in the axial plane (Fig. 72). The standard pressure angle is taken in the normal plane. The tooth flank profile is obtained when the tool is swivelled by the mean lead angle about the line of symmetry of the trapezoidal section such that this line of symmetry passes through the worm axis and is perpendicular to it.

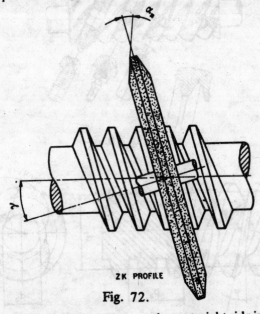

ZK PROFILE

Fig. 72.

Flanks of ZA, ZN and ZI worms have straight side in one plane or the other, simplifying the profile inspection and permitting machining in a lathe by straight sided tools, although only ZN worm allows good tool geometry and cutting conditions, irrespective of lead angle. Disc type milling cutter or grinding wheel requires a curved profile for producing these worms. On the other hand, in the case of ZK worms, these tools have a straight generating line, but the worm profile is not straight sided in any section. Thus it is easy to generate and inspect the tool profile but it is difficult to inspect the thread profile of finished worm. For lead angles less than $6°$ the profile *(ZK)* of worm produced by using a large grinding wheel with trapezoidal profile closely approximates *ZI* profile.

The shape of worm thread defines the tooth profile of worm wheel. The worm wheel is simply a gear element formed to be conjugate to the specified worm thread. Worm wheels are cut on hobbing machines and the hob is essentially a duplicate of mating worm except that it varies in tooth thickness and height to get proper root clearance and backlash in the mesh. In the case of small lot production, worm wheels are machined by fly-cutters. *ZN* type worm profile again allows good and simple tool geometry of fly cutters.

ZI worms can be manufactured with high precision. These worms can be accurately ground by the flat surface of the dish type wheel on a special grinding machine. This requirement of specialised manufacturing equipment has come in the way of wide-spread use of *ZI* worms. This is also the reason why *ZK* profile. is preferred as against *ZI* profile.

In many applications of worm gearing, it is neither necessary to harden and grind the worm, nor does the batch size justify the expense of special worm hob. In such cases, it is sufficient to finish machine the worm on a lathe with a single point tool and to machine the worm wheel on a hobbing machine with a flycutter. In such cases *ZN* worm tooth form is advantageous from the point of view of tool design.

Design considerations

SELECTION OF PARAMETERS (Table 80)
Module

Axial module is generally chosen from the standard series (refer Table 54). However considering the fact that worm and worm wheel cutting tools are to be specially made for each design of worm and worm wheel, designers can select intermediate values whenever needed. The common practice is to use a value of lead which is readily available on lathes or thread milling and grinding machines.

Lead Angle

Selection of lead angle depends on a number of factors. Efficiency of power transmission is determined by lead angle γ and coefficient of friction μ

Efficiency (worm driving wormwheel) $= (1 - \mu \ tan \ \gamma)/(1 + \mu \ cot \ \gamma)$

The efficiency increases with increasing values of γ and is maximum around $\gamma = 45^\circ$. Thus, higher lead angles are used for power transmitting worm gearing. However, high lead angles create manufacturing problems and require modification in tooth profile parameters like increase in pressure angle, reduction in dedendum and working depth. It is suggested not to use a lead angle of more than 6° per start. For instance if the lead angle is 30° there should be atleast 5 starts on the worm. If the number of starts are too few, the problem of designing tools and producing accurate curvature on worm threads and on the worm wheel teeth becomes too critical for good manufacturing practice. Maximum lead angles suitable for different pressure angles are as indicated below:

Pressure angle	Max. lead angle
$14\frac{1}{2}^\circ$	15° approximately max.
20°	25° approximately max.
25°	35° approximately max.
30°	45° max.

The drive will be irreversible or self locking if $tan^{-1} \mu \geq \gamma$

The exact lead angle at which a worm will be self locking depends on variables such as the surface finish, the kind of lubrication and the amount of vibration. Generally self locking occurs if the lead angle is less than 6°. If irreversibility is essential, it is recommended to use some form of a brake.

Pressure angle:

A pressure angle of 20° has been standardised for the general purpose application. For different applications, the pressure angles are as tabulated below:

Application	No. of starts	Pressure angle
Indexing mechanism	1 or 2	$14\frac{1}{2}^\circ$
Power transmission	1 or 2 3 or more	20° 25°
Fine pitch (instrument)	1 to 10	20°

Worm diameter:

Depending upon the desired lead angle and module, worm reference diameter gets fixed from the equation $d_1 = z_1 m_x/tan \gamma$. If the lead angle is not critical the diameter can be selected within a wide range depending upon wormshaft design and centre distance. Normally the reference diameter is expressed as a multiple of axial module $d_1 = q\, m_x$, where q is the diametral quotient and can vary from 6 to 20. If the worm is hollow, it is recommended that material below the thread be at least equal to depth of thread.

Worm gear correction

When the number of teeth on the worm gear is a limiting number then there will be undercutting of the teeth. The following values can be taken as the limit.

α_x	15^0	20^0	25^0
z_2 (Min.)	36	22	15

If standard pressure angle is in the normal section, the axial pressure angle increases with the lead angle and therefore z_2 (Min.) less than mentioned above are possible. For a given value of α_x, a smaller z_2 (Min.) is also possible by resorting to positive correction (i.e., shifting the cutter away from the worm wheel—increased centre distance). Formation of sharp teeth in end planes and decreased zone of contact however limit the amount of the positive unit correction that can be applied. For $\alpha_x = 20^0$ with $x = +0.3$, $z_2 = 16$ can be taken as the lower limit. The correction required (positive or negitive) to avoid undercutting for ZA worms ($z_2 > 16$) can be estimated from

$$x = 1 - \frac{z_2}{2}\, sin^2\, \alpha_x$$

Only the worm wheel is corrected while the size of the worm remains unaltered (Fig. 73.)

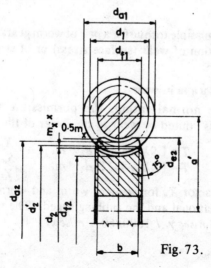

Fig. 73.

It is recommended to choose standardized sizes of worms as per IS 3734 (Table 79) which will in turn enable standardisation of hobs for worm wheels. If the centre distance is different from working centre distance, then it will be necessary to correct the wormwheel.

It is a good practice to keep $z_1 + z_2 > 40$ in power transmission worm gearing and not to exceed a positive correction $x = 0.5$ when worm is the driver.

Root interference (under cutting) in worm

For a given pressure angle of ZI worm the base circle goes on increasing with the increase in lead angle. If the base circle is larger than the working depth circle $(d_{f1} + 2c)$ there will be interference at the root of the worm. This can be avoided either by increasing the pressure angle or by reducing the tooth depth of the worm. In the later case root diameter and thread thickness of the worm are increased resulting in a decrease in the dedendum, while the tip diameter remains unaltered. The root diameter and dedendum of worm and addendum of wormwheel are calculated as follows:

$$tan\alpha_t = tan\alpha_n cosec\gamma$$

Base circle diameter $d_{b1} = d_1 cos\alpha_t = qm_x cos\alpha_t$

Root dia. of worm $d_{f1} = d_{b1} - 2c = m_x (qcos\alpha_t - 0.4cos\gamma)$

Dedendum of worm $h_{f1} = (d_1 - d_{f1})/2$
$$= 0.5\, m_x \left[q\, (1 - cos\alpha_t) + 0.4\, cos\gamma\right]$$

Addendum of wormwheel $h_{a2} = 0.5 m_x q\, (1 - cos\alpha_t)$

Load capacity

The permissible torque for a pair of worm gears is limited either by consideration of wear (surface stress) or of strength (bending stress).

Permissible torque in wear

For the normal rating, the permissible torque on the wormwheel is limited by wear to the lower of the two values.

$$T = 1.9 X_{w1}\, S_{w1}\, Y_z\, d_2^{1.8}\, m_x$$
$$T = 1.9 X_{w2}\, S_{w2}\, Y_z\, d_2^{1.8}\, m_x$$

X_w as per Fig. 75.
S_w From Table 81
Y_z from Table 82

Speed factor X_w for wear of worm and wormwheel, depend upon the rotational and the rubbing speed. (The rubbing speed is given by $\pi\, d_1 n_1 sec\, \gamma\, / \, 1000 \times 60$ m/sec.)

Permissible torque on the basis of strength

For the normal rating, the permissible torque on the wormwheel is limited by strength to the lower of the two values.

$$T = 1.8X_{s1} S_{s1} m_x l_r d_2 \cos \gamma \qquad X_s \text{ as per Fig. 74.}$$

$$T = 1.8X_{s2} S_{s2} m_x l_r d_2 \cos \gamma \qquad S_s \text{ from Table 81}$$

where $l_r = (d_{a1} + 2c) \sin^{-1}[b/(d_{a1} + 2c)]$

Lowest of the above four values of T determines the permissible torque on the wormwheel. This torque rating is based on a life of 26,000 hours.

These ratings for a worm and worm wheel pair are based on the assumption that the deflection of the worm shaft $\ngtr 0.003d_1$. Also assumed an adequate gear housing and lubrication.

Momentary overload should not exceed the value calculated from the equations:

for wear $T = 3.8S_{w2} Y_z d_2^{1.8} m_x$

and for strength $T = 4S_{s} d_2 l_r m_x \cos \gamma$

Influence of nature of sliding

The sliding contact between a worm and its gear can be separated into two types of action: *Approach* and *Recess*.

Approach is a sliding action where the gear tooth slides down the side of the worm tooth towards the worm central axis. This is a detrimental scraping that wears away the tooth surface.

Recess action is a sliding out action where the gear tooth slides up the worm tooth away from the worm central axis. The friction forces in recess action are lower than those in approach action, and they are in a direction that aids rotation. In addition, recess action tends to cold work the gear surfaces, improving contact and load capacity.

At the pitch line, where approach ends and recess begins, there is a reversal in the direction of sliding that tends to break down the oil film. This effect is eliminated with recess action. In units where the worm is always the driving member, recess action is the preferred operating mode.

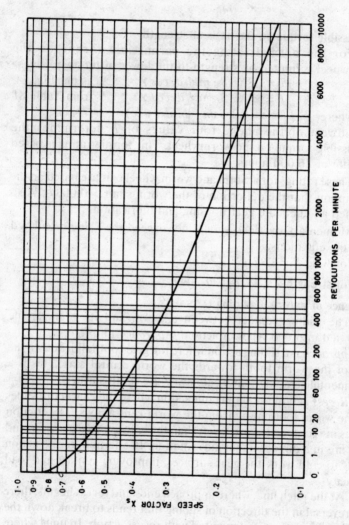

Fig. 74 Speed factor for worm gears for strength X_s

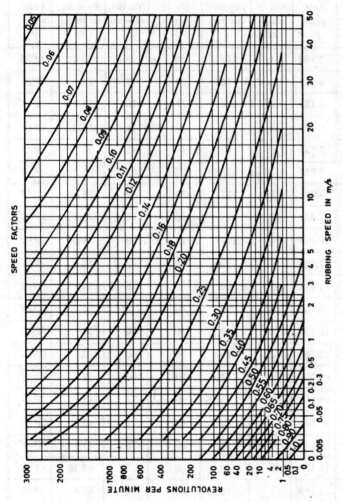

SPEED FACTORS

RUBBING SPEED IN m/s

REVOLUTIONS PER MINUTE

Fig. 75 Speed factor for worm gears for wear X_w.

Table 79 Dimensions of worms (Extract from IS 3734)

Axial module m_x	Axial pitch p_x	No. of starts z_1	Dimetral quotient q	Reference dia. of worm d_1	Tip dia. of worm d_{a1}	Root dia. of worm d_{f1}	Lead angle γ
1	3.142	1	16	16	18	13.6	3°35'
1.5	4.712	1	16	24	27	20.4	3°35'
2	6.283	1	(11)	22	26	17.2	5°12'
2	6.283	2	(11)	22	26	17.2	10°18'
2.5	7.854	1	16	40	45	34	3°35'
2.5	7.854	2	(11)	27.5	32.5	21.5	10°18'
3	9.425	1	(11)	33	39	25.8	5°12'
3	9.425	2	(11)	33	39	25.8	10°18'
3	9.425	4	(11)	33	39	25.8	19°59'
4	12.566	1	10	40	48	30.4	5°43'
4	12.566	2	10	40	48	30.4	11°19'
4	12.566	4	10	40	48	30.4	21°48'
5	15.708	1	10	50	60	38	5°43'
5	15.708	2	10	50	60	38	11°19'
5	15.708	4	10	50	60	38	21°48'
6	18.85	1	10	60	72	45.6	5°43'
6	18.85	2	10	60	72	45.6	11°19'
6	18.85	4	10	60	72	45.6	21°48'
6	18.85	6	10	60	72	49.36	30°58'
8	25.133	1	10	80	96	60.8	5°43'
8	25.133	2	10	80	96	60.8	11°19'
8	25.133	4	10	80	96	60.8	21°48'
8	25.133	4	8	64	80	48.52	26°34'

Bracketed values are non preferred.

Table 80 Equations for worm & wormwheel parameters

Type of profile		ZA	ZN	ZK	ZI
Standard module		m_x			
Pressure angle		α_x	α_n	α_n	α_n
Axial pitch	p_x	$p_x = \pi \cdot m_x$			
Lead	p_z	$p_z = z_1 \cdot p_x$			
Reference diameter of worm	d_1	$d_1 = q \cdot m_x$			
Relation between normal and axial module		$m_n = m_x \cdot \cos\gamma$			
Lead angle of worm	γ	$\tan\gamma = z_1 \cdot m_x / d_1$			
Addendum	h_a	$h_{a1} = m_x$ $h_{a2} = m_x(1+x)$ *			
Dedendum	h_f	$h_{f1} = 1 \cdot 2 m_x$ $h_{f2} = m_x(1 \cdot 2 - x)$ *			
Clearance	c	$0 \cdot 2 \, m_x$			
Total tooth depth	h	$2 \cdot 2 \, m_x$ *			
Centre distance	a	$\dfrac{d_1 + d_2}{2}$			
Working centre distance	a'	$\dfrac{d_1 + d_2'}{2}$			
Tip diameter of worm	d_{a1}	$d_{a1} = d_1 + 2h_{a1}$			
Root diamfter of worm	d_{f1} *	$d_{f1} = d_1 - 2h_{f1}$			
Pitch circle diameter of worm wheel	d_2	$d_2 = z_2 \cdot m_x$			
Working diamter of wormwheel	d_2'	$d_2' = d_2 + 2 \cdot x \cdot m_x$			
Throat diameter of wormwheel	d_{a2} *	$d_{a2} = d_2 + 2h_{a2}$			
Root diameter of wormwheel	d_{f2}	$d_{f2} = d_{a2} - 2(h_{a2} + h_{f2})$			
External diamfter of wormwheel	d_{e2}	$d_{e2} = (d_{a2} + m_x)\text{Max.}$ $= (d_{a2} + 0 \cdot 4 \, m_x)\text{Min} \cdot$			
Width of wormwheel	b	$b = 2m_x\sqrt{q+1}$			
Throat radius of wormwheel	r	$r = \dfrac{d_{f1}}{2} + c$			
Length of worm	l	$l = \sqrt{d_{a2}^2 - d_2^2}$			
Length of worm (without profile shifting)		$2m_x\sqrt{1 + z_2}$			

* For worms with *ZI* profile, check for root interference (page 314)

Table 81   Stress factors for worm gears S_s and S_w (Extract from IS 7443)

	Materials	IS reference	Bending stress factor S_s	Surface stress factor S_w when running with				
				A	B	C	D	E
A	Phosphor bronze centrifugally cast	IS 28-1958	7	—	0.85	0.85	0.92	1.55
	Phosphor bronze sand cast chilled	—	6.4	—	0.63	0.63	0.7	1.27
	Phosphor bronze sand cast	—	5	—	0.47	0.47	0.54	1.06
B	Grey cast iron	Grade 20 IS 210-1970	4.09	0.63	0.42	0.42	0.42	0.53
C	0.4% carbon steel normalized	C40 IS 1570-1961	14.1	1.1	0.7	—	—	—
D	0.55% carbon steel normalized	C55Mn75 IS 1570-1961	17.6	1.55	0.85	—	—	—
E	Carbon case hardening steel	C10, C14 IS 4432-1967	28.2	4.93	3.1	—	—	1.55
	3% nickel and nickel molybdenum case hardened steel	16Ni80Cr60 20Ni2Mo25 IS 4432-1967	33.11	5.41	3.1	—	—	1.55
	3½% nickel chromium	13Ni3Cr80 15Ni4Cr1 IS 1570-1961	35.22	6.19	3.1	—	—	1.55

Table 82   Worm gear zone factor Y_z

z_1 \ q	6	6.5	7	7.5	8	8.5	9	9.5	10	11	12	13	14	16	17	18	20
									Y_z								
1	1.045	1.048	1.052	1.065	1.084	1.107	1.128	1.137	1.143	1.16	1.202	1.26	1.318	1.374	1.402	1.437	1.508
2	0.991	1.028	1.055	1.099	1.144	1.183	1.214	1.223	1.231	1.25	1.28	1.32	1.36	1.418	1.447	1.49	1.575
3	0.822	0.89	0.969	1.109	1.209	1.26	1.305	1.333	1.35	1.365	1.393	1.422	1.442	1.502	1.532	1.58	1.674
4	0.826	0.883	0.981	1.098	1.204	1.301	1.38	1.428	1.46	1.49	1.515	1.545	1.57	1.634	1.666	1.71	1.798
5	0.947	0.991	1.05	1.122	1.216	1.315	1.417	1.49	1.55	1.61	1.632	1.652	1.675	1.735	1.765	1.805	1.886
6	1.132	1.145	1.172	1.22	1.287	1.35	1.438	1.521	1.588	1.675	1.694	1.714	1.733	1.789	1.818	1.854	1.928
7			1.316	1.34	1.37	1.405	1.452	1.54	1.614	1.704	1.725	1.74	1.76	1.817	1.846	1.88	1.95
8					1.437	1.462	1.5	1.557	1.623	1.715	1.738	1.753	1.778	1.838	1.868	1.898	1.96
9							1.573	1.604	1.648	1.72	1.743	1.767	1.79	1.85	1.88	1.91	1.96
10									1.68	1.728	1.748	1.773	1.798	1.858	1.888	1.92	1.97
11										1.732	1.753	1.777	1.802	1.862	1.892	1.924	1.98
12											1.76	1.78	1.806	1.866	1.895	1.927	1.987
13												1.784	1.806	1.867	1.898	1.931	1.992
14													1.811	1.871	1.9	1.933	1.998
																	2

1) The values are based on $b = 2m_x \sqrt{q + 1}$, symmetrical about the centre plane of the wheel.

2) For smaller face widths the value or Y_z must be reduced proportionately.

3) When it is necessary to obtain greater load capacity the worm wheel face width may be increased upto a maximum of 2.3 $m_x \sqrt{q + 1}$ and the zone factor increased proportionately.

4) The table applies to worm wheels having 30 teeth; variations in the number of teeth produce negligible changes in the value of Y_z

BEVEL GEARING

Bevel gears (Fig. 76) are most commonly used for transmitting power between intersecting axes at any angle. The pitch surfaces of the bevel gears are the surfaces of cones (Fig. 77) whose apices lie at the point of intersection of the two axes of rotation.

Crown gear is the basic member of bevel gears (Fig. 78). This may be compared to the basic rack which is a basic member of spur gear. The crown gear can be defined as a toothed disc having pitch surface (crown gear pitch plane) in the form of a flat circular surface perpendicular to its axis of rotation. The crown gear pitch plane has its centre at the intersection of bevel gear axes. A crown gear can be regarded as a bevel gear with a cone angle of 90^0 in which case the pitch cone surface and the pitch circle surfaces of bevel gears combine in the crown gear pitch plane. The relationship of bevel gear to crown gear is shown in Fig. 79.

All parameters of bevel gears cannot be inspected because of the complicated geometry. Only chordal tooth thickness, runout of pitch circle diameters and pitch can be measured. However, control of these parameters only is not sufficient for reliable operation, as the gears are very sensitive to assembly errors also.

In order to establish the correct position of bevel gears in its assembly, the contact pattern (bearing area) on the tooth flanks, achieved in the assembled conditions, is taken as the criterion. This takes care of manufacturing errors also.

As the contact area shifts and spreads under load, it may even move to the edges of the tooth under extreme conditions. This will result in excessive noise and wear and may even lead to tooth fracture. Such occurances can be reduced to a great extent by suitably controlling the tooth contact pattern during manufacture to suit given operating conditions. One of the methods adopted in manufacture to control the bearing area is by barrelling the tooth along the tooth length (Fig. 80) and sometimes along the tooth height also depending upon the type of bevel gear. Once this is done, achieving the exact position of contact area and shape, which is also called the localized tooth bearing, becomes easier. The localised tooth bearing permits larger tolerances in the mounting of gears in assembly.

The face width of bevel gears is generally limited to 1/3 to 1/4 of the cone distance. For higher widths, the thickness of tooth at the inner end becomes small and may break under load. The module as

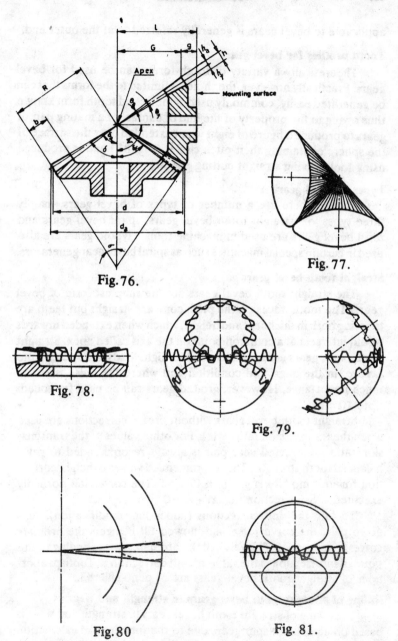

Fig. 76.

Fig. 77.

Fig. 78.

Fig. 79.

Fig. 80.

Fig. 81.

applicable to bevel gears is generally referred to at the outer end.

Tooth profiles for bevel gears

Theoretically a variety of tooth forms can be used for bevel gears. Practically however, the choice is limited to the forms that can be generated easily, commonly used is the octoid tooth form known thus owing to the property of the line of contact of a mating pair of gears to produce a figure of eight when extended over the surface of the sphere containing their pitch circles (Fig. 81) and is produced using tools having straight cutting edges.

Types of bevel gears

Though there are a number of types of bevel gears, mainly three types viz. straight tooth bevel gears, spiral bevel gears and zerol bevel gears are used in machine tools. Hypoid gears are also used in certain special machines such as spiral bevel gear generators.

Straight tooth bevel gears

The straight tooth bevel gears are the simplest form of bevel gears. The tooth traces on the pitch cone are straight but teeth are tapering both in thickness and height, which when extended inwards would intersect at a common point on the axis called apex. Straight tooth bevel gears are normally used for light loads and speeds upto $3m/sec$ in the unground condition and where the noise is not of much importance. However, ground gears can be used for speeds upto $10m/sec$.

Straight tooth bevel gears without profile corrections are used when the transmission ratio is 1:1. For other values of the transmission ratio a corrected gear pair is always recommended to get a balanced tooth strength. The recommended values of height correction for unit module is given in Fig. 82. The correction normally executed is S_0 correction i.e., $x_1 + x_2 = 0$.

To find the unit corrections (addendum modification) for a given pair, the curve for z_1 is followed till it meets the ordinate corresponding to z_2. The value of the ordinate corresponding to this point of intersection is the value of unit correction x. Tooth proportions of straight tooth bevel gears are as per Table 83.

Rating of straight tooth bevel gears in strength and wear

The rating of straight tooth bevel gears in strength and wear is based on equivalent spur gears due to the simplicity of calculation involved. This assumption has given satisfactory performance of gears in practice. The equivalent spur gear of a straight tooth bevel

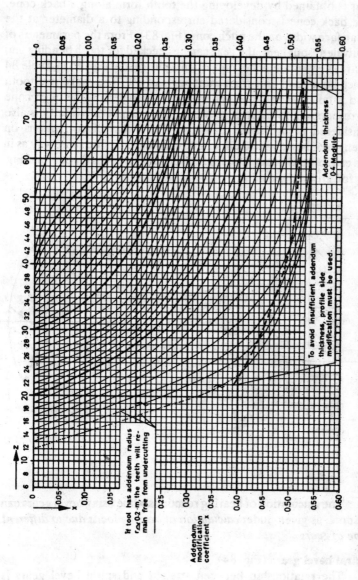

To avoid insufficient addendum thickness, profile side modification must be used.

If tool has addendum radius $r \approx 0.2 \cdot m$, the teeth will remain free from undercutting

Addendum thickness 0·4 Module

Addendum modification coefficient x

Fig. 82 Profile modifications for straight bevel gears with 90° shaft angle and 20° pressure angle (Extract from VDF News No. 34. 1968)

gear is obtained by developing the tooth form along a back cone. The back cone is considered corresponding to a diameter at the mean face width on the pitch cone (Fig. 83). From the parameters of equivalent spur gear, the gear can be checked for tooth bending and wear. The parameters of equivalent spur gear are given in Table 84 which hold good only for a shaft angle of 90°. The tangential tooth load is taken as the load acting at the diameter on the pitch cone corresponding to mean face width. The peripheral velocity is also considered at the same diameter. The effect of corrections on strength of bevel gear teeth can be considered in the same way as in the case of spur gears. Use Fig. 56 for generated (cut) gears and Fig. 57 for ground gears.

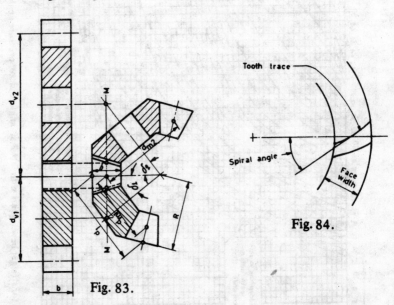

Fig. 83.

Fig. 84.

The calculation of bearing reactions while using bevel gears can be done as given under *calculation of bearing loads due to different type of gears*

Spiral bevel gears (Fig. 84).

The relationship between straight and spiral bevel gears is comparable with that of spur and helical gears. The curved and oblique teeth of spiral bevel gears result in a considerable amount of overlap, thus ensuring more than one tooth in contact at all times.

This leads to gradual engagement and continuous pitch line contact and results in larger load carrying capacity and smooth running compared to straight bevel gears of the same size.

Spiral bevel gears are suitable for transmission ratios as high as 1:10.

Minimum virtual number of teeth in a spiral bevel gear should be 14 to avoid undercutting. However with suitable addendum modification, smaller number of teeth can also be used.

The curved and oblique tooth form of spiral bevel gears gives rise to greater thrust loads than straight tooth bevel gears. Rolling thrust bearings are usually used to take up these forces as against the use of plain bearings in case of straight bevel gears.

The spiral angle of spiral bevel gear tooth is defined as an angle between the tangent to the tooth trace and generator of pitch cone (Fig. 84). This angle is different at different points along the tooth trace. Therefore, the value of spiral angle at the middle of face width of gear is taken as the value characterising the inclination of tooth trace. This is called mean spiral angle. Most commonly used spiral angle is 35°. A smaller spiral angle may result in undercut and a reduced contact ratio.

A right hand spiral bevel gear is one in which the tooth is inclined in the clockwise direction radially outwards through the midpoint of teeth when viewed at the face of the gear (Fig. 85a).

A left hand spiral bevel gear is one, in which the tooth is inclined in counter clockwise direction radially outwards through the midpoint of teeth when viewed at the face of the gear (Fig. 85b).

A right hand gear meshes with left hand pinion and a left hand gear meshes with right hand pinion.

Allowable surface speeds for spiral bevel gears can be taken as 11 *m* /sec. for unground gears and 35 *m*/sec. for ground gears.

Types of tooth traces

The tooth traces of spiral bevel gears when developed on the crown wheel pitch plane can be a part of a circle (Gleason method), extended involute (Klingelnberg method) or extended epicycloid (Oerlikon & Klingelnberg method). The advantage of using a circular arc is that the gears with this tooth trace can be ground in the same way as it is cut whereas gears with other types of tooth traces cannot be ground easily.

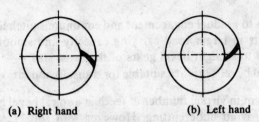

(a) **Right hand**   (b) **Left hand**

Fig. 85 Spiral bevel gears

Zerol bevel gears (Fig. 86)

Zerol bevel gears are spiral bevel gears with zero spiral angle. As these gears are manufactured on spiral bevel gear generators, localised tooth contact can be easily achieved. These gears produce the same thrust loads as straight tooth bevel gears and may be used in the same mountings.

Alowable surface velocity for unground zerol bevel gear can be taken as 5 *m/sec.* and for ground zerol gear upto 16 *m/sec.* Tooth proportions, strength and wear calculation for zerol bevel gears are same as those for straight tooth bevel gears.

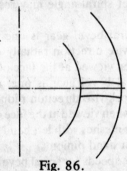

Fig. 86.

Accuracy grades of bevel gears

Depending upon the functional requirement of bevel gears, different accuracy grades have been formulated. Table 85 gives the application data, the method of manufacture, the permissible surface velocity and surface finish of tooth for different classes of accuracy.

Tolerances for bevel gears

For the purpose of simplicity, tolerance on bevel gears have been grouped under the following headings:

1) Tolerances on the gear blank and
2) Tolerances on the gear

Tolerances on the gear blank (Table 86)
1) Tolerance on tip diameter f_{da}
2) Tolerance on tip angle $f_{\delta a}$
3) Tolerance on the distance from tip to mounting surface f_g
4) Tolerance on back cone angle f_σ
5) Tolerance on the distance from cone apex to mounting surface f_i

Tolerance on the distance from cone apex to mounting surface is required to work out fixture and machine setting.

The following geometrical accuracies on gear blanks are also important. Table 87 gives the recommended values of:
1) runout of mounting surface f_x
2) runout of face and back cone angle f_{xa}

Tolerances on the gear
1) Adjacent pitch error f_t (Table 88)
2) Cumulative pitch error F_t (Table 89)
3) Tolerance on tooth thickness f_s (Table 90)
4) Minimum thinning of tooth Δs (Table 91)
5) Radial runout of tooth flanks f_r (Table 92)

Backlash (Table 93)

The minimum backlash values are fixed independent of the class of accuracy of the gear. The gearing is divided into four groups on the basis of backlash requirement. They are as follows:
1) Gearing with zero backlash j_{nz}
2) Gearing with reduced backlash j_{nr}
3) Gearing with standard backlash j_{ns}
4) Gearing with increased backlash j_{ni}

Wherever possible gearing with standard backlash is to be preferred.

Table 83 Tooth proportions of straight tooth bevel gears (for 90° shaft angle and $\alpha_n = 20°$) Refer to Fig. 76

Sl No.	Detail	Symbol	Pinion	Gear
1	Pitch circle diameter	d	$d_1 = z_1 m_n$	$d_2 = z_2 \cdot m_n$
2	Pitch cone angle	δ	$\tan \delta_1 = \dfrac{z_1}{z_2}$	$\tan \delta_2 = \dfrac{z_2}{z_1}$
3	Cone distance	R	$R = \dfrac{d_1}{2 \sin \delta_1}$	$R = \dfrac{d_2}{2 \sin \delta_2}$
4	Correction on pinion	x @	$x = \dfrac{14 - z_{v1}}{17}$ to avoid undercutting (for z_{v1} see Table 84) (For balanced design, correction values to be taken from Fig. 82)	
5	Addendum	h_a	$h_{a1} = m_n(1+x)$	$h_{a2} = m_n(1-x)$
6	Dedendum	h_f	$h_{f1} = m_n(1.1667^* - x)$	$h_{f2} = m_n(1.1667^* + x)$
7	Total tooth height	h	$h_1 = h_{a1} + h_{f1}$	$h_2 = h_{a2} + h_{f2}$
8	Addendum angle	θ_a	$\tan \theta_{a1} = \dfrac{h_{a1}}{R}$	$\tan \theta_{a2} = \dfrac{h_{a2}}{R}$
9	Dedendum angle	θ_f	$\tan \theta_{f1} = \dfrac{h_{f1}}{R}$	$\tan \theta_{f2} = \dfrac{h_{f2}}{R}$
10	Tip angle	δ_a	$\delta_{a1} = \delta_1 + \theta_{a1}$	$\delta_{a2} = \delta_2 + \theta_{a2}$
11	Root angle	δ_f	$\delta_{r1} = \delta_1 - \theta_{f1}$	$\delta_{f2} = \delta_2 - \theta_{f2}$
12	Tip diameter	d_a	$d_{a1} = d_1 + 2h_{a1} \cos \delta_1$	$d_{a2} = d_2 + 2h_{a2} \cos \delta_2$
13	Circular tooth thickness	s	$s = \dfrac{\pi m_n}{2} + 2 m_n x \tan \alpha_n$	

Table 83 Tooth proportions of straight tooth bevel gears (for 90° shaft angle and $\alpha_n = 20°$) (Concld.)

Sl. No.	Detail	Symbol	Pinion	Gear
14	Chordal tooth thickness	$\bar{s}$	$\bar{s}_1 = s\left[1 - \frac{1}{6}\left(\frac{s}{d_1}\right)^2\right]$	$\bar{s}_2 = s\left[1 - \frac{1}{6}\left(\frac{s}{d_2}\right)^2\right]$
15	Chordal tooth height	$\bar{h}_a$	$\bar{h}_{a1} = h_{a1} + \bar{s}_1^2 \cos \delta_1 / 4 d_1$	$\bar{h}_{a2} = h_{a2} + \bar{s}_2^2 \cos \delta_2 / 4 d_2$
16*	Clearance*	c	$0.1667 m_n$	$0.1667 m_n$
17	Face width	b	\multicolumn{2}{|c|}{$\frac{1}{3}$ to $\frac{1}{4} R$}
18	Module	m_n	\multicolumn{2}{|c|}{$m_n = \dfrac{b}{8 \text{ to } 10} = m_t$}
19	Distance from pitch apex to tip	G	$G_1 = \frac{d_1}{2} \cot \delta_1 - h_{a1} \sin \delta_1$	$G_2 = \frac{d_2}{2} \cot \delta_2 - h_{a2} \sin \delta_2$
20	Distance from pitch apex to reference surface	l	l_1 to be chosen	l_2 to be chosen
21	Distance from reference surface to tip	g	$g_1 = l_1 - G_1$	$g_2 = l_2 - G_2$
	For shaft angle $\gtrless$ 90° except pitch cone angle all other parameters are same as that of 90° shaft angle			
22	Pitch cone angle for $\Sigma \neq 90°$	δ	$\tan \delta_1 = \dfrac{\sin \Sigma}{\dfrac{z_2}{z_1} + \cos \Sigma}$	$\tan \delta_2 = \dfrac{\sin \Sigma}{\dfrac{z_1}{z_2} + \cos \Sigma}$

* Clearance $c = 0.188\, m + 0.050$ for Gleason system $= 0.2\, m$ as per IS 5037

@ If x is negative, use $x = 0$.

Table 84 Parameters of an equivalent spur gear for a given straight tooth bevel gear. (Valid only for $\Sigma = 90°$)

Sl No.	Details	Symbol	Pinion		Gear
1	Pressure angle	α_{nv}	α_n		α_n
2	Transmission ratio	i_v		$i_v = \dfrac{z_{v2}}{z_{v1}}$	
3	Number of teeth	z_v	$z_{v1} = \dfrac{z_1}{\cos \delta_1}$		$z_{v2} = \dfrac{z_2}{\cos \delta_2}$
4	Pitch circle diameter	d_v	$d_{v1} = \dfrac{d_{m1}}{\cos \delta_1}$		$d_{v2} = \dfrac{d_{m2}}{\cos \delta_2}$
5	Transverse module	m_{tv}		$\dfrac{d_{v1}}{z_{v1}}$	
6	Tooth width	b_v		$b_v = b$	
7	Equivalent tooth load	F_{tv}		$F_{tv} = F_t$	
8	Peripheral velocity	v		$v = \dfrac{\pi \cdot d_m n}{60 \times 1000}$	
9	Mean pitch diameter	d_m	$d_{m1} = d_1 - b \sin \delta_1$		$d_{m2} = d_2 - b \sin \delta_2$

Table 85 Bevel gear application data

Class of accuracy GOST 1758	Method of manufacture	Surface finish R_a in μm	Peripheral velocity	Application
6	Precision grinding and lapping (hardened)	0.4—0.8	More than 5 m/sec. for straight teeth. More than 10 m/sec. for spiral teeth	Mechanisms where very high speed and accurate gearing is required.
7	Grinding and lapping or shaving (hardened)	0.8—1.6	2 to 5 m/sec for straight teeth. 5 to 10 m/sec. for spiral teeth.	High speed gear boxes requiring continuous noise free running. Also for transmitting heavy loads and in indexing mechanisms.
8	Shaping, hardened, finished by lapping	1.6—3.2	Up to 2 m/sec. for straight teeth. Upto 5 m/sec. for spiral teeth.	High speed-low power or low speed high power transmissions. Standard gear boxes & other mehchanism.
9	Milling using formed cutter or shaping. Normally unhardened.	3.2—6.3	Upto 1 m/sec.	Low power drives and mechanisms not requiring continuous operation. e.g. hand drives

Table 86 Tolerances on bevel gear blank

Tolerance on	Symbol	Module		Diameter of blank 0-800 (for all classes of accuracy)
		over	upto	
Tip diameter μm	f_{da}	0.3	0.5	− 75
		0.5	1.25	− 100
		1.25	10	− 125
Distance from tip to mounting surface μm	f_g	—	0.5	− 25
		0.5	1.25	− 50
		1.25	10	− 75
Distance from apex to mounting surface μm	f_t	—	0.5	− 25
		0.5	1.25	− 50
		1.25	10	− 75
Tip angle $sec.$	$f_{\delta a}$	0.3	0.5	+ 30
		0.5	1.25	+ 15
		1.25	10	+ 8
Back cone angle $sec.$	f_σ	—	0.5	± 60
		0.5	1.25	± 30
		1.25	10	± 15

Table 87 Geometrical accuracies of bevel gear blank

Class of accuracy	Runout of face and back cone f_{xa}	Runout of mounting surface, f_x μm	
		upto diameter 50	above diameter 50
6		5	10
7	IT6	8	16
8		25	40
9			

Table 88 Adjacent pitch error f_i in μm

Class of accuracy	Module		Pitch circle diameter d				
	over	upto	upto 50	over 50 upto 100	over 100 upto 200	over 200 upto 400	over 400 —
6	1	3	16	16	17	18	20
	3	6	17	18	18	20	22
	6	10	21	22	22	24	24
7	1	3	22	22	24	26	28
	3	6	24	25	26	28	30
	6	10	30	30	32	32	35
8	1	3	30	32	32	35	38
	3	6	34	35	36	38	42
	6	10	40	42	44	45	50
9	1	3	42	44	45	50	55
	3	6	48	50	50	55	60
	6	10	58	60	60	65	70

Table 89 Cumulative pitch error F_t in μm

Class of accuracy	Module		Pitch circle diameter d					
	over	upto	upto 50	over 50 upto 80	over 80 upto 120	over 120 upto 200	over 200 upto 320	over 320 upto 500
6	1	10	30	36	42	48	55	70
7	1	10	45	55	65	75	90	110
8	1	10	70	90	105	120	140	180
9	1	10	110	140	170	200	240	280

Table 90 Tolerance on tooth thickness f_s in μm

type of back-lash	Radial runout of tooth flanks f_r in μms.												
	over 8 upto16	16 20	20 25	25 32	32 40	40 50	50 60	60 80	80 100	100 120	120 160	160 200	200 250 320
j_{nt}	36	38	42	48	55	65	70	85	100	120	150	180	220 280
j_{nr}	40	42	48	55	60	70	80	100	115	130	170	210	250 320
j_{ns}	45	48	52	60	70	80	90	110	130	150	190	240	280 340
j_{nl}	50	55	60	70	80	90	100	120	140	170	200	250	300 380

Table 91   Minimum thinning of tooth Δs in μm

Class of accuracy	Type of backlash	Module in mm over	Module in mm upto	Pitch circle diameter, d — upto 50	over 50 upto 80	over 80 upto 120	over 120 upto 200	over 200 upto 320	over 320 upto 500
6	jnz	1	2.5	12	16	20	25	32	40
		2.5	6	13	17	21	26	34	42
		6	10	—	18	22	28	36	45
	jnr	1	2.5	25	32	40	50	60	80
		2.5	6	26	34	42	52	65	85
		6	10	—	36	45	55	70	90
	jns	1	2.5	50	60	80	100	120	160
		2.5	6	52	65	85	105	130	170
		6	10	—	70	90	110	140	180
	jni	1	2.5	100	120	160	200	250	320
		2.5	6	105	130	170	210	250	340
		6	10	—	140	180	220	280	360
7	jnz	1	2.5	12	16	20	25	32	40
		2.5	6	13	17	21	26	34	42
		6	10	—	18	22	28	36	45
	jnr	1	2.5	25	32	40	50	60	80
		2.5	6	26	34	42	52	65	85
		6	10	—	36	45	55	70	90
	jns	1	2.5	50	60	80	100	120	160
		2.5	6	52	65	85	105	130	170
		6	10	—	70	90	110	140	180
	jni	1	2.5	100	120	160	200	250	320
		2.5	6	105	130	170	210	250	340
		6	10	—	140	180	220	280	360
8	jnr	1	2.5	25	32	40	50	60	80
		2.5	6	26	34	42	52	65	85
		6	10	—	36	45	55	70	90
	jns	1	2.5	50	60	80	100	120	160
		2.5	6	52	65	85	105	130	170
		6	10	—	70	90	110	140	180
	jni	1	2.5	100	120	160	200	250	320
		2.5	6	105	130	170	210	250	340
		6	10	—	140	180	220	280	360
9	jnr	1	2.5	25	32	40	50	60	80
		2.5	6	26	34	42	52	65	85
		6	10	—	36	45	55	70	90
	jns	1	2.5	50	60	80	100	120	160
		2.5	6	52	65	85	105	130	170
		6	10	—	70	90	110	140	180
	jni	1	2.5	100	120	160	200	250	320
		2.5	6	105	130	170	210	250	340
		6	10	—	140	180	220	280	360

jns is preferred type of meshing

Table 92 Radial runout of tooth flanks f_r in μm

Class of of accuracy	Module		Pitch cone angle		Pitch circle diameter d					
	over	upto	over	upto	over upto 50	50 80	80 120	120 200	200 320	320 500
6	1	10	—	20°	21	25	30	36	48	50
			20°	30°	24	28	34	40	48	55
			30°	40°	28	32	38	45	52	60
			40°	—	30	36	42	50	60	70
7	1	10	—	20°	34	40	48	55	65	80
			20°	30°	36	45	55	65	75	90
			30°	40°	45	52	60	70	80	95
			40°	—	48	55	65	80	95	110
8	1	10	—	20°	52	63	75	90	105	125
			20°	30°	60	70	90	105	120	140
			30°	40°	70	80	95	110	125	150
			40°	—	75	90	105	125	150	180
9	1	10	—	20°	85	100	120	140	170	200
			20°	30°	95	110	130	160	190	220
			30°	40°	110	130	150	180	210	240
			40°	—	120	140	170	200	240	280

Table 93 Minimum backlash j_n in μm

Type of backlash	Length of pitch cone generator R							
	over upto 50	50 80	80 120	120 200	200 320	320 500	500 800	800 1250
j_{nz}	0	0	0	0	0	0	0	0
j_{nr}	40	50	55	85	100	130	170	210
j_{ns}	85	100	130	170	210	260	340	420
j_{ni}	170	210	260	340	420	530	670	850

PLANETARY GEARING

Planetary gearing consists of a system of gears in which two co-axial gears are connected by a number of equally spaced gears or gear assemblies called planets mounted on a carrier which is also co-axial with the first two gears.

Planetary gearing is classified as
   i) simple planetary trains, and
   ii) compound planetary trains.

A simple planetary train has four basic elements namely (i) an external gear called sun wheel (ii) an internal gear called annulus (iii) a planet carrier and (iv) planets. The sun wheel, the annulus and the planet carrier are co-axial. Each one of these could be used as input, output or reaction member.

Compound planetary trains are obtained by combining two or more simple planetary trains such that two members of each train are connected to two members of another train. In case of compound planetary trains internal gears could be replaced by external gears.

Planetary trains could be used as (i) fixed ratio drives or as (ii) differential drives. In a fixed ratio drive, one of the three co-axial shafts is locked and the other two are used as input or output member as desired. In a differential drive all the three co-axial members are allowed to rotate with any one of them as the driver and the other two as followers or vice-versa.

Speed ratios for the combinations of planetary drives commonly used in machine tools are given in Tables 94, 95 and 96. Only theoritical range of transmission ratios are mentioned in the tables. The practical value depends on the space available, power transmission capacity, etc. If the transmission ratio, i, is preceded by a negative sign, it indicates that the driver and follower are rotating in opposite directions, otherwise they rotate in the same direction.

Planetary gears have the following advantages:
1. symmetrical and compact construction with possibility to obtain very high transmission ratios.
2. co-axial input and output shafts.
3. load sharing between several pinions.
4. low weight and efficient space utilization.
5. low flywheel effect.
6. low rolling and sliding velocities at tooth flanks.
7. silent running.

8. high efficiency.
9. reduced sensitivity to shock loading.
10. differential output speeds and torques.
11. complete balance of static forces within the gear train.
12. no resultant tooth loads on the bearings of co-axial shafts
    when spur gears are used.

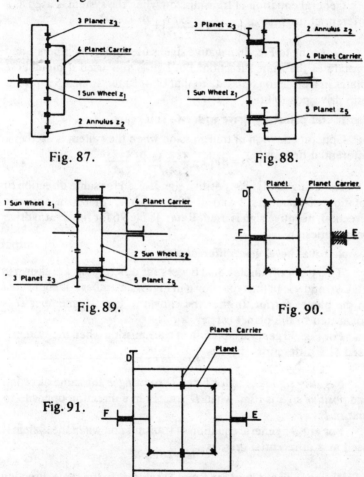

Fig. 87.

Fig. 88.

Fig. 89.

Fig. 90.

Fig. 91.

Simple planetary train

General equation of transmission when the system is used as a differential drive is $n_1 = \dfrac{n_4\,(z_1 + z_2)}{z_1} \pm \dfrac{n_2 . z_2}{z_1}$

Referring to Fig. 87 negative sign is used if members 2 and 4 rotate in the same direction. Positive sign is used when member 2 rotates in a direction opposite to that of 4. Refer Table 94 for fixed ratio drive possibilities.

Compound planetary drive with one sun gear and one internal gear

General equation of transmission when the system is used as a differential drive is

$$n_1 = n_4\left(1 + \frac{z_2.z_3}{z_1.z_5}\right) \pm n_2\frac{(z_2.z_3)}{z_1.z_5}$$

Referring to Fig. 88 negative sign is used when members 2 and 4 rotate in the same direction. Positive sign is used if member 2 rotates in a direction opposite to that of 4. Refer Table 95 for fixed ratio drive possibilities.

Compound planetary drive with two sun gears

General equation of transmission when the system is used as a differential drive is

$$n_1 = n_4\left(1 - \frac{z_2.z_3}{z_1.z_5}\right) \pm \frac{n_2\,z_2.z_3}{z_1.z_5}$$

Referring to Fig. 89, positive sign is used for same direction of rotation of members 2 and 4. If member 2 rotates in opposite direction, negative sign is used. Refer Table 96 for fixed ratio drive possibilities.

Simple equal bevel gear differentials

Two types of simple equal bevel gear differentials are shown in Figs. 90 and 91. In the type shown in Fig. 90, member F is connected to the planet carrier. In the type shown in Fig. 91, member D is connected to the planet carrier.

For Fig. 90 general equation of transmission when the system is used as a differential drive is

$$n_E = 2n_F \mp n_D$$

Negative sign is used for F and D rotating in the same direction and *positive sign* is used when D rotates in a direction opposite to that of F.

For Fig. 91 general equation of transmission when the system is used as a differential drive is

$$n_E = 2n_D \mp n_F$$

Negative sign is used for F and D rotating in the same direction and *positive sign* is used when F rotates in a direction opposite to that of D.

In the above equations by fixing any one of the members i.e., by having corresponding $n = 0$, the fixed drive relationship is obtained.

Table 94 Fixed ratio drive possibilities (Simple planetary train - Fig. 87)

	1	2	3	4	5	6
Drive no.	1	2	3	4	5	6
Fixed member	1	1	2	2	4	4
Driver	2	4	1	4	1	2
Driven	4	2	4	1	2	1
Transmission ratio $i = \dfrac{rpm \text{ of the driver}}{rpm \text{ of the driven}}$	$1+\dfrac{z_1}{z_2}$	$\dfrac{z_2}{z_1+z_2}$	$1+\dfrac{z_2}{z_1}$	$\dfrac{z_1}{z_1+z_2}$	$-\dfrac{z_2}{z_1}$	$-\dfrac{z_1}{z_2}$
Theoretical range of i	$1<i<2$	$0.5<i<1$	$2<i<\infty$	$0<i<0.5$	$-\infty<i<-1$	$-1<i<0$
Efficiency η	$1/\eta = \dfrac{1}{i}+\left(1-\dfrac{1}{i}\right)\dfrac{1}{\eta_F}$	$1/\eta = i+(1-i)\dfrac{1}{\eta_F}$	$1/\eta = \dfrac{1}{i}+\left(1-\dfrac{1}{i}\right)\eta_F$	$1/\eta = i+(1-i)\dfrac{1}{\eta_F}$	$\eta=\eta_F$	

Table 95 Fixed ratio drive possibilities (Compound planetary drive with one sun gear and one internal gear - Fig. 88)

Drive no.	1	2	3	4	5	6
Fixed member	1	1	2	2	4	4
Driver	2	4	1	4	1	2
Driven	4	2	4	1	2	1
Transmission ratio $i = \dfrac{rpm \text{ of the driver}}{rpm \text{ of the driven}}$	$1 + \dfrac{z_1 z_5}{z_2 z_3}$	$\dfrac{z_2 z_3}{z_1 z_5 + z_2 z_3}$	$1 + \dfrac{z_2 z_3}{z_1 z_5}$	$\dfrac{z_1 z_5}{z_1 z_5 + z_2 z_3}$	$-\dfrac{z_2 z_3}{z_1 z_5}$	$-\dfrac{z_1 z_5}{z_2 z_3}$
Theoretical range of i	$1 < i < 2$	$0 \cdot 5 < i < 1$	$2 < i < \infty$	$0 < i < 0 \cdot 5$	$-\infty < i < 0$	$-1 < i < 0$
Efficiency η	$1/\eta =$ $\dfrac{1}{i} + \left(1 - \dfrac{1}{i}\right)\dfrac{1}{\eta_F}$	$1/\eta =$ $i + (1-i)\dfrac{1}{\eta_F}$	$1/\eta =$ $\dfrac{1}{i} + \left(1 - \dfrac{1}{i}\right)\dfrac{1}{\eta_F}$	$1/\eta =$ $i + (1-i)\dfrac{1}{\eta_F}$	$\eta = \eta_F$	

Table 96 Fixed ratio drive possibilities (Compound planetary drive with two sun gears - Fig. 89)

Drive no.	1	2	3	4	5	6
Fixed member	1	1	2	2	4	4
Driver	2	4	1	4	1	2
Driven	4	2	4	1	2	1
Transmission ratio $i = \dfrac{rpm \text{ of the driver}}{rpm \text{ of the driven}}$	$1 - \dfrac{z_1 z_5}{z_2 z_3}$	$\dfrac{z_2 z_3}{z_2 z_3 - z_1 z_5}$	$1 - \dfrac{z_2 z_3}{z_1 z_5}$	$\dfrac{z_1 z_5}{z_1 z_5 - z_2 z_3}$	$\dfrac{z_2 z_3}{z_1 z_5}$	$\dfrac{z_1 z_5}{z_2 z_3}$
Theoretical range of i	$0 < i < 1$	$1 < i < \infty$	$-\infty < i < 0$	$-\infty < i < 0$	$0 < i < 1$	$1 < i < \infty$
Efficiency η	$1/\eta = \dfrac{1}{i} + \left(1 - \dfrac{1}{i}\right)\dfrac{1}{\eta_F}$	$1/\eta = i + (1-i)\dfrac{1}{\eta_F}$	$1/\eta = \dfrac{1}{i} + \left(1 + \dfrac{1}{i}\right)\dfrac{1}{\eta_F}$	$1/\eta = i + (1-i)\dfrac{1}{\eta_F}$	$\eta = \eta_F$	

CALCULATION OF BEARING LOADS DUE TO DIFFERENT TYPES OF GEARS

Spur and helical gears

In spur and helical gear transmission (Fig. 92) the bearings at A and B are subjected to radial loads caused by tangential and separating forces.

$$F = \sqrt{F_t^2 + F_s^2} = \frac{F_t}{cos\alpha'_t} \qquad \text{where } F_t = \frac{975000P}{n} \cdot \frac{2}{d}$$

$$cos\alpha'_t = \frac{a}{a'} \cdot cos\alpha_t \qquad \text{and } \alpha_t = tan^{-1}\left(\frac{tan\alpha_n}{cos\beta}\right)$$

In helical gears, in addition to the above forces the bearings are subjected to a radial force due to the moment M generated by the axial force F_x

$$M = F_x \cdot \frac{d'}{2} = F_t \frac{d'}{2} tan\beta'$$

The bearing load at A and B acting radially are

$$F_A = F_t \sqrt{\left(\frac{l_B}{l}\right)^2 + \left[\frac{l_B}{l}tan\alpha_t' \pm \frac{d'}{2l} tan\beta'\right]^2}$$

$$F_B = F_t \sqrt{\left(\frac{l_A}{l}\right)^2 + \left[\frac{l_A}{l}tan\alpha_t' \pm \frac{d'}{2l} tan\beta'\right]^2}$$

As regards $\frac{d'}{2l} tan\beta'$ + sign for F_A and − sign for F_B hold good for the direction of rotation and direction of helix of the driving gear as shown in Fig. 92. If the direction of rotation is opposite then, -sign for F_A and + sign for F_B will hold good. The axial force $F_x = F_t tan\beta'$ has to be taken up by any one of the two bearings. In Fig. 92, bearing A is designed to take both radial and axial force.

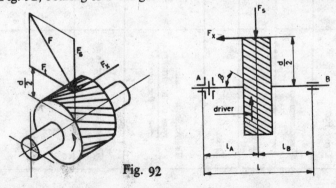

Fig. 92

Bevel gears

Seperating force & axial thrust

$$F_{s\,1,2} = F_t \left(\frac{tan\alpha_n \cdot cos\delta_{1,2}}{cos\beta} \pm tan\beta.sin\delta_{1,2} \right)$$

$$F_{x\,1,2} = F_t \left(\frac{tan\alpha_n.sin\delta_{1,2}}{cos\beta} \pm tan\beta.cos\delta_{1,2} \right)$$

F_x is considered +ve when it is directed away from cone apex

Tilting moment

$$M_{1,2} = F_{x\,1,2} \frac{d_{m\,1,2}}{2}$$

The direction of the tooth spiral is observed from the cone apex radially outwards.

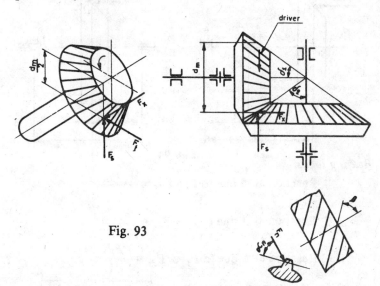

Fig. 93

$d_m = (d - bsin2\delta)$

M = tilting moment on shaft caused by axial force

F_t is calculated as in spur and helical gears.

The signs for the last term in the equations for F_s and F_x are governed by the following rules

Rule for	Direction of tooth spiral and rotation	
	Same	Opposite
Driver	$-$in equation F_s $+$ in equation F_x	$+$in equation F_s $-$ in equation F_x
Driven	$+$in equation F_s $-$ in equation F_x	$-$in equation F_s $+$ in equation F_x

Note: In Fig. 93 the direction of rotation and direction of tooth spiral are opposite.

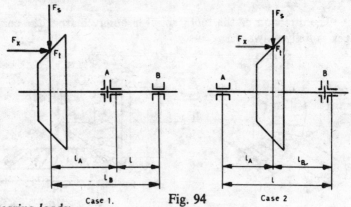

Case 1.   Fig. 94   Case 2

Bearing loads:

Reaction at A due to $F_s = F_{As} = \dfrac{F_s l_B}{l}$

Reaction at A due to $F_x = F_{Ax} = \dfrac{F_x d_m}{2l}$

Reaction at A due to $F_t = F_{At} = \dfrac{F_t l_B}{l}$

Resultant radial reaction at $A = F_A = \sqrt{F_{At}^2 + (F_{As} - F_{Ax})^2}$
(for case 1 and 2 of Fig. 94)

Similarly the reactions at B due to F_s, F_t and F_x can be calculated. If F_{Bs}, F_{Bt} and F_{Bx} are the corresponding reactions then the resultant

$$F_B = \sqrt{F_{Bt}^2 + (F_{Bs} \pm F_{Bx})^2} \quad \left(\begin{array}{l} - \text{ for case } 1 \\ + \text{ for case } 2 \end{array}\right)$$

Worm gears
(also applicable for crossed helical gears with a shaft angle 90°)

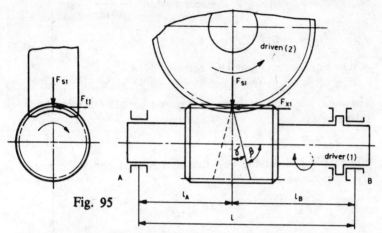

Fig. 95

Note: Suffixes 1 and 2 refer to worm (driver) and worm wheel (driven) respectively and $\gamma_1 = \beta_2 = (90° - \beta_1)$

Forces on the worm (gear 1)
$$F_{t1} = \frac{975000\ P}{n_1} \cdot \frac{2}{d_1}$$

Bearing reactions (Radial)

	At bearing A	At bearing B
due to F_{t1}	$F_{At} = F_{t1}\ \dfrac{l_B}{l}$	$F_{Bt} = F_{t1}\ \dfrac{l_B}{l}$
due to F_{s1}	$F_{As} = F_{s1}\ \dfrac{l_B}{l}$	$F_{Bs} = F_{s1}\ \dfrac{l_A}{l}$
due to F_{x1}	$F_{Ax} = F_{x1}\dfrac{d_1}{2l}$	$F_{Bx} = F_{x1}\dfrac{d_1}{2l}$

The resultant reaction at A $F_A = \sqrt{F_{At}^2 + (F_{As} \pm F_{Ax})^2}$

The resultant reaction at B $F_B = \sqrt{F_{Bt}^2 + (F_{Bs} \pm F_{Bx})^2}$

For the direction of rotation shown above positive sign for F_A and negative sign for F_B is used. For opposite direction of rotation negative sign for F_A and positive sign for F_B is used.

The axial force F_x can be taken up by any one of the bearings. In Fig.95 bearing B is designed to take both radial and axial force. The direction is opposite to the one shown if the driver rotates in the opposite direction.

$$F_{x1} = F_{t1} \cot(\gamma_1 \pm \phi) \quad \text{where } \phi = \tan^{-1} \frac{\mu}{\cos\alpha_n}$$

Values of μ as per Table 77

+ sign when worm (gear 1) is driving
− sign when worm gear (gear 2) is driving

The most appropriate form for crossed helical gears in terms of helix angle is

$$F_{x1} = F_{t1} \tan (\beta_1 \pm \phi)$$

When gear 1 is driving negative sign is used and when gear 2 is driving positive sign is used.

$$F_{s1} = F_{x1} \frac{\tan\alpha_n}{\cos\gamma_1}$$

Forces on worm wheel (gear 2)

$$F_{t2} = F_{x1}$$
$$F_{x2} = F_{t1}$$
$$F_{s1} = F_{s2}$$

INFORMATION FOR GIVING GEAR DATA ON DRAWINGS

To facilitate proper manufacture, inspection and assembly of different types of gears it is necessary to indicate not only the dimensions and tolerances but all other relevant data. While in general the dimensions and major tolerances are included in the dimensional drawing, all other tolerances and additional data are entered in a separate table. A typical table for this purpose is as shown in Table 97.

Table 97

Type of gear			
Heat treatment			
Tooth surface			
Class of accuracy			
Module	Normal	m_n	
	Axial	m_x	
	Transverse	m_t	
No. of teeth/starts		z	
Basic profile	Pressure angle	α	
	Addendum	h_a	
	Bottom clearance	c	
Addendum modification coefficient		x	
Shifting of base profile		xm	
Helix angle		β	
Hand of helix			
Lead angle		γ	
Lead		p_z	
Pitch circle diameter		d	
Theoretical tooth depth		h	
Base circle diameter		d_b	
Root diameter		d_f	
Inspection data (In normal plane)	Base tangent length	W_k	
	Chordal tooth thickness	$\bar{s}$	
	Chordal height	$\bar{h}_a$	
Allowable errors			
Mating gear/Worm	Drawing No		
	No. of teeth/starts	z	
	Centre distance	a'	
	Shaft angle	Σ	

LIFE CALCULATION OF BEARINGS

Nomenclature

C Basic dynamic load rating, kgf

C_o Basic static load rating, kgf

E Diameter of outer ring raceway, mm

f_a Factor for nature of loading (Table 98)

f_b Factor for bearing internal dimensions (Table 99)

F_a Actual axial load, kgf

$F_{a\,max}$ Maximum permissible axial load, kgf

F_r Actual radial load, kgf

L Nominal life, *millions of revolutions*

L_h Nominal life, *hours*

n Rotational speed, *rpm*

P Equivalent dynamic bearing load, kgf

P_o Equivalent static bearing load, kgf

S_o Static safety factor

X Radial factor for dynamic load

X_o Radial factor for static load

Y Axial factor for dynamic load

Y_o Axial factor for static load

ϵ An exponent $= 3$ for point contacts (ball bearings)
                   $= 10/3$ for line contacts (roller bearings)

The basic dynamic load rating C which can be found in the dimension tables of bearings is that constant stationary radial load which 90% of a sufficiently large group of apparently identical bearings will withstand for a minimum of one million revolutions before the first sign of bearing fatigue failure developes.

The relationship between the nominal life, the basic dynamic load rating and the bearing load is expressed by the equation:

$$L = (C/P)^{\epsilon}$$

For bearings operating at constant speed it is more convenient to deal with a life expressed in working hours and the following equation may be used.

$$L = (1000000/60n)\,(C/P)^{\epsilon}$$

Equivalent dynamic bearing load

The equivalent dynamic bearing load is defined as that hypothetical constant radial load, or axial load in case of thrust bearings, which, if applied, would have the same effect on the life of the rotating bearing as the actual loads.

Combined bearing load

Radial bearings are often subjected simultaneously to radial and axial loads. If the magnitude and direction of the resultant load is constant, the equivalent dynamic load is obtained from the following general equation:

$$P = XF_r + YF_a$$

Data required for finding out X and Y are given in the bearing tables.

Cylindrical.roller bearings incorporating one ring without flanges (types NU, N, NNK, CRL, CRM) and needle bearings can carry only radial loads, that is $P = F_r$. Other types of cylindrical roller bearings can carry limited axial loads (see under axial load of cylindrical roller bearings).

Axial load carrying capacity of cylindrical roller bearings

Cylindrical radial bearings are normally used to carry radial loads only. However, bearings with guiding flanges on both inner and outer rings (types NJ, NUP and NJ) with angle rings can also carry limited axial loads. The axial load carrying capacity in these cases is not governed by fatigue characteristics of the material, but is dependent upon the ability of sliding surfaces of the roller ends and guiding flanges to carry the load. Lubrication is therefore critical. It is not possible to exactly calculate the axial load that can be carried, but the following simplified equations based on tests and experience are a useful guide.

For Grease Lubrication

$$F_{a\,max} = f_a\,f_b\,E^2\,(2-nE/100000)$$

For oil lubrication

When $nE \leqslant 120000$, $F_{amax.} = f_af_bE^2\,(2-nE/100000)$

$$nE \geqslant 120000,\ F_{amax.} = f_af_bE^2\,(1-nE/600000)$$

Table 98 . Values for factor f_a

Nature of axial load	f_a
Constant, continuously acting ($F_a/F_r \leqslant 0.4$)	0.2
Fluctuating or intermittent	0.4
Shock loading	0.6

Table 99. Values for factor f_b

Bearing series	f_b
NJ 2, NJ 22, NUP 2, NUP 22	0.24
NJ 3,NJ 23,NUP 3,NUP 23,NJ 2 E, NJ 22 E,NUP 2 E,NUP 22 E	0.3
NJ 4,NUP 4	0.33
NJ 3 E,NJ 23 E,NUP 3 E,NUP 23 E	0.35

For cylindrical rollers to function satisfactorily under a constant axial load and to carry a simultaneously acting radial load, the ratio F_a/F_r should not exceed 0.4, since the sliding friction between the roller ends and guiding flanges causes higher running temperatures than would normally be the case with a radial load.

Theoretically, the axial load also has some effect on life of bearings, however, for all practical purposes, this can be ignored provided the ratio F_a/F_r does not exceed 0.13 for dimension series 02 and 03 or 0.2 for dimension series 22 and 23.

Thrust bearings

Thrust ball bearings can carry only axial loads. If loads act centrally the following equation is valid.

$$P = F_a$$

Spherical roller thrust bearings can carry a radial load F_r provided it does not exceed 55% of simultaneously acting axial load F_a. In such cases, the equivalent dynamic bearing load is given by

$$P = F_a + 1.2 \ F_r$$

Required life

The life expectancy of bearings used in machine tools depends on several varying factors and a definite rule cannot be applied. It depends upon the kind of machine tool, whether it is a special purpose machine working under constant working conditions or universal machine tool where the machining conditions are regularly varying. It also depends upon the position in which the bearing is used and the ease with which it can be replaced. The life expectancy of the bearing may coincide with the life expectancy of the machine or be only a part of it, for example it may be equal to the overhauling period decided by experience. However, the life expectancy of a bearing used in machine tool working continuously for 8 hours a day at constant load and speed is choosen between

20,000 to 30,000 hours. A value between 50,000 to 60,000 hours is selected for machines working continuously all the 24 hours of the day.

Very often the bearings used in machine tools are subjected to varying loads and work at varying speeds. In these circumstances, it is necessary to make an estimate of equivalent required life which will ultimately give the bearing a life expectancy as chosen above. For this it is necessary to estimate from previous experience with such machines the values of different loads that will act and periods as a percentage of total time these loads will act. With this data the expected life of bearing can be estimated as follows:

Let $P_1, P_2, \ldots P_Z$ be constant equivalent dynamic load components

$n_1, n_2, \ldots n_Z$ be the corresponding constant speed values (*rpm*)

$L_1, L_2, \ldots L_Z$ imaginary life in hours corresponding to loads $P_1, P_2, \ldots P_Z$ and speeds $n_1, n_2, \ldots n_Z$

$q_1, q_2, \ldots q_Z$ Corresponding percentage of total time for which the above loads act

L_h Overall life of bearing in hours

$$L_h = \cfrac{100}{\cfrac{q_1}{L_1} + \cfrac{q_2}{L_2} + \cdots \cfrac{q_Z}{L_Z}}$$

Static load carrying capacity

The static load carrying capacity of a rolling bearing is the load at which the permanent overall deformation of the rolling element and raceway at the highly stressed point is 0.0001 of the rolling element diameter.

Equivalent static bearing load

For bearings which are loaded statically, or when making only very small rotations or oscillatory movements, the permissible deformation is generally governed by the rated static capacity.

The equivalent static bearing load is defined as that constant radial load, (axial load in case of thrust bearings) which, if applied, would cause the same total deformation at the highly stressed contact as the actual load.

It is given by:

$$P_o = X_o F_r + Y_o F_a$$

All data required for calculating X_o and Y_o are included in the bearing tables. If calculated value of P_o is less than F_r (in the case of radial bearings) then $P_o = F_r$ is used. If calculated value of P_o is less than F_a (in the case of thrust bearings) then $P_o = F_a$ is used.

Static load rating

The required static load rating C_o of bearing is given by
$$C_o = S_o P_o$$
The following values of S_o may be chosen:

$S_o = 0.5$ where smooth, vibration free running is assured.

$S_o = 1$ for average working conditions and normal demands on quiet running.

$S_o = 2$ for shock loads and where high demands on quiet running exist.

$S_o \geqslant 2$ for spherical roller thrust bearings.

ELASTIC DEFORMATIONS IN ROLLING BEARINGS

When a bearing is loaded, its rings are displaced with respect to each other due to elastic deformation. The axial and radial elastic deformation of the bearings affect the rigidity of the spindle units as well as the overall static rigidity of the machine tool itself. Table 100 gives an idea of the axial and radial deformations to which the bearings are subjected to under load.

Nomenclature

D_w	Rolling element diameter, *mm*
F_a	Axial load, *kgf*
F_r	Radial load, *kgf*
i	Number of rows of rolling elements
l_a	Effective roller length, *mm*
Q	Maximum rolling element load, *kgf*
z	Number of rolling elements per row
$\propto$	Contact angle, *degrees*
$\propto_o$	Contact angle at no load, *degrees*
δ_a	Axial displacement, *μm*
δ_r	Radial displacement, *μm*
Δ_r	Radial clearance, *μm*

There is no equation for axial displacement of deep groove ball bearings, in Table 100, as it varies with the clearance in the bearing. Figure 96 shows the variation of contact angle with radial clearance. The dispersion of the curve is due to the small variation in ball groove dimension.

Figure 97 shows approximately the variation of axial displacement in deep groove ball bearings as a function of load and bearing clearance.

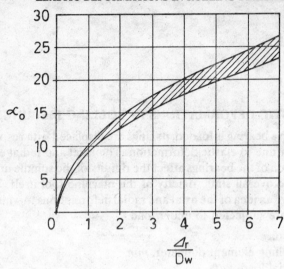

Fig. 96

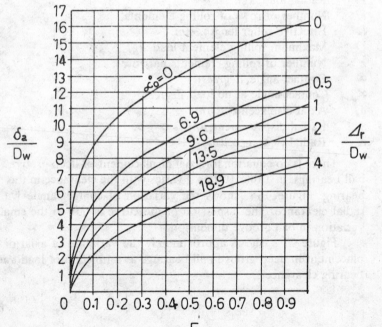

Fig. 97

Table 100 Elastic deformation in rolling bearings

Type of bearing	Loading conditions	
	$\delta_a = 0$ [1]	$\delta_r = 0$ [2]
Self-aligning ball bearings	$\delta_r = \dfrac{3 \cdot 2}{\cos\alpha}\sqrt[3]{\dfrac{Q^2}{D_w}}$	—
Deep groove ball bearings	$\delta_r = 2\sqrt[3]{\dfrac{Q^2}{D_w}}$	—
Angular contact ball bearings	$\delta_r = \dfrac{2}{\cos\alpha}\sqrt[3]{\dfrac{Q^2}{D_w}}$	$\delta_a = \dfrac{2}{\sin\alpha}\sqrt[3]{\dfrac{Q^2}{D_w}}$
Roller bearings with line contact for both raceways	$\delta_r = \dfrac{0 \cdot 6}{\cos\alpha}\dfrac{Q^{0.9}}{l_a^{0.8}}$	$\delta_a = \dfrac{0 \cdot 6}{\sin\alpha}\dfrac{Q^{0.9}}{l_a^{0.8}}$
Roller bearing with line contact for one raceway and point contact for the other	$\delta_r = \dfrac{1 \cdot 2}{\cos\alpha}\dfrac{Q^{0.75}}{l_a^{0.5}}$	$\delta_a = \dfrac{1 \cdot 2}{\sin\alpha}\dfrac{Q^{0.75}}{l_a^{0.5}}$
Thrust ball bearings	—	$\delta_a = \dfrac{2 \cdot 4}{\sin\alpha}\sqrt[3]{\dfrac{Q^2}{D_w}}$
Rolling element load	$Q = \dfrac{5 F_r}{iz\cos\alpha}$	$Q = \dfrac{F_a}{z\sin\alpha}$

1) In the case of bearings transmitting both axial and radial loads, the loading condition under which $\delta_a = 0$, occurs when $F_a \lesssim 1.25\, F_r\, \tan\alpha$. For other load conditions and data on dimensions of the bearing elements reference may be made to bearing manufacturers.

2) $\delta_r = 0$, when $F_r = 0$.

SPEED LIMITS FOR ROLLING BEARINGS

The limiting speed is dependent on various factors like bearing size and type, cage design, bearing clearance, method of lubrication, accuracy, type of arrangement of bearings and the bearing load. Because of these various factors that affect the limiting speed, it is not always possible to exactly determine the limiting speed. However, a good approximate estimation could be made assuming that a suitable method of lubrication giving correctly metered quantity and right kind of lubricant is used.

The limiting speed n_g in *rpm* is given by

Bearing outer diameter D smaller than 30 *mm*	Bearing outer diameter D 30 *mm* and larger
$n_g = \dfrac{3A}{D+30}$ (*rpm*)	$n_g = \dfrac{A}{D-10}$ (*rpm*)

where D is in *mm* and A is constant given in Table 101

If the recommended limiting speeds have to be exceeded, special precautions must be taken with regard to lubrication and cooling and a circulating oil system may be required. With very high speeds, it may be necessary to cool the circulating oil, or use oil mist or oil jet lubrication. Additionally, for some high speed applications bearings having special cages, greater radial internal clearance or manufactured to a higher degree of accuracy (Tolerance classes *P6*, *P5*, etc.) must be used.

Approximate values for the permissible bearing operating speeds under such conditions can be obtained by multiplying the appropriate limiting speeds for oil lubrication, listed in the bearing tables, by the relevant factor given in Table 102

The maximum values attainable recommended above can be further exceeded by using mist lubrications and taking certain other precautions. Definite data on this is not available.

Minimum preload for ball and roller thrust bearing

Ball and roller thrust bearing in high speed applications must always be preloaded with a certain minimum force F_a to prevent the sliding between the rolling elements and the tracks. The value of preload required can be calculated from the formula:

$$F_a \geqslant K \left(\frac{nC_o}{1000000} \right)^2 \qquad \qquad \ldots(4)$$

F_a = axial load in *kgf*
n = speed in rev/min

C_o = static load carrying capacity of bearing in kgf
K = a factor dependent on type of bearing
    0.08   for thrust ball bearings
    0.02   for angular contact thrust ball bearings
    0.004 for thrust roller bearings
    0.03   for angular contact ball bearings of series 72B and
    73B.

Table 101 Constant A for the calculation of limiting speed

Bearing Type	Constant A	
	Grease Lubrication	Oil Lubrication
Radial Bearings:		
Deep groove ball bearings, single row	500 000	630 000
Single row with seals	360 000	—
Double row	320 000	400 000
Magneto bearings	500 000	630 000
Angular contact ball bearings, single row	500 000	630 000
single row, paired	400 000	500 000
double row	360 000	450 000
Four-point bearings	400 000	500 000
Self-aligning ball bearings	500 000	630 000
Self-aligning ball bearings with extended inner ring	250 000	320 000
Cylindrical roller bearings, single row	500 000	630 000
double row	500 000	630 000
Taper roller bearings	320 000	400 000
Barrel roller bearings	220 000	280 000
Spherical roller bearings	200 000 to 320 000	250 000 to 400 000
Thrust bearings:		
Thrust ball bearings	140 000	200 000
Angular contact thrust ball bearings	220 000	320 000
Cylindrical roller thrust bearings	90 000	120 000
Spherical roller thrust bearings	—	280 000

The speed limits n_g apply to bearings of standard design provided the external load is about 10 per cent of the basic dynamic load rating C. It is further assumed that the load direction on radial bearings is radial and on thrust bearings axial.

Combined loading decreases the limiting speed. The reduced value is obtained by multiplying the limiting speed n_g by a correction factor which depends on the load angle (β) as expressed by

$$\tan \beta = \frac{F_a}{F_r} \quad \text{given in Fig. 98}$$

Table 102

Bearing type	Factor
Deep groove ball bearing	3
Self-aligning ball bearings	1.5
Angular contact ball bearings, single row	1.5
Cylindrical roller bearings, single row	2.2
Cylindrical roller bearings, double row	2.2
Thrust ball bearings	1.4
Spherical roller thrust bearings	2

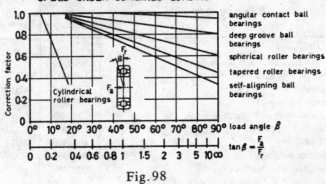

CORRECTION FACTOR FOR LIMITING
SPEED UNDER COMBINED LOADING

Fig. 98

BEARINGS

Calculation factors for bearings

DEEP GROOVE BALL BEARINGS

Equivalent bearing load

Dynamic $\qquad P = XF_r + YF_a$

Static $\qquad P_o = 0.6F_r + 0.5F_a$ when $P_o < F_r$, use $P_o = F_r$

Table 103

Calculation factors dynamic		$F_a/F_r \leqslant e$		$F_a/F_r > e$	
F_a/C_o	e	X	Y	X	Y
0.025	0.22	1	0	0.56	2
0.04	0.24	1	0	0.56	1.8
0.07	0.24	1	0	0.56	1.6
0.13	0.31	1	0	0.56	1.4
0.25	0.37	1	0	0.56	1.2
0.5	0.44	1	0	0.56	1

ANGULAR CONTACT BALL BEARINGS

Equivalent bearings load:

Dynamic $\qquad P = XF_r + YF_a$

Static $\qquad P_o = 0.5F_r + Y_oF_a$ when $P_o < F_r$ use $P_o = F_r$

Table 104

Calculation factors dynamic		$F_a/F_r \leqslant e$		$F_a/F_r > e$		Static
F_a/C_o	e	X	Y	X	Y	Y_o
Series 70C, 72C, 70CG and 72CG						
0.015	0.38	1	0	0.44	1.47	0.46
0.029	0.04	1	0	0.44	1.4	0.46
0.058	0.43	1	0	0.44	1.3	0.46
0.087	0.46	1	0	0.44	1.23	0.46
0.12	0.47	1	0	0.44	1.19	0.46
0.17	0.5	1	0	0.44	1.12	0.46
0.29	0.55	1	0	0.44	1.02	0.46
0.44	0.56	1	0	0.44	1	0.46
0.44	0.56	1	0	0.44	1	0.46
Series 72B and 72BG						
—	1.14	1	0	0.35	0.57	0.26

CYLINDRICAL ROLLER BEARINGS — SINGLE ROW

Equivalent bearing load

Dynamic $\qquad P = F_r$

Static $\qquad P_o = F_r$

NEEDLE ROLLER BEARINGS

Equivalent bearing load
Dynamic $P = F_r$
Static $P_o = F_r$

TAPER ROLLER BEARINGS

Equivalent bearing load
Dynamic $P = XF_r + YF_a$
Static $P_o = 0.5F_r + Y_oF_a$ when $P_o < F_r$, use $P_o = F_r$

Table 105

Calculation factors dynamic		$F_a/F_r \leq e$		$F_a/F_r > e$		Static
Designation	e	X	Y	X	Y	Y_o
32206	0.37	1	0	0.4	1.6	0.9
32207	0.37	1	0	0.4	1.6	0.9
32208	0.37	1	0	0.4	1.6	0.9
32209	0.4	1	0	0.4	1.5	0.8
32210	0.43	1	0	0.4	1.4	0.8
32211	0.4	1	0	0.4	1.5	0.8
32212	0.4	1	0	0.4	1.5	0.8
32213	0.4	1	0	0.4	1.5	0.8
32214	0.43	1	0	0.4	1.4	0.8
32215	0.43	1	0	0.4	1.4	0.8
32216	0.43	1	0	0.4	1.4	0.8
32217	0.43	1	0	0.4	1.4	0.8
32218	0.43	1	0	0.4	1.4	0.8
32219	0.43	1	0	0.4	1.4	0.8
32220	0.43	1	0	0.4	1.4	0.8
32221	0.43	1	0	0.4	1.4	0.8
32222	0.43	1	0	0.4	1.4	0.8
32224	0.43	1	0	0.4	1.4	0.8
32226	0.43	1	0	0.4	1.4	0.8
32228	0.43	1	0	0.4	1.4	0.8

THRUST BALL BEARINGS — SINGLE ROW

Equivalent bearing load
Dynamic $P = F_a$
Static $P_o = F_a$

DEEP GROOVE BALL BEARINGS

All dimensions in *mm*

Table 106 Dimension series 02

IS designation	Boundary dimensions				Abutment dimensions			Basic load rating *kgf*		Limiting speed *rpm* Lubrication		Equivalent *DIN & CSN* designation
	d	D	B	r Nom.	d_a Min.	D_a Max.	r_a Max.	Dynamic C	Static C_o	Grease	Oil	
10 BC 02	10	30	9	1	14	26	0.6	475	270	24000	30000	6200
12 BC 02	12	32	10	1	16	28	0.6	540	315	22000	28000	6201
15 BC 02	15	35	11	1	19	31	0.6	600	364	19000	24000	6202
17 BC 02	17	40	12	1	21	36	0.6	752	453	17000	20000	6203
20 BC 02	20	47	14	1.5	25	42	1	997	621	15000	18000	6204
25 BC 02	25	52	15	1.5	30	47	1	1088.6	707	12000	15000	6205
30 BC 02	30	62	16	1.5	35	57	1	1519.58	1016	10000	13000	6206
35 BC 02	35	72	17	2	41.5	65.5	1	1995.8	1406	9000	11000	6207
40 BC 02	40	80	18	2	46.5	73.5	1	2404	1655	8500	10000	6208
45 BC 02	45	85	19	2	51.5	78.5	1	2585	1882	7500	9000	6209
50 BC 02	50	90	20	2	56.5	83.5	1	2766	1995	7000	8500	6210
55 BC 02	55	100	21	2.5	63	92	1.5	3401	2540	6300	7500	6211
60 BC 02	60	110	22	2.5	68	102	1.5	3764	2857	6000	7000	6212
65 BC 02	65	120	23	2.5	73	112	1.5	4377	3469	5300	6300	6213
70 BC 02	70	125	24	2.5	78	117	1.5	4808	3855	5000	6000	6214
75 BC 02	75	130	25	2.5	83	122	1.5	5171	4150	4800	5600	6215
80 BC 02	80	140	26	3	89	131	2	5669	4535	4500	5300	6216
85 BC 02	85	150	28	3	94	141	2	6363	5352	4300	5000	6217
90 BC 02	90	160	30	3	99	151	2	7284	5914	3800	4500	6218

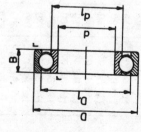

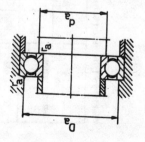

Table 106 Dimension series 02 (concld.)

DEEP GROOVE BALL BEARINGS (Contd.)

All dimensions in mm

IS designation	Boundary dimensions				Abutment dimensions			Basic load rating kgf		Limiting speed rpm Lubrication		Equivalent DIN & CSN designation
	d	D	B	r Nom.	d_a Min.	D_a Max.	r_a Max.	Dynamic C	Static C_o	Grease	Oil	
95 BC 02	95	170	32	3.5	106	159	2	8191	6831	3600	4300	6219
100 BC 02	100	180	34	3.5	111	169	2	9616	7983	3400	4000	6220
105 BC 02	105	190	36	3.5	116	179	2	10341	9071	3200	3800	6221
110 BC 02	110	200	38	3.5	121	189	2	11340	10160	3000	3600	6222
120 BC 02	120	215	40	3.5	131	204	2	11340	10160	2800	3400	6224
130 BC 02	130	230	40	4	143	217	2.5	12247	11340	2600	3200	6226
140 BC 02	140	250	42	4	153	237	2.5	12927	12473	2400	3000	6228
150 BC 02	150	270	45	4	163	257	2.5	13834	14061	2000	2600	6230
160 BC 02	160	290	48	4	173	277	2.5	14515	14741	1900	2400	6232
170 BC 02	170	310	52	5	186	294	3	16329	18144	1900	2400	6234
180 BC 02	180	320	52	5	196	304	3	17690	19958	1800	2200	6236
190 BC 02	190	340	55	5	206	324	3	19958	23587	1700	2000	6238
200 BC 02	200	360	58	5	216	344	3	21091	25401	1700	2000	6240
220 BC 02	220	400	65	5	236	384	3	23133	29710	1500	1800	6244
240 BC 02	240	440	72	5	256	424	3	28123	39235	1300	1600	6248

Table 107 Dimension series 03 DEEP GROOVE BALL BEARINGS (Contd.) All dimensions in *mm*

| IS designation | Boundary dimensions | | | | Abutment dimensions | | | Basic load rating kgf | | Limiting speed rpm | | Equivalent DIN & CSN designation |
	d	D	B	r Nom.	d_a Min.	D_a Max.	r_a Max.	Dynamic C	Static C_o	Lubrication Grease	Oil	
10 BC 03	10	35	11	1	14	31	0.6	640	380	20000	26000	6300
12 BC 03	12	37	12	1.5	17	32	1	765	475	19000	24000	6301
15 BC 03	15	42	13	1.5	20	37	1	900	550	17000	20000	6302
17 BC 03	17	47	14	1.5	22	42	1	1060	670	16000	19000	6303
20 BC 03	20	52	15	2	26.5	45.5	1	1250	800	13000	16000	6304
25 BC 03	25	62	17	2	31.5	55.5	1	1400	1080	11000	14000	6305
30 BC 03	30	72	19	2	36.5	65.5	1	1860	1600	9000	11000	6306
35 BC 03	35	80	21	2.5	43	72	1.5	2600	1830	8500	10000	6307
40 BC 03	40	90	23	2.5	48	82	1.5	3200	2280	7500	9000	6308
45 BC 03	45	100	25	2.5	53	92	1.5	4150	3050	6700	8000	6309
50 BC 03	50	110	27	3	59	101	2	4800	3650	6300	7500	6310
55 BC 03	55	120	29	3	64	111	2	5600	4250	5500	6700	6311
60 BC 03	60	130	31	3.5	71	119	2	6400	4900	5000	6000	6312
65 BC 03	65	140	33	3.5	76	129	2	7200	5700	4800	5600	6313
70 BC 03	70	150	35	3.5	81	139	2	8150	6400	4500	5300	6314
75 BC 03	75	160	37	3.5	86	149	2	9000	7350	4300	5000	6315
80 BC 03	80	170	39	3.5	91	159	2	9650	8150	3800	4500	6316
85 BC 03	85	180	41	4	98	167	2.5	10400	9150	3600	4300	6317
90 BC 03	90	190	43	4	103	177	2.5	11200	10000	3400	4000	6318
95 BC 03	95	200	45	4	108	187	2.5	12000	11200	3200	3800	6319
100 BC 03	100	215	47	4	113	202	2.5	13700	13400	3000	3600	6320
105 BC 03	105	225	49	4	118	212	2.5	14300	14600	2800	3400	6321
110 BC 03	110	240	50	4	123	227	2.5	16000	17000	2600	3200	6322
120 BC 03	120	260	55	4	133	247	2.5	17000	18300	2400	3000	6324

DEEP GROOVE BALL BEARINGS (Contd.)

Table 107 Dimension series 03 (concld.)

All dimensions in *mm*

IS designation	Boundary dimensions				Abutment dimensions			Basic load rating kgf		Limiting speed rpm Lubrication		Equivalent DIN & CSN designation
	d	D	B	r Nom.	d_a Min.	D_a Max.	r_a Max.	Dynamic C	Static C_o	Grease	Oil	
130 BC 03	130	280	58	5	146	264	3	18000	19600	2200	2800	6326
140 BC 03	140	300	62	5	156	284	3	20000	22800	2000	2600	6328
150 BC 03	150	320	65	5	166	304	3	21600	26000	1900	2400	6330

Table 108 Dimension series 04

All dimensions in *mm*

IS designation	Boundary dimensions				Abutment dimensions			Basic load rating kgf		Limiting speed rpm Lubrication		Equivalent DIN & CSN designation
	d	D	B	r Nom.	d_a Min.	D_a Max.	r_a Max.	Dynamic C	Static C_o	Grease	Oil	
17 BC 04	17	62	17	2	23.5	55.5	1	1800	1200	12000	15000	6403
20 BC 04	20	72	19	2	26.5	65.5	1	2400	1700	10000	13000	6404
25 BC 04	25	80	21	2.5	33	72	1.5	2800	2000	9000	11000	6405
30 BC 04	30	90	23	2.5	38	82	1.5	3400	2450	8500	10000	6406
35 BC 04	35	100	25	2.5	43	92	1.5	4300	3150	7000	8500	6407
40 BC 04	40	110	27	3	49	101	2	5000	3750	6700	8000	6408
45 BC 04	45	120	29	3	54	111	2	6000	4650	6000	7000	6409
50 BC 04	50	130	31	3.5	61	119	2	6800	5300	5300	6300	6410
55 BC 04	55	140	33	3.5	66	129	2	7800	6400	5000	6000	6411
60 BC 04	60	150	35	3.5	71	139	2	8500	7100	4800	5600	6412
65 BC 04	65	160	37	3.5	76	149	2	9300	8000	4500	5300	6413
70 BC 04	70	180	42	4	83	167	2.5	11200	10600	3800	4500	6414
75 BC 04	75	190	45	4	88	177	2.5	12000	11600	3600	4300	6415

Table 109 Dimension series 10   DEEP GROOVE BALL BEARINGS (Contd.)   All dimensions in mm

| IS designation | Boundary dimensions | | | | Abutment dimensions | | | Basic load rating kgf | | Limiting speed rpm | | Equivalent DIN & CSN designation |
	d	D	B	r Nom.	da Min.	Da Max.	ra Max.	Dynamic C	Static Co	Lubrication Grease	Oil	
10 BC 10	10	26	8	0.5	12	24	0.3	360	200	30000	36000	6000
12 BC 10	12	28	8	0.5	14	26	0.3	400	228	26000	32000	6001
15 BC 10	15	32	9	0.5	17	30	0.3	430	255	22000	28000	6002
17 BC 10	17	35	10	0.5	19	33	0.3	475	285	19000	24000	6003
20 BC 10	20	42	12	1	24	38	0.6	735	455	17000	20000	6004
25 BC 10	25	47	12	1	29	43	0.6	780	500	15000	18000	6005
30 BC 10	30	55	13	1.5	35	50	1	1040	695	12000	15000	6006
35 BC 10	35	62	14	1.5	40	57	1	1250	865	10000	13000	6007
40 BC 10	40	68	15	1.5	45	63	1	1320	950	9500	12000	6008
45 BC 10	45	75	16	1.5	50	70	1	1660	1250	9000	11000	6009
50 BC 10	50	80	16	1.5	55	75	1	1700	1340	8500	10000	6010
55 BC 10	55	90	18	2	61.5	83.5	1	2240	1730	7500	9000	6011
60 BC 10	60	95	18	2	66.5	88.5	1	2320	1860	6700	8000	6012
65 BC 10	65	100	18	2	71.5	93.5	1	2400	2000	6300	7500	6013
70 BC 10	70	110	20	2	76.5	103.5	1	3000	2500	6000	7000	6014
75 BC 10	75	115	20	2	81.5	108.5	1	3100	2650	5600	6700	6015
80 BC 10	80	125	22	2	86.5	118.5	1	3750	3200	5300	6300	6016
85 BC 10	85	130	22	2	91.5	123.5	1	3900	3400	5000	6000	6017
90 BC 10	90	140	24	2.5	98	132	1.5	4550	4000	4800	5600	6018
95 BC 10	95	145	24	2.5	103	137	1.5	4750	4250	4500	5300	6019
100 BC 10	100	150	24	2.5	108	142	1.5	4750	4250	4300	5000	6020
105 BC 10	105	160	26	3	114	151	2	5700	5200	4000	4800	6021
110 BC 10	110	170	28	3	119	161	2	6400	5850	3800	4500	6022

DEEP GROOVE BALL BEARINGS (Concld.)

All dimensions in *mm*

Table 109 Dimension series 10 (Concld.)

IS designation	Boundary dimensions				Abutment dimensions			Basic load rating kgf		Limiting speed rpm Lubrication		Equivalent DIN & CSN designation
	d	D	B	r Nom.	d_a Min.	D_a Max.	r_a Max.	Dynamic C	Static C_o	Grease	Oil	
120 BC 10	120	180	28	3	129	171	2	6700	6200	3400	4000	6024
130 BC 10	130	200	33	3	139	191	2	8300	8000	3200	3800	6026
140 BC 10	140	210	33	3	149	201	2	8650	8500	3000	3600	6028
150 BC 10	150	225	35	3.5	161	214	2	9800	9800	2600	3200	6030
160 BC 10	160	240	38	3.5	171	229	2	11200	11400	2400	3000	6032
170 BC 10	170	260	42	3.5	181	249	2	13200	13700	2200	2800	6034
180 BC 10	180	280	46	3.5	191	269	2	15000	16000	2000	2600	6036
190 BC 10	190	290	46	3.5	201	279	2	15300	17000	2000	2600	6038
200 BC 10	200	310	51	3.5	211	299	2	17000	19600	1900	2400	6040
220 BC 10	220	340	56	4	233	327	2.5	19200	23000	1800	2200	6044
240 BC 10	240	360	56	4	253	347	2.5	20000	25000	1700	2000	6048

SINGLE ROW ANGULAR CONTACT BALL BEARINGS

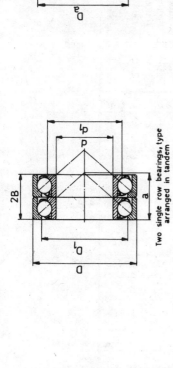

Basic design

Two single row bearings, type arranged in tandem

All dimensions in *mm*

Table 110 Dimension series 10

IS designation	Boundary dimensions					Abutment dimensions			Basic load rating kgf		Limiting speed rpm Lubrication		Equivalent DIN & CSN designation
	d	D	B	r Nom.	r_1	d_a Min.	D_a Max.	r_a Max.	Dynamic C	Static C_0	Grease	Oil	
10 BN 10	10	26	8	0.5	0.2	12	24	0.3	392	218	43000	55000	7000 C
12 BN 10	12	28	8	0.5	0.2	14	26	0.3	423	248	38000	50000	7001 C
15 BN 10	15	32	9	0.5	0.2	17	30	0.3	490	298	32000	43000	7002 C
17 BN 10	17	35	10	0.5	0.2	19	33	0.3	568	354	28000	38000	7003 C
20 BN 10	20	42	12	1	0.5	25	37	0.6	818	536	22000	32000	7004 C
25 BN 10	25	47	12	1	0.5	30	42	0.6	910	623	19000	28000	7005 C
30 BN 10	30	55	13	1.5	0.5	36	49	1	1180	865	17000	24000	7006 C

SINGLE ROW ANGULAR CONTACT BALL BEARINGS (Contd.)

Table 110 Dimension series 10 (concld.)

All dimensions in *mm*

IS designation	Boundary dimensions					Abutment dimensions			Basic load rating *kgf*		Limiting speed *rpm*		Equivalent *DIN&CSN* designation
											Lubrication		
	d	D	B	r Nom.	r_1	d_a Min.	D_a Max.	r_a Max.	Dynamic C	Static C_o	Grease	Oil	
35 BN 10	35	62	14	1.5	0.5	41	56	1	1475	1135	15000	20000	7007 C
40 BN 10	40	68	15	1.5	0.5	46	62	1	1545	1225	13000	18000	7008 C
45 BN 10	45	75	16	1.5	0.5	51	69	1	1885	1545	11000	16000	7009 C
50 BN 10	50	80	16	1.5	0.5	56	74	1	1950	1635	10000	15000	7010 C
55 BN 10	55	90	18	2	0.8	62	83	1	2725	2315	9500	14000	7011 C
60 BN 10	60	95	18	2	0.8	67	88	1	2770	2455	9000	13000	7012 C
65 BN 10	65	100	18	2	0.8	72	93	1	2815	2545	8500	12000	7013 C

Table 111 Dimension series 10   SINGLE ROW ANGULAR CONTACT BALL BEARINGS (Contd.)

All dimensions in *mm*

IS designation	Boundary dimensions					Abutment dimensions			Basic load rating kgf		Limiting speed rpm Lubrication		Equivalent DIN & CSN designation
	d	D	B	r Nom.	r_i	d_a Min.	D_a Max.	r_a Max.	Dynamic C	Static C_o	Grease	Oil	
10 BN 10*	10	26	16	0.5	0.2	12	24	0.3	622	436	34000	44000	2 x 7000 CG
12 BN 10*	12	28	16	0.5	0.2	14	26	0.3	695	490	30000	40000	2 x 7001 CG
15 BN 10*	15	32	18	0.5	0.2	17	30	0.3	800	595	24000	34000	2 x 7002 CG
17 BN 10*	17	35	20	0.5	0.2	19	33	0.3	910	710	20000	30000	2 x 7003 CG
20 BN 10*	20	42	24	1	0.5	25	37	0.6	1365	1072	18000	26000	2 x 7004 CG
25 BN 10*	25	47	24	1	0.5	30	42	0.6	1475	1245	16000	22000	2 x 7005 CG
30 BN 10*	30	55	26	1.5	0.5	36	49	1	1930	1725	14000	19000	2 x 7006 CG
35 BN 10*	35	62	28	1.5	0.5	41	56	1	2455	2270	11000	16000	2 x 7007 CG
40 BN 10*	40	68	30	1.5	0.5	46	62	1	2500	2455	10000	15000	2 x 7008 CG
45 BN 10*	45	75	32	1.5	0.5	51	69	1	3090	3090	9000	13000	2 x 7009 CG
50 BN 10*	50	80	32	1.5	0.5	56	74	1	3225	3270	8500	12000	2 x 7010 CG
55 BN 10*	55	90	36	2	0.8	62	83	1	4380	4640	8000	11000	2 x 7011 CG
60 BN 10*	60	95	36	2	0.8	67	88	1	4450	4915	7500	10000	2 x 7012 CG
65 BN 10*	65	100	36	2	0.8	72	93	1	4640	5100	7000	9500	2 x 7013 CG

Note: *The bearings are manufactured for paired mounting, face to face, back to back, or in tandem and the symbol D should be incorporated in the supplementary designation while ordering. For details refer IS 2398-1967. Identification code for rolling bearings.

Table 112 Dimension series 02.

SINGLE ROW ANGULAR CONTACT BALL BEARINGS (Contd.)

All dimensions in mm

IS designation	Boundary dimensions					Abutment dimensions			Basic load rating kgf		Limiting speed Lubrication		Equivalent DIN & CSN designation
	d	D	B	r nom	r_1	d_a Min.	D_a Max.	r_a Max.	Dynamic C	Static C_o	Grease	Oil	
10 BN 02	10	30	9	1	0.5	15	25	0.6	499	297	38000	50000	7200C
12 BN 02	12	32	10	1	0.5	17	27	0.6	576	347	34000	45000	7201C
15 BN 02	15	35	11	1	0.5	20	30	0.6	694	431	30000	40000	7202C
17 BN 02	17	40	12	1	0.5	22	35	0.6	844	544	24000	34000	7203C
20 BN 02	20	47	14	1.5	0.5	26	41	1	1134	771	20000	30000	7204C
25 BN 02	25	52	15	1.5	0.5	31	46	1	1293	907	18000	26000	7205C
30 BN 02	30	62	16	1.5	0.5	36	56	1	1769	1315	16000	22000	7206C
35 BN 02	35	72	17	2	0.5	42	65	1	2404	1769	13000	18000	7207C
40 BN 02	40	80	18	2	1	47	73	1	2858	2223	11000	16000	7208C
45 BN 02	45	85	19	2	1	52	78	1	3221	2540	10000	15000	7209C
50 BN 02	50	90	20	2	1	57	83	1	3334	2767	9500	14000	7210C
55 BN 02	55	100	21	2.5	1	64	91	1.5	4150	3470	9000	13000	7211C
60 BN 02	60	110	22	2.5	1	69	101	1.5	4990	4218	8000	11000	7212C
65 BN 02	65	120	23	2.5	1	74	111	1.5	5443	4627	8000	11000	7213C
70 BN 02	70	125	24	2.5	1	79	116	1.5	5851	5171	7500	10000	7214C
75 BN 02	75	130	25	2.5	1	84	121	1.5	6078	5443	7000	9500	7215C
80 BN 02	80	140	26	3	1.5	90	130	2	7394	6486	6700	9000	7216C
85 BN 02	85	150	28	3	1.5	95	140	2	7847	7076	6000	8000	7217C
90 BN 02	90	160	30	3	1.5	100	150	2	9616	8891	5600	7500	7218C
95 BN 02	95	170	32	3.5	2	107	158	2	10342	9616	5600	7500	7219C
100 BN 02	100	180	34	3.5	2	112	168	2	11567	10886	5300	7000	7220C

Table 113 Dimension series 02.

SINGLE ROW ANGULAR CONTACT BALL BEARINGS (Contd.)

All dimensions in mm

IS designation	Boundary dimensions					Abutment dimensions			Basic load rating kgf		Limiting speed rpm Lubrication		Equivalent DIN & CSN designation
	d	D	B	r Nom.	r₁	da Min.	Da Max.	ra Max.	Dynamic C	Static Co	Grease	Oil	
10 BN 02 *	10	30	18	1	0.5	15	25	0.6	816	594	30000	40000	2 x 7200 CG
12 BN 02 *	12	32	20	1	0.5	17	27	0.6	943	694	26000	36000	2 x 7201 CG
15 BN 02 *	15	35	22	1	0.5	20	30	0.6	1111	861	22000	32000	2 x 7202 CG
17 BN 02 *	17	40	24	1.5	0.5	22	35	0.6	1383	1089	19000	28000	2 x 7203 CG
20 BN 02 *	20	47	28	1.5	0.5	26	41	1	1837	1542	17000	24000	2 x 7204 CG
25 BN 02 *	25	52	30	1.5	0.5	31	46	1	2109	1814	15000	20000	2 x 7205 CG
30 BN 02 *	30	62	32	2	0.5	36	56	1	2903	2630	12000	17000	2 x 7206 CG
35 BN 02 *	35	72	34	2	0.5	42	65	1	3924	3538	10000	15000	2 x 7207 CG
40 BN 02 *	40	80	36	2	1	47	73	1	4626	4445	9000	13000	2 x 7208 CG
45 BN 02 *	45	85	38	2	1	52	78	1	5171	5080	8500	12000	2 x 7209 CG
50 BN 02 *	50	90	40	2	1	57	83	1	5443	5534	8000	11000	2 x 7210 CG
55 BN 02 *	55	100	42	2.5	1	64	91	1.5	6804	6940	7500	10000	2 x 7211 CG
60 BN 02 *	60	110	44	2.5	1	69	101	1.5	8165	8437	6700	9000	2 x 7212 CG
65 BN 02 *	65	120	46	2.5	1	74	111	1.5	8890	9253	6300	8500	2 x 7213 CG
70 BN 02 *	70	125	48	2.5	1	79	116	1.5	9616	10342	6000	8000	2 x 7214 CG
75 BN 02 *	75	130	50	2.5	1	84	121	1.5	10161	10886	5600	7500	2 x 7215 CG
80 BN 02 *	80	140	52	3	1.5	90	130	2	11794	12973	5300	7000	2 x 7216 CG
85 BN 02 *	85	150	56	3	1.5	95	140	2	12701	14152	5000	6700	2 x 7217 CG
90 BN 02 *	90	160	60	3	1.5	100	150	2	15422	17781	4800	6300	2 x 7218 CG
95 BN 02 *	95	170	64	3.5	2	107	158	2	16556	19233	4500	6000	2 x 7219 CG
100 BN 02 *	100	180	68	3.5	2	112	168	2	18824	21773	4300	5600	2 x 7220 CG

Note: *The bearings are manufactured for paired mounting, face to face, back to back or in tandem and symbol D should be incorporated in the supplimentary designation while ordering. For details refer IS 2398-1967. *Identification code for rolling bearings.*

SINGLE ROW ANGULAR CONTACT BALL BEARINGS (Contd.)

Table 114 Dimension series 02

All dimensions in mm

IS designation	Boundary dimensions					Abutment dimension			Basic load rating kgf		Limiting speed rpm Lubrication		Equivalent DIN & CSN designation
	d	D	B	r nom.	r_1	d_a Min.	D_a Max.	r_a Max.	Dynamic C	Static C_o	Grease	Oil	
10 BT 02	10	30	9	1	0.5	15	25	0.6	392	215	19000	28000	7200 B
12 BT 02	12	32	10	1	0.5	17	27	0.6	544	308	17000	24000	7201 B
15 BT 02	15	35	11	1	0.5	20	30	0.6	621	376	16000	22000	7202 B
17 BT 02	17	40	12	1	0.8	22	35	0.6	785	472	14000	19000	7203 B
20 BT 02	20	47	14	1.5	0.8	26	41	1	1034	649	11000	16000	7204 B
25 BT 02	25	52	15	1.5	0.8	31	46	1	1157	785	9500	14000	7205 B
30 BT 02	30	62	16	1.5	0.8	36	56	1	1565	1111	8500	12000	7206 B
35 BT 02	35	72	17	2	1	42	65	1	2109	1520	7500	10000	7207 B
40 BT 02	40	80	18	2	1	47	73	1	2495	1882	6700	9000	7208 B
45 BT 02	45	85	19	2	1	52	78	1	2812	2155	6300	8500	7209 B
50 BT 02	50	90	20	2	1	57	83	1	2903	2359	5600	7500	7210 B
55 BT 02	55	100	21	2.5	1.2	64	91	1.5	3697	2971	5300	7000	7211 B
60 BT 02	60	110	22	2.5	1.2	69	101	1.5	4377	3697	4800	6300	7212 B
65 BT 02	65	120	23	2.5	1.2	74	111	1.5	4990	4300	4300	5600	7213 B
70 BT 02	70	125	24	2.5	1.2	79	116	1.5	5352	4717	4300	5600	7214 B
75 BT 02	75	130	25	2.5	1.2	84	121	1.5	5534	4990	4000	5300	7215 B
80 BT 02	80	140	26	3	1.5	90	130	2	6214	5670	3600	4800	7216 B
85 BT 02	85	150	28	3	1.5	95	140	2	7076	6486	3400	4500	7217 B
90 BT 02	90	160	30	3	1.5	100	150	2	8301	7711	3200	4300	7218 B
95 BT 02	95	170	32	3.5	2	107	158	2	9435	8754	3000	4000	7219 B
100 BT 02	100	180	34	3.5	2	112	168	2	10161	9253	2800	3800	7220 B
110 BT 02	110	200	38	3.5	2	122	188	2	12020	11567	2400	3400	7222 B
120 BT 02	120	215	40	3.5	2	132	203	2	12928	13154	2200	3200	7224 B
130 BT 02	130	230	40	4	2	144	216	2.5	13835	14515	1900	2800	7226 B
140 BT 02	140	250	42	4	2	154	236	2.5	14288	15422	1800	2600	7228 B

All dimensions in *mm*

SINGLE ROW ANGULAR CONTACT BALL BEARINGS (Concld.)

Table 115 Dimension series 02

IS designation	Boundary dimensions				Abutment dimensions				Basic load rating kgf		Limiting speed rpm		Equivalent DIN & CSN designation
	d	D	B	r Nom.	r_1	d_a Min.	D_a Max.	r_a Max.	Dynamic C	Static C_o	Grease	Oil	
											Lubrication		
15 BT 02 *	15	35	22	1	0.5	20	30	0.6	1016	752	12000	17000	2 x 7202 BG
17 BT 02 *	17	40	24	1	0.8	22	35	0.6	1270	943	10000	15000	2 x 7203 BG
20 BT 02 *	20	47	28	1.5	0.8	26	41	1	1656	1297	9000	13000	2 x 7204 BG
25 BT 02 *	25	52	30	1.5	0.8	31	46	1	1837	1569	8000	11000	2 x 7205 BG
30 BT 02 *	30	62	32	1.5	0.8	36	56	1	2586	2223	7000	9500	2 x 7206 BG
35 BT 02 *	35	72	34	2	1	42	65	1	3470	3039	6000	8000	2 x 7207 BG
40 BT 02 *	40	80	36	2	1	47	73	1	4082	3765	5300	7000	2 x 7208 BG
45 BT 02 *	45	85	38	2	1	52	78	1	4536	4309	5000	6700	2 x 7209 BG
50 BT 02 *	50	90	40	2	1	57	83	1	4717	4717	4500	6000	2 x 7210 BG
55 BT 02 *	55	100	42	2.5	1.2	64	91	1.5	5761	5942	4300	5600	2 x 7211 BG
60 BT 02 *	60	110	44	2.5	1.2	69	101	1.5	7076	7394	3800	5000	2 x 7212 BG
65 BT 02 *	65	120	46	2.5	1.2	74	111	1.5	7983	8618	3400	4500	2 x 7213 BG
70 BT 02 *	70	125	48	2.5	1.2	79	116	1.5	8754	9435	3200	4300	2 x 7214 BG
75 BT 02 *	75	130	50	2.5	1.2	84	121	1.5	8891	9979	3200	4300	2 x 7215 BG
80 BT 02 *	80	140	52	3	1.5	90	130	2	10161	11340	2800	3800	2 x 7216 BG
85 BT 02 *	85	150	56	3	1.5	95	140	2	11340	12973	2600	3800	2 x 7217 BG
90 BT 02 *	90	160	60	3	1.5	100	150	2	13608	15422	2400	3400	2 x 7218 BG
95 BT 02 *	95	170	64	3.5	2	107	158	2	15196	17509	2200	3200	2 x 7219 BG
100 BT 02 *	100	180	68	3.5	2	112	168	2	16330	18507	2000	3000	2 x 7220 BG
110 BT 02 *	110	200	76	3.5	2	122	188	2	19278	23134	1900	2800	2 x 7222 BG
120 BT 02 *	120	215	80	3.5	2	132	203	2	20639	26309	1700	2400	2 x 7224 BG
130 BT 02 *	130	230	80	4	2	144	216	2.5	22226	29030	1700	2400	2 x 7226 BG
140 BT 02 *	140	250	84	4	2	154	236	2.5	23134	30845	1600	2200	2 x 7228 BG

Note: * The bearings are manufactured for paired mounting, face to face, back to back, or in tandem and symbol *D* should be incorporated in the supplementary designation while ordering. For details refer IS 2398-1967, *Identification code for rolling bearings.*

SINGLE THRUST BALL BEARINGS

Table 116   Dimension series 14   All dimensions in *mm*

IS designation	d	D	H	d₁	r	d_a Min.	D_a Max.	r_a Max.	Dynamic C kgf	Static C_o kgf	Grease	Oil	SKF designation
25 TA 14	25	60	24	27	1.5	41	36	1	4400	7350	4000		51405
30 TA 14	30	70	28	32	1.5	54	46	1	5700	10400	2200	3200	51408
35 TA 14	35	80	32	37	2	62	53	1	6800	12700	1800	2600	51407
40 TA 14	40	90	36	42	2	70	60	1	8800	17000	1700	2400	51408
45 TA 14	45	100	39	47	2	78	67	1	10200	20000	1600	2200	51409
50 TA 14	50	110	43	52	2.5	86	74	1.5	12500	25500	1500	2000	51410
55 TA 14	55	120	48	57	2.5	94	81	1.5	14000	30000	1300	1800	51411
60 TA 14	60	130	51	62	2.5	102	88	1.5	16000	36000	1200	1700	51412
65 TA 14	65	140	56	68	3	110	95	2	18300	40500	1000	1500	51413
70 TA 14	70	150	60	73	3	118	102	2	19600	45500	950	1400	51414
75 TA 14	75	160	65	78	3	125	110	2	21200	51000	900	1300	51415
80 TA 14	80	170	68	83	3.5	133	117	2	22800	56000	850	1200	51416
85 TA 14	85	180	72	88	3.5	141	124	2	24000	62000	850	1200	51417
90 TA 14	90	190	77	93	3.5	149	131	2	25500	68000	800	1100	51418
100 TA 14	100	210	85	103	4	165	145	2.5	31000	88000	700	950	51420

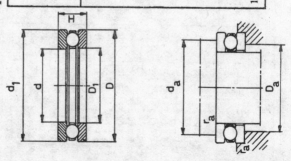

DOUBLE THRUST BALL BEARINGS

All dimensions in *mm*

Table 117 Dimension series 22

IS designation	d	D	H	d₁	d₂ Min.	a	r	r₁	Dₐ Max.	rₐ Min.	Basic load rating kgf		Limiting speed rpm	SKF designation
											Dynamic C	Static C₀		
25 TDC 22	25	47	28	20	25.2	7	1	0.5	34	0.6	2160	4150	6000	52205
30 TDC 22	30	52	29	25	30.2	7	1	0.5	39	0.6	2280	4800	6000	52206
35 TDC 22	35	62	34	30	35.2	8	1.5	0.5	46	1	3050	6400	5000	52207
40 TDC 22	40	68	36	30	40.2	9	1.5	1	51	1	3450	7650	5000	52208
45 TDC 22	45	73	37	35	45.2	9	1.5	1	56	1	3650	8650	4000	52209
50 TDC 22	50	78	39	40	50.2	9	1.5	1	61	1	3750	9150	4000	52210
55 TDC 22	55	90	45	45	55.2	10	1.5	1	69	1	5500	13200	3000	52211
60 TDC 22	60	95	46	50	60.2	10	1.5	1	74	1	5700	14600	3000	52212
65 TDC 22	65	100	47	55	65.2	10	1.5	1	79	1	5850	15600	2500	52213
70 TDC 22	70	105	47	55	70.2	10	1.5	1.5	84	1	6000	16300	2500	52214
75 TDC 22	75	110	47	60	75.2	10	1.5	1.5	89	1	6100	17300	2500	52215
80 TDC 22	80	115	48	65	80.2	10	1.5	1.5	94	1	6200	18000	2000	52216
85 TDC 22	85	125	55	70	85.2	12	1.5	1.5	101	1	7500	22000	2000	52217
90 TDC 22	90	135	62	75	90.2	14	2	1.5	108	1	9150	27000	2000	52218
100 TDC 22	100	150	67	85	100.2	15	2	1.5	120	1	11400	34000	1600	52220

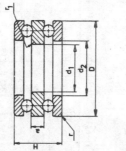

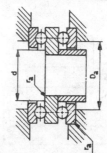

SINGLE ROW CYLINDRICAL ROLLER BEARINGS

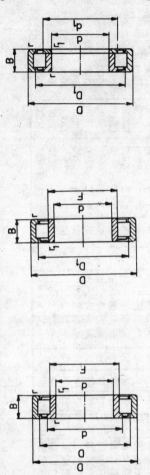

Type N   Type NJ   Type NU   Type NUP

Table118   Dimension series 02

All dimensions in mm

IS designation				d	D	B	r	r₁	E	Basic load rating kgf Dynamic C	Static Co	Limiting speed rpm Lubrication Grease	Oil	Equivalent SKF designation Type RN (N)	Type RJ (NJ)	Type RU (NU)	Type RT (NUP)
Type RN	Type RJ	Type RU	Type RT														
15 RN 02	15 RJ 02	15 RU 02	—	15	35	11	1	0.5	29.3	830	430	19000	24000	N 202	NJ 202	NU 202	—
17 RN 02	17 RJ 02	17 RU 02	—	17	40	12	1	0.5	33.9	965	520	17000	20000	N 203	NJ 203	NU 203	—
20 RN 02	20 RJ 02	20 RU 02	20 RT 02	20	47	14	1.5	1	40	1040	695	15000	18000	N 204	NJ 204	NU 204	NUP 204
25 RN 02	25 RJ 02	25 RU 02	25 RT 02	25	52	15	1.5	1	45	1200	850	12000	13000	N 205	NJ 205	NU 205	NUP 205
30 RN 02	30 RJ 02	30 RU 02	30 RT 02	30	62	16	1.5	1	53.5	1560	1160	10000	11000	N 206	NJ 206	NU 206	NUP 206
35 RN 02	35 RJ 02	35 RU 02	35 RT 02	35	72	17	2	1	61.8	2360	1700	9000	10000	N 207	NJ 207	NU 207	NUP 207
40 RN 02	40 RJ 02	40 RU 02	40 RT 02	40	80	18	2	2	70	3100	2300	8500	9000	N 208	NJ 208	NU 208	NUP 208
45 RN 02	45 RJ 02	45 RU 02	45 RT 02	45	85	19	2	2	75	3250	2500	7500	8500	N 209	NJ 209	NU 209	NUP 209
50 RN 02	50 RJ 02	50 RU 02	50 RT 02	50	90	20	2	2	80.4	3400	2700	7000	—	N 210	NJ 210	NU 210	NUP 210

SINGLE ROW CYLINDRICAL ROLLER BEARINGS (Concld.)

Table 118 Dimension series 02 (Concld.)

All dimensions in mm

IS designation										Basic load rating kgf		Limiting speed rpm Lubrication		Equivalent SKF designation			
Type RN	Type RJ	Type RU	Type RT	d	D	B	r	r₁	E	Dynamic C	Static C₀	Grease	Oil	Type RN (N)	Type RJ (NJ)	Type RU (NU)	Type RT (NUP)
55 RN 02	55 RJ 02	55 RU 02	55 RT 02	55	100	21	2.5	2	88.5	4150	3250	6300	7500	N 211	NJ 211	NU 211	NUP 211
60 RN 02	60 RJ 02	60 RU 02	60 RT 02	60	110	22	2.5	2.5	97.5	4800	4000	5600	6700	N 212	NJ 212	NU 212	NUP 212
65 RN 02	65 RJ 02	65 RU 02	65 RT 02	65	120	23	2.5	2.5	105.6	5700	4750	5300	6300	N 213	NJ 213	NU 213	NUP 213
70 RN 02	70 RJ 02	70 RU 02	70 RT 02	70	125	24	2.5	2.5	110.5	6000	5000	5000	6000	N 214	NJ 214	NU 214	NUP 214
75 RN 02	75 RJ 02	75 RU 02	75 RT 02	75	130	25	2.5	2.5	116.5	6950	5850	4800	5600	N 215	NJ 215	NU 215	NUP 215
80 RN 02	80 RJ 02	80 RU 02	80 RT 02	80	140	26	3	3	125.3	7800	6800	4500	5300	N 216	NJ 216	NU 216	NUP 216
85 RN 02	85 RJ 02	85 RU 02	85 RT 02	85	150	28	3	3	133.8	9000	7800	4300	5000	N 217	NJ 217	NU 217	NUP 217
90 RN 02	90 RJ 02	90 RU 02	90 RT 02	90	160	30	3	3	143	11000	9300	3800	4500	N 218	NJ 218	NU 218	NUP 218
95 RN 02	95 RJ 02	95 RU 02	95 RT 02	95	170	32	3.5	3.5	151.5	12500	11000	3600	4300	N 219	NJ 219	NU 219	NUP 219
100 RN 02	100 RJ 02	100 RU 02	100 RT 02	100	180	34	3.5	3.5	160	14000	12200	3400	4000	N 220	NJ 220	NU 220	NUP 220
105 RN 02	105 RJ 02	105 RU 02	105 RT 02	105	190	36	3.5	3.5	168.8	18371	14062	3200	3800	N 221	NJ 221	NU 221	NUP 222
110 RN 02	110 RJ 02	110 RU 02	110 RT 02	110	200	38	3.5	3.5	178.5	22226	16556	3000	3600	N 222	NJ 222	NU 222	NUP 222
120 RN 02	120 RJ 02	120 RU 02	120 RT 02	120	215	40	3.5	3.5	191.5	24041	18371	2800	3400	N 224	NJ 224	NU 224	NUP 224
130 RN 02	130 RJ 02	130 RU 02	130 RT 02	130	230	40	4	4	204	25855	20639	2600	3200	N 226	NJ 226	NU 226	NUP 226
—	140 RJ 02	140 RU 02	140 RT 02	140	250	42	4	4	221	29711	24041	2400	3000	—	NJ 228	NU 228	NUP 228
—	150 RJ 02	150 RU 02	150 RT 02	150	270	45	4	4	238	34700	28123	2000	2600	—	NJ 230	NU 230	NUP 230

ABUTMENT DIMENSIONS FOR SINGLE ROW CYLINDRICAL ROLLER BEARINGS

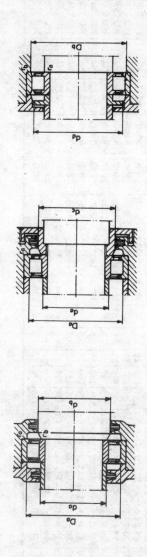

All dimensions in *mm*

Table 119 Dimension series 02

Type RN	Type RJ	Type RU	Type RT	d_a Max.	D_a Max.	d_b Min.	D_b Min.	d_c Min.	d_d Max.	r_a Max.	r_b Max.	Type RN (N)	Type RJ (NJ)	Type RU (NU)	Type RT (NUP)
15 RN 02	15 RJ 02	15 RU 02	—	18	31	21	31	24	28	0.6	0.3	N 202	NJ 202	NU 202	—
17 RN 02	17 RJ 02	17 RU 02	—	21	36	25	36	27	32	0.6	0.3	N 203	NJ 203	NU 203	—
20 RN 02	20 RJ 02	20 RU 02	20 RT 02	26	42	29	42	32	38	1	0.6	N 204	NJ 204	NU 204	NUP 204
25 RN 02	25 RJ 02	25 RU 02	25 RT 02	31	47	34	47	37	43	1	0.6	N 205	NJ 205	NU 205	NUP 205
30 RN 02	30 RJ 02	30 RU 02	30 RT 02	37	57	40	56	44	52	1	0.6	N 206	NJ 206	NU 206	NUP 206
35 RN 02	35 RJ 02	35 RU 02	35 RT 02	43	65.5	46	64	50	60	1	0.6	N 207	NJ 207	NU 207	NUP 207
40 RN 02	40 RJ 02	40 RU 02	40 RT 02	49	73.5	52	72	56	68	1	1	N 208	NJ 208	NU 208	NUP 208
45 RN 02	45 RJ 02	45 RU 02	45 RT 02	54	78.5	57	77	61	73	1	1	N 209	NJ 209	NU 209	NUP 209

IS designation | Equivalent *SKF* designation

ABUTMENT DIMENSIONS FOR SINGLE ROW CYLINDRICAL ROLLER BEARINGS (Concld.)

Table 119 Dimension series 02

All dimensions in *mm*

IS designation				da Max.	Da Max.	db Min.	Db Min.	dc Min.	dd Max.	ra Max.	rb Max.	Equivalent SKF designation			
Type RN	Type RJ	Type RU	Type RT									Type RN (N)	Type RJ (NJ)	Type RU (NU)	Type RT (NUP)
50 RN 02	50 RJ 02	50 RU 02	50 RT 02	58	83.5	62	83	67	78	1	1	N 210	NJ 210	NU 210	NUP 210
55 RN 02	55 RJ 02	55 RU 02	55 RT 02	65	92	68	91	73	86	1.5	1	N 211	NJ 211	NU 211	NUP 211
60 RN 02	60 RJ 02	60 RU 02	60 RT 02	71	102	75	100	80	95	1.5	1.5	N 212	NJ 212	NU 212	NUP 212
65 RN 02	65 RJ 02	65 RU 02	65 RT 02	77	112	81	108	87	103	1.5	1.5	N 213	NJ 213	NU 213	NUP 213
70 RN 02	70 RJ 02	70 RU 02	70 RT 02	82	117	86	114	92	108	1.5	1.5	N 214	NJ 214	NU 214	NUP 214
75 RN 02	75 RJ 02	75 RU 02	75 RT 02	87	122	90	120	96	114	1.5	1.5	N 215	NJ 215	NU 215	NUP 215
80 RN 02	80 RJ 02	80 RU 02	80 RT 02	94	131	97	128	104	123	2	2	N 216	NJ 216	NU 216	NUP 216
85 RN 02	85 RJ 02	85 RU 02	85 RT 02	99	141	104	137	110	—	2	2	N 217	NJ 217	NU 217	NUP 217
90 RN 02	90 RJ 02	90 RU 02	90 RT 02	105	151	109	146	116	140	2	2	N 218	NJ 218	NU 218	NUP 218
95 RN 02	95 RJ 02	95 RU 02	95 RT 02	111	159	116	155	123	149	2	2	N 219	NJ 219	NU 219	NUP 219
100 RN 02	100 RJ 02	100 RU 02	100 RT 02	117	169	122	164	130	157	2	2	N 220	NJ 220	NU 220	NUP 220
105 RN 02	105 RJ 02	105 RU 02	105 RT 02	124	179	129	173	137	166	2	2	N 221	NJ 221	NU 221	NUP 221
110 RN 02	110 RJ 02	110 RU 02	110 RT 02	130	189	135	182	144	175	2	2	N 222	NJ 222	NU 222	NUP 222
120 RN 02	120 RJ 02	120 RU 02	120 RT 02	141	204	146	196	156	188	2	2	N 224	NJ 224	NU 224	NUP 224
130 RN 02	130 RJ 02	130 RU 02	130 RT 02	151	217	158	208	168	201	2.5	2.5	N 226	NJ 226	NU 226	NUP 226
	140 RJ 02	140 RU 02	140 RT 02	166	237	171	—	182	—	2.5	2.5	—	NJ 228	NU 228	NUP 228
	150 RJ 02	150 RU 02	150 RT 02	179	257	184	—	196	—	2.5	2.5	—	NJ 230	NU 230	NUP 230

DOUBLE ROW ROLLER BEARINGS

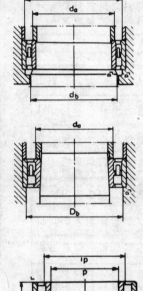

All dimensions in *mm*

Table 120 Dimension series 30

| IS designation | Boundary dimensions | | | | Abutment dimensions | | | Basic load rating kgf | | Limiting speed rpm | | Equivalent DIN & CSN designation |
	d	D	B	r Nom.	d_a Min.	D_a Max.	r_a	Dynamic C	Static C_o	Grease	Oil	
30 RD 30 K	30	55	19	1.5	35	50	1	2540	1814	11000	14000	NN 3006 K
35 RD 30 K	35	62	20	1.5	40	57	1	3266	2495	9500	12000	NN 3007 K
40 RD 40 K	40	68	21	1.5	45	63	1	3765	2858	9000	11000	NN 3008 K
45 RD 30 K	45	75	23	1.5	50	70	1	4218	3402	8000	9500	NN 3009 K
50 RD 30 K	50	80	23	1.5	55	75	1	4536	3765	7500	9000	NN 3010 K
55 RD 30 K	55	90	26	2	61.5	83.5	1	5988	4990	6700	8000	NN 3011 K
60 RD 30 K	60	95	26	2	66.5	88.5	1	6350	5443	6300	7500	NN 3012 K
65 RD 30 K	65	100	26	2	71.5	93.5	1	6486	5670	6000	7000	NN 3013 K
70 RD 30 K	70	110	30	2	76.5	103.5	1	8437	7530	5300	6300	NN 3014 K
75 RD 30 K	75	115	30	2	81.5	108.5	1	8437	7530	5000	6000	NN 3015 K
80 RD 30 K	80	125	34	2	86.5	118.5	1	10342	9435	4800	5600	NN 3016 K

DOUBLE ROW ROLLER BEARINGS (Concld.)

Table 120 Dimension series 30 (concld.)

All dimensions in *mm*

IS designation	Boundary dimensions				Abutment dimensions			Basic load rating *kgf*		Limiting speed *rpm* Lubrication		Equivalent *DIN & CSN* designation
	d	D	B	r Nom.	da Min.	Da Max.	ra	Dynamic C	Static Co	Grease	Oil	
85 RD 30 K	85	130	34	2	91.5	123.5	1	10704	10161	4500	5300	NN 3017 K
90 RD 30 K	90	140	37	2.5	98	132	1.5	12474	11794	4300	5000	NN 3018 K
95 RD 30 K	95	145	37	2.5	103	137	1.5	12928	12247	4000	4800	NN 3019 K
100 RD 30 K	100	150	37	2.5	108	142	1.5	13608	13154	3800	4500	NN 3020 K
105 RD 30 K	105	160	41	3	114	151	2	16556	16103	3600	4300	NN 3021 K
110 RD 30 K	110	170	45	3	119	161	2	19958	19278	3400	4000	NN 3022 K
120 RD 30 K	120	180	46	3	129	171	2	20639	20639	3200	3800	NN 3024 K
130 RD 30 K	130	200	52	3	139	191	2	25855	25402	2800	3400	NN 3026 K
140 RD 30 K	140	210	53	3	149	201	2	27216	27670	2600	3200	NN 3028 K
150 RD 30 K	150	225	56	3.5	161	214	2	29711	30391	2400	3000	NN 3030 K
160 RD 30 K	160	240	60	3.5	171	225	2	33340	34700	2200	2800	NN 3032 K
170 RD 30 K	170	260	67	3.5	181	249	2	40824	43090	2000	2600	NN 3034 K
180 RD 30 K	180	280	74	3.5	191	269	2	50803	53525	1900	2400	NN 3036 K
190 RD 30 K	190	290	75	3.5	201	275	2	53525	57607	1900	2400	NN 3038 K
200 RD 30 K	200	310	82	3.5	211	299	2	58514	62143	1800	2200	NN 3040 K
220 RD 30 K	220	340	90	4	233	327	2.5	73937	79834	1700	2000	NN 3044 K
240 RD 30 K	240	360	92	4	253	347	2.5	78473	86184	1500	1800	NN 3048 K
260 RD 30 K	260	400	104	5	276	384	3	94349	103420	1300	1600	NN 3052 K
280 RD 30 K	280	430	106	5	296	404	3	97978	111132	1200	1500	NN 3056 K
300 RD 30 K	300	460	118	5	316	444	3	117936	136080	1100	1400	NN 3060 K
320 RD 30 K	320	480	121	5	336	464	3	122472	140616	1000	1300	NN 3064 K
340 RD 30 K	340	520	133	6	360	500	4	147420	172368	950	1200	NN 3068 K
360 RD 30 K	360	540	134	6	380	520	4	154224	188244	950	1200	NN 3072 K

RING GAUGES FOR NN30K BEARINGS

Table 121   Dimensioning and inspection of tapered shaft seatings for NN 30K bearings

All dimensions in *mm*.

Ring gauge No.	Bearing No. SP & UP classes	Ring gauge dimensions			Dimensioning of bearing seating				Tol. on	
		d	D	B	d_1	a^*	b	c	c	d_2
GRA 3006	NN 3006 K	30	52	19	30.1	4	24	6.2	0.1	32
GRA 3007	NN 3007 K	35	57	20	35.1	6	25	6.2	0.1	37
GRA 3008	NN 3008 K	40	62	21	40.1	6	28	8.2	0.1	42
GRA 3009	NN 3009 K	45	67	23	45.1	8	30	8.2	0.1	47
GRA 3010	NN 3010 K	50	72	23	50.1	8	30	8.2	0.1	52
GRA 3011	NN 3011 K	55	77	26	55.15	8	32.5	8.3	0.12	57
GRA 3012	NN 3012 K	60	82	26	60.15	8	34.5	10.3	0.12	62
GRA 3013	NN 3013 K	65	88	26	65.15	10	34.5	10.3	0.12	67
GRA 3014	NN 3014 K	70	95	30	70.15	10	38.5	10.3	0.12	73
GRA 3015	NN 3015 K	75	100	30	75.15	10	38.5	10.3	0.12	78
GRA 3016	NN 3016 K	80	105	34	80.15	10	44.5	12.3	0.12	83
GRA 3017	NN 3017 K	85	112	34	85.2	12	44	12.4	0.15	88
GRA 3018	NN 3018 K	90	120	37	90.2	12	47	12.4	0.15	93
GRA 3019	NN 3019 K	95	128	37	95.2	12	47	12.4	0.15	98
GRA 3020	NN 3020 K	100	135	37	100.2	12	47	12.4	0.15	103

Notes:

* To allow for final adjustment when mounting, intial width of distance ring should be $a+0.2$ *mm*. Final width should be determined during mounting when the desired internal clearance in the bearing is obtained.

All measurements must be made from the polished surface at the large end of the taper bore on the ring gauge.

NN 30 K BEARING

The permissible axial runout of shoulders, and errors in parallelism of distance rings and between gauge reference face and shoulder are as follows:

Class of NN 30 K bearings	Values in μ m			
	Diameter range in *mm* (OD)			
	13-50	50-120	120-180	>180
SP	3	4	5	7
UP	2	3	4	5

TAPER ROLLER BEARINGS

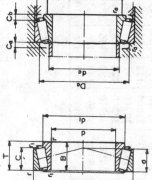

Table 122 Dimension series 22

All dimensions in *mm*

IS designation	d	D	T	B	r	r₁	Basic load rating Dynamic C kgf	Basic load rating Static C₀ kgf	db Min.	da Max.	Da Min.	Db Min.	ra Max.	Limiting speed rpm Lubrication Grease	Limiting speed rpm Lubrication Oil	SKF designation
30 KB 22	30	62	21.25	20	1.5	0.5	3200	2750	36	37	52	58	1	6300	8500	32206
35 KB 22	35	72	24.25	23	2	0.8	4300	3650	42	43	61	67	1	5300	7000	32207
40 KB 22	40	80	24.75	23	2	0.8	4750	4050	47	48	68	75	1	4800	6300	32208
45 KB 22	45	85	24.75	23	2	0.8	5200	4650	52	53	73	80	1	4500	6000	32209
50 KB 22	50	90	24.75	25	2	0.8	5300	4800	57	58	78	85	1	4300	5600	32210
55 KB 22	55	100	26.75	25	2.5	0.8	6700	6300	63	64	87	95	1.5	3800	5000	32211
60 KB 22	60	110	29.75	28	2.5	0.8	8000	7650	69	69	95	104	1.5	3400	4500	32212
65 KB 22	65	120	32.75	31	2.5	0.8	9800	9300	74	76	104	115	1.5	3000	4000	32213
70 KB 22	70	125	33.25	31	2.5	0.8	9800	9300	79	80	108	119	1.5	2800	3800	32214
75 KB 22	75	130	33.25	31	2.5	0.8	10400	10200	84	85	114	125	1.5	2600	3600	32215
80 KB 22	80	140	35.25	33	3	1	12000	11600	90	90	122	134	2	2400	3400	32216

TAPER ROLLER BEARINGS (Contd.)

Table 122 Dimension series 22 (concld.)

All dimensions in *mm*

IS designation	d	D	T	B	r	r_1	Basic load rating Dynamic C kgf	Static C_o kgf	db Min.	da Max.	Da Min	Db Min.	ra Max.	Limiting speed rpm Lubrication Grease	Oil	SKF designation
85 KB 22	85	150	38.5	36	3	1	13700	13700	95	96	130	142	2	2200	3200	32217
90 KB 22	90	160	42.5	40	3	1	16300	16600	100	102	138	152	2	2000	3000	32218
95 KB 22	95	170	45.5	43	3.5	1.2	18600	18600	107	108	145	161	2	1900	2800	32219
100 KB 22	100	180	49	46	3.5	1.2	20800	21200	112	114	154	171	2	1800	2600	32220
105 KB 22	105	190	53	50	3.5	1.2	31525	29711	117	120	161	180	2	1800	2600	32221
110 KB 22	110	200	56	53	3.5	1.2	34700	33340	122	126	170	190	2	1700	2400	32222
120 KB 22	120	215	61.5	58	3.5	1.2	36288	34700	132	136	181	204	2	1600	2200	32224
130 KB 22	130	230	67.75	64	4	1.5	43092	41504	144	146	193	219	2.5	1500	2000	32226
140 KB 22	140	250	71.75	68	4	1.5	49896	48989	154	158	210	238	2.5	1400	1900	32228

NEEDLE BEARINGS

Table 123   Light series

All dimensions in *mm*

Needle bearing with inner ring	Needle bearing without inner ring	d Nom.	di Nom.	D Nom.	B Nom.	r	Basic load rating		Limiting speed rpm	Equivalent NRB designation	
							Dynamic C kgf	Static Co kgf		Needle bearing with inner ring	Needle bearing without inner ring
NEA 1012	NES 1012	12	17.6	28	15	1	1150	960	21600	Na 1012	Na 1012 S/Bi
NEA 1015	NES 1015	15	20.8	32	15	1	1285-	1110	18300	Na 1015	Na 1015 S/Bi
NEA 1017	NES 1017	17	23.9	35	15	1	1420	1250	15900	Na 1017	Na 1017 S/Bi
NEA 1020	NES 1020	20	28.7	42	18	1	2000	1870	13200	Na 1020	Na 1020 S/Bi
NEA 1025	NES 1025	25	33.5	47	18	1	2200	2160	11100	Na 1025	Na 1025 S/Bi
NEA 1030	NES 1030	30	38.2	52	18	1	2440	2440	10000	Na 1030	Na 1030 S/Bi
NEA 1035	NES 1035	35	44	58	18	1	2700	2790	8600	Na 1035	Na 1035 S/Bi
NEA 1040	NES 1040	40	49.7	65	18	1.5	2930	3130	7600	Na 1040	Na 1040 S/Bi
NEA 1045	NES 1045	45	55.4	72	18	1.5	3160	3470	6900	Na 1045	Na 1045 S/Bi
NEA 1050	NES 1050	50	62.1	80	20	2	3420	3860	6100	Na 1050	Na 1050 S/Bi
NEA 1055	NES 1055	55	68.8	85	20	2	3690	4250	5500	Na 1055	Na 1055 S/Bi
NEA 1060	NES 1060	60	72.6	90	20	2	3820	4460	5200	Na 1060	Na 1060 S/Bi
NEA 1065	NES 1065	65	78.3	95	20	2	4240	5050	4900	Na 1065	Na 1065 S/Bi
NEA 1070	NES 1070	70	83.1	100	20	2	4420	5390	4500	Na 1070	Na 1070 S/Bi
NEA 1075	NES 1075	75	88	110	24	2	6600	8200	4300	Na 1075	Na 1075 S/Bi
NEA 1080	NES 1080	80	96	115	24	2	7000	8850	4000	Na 1080	Na 1080 S/Bi

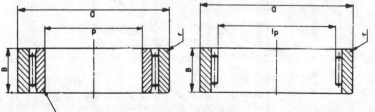

Table 124 Medium series

NEEDLE BEARINGS (Contd.)

All dimensions in *mm*

Needle bearing with inner ring	Needle bearing without inner ring	d Nom.	di Nom.	D Nom.	B Nom.	r	Basic load rating		Limiting speed rpm	Equivalent NRB designation	
							Dynamic C kgf	Static Co kgf		Needle bearing with inner ring	Needle bearing without inner ring
NEA 2015	NES 2015	15	22.1	35	22	1	2480	2200	17200	Na 2015	Na 2015 S/Bi
NEA 2020	NES 2020	20	28.7	42	22	1	2980	2680	13200	Na 2020	Na 2020 S/Bi
NEA 2025	NES 2025	25	33.5	47	22	1	3220	3220	11100	Na 2025	Na 2025 S/Bi
NEA 2030	NES 2030	30	38.2	52	22	1	3640	3640	10000	Na 2030	Na 2030 S/Bi
NEA 2035	NES 2035	35	44	58	22	1	4000	4150	8600	Na 2035	Na 2035 S/Bi
NEA 2040	NES 2040	40	49.7	62	22	1.5	4350	4650	9600	Na 2040	Na 2040 S/Bi
NEA 2045	NES 2045	45	55.4	72	22	1.5	4750	5150	6900	Na 2045	Na 2045 S/Bi
NEA 2050	NES 2050	50	62.1	80	28	2	6500	7600	6100	Na 2050	Na 2050 S/Bi
NEA 2055	NES 2055	55	68.8	85	28	2	7250	8400	5500	Na 2055	Na 2055 S/Bi
NEA 2060	NES 2060	60	72.6	90	28	2	7550	8600	5200	Na 2060	Na 2060 S/Bi
NEA 2065	NES 2065	65	78.3	95	28	2	8200	9750	4900	Na 2065	Na 2065 S/Bi
NEA 2070	NES 2070	70	83.1	100	28	2	8500	10300	4500	Na 2070	Na 2070 S/Bi
NEA 2075	NES 2075	75	88	110	32	2	10800	13400	4300	Na 2075	Na 2075 S/Bi
NEA 2080	NES 2080	80	96	115	32	2	11500	14600	4000	Na 2080	Na 2080 S/Bi
NEA 2085	NES 2085	85	99.5	120	32	2	11800	15100	3800	Na 2085	Na 2085 S/Bi
NEA 2090	NES 2090	90	104.7	125	32	2	12500	15950	3600	Na 2090	Na 2090 S/Bi
NEA 2095	NES 2095	95	109.1	130	32	2	12100	16600	3500	Na 2095	Na 2095 S/Bi
NEA 2100	NES 2100	100	114.7	135	32	2	12800	17400	3300	Na 2100	Na 2100 S/Bi
NEA 2105	NES 2105	105	119.2	140	32	2	13400	18100	3200	Na 2105	Na 2105 S/Bi
NEA 2110	NES 2110	110	124.7	145	34	2	13900	18900	3000	Na 2110	Na 2110 S/Bi

All dimensions in *mm*

NEEDLE BEARINGS (Contd.)

Table 124   Medium series (Concld.)

Needle bearing with inner ring	Needle bearing without inner ring	d Nom.	di Nom.	D Nom.	B Nom.	r	Basic load rating		Limiting speed rpm	Equivalent NRB designation	
							Dynamic C kgf	Static Co kgf		Needle bearing with inner ring	Needle bearing without inner ring
NEA 2115	NES 2115	115	132.5	155	34	2	14400	20000	2900	Na 2115	Na 2115 S/Bi
NEA 2120	NES 2120	120	137	160	34	2	14800	20600	2800	Na 2120	Na 2120 S/Bi
NEA 2125	NES 2125	125	143.5	165	34	2	15200	21600	2700	Na 2125	Na 2125 S/Bi
NEA 2130	NES 2130	130	148	170	34	2	15500	22100	2600	Na 2130	Na 2130 S/Bi
NEA 2140	NES 2140	140	158	180	36	2	16300	23800	2400	Na 2140	Na 2140 S/Bi
NEA 2150	NES 2150	150	170.5	195	36	2	17200	25300	2200	Na 2150	Na 2150 S/Bi
NEA 2160	NES 2160	160	179.3	205	36	2	17800	26800	2100	Na 2160	Na 2160 S/Bi
NEA 2170	NES 2170	170	193.8	220	42	3	24300	37500	2000	Na 2170	Na 2170 S/Bi
NEA 2180	NES 2180	180	202.6	230	42	3	25100	39200	1900	Na 2180	Na 2180 S/Bi
NEA 2190	NES 2190	190	216	240	42	3	26200	41700	1800	Na 2190	Na 2190 S/Bi
NEA 2200	NES 2200	200	224.1	255	42	3	26800	43200	1700	Na 2200	Na 2200 S/Bi
NEA 2210	NES 2210	210	236	265	42	3	28800	45500	1600	Na 2210	Na 2210 S/Bi
NEA 2220	NES 2220	220	248.4	280	49	3	34400	56800	1500	Na 2220	Na 2220 S/Bi
NEA 2230	NES 2230	230	258.4	290	49	3	35300	59000	1500	Na 2230	Na 2230 S/Bi
NEA 2240	NES 2240	240	269.6	300	49	3	36400	61500	1400	Na 2240	Na 2240 S/Bi
NEA 2250	NES 2250	250	281.9	315	49	3	37500	64500	1300	Na 2250	Na 2250 S/Bi

All dimensions in *mm*

Table 125 Heavy series

NEEDLE BEARINGS (Contd.)

Needle bearing with inner ring	Needle bearing without inner ring	d Nom.	di Nom.	D Nom.	B Nom.	r	Basic load rating		Limiting speed rpm	Equivalent NRB designation	
							Dynamic C kgf	Static Co kgf		Needle bearing with inner ring	Needle bearing without inner ring
NEA 3030	NES 3030	30	44	62	30	1	3650	6950	8600	Na 3030	Na 3030 S/Bi
NEA 3035	NES 3035	35	49.7	72	36	1	9350	10000	7600	Na 3035	Na 3035 S/Bi
NEA 3040	NES 3040	40	55.4	80	36	1.5	10500	11100	6900	Na 3040	Na 3040 S/Bi
NEA 3045	NES 3045	45	62.1	85	38	1.5	10900	12400	6100	Na 3045	Na 3045 S/Bi
NEA 3050	NES 3050	50	68.8	90	38	2	11800	13700	5500	Na 3050	Na 3050 S/Bi
NEA 3055	NES 3055	55	72.6	95	38	2	12200	14400	5200	Na 3055	Na 3055 S/Bi
NEA 3060	NES 3060	60	78.3	100	38	2	12900	15400	4900	Na 3060	Na 3060 S/Bi
NEA 3065	NES 3065	65	83.1	105	38	2	13400	16400	4500	Na 3065	Na 3065 S/Bi
NEA 3070	NES 3070	70	88	110	38	2	14000	17300	4300	Na 3070	Na 3070 S/Bi
NEA 3075	NES 3075	75	96	120	38	2	14300	18800	4000	Na 3075	Na 3075 S/Bi
NEA 3080	NES 3080	80	99.5	125	38	2	15200	19400	3800	Na 3080	Na 3080 S/Bi
NEA 3085	NES 3085	85	104.7	130	43	2	15700	20200	3600	Na 3085	Na 3085 S/Bi
NEA 3090	NES 3090	90	109.1	135	43	2	19300	25400	3500	Na 3090	Na 3090 S/Bi
NEA 3095	NES 3095	95	114.7	140	43	2	20000	26800	3300	Na 3095	Na 3095 S/Bi
NEA 3100	NES 3100	100	119.2	145	45	2	20500	27600	3200	Na 3100	Na 3100 S/Bi
NEA 3105	NES 3105	105	124.7	150	45	2	21200	28900	3000	Na 3105	Na 3105 S/Bi
NEA 3110	NES 3110	110	132.5	160	45	2	22000	30600	2900	Na 3110	Na 3110 S/Bi
NEA 3115	NES 3115	115	137	165	45	2	22600	31800	2800	Na 3115	Na 3115 S/Bi
NEA 3120	NES 3120	120	143.5	170	45	2	23300	33000	2700	Na 3120	Na 3120 S/Bi
NEA 3125	NES 3125	125	152.8	185	52	2	27900	40500	2500	Na 3125	Na 3125 S/Bi
NEA 3130	NES 3130	130	158	190	52	2	28600	41800	2400	Na 3130	Na 3130 S/Bi

NEEDLE BEARINGS (Concld.)

Table 125 Heavy series (Concld.)

All dimensions in mm

Needle bearing with inner ring	Needle bearing without inner ring	d Nom.	di Nom.	D Nom.	B Nom.	r	Basic load rating		Limiting speed rpm	Equivalent NRB designation	
							Dynamic C kgf	Static Co kgf		Needle bearing with inner ring	Needle bearing without inner ring
NEA 3140	NES 3140	140	170.5	205	52	2	30000	45000	2200	Na 3140	Na 3140 S/Bi
NEA 3150	NES 3150	150	179.3	215	52	2	31400	47300	2100	Na 3150	Na 3150 S/Bi
NEA 3160	NES 3160	160	193.8	230	57	3	37600	58000	2000	Na 3160	Na 3160 S/Bi
NEA 3170	NES 3170	170	202.6	245	57	3	38800	60500	1900	Na 3170	Na 3170 S/Bi
NEA 3180	NES 3180	180	216	255	57	3	40500	64500	1800	Na 3180	Na 3180 S/Bi
NEA 3190	NES 3190	190	224.1	265	57	3	—	—	1700	Na 3190	Na 3190 S/Bi
NEA 3200	NES 3200	200	236	280	57	3	43000	70500	1600	Na 3200	Na 3200 S/Bi
NEA 3210	NES 3210	210	248.4	290	64	3	50000	83000	1500	Na 3210	Na 3210 S/Bi
NEA 3220	NES 3220	220	258.4	300	64	3	51400	86400	1500	Na 3220	Na 3220 S/Bi
NEA 3230	NES 3230	230	269.6	315	64	3	52800	90000	1400	Na 3230	Na 3230 S/Bi
NEA 3240	NES 3240	240	281.5	325	64	3	54500	94000	1300	Na 3240	Na 3240 S/Bi
NEA 3250	NES 3250	250	290.9	340	74	3	68000	118000	1300	Na 3250	Na 3250 S/Bi

NEEDLE BUSHES, FULL COMPLEMENT, RETAINED NEEDLES

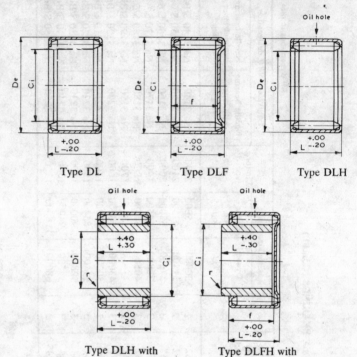

Type DL   Type DLF   Type DLH

Type DLH with
cylindrical inner ring

Type DLFH with
cylindrical inner ring

Table 126 Types *DL; DLH, DLF, DLFH*   All dimensions in *mm*

NRB designation								Basic load rating*		
DL DLH DLF DLFH	DL/... DLH/.... DLF/.... DLFH/..	D_i	C_i	D_e	L	f	−	Dynamic C kgf	Static C_o kgf	Limiting speed rpm
610	—	—	6	12	10	7.8		309	274	50000
810	—	—	8	14	10	7.8		386	347	37500
DLC 910	—	—	9	13	10	—		570	480	33000
914 12	—	—	9	14	12	9.8		630	562	33000
1012	—	—	10	16	12	9.8		640	593	30000
1210	—	—	12	18	10	—		525	496	25000
1212	1212/08	8	12	18	12	9.8		735	699	25000
1212	1212/09	9	12	18	12	—		—	—	—
1312	1312/09	9	13	19	12	9.8		766	737	23000
1412	1412/10	10	14	20	12	9.8		825	797	21500
1412	1412/11	11	14	20	12	—		—	—	—
1512	1512/11	11	15	21	12	9.8		855	840	20000

* For shaft raceway hardness of 60 *HRC*

NEEDLE BUSHES, FULL COMPLIMENT, GREASE RETAINED NEEDLES

Table 126 Types *DL, DLH, DLF, DLFH* (Concld.)   All dimensions in *mm*

DL DLH DLF DLFH	DL/.... DLH/:.. DLF/... DLFH/..	D_i	C_i	D_e	L	f	Basic load rating Dynamic C kgf	Static C_o kgf	Limiting Speed rpm
1512	1512/12	12	15	21	12	—	—	—	—
DLC 1516	—	—	15	21	16	—	1200	1180	20000
1612	1612/12	12	16	22	12	9.8	912	902	18500
1712	1712/13	13	17	23	12	9.8	945	940	17500
1812	1812/13	13	18	24	12	9.8	992	1000	16500
1816	1816/13	13	18	24	16	13.8	1555	1575	16500
2012	2012/15	15	20	26	12	9.8	1070	1100	15000
2016	2016/15	15	20	26	16	13.8	1680	1730	15000
2216	2216/17	17	22	28	16	13.8	1795	1890	13500
2516	2516/20	20	25	33	16	13.8	1680	1890	12000
2520	2520/20	20	25	33	20	17.8	2390	2690	12000
2820	2820/23	23	28	36	20	17.8	2590	2970	11000
3016	3016/25	25	30	38	16	13.8	1900	2210	10000
3020	3020/25	25	30	38	20	17.8	2720	3160	10000
3025	—	—	30	38	25	—	3720	4340	10000
3542 16P	—	—	35	42	16	—	2240	2690	8500
3516	3516/30	30	35	43	16	13.8	2120	2560	8500
3520	3520/30	30	35	43	20	17.8	3020	3640	8500
3520	3520/28	28	35	43	20	—	—	—	—
4016	4016/35	35	40	48	16	13.8	2330	2900	7500
4020	4020/35	35	40	48	20	17.8	3210	4120	7500
4416	4416/40	40	44	52	16	13.8	2500	3160	6800
4516(P)	4516/40	40	45	52	16	—	2750	3450	6500
4716	—	—	47	55	16	—	2620	3370	6400
5020	5020/45	45	50	58	20	—	3900	5030	6000
5520	5520/50	50	55	63	20	17.8	4025	5550	5500

NEEDLE BUSHES, FULL COMPLIMENT, GREASE RETAINED NEEDLES

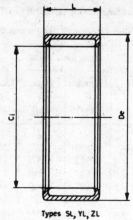

Types SL, YL, ZL   Types CN, CNS

Table 127 Open end types *SL, YL & ZL*   All dimensions in *mm*

NRB designation	C_i	D_c	L -0.1 -0.2	Basic load rating Dynamic C kgf	Basic load rating Static C_o kgf	Limiting speed rpm
SL 20429	10	16	8.95	610	510	30000
SL 20017	12	18	9.95	810	690	25000
SL 122012	12	20	12	1020	930	25000
SL 20018	16	22	12.15	1300	1175	18500
SL 20019	18	24	12.8	1520	1410	16500
SL 1816 P	18	24	16	2000	1440	16500
SL 20020	22	28	14.8	2100	2010	13500

Table 128 Closed end types *CN & CNS*   All dimensions in *mm*

NRB designation	C_i	D_a	L 0 -0.2	L_1 0 -0.2	f 0 +0.3	Basic load rating Dynamic C kgf	Basic load rating Static C_o kgf	Limiting speed rpm
CNS 1009*	10	16	8.95	10.15	7.4	610	510	30000
CN 1210	12	18	9.95	10.7	8.4	110	695	25000
CN 1612	16	22	12.15	13.1	10.6	1300	1180	18500
CN 1813	18	24	12.8	13.8	11.25	1520	1410	16500
CN 2213	22	28	12.8	13.8	11.25	1740	1680	13500
CN 2215	22	28	14.8	15.8	13.25	2100	2010	13500

* This bearing is not designed to support axial loads due to its configuration and should be positioned as far apart as possible at the ends of the shaft.

Table 125 a Shaft and housing tolerances (For *Na* & *Na..... S/Bi* series)

Condition	Load direction	Shaft dimension *Di*				Shaft dimension *Ci*	Housing dimension *De*	Ovality+ conicity
		< 80	85 – 130	140 – 220	≥ 230			
Shaft revolving housing stationary	Fixed	k5	m5	n6	p6	h5	J7	for *Di* and *De* less than half the tolerance, for *Ci* less than quarter of the tolerance
	Revolving with shaft	h5	h5	h6	h6	g5	M7	
	Indeterminate	k5	m5	n6	p6	g5	M7	
Shaft stationary housing revolving	Fixed	h5	h5	h6	h6	g5	M7	
	Revolving with housing	k5	m5	n6	p6	h5	J7	
	Indeterminate	k5	m5	n6	p6	g5	M7	
Shaft and housing revolving	Any	k5	m5	n6	p6	g5	M7	
Oscillatory motion	Any	k5	k5	m6	m6	h5	M7	

— Housing tolerances given above are for rigid steel or CI housings.
— The outer ring must be a tighter fit in thin walled or non-ferrous metal housings.
— If light alloy housings reach a considerably higher (or lower) than 20º C the housing fits must be adjusted to allow for differential expansion (or contraction).

SHAFT AND HOUSING TOLERANCES

(For Needle bushes, full complement, retained needles)

Table 126 a Shaft tolerance

	Dimension *Di*	Dimension *Ci*
Continuous rotation	k6	h6
Oscillatory motion	m6	j6
	Ovality+ conicity less than half the tolerance	Ovality+ conicity less than quarter of the tolerance

Table 126 b Housing Tolerances — Dimension *De*

Steel H7	Type DL and derivatives (except type DL ..... P)	M7	Non-ferrous metal(1)
Or			Or
Cast iron N7	Type DL ..... P and DLC	R7	Thin walled steel

SHAFT AND HOUSING TOLERANCES

(For Needle bushes, full complement, grease-retained needles)

Table 128 a Shaft tolerances - Dimension *Ci*

Continuous rotation	h6	All bushes
Oscillatory motion	j6	Types SL, YL & ZL
	k6	Types CN and CNS

Ovality+conicity less than quarter of the tolerance.

Table 128 b Housing tolerances - Dimension *De*

Bush	Housing	
	Steel or Cast iron	Non-ferrous metal(1) or thin walled steel
CN & CNS	F7	H7
SL 20429	F7	H7
SL 20017	F7	H7
SL 122012	H7	M7
SL 20018	+0.045/+0.024	+0.024/+0.003
SL 20019	+0.043/+0.022	H7
SL 1816 P	N7	R7
SL 20020	+0.035/+0.014	+0.014/—0.007

Ovality+conicity less than half the tolerance

(1) If the non-ferrous metal housing reaches a temperature considerably higher (or lower) than 20°C, differential expansion (or contraction) must be taken into account and the tolerance adjusted accordingly.

NEEDLE CAGES

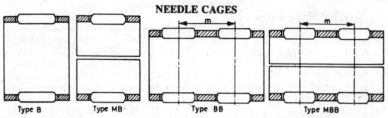

Type B   Type MB   Type BB   Type MBB

Table 129 Type *B*   All dimensions in *mm*

NRB designation	C_i	C_e	L	Basic load rating Dynamic C kgf	Static C_o kgf	Limiting speed rpm
B 698	6	9	8	288	175	65000
B 61013	6	10	13	473	305	65000
B 81110	8	11	10	418	287	50000
B 91210	9	12	10	440	310	44000
B 10139	10	13	9	336	222	40000
B 101313	10	13	13	530	403	40000
B 121513	12	15	13	580	461	33000
B 50386	12	15	13	580	461	33000
B 50255	12	15	13.7	575	456	33000
B 50114	13	16	14	575	461	31000
B 141813	14	18	13	790	605	28000
B 50160	14.4	20.4	19.9	1400	1030	28000
B 50320	14.8	19.8	10	585	368	27000
B 151913	15	19	13	820	645	27000
B 5019015 x 19x 20				970	785	27000
B 50113	15.2	22.21	12	1010	645	26000
B 162013	16	20	13	850	685	25000
B 5079216 x 20 x 20				1155	1010	25000
B 50139	16	21	10	680	453	25000
B 16221216 x 22 x 12				990	650	25000
B 162217.2	16	22	17.2	1160	815	25000
B 172115	17	21	15	1030	885	23000
B 172717	17	27	17	1780	1365	23000
B 182213	18	22	13	910	760	22000
B 5078718 x 22 x 21.6				1400	1290	22000
B 182420	18	24	20	1650	1310	22000
B 182421	18	24	21	1650	1310	22000
B 182616	18	26	16	1660	1150	22000
B 182816	18	28	16	1950	1310	22000
B 192313	19	23	13	940	795	21000
B 202413	20	24	13	965	835	20000
B 202417	20	24	17	1290	1210	20000
B 202613.6	20	26	13.6	1080	770	20000
B 222613	22	26	13	985	875	18000
B 50650	22	26	14	985	875	18000
B 222617	22	26	17	1320	1270	18000
B 5074322 x 29 x 15.6				1730	1310	18000

NEEDLE CAGES (Contd.)

Table 129 Type *B* (contd.)   All dimensions in *mm*

NRB designation	C_i	C_e	L	Basic load rating Dynamic C kgf	Static C_o kgf	Limiting speed rpm
B 5079322.9x28.9x13.7				1375	1040	17000
B 50010	23.95	27.92	19.2	1200	1150	17000
B 5079424 x 28 x 10				780	660	16000
B 50468	24.8	33.8	25.2	3180	2580	16000
B 252913	25	29	13	1060	985	16000
B 50173	25	29	16	1100	1020	16000
B 252917	25	29	17	1420	1440	16000
B 50008	25	29	18.4	1100	1020	16000
B 253020	25	30	20	1840	1750	16000
B 253124	25	31	24	2350	2160	16000
B 283317	28	33	17	1770	1710	14000
B 283327	28	33	27	2780	3060	14000
B 283825.2	28	38	25.2	3890	3230	14000
B 284421	28	44	21	4110	2850	14000
B 50105	29.5	33.5	18.4	1140	1120	13500
B 303517	30	35	17	1650	1820	13000
B 303527	30	35	27	2830	3180	13000
B 304030	30	40	30	4755	4150	13300
B 323717	32	37	17	1890	1920	12500
B 323727	32	37	27	2960	3430	12500
B 50174	35	40	13	1240	1140	11500
B 354017	35	40	17	1960	2050	11500
B 354027	35	40	27	3070	3670	11500
B 50329	35	42	16	2240	1990	11500
B 374227	37	42	27	3190	3920	11000
B 404517	40	45	17	2100	2330	10000
B 404527	40	45	27	3290	4160	10000
B 424717	42	47	17	2170	2460	9500
B 424727	42	47	27	3400	4400	9500
B 425430.7	42	54	30.7	5500	4800	9500
B 454919.2	45	49	19.2	1840	2320	8900
B 455017	45	50	17	2230	2600	9000
B 455021	45	50	21	2745	3430	9000
B 455027	45	50	27	3500	4650	9000
B 455218	45	52	18	2630	2600	9000
B 455221	45	52	21	4450	4750	9000
B 455320	45	53	20	4450	4750	9000
B 475217	47	52	17	2290	2740	8500
B 475227	47	52	27	3600	4890	8500
B 485317	48	53	17	2280	2740	8500
B 505520	50	55	20	2490	3090	8000
B 505530	50	55	30	3650	5050	8000
B 505623	50	56	23	3110	3640	8000

NEEDLE CAGES (Contd.)

Table 129 Types *B, BB* (contd.) · All dimensions in *mm*

| NRB designation | C_i | C_e | L | Basic load rating | | Limiting speed rpm˙ |
				Dynamic C kgf	Static C_o kgf	
B 505825	50	58	25	4390	4730	8000
B 556020	55	60	20	2580	3340	7500
B 556030	55	60	30	3550	5000	7500
B 556324.9	55	63	24.9	4280	4680	7500
B 606520	60	65	20	2720	3660	6700
B 606530	60	65	30	3980	6000	6700
B 606825	60	68	25	4820	5600	6700
B 657020	65	70	20	2800	3900	6000
B 657023	65	70	23	3480	6100	6000
B 657030	65	70	30	4100	6400	6000
B 657323	65	73	23	4600	5475	6000
B 657325	65	73	25	5050	6100	6000
B 707620	70	76	20	3270	4310	5700
B 707630	70	76	30	4920	7300	5700
B 707825	70	78	25	5200	6400	5700
B 737920	73	79	20	3350	4500	5500
B 758330	75	83	30	6100	8050	5300
B 808620	80	86	20	3530	4970	5000
B 808830	80	88	30	6400	8700	5000
B 859220	85	92	20	3650	4700	4700
B 859330	85	93	30	6500	9100	4700
B 909830	90	98	30	6700	9500	4500
BB 50378	16	20	44.7	1910	1910	25000
BB 293230	29	32	30	1850	2540	13800
BB 293330	29	33	30	2200	2610	13800
BB 303526.5	30	35	26.5	2300	2450	13000
BB 303532	30	35	32	2700	3000	13000
BB 50272	30.4	40.4	28.4	3870	3230	13000
BB 50033	32	35	35.8	1500	1970	12500
BB 323732	32	37	32	2830	3230	12500
BB 323758	32	37	58	4500	5850	12500
BB 323827.6	32	38	27.6	2550	2530	12500
BB 333730.7	33	37	30.7	2360	2990	12000
BB 343945	34	39	45	3840	4860	12000
BB 354053.8	35	40	53.8	4730	6400	11500
BB 50185	36	39	36.9	1840	2670	11000
BB 374217	37	42	17	2030	2190	11000
BB 374227.5	37	42	27.5	2410	2720	11000
BB 50342	40.2	44.2	25.8	2200	2880	10000
BB 50621	40.2	44.2	32.4	2850	3980	10000
BB 50808	41.9	46.9	32	2770	3350	9500
BB 50347	42	46	51.4	3200	4720	9500
BB 424734	42	47	34	2970	3690	9500
BB 50253	43	46	51.4	2280	3750	9500

NEEDLE CAGES (Concld.)

Table 129 Types *BB, MB, MBB* (Concld.)   All dimensions in *mm*

NRB designation	Ci	Ce	L	Basic load rating		Limiting speed rpm
				Dynamic C kgf	Static Co kgf	
BB 434834	43	48	34	3780	5080	9500
BB 50186	45	49	36.6	2410	3310	9000
BB 455027	45	50	27	3100	3200	9000
BB 505736	50	57	36	4260	4950	8000
BB 515633	51	56	33	3250	4380	8000
BB 586440	58	64	40	5250	7500	6900
BB 586451	58	64	51	6100	9100	6900
BB 606632.9	60	66	32.9	4560	6300	6700
BB 50227	60	66	37.9	4560	6300	6700
BB 50162	60	66	39.9	4560	6300	6700
BB 50790	60	68	30	4935	5650	6700
BB 50791	60	68	34	5660	6750	6700
BB 657055.9	65	70	55.6	7600	11700	6000
BB 687434.6	68	74	34.9	4780	6950	5900
BB 50181	73	79	39.9	5050	7600	5500
BB 758346	75	83	46	7150	10600	5300
MB 50265	19	23	17.2	900	745	21000
MB 283317	28	33	17	1710	1650	14000
MB 323714.2	32	37	14.2	1250	1130	12500
MB 50629	32	37	20	1880	1905	12500
MB 50133	39	44	24	2450	2830	10200
MBB 71018	7	10	18	445	305	57000
MBB 151922.2	15	19	22.2	1000	820	27000
MBB 283224	28	32	24	1370	1410	14000
MBB 50263	31	36	30	2160	2250	13000
MBB 50441	35	40	32	2780	2230	11500
MBB 354034.9	35	40	34.9	3000	3570	11500
MBB 50080	35	42	36	3360	3350	11500
MBB 374225.5	37	42	25.5	2350	2630	11000
MBB 374227.5	37	42	27.5	2350	2630	11000
MBB 374228	37	42	28	2350	2630	11000
MBB 50622	40.2	44.2	32.4	2790	3880	10000
MBB 50361	42	46	51.4	3790	5850	9500
MBB 50633	42.5	46.5	51.4	3140	4590	9500
MBB 50635	43	47	51.4	3150	4600	9500
MBB 556054	55	60	54	4250	6350	7500

Note: Shaft: *h*5 Housing:*G*6 on Dim. *L: H*11
Ovality and conicity of shaft and of housing less than quarter of shaft and housing tolerances

RADIAL INTERNAL CLEARANCES IN ANTIFRICTION BEARINGS

Table 130 Deep groove ball bearings with Cylindrical bore

Bearing bore diameter d mm		Group 2(C2)				Group 0 (C0) (Normal group)				Group 3 (C3)				Group 4 (C4)			
		Manufacturing limits		Acceptance limits		Manufacturing limits		Acceptance limits		Manufacturing limits		Acceptance limits		Manufacturing limits		Acceptance limits	
Over	Upto	Min.	Max.	Min.	Max.	Min	Max.	Min.	Max.	Min.	Max.	Min	Max.	Min.	Max.	Min.	Max.
10	18	0	8	–	9	5	15	3	18	13	23	11	25	20	30	18	33
18	24	0	9	–	10	7	17	5	20	15	25	13	28	23	33	20	36
24	30	0	10	–	11	8	18	5	20	15	25	13	28	25	38	23	41
30	40	0	10	–	11	8	18	6	20	18	30	15	33	30	43	28	46
40	50	0	10	–	11	8	20	6	23	20	33	18	36	33	48	30	51
50	65	3	13	–	15	10	25	8	28	25	41	25	43	41	58	38	61
65	80	3	13	–	15	13	28	10	30	28	48	25	51	48	69	46	71
80	100	3	15	–	18	15	33	12	36	33	56	30	58	56	81	53	84
100	120	3	18	–	20	18	38	15	41	38	63	36	66	63	94	61	97
120	140	3	20	–	23	20	46	18	48	46	76	41	81	76	109	71	114
140	160	3	20	–	23	20	51	18	53	51	86	46	91	86	124	81	130
160	180	3	23	–	25	23	58	20	61	58	97	53	102	97	140	91	147
180	200	3	28	–	30	28	69	25	71	69	112	63	117	112	157	107	163
200	225	4	32	–	–	32	82	–	–	82	132	–	–	132	187	–	–
225	250	4	36	–	–	36	92	–	–	92	152	–	–	152	217	–	–

RADIAL INTERNAL CLEARANCES IN ANTIFRICTION BEARINGS (Contd.)

Table 131 Cylindrical roller bearings with cylindrical bore

Deviations in microns

Bearing bore diameter d mm		Group 1 (C1) Limits		Group 2 (C2) Inter-changeable		Group 2 (C2) Matched		Group 0 (C0) (Normal group) Inter-changeable		Group 0 (C0) (Normal group) Matched		Group 3 (C3) Inter-changeable		Group 3 (C3) Matched		Group 4 (C4) Inter-changeable		Group 4 (C4) Matched	
Over	Upto	Min.	Max.	Min.	Max.	Min.	Max.	Min.	Max.	Min.	Max.	Min.	Max.	Min.	Max.	Min.	Max.	Min.	Max.
10	18	—	—	0	30	10	20	10	40	20	30	25	55	35	45	35	65	45	55
18	24	5	15	0	30	10	20	10	40	20	30	25	55	35	45	35	65	45	55
24	30	5	15	0	30	10	25	10	45	25	35	30	65	40	50	40	70	50	60
30	40	5	15	5	35	12	25	15	50	25	40	35	70	45	55	45	80	55	70
40	50	5	18	5	40	15	30	20	55	30	45	40	75	50	65	55	90	65	80
50	65	5	20	5	45	15	35	20	65	35	50	45	90	55	75	65	105	75	90
65	80	5	25	10	55	20	40	25	75	40	60	55	105	70	90	75	125	90	110
80	100	10	30	10	60	25	45	30	80	45	70	65	115	80	105	90	140	105	125
100	120	10	30	10	65	25	50	35	90	50	80	80	135	95	120	105	160	120	145
120	140	10	35	15	75	30	60	40	105	60	90	90	155	105	135	115	180	135	160
140	160	10	35	20	80	35	65	50	115	65	100	100	165	115	150	130	195	150	180
160	180	10	40	25	85	35	75	60	125	75	110	110	175	125	165	150	215	165	200
180	200	15	45	30	95	40	80	65	135	80	120	125	195	140	180	165	235	180	220
200	225	15	50	40	105	45	90	75	150	90	135	140	215	155	200	180	255	200	240
225	250	15	50	50	115	50	100	90	165	100	150	155	230	170	215	205	280	215	265

RADIAL INTERNAL CLEARANCES IN ANTIFRICTION BEARINGS (Concld.)

Table 132 Cylindrical roller bearings with taper bore

Deviations in microns

Bearing bore diameter d mm		Group 1 (C1) Limits		Group 2 (C2)				Group 0 (C0) (Normal group)				Group 3 (C3)				Group 4 (C4)			
Over	Upto			Inter-changeable		Matched		Inter-changeable		Matched		Inter-changeable		Matched		Inter-changeable		Matched	
		Min.	Max.	Min.	Max.	Min.	Max.	Min.	Max.	Min.	Max.	Min.	Max.	Min.	Max.	Min.	Max.	Min.	Max.
—	18	—	—	10	40	20	30	25	55	35	45	35	65	45	55	45	75	55	65
18	24	10	20	10	40	20	30	25	55	35	45	35	65	45	55	45	75	55	65
24	30	15	25	10	45	25	35	30	65	40	50	40	70	50	60	50	85	60	70
30	40	15	25	15	50	25	40	35	70	45	55	45	80	55	70	60	95	70	80
40	50	17	30	20	55	30	45	40	75	50	65	45	90	65	80	70	105	80	95
50	65	20	30	20	65	35	50	45	90	55	75	55	105	75	90	80	125	90	110
65	80	25	40	25	75	40	60	55	105	70	90	65	125	90	110	95	145	110	130
80	100	35	55	30	80	45	70	65	115	80	105	75	140	105	125	110	160	125	150
100	120	40	60	35	90	50	80	80	135	95	120	90	160	120	145	130	185	145	170
120	140	45	70	40	105	60	90	90	155	105	135	105	180	135	160	145	210	160	190
140	160	50	75	50	115	65	100	100	165	115	150	115	195	150	180	165	230	180	215
160	180	55	85	60	125	75	110	110	175	125	165	130	215	165	200	190	255	200	240
180	200	60	90	65	135	80	120	125	195	140	180	150	235	180	220	205	275	220	260
200	225	60	95	75	150	90	135	140	215	155	200	165	255	200	240	225	300	240	285
225	250	65	100	90	165	100	150	155	230	170	215	180	280	215	265	255	330	265	315

TOLERANCES FOR ANTIFRICTION BEARINGS (Ball and roller bearings except taper roller bearings)

Table 133   Normal tolerance class (PO)

Deviations in microns

Inner ring

Nominal bore diameter d mm		Cylindrical bore diameter, deviations				Width B		Variation in individual ring	Radial runout
		d_m		d		Deviations			
Over	Upto	Max.	Min.	Max.	Min.	Max.	Min.	Max.	Max.
10	18	0	-8	3	-11	0	-120	20	10
18	30	0	-10	3	-13	0	-120	20	13
30	50	0	-12	3	-15	0	-120	20	15
50	80	0	-15	4	-19	0	-150	25	20
80	120	0	-20	5	-25	0	-200	25	25
120	180	0	-25	6	-31	0	-250	30	30
180	250	0	-30	8	-38	0	-300	30	40

Outer ring

Nominal outside diameter D mm		Outside diameter deviations				Width B deviations		Radial runout
		D_m		D				
Over	Upto	Max.	Min.	Max.	Min.	Max.	Min.	Max.
18	30	0	-9	2	-11	0	-120	15
30	50	0	-11	3	-14	0	-120	20
50	80	0	-13	4	-17	0	-120	25
80	120	0	-15	5	-20	0	-150	35
120	150	0	-18	6	-24	0	-200	40
150	180	0	-25	7	-32	0	-250	45
180	250	0	-30	8	-38	0	-300	50
250	315	0	-35	9	-44	0	-350	60
315	400	0	-40	10	-50	0	-400	70

TOLERANCES FOR ANTIFRICTION BEARINGS (Ball and roller bearings except taper roller bearings) (Contd.)

Table 134 Tolerance class 6 (*P*6)

Deviations in microns

Inner ring

Nominal bore diameter d mm		Cylindrical bore diameter deviations				width B			Radial runout
		d_m		d		Deviations		Variation in individual ring	
Over	Upto	Max.	Min.	Max.	Min.	Max.	Min.	Max.	Max.
10	18	0	-7	1	-8	0	-120	20	7
18	30	0	-8	1	-9	0	-120	20	8
30	50	0	-10	1	-11	0	-120	20	10
50	80	0	-12	2	-14	0	-150	25	10
80	120	0	-15	3	-18	0	-200	25	13
120	180	0	-18	3	-21	0	-250	30	18
180	250	0	-22	4	-26	0	-300	30	20

Outer ring

Nominal outside diameter D mm		Outside diameter deviations				Width B Deviations		Radial runout
		D_m		D				
Over	Upto	Max.	Min.	Max.	Min.	Max.	Min.	Max.
18	30	0	-8	1	-9	0	-120	9
30	50	0	-9	2	-11	0	-120.	10
50	80	0	-11	2	-13	0	-120	13
80	120	0	-13	2	-15	0	-150	18
120	150	0	-15	3	-18	0	-200	20
150	180	0	-18	3	-21	0	-250	23
180	250	0	-20	4	-24	0	-300	25
250	315	0	-25	4	-29	0	-350	30
315	400	0	-28	5	-33	0	-400	35
400	500	0	-33	5	-38	0	-450	40
500	630	0	-38	7	-45	0	-500	50

TOLERANCES FOR ANTIFRICTION BEARINGS (Ball and roller bearings except taper roller bearings)

Table 135 Tolerance class 5 (P5)　　　　Deviations in microns.

Inner ring

Nominal bore diameter d mm		Cylindrical bore diameter deviations				Width B Deviations		Variation in individual ring	Runout		
		d_m		d					Radial	Ref. side with bore	Groove with ref. side
Over	Upto	Max.	Min.	Max.	Min.	Max.	Min.	Max.	Max.	Max.	Max.
10	18	0	-5	0	-5	0	-80	5	3.5	7	7
18	30	0	-6	0	-6	0	-120	5	4	8	8
30	50	0	-8	0	-8	0	-120	5	5	8	8
50	80	0	-9	0	-9	0	-150	6	5	8	8
80	120	0	-10	0	-10	0	-200	7	6	9	9
120	180	0	-13	0	-13	0	-250	8	8	10	10
180	250	0	-15	0	-15	0	-300	10	10	11	13

Outer ring

Nominal outside diameter D mm		Outside diameter Deviations				Width B Deviations		Variation in individual ring	Runout		
		D_m		D					Radial	OD with ref. side	Groove with ref. side
Over	Upto	Max.	Min.	Max.	Min.	Max.	Min.	Max.	Max.	Max.	Max.
18	30	0	-6	0	-6	0	-80	5	6	8	8
30	50	0	-7	0	-7	0	-120	5	7	8	8
50	80	0	-9	0	-9	0	-120	6	8	8	10
80	120	0	-10	0	-10	0	-150	8	10	9	11
120	150	0	-11	0	-11	0	-200	8	11	10	13
150	180	0	-13	0	-13	0	-250	8	13	10	14
180	250	0	-15	0	-15	0	-300	10	15	11	15
250	315	0	-18	0	-18	0	-350	11	18	13	18
315	400	0	-20	0	-20	0	-400	13	20	13	20
400	500	0	-23	0	-23	0	-450	15	23	15	23
500	630	0	-28	0	-28	0	-500	18	25	18	25

TOLERANCES FOR ANTIFRICTION BEARINGS (Ball and roller bearings except taper roller bearings) (Contd.)

Table 136 Tolerance class 4 ($P4$)

Deviations in microns

Inner ring

Nominal bore diameter d mm		Cylindrical bore diameter deviations				Width B Deviations		Variation in individual ring	Runout		
		d_m		d					Radial	Ref. side with bore	Groove with ref. side
Over	Upto	Max.	Min.	Max.	Min.	Max.	Min.	Max.	Max.	Max.	Max.
10	18	0	- 4	0	- 4	0	- 80	2.5	2.5	3	3
18	30	0	- 5	0	- 5	0	-120	2.5	3	4	4
30	50	0	- 6	0	- 6	0	-120	3	4	4	4
50	80	0	- 7	0	- 7	0	-150	4	4	5	4
80	120	0	- 8	0	- 8	0	-200	4	5	5	5
120	180	0	-10	0	-10	0	-250	5	6	6	7
180	250	0	-12	0	-12	0	-300	6	8	7	8

Outer ring

Nominal outside diameter D mm		Outside diameter deviations				Width B Deviations		Variation in individual ring	Runout		
		D_m		D					Radial	OD with ref. side	Groove with ref. side
Over	Upto	Max.	Min.	Max.	Min.	Max.	Min.	Max.	Max.	Max.	Max.
18	30	0	- 5	0	- 5	0	- 80	2.5	4	4	5
30	50	0	- 6	0	- 6	0	-120	2.5	5	4	5
50	80	0	- 7	0	- 7	0	-120	3	5	4	5
80	120	0	- 8	0	- 8	0	-150	4	6	5	6
120	150	0	- 9	0	- 9	0	-200	5	7	5	7
150	180	0	-10	0	-10	0	-250	5	8	5	8
180	250	0	-11	0	-11	0	-300	7	10	7	10
250	315	0	-13	0	-13	0	-350	7	11	8	10
315	400	0	-15	0	-15	0	-400	8	13	10	13

TOLERANCES FOR ANTIFRICTION BEARINGS (Ball and roller bearings except taper roller bearings) (Contd.)

Table 137 Tolerance class special precision (SP)

Deviations in microns

Inner ring

Nominal bore diameter d mm		Cylindrical bore diameter			Width B			Runout	
Over	Upto	Deviations		Variations	Deviations		Variations	Radial	Reference side with bore
		Max.	Min.	Max.	Max.	Min.	Max.	Max.	Max.
18	30	0	-6	3	0	-100	5	3	8
30	50	0	-8	4	0	-120	5	4	8
50	80	0	-9	5	0	-150	6	4	8
80	120	0	-10	5	0	-200	7	5	9
120	180	0	-13	7	0	-250	8	6	10
180	250	0	-15	8	0	-300	10	8	11

Outer ring

Nominal outside diameter D mm		Outside diameter			Width B		Runout	
Over	Upto	Deviations		Variations	Deviations	Variations	Radial	OD with ref. side
		Max.	Min.	Max.			Max.	Max.
30	50	0	-7	4	Values are identical to those of inner ring of same bearing		5	8
50	80	0	-9	5			5	8
80	120	0	-10	5			6	9
120	150	0	-11	6			7	10
150	180	0	-13	7			8	10
180	250	0	-15	8			10	11
250	315	0	-18	9			11	13
315	400	0	-20	10			13	13
400	500	0	-23	12			15	15
500	630	0	-28	14			17	18

TOLERANCES FOR ANTIFRICTION BEARINGS (Ball and roller bearings except taper roller bearings) (Concld.)

Deviations in microns

Table 138. Tolerance class ultra precision (UP)

Inner ring

Nominal bore diameter d mm Over	Upto	Cylindrical bore diameter Deviations Max	Min	Variations Max	Width B Deviations Max	Min	Variation Radial Max	Runout Radial Max	Ref. side with bore Max	Groove with ref. side Max
10	18	0	-4	2	0	-25	1.5	1.5	2	2
18	30	0	-5	2.5	0	-25	1.5	1.5	3	3
30	50	0	-6	3	0	-30	2	2	3	3
50	80	0	-7	3.5	0	-40	3	2	4	3
80	120	0	-8	4	0	-50	3	3	4	3
120	180	0	-10	5	0	-60	4	3	5	6
180	250	0	-12	6	0	-75	5	4	6	7

Outer ring

Nominal outside diameter D mm Over	Upto	Outside diameter Deviations Max	Min	Variation Max	Width B (deviation / Variation / Max)	Runout Radial Max	OD with ref. side Max	Groove with ref. side Max
18	30	0	-4	2	Values are identical to those of inner ring of same bearing	2	2	4
30	50	0	-5	3		3	2	4
50	80	0	-6	3		3	2	4
80	120	0	-7	3.5		3	3	5
120	150	0	-8	4		4	3	6
150	180	0	-9	4.5		4	3	7
180	250	0	-10	5		5	4	9
250	315	0	-12	6		6	4	9

TOLERANCES FOR ANTIFRICTION BEARINGS (Taper roller bearings)

Table 139 Tolerance class normal (P0)

Deviations in microns

Inner ring and assembled bearing

Nominal bore diameter d mm		Inner ring							Assembled bearing width T	
		Cylindrical bore diameter				Width B		Radial runout		
		dm		d						
Over	Upto	Max.	Min.	Max.	Min.	Max.	Min.	Max.	Max.	Min.
10	18	0	-8	3	-11	0	-200	15	200	0
18	30	0	-10	3	-13	0	-200	18	200	0
30	50	0	-12	3	-15	0	-240	20	200	0
50	80	0	-15	4	-19	0	-300	25	200	0
80	120	0	-20	5	-25	0	-400	30	200	-200
120	180	0	-25	6	-31	0	-500	35	350	-250
180	250	0	-30	8	-38	0	-600		350	-250

Outer ring

Nominal outside diameter D mm		Outside diameter				Radial runout
		Dm		D		
Over	Upto	Max.	Min.	Max.	Min.	Max.
18	30	0	-9	2	-11	18
30	50	0	-11	3	-14	20
50	80	0	-13	4	-17	25
80	120	0	-15	5	-20	35
120	150	0	-18	6	-24	40
150	180	0	-25	7	-32	45
180	250	0	-30	8	-38	50
250	315	0	-35	9	-44	60
315	400	0	-40	10	-50	70
400	500	0	-45	12	-57	80
500	630	0	-50	14	-64	100

TOLERANCES FOR ANTIFRICTION BEARINGS (Taper roller bearings) (Contd.)

Table 140 Tolerance class 6 (P6)

Deviations in microns

Inner ring and assembled bearing

| Nominal bore diameter d mm | | Cylindrical bore diameter | | | | Width B | | Radial runout | Assembled bearing width T | |
| | | d_m | | d | | | | | | |
Over	Upto	Max.	Min.	Max.	Min.	Max.	Min.	Max.	Max.	Min.
10	18	0	- 7	1	- 8	0	-200	7	200	0
18	30	0	- 8	1	- 9	0	-200	8	200	0
30	50	0	-10	1	-11	0	-240	10	200	0
50	80	0	-12	2	-14	0	-300	10	200	0
80	120	0	-15	3	-18	0	-400	13	200	-200
120	180	0	-18	3	-21	0	-500	18	350	-250
180	250	0	-22	4	-26	0	-600	20	350	-250

Outer ring

| Nominal outside diameter D mm | | Outside diameter | | | | Radial runout | |
| | | D_m | | D | | | |
Over	Upto	Max.	Min.	Max.	Min.	Max.	Min.
18	30	0	- 8	1	- 9	9	- 9
30	50	0	- 9	2	-11	10	-11
50	80	0	-11	2	-13	13	-13
80	120	0	-13	2	-15	18	-15
120	150	0	-15	3	-18	20	-18
150	180	0	-18	3	-21	23	-21
180	250	0	-20	4	-24	25	-24
250	315	0	-25	4	-29	30	-29
315	400	0	-28	5	-33	35	-33

TOLERANCES FOR ANTIFRICTION BEARINGS (Taper roller bearings) (Concld.)

Table 141 Tolerance class 5 (P5)

Deviations in microns

Inner ring and assembled bearing

Nominal bore diameter d mm		Inner ring				Width B		Runout		Assembled bearing width T	
		Cylindrical bore diameter						Radial	Thrust Side with bore		
		d_m		d							
Over	Upto	Max.	Min.	Max.	Min.	Max.	Min.	Max.	Max.	Max.	Min.
10	18	0	-7	1	-8	0	-200	3.5	7	200	0
18	30	0	-8	1	-9	0	-200	4	8	200	0
30	50	0	-10	1	-11	0	-240	5	8	200	0
50	80	0	-12	2	-14	0	-300	5	8	200	0
80	120	0	-15	3	-18	0	-400	6	9	200	-200
120	180	0	-18	3	-21	0	-500	8	10	350	-250
180	250	0	-22	4	-26	0	-600	10	11	350	-250

Outer ring

Nominal outside diameter D mm		Outside diameter				Runout	
		D_m		D		Radial	Outside surface with thrust side
Over	Upto	Max.	Min.	Max.	Min.	Max.	Max.
18	30	0	- 8	1	- 9	6	8
30	50	0	- 9	2	-11	7	8
50	80	0	-11	2	-13	8	8
80	120	0	-13	2	-15	10	9
120	150	0	-15	3	-18	11	10
150	180	0	-18	3	-21	13	10
180	250	0	-20	4	-24	15	11
250	315	0	-25	4	-29	18	13
315	400	0	-28	5	-33	20	13

TOLERANCES FOR ANTIFRICTION BEARINGS (Thrust bearings)

Table 142 Tolerance class normal (*PO*) Deviations in microns

Nominal bore diameter *d* mm		Shaft washer Bore diameter Deviations		All washers Raceway runout with face of washer	Outside diameter *D* mm		Housing washer Outside diameter *D* deviations	
Over	Upto	Min.	Max.	Max.	Over	Upto	Max.	Min.
—	18	-8	6	10	—	18	0	-11
18	30	-10	8	10	18	30	0	-13
30	50	-12	10	10	30	50	0	-16
50	80	-15	13	10	50	80	0	-19
80	120	-20	16	15	80	120	0	-22
120	180	-25	18	15	120	180	0	-25
180	250	-30	22	20	180	250	0	-30
					250	315	0	-35
					315	400	0	-40
					400	500	0	-45

Table 143 Tolerance classes 6 & 5 (*P6* & *P5*) Deviation in microns

Nominal bore diameter *d* mm		Shaft washer Bore diameter *d* deviations		All washers Raceway runout with seat face of washer Max.		Outside diameter *D* mm		Housing washer Outside diameter *D* deviations	
Over	Upto	Max.	Min.	P6	P5	Over	Upto	Max.	Min.
—	18	0	- 8	5	3	—	18	—	-11
18	30	0	-10	5	3	18	30	0	-13
30	50	0	-12	6	3	30	50	0	-16
50	80	0	-15	7	4	50	80	0	-19
80	120	0	-20	8	4	80	120	0	-22
120	180	0	-25	9	5	120	180	0	-25
180	250	0	-30	10	5	180	250	0	-30
						250	315	0	-35
						315	400	0	-40
						400	500	0	-45

TOLERANCES FOR ANTIFRICTION BEARINGS (Thrust bearings) (Concld.)

Table 144 Tolerance class 4 (P4) — Deviations in microns

Nominal bore diameter d mm		Shaft washer Bore diameter d deviations		All washers Raceway runout with seat face of washer	Outside diameter D mm		Housing washer Outside diameter D deviations	
Over	Upto	Max.	Min.	Max.	Over	Upto	Max.	Min.
—	18	0	- 7	2	—	18	0	- 7
18	30	0	- 8	2	18	30	0	- 8
30	50	0	-10	2	30	50	0	- 9
50	80	0	-12	3	50	80	0	-11
80	120	0	-15	3	80	120	0	-13
120	180	0	-18	4	120	180	0	-15
180	250	0	-22	4	180	250	0	-20
					250	315	0	-25
					315	400	0	-28
					400	500	0	-33

Table 145 Tolerance class special precission (SP) — Deviations in microns

Nominal bore diameter d mm		Shaft washer Bore diameter deviations dm			d	Raceway runout with seat face	Height of bearing		Width of housing washer		Housing washer Outside diameter D		Outside diameter deviations	
Over	Upto	Max.	Min.	Max.	Min.	Max.	Max.	Min.	Max.	Min.	Over	Upto	Max.	Min.
10	18	0	- 7	1	- 8	3	0	- 40	0	-25	30	50	-20	-27
18	30	0	- 8	1	- 9	3	0	- 40	0	-25	50	80	-24	-33
30	50	0	-10	1	-11	3	0	- 50	0	-30	80	120	-28	-38
50	80	0	-12	2	-14	4	0	- 60	0	-30	120	150	-33	-44
80	120	0	-15	3	-18	4	0	- 70	0	-30	150	180	-33	-46
120	180	0	-18	3	-21	5	0	- 80	0	-30	180	250	-37	-52
180	250	0	-22	4	-26	5	0	-100	-0	-30	250	315	-41	-59

Table 146 Tolerance class ultra precision (UP) — Deviations in microns

Nominal bore diameter d mm		Shaft washer Bore diameter d deviations		Raceway runout with seat face	Height of bearings		Width of housing washer		Housing washer Outside diameter D		Outside diameter deviations	
Over	Upto	Max.	Min.	Max.	Max.	Min.	Max.	Min.	Over	Upto	Max.	Min.
10	18	0	- 5	1.5	0	- 40	0	-25	30	50	-20	-27
18	30	0	- 6	1.5	0	- 40	0	-25	50	80	-24	-33
30	50	0	- 8	1.5	0	- 50	0	-30	80	120	-28	-38
50	80	0	- 9	2	0	- 60	0	-30	120	150	-33	-44
80	120	0	-10	2	0	- 70	0	-30	150	180	-33	-46
120	180	0	-13	3	0	- 80	0	-30	180	250	-37	-52
180	250	0	-15	3	0	-100	0	-30	250	315	-41	-59

Table 147

TOLERANCES FOR MOUNTING OF MACHINE TOOL SPINDLE BEARINGS

	Bearing type	Diameter range mm	Shafts * P6	Shafts * P5, SP	Shafts * UP	Requirement	Housing P6	Housing P5, SP	Housing UP
Radial bearings	Radial ball bearing	≤ 18	h5	h4	h3	Free bearings, normal and light loads	J6	J$_s$5	J$_s$4
		(18) - 100	j5	j4	js3	Rotating outer ring load	M6	M5	M4
		100	k5	k4	-	Normal and light load	K6	K5	K4
	Cylindrical roller bearings **	≤ 40	j5	js4	-	Heavily loaded and rotating	M6	M5	M4
		(40) - 140	k5	k4	-	Outer ring — Adjustable	J6	J$_s$5	-
		(140) - 200	m5	m5	-	Located	K6	K5	-
	Taper roller bearings	≤ 40	j6	j5	-	Rotating outer ring load	M6	M5	-
		(40) - 140	k6	k6	-				
		(140) - 200	m6	m5	-				
Thrust bearings	Angular contact thrust ball bearings	All diameters	-	h5	h4	Bearings have common housing seating with cylindrical roller bearings *NN30K* or *NU49K*	K6	K5	K4
	Thrust ball bearings	All diameter	h6	h5	-		H8	H7	-

Only applicable to cylindrical bore bearings   ** Tapered bearing seating checked with ring gauge

j6 or j5 fit to be selected for adjustable lightly loaded bearings.

Tighter fit than that obtained with given tolerances should not be selected although this may be recommended for cylindrical roller bearings.

Table 148

TOLERANCE FOR MOUNTING OF NORMAL CLASS BEARINGS

Housing seatings for Radial Bearings

Conditions			Examples	Tolerance	Remarks
Split or solid housing	Stationary outer ring load	All loads	Bearing arrangements generally; to be used only for hand drive	H 7	Outer ring *easily* displaced
Solid housing		Very true running and great rigidity under variable load.	Roller bearings for machine tool main spindles D>125 mm	N6	Outer ring cannot be displaced
			D≤125 mm	M6	
	Bearing arrangement extra accurate.	Very true running under light loads of indeterminate direction	Ball bearings at the work end of grinding spindles; locating bearings in high-speed centrifugal compressors	K6	Outer ring cannot as a rule, be displaced
		Very true running; axial mobility of outer ring desirable	Ball bearings at drive end of grinding spindles; axially free bearings in high-speed centrifugal compressors	J 6	Outer ring can be displaced

Note:

(1) This table applies to cast-iron or steel housing. For light alloy housing, limits are generally used that give a somewhat tighter fit than is obtained with the limits specified above. The tolerance given in this table does not apply to seatings for taper roller bearings with inch dimensions.

(2) The information about axial mobility of the outer ring indicates whether the fit is suitable for non-separable bearings required to have axial freedom.

Table 148

TOLERANCES FOR MOUNTING OF NORMAL CLASS BEARINGS (Contd.)

(contd.)

		Shaft seatings for radial bearings				
		Shaft dia. in milimeters				
		Bearing with Cylindrical bore				
Conditions	Examples	Ball bearing	Cylindrical & taper roller bearings	Spherical roller bearings	Tolerance	Ramarks
Stationary Inner ring load						
Easy axial displacement desirable	Shafts	All diameters			g6	By light load is understood such loads as are, as a rule, not greater than 6-7% of the basic dynamic capacity C. For very accurate bearing arrangements $j5$, $k5$, & $m5$ are used instead of $j6$, $k6$ & $m6$.
Easy axial displacement unnecessary	Tension pulleys, rope sheaves	All diameters			n6	
Rotating inner ring load or direction of loading indeterminate						
Light and variable loads.	Electrical apparatus, machine tools, Pumps,	≤18	—	—	h5	
		(18)-100	≤40	≤40	j6	
		(100)-200	(40)-140	(40)-100	k6	
		—	(140)-200	(100)-200	m6	
Normal and heavy loads	Bearing arrangements generally; electric motors pumps; gearing, wood working machine	≤18	—	—	j5	For taper roller bearings $k6$ or $m6$ can be used as a rule instead of $k5$ or $m5$, for in the fitting of bearings of this type there is no need to take the decrease in bearing slackness into account.
		(18)-100	≤40	≤40	k5	
		(100)-140	(40)-100	(40)-65	m5	
		(140)-200	(100)-140	(65)-100	m6	
		(200)-280	(140)-200	(100)-140	n6	
Purely axial load	Bearing arrangements of all kinds	All diameters			j6	

TOLERANCES FOR MOUNTING OF NORMAL CLASS BEARINGS (Concld.)

Table 148 (Concld.)

Seatings for thrust bearings			
Conditions		**Tolerance**	**Remarks**
Purely axial load	Thrust ball bearings	H8	Where great accuracy is unnecessary, the housing washer or seating ring is fitted so as to have radial play
	Spherical roller thrust bearings where another bearing takes care of the radial location	—	Housing wahser fitted so as to have radial play
Combined (radial and axial) load on the spherical roller thrust bearings	Stationary load on housing washer or direction of loading indeterminate	J7	
	Rotating load on housing washer	K7	General
		M7	Comparatively heavy radial load

Shaft seatings for thrust bearings			
Conditions			**Tolerance**
Purely axial load	Thrust ball bearings	All diameters	j6
	Spherical roller thrust bearings	All diameters	j6
Combined (radial and axial) load on spherical roller thrust bearings	Stationary load on shaft washers.	All diameters	j6
	Rotating load on shaft washer, or direction of loading indeterminate.	$d \leqslant 200\ mm$ $d = (200)\text{-}400\ mm$ $d > 400\ mm$	k6

Table 149

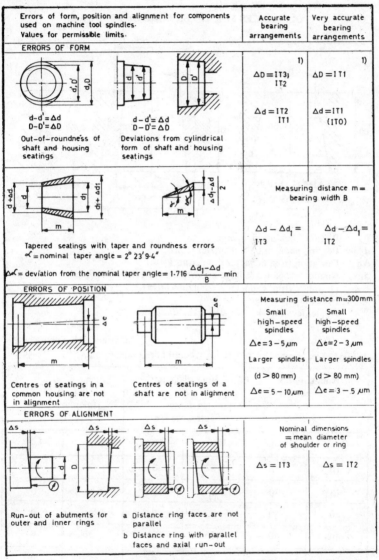

Errors of form, position and alignment for components used on machine tool spindles. Values for permissible limits.	Accurate bearing arrangements	Very accurate bearing arrangements
ERRORS OF FORM		
d − d′ = Δd D − D′ = ΔD Out−of−roundness of shaft and housing seatings	d − d′ = Δd D − D′ = ΔD Deviations from cylindrical form of shaft and housing seatings	1) $\Delta D = IT3;$ IT2 $\Delta d = IT2$ IT1	1) $\Delta D = IT1$ $\Delta d = IT1$ (IT0)

Tapered seatings with taper and roundness errors
α = nominal taper angle = 2° 23′ 9·4″

$\Delta\alpha$ = deviation from the nominal taper angle = $1.716 \dfrac{\Delta d_1 - \Delta d}{B}$ min

Measuring distance m = bearing width B

| $\Delta d - \Delta d_1 =$ IT3 | $\Delta d - \Delta d_1 =$ IT2 |

ERRORS OF POSITION

Centres of seatings in a common housing are not in alignment

Centres of seatings of a shaft are not in alignment

Measuring distance m = 300 mm

| Small high−speed spindles $\Delta e = 3 - 5\,\mu m$ Larger spindles (d > 80 mm) $\Delta e = 5 - 10\,\mu m$ | Small high−speed spindles $\Delta e = 2 - 3\,\mu m$ Larger spindles (d > 80 mm) $\Delta e = 3 - 5\,\mu m$ |

ERRORS OF ALIGNMENT

Run−out of abutments for outer and inner rings

a Distance ring faces are not parallel
b Distance ring with parallel faces and axial run−out

Nominal dimensions = mean diameter of shoulder or ring

| $\Delta s = IT3$ | $\Delta s = IT2$ |

1) IT3 is preferred when shaft or housing tolerence is of the 6th grade
IT2 is preferred when shaft or housing tolerence is of the 5th grade
IT1 is preferred when shaft or housing tolerence is of the 4th grade
IT0 is preferred when shaft or housing tolerence is of the 3rd grade

TOLERANCES FOR TAPER BORE OF BEARINGS

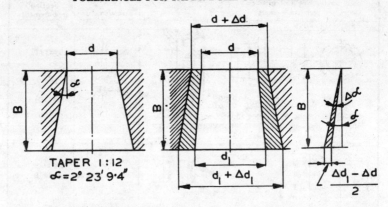

TAPER 1:12
$\alpha = 2° 23' 9.4''$

B	Width of bearing ring, mm
d	Bearing nominal bore diameter, mm
Δd	Deviation from diameter d, microns (μm)
d_1	Nominal diameter at large end of bore $= d + 0.083333B$, mm
Δd_1	Deviation from diameter d_1, microns (μm)
α	Nominal taper angle $2° 23' 9.4''$
$\Delta \alpha$	Angle deviation $= 1.716(\Delta d_1 - \Delta d)/B$ *minutes*

Deviations in microns (μm)

Table 150   Normal tolerances for tapered bore, Taper 1:12

Nominal bore diameter d mm		Bore diameter d deviations (Tolerance $H8$)		$(\Delta d_1 - \Delta d)$ (tolerance grade $IT7$)	
Over	Upto	High	Low	High	Low
10	18	27	0	18	0
18	30	33	0	21	0
30	50	39	0	25	0
50	80	46	0	30	0
80	120	54	0	35	0
120	180	63	0	40	0
180	250	72	0	46	0

BEARING MOUNTING ARRANGEMENTS

The following basic principles should be satisfied for satisfactory performance of bearing mounting arrangements:

1. The shaft or other rotating component must be axially located in both directions.
2. The bearing arrangement must be able to accommodate relative axial expansion between the shaft and the housing.
3. Proper fits on shaft and housing demanded by the mounting requirements, conditions of load and rotation must be provided.
4. Proper lubrication systems to provide lubricant circulation or retention should be used.
5. Actual load conditions and load carrying capacity of bearings should be compatible.

Typical bearing mounting arrangements and a brief description of these arrangements are given in Table 151.

Pre-loading of bearings

Preloading of bearings is a method which is adopted to reduce bearing deflection under the working loads. It makes use of the fact that the rate of deflection in a bearing reduces with increasing load. The normal procedure, in practice, is to adjust two bearings or two rows of rolling elements against one another by applying a permanent axial load. Preloading can be applied to radial, angular contact and thrust ball bearings as well as tapered roller bearings. Sometimes two rows of rollers in a spherical roller bearing are also preloaded against one another by making the outer ring in two halves, circumferentially split, and adjusting the two halves against one another.

Preloading can also be applied to single bearings by radial expansion of inner ring or contraction of outer ring during the fitting process. A typical example of this is double row cylindrical roller bearing (NNK type) used on machine tool spindles. These bearings have a tapered bore (taper 1:12) provided at the inner ring and the ring is forced up the corresponding tapered seating on the spindle. Great care must be taken with this method of preloading as it is easy to set up excessive preload and cause overload and over heating in the bearing.

Angular contact and some times deep groove ball bearings are supplied by bearing manufactures with their rings specially face ground such that when mounted in pairs they result in a

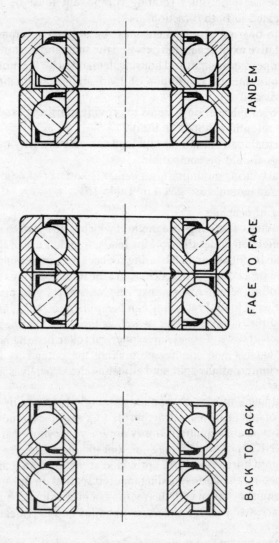

Fig. 99   Paired bearing arrangements

TANDEM

FACE TO FACE

BACK TO BACK

predetermined preload or end movement. They can be arranged in face to face, back to back or in tandem combinations (see Fig. 99).

Many times single bearings are used and are paired by using two rings one between the outer rings and second between the inner rings, the widths of these two rings are controlled to give desired preload and kind of mounting.

Angular contact ball bearings and tapered roller bearings are also widely used in pairs, with the individual bearings spaced from one another. In these cases the necessary adjustment of preload or end play is obtained by the use of shims, screwed end covers, shaft nuts, etc. The normal procedure is to first take up the free end movement in the assembly and then either to measure the shim thickness required or calculate the required final adjustment of the screw member, to give the desired preload or end movement. Where a screw thread adjustment is adopted it is also necessary to incorporate an efficient locking arrangement.

Taper roller bearings have a small deflection rate and final adjustment of preloading, if required, has to be done very carefully. It is therefore, common practice in case of taper roller bearings to assess the amount of preload by measuring the friction torque in the assembly when it is rotated.

Excessive friction torque can result in overloading and overheating of the bearings resulting in premature failure of bearings. The desired amount of preload, required to ensure sufficient rigidity without overheating can only be determined by actual running tests. Previous experience with established applications of similar designs provide useful guidance.

Spindle bearing arrangements

The quality of performance of machine tool is to a large extent influenced by the relationship maintained between the tool and the work piece. The spindle provides the drive to, either the work piece or tool depending upon the kind of machine. The performance of bearing arrangement supporting the spindle is one of the principal contributors in maintaining the relationship between the tool and the workpiece. The conditions required for a good spindle bearing arrangement can be summerised as follows:

1. Maximum rigidity, that is, minimum deflection under load. This is obtained by using adequately rigid spindle and use of bearings permitting minimum deflections (refer *Spindles* chapter).

2. Greatest running accuracy under all working conditions.

This calls for selection of bearings of proper precision class. Normal accuracy bearings are generally suitable for machine tools meant for rough turning or rough grinding operations, where accuracy is not of prime importance. Bearings of precision class P5, P4, P2, SP (special precision) or UP (ultra precision) should be selected for machine tool spindles used for finishing depending upon the requirements.

3. Minimum bearing clearance, which can be adjusted within certain limits or completely eliminated.

Bearing clearance must be adopted to the working conditions. It is desirable to adjust the bearing internal clearance to very close limits. For good results it may be necessary to preload the bearings. Different amount of adjustment may be needed for otherwise identical machines.

4. Low working temperature, constant over the complete speed range

To maintain a minimum temperature increase during operation at high speeds, the bearing friction must be a minimum. Thermal expansion caused by temperature rise may affect machining accuracy and must be taken into account when the accuracy of machine tools is being considered.

5. Low sectional height bearings permitting the use of robust spindles where space limitations exist.

6. Proper and adequate lubrication.

7. Proper sealing methods, to ensure that the lubricant is retained and to prevent dust or external matter from entering into areas which are liable to be affected from these.

8. Long bearing service life.

To meet these stringent demands on bearings required for machine tool spindles, bearings made to different tolerance classes, different clearances are available. Also certain manufacturers have made available some special bearings particularly suited for machine tool spindle applications. A proper and judicious selection of the type of bearing dependent on factors discussed above must be made.

Typical mounting arrangements of spindle bearings are shown in Figs. 100 to 111.

Table 151

Application	Figure	Loading conditions	Drive conditions	Axial location	Relative axial expansion	Seating fits	Remarks
Two single row radial ball bearings		Radial load at both bearings, (axial load in either direction on locating bearing)	Shaft rotating bearing housing stationary, direction of radial load constant on outer ring	In both directions by one bearing with both rings clamped	At the bearing at other end. Outer rings ring free axially	Inner rings interference. Outer rings sliding	Bearing with snap ring groove can be used alternatively for locating
Two single row radial ball bearings		Radial load at both bearings	Shaft rotating housing stationary. Direction of radial load constant on outer ring	By one bearing in each direction	Axial clearance a provided between outer ring and end cover abutment face	Inner ring interference outer sliding	Value of clearance $a=(12 \times 10^{-6}t +0.15)mm$ where t=maximum possible temperature rise in shaft in $^\circ C$.
Two single row radial ball bearings with rotating housing		Radial load at both the bearings axial load in either direction on locating bearing	Housing rotating, shaft stationary, direction of radial load constant on inner rings	In both the directions by one bearing with both rings clamped	At the bearing at opposite end, ring clamped free end ways	Inner ring sliding outer ring interference, inner fence	

Table 151 *(contd.)*

Application	Figure	Loading conditions	Drive conditions	Axial location	Relative axial expansion	Seating fits	Remarks
Matched pair (face to face) taper roller bearing and one cylindrical roller bearing		Radial load at both ends, axial load in both directions	Shaft rotating, housing stationary. Radial load direction constant on outer ring	In both directions by tapered roller bearings	Accommodated on inner ring of the roller bearing	Inner ring interference outer ring sliding	Used where close control of end movement is essential and for heavy axial loads
Matched pair (face to face) angular contact ball bearing preloaded by spring		Radial load at both ends, axial load in both directions	Shaft rotating, housing stationary. Radial load directions constant on outer ring	In both directions by paired angular contact ball bearing at left end	At outer ring of deep groove ball bearing at right end	Inner ring interference outer rings sliding	This provides very accurate and silent running bearing arrangement
Single row angular contact ball bearings (face to face mounting)		Radial load at both ends axial load in one direction at each end	Shaft rotating, housing stationary. Direction of radial load constant on outer ring	By one bearing in each direction	Allowance must be made during assembly for adjustment by shims between flange face and housing.	Inner rings interference as outer rings sliding	Rise in temperature has a tendency to reduce end play of the bearing system

Table 151

(Concld.)

Application	Figure	Loading conditions	Drive conditions	Axial location	Relative axial expansion	Seating fits	Remarks
Two single row angular contact ball bearings (back to back mounting) Matched pair		Radial load in both positions, axial load in one direction at each position	Housing rotating, shaft stationary, direction of radial load constant on inner ring	By one bearing in each direction	Adjusted by shaft nuts	Inner ring sliding outer rings interference	
Two tapered roller bearings (back to back)		Radial and axial load in one direction at each position	Shaft rotating, housing stationary, constant direction of radial load on outer rings	By one bearing in each direction	Adjusted by shaft nuts	Inner rings interference, outer rings sliding	
Two tapered roller bearings (face to face mounting)		Radial and axial load in one direction at each position	Shaft rotating, housing stationary, constant direction of radial load on outer ring	By one bearing in each direction	Adjusted by shims between flange and housing faces	Inner ring interference outer ring sliding	

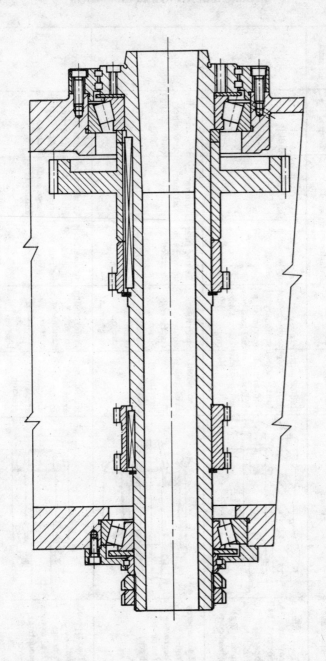

Fig. 100 Lathe spindle

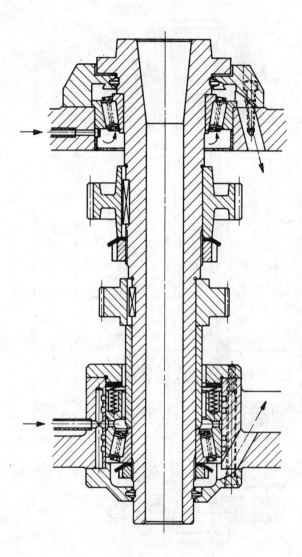

Fig. 101   Medium powered lathe spindle with Gamet P type bearing

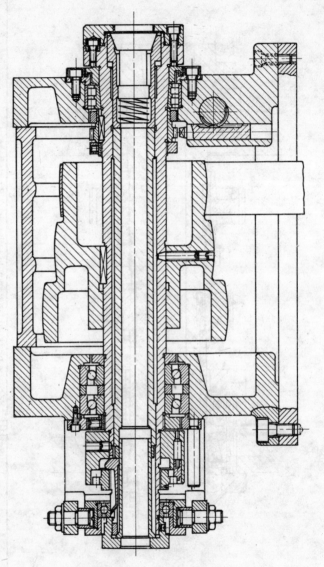

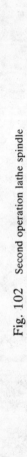

Fig. 102   Second operation lathe spindle

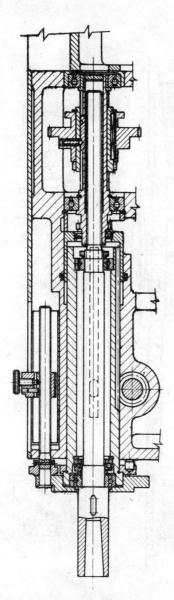

Fig. 103   Drilling machine spindle

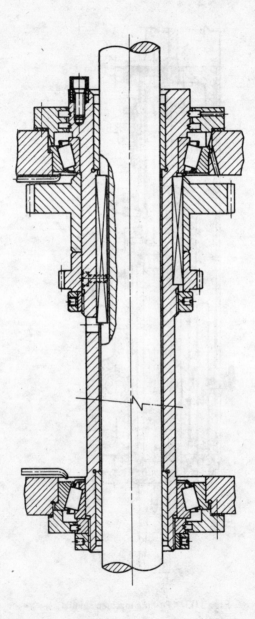

Fig. 104   Boring machine spindle

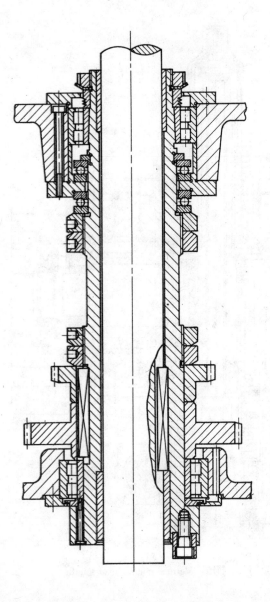

Fig. 105   Boring machine spindle

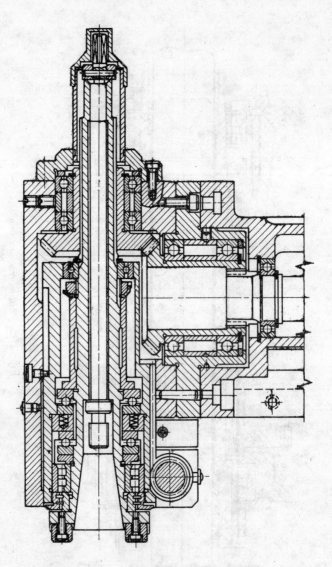

Fig. 106   Vertical milling head spindle

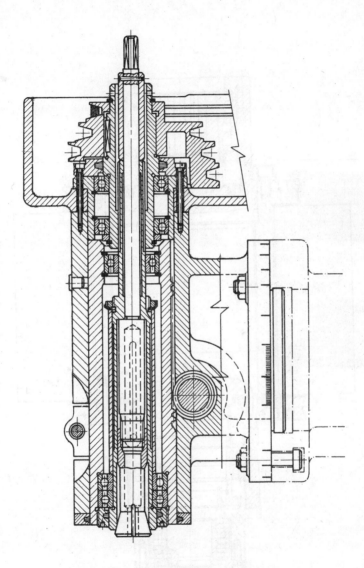

Fig. 107   High speed milling head spindle

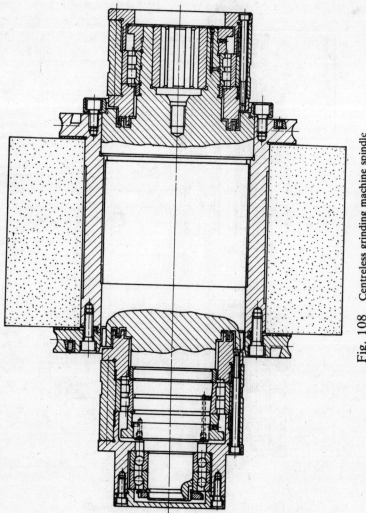

Fig. 108   Centreless grinding machine spindle

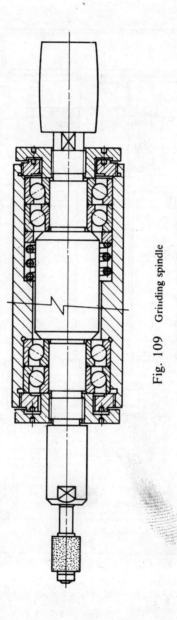

Fig. 109 Grinding spindle

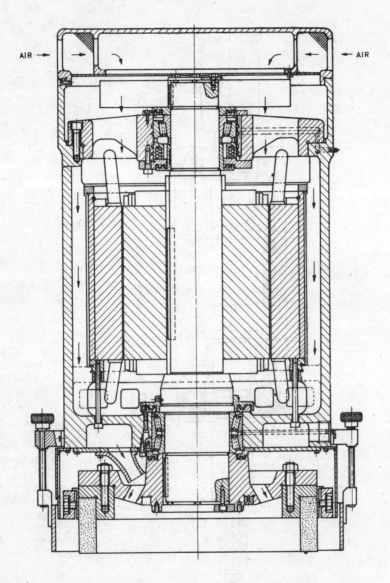

AIR →   ← AIR

Fig. 110   Vertical surface grinder spindle

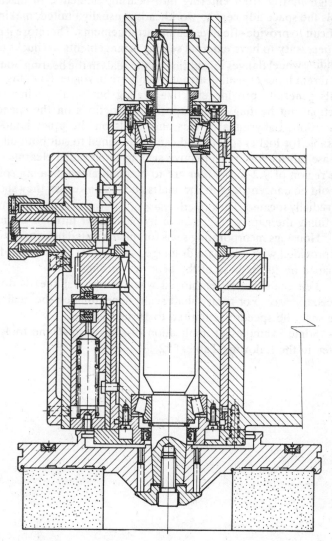

Fig. 111   Twist drill grinder spindle

SEALING OF ROLLING BEARINGS

Seals are used to retain lubricant and to prevent dirt and other foreign matter from entering into bearing elements. In machine tools the space adjacent to workheads is usually limited, making it difficult to provide effective sealing arrangements. Therefore it may be necessary to have external sealing arrangements in chucks or in grinding wheel flanges in addition to the seals in the bearing housing. Different types of sealing arrangements are in vogue. Rubbing type seals generally provide effective sealing but result in increased friction and heating, thus imposing restrictions on the speed of operation. Labyrinths or clearance seals on the other hand are suitable for higher speeds and can be designed to suit both oil and grease lubrication. For effective sealing small radial clearances in the region of 0.2 to 0.3 *mm* are to be maintained. Sealing collars should be concentric with the shafts and if screwed on they should be radially located. The effectiveness of labyrinths can be improved by filling the gaps with grease or by passing air through them.

Housings incorporating seals for oil lubricated bearings should be provided with adequate drainage passages for the return of the escaped oil to the base of the bearing housing.

Felt seals can be employed when peripheral speeds do not exceed 8*m/sec*. For rotary shaft seals, depending on the quality of the seal, the speeds are limited to 10 to 15*m/sec*.

Some examples of application of seals in machine tools are given in the following Table. 152

Table 152

Figure	Method of lubrication	Type of seal and application
	Grease or Oil mist	*Labyrinth seals:* This type of seal is frictionless and suitable for high speed spindles. The sealing collar should be located on the shaft and dynamically balanced.
	Oil, small quantities	*Labyrinth seals with oil drainage grooves:* This type of seal is suitable for most types of spindles. It can be reinforced with external flinger ring or collar if spindle is exposed to swarf or coolant.
	Grease	*Reinforced labyrinth seals:* Under difficult working conditions the labyrinth should be reinforced with a rubbing sealing collar of oil resistant material. The collar must only be in light contact with the spindle so that friction is small. This seal is suitable for slow and medium speed spindles where coolant or cutting fluid may spill over the housings.
	Grease or oil mist	*Felt seals:* Normally used where rubbing speed is less than 8 *m/sec.* suitable for self aligning bearings.
	Free oil Circulation	*Gap seals with oil grooves:* This type of seal is used when running conditions necessitate liberal oil circulation to cool the bearings. It has good sealing properties against ingress of foreign matter; a drainage groove returns escaped oil. If coolant is used during machining the seal should be supplemented with an external flinger or collar.

Table 152

(Concld.)

Figure	Method of lubrication	Type of seal and application
	Oil or Grease	*Rubbing seals:* This type is used where cutting fluid spilis over the spindle nose. The rubbing seal, which is shielded by a swarf guard, is mounted so that it gives maximum protection to the bearing. The spindle speed must not exceed the maximum permissable surface speed for this type of seal.
For Vertical spindle	Oil or Grease	*Labyrinth seals:* This type of seal is used where conditions are onerous and where it is essential to prevent ingress of foreign matter as with vertical spindles. The main task of the inner labyrinth of the seal is to retain grease in the bearing housing.
	Oil	*Labyrinths with oil flingers:* This type incorporates an oil flinger to throw off the oil which has passed through the bearing. The oil is collected in the housing and led away.
For Vertical spindle	Grease	*Gap seals:* This type of seal is generally used on smaller sizes of drilling spindle.

HYDRODYNAMIC BEARINGS

Nomenclature

A_o Auxiliary quantity used in equation 11, $mm^2 \cdot {}^{\circ}C/kgf. \ sec.$

B Auxiliary quantity used in equation 7, $m/sec. \ {}^{\circ}C$

c_h Specific heat of oil, $kcal/kgf \ {}^{\circ}C$

d Diameter of journal, mm

D Diameter of bearing, mm

E Modulus of elasticity, kgf/mm^2

f_b Allowable stress in bending for the material of the journal, kgf/mm^2

f_c Allowable contact stress for softer material 0.03 to 0.05 times $HB, kgf/mm^2$

h_o Minimum thickness of oil film, mm

H Heat generated, $kcal/sec.$

H_1 Quantity of heat dissipated by oil flowing out of bearing, $kcal/sec.$

H_2 Quantity of heat dissipated through the bearing metal into surroundings, $kcal/sec.$

l Length of the journal, mm

L Length between bearings, mm

n Speed, rpm

p Mean specific pressure, kgf/mm^2

P Radial load on the bearing, kgf

P_{all} Allowable load for the bearing, kgf

P_a Pressure distribution along the axis

P_r Pressure distribution around the periphery

Q Rate of flow of oil in the bearing, m^3/sec

R_1 Surface roughness value of journal (peak to valley), mm

R_2 Surface roughness value of bearing (peak to valley), mm

t_{in} Inlet temperature of oil, $^{\circ}C$

t_m Mean bearing temperature, $^{\circ}C$

v Velocity of journal, $m/sec.$

X Eccentricity ratio

y Deflection of journal due to bending, mm

α Coefficient of heat transfer, $kcal/m^2sec.{}^{\circ}C$

μ Coefficient of friction

η Dynamic viscosity of oil, $kgf.sec/mm^2$

η_t Dynamic viscosity of oil at $t_m, kgf \ sec / mm^2$

ϱ Specific weight of oil, kgf/m^3

ϕ Sommerfeld number
ψ Clearance ratio

For machine tools of normal accuracy rolling element bearings are most commonly used, because of their commercial availability, ease of assembly and maintenance. However for precision and high precision machine tools, and for high speed spindles fluid film bearings are widely used to ensure a high rotational accuracy of the spindle and maintain the precision for a long time. Fluid film bearings also possess good damping and heat dissipation properties. In full fluid film lubrication under normal running conditions there is no metal to metal contact. Such a bearing is said to be operating under hydrodynamic conditions. However a low velocity of the moving surface, an insufficient quantity of oil delivered to the bearing or an increased bearing load may result in a thin film of lubricant causing partial metal to metal contact during running. Such a lubrication is known as *Boundary lubrication*. Bearings operated under such conditions are known as plain sleeve bearings.

The plain full journal bearing (360º contact with journal) is most commonly used in industry. Compared to radial antifriction bearings, it has the following advantages:

1) requires less radial space,
2) silent in operation, good damping,.
3) less sensitivity to injury by foreign matter,
4) better suited to unanticipated overload and shock conditions and

5) generally lower initial cost.

Design of simple sleeve bearings

Simple sleeve bearings are used in machine tools only for low speed and lightly loaded applications such as hand drives, low speed oscillating drives, etc. They also find use in certain high load applications such as eccentric press. In the design of sleeve bearings the following factors are considered.

Table 153

Velocity	Peripheral velocity of journal not to exceed 0.1 m/sec.
Lubrication	At least intermittent lubrication either by oil or grease to be provided.
Contact stress	Contact stress should be within permissible values $\sqrt{\dfrac{\psi\, PE}{2.88 ld}} < f_c$

Table 153   (Concld.)

Strength	Diameter of journal depends upon the diameter of the spindle or shaft, of which the journal constitutes a part. The spindle diameter is calculated from the required strength or stiffness consideration. However, if a reduction in diameter is made at the journal, an additional check may be made for strength from the equation $$d = \sqrt[3]{\frac{5\,Pl}{f_b}}$$
Mean specific pressure	$p = \dfrac{P}{l.D} < \dfrac{1}{10}$ of the values given in Table 160
Ratio $\dfrac{l}{d}$	An increase of length l decreases the mean pressure and increases the sensitivity of entire bearing to shaft misalignment and temperature rise. This may rupture the oil film at the ends causing undue wear, seizure, etc. Recommended values of l/d for machine tool bearings vary from 0.8 to 1.2. However higher values upto 2 can be used to increase the load carrying capacity.

Design of hydrodynamic bearings

As the journal rotates in the bearing, it carries with it concentric layers of oil. The journal acts as a pump feeding oil into the clearance, which narrows gradually in the direction of rotation and causes the pressure to rise. A state of equilibrium sets in, when the hydrodynamic load developed balances the external load (see Fig. 112).

Load carrying capacity of journal bearings operating under hydro-dynamic conditions depends upon certain factors such as surface finish, wiping and l/d ratio. Requirement for these factors are given in Table 154.

Table 154

Surface finish	Magnitude of h_o–the minimum film thickness of oil–should be such that irregularities left after machining of the surfaces of journal and bearing should not rupture the oil film. Factor of safety = h_o/h critical = 1.2 to 1.5 where, h critical = $R_1 + R_2 + y$ R_1 and R_2 are obtained by multiplying the corresponding values of R_a by 4. Values of y are obtained from equation in Fig. 113.
Wiping	At high surface speeds and high loads, the bearing may become plastic and *wipe* under local high temperature conditions in the oil film–Wiping is important in I. C. engines, steam turbines, etc. However in machine tools, calculation for wiping is not necessary.
l/d ratio	To be preferably between 0.4 and 1.

Load carrying capacity

$$P = \frac{1000\,\eta\,vl}{\psi^2}\;\phi \quad \ldots(1)$$

Recommended values of ψ are given in Fig. 114. For a given size, the load capacity can be increased by increasing the viscosity of oil η or by decreasing the clearance ratio ψ. During running, the viscosity does not remain constant due to the heat generated. As the temperature increases the oil film thickness reduces with a consequent reduction in the load carrying capacity.

In equation (1) ϕ is a non-dimensional number known as the *Sommerfeld number*. It is very difficult to calculate ϕ in a general form. However by introducing an additional assumption that the hydrodynamic oil film is confined to an arc of $180°$ of the clearance between the journal and bearing, it is possible to get an equation of the form $\phi = f(x)$ for each value of l/d as:-

$$\phi = \frac{\psi^2 P}{1000\,\eta\,vl} = \frac{19.11 p \psi^2}{\eta n} = \frac{1.02}{h_o/d\psi\,[1 + 4.62\,(d/l)^2\,(0.026 + h_o/d\psi)]} \quad \ldots(2)$$

Values of ϕ for various ratios of l/d and $X = 1 - \dfrac{2\,h_o}{d\psi}$ are tabulated in Table 155.

Calculation for temperature rise

Heat generated in the bearing causes as increase in the oil temperature which in turn affects the oil film thickness. The heat generated should therefore be checked to be within the limits by properly selecting the oil and the dimensions of the bearing. The oil viscosity η should be taken for the mean temperature of oil in the bearing. t_m should not be more than $70°-80°C$ in machine tools. When the bearing operating conditions are stable, the quantity of heat H produced by friction of oil layers in the clearance is equal to the amount of heat withdrawn from the bearing.

$$H = H_1 + H_2 \quad \ldots(3)$$

$$H = \frac{Pv\mu}{427} \quad \ldots(4)$$

$$H_1 = c_h pQ\,(t_m - t_{in}) \quad \ldots(5)$$

$$H_2 = \frac{\alpha\,\pi dl\,(t_m - t_{in})}{10^6} \quad \ldots(6)$$

Table 155 Values of ϕ

l/d	0.33	0.4	0.5	0.6	0.65	0.7	0.75	0.8	0.83	0.9	0.925	0.95	0.975	0.99
							value of ϕ							
1.5	1.37	1.76	2.47	3.5	4.16	5.1	6.42	8.32	10.3	17.9	23.8	37.3	76.7	196.4
1.3	1.2	1.55	2.2	3.17	3.8	4.68	5.96	7.79	9.68	17.2	22.9	36.2	75.2	193.9
1.2	1.11	1.43	2.05	2.97	3.58	4.44	5.68	7.47	9.3	16.66	22.34	35.4	74.1	192
1.1	1	1.3	1.89	2.75	3.33	4.15	5.34	7.08	8.86	16	21.6	34.4	72.7	190
1	0.896	1.17	1.7	2.51	3.05	3.83	4.96	6.61	8.35	15.31	20.7	33.4	71.1	187
0.9	0.781	1.03	1.51	2.24	2.75	3.47	4.54	6.1	7.75	14.43	19.74	32	68.95	183.4
0.8	0.662	0.874	1.3	1.95	2.4	3.06	4.03	5.48	7.02	13.35	18.4	30.2	66.1	178.7
0.7	0.54	0.716	1.07	1.64	2.03	2.61	3.47	4.78	6.18	12	16.8	27.85	66.2	171.8
0.6	0.421	0.562	0.85	1.31	1.64	2.13	2.87	3.98	5.21	10.42	14.75	24.95	57.3	162.5
0.5	0.31	0.416	0.636	0.99	1.25	1.64	2.22	3.15	4.16	8.52	12.33	21.28	50.6	149
0.4	0.208	0.281	0.433	0.705	0.867	1.14	1.57	2.26	3.02	6.43	9.55	16.72	41.65	129.4

X

Noting that $c_h \rho \approx 405$ $kcal/m^3$ ^oC for all grades of oil

$$(t_m - t_{in}) = 5.78 \quad \frac{\mu p/\psi}{\dfrac{Q \times 10^6}{\psi v l d} + \dfrac{B}{\psi v}} \quad ......(7)$$

Where $B = \alpha \pi /405$ is the coefficient which depends on the type of bearing as shown in Table 156

Table 156   Values of α and B

Type of bearing	$\dfrac{kcal}{m^2 \ sec \ ^oC}$	$B = \alpha \pi /405$
Light bearings with difficult heat dissipation (eg., in case of high ambient temperature).	0.013	1×10^{-4}
Standard bearings operating under medium conditions.	0.018	1.4×10^{-4}
Heavy bearings with intensive heat dissipation (eg., cooled by water or air).	0.033	2.5×10^{-4}

The ratios μ/ψ and $Q.10^6/\psi v l d$ can be calculated from the formulae 8 and 9. Calculated values are tabulated in Table 157.

$$\mu/\psi = 0.150 + 1.92 \ (1.119 - X) \left[(1 + 2.31 \ (d/l)^2 - (1.052 + X)\right]..(8)$$

$$\frac{Q \times 10^6}{\psi v l d} \left[0.285 \ (0.2035 + X) \ \frac{0.072 \ (l/d)^2 - 1.05 + X}{0.433 \ (l/d)^2 + 1.05 - X}\right] ......(9)$$

If there is insufficient heat dissipation through the bearing due to a high ambient temperature or a low heat conductivity of the shell made of a nonmetallic material, then use

$$(t_m - t_{in}) = 5.78 \ \frac{p\mu/\psi}{Q 10^6/\psi v l d} \quad ....(10)$$

After obtaining $(t_m - t_{in})$ from either (7) or (10) and using the equation $t_m = t_{in} + A\eta - (11)$ (where A_o and t_{in} are constants) a straight line graph of t_m versus η is constructed as shown in Fig. 115.

The intersection point of the straight line with the characteristic curve of the oil under consideration will give both the oil mean temperature t_m and the corresponding viscosity η_t

Rate of circulation of oil through the bearing is calculated from Table 157 for known values of l, d, ψ, v and X.

Note: In the above analysis only single wedge bearings are considered, in which case the journal takes different positions at different speeds relative to bearing sleeve and the bearing is optimised only for one speed. If a stable position is necessary multiwedge type of the bearings are to be used. Table 160 indicates different materials used in the construction of hydrodynamic bearings.

Table 157 Values of μ/ψ and $(Q\,10^6/\psi\,vld)$

$\frac{l}{d}$	Ratio	X							
		0.33	0.5	0.7	0.8	0.9	0.95	0.975	0.99
1.5	μ/ψ	2.87	1.92	1.24	0.97	0.67	0.48	0.35	0.22
	$Q/\psi vld$	0.0697	0.0896	0.1038	0.1009	0.0888	0.077	0.067	0.0576
1.3	μ/ψ	3.26	2.13	1.33	1.02	0.7	0.49	0.35	0.22
	$Q/\psi vld$	0.0775	0.1006	0.1171	0.1134	0.0986	0.093	0.0718	0.0605
1	μ/ψ	4.34	2.71	1.58	1.16	0.76	0.52	0.36	0.23
	$Q/\psi vld$	0.0917	0.1215	0.144	0.1415	0.122	0.1017	0.0843	0.0686
0.9	μ/ψ	4.96	3.04	1.72	1.24	0.79	0.54	0.37	0.23
	$Q/\psi vld$	0.0968	0.1295	0.1563	0.1537	0.1334	0.1108	0.0909	0.0729
0.8	μ/ψ	4.83	3.5	1.92	1.35	0.84	0.56	0.38	0.24
	$Q/\psi vld$	0.1023	0.1382	0.1695	0.1685	0.147	0.122	0.0997	0.0784
0.7	μ/ψ	7.1	4.18	2.21	1.52	0.91	0.6	0.4	0.24
	$Q/\psi vld$	0.1078	0.1474	0.184	0.1852	0.164	0.1367	0.1115	0.0864
0.6	μ/ψ	9.09	5.22	2.66	1.78	1.02	0.65	0.42	0.26
	$Q/\psi vld$	0.1132	0.1566	0.1995	0.204	0.1842	0.1552	0.1267	0.0974
0.5	μ/ψ	12.3	6.94	3.39	2.2	1.2	0.73	0.46	0.27
	$Q/\psi vld$	0.1183	0.1656	0.2152	0.224	0.208	0.1785	0.1473	0.1133
0.4	μ/ψ	18.3	10.1	4.47	2.98	1.54	0.89	0.53	0.13
	$Q/\psi vld$	0.123	0.174	0.232	0.245	0.235	0.207	0.175	0.136

Table 158 Table for quick calculation of bearings

Parameters	Unit	Formulae	Remarks
d	mm		
n	rpm		given
P	kgf		
v	m/sec	$v = \dfrac{\pi dn}{60,000}$	calculate
l	mm		assume
p	kgf/mm^2	$p = \dfrac{P}{l.d}$ See Table 160 for allowable values	calculate
l/d			calculate
h_o	mm	Refer Table 154	
ψ	—	Refer Fig. 114	
η	$kgf \cdot sec/mm^2$		assume
X	—	$X = 1 - \dfrac{2\,h_o}{d\,\psi}$	calculate
$D\text{-}d$	mm	$D - d = \psi.d$	
ϕ	—	Refer Table 155 for known values of X and l/d	
P_{all}	kgf	$P_{all} = \dfrac{\phi \eta n \cdot l.d}{19.11 \psi^2}$; Margin $= \dfrac{P_{all}}{P} \geqslant 1 \cdot 25$	
Q	m^3/sec	Refer Table 157	calculate
$t_m - t_{in}$	^oC	Refer equation 7 or 10; Max. Temperature not to exceed 70-80°	

Table 159 Table for checking calculations of bearings

Parameter	Unit	Formulae		Remarks
d	mm	d_{max}		
		d_{min}		
D	mm	D_{max}		
		D_{min}		given
P	kgf			
l	mm			
n	rpm			
η	$kgf \cdot sec/mm^2$			
ψ_{max}	–	$\psi_{max} = \dfrac{D_{max} - d_{min}}{d}$		
ψ_{min}	–	$\psi_{min} = \dfrac{D_{min} - d_{max}}{d}$		
l/d	–			calculate
p	kgf/mm^2	$p = \dfrac{P}{l.d}$ See Table 160 for allowable values		
ϕ	–	$\phi = \dfrac{19.11\, p\psi^2}{n.\eta}$	$\phi_{max} =$	
			$\phi_{min} =$	
X	–	X should be within 0.7 to 0.96		
$t_m - t_{in}$	$°C$	Refer equation 7 or 10 Max. temperature should not exceed 70-80°C		
η_t	$kgf\ sec/mm^2$	Refer viscosity versus temperature characteristics for the oil under consideration		
P	kgf	$\psi_{min} =$ $P_{max} =$ $\psi_{max} =$ $P_{min} =$	using $P = \dfrac{\phi \eta_t\, n.i.d}{19\cdot11\psi^2}$ margin $= \dfrac{P_{min}}{P} =$	calculate
Q	m^3/sec	Refer Table 157		

Table 160 Materials for hydrodynamic bearings

Sl No.	Material	Applications	Max. Mean specific pressure kgf/mm^2	Modulus of elasticity E kgf/mm^2	Hardness HB
1	Aluminium bronze	Resistant to wear, used in heavy duty bearing requiring high strength and good impact resistance	4	11250	202
2	Phosphor bronze	Excellent corrosion resistance, with moderate strength. Very good bearing material when working under reasonable to good lubrication condition. Used for plain bearing bushes running against steel shafts where lubrication is adequate and shaft misalignment is minimum.	6.1	11200	60
3	Cast bronze	Very good running properties. Used for heavily loaded bearings.	8	12000	100-120
4	Tin bronze		10	10000	40-110
5	Leaded bronze	Less sensitive to impact loads, less wear.	7.5	8000	19-60
6	White metals	Good antiseizure properties. Sensitive to impact loads, used for overlays for lightly loaded bearings.	3	3100-6300	6-30
7	Low tin lead alloy	Better resistance to impact loads compared to white metals, higher wear resistance.	2	1600-3100	18-36
8	Aluminium alloys	Higher wear resistance. Sensitive to edge loading.	4	6900-7500	40-180

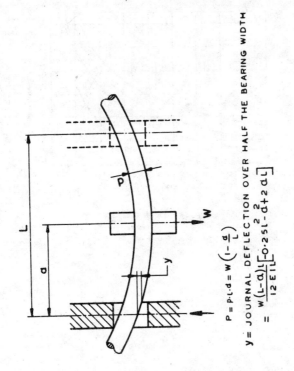

$$P = p \cdot L \cdot d = w \left(1 - \frac{d}{L}\right)$$

y = JOURNAL DEFLECTION OVER HALF THE BEARING WIDTH

$$= \frac{w(L-a)}{12 E I L}\left[0.25 L^2 - d^2 + 2 a L\right]$$

Fig. 113 (a) Deflection of journal due to bending

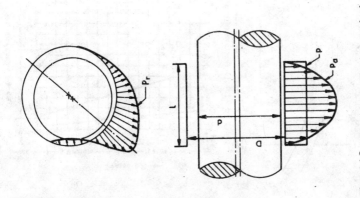

Fig. 112 Pressure distribution on the bearing

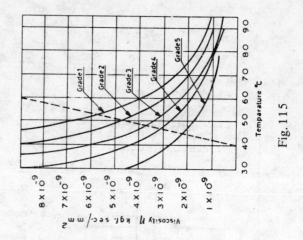

Fig. 115

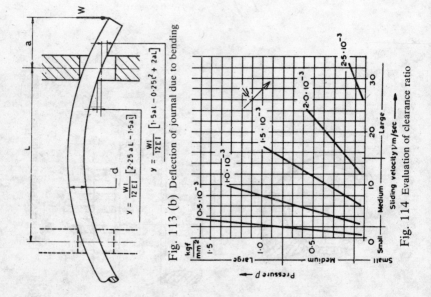

$$y = \frac{Wl}{12EI}\left[2 \cdot 25\,al - 1 \cdot 5al\right]$$

$$y = \frac{Wl}{12EI}\left[1 \cdot 5al - 0 \cdot 25l^2 + 2al\right]$$

Fig. 113 (b) Deflection of journal due to bending

Fig. 114 Evaluation of clearance ratio

HELICAL COMPRESSION SPRING

Nomenclature

A	Average work done, $kgf.mm$
C	Spring rate, kgf/mm
d	Wire diameter, mm
D_i	Inside diameter, mm
D_m	Mean diameter, mm
D_o	Outside diameter, mm
f	Deflection corresponding to load F, mm
$f_{1,2..n}$	Deflections corresponding to loads $F_{1,2..n}$ and lengths $L_{1,2..n}$, mm
f_s	Deflection corresponding to solid length, mm
f''	Deflection per coil, mm
F	Load on the spring corresponding to length L, kgf
$F_{1,2..n}$	Loads on the spring corresponding to lengths $L_{1,2..n}$, kgf
F_s	Load corresponding to solid length, kgf
g	Acceleration due to gravity, mm/sec^2
G	Modulus of rigidity, kgf/mm^2
h	Difference between two spring lengths, mm
i_f	Number of working coils
i_t	Total number of coils
k	Stress concentration factor
l	Developed length of wire, mm
L_o	Free length of spring, mm
$L_{1,2..n}$	Spring lengths corresponding to loads $F_{1,2..n}$, mm
L_s	Solid length of spring, mm
n_R	Natural frequency of spring guided at the ends, Hz
w	Spring index (coil ratio)
x	Clearance factor
y	Specific weight of spring material, kgf/mm^3
σ_u	Tensile strength, kgf/mm^2
Σa	Sum of minimum clearances between working coils at maximum permissible load, mm
τ	Shear stress at load F, kgf/mm^2
τ_a	Maximum allowable shear stress, kgf/mm^2
τ_k	Corrected shear stress at load F, kgf/mm^2
$\tau_{k1,k2..kn}$	Corrected shear stress at loads $F_{1,2..n}$, kgf/mm^2
τ_{kh}	Corrected shear stress amplitude in the case of springs subjected to variable loading, kgf/mm^2

τ_{kH}	Stress amplitude for fatigue limit, kgf/mm^2
τ_{kL}	Lower stress range for fatigue limit, kgf/mm^2
τ_{kU}	Upper stress range for fatigue limit, kgf/mm^2
τ_s	Actual shear stress at solid length, kgf/mm^2
τ_{sa}	Maximum allowable shear stress at solid length, kgf/mm^2

Helical compression or tension springs are made of bar stock or wire coiled into helical form, the load being applied along the helix axis.

Spring design

The design of helical compression spring is based on the maximum allowable shear stress τ_a Table 161 gives the τ_a values for cold coiled compression springs made from cold drawn, patented unalloyed spring steel wires of grade 1, 2 and 3 of IS 4454. Diameter range from 0.4 to 10mm of grade 2 is normally used for statically or infrequently loaded springs. The theoretical characteristic curve of a compression spring is as shown in Fig. 116

STRESS CONCENTRATION FACTOR k

Owing to the curvature, the actual shear stress on the inside of the coil varies for different coil diameters. To take this effect into consideration a stress concentration factor k is used. Values of k are given in Table 162 . This factor need not be considered for springs with static or infrequently varying loads.

FORMATION OF END COILS

Four different types of end formations are used (ref Fig. 117). To ensure uniform axial loading of working coils, closed and tapered, ground ends (also known as squared and ground ends) are generally used, for $d \geqslant 0.5\ mm$. The end coils are usually offset by 180^0 i.e., the working coils are always an odd multiple of ½.

MINIMUM AGGREGATE CLEARANCE Σa

This is the minimum sum of clearance spaces between the working coils, when the spring is subjected to the maximum permissible load F_n. The value of Σa is calculated with the help of Fig.118

RESISTANCE TO BUCKLING

Buckling will occur if the maximum relative deflection given by the graphs in Fig. 119 is exceeded at the relevant degree of slenderness. If the spring is vulnerable to buckling, it should be guided either in a bore or on a mandrel.

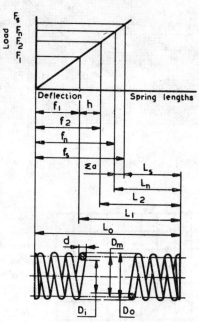

Fig. 116   Theoretical characteristic curve for compression springs

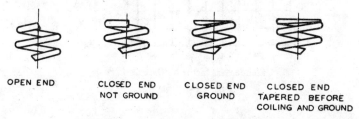

OPEN END   CLOSED END   CLOSED END   CLOSED END
                NOT GROUND   GROUND   TAPERED BEFORE
                                                 COILING AND GROUND

Fig. 117 Coil ends for compression springs

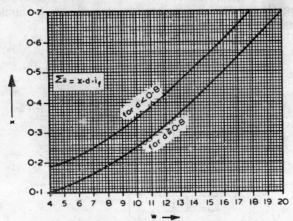

Fig. 118 Values of x against spring index w

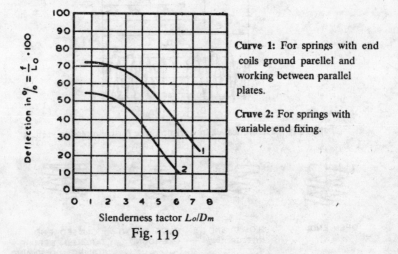

Slenderness factor L_0/D_m

Fig. 119

Curve 1: For springs with end coils ground parellel and working between parallel plates.

Cruve 2: For springs with variable end fixing.

Selection of the spring parameters

The requirements of the spring are analysed as follows:

a) Load–minimum and maximum loads, required stiffness, specific deflection, etc.

b) Space–minimum inside diameter, maximum outside diameter, spring lengths at different loads, stroke, etc.

c) Nature of load–whether static or infrequently varying ($<10^4$ load cycles during the life of equipment) or fatigue loading.

After ascertaining the above requirements a combination of d and D_m to satisfy the load and space availability is selected from Table 163 which indicates the maximum allowable load F_n and the deflection per coil f'' under this load. For static and infrequently varying loads, F_n and f'' can be multiplied by the stress concentration factor k.

Then the number of working coils, solid length, free length, etc., are calculated from the equations given in Table 164.

It is further checked that the theoretical stress τ_s at solid length, should not exceed max. permissible shear stress τ_a by more than 12%.

For springs subjected to fatigue loading the permissible shear stress is taken from the fatigue strength diagram (Fig. 120). The design of the spring is based upon the stress amplitude $\tau_{kh} = (\tau_{k2} - \tau_{k1})$ to lie within the stress amplitude τ_{kH} for the fatigue limit and $\tau_{k1} = \tau_{kL}$ and $\tau_{k2} \leqslant \tau_{kU}$ where τ_{kL} and τ_{kU} are lower and upper limits of the stress range for fatigue limit. When the springs are shot peened the stress range τ_{kH} increases as shown in Fig. 120

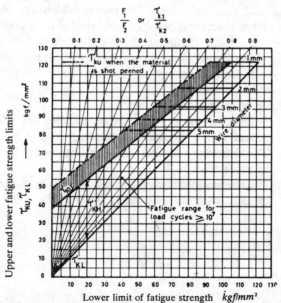

Fig. 120 Fatigue strength for cold coiled compression springs of cold drawn patented spring steel wire of grade 3 part I of IS 4454

The following points should be noted while a spring under fatigue loading is being designed.

(i) Ratio of F_1/F_2 should be high, i.e., the load range should be as low as possible.

(ii) Number of working coils and spring index should be as large as possible, i.e., the spring should be very soft.

(iii) Grade 3 wire material of IS 4454 should be used for springs subjected to severe dynamic or fatigue loading.

Materials and allowable stresses

High carbon, high strength steels are commonly used for helical compression springs. The physical properties of the wires, e.g., tensile strength, etc., are specified in IS 4454. The chemical composition for cold drawn patented steel wire is similar to that of high carbon steels, i.e. $C80$ and $C85$. Some deviation from the specified chemical composition is permissible provided the physical properties satisfy the specifications.

Table 165 lists the commonly used spring materials.

Tolerance and testing

Various tolerances can be specified on springs, but it is recommended that only those tolerances which are absolutely necessary from functional point of view should be specified.

Tolerance on (i) wire diameter, (ii) coil diameters, (iii) unloaded length or free length, (iv) axial loads, (v) spring rate, (vi) straightness & parrallelism and (vii) number of coils can be specified when functionally required.

Testing of springs consists of:

i) Static load test at maximum permissible load F_n

ii) Actual load to compress the spring to solid length
This should not exceed the theoritical load at solid length by more than 50%

iii) Theoretical characterstics or spring rate should be checked over the load range from $0.3\ F_n$ to $0.7\ F_n$.

Table 161 Maximum permissible values of Torsional shear stress τ_a kgf/mm^2

d mm	Grade 1	Grade 2	Grade 3
0.4	86.5	104	124
0.5	85	102.5	122
0.63	83.5	101	119.5
0.8	82	99.5	116.5
1	80	97	114
1.25	78	94.5	109
1.6	75	91	106
2	72.5	87.5	101.5
2.5	70	83.5	96.5
3.2	67	79	91.5
4	64	75.5	86.5
5	60.5	71	81.5
6.3	57	67	76.5
8	53.5	62	71.5
10	50	57.5	66

Table 162 Stress concentration factor k

D_m/d	3.2	4	5	6.3	8	10	12.5	16	20
k	1.5	1.38	1.29	1.22	1.17	1.13	1.11	1.08	1.06

Table 163   Selection chart for helical compression springs (material grade 2, IS 4454, Part 1)

D_m		d = 0.32	0.40	0.5	0.63	0.80	1.00	1.25	1.6	2.0	2.5	3.2	4.0	5.0	6.3	8
	τ_a	104.5	104	102.5	101	99.5	97	94.5	91	87.5	83.5	79	75.5	71	67	62
4	F_n	0.304	0.576	1.073	2.207	3.865	6.878	11.998	21.346							
	f^H	1.788	1.387	1.059	0.794	0.582	0.424	0.303	0.201							
5	F_n	0.248	0.472	0.886	1.690	3.260	5.887	10.469	19.180							
	f^H	2.851	2.224	1.709	1.292	0.959	0.709	0.517	0.353							
6.3	F_n	0.200	0.383	0.722	1.387	2.702	4.935	8.909	16.689	28.690						
	f^H	4.602	3.605	2.786	2.122	1.590	1.189	0.879	0.614	0.432						
8	F_n		0.307	0.581	1.122	2.203	4.061	7.419	14.138	24.816	42.407					
	f^H		5.914	4.590	3.516	2.654	2.004	1.500	1.065	0.765	0.536					
10	F_n			0.472	0.916	1.808	3.356	6.182	11.927	21.241	37.003	66.603				
	f^H			7.286	5.605	4.255	3.234	2.441	1.754	1.280	0.913	0.612				
12.5	F_n				0.745	1.476	2.754	5.108	9.949	17.919	31.673	58.254	99.456			
	f^H				8.899	6.786	5.185	3.939	2.858	2.108	1.526	1.046	0.731			
16	F_n					1.174	2.200	4.104	8.059	14.654	26.221	49.096	85.652	144.236		
	f^H					11.317	8.687	6.637	4.855	3.616	2.650	1.848	1.321	0.911		
20	F_n						1.788	3.349	6.615	12.108	21.850	41.416	73.313	125.855	217.048	
	f^H						13.791	10.579	7.783	5.835	4.313	3.046	2.208	1.553	1.062	
25	F_n							2.722	5.401	9.938	18.054	34.548	61.845	107.724	189.539	326.691
	f^H							16.794	12.412	9.355	6.961	4.962	3.638	2.596	1.812	1.201
32	F_n								4.295	7.940	14.506	27.986	50.577	89.182	159.531	281.347
	f^H								20.700	15.672	11.279	8.429	6.240	4.507	3.198	2.169
40	F_n									6.453	11.838	22.971	41.791	74.317	134.457	240.817
	f^H									24.880	18.695	13.513	10.070	7.335	5.265	3.627
50	F_n										9.622	18.755	34.301	61.406	112.083	203.147
	f^H										29.678	21.550	16.143	11.837	8.572	5.975
63	F_n											15.137	27.803	50.042	91.996	168.338
	f^H											34.791	26.174	19.297	14.075	9.905

Table 164 Design formulae for helical compression spring (Ref. Fig. 116)

Sl. No.	Nomenclature	Symbol	Unit	Equation
1	Wire diameter	d	mm	$\sqrt[3]{\dfrac{8F\,D_m\,k}{\pi\,\tau_k}}$
2	Mean diameter	D_m	mm	$\dfrac{\pi d^3 \tau_k}{8F\,k}$
3	Outside diameter	D_o	mm	$D_m + d$
4	Inside diameter	D_i	mm	$D_m - d$
5	Hole diamter	For guided springs	mm	$1{\cdot}05\,D_0$ (minimum)
6	Mandrel diamter		mm	$0{\cdot}95\,D_i$ (maximum)
7	Condition for guiding of spring	–	–	$\dfrac{L_o}{D_m} > 2{\cdot}6$
8	Spring Index or coil ratio	w	–	$\dfrac{D_m}{d}$ $(3 < w < 15)$
9	Stress concentration factor	k	–	$\dfrac{4w+2}{4w-3}$ (see Table 162)
10	Spring rate	C	kgf/mm	$\dfrac{G\,d^4}{8D_m^3 i_f}$
11	Specific deflection	–	mm/kgf	$\dfrac{1}{C}$
12	Number of working coils	i_f	–	$\dfrac{G d^4 f}{8 D_m^3 F}$
13	Load on the spring	F	kgf	$\dfrac{\pi d^3 \tau}{8 D_m}$
14	Deflection of the spring corresponding to load F	f	mm	$\dfrac{8 D_m^3 i_f F}{G d^4}$
15	Ultimate tensile strength of wire	σ_u	kgf/mm²	Selected from IS 4454

Table 164   Design formulae for helical compression spring (Contd.)

Sl. No.	Nomenclature	Symbol	Unit	Equation
16	Maximum allowable shear stress	τ_a	kgf/mm^2	$0{\cdot}5\sigma_u$ (from Table 161)
17	Modulus of rigidity of spring	G	kgf/mm^2	selected from Table 165
18	Uncorrected shear stress at load F	τ	kgf/mm^2	$\dfrac{8D_m F}{\pi d^3}$
19	Corrected shear stress at load F	τ_k	kgf/mm^2	$k \cdot \tau$
20	Average work done	A	$kgf\,mm$	$0{\cdot}5F.f$
21	Solid length of spring (all coils closed)	L_s	mm	$i_t{\cdot}d$
22	Total number of coils	i_t	—	$i_f +$ end coils
23	Sum of minimum clearance between working coils at max. premissible load	Σa	mm	$x.d.i_f$ (ref. Fig.118 for x)
24	Minimum permissible test length	L_n	mm	$L_s + \Sigma a$
25	Max. deflection corresponding to max. load F_n	f_n	mm	$\dfrac{F_n}{C}$
26	Free length of the spring	L_o	mm	$L_n + f_n$
27	Load on the spring at any length L	F	kgf	$(L_o - L)C$
28	Deflection/unit coil	f''	mm	f/i_f
29	Length of spring at any load F	L	mm	$L_o - \dfrac{F}{C}$
30	Developed length of wire	l	mm	$i_t\sqrt{\pi^2(D_m - d)^2 + (\Sigma a)^2}$
31	Natural frequency of spring guided at ends	n_R	Hz	$\dfrac{1}{2\pi} \cdot \dfrac{d}{i_f D_m^2}\left(\dfrac{Gg}{2y}\right)^{\frac{1}{2}}$
32	Acceleration due to gravity	g	mm/sec^2	9800
33	Specific weight of the material of spring	y	kgf/mm^3	7.85×10^{-6} for steel

Table 164 Design formulae for helical compression spring (Concld.)

Sl. No	Nomenclature	Symbol	Unit	Equations
34	Max. allowable shear stress at load corresponding to solid length	τ_{sa}	kgf/mm^2	$1\cdot12\tau_a$
35	Actual shear stress at solid length	τ_s	kgf/mm^2	$\tau_a\left(1+\dfrac{\Sigma a}{f_n}\right)$

Table 165 Commonly used spring materials and their modulus of rigidity

Sl. No.	Material		Applications	G kgf/mm^2
1	Cold drawn, patented, unalloyed spring steel wires of grade 1, 2, 3 & 4 of Part I IS 4454.		General engineering use for Helical springs	8300
2	Oil hardened and tempered unalloyed spring steel wire Part II IS 4454		Springs for Dynamic loading	8000
3	Oil hardened and tempered alloy spring steel wire Part II, IS 4454.		Springs working at moderately elevated temperature	8000
4	Stainless steel grade 18/8	Tempered	Corrosive atmospheres and moderately elevated temperature	7700
		untempered		7300
5	Stainless steel grade 17/7	Tempered		8100
		Untempered		7700
6	Hard drawn brass wire		Corrosive atmosphere	3500

HELICAL TENSION SPRINGS

Nomenclature

A	Work done by the spring, $kgf\,mm$
C	Spring rate, kgf/mm
d	Wire diameter, mm
D_i	Inside diameter of the spring, mm
D_m	Mean diameter of the spring, mm
D_o	Outside diameter of the spring, mm
$f_{1,2,.n}$	Deflections corresponding to total loads $F_{1,2,.n}$, mm
$F_{1,2,.n}$	Total loads on the spring (including initial tension) corresponding to deflections $f_{1,2,.n}$, kgf
F_o	Initial tension on the spring, kgf
G	Modulus of rigidity, kgf/mm^2
h	Difference between two spring lengths, mm
i_f	No. of working coils
i_t	Total no. of coils
k	Stress correction factor
$L_{1,2,.n}$	Lengths of the springs corresponding to deflections $f_{1,2,.n}$, mm
L_H	Distance of inside edge of hook from body of the spring, mm
L_K	Body length of the spring, mm
m	Hook opening, mm
w	Spring index (coil ratio)
σ_u	Tensile strength of wire material, kgf/mm^2
τ	Shear stress (uncorrected), kgf/mm^2
τ_a	Allowable shear stress, kgf/mm^2
τ_k	Corrected shear stress, kgf/mm^2
τ_{oa}	Allowable shear stress for initial tension, kgf/mm^2

The design and calculation of tension springs are similar to that of compression springs except that the maximum permissible shear stress is modified to a smaller value. Tension springs are usually wound with initial tension, i.e., coils pressed against each other in cold coiled condition. The initial tension obtainable in this way is governed primarily by the quality of the wire, the diameter of the wire, the spring index and the manufacturing method. A small amount of initial tension will always be present since it is very difficult to wind a spring without any initial tension. However hot coiled tension springs cannot be made with any initial tension, since the heat treatment applied causes air gaps to occur between coils which removes the initial tension.

Several end formations are used, the commonly used being machine formed half hooks or full loop as shown in Figs. 122 to 127.

Table 164 of compression springs can be used for tension spring design also, by reducing the values of F and f to 90% of the values given in the table. The general parameters and tension spring characteristic curve are given in Fig. 121

It is recommended to avoid tension springs subjected to fatigue loading as the failure usually occurs at the hooks owing to excessive stresses developed which are very difficult to evaluate. If it is not at all possible to avoid the use of tension springs subjected to fatigue loading, then the end supports have to be modified as shown in Figs. 128 and 129.

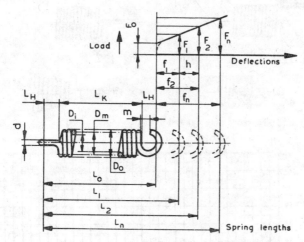

Fig. 121   Theoretical characteristic curve for tension springs

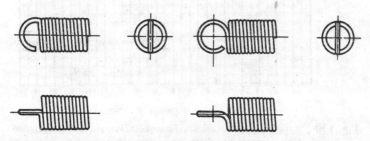

Fig. 122 $L_H = 0.55$ to $0.8\ D_i$ Fig. 123 $L_H = 0.8$ to $1.1\ D_i$

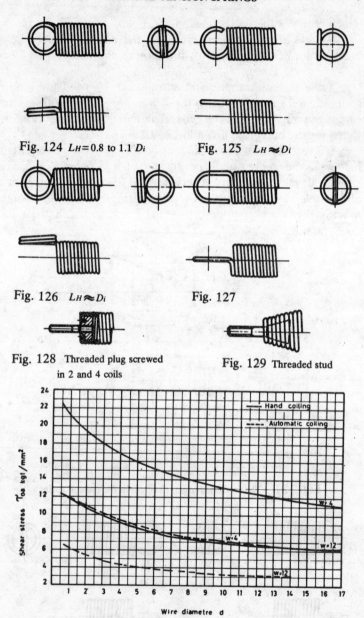

Fig. 124 $L_H = 0.8$ to 1.1 D_i Fig. 125 $L_H \approx D_i$

Fig. 126 $L_H \approx D_i$ Fig. 127

Fig. 128   Threaded plug screwed
in 2 and 4 coils

Fig. 129   Threaded stud

Fig. 130   Approximate values of allowable initial shear stress τ_{oa} for cold coiled
tension springs made from hand drawn patented spring steel wire of
grade 2 of part 1 of IS 4454

Table 166 Design formulae for Tension Springs

Sl. No.	Nomenclature	Symbol	Unit	Equation
1	Wire diamter	d	mm	$\sqrt[3]{\dfrac{8k(F-F_o)D_m}{\pi\,\tau_k}}$
2	Mean coil diamter	D_m	mm	$\dfrac{\pi d^3 \tau_k}{8(F-F_o)k}$
3	Outside diamter of spring	D_o	mm	$D_m + d$
4	Inside diamter of spring	D_i	mm	$D_m - d$
5	Spring Index or coil ratio	w	—	$\dfrac{D_m}{d}$
6	Stress correction factor	k	—	$\dfrac{4w+2}{4w-3}$
7	Spring rate	C	kgf/mm	$\dfrac{Gd^4}{8D_m^{\,3}i_f}$
8	Specific deflection	—	mm/kgf	$\dfrac{1}{C}$
9	Number of working coils	i_f	—	$\dfrac{Gd^4 f}{8D_m^{\,3}(F-F_o)}$
10	Load on the spring	$F-F_o$	kgf	$\dfrac{\pi d^3 \tau}{8D_m}$
11	Deflection of the spring	f	mm	$\dfrac{8D_m^{\,3}(F-F_o)i_f}{G\,d^4}$
12	Ultimate tensile strength	σ_u	kgf/mm²	Selected from IS 4454
13	Allowable shear stress	τ_a	kgf/mm²	90% of the corresponding values from Table 161
14	Uncorrected shear stress	τ	kgf/mm²	$\dfrac{8D_m(F-F_o)}{\pi\cdot d^3}$

Table 166 Design formulae for tension springs (contd.)

Sl. No.	Nomenclature	Symbol	Unit	Equation
15	Corrected shear stress	τ_k	kgf/mm^2	$k \cdot \tau \leqslant \tau_a$
16	Work done by spring	A	$kgf\,mm$	$0.5f(F + F_o)$
17	Distance of inside edge of hook from body of the spring	L_H	mm	Ref. Figs. 122 to 127 (lower values preferred)
18	Body length (all coils closed)	L_K	mm	$(i_f + 1)d$
19	Free length of spring	L_O	mm	$L_K + 2L_H$ (for equal loops)
20	Maximum permissible test length	L_n	mm	$L_O + \dfrac{(F_n - F_o)}{C}$
21	Maximum permissible test load	$(F_n - F_o)$	kgf	$\dfrac{\pi d^3 \tau_a}{8 D_m k}$
22	Allowable shear stress for initial tensioning of the spring (for cold coiled tension springs only)	τ_{oa}	kgf/mm^2	Selected from Fig. 130
23	Maximum permissible initial tension	F_o	kgf	$\dfrac{\pi d^3 \tau_{oa}}{8 D_m k}$
24	Total number of coils	i_t	—	i_f + no. of end coils
25	Modulus of rigidity	G	kgf/mm^2	Selected from Table 165
26	Hook opening	m	mm	If provided, $\not< 2d$

HELICAL TORSION SPRINGS

Nomenclature

a	Clearance between consecutive coils, mm
A	Work done on the spring, $kgf.mm$
C	Spring rate, $kgf.mm/deg.$
d	Wire diameter, mm
D_i	Inside diameter of the coil, mm
$D_{i\alpha}$	Inside diameter of the coil corresponding to a deflection α, mm
D_m	Mean diameter of the coil, mm
D_o	Outside diameter of the coil, mm
$D_{o\alpha}$	Outside diameter of the coil corresponding to a deflection α, mm
E	Modulus of elasticity, kgf/mm^2
$F_{1,2,.n}$	Force at a distance R from the coil centre corresponding to deflections $\alpha_{1.2..n}$, kgf
i_f	No. of active coils
k	Stress concentration factor
l	Developed length of spring (active coils only), mm
L_{KO}	Axial length of spring body under free condition, mm
$M_{1,2,.n}$	Torques on the spring at deflections $\alpha_{1,2,n}$, $kgf.mm$
r	Inside radius of the wire at the bend for springs with radial ends, mm
R	Distance of the point of application of force from coil centre, mm
w	Spring index
$\alpha_{1,2,.n}$	Angular deflection of the spring corresponding to moments $M_{1.2..n}$, $deg.$
α'	Total angular deflection of the spring, $deg.$
α_h	Stroke of the spring (for working torques), $deg.$
β	Additional angular deflection when the loading end is not fixed, $deg.$
σ_b	Bending stress induced (without considering stress concentration effect), kgf/mm^2
σ_{ba}	Maximum allowable bending stress, kgf/mm^2
σ_{bnk}	Actual stress under maximum torque M_n, kgf/mm^2
σ_u	Tensile strength of the wire material, kgf/mm^2

Helical torsion springs are similar in form to helical compression springs. These are made of bar stock or wire coiled into

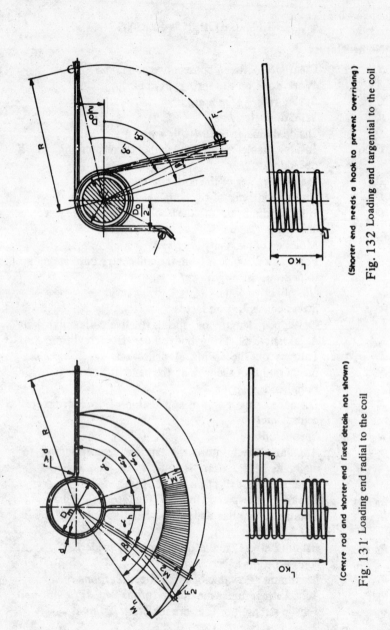

(Shorter end needs a hook to prevent overriding)

Fig. 132 Loading end targential to the coil

(Centre rod and shorter end fixed details not shown)

Fig. 131' Loading end radial to the coil

helical form. The ends are formed in such a way (Figs.131 & 132) that the spring may be subjected to a torque about the coil axis instead of an axial load. The mode of stressing of helical torsion spring is bending, in contrast to the helical compression spring where the stress is torsional shear. It is advisable to load the torsion spring.in such a way that the spring tends to wind up causing a decrease in the diameter. In the design of torsion spring, care must be taken to provide sufficient clearance between the arbor rod about which the spring is wound and the inner diameter of the spring. As a guide, arbor diameter $\approx$ (0.8 to 0.9) D_i and hole diameter $\approx$ (1.1 to 1.2) D_o. Reference should be made to equations 18 & 19 of Table 167 also.

Due to manufacturing reasons torsion springs are usually made of round wires only. However, when maximum energy storage is required in a given space the use of square or rectangular sections may be advisable.

Note: The maximum allowable bending stress σ_{ba} given in the table holds good for statically or infrequently loaded springs.

Table 167 Design formulae for helical torsion springs for round wire (cold coiled) Ref. Figs. 131 & 132

Sl. No.	Nomenclature	Symbol	Unit	Equation
1	Torque on the spring	M	$kgf\,mm$	$\dfrac{\pi d^3 \sigma_b}{32}$ $\quad FR \approx \dfrac{E d^4 \alpha}{1167 l}$ $E = 21 \times 10^3\ kgf/mm^2$ for hard drawn patented steel wire as per IS 4454
2	Wire diamter	d	mm	A standard size is selected
3	Max. allowable bending stress	σ_{ba}	kgf/mm^2	$0.7\,\sigma_u$ $\quad [= 1.4\,\tau_a$ τ_a from Table 161
4	Mean diamter of the coil	D_m	mm	Chosen to suit the requirements
5	Spring index	w	—	$\dfrac{D_m}{d}$ $\quad (3 < w < 15)$

Table 167 Design formulae for helical torsion springs of round wire   (contd.)

Sl. No.	Nomenclature	Symbol	Unit	Equation
6	Stress concentration factor	k	—	$\dfrac{4w^2 - w - 1}{4w^2 - 4w}$ if $w > \left(\dfrac{2r}{d}+1\right)$ then substitute w by $\dfrac{2r}{d}+1$ to get k for the case in Fig. 131
7	Angular deflection due to torque M	α	deg	$\approx \dfrac{1167M\,l}{E\,d^4}$
8	Force at the loading end at a radius R	F	kgf	$\dfrac{M}{R}$
9	Actual stress under maximum torque M_n	σ_{bnk}	kgf/mm^2	$\approx \dfrac{10 \cdot 19 M_n\,k}{d^3} \leqslant \sigma_{ba}$
10	No. of active coils	i_f	—	$\approx \dfrac{d^4.E.\alpha}{3667\,D_m M}$ if $(a+d) \leqslant \dfrac{D_m}{4}$ $\approx \dfrac{d^4.E.\alpha}{1167.M\sqrt{(\pi D_m)^2+(a+d)^2}}$ if $(a+d) > \dfrac{D_m}{4}$
11	Spring rate	C	$\dfrac{kgf\,mm}{deg}$	$\approx \dfrac{d^4.E}{1167.l}$
12	Clearance between consecutive coils	a	mm	$0 \cdot 1$ to $0 \cdot 5$
13	Developed length of spring (of active coils only)	l	mm	$\pi D_m\,i_f$ if $(a+d) \leqslant \dfrac{D_m}{4}$ $i_f\sqrt{(\pi D_m)^2+(a+d)^2}$ if $(a+d) > \dfrac{D_m}{4}$

Table 167 Design formulae for helical torsion springs of round wire (Concld.)

Sl. No.	Nomenclature	Symbol	Unit	Equation
14	Axial length of the spring body under free condition	L_{KO}	mm	$i_f(a+d)+d$
15	Additional angular deflection when the loading end is not fixed	β	deg	$\approx \dfrac{48.68\,F(2R-D_m)^3}{E.R.d^4}$ When the end is bent to be radial to the coil (Fig. 131 $r>d$) $\approx \dfrac{97.37\,F(4R^2-D_m^2)}{Ed^4}$ when the end is tangential to the coil (Fig. 132)
16	Total angular deflection of the loading end	α'	deg	$\alpha + \beta$
17	Work done on the spring	A	kgf mm	$\approx \dfrac{M\alpha}{114\cdot 6}$
18	Inside diamter after deflection	$D_{i\alpha}$	mm	$\dfrac{D_m\,i_f}{i_f+\dfrac{\alpha}{360}}-d$
19	Outside diamter after deflection	$D_{o\alpha}$	mm	$\dfrac{D_m\,i_f}{i_f+\dfrac{\alpha}{360}}+d$
20	Stroke of the spring for working torques M_1 to M_2	α_h	deg	$\alpha_1 - \alpha_2$

FLAT SPIRAL SPRINGS

Nomenclature

b	Width of the spring strip, mm
d	Diameter of the arbor, mm
D	Diameter (inner) of the casing, mm
E	Modulus of elasticity, kgf/mm^2
h	Thickness of the spring strip, mm
k	Stress concentration factor
l	Developed active length of the spring, mm
M_n	Max. bending moment corresponding to max. torque condition, $kgf.mm$
M_n'	Max. bending moment for finding the thickness of the spring strip, $kgf.mm$
N_n	Total no. of turns of the spring corresponding to fully wound condition
N_o	Absolute no. of turns of the unwound spring (free state)
$\triangle N$	No. of turns during working stroke
$\triangle N_1$	No. of turns from free state to initial working torque
$\triangle N_n$	No. of turns from free state to fully wound condition
$T_{1,2,\&n}$	Torques corresponding to initial and final working strokes and to fully wound condition, $kgf.mm$
α	Angle of rotation during working stroke, $deg.$
σ_{ba}	Maximum allowable bending stress, kgf/mm^2
σ_{bn}	Bending stress induced under fully wound condition, kgf/mm^2

Spiral springs consist essentially a flat strip wound to form a spiral shape (Fig. 133 or 134). Usually the inner end is clamped to the arbor while the outer end may be either pinned or clamped. Common applications of flat spiral springs are found in drilling quill movement control, counterbalancing, etc. The springs are generally subjected to bending stress. In practice, to avoid excessive stress, the arbor diameter is usually made about 15 to 25 times the spring strip thickness.

The formulae given in Table 168 are based upon the following assumptions.

1) The portion of the coil connected to the arbor and casing is not

included in the developed length.

2) The spring characteristics are linear.

3) In the free state the spring is in contact with the casing and the coils do not touch one another.

4) When the spring is wound completely the coils touch one another.

5) Total sectional area of the wound spring occupies approximately half the available space between arbor and casing.

Table 168 Design formulae for Flat Spiral Springs

Sl No	Nomenclature	Symbol	Unit	Inner End Clamped	
				Case 1: outer End Clamped	Case 2: Outer End Hinged
1	Arrangement of the spring and positions(x) of max. bending stresses.	—	—	Fig. 133	Fig. 134
2	Nature of bending moment	—	—	M constant	M variable
3	Characteristic of the spring	—	—	Fig. 135	
4	Diamter of the arbor	d	mm	Chosen to suit the requirements	
5	Initial working torque on the spring in the preloaded condition	T_1	$kgf\,mm$	Chosen to suit the requirements	
6	Final working torque on the spring	T_2	$kgf\,mm$	1.2 to $1.5T_1$	
7	Max. torque on the spring when it is fully wound	T_n	$kgf\,mm$	1.2 to 1.5 T_2	
8	Angle of rotation during working stroke	α	deg	Chosen to suit the requirements	
9	No. of turns during working stroke	ΔN	—	$\alpha/360$	

Table 168 Design formulae for Flat Spiral Springs (Contd.)

Sl No.	Nomenclature	Symbol	Unit	Inner End Clamped	
				Case 1: Outer End Clamped	Case 2: Outer End Hinged
10	No. of turns from free state of max. torque condition	ΔN_n	—	$\left(\dfrac{T_n}{T_2-T_1}\right)\cdot\Delta N$	
11	No. of turns during preloaded condition	ΔN_1	—	$\dfrac{T_1}{T_n}\cdot\Delta N_n$	
12	Max. bending moment on the spring corresponding to max. torque condition	M_n	kgf mm	T_n	$2T_n$
13	Max. bending moment for finding the thickness of spring strip.	M_n'	kgf mm	$k.M_n$ from Table 169	M_n
14	Max. bending stress induced	σ_{bn}	kgf/mm²	$\dfrac{6M_n'}{bh^2}$	
15	Max. allowable bending stress	σ_{ba}	kgf/mm²	$160.h^{-0.272}$ for srping steel	
16	Thickness of spring strip	h	mm	$\left[\dfrac{6M_n'}{160\dfrac{b}{h}}\right]^{\frac{1}{2.728}}$ Ref. Fig. 136	
17	Width of the spring strip	b	mm	$\dfrac{6M_n'}{160h^{1.728}}$	
18	Developed length of the active spring	l	mm	$\dfrac{\Delta N_n \pi E\, bh^3}{6T_n}$	$\dfrac{\Delta N_n \pi E\, bh^3}{7.5\,T_n}$
19	Total No. of turns of the wound spring at T_n	N_n	—	$\dfrac{\sqrt{\dfrac{4}{\pi}\cdot lh+d^2}-d}{2h}$	
20	Absolute no. of turns for the unwound spring	N_o	—	$N_n-\Delta N_n$	
21	Inner diamter of the casing	D	mm	$\dfrac{l}{\pi N_o}+N_o h$	

Table 169 Stress concentration factor k

r/h	1	1.5	2	2.5	3	3.5	4	4.5	5
k	1.53	1.3	1.23	1.18	1.13	1.1	1.09	1.08	1.075

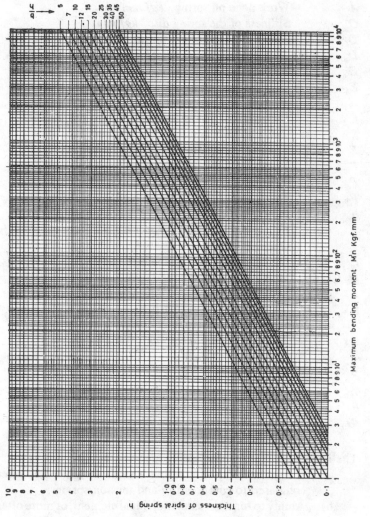

Fig. 136 (For spring steel only)

DISC SPRINGS

Nomenclature

A	Work done on spring, $kgf.mm$
C	Stiffness of the spring disc, kgf/mm
D_i	Inside diameter of the spring disc, mm
D_o	Outside diameter of the spring disc, mm
E	Modulus of elasticity, kgf/mm^2
E_n	Energy stored in a pack when compressed to flat position, $kgf.mm$
f	Deflection of spring disc under a load F, mm
f_t	Deflection of compound spring, mm
F	Load on spring disc (for a deflection f), kgf
F_R	Frictional force of a compound spring, kgf
F_S	Load required to flatten a spring disc, kgf
F_T	Total load applied on a compound spring, kgf
h	Free height of cone frustum of unloaded disc, mm
i	No. of discs or packs facing alternate directions in the compound spring
l_o	Overall height of a spring disc, mm
L_o	Free length of compound spring, mm
L_s	Length of compound spring when fully flattened, mm
n	No. of discs facing in the same direction in the pack
r	Ratio of outer to inner diameters of the spring disc,
$s(s')$	Thickness of the disc, mm
α	Coefficient for load
β	Coefficient for stress
γ	Coefficient for stress
μ	Poisson's ratio
σ_I	Stress induced at the inside upper edge of the disc, kgf/mm^2

The main features of a disc spring are:

a) A high load capacity within a small design space, when compared to compression and tension springs,

b) Ability to function as an individual element, or more often in groups of different combinations creating disc spring stacks operating over a central guide rod,

c) Ability to provide varied spring rates for a design requirement simply by changing the stacking configuration (Figs. 140, 141 and 142)

d) Greater security of operation since the failure of one disc element within a stack does not totally lead to failure of the entire assembly,

e) The stack can be selected to operate within known stress limits, thus giving predictable performance.

However disc springs should always have a minimum preload value if subjected to fatigue loading. Recommended value corresponds to 10% to 15% of the total available deflection. Otherwise cracking appears at the top inner edge of the spring causing premature failure.

The spring characteristics of various disc springs are as shown in Fig. 139. It can be seen that

i) for $h/s \leqslant 0.75$, approximate linear characteristic is obtained,

ii) for $0.75 \leqslant h/s \leqslant 1.4$, spring stiffness decreases,

iii) for $h/s > 1.4$, the force is maximum between $f/h = 0$ to $f/h = 1$. Then stiffness changes from positive to negative.

Discs can be combined in series or in parallel or suitably compounded depending upon the requirement as shown in Figs. 140, 141 and 142. Friction force F_R to be added to F taking into account the sign.

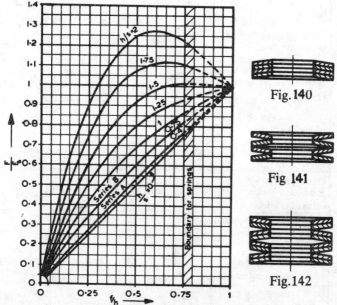

Fig. 140

Fig 141

Fig. 142

Fig. 139 Characteristic curves for disc springs

Table 170 Design formulae for Disc Springs or Belleville Washers

Sl No.	Nomenclature	Symbol	Unit	Equation
1	Outside diamter of spring	D_o	mm	
2	Inside diamter of spring	D_i	mm	
3	Thickness of the disc.	s	mm	Refer to Figs.137 and 138
4	Free height of cone frustum of unloaded disc.	h	mm	
5	Overall height	l_o	mm	$h+s$ (Refer to Figs.137 and 138)
6	Max. permissible deflection of individual disc.	f	mm	$0.75h$
7	Deflection of compound spring.	f_t	mm	$i.f$
8	Load on each disc (to produce deflection f)	F	kgf	$\dfrac{4E}{(1-\mu^2)} \cdot \dfrac{s^4}{\alpha D_o^2}\dfrac{f}{s} \times \left[\left(\dfrac{h}{s}-\dfrac{f}{s}\right)\left(\dfrac{h}{s}-\dfrac{f}{2s}\right)+1\right]$
9	Load required to flatten the disc	F_s	kgf	$\dfrac{4E}{(1-\mu^2)}\dfrac{s^3.h}{\alpha D_o^2}$
10	No. of discs facing in the same direction in the pack.	n	—	$\dfrac{F_t}{F}$
11	No. of discs or packs facing alternate directions in the compound spring.	i	—	$\dfrac{f_t}{f}$
12	Free length of the compound spring without any load	L_o	mm	$i\ [l_o+(n-1)s]$
13	Length of the compound spring or pack when fully flattened.	L_s	kgf	$n.i.s$
14	Frictional force of a compound spring or pack	F_R	kgf	(0.02 to 0.03) nF-Loading cycle (−0.02 to −0.03) nF-unloading cycle
15	Total load applied on compound spring or pack	F_T	kgf	$n\ F$

Table 170 Design formulae for Disc Springs or Belleville Washers (contd.)

Sl No.	Nomenclature	Symbol	Unit	Equation
16	Energy stored in a pack when compressed to flat position	E_n	kgf mm	$\dfrac{4E}{(1-\mu^2)}\ \dfrac{n}{\alpha D_o^2}\left[\dfrac{sh^4}{8}+\dfrac{s^3h^2}{2}\right]$
17	Stiffness of the spring disc	C	kgf/mm	$\dfrac{4E}{(1-\mu^2)}\ \dfrac{s^3}{\alpha D_o^2}\times$ $\left[\left(\dfrac{h}{s}\right)^2-3\dfrac{h}{s}\cdot\dfrac{f}{s}+1\cdot5\left(\dfrac{f}{s}\right)^2+1\right]$
18	Compressive stress induced on the inside upper edge of disc Ref. Figs. 137, 138	σ_I	kgf/mm²	$\dfrac{4E}{(1-\mu^2)}\ \dfrac{s^2}{\alpha D_o^2\cdot s}\ \dfrac{f}{}\times$ $\left[\beta\left(\dfrac{h}{s}-\dfrac{f}{2s}\right)+\gamma\right]<\sigma_I\ \text{max.}$ Table 171
19	Modulus of elasticity of material of spring	E	kgf/mm²	21000 (for spring steels)
20	Poisson's ratio	μ	—	for spring steel $\mu=0.3$
21	Ratio of outer to inner diamters	r	—	D_o/D_i (Recommended value from 1.5 to 3.5
22	Coefficient for load	α	—	
23	Coefficient for stress	β	—	Refer to Table 172
24	Coefficient for stress	γ	—	
25	Work done on spring for producing a deflection f	A	kgf mm	$\dfrac{2E}{(1-\mu^2)}\ \dfrac{s^5}{\alpha D_o^2}\left(\dfrac{f}{s}\right)^2$ $\left[\left(\dfrac{h}{s}-\dfrac{f}{2s}\right)^2+1\right]$

Table 171 Permissible values of stress at the inside upper edge of the disc for static loading

For producing a deflection $f\approx0.75\,h$	$\sigma_I\ \text{max}=200$ to $240\,kgf/mm^2$
For flattening the spring $f=h$	$\sigma_I\ \text{max}=260$ to $300\,kgf/mm^2$

Table 172 Coefficients $\alpha\ \beta$ & γ

r	α	β	γ	r	α	β	γ	r	α	β	γ
1.2	0.29	1.02	1.05	2.2	0.73	1.26	1.45	3.2	0.79	1.46	1.81
1.3	0.39	1.04	1.09	2.3	0.74	1.29	1.49	3.4	0 80	1.50	1.87
1.4	0.46	1.07	1.14	2.4	0.75	1.31	1.53	3.6	0.80	1.54	1.94
1.5	0.53	1.10	1.18	2.5	0.76	1.33	1.56	3.8	0.80	1.57	2.00
1.6	0.57	1.12	1.22	2.6	0.77	1.35	1.60	4.0	0.80	1.60	2.07
1.7	0.61	1.15	1.26	2.7	0.78	1.37	1.63	4.2	0.80	1.64	2.13
1.8	0.65	1.17	1.30	2.8	0.78	1.39	1.67	4.4	0.80	1.67	2.19
1.9	0.67	1.20	1.34	2.9	0.78	1.41	1.70	4.6	0.80	1.70	2.25
2.0	0.69	1.22	1.38	3.0	0.78	1.43	1.74	4.8	0.79	1.73	2.31
2.1	0.71	1.24	1.42	3.1	0.79	1.45	1.77	5.0	0 79	1.76	2.37

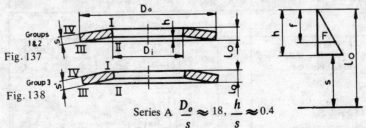

Fig. 137

Groups 1&2

Fig. 138

Group 3

Series A $\dfrac{D_o}{s} \approx 18, \ \dfrac{h}{s} \approx 0.4$

Table 173   All dimensions in *mm*

Group	D_o h_{12}	D_i H_{12}	s	h	l_o	at $f \approx 0.75h$			
						F kgf	f	$l_o - f$	σ 1) kgf/mm²
1	8	4.2	0.4	0.2	0.6	21	0.15	0.45	125*
	10	5.2	0.5	0.25	0.75	34	0.19	0.56	125*
	12.5	6.2	0.7	0.3	1	67	0.22	0.78	142*
	14	7.2	0.8	0.3	1.1	81	0.22	0.88	134*
	16	8.2	0.9	0.35	1.25	103·	0.26	0.99	
2	18	9.2	1	0.4	1.4	128	0.3	1.1	133*
	20	10.2	1.1	0.45	1.55	155	0.34	1.21	132*
	22.5	11.2	1.25	50.5	1.75	195	0.37	1.38	133*
	25	12.2	1.5	0.55	2.05	298	0.41	1.64	146*
	28	14.2		0.65	2.15	290	0.49	1.66	131*
	31.5	16.3	1.75	0.7	2.45	398	0.52	1.93	133*
	35.5	18.3	2	0.8	2.8	528	0.6	2.2	137*
	40	20.4	2.25	0.9	3.15	660	0.67	2.48	136*
	45	22.4	2.5	1	3.5	790	0.75	2.75	133*
	50	25.4	3	1.1	4.1	1220	0.82	3.28	146*
	56	28.5	3	1.3	4.3	1150	0.97	3.33	131*
	63	31	3.5	1.4	4.9	1530	1.05	3.85	133*
3	71	36	4	1.6	5.6	2100	1.2	4.4	124
	80	41	5	1.7	6.7	3500	1.28	5.42	127
	90	46		2	7	3200	1.5	5.5	122
	100	51	6	2.2	8.2	4900	1.65	6.55	127
	112	57		2.5	8.5	4500	1.88	6.62	117
	125	64	8	2.6	10.6	8800	1.95	8.65	126
	140	72		3.2	11.2	8700	2.4	8.8	130
	160	82	10	3.5	13.5	14000	2.6	10.9	132
	180	92		4	14	12800	3	11	122
	200	102	12	4.2	16.2	18700	3.15	13.05	120
	225	112		5	17	17500	3.75	13.25	115
	250	127	14	5.6	19.6	25000	4.2	15.4	123

Note: 1)   The figures listed are in each case the maximum theoretical tensile stresses at the underside of the individual disc. For the values denoted by * this maximum tensile stress is evaluated for position II whilst for others; it applies to position III

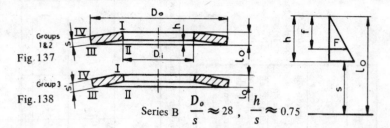

Fig. 137 Groups 1 & 2

Fig. 138 Group 3

Series B $\dfrac{D_o}{s} \approx 28$, $\dfrac{h}{s} \approx 0.75$

Table 174

All dimensions in *mm*

Group	D_o h_{12}	D_i H_{12}	s	h	l_0	at $f \approx 0.75h$			
						F kgf	f	$l_0 - f$	σ 1) kgf/mm²
	8	4.2	0.3	0.25	0.55	12	0.19	0.36	135
	10	5.2	0.4	0.3	0.7	21	0.22	0.48	131
1	12.5	6.2		0.35	0.85	30	0.26	0.59	114
	14	7.2	0.5	0.4	0.9	28	0.3	0.6	113
	16	8.2	0.6	0.45	1.05	42	0.34	0.71	114
	18	9.2	0.7	0.5	1.2	58	0.37	0.83	
	20	10.2	0.8	0.55	1.35	76	0.41	0.94	115
	22.5	11.2		0.65	1.45	72	0.49	0.96	111
	25	12.2	0.9	0.7	1.6	88	0.52	1.08	105
	28	14.2	1	0.8	1.8	113	0.6	1.2	112
	31.5	16.3	1.25	0.9	2.15	194	0.67	1.48	122
2	35.5	18.3		1	2.25	173	0.75	1.5	110
	40	20.4	1.5	1.15	2.65	267	0.86	1.79	117
	45	22.4	1.75	1.3	3.05	372	0.97	2.08	118
	50	25.4	2	1.4	3.4	485	1.05	2.35	117
	56	28.5		1.6	3.6	452	1.2	2.4	112
	63	31	2.5	1.75	4.25	730	1.31	2.94	
	71	36		2	4.5	690	1.5	3	108
	80	41	3	2.3	5.3	1070	1.72	3.58	117
	90	46	3.5	2.5	6	1450	1.88	4.12	114
	100	51		2.8	6.3	1330	2.1	4.2	108
	112	57	4	3.2	7.2	1830	2.4	4.8	199
3	125	64	5	3.5	8.5	3100	2.65	5.85	127
	140	72		4	9	2850	3	6	122
	160	82	6	4.5	10.5	4200	3.4	7.1	
	180	92		5.1	11.1	3800	3.8	7.3	115
	200	102	8	5.6	13.6	7800	4.2	9.4	125
	225	112		6.5	14.5	7200	4.85	9.65	119
	250	127	10	7	17	12200	5.25	11.75	126

Note: 1)   Maximum theoretical tensile stress evaluated for position III

Table 175 Recommended materials and method of manufacture for disc springs

IS*	DIN	Group No.	Thickness of Disc	Condition of material	Method of manufacture	Remarks
2507-1965 C75, C70	Ck67 as per DIN 17222	1	s<1	Cold rolled strip steel according to DIN 1544, close Gr. tol. (F)	Cold formed	—
50Cr1V23 as per IS 2507-1965	67SiCr5 or 50CrV4 as per DIN 17221 and 17222	2	4>s⩾1	-do-	Cold formed	Inside and outside dia machined. Edges radiused at inside dia.
		3	14⩾s⩾4	Hot rolled strip steel according to DIN 1016, rolled wide flat steel according to DIN 59200 drop forged blanks	Hot formed	Spring machined all over. Edges radiused at inside and outside dia.., provided with seating faces and reduced disc thickness

*Nearest equivalent to DIN

Table 176 Reduced disc thickness for disc springs for group 3 ($s' \approx 0.94\,s$)

s mm	4	5	6	8	10	12	14
s' mm	3 75	4.7	5.6	7.5	9.4	11.25	13.1

Table 177 Permissible variation for disc thickness and free height

Group	Disc thickness	Permissible variation for s in mm +	Permissible variation for s in mm −	Permissible variation for l_0 in mm +	Permissible variation for l_0 in mm −
1	0.3,0.4	0.02	0.03	0.075	0.025
	0.5,0.6	0.025	0.035		
	0.7,0.8	0.03	0.04		
	0.9	0.035	0.045		
2	1	0.035	0.045	0.1	0.05
	1.1,1.25	0.04	0.05		
	1.5,1.75	0.045	0.055		
	2	0.05	0.06		
	2.25	0.055	0.065	0.15	0.05
	2.5	0.055	0.065		
	3,3.5	0.065	0.075		
3	4 to 14	0.08	0.08	0.2	0.2

Table 178

Group	Permissible variation (using a suitable lubricant) in spring force F at $f \approx 0.75\,h$ in %
1	+25 to −7.5
2	+15 to −7.5
3	+5

Clearance between the guiding pin and the inside diameter

The inside diameter of the disc spring reduces when the discs are subjected to an axial load and hence a certain amount of clearance must be provided between the centrally located guiding pin and the inner diameter D_i when the inside diameter is used for locating purposes. The clearance values indicated in Table 179 are recommended in such cases.

Table 179

Inside diameter D_i in mm	Approx. clearance mm
4.2 to 14.2	0.2
16.3 to 18.3	0.3
20.4 to 25.4	0.4
28.5	0.5
31 to 64	1
72 to 127	2

Note: The clearance values as recommended in Table 179 have been taken into account while taking the inside diameter D_i in Tables 173 and 174. Hence locating pins of standard diameters can be used for central pin.

MACHINE TOOL DESIGN

ISO SYSTEM OF TOLERANCES

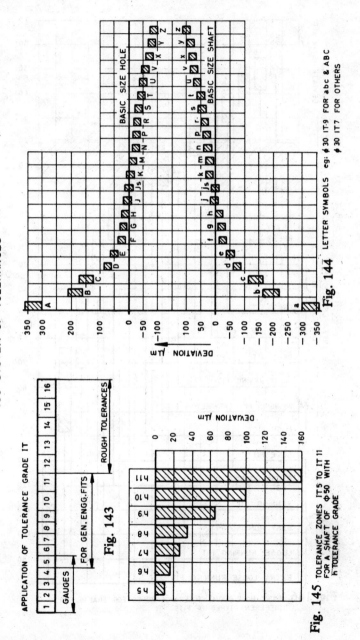

Fig. 144 LETTER SYMBOLS eg: φ 30 IT·9 FOR abc & ABC
φ 30 IT7 FOR OTHERS

APPLICATION OF TOLERANCE GRADE IT

Fig. 143

Fig. 145 TOLERANCE ZONES IT5 TO IT 11
FOR A SHAFT OF φ50 WITH
h TOLERANCE GRADE

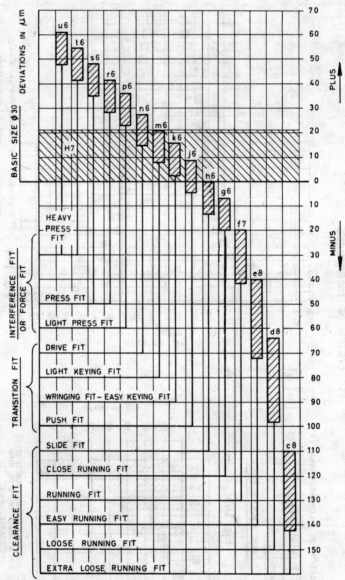

Fig. 146   BASIC HOLE φ30H7 TOLERANCE ZONES FOR SHAFTS FOR
DIFFERENT TYPES OF FITS

Table 180 List of recommended fits

BASIC HOLE	\- SHAFTS FOR \-																
	RUNNING OR CLEARANCE FIT								TRANSITION FIT				INTERFERENCE FIT				
	a	b	c	d	e	f	g	h	j	k	m	n	p	r	s	t	u
H6					e7												
						f6											
							g5	h5	j5	k5	m5	n5	p5	r5	s5	t5	u5
H7	a9																
		b8	c8	d8	e8												
						f7											
							g6	h6	j6	k6	m6	n6	p6	r6	s6	t6	u6
H8				d9	e9	f9		h9									
					e8	f8		h8									
								h7	j7	k7	m7	n7	p7	r7	s7	t7	u7
H10								h10									
	a11	b11	c11	d11				h11									
H12								h12									

BASIC SHAFT	\- HOLES FOR \-																
	RUNNING OR CLEARANCE FIT								TRANSITION FIT				INTERFERENCE FIT				
	A	B	C	D	E	F	G	H	J	K	M	N	P	R	S	T	U
h5					E7												
						F6	G6	H6	J6	K6	M6	N6	P6	R6	S6	T6	U6
h6	A9			D9													
		B8	C8	D8	E8												
						F7	G7	H7	J7	K7	M7	N7	P7	R7	S7	T7	U7
h7	A9	B9															
			C8					H8	J8	K8	M8	N8					
h8		B9	C9														
					E8	F8		H8									
h9			C10	D10													
					E9			H8									
						F8		H9									
h10								H10									
h11	A11	B11	C11	D11				H11									
h12								H12									

▼ : FIRST PREFERENCE
▽ : SECOND PREFERENCE
OTHERS: THIRD PREFERENCE

ALLOWABLE DEVIATIONS FOR DIMENSIONS WITHOUT SPECIFIED TOLERANCES (As per IS 2102-1969)

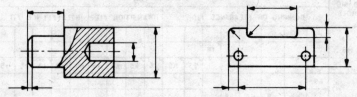

Table 181 Allowable deviations for linear dimensions   All dimensions in *mm*

Deviation ± for dimension									
Over	0.5	3	6	30	120	315	1000	2000	4000
Upto	3	6	30	120	315	1000	2000	4000	8000
For machined parts ‡	0.1	0.1	0.2	0.3	0.5	0.8	1.2	2	3
	0.2*	0.5*	1*	2*	4*	—	—	—	—
For sheet metal work ††	—	0.5	1	1.5	2	3	4	6	8
	—	1*	2*	4*	8*	—	—	—	—

* Applicable for radii and chamfers

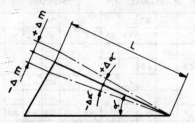

Table 182 Allowable deviations for angular dimensions   All dimensions in *mm*

Deviation ± for length *L* of shorter side of angle.					
		Over	—	10	50	120
		Upto	10	50	120	—
For machined parts	Δm		0.1	0.2	0.6	0.8
	$\Delta \alpha$		1°	30	20	10
For sheet metal work	Δm		0.3	0.8	1.8	2.4
	$\Delta \alpha$		3°	2°	1°	30

‡   For machined parts, allowable deviations according to medium grade of IS: 2102 - 1969

††   For sheet metal work, allowable deviations according to extra coarse grade of IS 2102 - 1969

Table 185   LIMITS AND FITS

Typical applications	*Symbol and grade	Diameter limit												
		Over 1 Upto 3	3 6	6 10	10 18	18 30	30 50	50 80	80 120	120 180	180 250	250 315	315 400	400 500
Major dia. of external splines	a 11	−270 −330	−270 −345	−280 −370	−290 −400	−300 −430								
Width of T nuts	a13	−270 −410	−270 −450	−280 −500	−290 −560	−300 −630								
	f6	−6 −12	−10 −18	−13 −22	−16 −27	−20 −33	−25 −41	−30 −49	−36 −58	−43 −68	−50 −79	−56 −88	−62 −98	−68 −108
	h7	0 −10	0 −12	0 −15	0 −18	0 −21	0 −25	0 −30	0 −35	0 −40	0 −46	0 −52	0 −57	0 −63
Socket head dia.	h12	0 −100	0 −120	0 −150	0 −180	0 −210	0 −250	0 −300	0 −350	0 −400	0 −460	0 −520	0 −570	0 −630
Socket head height	h13	0 −140	0 −180	0 −220	0 −270	0 −330	0 −390	0 −460	0 −540	0 −630	0 −720	0 −810	0 −890	0 −970
Slotted and castle nut dimensions	h14	0 −250	0 −300	0 −360	0 −430	0 −520	0 −620	0 −740	0 −870	0 −1000	0 −1150	0 −1300	0 −1400	0 −1550
	h15	0 −400	0 −480	0 −580	0 −700	0 −840	0 −1000	0 −1200	0 −1400	0 −1600	0 −1850	0 −2100	0 −2300	0 −2500
	h16	0 −600	0 −750	0 −900	0 −1100	0 −1300	0 −1600	0 −1900	0 −2200	0 −2500	0 −2900	0 −3200	0 −3600	0 −4000
Symmetrical length tolerances	js9	+12.5 −12.5	+15 −15	+18 −18	+21.5 −21.5	+26 −26	+31 −31	+37 −37	+43.5 −43.5	+50 −50	+57.5 −57.5	+65 −65	+70 −70	+77.5 −77.5
	js14	+125 −125	+150 −150	+180 −180	+215 −215	+260 −260	+310 −310	+370 −370	+435 −435	+500 −500	+575 −575	+650 −650	+700 −700	+775 −775
	js15	+200 −200	+240 −240	+290 −290	+350 −350	+420 −420	+500 −500	+600 −600	+700 −700	+800 −800	+925 −925	+1050 −1050	+1150 −1150	+1250 −1250
Hydraulic fittings	A11	+330 +270	+345 +270	+370 +280	+400 +290	+430 +300								
	B11	+200 +140	+215 +140	+240 +150	+260 +150	+290 +160								
Key-way width in spacer rings	C11	+120 +60	+145 +70	+170 +80	+205 +95	+240 +110								
Width of internal splines	D9	+45 +20	+60 +30	+76 +40	+93 +50	+117 +65	+142 +80	+174 +100	+207 +120	+245 +145	+285 +170	+320 +190	+350 +210	+385 +230
Hexagonal socket width	D12	+120 +20	+150 +30	+190 +40	+230 +50	+275 +65	+330 +80	+400 +100	+470 +120	+545 +145	+630 +170	+710 +190	+780 +210	+860 +230
Circlips	H10	+40 0	+48 0	+58 0	+70 0	+84 0	+100 0	+120 0	+140 0	+160 0	+185 0	+210 0	+230 0	+250 0
Counterbore diameter	H12	+100 0	+120 0	+150 0	+180 0	+210 0	+250 0	+300 0	+350 0	+400 0	+460 0	+520 0	+570 0	+630 0
Major dia. of internal splines	H13	140 0	180 0	220 0	270 0	330 0	390 0	460 0	540 0	630 0	720 0	810 0	890 0	970 0
Key-way width	Js9	+12.5 −12.5	+15 −15	+18 −18	+21.5 −21.5	+26 −26	+31 −31	+37 −37	+43.5 −43.5	+50 −50	+57.5 −57.5	+65 −65	+70 −70	+77.5 −77.5
Symmetrical tolerances in slots	Js14	+125 −125	+150 −150	+180 −180	+215 −215	+260 −260	+310 −310	+370 −370	+435 −435	+500 −500	+575 −575	+650 −650	+700 −700	+775 −775
	Js16	+300 −300	+375 −375	+450 −450	+550 −550	+650 −650	+800 −800	+950 −950	+1100 −1100	+1250 −1250	+1450 −1450	+1600 −1600	+1800 −1800	+2000 −2000

*NOT TO BE SPECIFIED FOR GENERAL LIMITS AND FITS.

Table 186

Diameter steps in mm	Value of tolerance in microns (1 micron = 0.001 mm) Tolerance grades IT																	
	01	0	1	2	3	4	5	6	7	8	9	10	11	12	13	14*	15*	16*
Upto 3	0.3	0.5	0.8	1.2	2	3	4	6	10	14	25	40	60	100	140	250	400	600
Over 3 Upto 6	0.4	0.6	1	1.5	2.5	4	5	8	12	18	30	48	75	120	180	300	480	750
Over 6 Upto 10	0.4	0.6	1	1.5	2.5	4	6	9	15	22	36	58	90	150	220	360	580	900
Over 10 Upto 18	0.5	0.8	1.2	2	3	5	8	11	18	27	43	70	110	180	270	430	700	1100
Over 18 Upto 30	0.6	1	1.5	2.5	4	6	9	13	21	33	52	84	130	210	330	520	840	1300
Over 30 Upto 50	0.6	1	1.5	2.5	4	7	11	16	25	39	62	100	160	250	390	620	1000	1600
Over 50 Upto 80	0.8	1.2	2	3	5	8	13	19	30	46	74	120	190	300	460	740	1200	1900
Over 80 Upto 120	1	1.5	2.5	4	6	10	15	22	35	54	87	140	220	350	540	870	1400	2200
Over 120 Upto 180	1.2	2	3.5	5	8	12	18	25	40	63	100	160	250	400	630	1000	1600	2500
Over 180 Upto 250	2	3	4.5	7	10	14	20	29	46	72	115	185	290	460	720	1150	1850	2900
Over 250 Upto 315	2.5	4	6	8	12	16	23	32	52	81	130	210	320	520	810	1300	2100	3200
Over 315 Upto 400	3	5	7	9	13	18	25	36	57	89	140	230	360	570	890	1400	2300	3600
Over 400 Upto 500	4	6	8	10	15	20	27	40	63	97	155	250	400	630	970	1550	2500	4000

*Upto 1mm, Grades 14 to 16 are not provided.

SELECTED SHAFT AND HOLE TOLERANCES & FITS

SELECTED FITS

Table 187 Basic hole system

Hole	Shaft											
H6				g5	h5	j5	k5	m5				
H7	d8		f7	g6	h6	j6	k6	m6	n6	p6	r6	s6
H8		e8	f8		h8	(j7)	(k7)	(m7)	(n7)			
H11	d11				h11							

Table 188 Basic shaft system

Shaft	Hole								
h5					H6	J6	K6	M6	N6
h6				G7	H7	(J7)	K7	(M7)	N7
h8		E8	F8		H8				
h11	D11				H11				

► Preferred fits

Factors which influence selection of fits

PROBABLE SIZE OF ACTUAL DIMENSION

According to the principle of statistical probability, it is required to select fits such that clearance or interference of both parts occur within the middle two-thirds of tolerance zone.

Working temperature difference of shafts and flanges (holes), especially when the temperature difference is high.

DEFORMATION OF MATERIAL

If changes of dimensions due to temperature are not considered, then it is necessary in case of heavy force fits to calculate the stress after pressing so that the elastic limit is not exceeded.

The choice of fits considerably depends on the material of mating parts, the workmanship, running conditions, types of lubrication, etc. Therefore, all such factors should be considered before deciding fits.

The values of surface roughness indicated for the mating parts are for guidance only.

Table 189 Selected shaft and hole tolerances and fits

Type of fit	Fit / Surface roughness	Uses	Examples
Clearance fit	H7/d8 1·6 / 1·6	This combination of hole and shaft provides a loose running fit which in general is suitable for applications as gland seals, loose pulleys, idle gears, etc.	H7/d8
	H8/e8 1·6 / 1·6	This combination is commonly used as an easy running fit for bearings where an appreciable clearance is permissible, e.g., for widely separated bearings or several bearings in line.	H8/e8
	H8/f8 0·8 / 0·8	This is the most commonly used normal running fit which is relatively easy to produce. Typical applications are shaft bearings in gear boxes, gears running on fixed shafts, pump driving shaft bearings, etc. The combination H7/f7 satisfies most of the requirements of running fit.	H7/s6 H8/f8 H7/f7
	H7/f7 0·8 / 0·8		H7/p6 H7/f7
	H7/g6 0·8 / 0·4	The amount of clearance provided by this fit is small and it is not recommended for a continuously running bearing except in precision application where shaft loadings are light and conditions suitable. It may be used in high precision machine tools and measuring instruments as a precision sliding fit or as location fit or spigot fit	H7/g6
	H6/g5 0·4 / 0·2		H6/g5
Transition fit	H11/h11 3·2 / 1·6	For parts where cold drawn material can be used to advantage. Rough machines, transport equipment, cranes, lifts, etc.	H11/h11

Table 189   Selected shaft and hole tolerances and fits (Contd.)

Type of fit	Fit / Surface roughness	Uses	Examples
Transition fit	H8 / h8 0·8 / 0·8 H7 / h6 0·8 / 0·4 H6 / h5 0·2 / 0·1	Though the upper shaft limit of this fit is zero, a slight clearance will usually be present. The fit is widely used for non-running assemblies such as spigots and location fits and is the closest available clearance fit. Finer grades may be used to give a precision sliding fit where conditions are suitable e. g. where no substantial temperature differences are encountered.	H8/h8 H8/h8 H7/h6 H6/h5
	H8 / j7 0·8 / 0·8 H7 / j6 0·8 / 0·8 H6 / j5 0·8 / 0·8	This fit averages a slight clearance although a small amount of interference can occur under extreme conditions. Recommended as a location fit, in cases where the clearances should be kept small and the interference also kept to a minimum in order to ensure both accurate location and easy assembly and dismantling. Typical applications are coupling spigots and recesses, gear rings fitted to hubs, etc.	H8/j7 H7/j6 H6/j5
	H7 / k6 0·8 / 0·8 H6 / k5 0·8 / 0·8	This is a true transition fit averaging virtually no clearance and is recommended for location fit where a slight interference can be tolerated with the object, e. g. of eliminating vibration. The parts can be comparatively easily assembled or dismantled by hand or by using hammer (couplings, handwheels, gears on shafts, levers, handles, etc.)	H7/k6 H6/k5

Table 189 Slected shaft and hole tolerances and fits (Concld.)

Type of fit	Fit / Surface roughness	Uses	Examples
Transition fit	H6 / m5 0·8// 0·8/	This transition fit averages a slight interference. It is suitable for general light keying fits where accurate location with freedom from play is necessary (gear wheels, pulleys, fly wheels, etc.)	H6/m5
	H7 / n6 0·8// 0·8/	Averages some interference and gives clearance only when the minimum metal extreme of sizes is approached. For parts which should be rigidly connected and which can be assembled and dismantled only by greater force using assembly fixtures.	H7/n6
Interference fit	H7 / p6 0·8// 0·8/	Provides the first interference fit. Amount of interference, though small, is sufficient to give on non ferrous parts a press fit which can be easily dismantled and assembled when required without over straining the parts.	H7/p6
	H7 / r6 0·8// 0·8/	Provides a medium drive fit on ferrous parts and a light drive fit on non ferrous parts,which can be easily dismantled when required. Commonly used for applications such as press-in bearing bushes, sleeves, seatings, gear wheels, etc.	H7/r6
	H7 / s6 0·8// 0·8/	Provides a heavy drive fit. Commonly used for permanent or semi-permanent assembly of steel & cast iron members which have to carry high load & for such applications, as bearing bushes in housings. In larger sizes, amount of interference may necessitate assembly by shrinking to avoid damage to fitting surfaces.	H7/s6

TOLERANCES OF FORM AND POSITION
GUIDE FOR SELECTION FOR MACHINE TOOL PARTS

The table gives a guidline for the selection of tolerance values of form and position for a few critical components of light and medium size machine tools and also give a range of values of errors of form obtainable normally and with difficulty by various methods of machining. However the values are to be selected in the light of required function, economy of manufacture, class of accuracy of machine, etc. Also it is not necessary to prescribe tolerances for all errors of form and position, as shown in the examples and the designer must exercise his own choice in the light of functional requirements. The tolerance values given are applicable not only to parts indicated in the sketches, but also to parts having similar functions.

Table 190 Tolerances of form and position

MOUNTING SURFACES FOR JOURNAL BEARINGS OF SPINDLES						
	Kind of error	Kind of machine tool/Feature		Tolerance values in μm		
				Class of accuracy		
				Normal	Precision	High precision
	O	Lathes	$D \leqslant 100$	3	2	1
			$D > 100$	3	2	2
		Grinding machines	$D \leqslant 100$	1	0.8	0.5
			$D > 100$	1	0.8	0.8
	⟡	Lathes	$D \leqslant 100$	5	3	2
			$D > 100$	5	3	3
		Grinding machines	$D \leqslant 100$	2	1.5	0.8
			$D > 100$	2	1.5	1

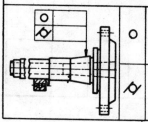

MOUNTING SURFACES FOR ANTIFRICTION BEARINGS (CYLINDRICAL & TAPERED)		
	O	Permissible values of circularity for the taper surface for mounting *NN30K* bearings will be same as those indicated for *P2* and *P5* class bearings by the bearing manufactures.
	⟡	The same values will hold good for other types of bearings also and for cylindrical mounting surface.

Table 190 Tolerances of form and position (Contd.)

MOUNTING SURFACES FOR BEARINGS OTHER THAN MAIN BEARINGS (AUXILIARY AXIAL AND RADIAL BEARINGS), PULLEYS, GEARWHEELS, PRECISION BUSHES, ETC.

	Kind of error	Kind of machine tool/Feature	Tolerance values in μm		
			Class of accuracy		
			Normal	Precision	High precision
	O	Lathes	8	5	3
		Milling machines	10	8	–
		Boring machines	8	5	3
		Grinding machines	5	3	2
	@	Lathes	8	5	5
		Milling machines	10	8	–
		Boring machines	10	5	5
		Grinding machines	5	5	3
		Lathes	10	8	5
		Milling machines	15	8	–
		Boring machines	10	8	5
		Grinding machines	10	5	3

@ Not required for surfaces other than bearings.

IMPORTANT AXIAL BEARING SURFACES (BEARING SURFACES PREVENTING AXIAL MOVEMENT).

Facial bearing surface		Lathes	5	3	3
	Non Sliding	Milling machines	10	8	–
		Boring machines	5	3	3
		Grinding machines	3	2.5	2
	Sliding	Lathes	5	3	3
		Milling machines	10	5	–
		Boring machines	5	3	2
		Grinding machines	3	2	1

TAPERED END OF GRINDING SPINDLES

O	–	2	1.5	0.8
	–	3	2	1

Table 190   Tolerances of form and position (Contd.)

SPINDLE NOSES OF LATHES, MILLING MACHINES, BORING MACHINES, ETC.

	Kind of error	Kind of machine tool/Feature	Tolerances value in μm		
			Class of accuracy		
			Normal	Precision	High precision
	O	Taper bore	3	3	2
		External taper	5	3	2
	⟋	Taper bore	5	5	3
		External taper	8	5	3
		Facial surface	5	5	3
	▱	Facial surface	5	5	3

SPINDLE NOSES OF MILLING MACHINES WITH STEEP TAPER BORE

	O	$D \leqslant 40$	5	3	–
		$D > 40$	5	5	–
	⟋	$D \leqslant 40$	8	5	–
		$D > 40$	10	8	–

SPINDLE NOSES OF BENCH AND RADIAL DRILLING MACHINES

	O	Bench drilling m/c.	10	–	–
		Radial drilling m/c.	8	–	–
	⟋	Bench drilling m/c.	15	–	–
		Radial drilling m/c.	10	–	–

HOUSING BUSH FOR SPINDLE BEARINGS

	O	–	10	5	2
	Radial	–	15	8	3
	Facial	–	10	5	3
	//	–	8	5	2

SPACERS, ETC.

	//	–	8	5	2

Table 190   Tolerances of form and position (Contd.)

INSIDE SURFACES OF SLEEVES FOR MOUNTING OF BEARINGS

	Kind of error	Kind of machine tool	Tolerance values in μm		
			Class of accuracy		
			Normal	Precision	High precision
	O	Drilling machines	8	–	–
		Boring machines	5	3	2
		Milling machines	8	5	–
	↗ Radial	Drilling machines	10	–	–
		Boring machines	8	5	3
		Milling machines	10	8	–
	↗ Facial	Drilling machines	8	–	–
		Boring machines	5	3	2
		Milling machines	8	5	–

EXTERNAL CYLINDRICAL SURFACES OF SLEEVES OF MAIN SPINDLES OF MACHINE TOOLS

	Kind of error	Kind of machine tool	Normal	Precision	High precision
	O	Drilling machines	5	–	–
		Boring machines	5	3	2
		Milling machines	8	5	–
	⌀	Drilling machines	8	–	–
		Boring machines	5	3	3
		Milling machines	8	5	–

EXTERNAL CYLINDRICAL SURFACES OF TAILSTOCK SLEEVE

	Kind of error	Kind of machine tool	Normal	Precision	High precision
	O	Lathes	5	3	2
		Grinding machines	2	2	1
	⌀	Lathes	5	5	3
		Grinding machines	3	3	2

TAPERED BORE

	Kind of error	Kind of machine tool	Normal	Precision	High precision
	O	Lathes	5	3	2
		Grinding machines	3	2	1
	↗	Lathes	5	5	3
		Grinding machines	3	3	2

Table 190 Tolerances of form and position (Contd.)

MOUNTING SURFACES FOR JOURNAL BEARINGS

Type		Diameter mm		Tolerance values in μm			
				O	⬦	✦	□
		Over	Upto	X_1	X_2	X_3	X_4
For peripheral speeds	≤10 m/sec.	10	50	2	3	8	8
		50	120	2	3	8	8
		120	250	3	5	10	10
	>10 m/sec.	10	50	1	2	5	5
		50	120	1	2	5	5
		120	250	1.5	3	8	8

MOUNTING SURFACES FOR ANTIFRICTION BEARINGS

Type		Diameter mm		O	⬦	✦	□
		Over	Upto	X_1	X_2	X_3	X_4
For normal mounting P0 & P6		10	50	2	3	5	5
		50	120	5	5	8	8
		120	250	5	5	10	10
For precision mounting P5 & P4		10	50	1	2	3	3
		50	120	2	3	5	5
		120	250	3	3	8	8

MOUNTING SURFACES FOR GEARWHEELS, PULLEYS, ETC.

Type		Diameter, mm		Tolerance values in μm				
				O	⬦	—	✦	
		Over	Upto	X_1	X_2	X_3	X_4	X_5
For peripheral speeds	≤6 m/sec.	10	50	3	5	3	4	10
		50	120	3	5	3	4	15
		120	250	3	5	5	5	20
	>6 m/sec.	10	50	2	3	2	2.5	8
		50	120	2	3	2	2.5	10
		120	250	2	3	3	4	15

Table 190 Tolerances of form and position (Contd.)

MOUNTING SURFACES FOR HIGH PRECISION ROTATING PARTS (e.g. BEARING BUSH)

Type	D, mm		Tolerance values in μm					
			O	∿	□	/		
	Over	Upto	X_1	X_2	X_3	X_4	X_5	X_6
For peripheral speeds ⩽10 m/sec.	10	50	2	3	3	5	8	5
	50	120	3	3	5	8	10	8
	120	250	3	5	8	10	15	10
For peripheral speeds >10 m/sec.	10	50	1.5	1.5	2	3	5	3
	50	120	2	2	3	5	8	5
	120	250	2	3	5	8	10	8

BEARING SURFACE FOR ROTATING OR SLIDING PARTS (e.g. BUSH)

Accuracy	D, mm		Tolerance values in μm				
			O	∿	/		
	Over	Upto	X_1	X_2	X_3	X_4	X_5
Normal	10	50	3	3	5	10	10
	50	120	5	5	8	15	10
	120	250	5	8	10	20	15
Precision	10	50	2	2·	3	8	8
	50	120	3	3	5	10	8
	120	250	3	5	8	15	10

FUNCTIONAL CYLINDRICAL SURFACES (e.g. PISTON)

Pressure kgf/cm^2	D, mm		Tolerance values in μm		
	Over	Upto	−	O	∿
⩽63	10	50	5	3	5
	50	120	8	5	8
	120	250	10	8	10
>63	10	50	5	2	3
	50	120	8	3	5
	120	250	10	5	8

Table 190 Tolerances of form and position (Contd.)

FACIAL SURFACES FOR CLAMPING (e.g. NUT)

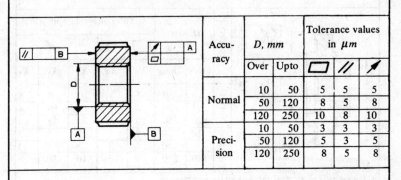

Accuracy	D, mm		Tolerance values in µm		
	Over	Upto	▭	//	⟋
Normal	10	50	5	5	5
	50	120	8	5	8
	120	250	10	8	10
Precision	10	50	3	3	3
	50	120	5	3	5
	120	250	8	5	8

SURFACE FOR LOCATING THE POSITION OF ANOTHER PART (e.g. HUB)

Accuracy	D, mm		Tolerance values in µm							
	Over	Upto	X_1	X_2	X_3	X_4	X_5	X_6	X_7	X_8
Normal	10	50	8	2	3	3	5	5	10	8
	50	120	8	2.5	5	5	8	5	10	8
	120	250	10	3	5	8	8	8	10	10
Precision	10	50	5	1	2	2	3	3	5	5
	50	120	5	1.5	3	3	5	3	5	5
	120	250	8	2	3	5	5	5	5	8

SURFACE FOR LOCATING THE POSITION OF ANOTHER PART (e.g. SLEEVE)

Accuracy	D, mm		Tolerance values in µm					
	Over	Upto	X_1	X_2	X_3	X_4	X_5	X_6
Normal	10	50	8	3	3	5	8	8
	50	120	8	3	5	8	10	8
	120	250	10	5	8	10	15	10
Precision	10	50	5	2	2	3	5	5
	50	120	5	2	3	5	8	5
	120	250	8	3	5	8	10	8

Table 190 Tolerances of form and position (Contd.)

SURFACE FOR LOCATING THE POSTION OF ANOTHER PART ENSURING COAXIALITY (e.g. FLANGE)

Accuracy	D, mm		Tolerance values in μm				
			$\Box$	$\bigcirc$		$\nearrow$	
	Over	Upto	X_1	X_2	X_3	X_4	X_5
Normal	10	50	5	3	10	8	5
	50	120	8	5	10	10	8
	120	250	10	5	15	15	10
Precision	10	50	3	2	8	5	3
	50	120	5	3	8	8	5
	120	250	8	3	10	10	8

HEADSTOCK BODY OF MACHINE TOOLS

Kind of error	Type of mounting	Kind of machine tool	Tolerance values in μm		
			Class of accuracy		
			Normal	Precision	High precision
$\bigcirc$	Bearings mounted directly in the body	Lathes	8	5	3
		Boring m/cs	8	5	3
		Grinding m/cs	5	3	3
$\nearrow$		Lathes	10	8	5
		Boring m/cs	10	8	5
		Grinding m/cs	8	5	5
$\bigodot$		Lathes	20	15	8
		Boring m/cs	15	10	5
		Grinding m/cs	5	3	2
$\bigcirc$	Bearings mounted in a bush fitted to the body	Lathes	10	8	5
		Boring m/cs	10	8	5
		Grinding m/cs	8	5	5
$\nearrow$		Lathes	10	10	8
		Boring m/cs	10	10	8
		Grinding m/cs	10	8	8
$\bigodot$		Lathes	20	15	8
		Boring m/cs	15	10	5
		Grinding m/cs	8	5	3

Table 190 Tolerances of form and position (Contd.)

HEADSTOCK BODY OF MACHINE TOOLS

Kind of error	Type of mounting	Kind of machine tool	Tolerance values in μm Class of accuracy		
			Normal	Precision	High Precision
□	Base scraped	Lathes	30	30	20
		Boring m/cs	30	30	20
		Grinding m/cs	20	15	10
//		Lathes	15	10	10
		Boring m/cs	15	10	10
		Grinding m/cs	10	8	8
□	Base milled or planed	Lathes	—	30	—
		Boring m/cs	—	30	—
		Grinding m/cs	—	20	—
//		Lathes	30	20	20
		Boring m/cs	30	20	20
		Grinding m/cs	20	15	10

GEAR BOX BODY

Type: a) For transfer of main motion, high torques, etc.

b) For transfer of auxiliary motions, drive for lead screw, indexing mechanism, etc.

Kind of error		Speed rpm	Tolerance values in μm Axial distance of bearings, l mm					
			100	160	250	400	630	> 630
//	a	≤ 1400	20	25	30	40	50	60
		> 1400 ≤ 2800	10	15	20	25	30	40
		> 2800	8	10	10	15	20	25
	b	≤ 1400	8	10	10	15	20	25
		> 1400	5	8	8	10	15	20

Kind of error		Speed rpm	D, mm					
			≤25	40	63	100	120	160
◎	a	≤ 1400	10	10	15	15	20	20
		> 1400 ≤ 2800	8	8	10	10	10	15
		> 2800	5	5	5	5	8	8
	b	≤ 1400	3	3	3	5	8	8
		> 1400	3	5	5	8	8	10
○	a, b	—	3	5	5	8	8	10
⌀	a, b	—	5	8	8	10	10	10

Table 190 Tolerances of form and position (Contd.)

SPUR AND HELICAL GEAR BLANKS

Kind of error	Module	D mm Over	D mm Upto	3	4	5	6	7	8
Facial	1–8	25	40	5	5	8	8	8	8
		40	63	8	8	8	8	8	10
		63	100	8	8	10	10	10	10
		100	160	10	10	10	10	15	15
		160	250	10	15	15	15	20	20
Radial	1–4	25	40	5	8	10	15	20	30
	1–2	40	63	5	8	10	15	20	30
	2–4			5	10	10	15	25	40
	1–2	63	100	5	8	10	15	20	30
	2–4			5	10	15	20	25	40
	4–8			8	10	15	20	30	40
	1–2	100	160	5	10	10	20	25	40
	2–4			8	10	15	20	25	40
	4–8			8	10	15	20	30	50
	1–4	160	250	8	10	15	20	30	40
	4–8			10	15	20	25	40	50

(Tolerance values in μm — Class of accuracy ISO 3, 4, 5, 6, 7, 8)

HYDRAULIC ELEMENTS

BODY WITHOUT SLEEVES

Type	Nominal bore N_b, mm	Nominal pressure N_p kgf/cm²	○	⟋	◎
Regulating — Pressure regulators, hydraulic valves, etc.,	≤16	≤63	2	3	–
	≤16	>63	1.5	2	–
	>16	≤63	3	5	–
	≤32	>63	2	5	–
	>32	≤63	5	8	–
		>63	3	5	–
Distributing — Reversible systems with valves operating due to the movement of piston, etc.	≤16	≤63	5	5	–
		>63	3		–
	>16	≤63	5	5	–
	≤32	>63	3		–
	>32	All values	8	8	–

Table 190 Tolerances of form and position (Contd.)

BORES IN THE BODY FOR PRESS FITTING THE SLEEVES

Type		Nominal bore N_b, mm	Nominal pressure N_p kgf/cm²	$\bigcirc$	$\oslash$	$\circledcirc$
Regulating	Sleeves ground after assembly	≤ 32	≤ 63	8	10	16
			> 63	5		
		> 32	≤ 63	10	15	
			≥ 63	8	15	
Distributing	Sleeves honed after assembly	≤ 32	≤ 63	8	10	10
			> 63	5		
		> 32	≤ 63	10	10	
			> 63	8	10	

SPACERS & GUIDE RINGS //

‡ Not for spacer rings

Type		Nominal bore	Nominal pressure	$\bigcirc$	$\oslash$	$/\!/$
Regulating	Ground after assembly	≤ 16	≤ 63	5	5	1
			> 63			0.5
		> 16 ≤ 32	≤ 63	5	5	2
			> 63			1
		> 32	All values	15	8	2
	Honed after assembly	≤ 16	≤ 63	5	5	1
			> 63	3	3	0.5
		> 16	≤ 63	5	5	2
		≤ 32	> 63	3	3	1
		> 32	All values	5	5	2

Type		Nominal bore N_b, mm	Nominal pressure N_p kgf/cm²	$\oslash$ X_1	$\nearrow$ X_2	$/\!/$ X_3	X_4
Distributing	Ground after assembly	≤ 16	All values	8	5	15	3
		> 16 ≤ 32	All values				
		> 32	All values	10	8		5
	Honed after assembly	≤ 16	≤ 63	5	5	5	3
			> 63	3	3		
		> 16	≤ 63	5	5		
		≤ 32	> 63	3	3		
		> 32	All values	5	5	8	5

Table 190   Tolerances of form and position (Contd.)

SPOOL VALVES SLEEVE

Type		Nominal bore N_b, mm	Nominal pressure N_p kgf/cm²	Tolerance values μm			
				X_1	X_2	X_3	X_4
Regulating	Ground after assembly	≤16	≤63	5	5	10	3
		≤16	>63				2
		>16	≤63				5
		≤32	>63				3
		>32	All values	8	8	15	5
	Honed after assembly	≤16	≤63	5	5	5	3
		≤16	>63	3	3		2
		>16	≤63	5	5		5
		≤32	>63	3	3		3
		>32	All values	5	5	8	5
Distributing	Ground after assembly	≤16	All values	8	5	15	10
		>16					20
		≤32					20
		>32		10	8		50
	Honed after assembly	≤16	≤63	5	5	5	10
			>63	3	3		
		>16	≤63	5	5		20
		≤32	>63	3	3		
		>32	All values	5	5	8	50

BODY WITH SPACERS AND GUIDERINGS FOR PRECISE REGULATION

SLEEVE   BODY

Type	Nominal bore N_b, mm	Nominal Pressure N_p kgf/cm²	Tolerance values μm		
			O	⌀	⟋
Regulating	≤16	≤63	2	3	
		>63 ≤160	1.5	2	
		>160	0.8		—
	>16	≤63	3	5	
	≤32	>63	2	5	
	>32	<63	5	8	
		>63	3	5	

Table 190 Tolerances of form and position (Concld.)

BODY WITH SINGLE LONG SLEEVE ONLY

Type	Nominal bore N_b, mm	Nominal pressure N_p kgf/cm²	Tolerance values, μm ⌀	⌂	↗
Regulating	≤ 16	≤ 63	2	3	5
Regulating	≤ 16	> 63	1.5	2	3
Regulating	> 16 ≤ 32	≤ 63	3	5	5
Regulating	> 16 ≤ 32	> 63	2		
Regulating	> 32	≤ 63	5	8	8
Regulating	> 32	> 63	3	5	5
Distributing	≤ 16	≤ 63	5	5	10
Distributing	≤ 16	> 63	3		
Distributing	> 16 ≤ 32	≤ 63	5	5	20
Distributing	> 16 ≤ 32	> 63	3		
Distributing	> 32	All values	8	8	50

REGULATING SPOOL VALVE AND SLEEVE

a—Facial surface regulating the flow and
pressure of liquid
a'—Free surface
b—Cylindrical surface

Nominal bore N_b mm	Nominal pressure N_p kgf/cm²	Tolerance values μm ⌀ X_1	⌂ X_2	↗ X_3	X_4	X_5
≤ 16	≤ 63	2	2	2	1.5	5
≤ 16	> 63 ≤ 160	1.5	1.5	1.5	1	5
≤ 16	> 160					
> 16 ≤ 32	≤ 63	2	2	3	1.5	10
> 16 ≤ 32	> 63	1.5		2.5	1	
> 32	≤ 63	3	3	5	3	20
> 32	> 63	2		3	2	

DISTRIBUTING VALVE

Nominal bore	Nominal pressure	⌀ X_1	⌂ X_2	↗ X_3	X_4	X_5
≤ 16	≤ 63	3	3	3	8	8
≤ 16	> 63	2				
> 16 ≤ 32	≤ 63	3		5	10	10
> 16 ≤ 32	> 63	2				
> 32	≤ 63	5	5	8	20	20
> 32	> 63	3				

Table 190a Tolerances of form and position obtainable from manufacturing processes

Cylindricity	Method of machining		Diameter mm	Values of error, μm
				Columns: 0.2, 0.25, 0.3, 0.5, 0.8, 1, 1.5, 2, 2.5, 3, 5, 8, 10, 15, 20, 30, 40, 50, 60
⌀	Turning	Shaft	36	
			50	
			70	
			100	
		Bore	36	
			50	
			70	
			100	
	Boring		36	
			50	
			70	
			100	
	Reaming		36	
			50	
			70	
			100	
	Grinding	Centreless	36	
			50	
			70	
			100	
		Internal	36	
			50	
			70	
			100	
		Cylindrical	36	
			50	
			70	
			100	
	Honing		36	
			50	
			70	
			100	
	Super-finishing		36	
			50	
			70	
			100	

////// Values achieved with difficulty   ✕✕✕✕ Values normally achieved

Table 190a Tolerances of form and position obtainable from manufacturing processes (Contd.)

Legend: ▨ = Values achieved with difficulty ▩ = Values normally achieved

Circularity	Method of machining	Diameter mm over	upto	0·5	0·8	1	1·5	2	2·5	3	5	8	10	15	20	30	40	50	60	80	100	120
◯	Turning –Shaft	0	18								▨	▩	▩	▩								
		18	50								▨	▩	▩	▩								
		50	120								▨	▩	▩	▩								
		120	250									▨	▩	▩	▩							
		250	500									▨	▩	▩								
	Turning –Bore	0	18									▨	▩	▩								
		18	50									▨	▩	▩								
		50	120									▨	▩	▩								
		120	250											▨	▩							
		250	500											▩								
	Drilling –Twist drill	0	18													▩	▩					
		18	50														▨	▩				
		50	120																▨	▩	▩	
	Drilling –Gun drill	0	18								▨	▨	▩									
		18	50								▨	▨	▩									
		50	120									▨	▩	▩								
	Reaming	0	18								▨	▩	▩									
		18	50								▨	▩	▩									
		50	120									▨	▩	▩								
	Boring	0	18							▨	▩	▩										
		18	50							▨	▩	▩										
		50	120								▨	▩	▩									
		120	250								▨	▩	▩									
	Internal grinding	0	18				▨	▨	▨	▨	▩	▩										
		18	50		▨	▨	▨	▨	▨	▨	▩	▩										
		50	120		▨	▨	▨	▨	▨	▨	▩	▩										
		120	250			▨	▨	▨	▨	▨	▩											
		250	500				▨	▨	▨	▨	▩	▩										

▨ Values achieved with difficulty   ▩ Values normally achieved

Table 190a Tolerances of form and position obtainable from manufacturing processes (Concld.)

Flatness and straightness	Method of machining	Length of surface, mm	Values of error, μm
▱	Milling	750	
		1500	
		5000	
	Planing	750	
		1500	
		3000	
		5000	
│	Surface grinding	500	
		750	
		3000	
		5000	
	Scraping	750	
		1500	
		3000	
		5000	

Error scale columns (μm): 2, 2·5, 4, 5, 8, 10, 15, 20, 25, 30, 40, 50, 60, 80, 100, 140, 150, 200

Circularity	Method of machining	Diameter mm over	upto	Values of error, μm
○	Centreless grinding	0	18	
		18	50	
		50	120	
		120	250	
	Cylindrical grinding	0	18	
		18	120	
		120	250	
		250	500	
	Honing	0	18	
		18	50	
		50	120	
		120	250	
		250	500	
	Lapping	0	18	
		18	50	
		50	120	
		120	250	
	Super-finishing	0	18	
		18	120	
		120	250	

Error scale columns (μm): 0·1, 0·15, 0·2, 0·25, 0·3, 0·5, 0·6, 1, 1·5, 2, 2·5, 3, 5, 8, 10, 15, 20, 30, 40

⬚ Values achieved with difficulty   ▨ Values normally achieved

GUIDE FOR SELECTION OF SURFACE ROUGHNESS OF MACHINE TOOL PARTS

The following table recommends the specification of surface roughness for functional surfaces of machine tool parts. These values given are only a guide and the designer can make his own selection depending upon the function of the surface. However, from the point of production economy, it is better not to specify values finer than that are really necessary for satisfactory functioning of the component. The values of surface roughness recommended in the tables have been established from the point of production processes and type of machining of surface in practice.

Table 191 Guide for selection of surface roughness

SURFACES WITH ROTARY MOTION				Surface roughness R_a in μm	
Examples		Workmanship		a—Shaft	b—Hole
Shafts rotating in bush bearings		High precision		0.05 0.2	0.1 0.4
		Upto φ 50	Precision		0.4
			Normal	0.4	0.8
		Above φ 50	Precision		
			Normal	0.8	1.6
Main spindle of machine tools	Precision and high precisior machine tools	Upto φ180		0.025	0.1
		Above φ180		0.05	
	Normal accuracy machine tools			0.1 (0.05)	0.2

Table 191 Guide for selection of surface roughness (Contd.)

Examples		Workmanship	Surface roughness R_a in μm	
			a Shaft	b Hole
	Tailstock sleeve in tailstock body, ram moving in guideways or bushing	Precision	0.2	0.4
		Normal	0.4	
	Tapered shanks and sleeves	Precision	0.4	0.8
		Normal	0.8 (1.6)	1.6 (3.2)
	Shanks of clamping centres, bore of tailstock sleeve, etc.	Precision and high precision machine tools	0.2	0.4
		Normal accuracy machine tools	0.4	0.8
	Taper bores of spindles, mandrels, reduction sleeves, external taper of spindle, etc.	Precision and high precision machine tools	0.2	0.4
		Normal accuracy machine tools	0.4	0.8
DISMOUNTABLE SURFACES				
	Keys and keyways	Keys for tight fit (when milled)	0.8 (1.6)	
		Keyways in shafts and hubs — Precision	1.6	3.2
		Keyways in shafts and hubs — Normal	3.2	6.3
	Surfaces which do not transmit force or torque	Precision	0.8	1.6
		Normal	1.6	3.2

Table 191 Guide for selection of surface roughness (Contd.)

MOUNTING SURFACES FOR ANTIFRICTION BEARINGS

Examples	Workmanship	Surface roughness R_a in μm	
		a Shaft	*b* Hole
Mounting allowing axial movement of bearing to compensate shaft expansion or allowing axial adjustment	Precision	0.4	0.8
	Normal	0.8	1.6

SURFACES NOT DISMOUNTABLE

Shaft or bush press fitted into bodies in cold condition	Upto dia. 50	High precision	0.1	0.2
		Precision	0.4	0.8
		Normal	—	—
	Over dia. 50 Upto dia. 120 Over dia. 120		0.8	1.6
Bush press fitted into bodies while hot	Upto dia. 250		1.6	
	Over dia. 250			3.2

SPLINED SHAFTS AND HUBS

Examples	Workmanship	Surface roughness R_a in μm					
		Shaft			Hub		
		a	*b*	*c*	a^I	b^I	c^I
	Internal centring — Shaft sliding in hub	3.2	0.8	0.8	3.2	0.8	1.6
	Shaft fixed in hub	3.2	1.6	1.6	3.2	1.6	1.6

Table 191 Guide for selection of surface roughness (Contd.)

SPLINED SHAFTS AND HUBS

Examples		Workmanship	Surface roughness R_a in μm					
			Shaft			Hub		
			a	b	c	a'	b'	c'
	Centring onside of spline	Shaft sliding in hub	0.8	6.3	6.3	1.6	6.3	6.3
		Shaft fixed in hub	0.8	6.3	6.3	1.6	6.3	6.3

BEDS OF MACHINE TOOLS AND DOVETAIL GUIDEWAYS

Examples	Workmanship		Surface roughness R_a in μm
	Beds of machine tools, guideways, guide gibs (ground), etc.	Guide surfaces located horizontally — With continuous movement	0.4
		With occasional movement	0.8
	Dovetail guide surfaces of cross slides, adjustable gibs (ground)	Guide surfaces located vertically (columns)	0.8

SLIDING FACIAL BEARING SURFACES

Examples	Workmanship		Surface roughness R_a in μm
	Facial bearing surfaces which takeup axial forces	High precision	0.1
		Normal — Peripheral speed above 0,5 m/sec.	0.2
		Peripheral speed upto 0.5 m/sec.	0.4
	Sliding thrust bearings, side surfaces of gears in gear pumps	For occasional slow movement	0.8 (1.6)
		Where accuracy is not required	3.2

Table 191 Guide for selection of surface roughness (Contd.)

SURFACES WITH ROLLING MOTION			
Examples		Workmanship	Surface roughness R_a in μm
	Guideway of outer race	Precision	0.1
		Normal	0.2
	Guideway of needle rollers on shaft	Precision	0.05
		Normal	0.2
	Needle rollers	Precision	0.05
		Normal	0.1
	Surfaces in contact with rolling elements	Precision	0.05
		Normal	0.1–0.2
	Rolling elements	Balls	0.025
		Rollers	0.05
CONTACT SURFACES			
	Contact surfaces of indexing type tool post, clamping surfaces of the table of grinding machine	—	0.8
	Joining surfaces, planed and shaped	Precision	1.6–3.2
		Normal	6.3
	Surfaces for connecting machine parts, flanges of pipe fittings, etc.	Where accuracy is not required	12.5
	Bearing surfaces of gear boxes, covers, etc.	—	1.6
	Base surfaces of motors, brackets, etc.	—	6.3
FACIAL BEARING SURFACES			
	Faces of distance rings, bearing surfaces of washers, etc.	Precision	1.6
		Normal	3.2

Table 191 Guide for selection of surface roughness (Contd.)

FACIAL BEARING SURFACES

Examples		Workmanship	Surface roughness R_a in μm
	Faces of bushes, shoulders of shaft, facial bearing surfaces of antifriction bearings	High precision	0.4
		Precision	0.8
		Normal	**1.6**
	Functional surfaces of stops	Normal	1.6
		High Precision	0.4

TOOTH FLANKS OF WORM AND WORMWHEELS

Examples	Class of accuracy ISO		Workmanship	Peripheral speed in m/s	Surface roughness R_a in μm
	4	a	Worms most carefully ground and lapped	—	0.2
		b	Worm wheels lapped	—	
	6	a	Worms ground and lapped	Above 5 m/s	0.4
		b	Wormwheels scraped		
	7	a	Worms ground (accurately machined on lathe)	Below 5 m/s	0.8
		b	Wormwheels milled with profile ground tools		
	8	a	Worms machined on lathe or milled	Below 3 m/s	3.2
		b	Worm wheel milled with normal tools		
	9	a	Worms milled	Below 1 m/s	6.3
		b	Wormwheel milled with fly cutter		

Table 191 Guide for selection of surface roughness (Contd.)

FLANKS OF LEAD SCREWS

Examples	Workmanship	Screw	Nut
	Precision	0.4	0.8
	Normal	0.8	1.6

BELT TRANSMISSION

Examples		Workmanship	Surface roughness R_a in μm
	Peripheral surfaces of pulleys	Peripheral speed upto 10 m/sec.	6.3–3.2
		Peripheral speed over 10 m/sec.	0.8–1.6
	Grooves for V-belts	Peripheral speed upto 10 m/sec.	1.6
		Peripheral speed over 10 m/sec.	0.8

CAMS, TEMPLATES

Examples				Surface roughness
	Functional surfaces of templates	Tracer		0.4 (0.2)
		Template		0.4
	Functional surfaces of facial and peripheral cams	Follower (roller)	Precision	0.4
			Normal	1.6
		Cam	Precision	0.4
			Normal	1.6
	Follwer in cam grooves, grooves for shifter rings	Follower		0.4–0.8
		Grooves		0.8–1.6

INFLUENCING VISUAL PROPERTIES

Examples			
	Surfaces for graduations and dials	High precision surface for optical reading	0.012
		Precision surface for optical reading	0.05
		Dials for fine setting	0.8
		Dials for rough setting	1.6

Table 191 Guide for selection of surface roughness (Concld.)

SURFACES OF ROTATING PARTS MACHINED FOR MECHANICAL BALANCING			
Examples		**Workmanship**	**Surface roughness R_a in μm**
	Pulleys, side surfaces of gears	—	3.2–12.5
SURFACES OF CONTROL ELEMENTS			
	Surfaces of control elements (levers, cranks, handwheels, etc.)	Before polishing	1.6
		After polishing	0.8
MECHANICALLY MACHINED FROM PRODUCTION POINT OF VIEW			
	Bases of machines	—	12.5–25
	Machine tool tables, angle blocks with T-slots for clamping, etc.	—	0.8–1.6
Machined surfaces for painting			1.6–3.2
Machined surfaces meant for chrome plating, nickel plating and anodising (before polishing)			0.2–0.4

Table 1·2 Surface roughness obtainable from manufacturing processess
(R_a in μm)

Manufacturing process	0·012	0·025	0·05	0·1	0·2	0·4	0·8	1·6	3·2	6·3	12·5	25	50
Shaping – Casting													
Sand casting													
Permanent mould casting													
Die casting													
Shaping – Forging													
Forging													
Rolled surface													
Extrusion													
Machining – General													
Turning													
Boring													
Cylindrical grinding													
Centreless grinding													
Internal grinding													
Surface grinding													
Planing and shaping													
Broaching													
Drilling													
Reaming													
Face milling													
Shell milling													
Lapping													
Honing													
Super finishing													
Machining – Gears													
Lapping													
Grinding													
Grinding-criss cross													
Shaving													
Planing													
Shaping													
Hobbing													
Milling with formed cutters													
Milling of spiral bevel gears													
Machining – Manual													
Filing													
Hand polishing with emery paper													
Chipping													
Cutting with hacksaw													
Cutting with bandsaw													
Flame cutting													
Surface processing													
Grinding with abrasive belt													
Buffing with cloth													
Buffing with fibre wheel													
Shot blasting													
Tumbling													

Legend: ▨ Achieved with difficulty   ▩ Achieved normally   ◹ Roughing.

Table 193 Equivalent surface roughness symbols

Roughness Values Ra μm	Roughness grade number	Roughness grade symbol
50	N12	$\sim$
25	N11	$\triangledown$
12.5	N 10	
6.3	N 9	$\triangledown\triangledown$
3.2	N 8	
1.6	N 7	
0.8	N 6	$\triangledown\triangledown\triangledown$
0.4	N 5	
0.2	N 4	
0.1	N 3	$\triangledown\triangledown\triangledown\triangledown$
0.05	N 2	
0.025	N 1	

Table 194 Sizes for machine tool tables (As per IS 2642 - 1974) (With straight T-slots)

a	p	Table width W for No. of T-slots											
		2	3	4	5	6	7	8	9	10	11	12	13
10	40	80 (11)	125 (13.5)	160 (11)	200 (11)	250 (16)	280 (11)						
10, 12	50	100 (14.5)	160 (19.5)	200 (14.5)	250 (14.5)	320 (24.5)	360 (19.5)	400 (14.5)	450 (14.5)				
10, 12	63	125 (18.5)	200 (24.5)	250 (18)	320 (21.5)	400 (30)	450 (23.5)	500 (17)	560 (15.6)	630 (19)	710 (27.5)		
12, 14, 18	80	160 (24)	250 (29)	320 (24)	400 (24)	500 (34)	560 (24)	630 (19)	710 (19)	800 (24)	900 (34)	1000 (44)	1120 (64)
14, 18, 22	100	200 (30)	320 (40)	400 (30)	500 (30)	630 (45)	710 (35)	800 (30)	900 (30)	1000 (30)	1120 (40)	1250 (55)	1400 (80)
18, 22, 28	125	250 (37.5)	400 (50)	500 (37.5)	630 (40)	800 (62.5)	900 (50)	1000 (37.5)	1120 (35)	1250 (37.5)	1400 (50)	1600 (87.5)	1800 (125)
22, 28, 36	160	320 (50)	500 (60)	630 (45)	800 (50)	1000 (70)	1120 (50)	1250 (35)	1400 (30)	1600 (50)	1800 (70)	2000 (90)	2240 (130)
28, 36, 42	200	400 (64)	630 (79)	800 (64)	1000 (64)	1250 (89)	1400 (64)	1600 (64)	1800 (64)	2000 (64)	2240 (84)	2500 (114)	2800 (164)
36, 42	250	500 (82.5)	800 (107.5)	1000 (82.5)	1250 (82.5)	1600 (132.5)	1800 (107.5)	2000 (82.5)	2240 (77.5)	2500 (82.5)	2800 (107.5)	3150 (157.5)	3550 (232.5)
42	320			1250 (97.5)	1600 (112.5)	2000 (152.5)	2240 (112.5)	2500 (82.5)	2800 (72.5)	3150 (87.5)	3550 (127.5)	4000 (192.5)	4500 (282.5)

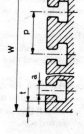

T-slots are generally arranged symmetric about a median slot. However when even number of T-slots are provided a reference T-slot is clearly marked on both the end faces.

Values of resulting edge spacing t corresponding to the maximum value of slot width a for each value of pitch p are given in brackets.

QUALITY OF SCRAPED SURFACES

Table 195   Quality of scraped surfaces - classification

Quality class	Number of contact surfaces	Surface roughness R_a m	Approximate distribution of contact surfaces in 25 x 25 mm square	Applications
1	From 24 Upto 32	≈ 0.2		Precision measuring machines, instruments & gauges.
2	From 13 Upto 23	≈ 0.4		Workshop inspection instruments and tools (surface plate, straight edge etc.) Exceptionally precision guideways of machines (m/c tools, slides of hydraulic equipment), bearing bushes.
3	From 9 Upto 12	≈ 0.8		Workshop jigs & fixtures and instruments. Guideways of machine tools and bearing bushes.
4	From 6 Upto 8	≈ 1.6		Guideways of heavy machine tools, surface of scraped tables.
5	From 3 Upto 5	≈ 3.2		Guideways of heavy machinery, contact surfaces of rotary table, covers of gear boxes, etc.

Notes:
1. Contact surfaces are checked by colour. The size of the surface plate used must be proportional to the size of the scraped surface.
2. The surface roughness of machined surface before scraping must not be coarser than $R_a = 3.2$ μm.
3. In cases of narrow or interrupted surfaces, where the determination of the number of contact surfaces on 25 x 25 mm square is not reliable, they can be determined on a surface of different size and then recalculated to the surface of 25 x 25 mm square.
4. In fourth and fifth quality class of scraped surface, the contact surfaces which protrude half or more of their own area beyond the periphery of measuring square are counted as half surfaces. If they protrude less than half their area, they are counted as whole surfaces.

GUIDEWAYS FOR MACHINE TOOLS

A guideway constrains either the tool or work to move in a defined path, usually either a straight line or a circle. Principal characteristics of the guideways are accuracy of travel, durability and rigidity. Normally friction guideways are used in machine tools, however antifriction as well as hydrostatic guideways are also resorted to in precision machine tools. Different types of friction guideways normally used in light to medium duty machine tools are dealt with in this chapter. Standardised dimensions in *mms* related to preferred guideway heights and distances between the guideways are given for flat, prismatic and dovetail guideways. Gibs, both headed and enclosed type are normally housed in shorter members are used for the purpose of clearance adjustment. They are either straight or have a taper of 1 : 50 or 1 : 100. For taper gibs, care should be taken to see that maximum load is applied to the thick end of gib. Whenever the length of the slide is too long as compared to the thickness and height of the guideway, two separate gibs may be provided one from each end. In dovetail guideways the contact surfaces may be either top and sides (Type TS) or the bottom and sides (Type BS) with clearance in the bottom or in the top respectively. In dovetail guideways with tapered gibs, the gib may be housed in the internal (Types TSI and BSI) or the external guideway (Types TSE and BSE).

Material and hardness requirements are covered in section on *Materials and heat treatment.*

For methods of lubrication of guideways and for details of types of lubrication grooves reference to be made to section on *Lubricants and lubrication.*

Recommendations regarding the quality of scraping are given in page 524.

Checking calculation of guideways

To make the guideways wear resistant, it is essential that the pressure distribution is uniform and the mean pressure should not be more than the values obtained by tests on machine tools. In normal calculations, it is assumed that the pressure distribution follows a linear law along the length of the guideway with a uniform pressure distribution along the width, Fig 147. This assumption is valid only when the rigidity of guideways on beds, columns, carriages, tables or other parts of machine tools considered as

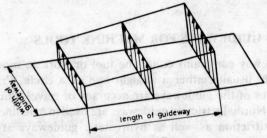

Fig. 147

beams or as thick plates, is many times higher than the rigidity of the contacting surface layers of these guideways. This condition is fulfilled in most of the modern machine tools and the above assumption about pressure distribution is valid.

MAXIMUM PERMISSIBLE PRESSURES ON GUIDEWAY SURFACES

For each contacting surface, the maximum pressure is calculated and compared with the maximum permissible value of pressure as obtained from experimental investigations are given in Table 196.

Table 196 Maximum permissible pressure (p Max.) for cast iron slideways (cast iron sliding on cast iron).

Condition of operation	p Max. kgf/mm^2
At low sliding speeds, in the order of the rates of feed (lathes and milling machines)	0.25 to 0.3
At high sliding speeds, in the order of the cutting speeds (planers, shapers and slotters)	0.08
For heavy machine tools – for slow sliding speeds	0.1
For heavy machine tools – for high sliding speeds	0.04
Slideways of grinding machines	0.005 to 0.008

Notes: 1. For special purpose machine tools, operating at constant heavy feeds and high speeds, the specified values of p Max. will be about 75% of the values indicated.

2. For steel on cast iron guideways, p Max. values are about the same as for cast iron on cast iron.

3. In case of steel ways on steel ways, the permissible values can be increased by 20 to 30 percent.

4. If checking calculations are limited to a determination of only mean pressures, it is recommended that the permissible mean value p_m be taken one half of p Max. values.

Table 197 Survey of types of profiles

Guideways		Applications
External	Internal	
Prismatic symmetric	Prismatic symmetric	For very accurate movement of parts. Self aligning in wear. External type ensures good removal of chips. Internal type retains lubricant.
Prismatic unsymmetric	Prismatic unsymmetric	Have same characteristics as symmetric guideways. Used in such cases where there is an unequal distribution of pressure on guideways. The internal type is a basic type for guideways of parts with rotary movement.
Flat	Flat	For normal accuracy requirements. Setting with straight and tapered gibs. Requires good workmanship and proper protection against chips.
Dovetail	Dovetail	Used in such cases where the height of the guideways is comparatively small. Not suited for cases where the forces try to pull out the guides.
Circular	Circular	Used mainly for axial loading. Easy for manufacture.

Table 198 Recommended distance between guideways (for flat and prismatic guideways)

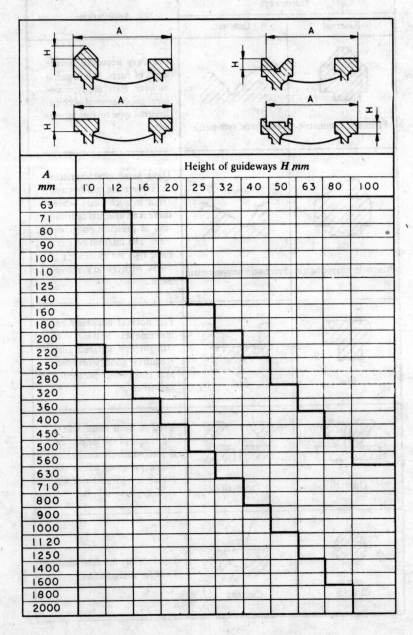

A mm	10	12	16	20	25	32	40	50	63	80	100
					Height of guideways H mm						
63											
71											
80											
90											
100											
110											
125											
140											
160											
180											
200											
220											
250											
280											
320											
360											
400											
450											
500											
560											
630											
710											
800											
900											
1000											
1120											
1250											
1400											
1600											
1800											
2000											

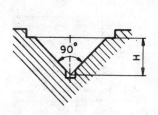 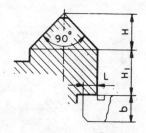

Table 199 Prismatic symmetrical guideways (triangular)

All dimensions in *mm*

H	10	12	16	20	25	32	40	50	63	80	100
	10	14	18	22	28	36	45	56	71	90	110
H_1	12	16	20	25	32	40	50	63	80	100	125
	14	18	22	28	36	45	56	71	90	110	140
Min.* L	5	6	8	10	12	16	20	25	32	40	50
	8	10	12	16	20	25	32	40	50	63	80
b	6	8	10	12	16	20	25	30	40	50	60

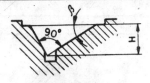

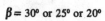

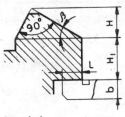

$\beta = 30°$ or $25°$ or $20°$

Table 200 Prismatic unsymmetric guideways (triangular)

All dimensions in *mm*

H	20	25	32	40	50	63	80	100	125	160	200
	22	28	36	45	56	71	90	110	–	–	–
H_1	25	32	40	50	63	80	100	125	–	–	–
	28	36	45	56	71	90	110	140	–	–	–
Min.* L	10	12	16	20	25	32	40	50	60	80	100
	16	20	25	32	40	50	63	80	–	–	–
b	12	16	20	25	30	40	50	60	–	–	–

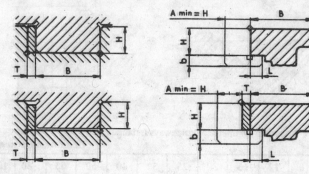

Table 201 Flat guideways   All dimensions in *mm*

H	10	12	16	20	25	32	40	50	63	80	100
	16	20	25	32	40	50	63	80	100	125	160
	20	25	32	40	50	63	80	100	125	160	200
B	25	32	40	50	63	80	100	125	160	200	250
	32	40	50	63	80	100	125	160	200	250	320
	40	50	63	80	100	125	160	200	250	320	400
T	3	4	5	5	6	8	10	12	16	20	25
Min.* L	5	6	8	10	12	16	20	25	32	40	50
	8	10	12	16	20	25	32	40	50	63	80
b	6	8	10	12*	16	20	25	30	40	50	60

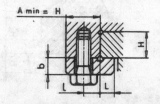

Table 202 Guide strips for flat guideways (without gibs)   All dimensions in *mm*

H	10	12	16	20	25	32	40	50	63	80	100
d	M5	M6	M6	M8	M10	M12	M12	M16	M16	M20	M24
Min.* L	5	6	8	10	12	16	20	25	32	40	50
	8	10	12	16	20	25	32	40	50	63	80
b	6	8	10	12	16	20	25	30	40	50	60
Min.* L_2	15	18	24	30	37	48	60	75	95	120	150
	18	22	28	36	45	57	72	90	113	143	180
l	4	5	5	6	8	10	10	12	12	15	18

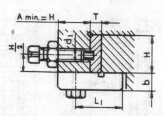
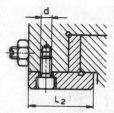

Table 203 Guide strip for flat guideways (with straight gibs)   All dimensions in *mm*

H	10	12	16	20	25	32	40	50	63	80	100
b	6	8	10	12	16	20	25	30	40	50	60
d	M6	M6	M6	M8	M10	M12	M12	M16	M16	M20	M24
d_1	M4	M5	M6	M8	M10	M12	M12	M16	M16	M20	M20
T	3	4	5	5	6	8	10	12	16	20	25
Min.* L_1	12	15	18	21	26	34	40	49	60	75	93
	15	19	22	27	34	43	52	64	78	98	123
Min.* L_2	18	22	29	35	43	56	70	87	111	140	175
	21	26	33	41	51	65	82	102	129	163	205

* Min. values are used where unit pressure on surface L is very small

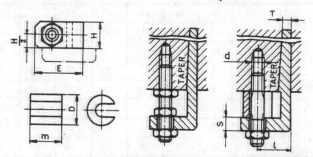

Table 204 Headed type taper gibs for flat guideways   All dimensions in *mm*

H	16	20	25	32	40	50	63	80	100
d	M8	M10	M10	M12	M12	M16	M16	M20	M20
T	5	5	6	8	10	12	16	20	25
l	20	22	24	28	30	38	46	58	65
E	30	32	35	40	42	55	62	78	85
S	8	10	10	12	12	16	16	20	20
Taper	1 : 50 for $L/H < 10$; 1 : 100 for $L/H > 10$ where L = Length of slide								
D	18	20	20	22	22	28	28	35	35
m	20	25	25	32	32	40	40	50	50

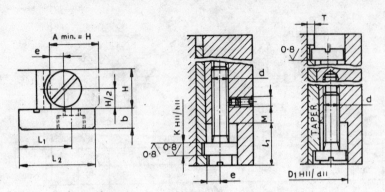

Set up of enclosed type taper gibs for flat guideways

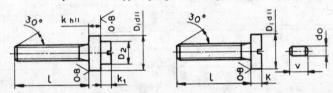

Table 205 Screws for gibs and locking inserts   All dimensions in *mm*

H	16	20	25	32	40	50	63	80	100
d	M6	M8	M8	M10	M12	M16	M16	M20	M24
D_1	14	18	18	25	32	40	40	50	63
l_1	25	25	25	30	35	40	40	45	50
K	5	6	6	8	8	10	10	12	12
L_1	24	30	34	42	56	68	80	100	125
L_2	32	42	50	62	80	100	120	150	190
b	10	12	16	20	25	30	40	50	60
e	5	6	6	8	10	12	12	14	16
M	M5	M6	M6	M8	M10	M12	M12	M16	M16
T	5	5	6	8	10	12	16	20	25
Taper	1 : 50 for L/H < 10, 1 : 100 for L/H ≥ 10 Where L is length of slide									
l	30	35	35	40	45	55	55	65	70
D_2	9	10	10	14	18	22	22	26	30
K_1	4	5	5	6	6	6	6	8	8
d_0	3.5	4.2	4.2	5.8	7.5	9	9	12	12
V	Length to suit dimension A and length of screw									

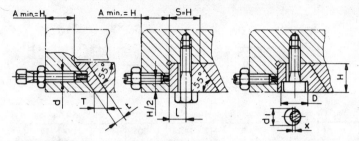

Table 206 Set up of straight gibs for dovetail guideways All dimensions in *mm*

H	12	16	20	25	32	40	50	63	80
T	6	8	10	12	16	20	25	–	–
t*	4.915	6.553	8.192	9.83	13.106	16.383	20.479	–	–
d	M5	M6	M8	M10	M12	M12	M16	M16	M20
l	–	9	12	14	16	20	25	32	40
D	–	14	18	20	22	22	30	30	35
d₁	–	8	10	12	14	14	18	18	23
x	–	1	1	1	1	1	1	1	1.5

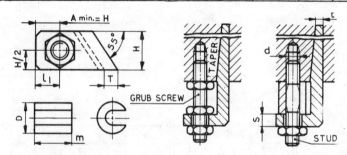

Table 207 Headed type taper gibs for dovetail guideways** All dimensions in *mm*

H	16	20	25	32	40	50	63	80
T	5	5	6	8	10	12	16	20
t*	4.096	4.096	4.915	6.553	8.192	9.83	13.106	16.385
S	8	10	10	12	12	16	16	20
d	M8	M10	M10	M12	M12	M16	M16	M20
l₁	10	12	16	20	20	25	25	32
Taper	1 : 50 for L/H<10, 1 : 100 for L/H≥10 where L= Length of slide							
D	18	20	20	22	22	28	28	35
m	20	25	25	32	32	40	40	50

* Theoretical values of gib thickness without the dovetail guideway tolerance and grinding allowance.
** Taper normal to gib thickness. 1 : 61.039 for 1 : 50 and 1 : 122.078 for 1 : 100

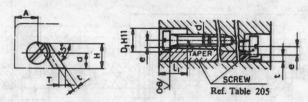

Ref. Table 205

Table 208 Set up of taper gibs for dovetail guideways-enclosed type   **All dimensions in** *mm*

H	6	8	10	12	16	20	25	32	40	50
a	8	8	8	10	10	12	15	20	25	30
A	12	12	16	16	16	20	25	32	40	50
d	M5	M5	M5	M6	M6	M8	M8	M10	M12	M16
D_1	12	12	12	14	14	18	18	25	32	40
e	4	4	4	5	5	6	6	8	10	12
l_1	15	15	15	25	25	25	25	30	35	40
Taper 1 : 50 for $L/H < 10$; 1 : 100 for $L/H \geq 10$ where L = Length of slide										
T	3	3	4	4	5	5	6	8	10	12
t^*	2.457	2.457	3.277	3.277	4.096	4.096	4.915	6.553	8.192	9.83

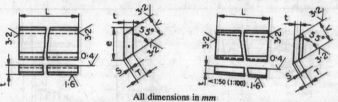

All dimensions in *mm*

Table 209 Straight gibs for dovetail guideways   **Table 210** Tapered gibs for dovetail guideways

V	T	t^*	e	S	V	T	t^*	S
6	3	2.457	9.045	1	6	3	2.457	1
8	4	3.277	12.061	1	8	3	2.457	1
10	5	4.096	15.076	1	10	4	3.277	1
12	6	4.915	18.091	1	12	4	3.277	1
16	8	6.553	24.121	1	16	5	4.096	1
20	10	8.192	30.151	1.5	20	5	4.096	1.5
25	12	9.83	37.402	2	25	6	4.915	2
32	16	13.106	48.242	2	32	8	6.553	2
40	20	16.383	60.303	3	40	10	8.192	3
50	25	20.479	75.378	3	50	12	9.83	3
63	30	24.574	94.116	3	63	16	13.106	3

Theoritical values of gib thickness without the dovetail guideway tolerance and grinding allowance.

Table 211 Recommended combinations of width A and height V or V_1 of dovetail guide ways .

All dimensions in *mm*

A \ V	6	8	10	12	16	20	25	32	40	50	63	80
V_1	6.5	8.5	10.5	12.5	16.5	20.5	25.5	33	41	51	64	81
32												
36												
40												
45												
50												
56												
63												
70												
80												
90												
100												
110												
125												
140												
160												
180												
200												
220												
250												
280												
315												
355												
400												
450												
500												
560												
630												
710												
800												

INSPECTION OF DOVETAIL GUIDEWAYS

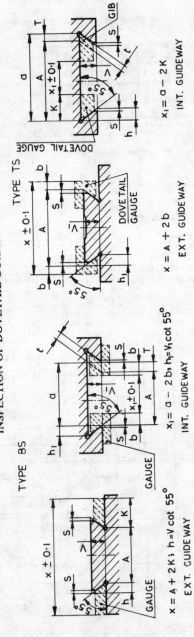

TYPE BS

$x = A + 2K;\ h = V\cot 55°$
EXT. GUIDEWAY

$x_1 = a - 2b;\ h_1 = V_1 \cot 55°$
INT. GUIDEWAY

TYPE TS

$x = A + 2b$
EXT. GUIDEWAY

$x_1 = a - 2K$
INT. GUIDEWAY

Table 212 Measurement of dovetail guideways with straight gibs

V	6	8	10	12	16	20	25	32	40	50	63	80
$2K$	20	20	30	30	50	50	60	60	100	100	160	160
h	4.201	5.602	7.002	8.403	11.203	14.004	17.505	22.407	28.009	35.011	44.113	56.017
$2b$	20	20	30	30	40	50	50	60	80	80	100	100
h_1	4.551	5.952	7.352	8.753	11.553	14.564	18.065	23.107	28.709	35.711	44.814	56.717

All dimensions in *mm*

Tolerance on angle 55° is ±0.03/100. For dimensions T, t and S Refer Table 209

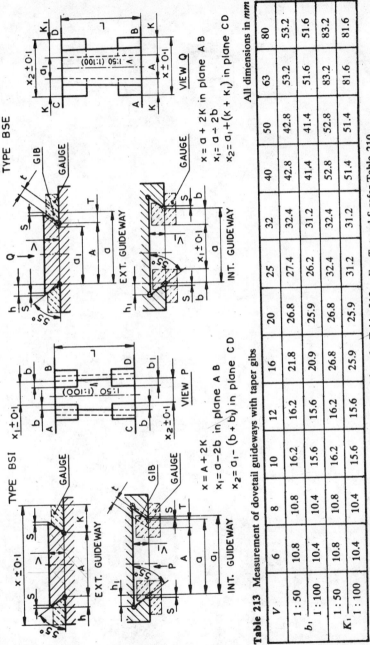

TYPE BSE

$x = a + 2K$ in plane A B
$x_1 = a + 2b$
$x_2 = a_1 + (k + k_1)$ in plane C D

TYPE BSI

$x = A + 2K$
$x_1 = a - 2b$ in plane A B
$x_2 = a_1 - (b + b_1)$ in plane C D

All dimensions in mm

Table 213 Measurement of dovetail guideways with taper gibs

V		6	8	10	12	16	20	25	32	40	50	63	80
b_1	1:50	10.8	10.8	16.2	16.2	21.8	26.8	27.4	32.4	42.8	42.8	53.2	53.2
	1:100	10.4	10.4	15.6	15.6	20.9	25.9	26.2	31.2	41.4	41.4	51.6	51.6
K_1	1:50	10.8	10.8	16.2	16.2	26.8	26.8	32.4	32.4	52.8	52.8	83.2	83.2
	1:100	10.4	10.4	15.6	15.6	25.9	25.9	31.2	31.2	51.4	51.4	81.6	81.6

For T, t and S refer Table 210

Tolerance on angle 55° is ±0.03/100. For 2K, 2b, h and h_1 refer Table 212

Measurement of dovetail guideways with taper gibs

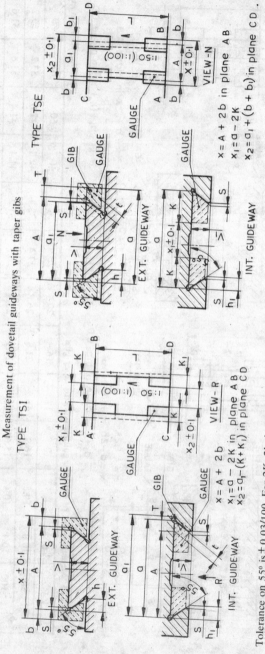

TYPE TSE

VIEW-N

$x = A + 2b$ in plane AB.

$x_1 = a - 2K$

$x_2 = a_1 + (b + b_1)$ in plane CD.

TYPE TSI

VIEW-R

$x = A + 2b$

$x_1 = a - 2K$ in plane AB

$x_2 = a_1 - (K + K_1)$ in plane CD

Tolerance on 55° is ± 0.03/100. For 2K, 2b, h and h_1 refer Table 212

For K_1 and b_1 refer Table 213

For T, t and S refer Table 210

MACHINE TOOL SPINDLES

Nomenclature

P Load, *kgf*

δ Deflection at the point of application of load P, *mm*

δ_1 Deflection due to radial yielding of the bearings, *mm*

δ_2 Deflection due to elastic bending of the spindle, *mm*

a Length of the overhanging portion of spindle, *mm*

S $=P/\delta=$Overall stiffness of the spindle unit, *kgf/mm*

S_A Radial stiffness of the bearing near the load point, *kgf/mm*.

S_B Radial stiffness of the bearing away from the load point, *kgf/mm*

I_a Moment of inertia of overhang portion of the spindle, mm^4

I_L Moment of inertia of spindle section between bearings, mm^4

E Modulus of elasticity of spindle material, kgf/mm^2

L Bearing span, *mm*

L_0 Static optimum bearing span, *mm*

Q Trial value for iterative determination of L_0

Rigidity calculations

A spindle provides drive to either the workpiece or the tool depending upon the type of machine tool.

The accuracy with which a spindle runs is affected by the elastic deformation of the spindle, its bearings, housing and other components of the arrangement. The stiffness of the bearings, the spindle diameter, all have an influence on the overall stiffness of the spindle system.

While the stiffness of the housing is an important factor affecting the overall stiffness of the spindle system, normally the housings when compared to other elements of the spindle arrangements are quite stiff and hence its effect is not discussed here.

The overhang has a great influence on the stiffness of the spindle, lesser the overhang the better it is. The influence of front bearing stiffness on overall stiffness is quite considerable. Hence a bearing with higher stiffness should be located at the front.

Ignoring the effects of housing deformation on the spindle, the total deflection of the spindle unit is due to the elastic deformation δ_2 of the spindle itself together with δ_1 the deflection caused by elastic deformation of the bearings. The total deflection of the bearing system due to load P at the point of application of load is given by

$$\delta = \delta_1 + \delta_2 = P\left[\frac{1}{S_A}\left(\frac{a+L}{L}\right)^2 + \frac{1}{S_B}\left(\frac{a}{L}\right)^2 + \frac{a^2}{3E}\left(\frac{L}{I_L} + \frac{a}{I_a}\right)\right]$$

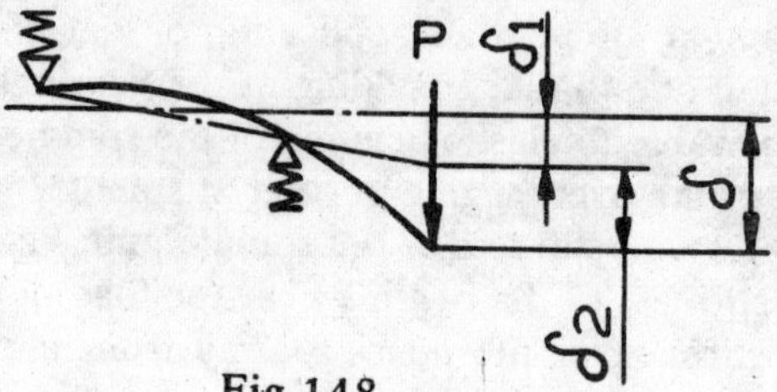

Fig.148

In the above equation if it is assumed that the bearings, the diameter of spindle and the overhang of the spindle are fixed, it can be seen that the stiffness becomes a function of bearing span only. Thus the optimum distance between the bearings, based on considerations of static deflection which gives the maximum stiffness to the spindle system, is given by

$$L_o \approx \left[6EI_L\left(\frac{1}{S_A} + \frac{1}{S_B}\right) + \left(\frac{6EI_L}{aS_A}\right)Q\right]^{1/3}$$

This equation for optimum bearing span could be solved by iteration by first taking $Q = 4a$ and then equal to the derived value of L_o for subsequent iterations.

It has been observed that the effect of change in stiffness of the spindle system is less when the bearing span exceeds the optimum, than when it is less than the optimum. An increase of about 20 percent on the bearing span reduces the spindle stiffness only by about 4 per cent. It is recommended to maintain the bearing span within these limits. However, it is not always possible to maintain the bearing span within these limits, because of other design considerations. In such cases a third bearing will have to be used despite the system becoming a statically indeterminate structure combined with the increased misalignment of the bores.

The journal bearings, both hydrodynamic and hydrostatic, as well as all types of antifriction bearings find considerable application in machine tools.

Spindle bearing arrangements using different types of antifriction bearings commonly used in machine tools are discussed in chapter *Bearing mounting arrangements.*

The hydrodynamic and hydrostatic bearings wherever used are to be specially designed and developed for a particular application. Design details of hydrodynamic bearings are discussed in chapter *Hydrodynamic bearings*

Different types of spindle noses adopted in machine tools are covered in pages 542 to 588.

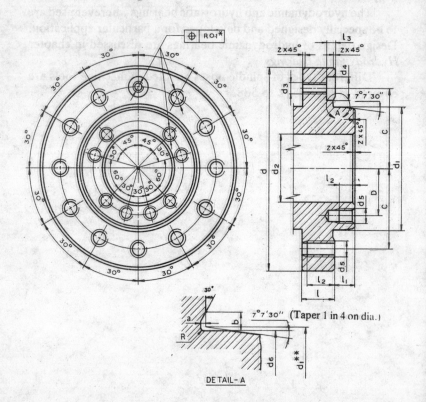

DETAIL-A

Fig. 149   Lathe spindle nose type A_1 size No. 5 to 11

Table 214 Dimensions for Lathe Spindle Noses Type A_1
(as per IS 2582 Part I -1972)

All dimensions in *mm*

Size No.	5	6	8	11
d	133	165	210	280
d_1	82.563 $\left(\begin{array}{c}+0.01\\0\end{array}\right)$	106.375 $\left(\begin{array}{c}+0.01\\0\end{array}\right)$	139.719 $\left(\begin{array}{c}+0.012\\0\end{array}\right)$	196.869 $\left(\begin{array}{c}+0.014\\0\end{array}\right)$
d_2 Max.	40 or Morse 5	56 or Morse 6	80 or Metric 80	125 or Metric 120
d_3	M6	M8	M8	M10
d_4 $H8$	15.9	19.05	23.8	28.6
d_5	M10	M12	M16	M18
d_6	82	106	139	196
a	1	1	1	1
b	6	8	8	8
C	52.4	66.7	85.7	117.5
D	30.95	41.3	55.55	82.55
$l_1\left[\begin{array}{c}0\\-0.025\end{array}\right]$	14.288	15.875	17.462	19.05
l_2	19	22	25	32
l_3	6	8	10	12
l	22	25	28	35
R	1	1	1	1
Z	1	1.6	1.6	1.6

Notes

1. Type A_1 spindle nose is provided with two rows of tapped holes and a driving button. The inner row of tapped holes provides means for attaching certain sizes of scroll chucks by the use of hexagon socket head cap screws which pass inside the scroll plate between the chuck jaws. The outer row of tapped holes provides means for mounting face plates, fixtures and other chucks by the use of hexagon socket head cap screws.

2. *Tolerance on C, D and angular dimensions is controlled by this position tolerance which is the permissible deviation of the hole centres with respect to their theoretical positions.

3. **Dimension d_1 is taken at the theoretical point of intersection between generating line of the cone and the face of the flange.

4. Tolerance on untoleranced dimensions : ± 0.4

5. IS 2582 (Part I)-1972 specifies dimensions for size Nos. 15 to 28 also.

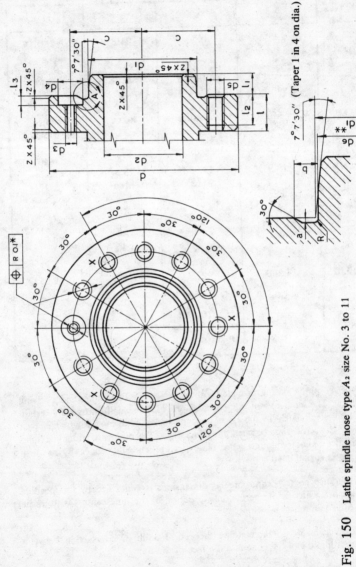

Fig. 150   Lathe spindle nose type A₂ size No. 3 to 11

Table 215 Dimensions for lathe spindle noses type A_2
(as per IS 2582 Part I -1972)　　　All dimensions in *mm*

Size No.	3	4	5	6	8	11
d	92	108	133	165	210	280
d_1	$53.975 \left(\begin{matrix} +0.008 \\ 0 \end{matrix} \right)$	$63.513 \left(\begin{matrix} +0.008 \\ 0 \end{matrix} \right)$	$82.563 \left(\begin{matrix} +0.01 \\ 0 \end{matrix} \right)$	$106.375 \left(\begin{matrix} +0.01 \\ 0 \end{matrix} \right)$	$139.719 \left(\begin{matrix} +0.012 \\ 0 \end{matrix} \right)$	$196.869 \left(\begin{matrix} +0.014 \\ 0 \end{matrix} \right)$
d_2 Max.	32 or Morse 4	40 or Morse 4	50 or Morse 5	71 or Metric 80	100 or Metric 100	150 or Metric 160
d_3	—	M6	M6	M8	M8	M10
d_4 $H8$	—	14.25	15.9	19.05	23.8	28.6
d_5	M10	M10	M10	M12	M16	M18
d_6	53.5	63	82	106	139	196
a	1	1	1	1	1	1
b	5	6	6	8	8	8
C	35.3	41.3	52.4	66.7	85.7	117.5
l_1	11	11	13	14	16	18
l_2	14	17	19	22	25	32
l_3	—	5	6	8	10	12
l	16	20	22	25	28	35
R	1	1	1	1	1	1
Z	1	1	1	1.6	1.6	1.6

Notes:

1. Type A_2 spindle nose is similar to Type A_1 except that the holes in the inner row are omitted. Type A_2 spindle nose may be used when the bore in the spindle nose is so large that there is no space left for providing the inner row of holes for screws and where the nature of job is such that it does not require the inner row of holes for screws for holding it.
   Type A_2 provides for mounting face plates, chucks and similar fixtures by the use of hexagon socket head cap screws.

2. For size 3, only the three holes marked X are provided.

3. *Tolerance on C and angular dimensions is controlled by this position tolerance which is the permissible deviation of the hole centres with respect to their theoretical positions.

4. **Dimension d_1 is taken at the theoretical point of intersection between the generating line of the cone and the face of the flange.

5. Tolerance on untoleranced dimensions : ± 0.4

6. IS 2582 (Part I)-1972 specifies dimensions for size Nos. 15 to 28 also.

19 (4

Table 217 Dimensions for driving button for lathe spindle noses

Nominal size	4	5	6	8	11	15	20
$D\,h8$	14.25	15.9	19.05	23.8	28.6	34.9	41.3
$D_1\,H12$	10.4	10.4	13.5	13.5	16.5	18.5	18.5
D_2	–	–	–	–	–	16	16
$d\,H12$	6.4	6.4	8.4	8.4	10.5	13	13
H	10	11	13	16	20	20	24
h	7	7	9	9	11	13	13
k	1	1	1.6	1.6	1.6	2	2
Size of screws (as per IS 2269)	M6 x 14	M6 x 14	M8 x 20	M8 x 20	M10 x 25	M12 x 25	M12 x 30

All dimensions in *mm*

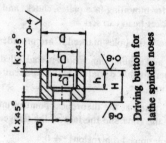

Driving button for lathe spindle noses

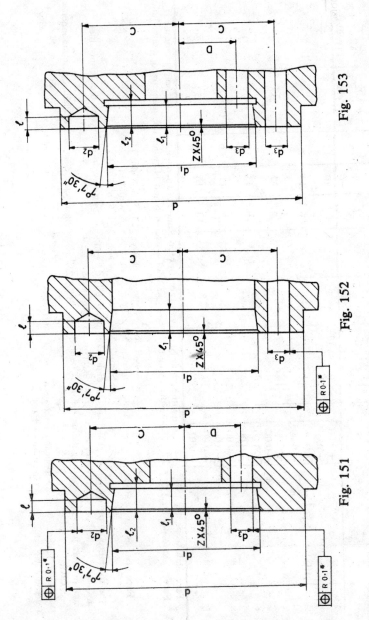

Fig. 153

Fig. 152

Fig. 151

Face plates for type A lathe spindle noses

Table 216 Dimensions for face plates for lathe spindle noses type A (as per IS 2582 Part 1-1972)

All dimensions in mm

Nominal size	d	d_1	$d_2 \begin{bmatrix} +0.1 \\ 0 \end{bmatrix}$	d_3	C	D	l	l_1	z	$l_2 \begin{matrix}+0.025\\0\end{matrix}$ for Type A_1	Min. for @ Type A_2
3	92	53.975 $\begin{matrix}+0.003\\-0.005\end{matrix}$	—	12	35.3	—	—	10	1	—	—
4	108	63.513 $\begin{matrix}+0.003\\-0.005\end{matrix}$	14.7	12	41.3	—	6.5	10	1	—	—
5	133	82.563 $\begin{matrix}+0.004\\-0.006\end{matrix}$	16.3	12	52.4	30.95	6.5	12	1	14.288	15
6	165	106.375 $\begin{matrix}+0.004\\-0.006\end{matrix}$	19.45	14	66.7	41.3	6.5	13	1	15.875	16
8	210	139.719 $\begin{matrix}+0.004\\-0.008\end{matrix}$	24.2	18	85.7	55.55	8	14	1.6	17.462	18
11	280	196.869 $\begin{matrix}+0.004\\-0.01\end{matrix}$	29.4	20	117.5	82.55	10	16	1.6	19.05	20

1. a) Face plate as per Fig.151   Used with Spindle Nose Type A_1 of sizes 5, 6, 8 and 11 (when the outer row of holes is not used)
   b) Face plate as per Fig.152   Used with i) Spindle Nose Type A_1 of sizes 5, 6, 8 and 11 (when the inner row of tapped holes is not used)
   ii) Spindle Nose Type A_2 of sizes 3, 4, 5, 6, 8 and 11
   c) Face plate as per Fig.153   Used with i) Spindle Nose Type A_1 of sizes 5, 6, 8 and 11
   ii) Spindle Nose Type A_2 of sizes 3, 4, 5, 6, 8 and 11

2. *The tolerance on dimensions C, D and angular dimensions is controlled by this position tolerance which is the permissible deviation of the hole centres with respect to their theoretical positions.

3. @ and possibly for type A_1 also, if the face plate is rigid enough not to risk bending when the screws are clamped on the inner bolt circle

4. The mounting dimensions are applicable for direct mounting lathe chucks also.

5. Tolerance on untoleranced dimensions : ±0.4

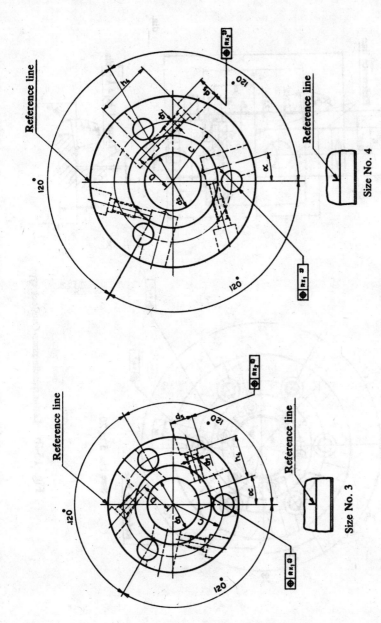

Fig. 154a Lathe spindle nose–Camlock type

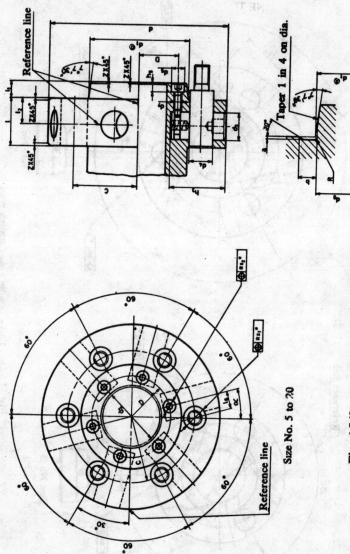

Size No. 5 to 20

Fig. 154b   Lathe spindle nose–Camlock type

Table 218 Dimensions for camlock type lathe spindle noses (as per IS 2582 part II — (1972))

Size No.	3	4	5	6	8	11	15	20
d	92	117	146	181	225	298	403	546
d_1	53.975 [+0.008/0]	63.513 [+0.008/0]	82.563 [+0.01/0]	106.375 [+0.01/0]	139.719 [+0.012/0]	196.869 [+0.014/0]	285.775 [+0.016/0]	412.775 [+0.02/0]
a_2 Max.	32 or Morse 4	38 or Morse 4	45 or Morse 5	65 or Morse 6	85 or Metric 80	135 or Metric 120	210	330
d_3 H8	19	19	22	26	29	32	35	42
d_4 [+0.05/0]	15.1	16.7	19.8	23	26.2	31	35.7	42.1
d_5	15.5	15.5	10.5	13.5	13.5	13.5	16.5	16.5
d_6	53.5	63	82	106	139	196	285	412
d_7	M8	M8	M6	M8	M8	M8	M10	M10
C	35.3	41.3	52.4	66.7	85.7	117.5	166.1	231.8
D	22.6	27	32.5	41	57	86	129	190
$\propto$	18° 18′ 6″	15° 36′	14° 55′	13° 46′	12° 18′	10° 30′	8° 35′	7° 5′
x_1	0.05	0.075	0.075	0.075	0.075	0.075	0.075	0.075
x_2	0.05	0.05	0.1	0.1	0.1	0.1	0.1	0.1
a	1	1	1	1	1	1	1	1
b	5	6	6	8	8	8	8	9
R	1	1	1	1	1	1	1	1
Z	1	1	1	1.6	1.6	1.6	2	2
l_1	11	11	13	14	16	18	19	21
l_2	17.5	17.5	20.6	23.8	27	31.8	36.5	42.9
h_3 [+0.2/0]	30	36	46	57	64	75	84	94
h_4 [±0.2]	—	40	—	—	—	—	—	—
h_5	—	—	7	9	9	9	11	11
l_6 [±0.1]	11.1	11.1	13.5	15.9	18.25	21.45	24.6	28.6
l Min.	32	34	38	45	50	60	70	82

Notes:

1.* Tolerance on C, D and angular dimensions is controlled by this position tolerance which is the permissible deviation of the hole centres with respect to their theoretical positions.

2.@ Dimension d_1 is taken at the theoretical point of intersection between the generating line of the cone and the face of the flange.

3. Tolerance on untoleranced dimensions: ± 0.4

All dimensions in *mm*

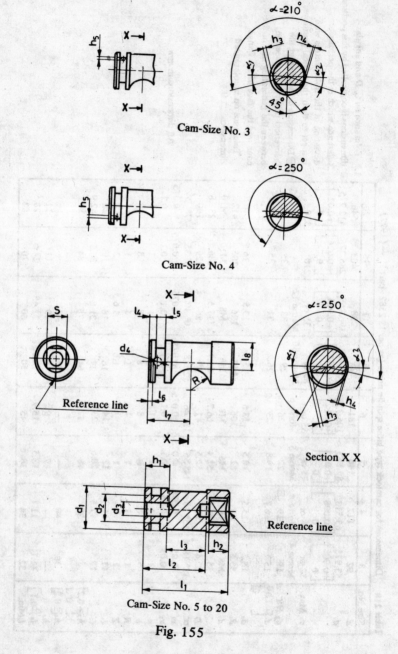

Cam-Size No. 3

Cam-Size No. 4

Reference line

Section X X

Reference line

Cam-Size No. 5 to 20

Fig. 155

Table 219 Dimensions for cams for camlock type lathe spindle noses
(as per IS 2582 Part II - 1972)   All dimensions in mm

Size No.	3	4	5	6	8	11	15	20
$d_1 e8$	19	19	22	26	29	32	35	42
d_2	13±0.2	13±0.2	14	17	21	24	27	33
d_3	—	—	7	10	10	10	10	10
$d_4 \pm 0.05$	—	—	4.5	6	6	6	8	8
$l_1 \begin{bmatrix} 0 \\ -0.1 \end{bmatrix}$	26.5	35	45	56	63	73	82	92
l_2	21.4	26.5	35	43	49	59	62	69
l_3 Min.	13	17	22	25	28	32	37	43
$l_4 \pm 0.1$	2.2	2.2	3	4.2	5.3	8.7	6	6
$l_5 \pm 0.05$ ⎱	3.6	3.6	—	—	—	—	—	—
$l_5 \pm 0.1$ ⎰	—	—	5	6.5	6.5	6.5	8.5	8.5
$l_6 \pm 0.1$	—	—	2	2.85	3.95	7.35	5.2	5.2
$l_7 \pm 0.2$	14.9	16.7	22.4	30.2	33.2	39.5	43.6	48.4
$l_8 \begin{bmatrix} 0 \\ -0.2 \end{bmatrix}$	13.4	11.9	14.2	16.7	18.9	21.2	23.5	27.8
R	7.5	9.5	11.1	12.7	14.2	16.7	19	22.2
$S D12$	8	10	11	12	14	17	17	22
h_1	—	—	13	15	15	15	15	15
h_2	8	9	11	12	14	16	16	20
$h_3 \begin{bmatrix} +0.3 \\ 0 \end{bmatrix}$	1.65	1.6	1.45	2.56	2.46	2.44	2.35	3.1
$h_4 \begin{bmatrix} +0.1 \\ 0 \end{bmatrix}$	0.15	0.15	0	0.45	0.36	0.28	0.2	0.5
h_5	1.2	1.2	—	—	—	—	—	—
$\propto_1$	15°	15°	15°	20°	20°	20°	20°	20°
$\propto_2$	15°	10°	10°	10°	10°	15°	15°	15°
Slope on $\propto$	1.6	1.9	1.9	2.64	2.64	2.64	2.64	3.18

Notes: 1.   * See tolerances on h_3 and h_4
2.   Tolerance on untoleranced dimensions : ±0.4

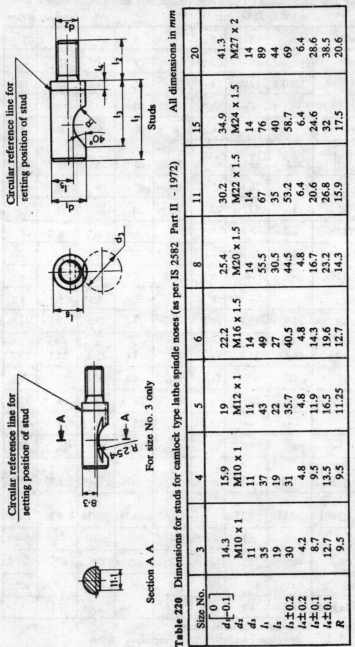

Circular reference line for setting position of stud

Studs

Circular reference line for setting position of stud

Section A A　　For size No. 3 only

Table 220　Dimensions for studs for camlock type lathe spindle noses (as per IS 2582 Part II – 1972)

All dimensions in mm

Size No.	3	4	5	6	8	11	15	20
$d_1 \begin{bmatrix} 0 \\ -0.1 \end{bmatrix}$	14.3	15.9	19	22.2	25.4	30.2	34.9	41.3
d_2	M10 x 1	M10 x 1	M12 x 1	M16 x 1.5	M20 x 1.5	M22 x 1.5	M24 x 1.5	M27 x 2
d_3	11	11	11	14	14	14	14	14
l_1	35	37	43	49	55.5	67	76	89
l_2	19	19	22	27	30.5	35	40	44
$l_3 \pm 0.2$	30	31	35.7	40.5	44.5	53.2	58.7	69
$l_4 \pm 0.2$	4.2	4.8	4.8	4.8	4.8	6.4	6.4	6.4
$l_5 \pm 0.1$	8.7	9.5	11.9	14.3	16.7	20.6	24.6	28.6
$l_6 \pm 0.1$	12.7	13.5	16.5	19.6	23.2	26.8	32	38.5
R	9.5	9.5	11.25	12.7	14.3	15.9	17.5	20.6

Note: Tolerance on untoleranced dimensions ± 0.4

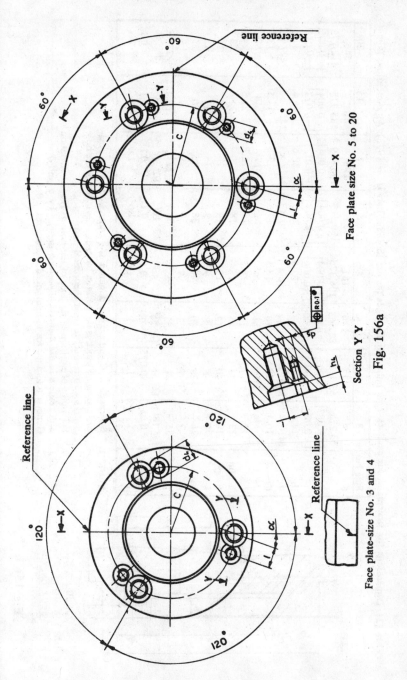

Face plate size No. 5 to 20

Face plate-size No. 3 and 4

Section Y Y

Fig. 156a

Table 221 Dimensions for face plates for camlock type lathe spindle noses (as per IS 2582 Part II — 1972)   All dimensions in mm

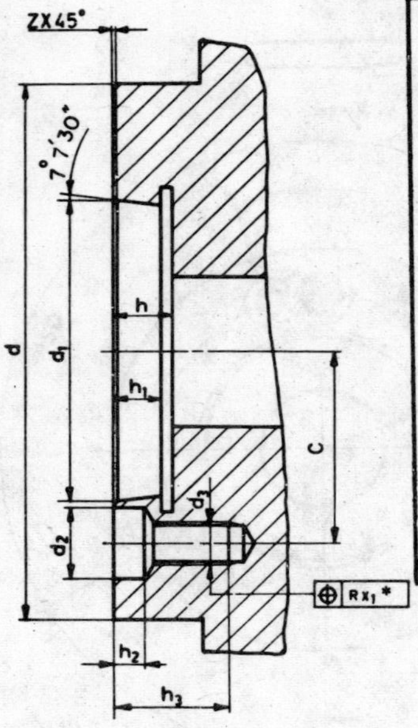

Size No.	3	4	5	6	8	11	15	20
d	92	117	146	181	225	298	403	546
d_1	53.975 $\begin{bmatrix}+0.003\\-0.005\end{bmatrix}$	63.513 $\begin{bmatrix}+0.003\\-0.005\end{bmatrix}$	82.563 $\begin{bmatrix}+0.004\\-0.006\end{bmatrix}$	106.375 $\begin{bmatrix}+0.004\\-0.006\end{bmatrix}$	139.719 $\begin{bmatrix}+0.004\\-0.008\end{bmatrix}$	196.869 $\begin{bmatrix}+0.004\\-0.01\end{bmatrix}$	285.775 $\begin{bmatrix}+0.004\\-0.012\end{bmatrix}$	412.775 $\begin{bmatrix}+0.005\\-0.015\end{bmatrix}$
d_2	14.6	16.2	19.4	22.6	25.8	30.6	35.4	41.6
d_3	M10 x 1	M10 x 1	M12 x 1	M16 x 1.5	M20 x 1.5	M22 x 1.5	M24 x 1.5	M27 x 2
d_4	10.5	10.5	10.5	13.5	13.5	13.5	13.5	13.5
d_5	M6	M6	M6	M8	M8	M8	M8	M8
l	11	11	12.5	15.5	17.5	18.7	21.5	24.8
h Min.	13	13	15	16	18	20	21	23
h_1	10	10	12	13	14	16	17	19
h_2	7	8	8	9.5	9.5	13	13	13
h_3	26	28	30	35	38	45	50	55
h_4	7	7	7	9	9	9	9	9
C	35.3	41.3	52.4	66.7	85.7	117.5	165.1	231.8
α	18° 18' 6"	15° 36'	14° 55'	13° 46'	12° 18'	10° 30'	8° 35'	7° 5'
x_1	0.05	0.075	0.1	0.1	0.1	0.1	0.1	0.1
z	1	1	1	1.6	1.6	1.6	2	2

Notes: 1. * Tolerance on C and angular dimensions is controlled by this position tolerance which is the permissible radial deviation of the hole centres with respect to their theoretical positions.

     @ Tolerance of position of the axis of the hole d_5 (radial deviation with respect to the theoretical position defined by α and l)

     2. General tolerance for untoleranced dimensions: ± 0.4

     3.

SECTION X X

Fig. 156 b

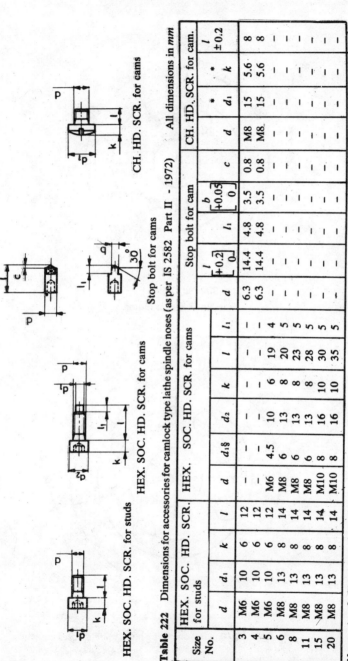

HEX. SOC. HD. SCR. for studs

HEX. SOC. HD. SCR. for cams

Stop bolt for cams

CH. HD. SCR. for cams

Table 222 Dimensions for accessories for camlock type lathe spindle noses (as per IS 2582 Part II - 1972)   All dimensions in *mm*

Size No.	HEX. SOC. HD. SCR. for studs				HEX. SOC. HD. SCR. for cams						Stop bolt for cam					CH. HD. SCR. for cam			
	d	d_1	k	l	d	d_1§	d_2	k	l	l_1	d	$l\,{}^{+0.2}_{0}$	l_1	$b\,{}^{+0.05}_{0}$	c	d	d_1*	k*	l ±0.2
3	M6	10	6	12	—	—	—	—	—	—	6.3	14.4	4.8	3.5	0.8	M8	15	5.6	8
4	M6	10	6	12	—	—	—	—	—	—	6.3	14.4	4.8	3.5	0.8	M8	15	5.6	8
5	M6	10	6	12	M6	4.5	10	6	19	4	—	—	—	—	—	—	—	—	—
6	M8	13	8	14	M8	6	13	8	20	5	—	—	—	—	—	—	—	—	—
8	M8	13	8	14	M8	6	13	8	23	5	—	—	—	—	—	—	—	—	—
11	M8	13	8	14	M8	6	13	8	28	5	—	—	—	—	—	—	—	—	—
15	M8	13	8	14	M10	8	16	10	30	5	—	—	—	—	—	—	—	—	—
20	M8	13	8	14	M10	8	16	10	35	5	—	—	—	—	—	—	—	—	—

Notes: 1. § The values indicated for d_1 are maximum.   2. * The values indicated for d_1 and k are maximum values.

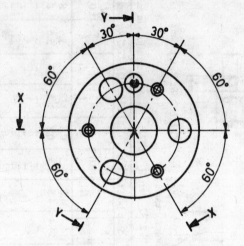

Size No. 3 and 4
Size No. 3 has no button

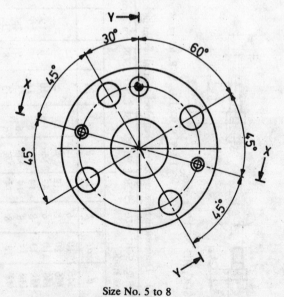

Size No. 5 to 8

Fig. 157a   Bayonet type lathe spindle noses

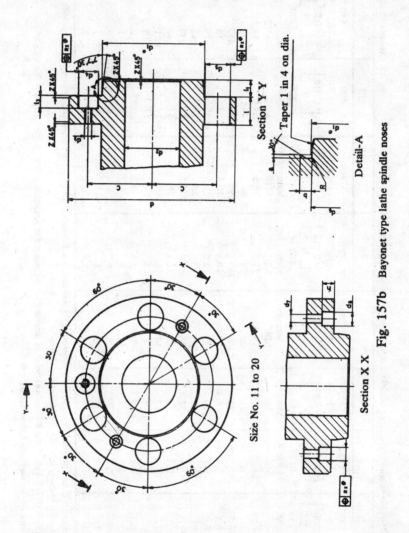

Section Y-Y

Detail-A

Taper 1 in 4 on dia.

Section X X

Size No. 11 to 20

Fig. 157b   Bayonet type lathe spindle noses

Table 223 Dimensions for bayonet type lathe spindle noses (as per IS 2582 Part III - 1972)

All dimensions in *mm*

Size No.	3	4	5	6	8	11	15	20
d	102	112	135	170	220	290	400	540
d_1*	$53.975^{+0.008}_{0}$	$63.513^{+0.008}_{0}$	$82.563^{+0.01}_{0}$	$106.375^{+0.01}_{0}$	$139.719^{+0.012}_{0}$	$196.869^{+0.014}_{0}$	$285.775^{+0.016}_{0}$	$412.775^{+0.02}_{0}$
d_2 Max.	32 or Morse 4	40 or Morse 4	50 or Morse 5	70 or Morse 6	100 or Metric 100	150 or Metric 140	220 —	300 —
d_3	—	M6	M6	M8	M8	M10	M12	M12
d_4 H8		14.25	15.9	19.05	23.8	28.6	34.9	41.3
d_5	21	21	21	23	29	36	43	43
d_6	53.5	63	82	106	139	196	285	412
d_7	6.4	6.4	6.4	8.4	10.5	10.5	13	13
d_8	10.4	10.4	10.4	13.5	16.5	16.5	19	19
h	10	10	10	11	12	13	15	15
C	37.5	42.5	52.4	66.7	85.7	117.5	165.1	231.8
l	16	20	22	25	28	35	42	48
l_1	11	11	13	14	16	18	19	21
l_3	—	5	6	8	10	12	12	16
x	0.1	0.1	0.1	0.1	0.1	0.1	0.15	0.15
a	1	1	1	1	1	1	1	1
b	5	6	6	8	8	8	8	9
R	1	1	1	1	1	1	1	1
Z	1	1	1	1.6	1.6	1.6	2	2

Notes: 1. Tolerance on C and angular dimensions is controlled by this position tolerance which is the permissible deviation of the hole centres with respect to their theoretical positions.

    2*. Dimension d_1 is taken at the theoretical point of intersection between the generating line of the cone and the face of the flange.

    3. Tolerance on untoleranced dimensions: ±0.4

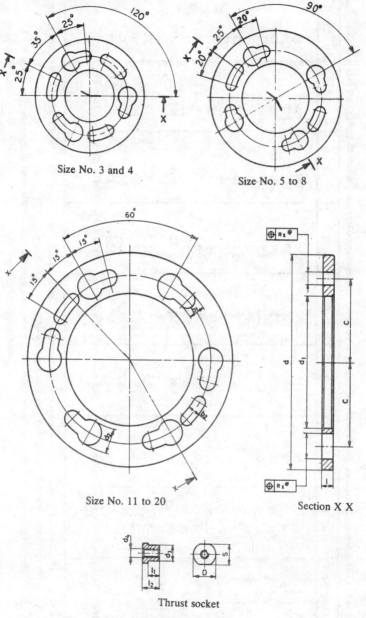

Size No. 3 and 4

Size No. 5 to 8

Size No. 11 to 20

Section X X

Thrust socket

Fig. 158 Bayonet disc

Table 224   Dimensions for bayonet discs and thrust sockets for bayonet type lathe spindle noses (as per IS 2582 (Part III) — 1972)

Size No.	3	4	5	6	8	11	15	20
d	110	120	145	180	230	300	410	550
d_1H8^*	50	60	80	100	130	185	270	400
d_2	21	21	21	23	29	36	43	43
$C\ {}^{\ 0}_{-0.1}$	37.5	42.5	52.4	66.7	85.7	117.5	165.1	231.8
$l\ {}^{\ 0}_{-0.1}$	5	6	8	10	12	16	18	22
b_1	11.5	11.5	11.5	14	18	23	27	27
b_2	11.5	11.5	11.5	14	18	18	23	23
x	0.1	0.1	0.1	0.1	0.1	0.1	0.15	0.15
d_3	11	11	11	13	17	17	22	22
d_4	M6	M6	M6	M8	M10	M10	M12	M12
$l_1\ {}^{+0.2}_{\ \ 0}$	5.2	6.2	8.2	10.2	12.2	16.2	18.3	22.3
l_2	8	9	12	15	18	22	26	30
D	16	16	16	19	25	25	32	32
S	14	14	14	17	22	22	27	27
HEX. SOC. HD. SCR.	M6 x 16	M6 x 20	M6 x 25	M8 x 30	M10 x 35	M10 x 45	M12 x 55	M12 x 65

Notes: 1.   Tolerance on C and angular dimensions is controlled by this position tolerance which is the permissible deviation of the hole centres with respect to their theoretical positions.

2*.   The given boring diameters are maximum values. The tolerance on the spindle diameter of the same dimension is $f\,7$.

All dimensions in *mm*

Table 225 Dimensions for face plates for bayonet type lathe spindle noses
(as per IS 2582 Part III - 1972)

All dimensions in *mm*

Size No.	3	4	5	6	8	11	15	20
d	102	112	135	170	220	290	400	540
d_1	53.975	63.513	82.563	106.375	139.719	196.869	285.775	412.775
	$\begin{bmatrix}+0.003\\-0.005\end{bmatrix}$	$\begin{bmatrix}+0.003\\-0.005\end{bmatrix}$	$\begin{bmatrix}+0.004\\-0.006\end{bmatrix}$	$\begin{bmatrix}+0.004\\-0.006\end{bmatrix}$	$\begin{bmatrix}+0.004\\-0.008\end{bmatrix}$	$\begin{bmatrix}+0.004\\-0.01\end{bmatrix}$	$\begin{bmatrix}+0.004\\-0.012\end{bmatrix}$	$\begin{bmatrix}+0.005\\-0.015\end{bmatrix}$
$d_2 \begin{bmatrix}+0.1\\0\end{bmatrix}$	—	14.7	16.3	19.45	24.25	29.4	35.7	42.1
d_3	M10	M10	M10	M12	M16	M20	M24	M24
d_4 Max.	51.5	61	79.6	103.2	136.2	192.9	281.5	408
C	37.5	42.5	52.4	66.7	85.7	117.5	165.1	231.8
x	0.1	0.1	0.1	0.1	0.1	0.1	0.15	0.15
l	—	6.5	6.5	6.5	8	10	10	10
l_1	10	10	12	13	14	16	17	19
h_1	15	15	15	18	24	30	36	36
h_2	18	18	18	22	28	34	40	40
Size of stud	M10 x 34	M10 x 39	M10 x 43	M12 x 50	M16 x 60	M20 x 75	M24 x 90	M24 x 100
Z	1	1	1	1.6	1.6	1.6	2	2

Notes: 1. Tolerance on C and angular dimensions is controlled by this position tolerance which is the permissible radial deviation of the hole centres with respect to their theoretical positions.

2. Tolerance on untoleranced dimensions: ± 0.4

3. Dimension d_1 is taken at the theoretical point or intersection between the generating line of the cone and the face of the flange.

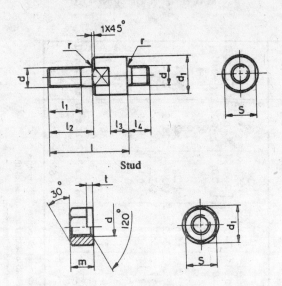

Stud

Nut with collar

Table 226   Dimensions for studs and nuts with collar for bayonet type lathe spindle noses (as per IS 2582 (Part III)–1972)   All dimensions in *mm*

Size No.	3	4	5	6	8	11	15	20
d	M10	M10	M10	M12	M16	M20	M24	M24
d_1 $h11$	19.5	19.5	19.5	21.5	27	34	41	41
l_1	18	18	18	20	25	30	36	36
l_2	20	22	24	28	35	44	52	56
l_3	5	8	10	12	12	15	20	26
l_4	12	12	12	15	20	25	30	30
l	34	39	43	50	60	75	90	100
S	17	17	17	19	24	30	36	36
r	0.6	0.6	0.6	1	1	1	1.6	1.6
m	12	12	12	14	18	22	27	27
t	3	3	3	3	3	4	4	4

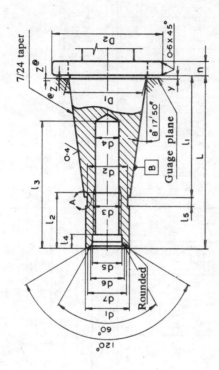

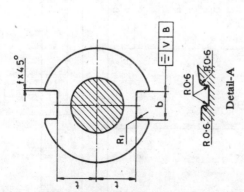

Fig. 159 Self-release 7/24 tapers for milling arbor and tool shank

Table 227 Dimensions for self-release 7/24 tapers for milling arbor and tool shanks (as per IS 2340—1972)　　All dimensions in mm

| Self-release 7/24 taper No. | External taper | | | | | | | | | Collar and driving slot | | | | | | |
	D_1 basic	d_1	Tolerance on d_1 $a10$	d_2	L Max.	l_1	l_5	y	Z@	D_2	n	b H12	t Max.	V	R_1	f
30	31.75	17.4	-0.29 / -0.36	16.5	70	50	3	1.6	0.4	50	8	16.1	16.2	0.12	1.6	1.6
40	44.45	25.3	-0.3 / -0.38	24	95	67	5	1.6	0.4	63	10	16.1	22.5	0.12	1.6	1.6
45	57.15	32.4	-0.31 / -0.41	30	110	86	6	3.2	0.4	80	10	19.3	29	0.12	1.6	1.6
50	69.85	39.6	-0.31 / -0.41	38	130	105	8	3.2	0.4	100	12	25.7	35.3	0.2	2	2
55	88.9	50.8	-0.34 / -0.46	48	168	130	9	3.2	0.4	130	14	25.7	45	0.2	2	2
60	107.95	60.2	-0.34 / -0.46	58	210	165	10	3.2	0.4	160	16	25.7	60	0.2	2	2

| Self-release 7/24 taper No. | Internal details | | | | | | | |
	d_3	d_4	d_5	d_6	d_7	l_2	l_3 Min.	l_4
30	M12	10.25	12.5	15.5	16	24	50	6
40	M16	14	17	19.5	23	30	70	8
45	M20	17.5	21	24	30	38	70	10
50	M24	21	25	28	36	45	90	11
55	M24	21	25	28	36	45	90	11
60	M30	26.5	31	35	45	56	110 (160)*	14

1. * Applicable for reduction sleeves
2. @Z is the maximum permissible deviation of position of diameter D_1 from guage plane.

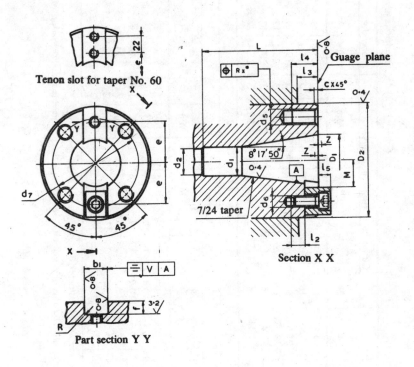

Tenon slot for taper No. 60

Section X X

Part section Y Y

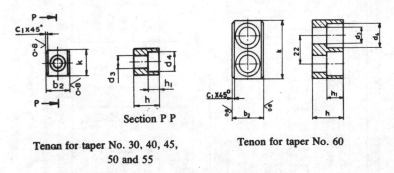

Section P P

Tenon for taper No. 30, 40, 45, 50 and 55

Tenon for taper No. 60

Fig. 160 Self-release 7/24 tapers for spindle noses for milling machines

Table 228 Dimensions for self-release 7/24 tapers for spindle noses for milling machines (as per IS 6681—1972) All dimensions in *mm*

Self-release 7/24 taper No.	Internal details						Cutter head insertion details						Tenon slot	
	D_1 basic	d_1 H12	d_2 Min.	L Min.	Z ‡	D_2 h5	l_3 Min.	d_7	x^*	d_5	l_4 Min.	C	b_1 M6	f Min.
30	31.75	17.4	17	73	0.4	69.832	12.5	54	0.075	M10	16	2	15.9	8
40	44.45	25.3	17	100	0.4	88.882	16	66.7	0.075	M12	20	2	15.9	8
45	57.15	32.4	21	120	0.4	101.6	18	80	0.075	M12	25	2	19	9.5
50	69.85	39.6	27	140	0.4	128.57	19	101.6	0.1	M16	25	3	25.4	12.5
55	88.9	50.4	27	178	0.4	152.4	25	120.6	0.1	M20	30	3	25.4	12.5
60	107.95	60.2	35	220	0.4	221.44	38	177.8	0.1	M20	30	3	25.4	12.5

Self-release 7/24 taper No.	Tenon slot						Tenon								
	e^{**} ±0.2	d_6	l_2	R	V	b_2 h5	d_3	d_4	h	h_1	k Max.	l_5 Max.	M Min.	C_1	HEX. SOC. HD. SCR
30	24.75	M6	9	1.6	0.06	15.9	6.4	10.4	16	7	16.5	8	16.5	1.6	M6 x 16
40	32.75	M6	9	1.6	0.06	15.9	6.4	10.4	16	7	19.5	8	23	1.6	M6 x 16
45	39.75	M6	9	1.6	0.06	19	6.4	10.4	19	7	19.5	9.5	30	1.6	M6 x 20
50	49.25	M12	18	2	0.08	25.4	13	19	25	13.5	26.5	12.5	36	2	M12 x 25
55	61.25	M12	18	2	0.08	25.4	13	19	25	13.5	26.5	12.5	48	2	M12 x 25
60	72.75	M12	18	2	0.08	25.4	13	19	25	13.5	45.5	12.5	61	2	M12 x 25

Notes: 1. ‡Z is the maximum permissible deviation of position of diameter D_1 from guage plane.

     2. *Tolerance on d_7 and angular dimensions is controlled by this position tolerance which is the permissible radial deviation of hole centres with respect to their theoretical positions.

     3. **Not mentioned in IS 6681-1972. Calculated from the values of M and k

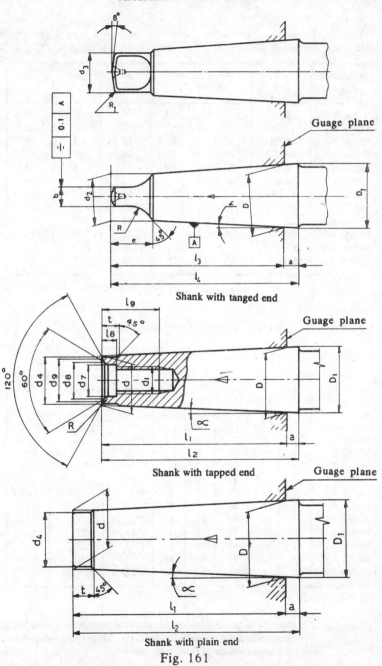

Shank with tanged end

Shank with tapped end

Shank with plain end

Fig. 161

Table 229 Dimensions for external self-holding tapers with tanged end shanks (as per IS 1715-1973)

All dimensions in *mm*

Designation of taper	D	D_1*	d_2*	d_3 Max.	l_3	l_4 Max.
Morse 0	9.045	9.2	6.1	6	56.5	59.5
Morse 1	12.065	12.2	9	8.7	62	65.5
Morse 2	17.78	18	14	13.5	75	80
Morse 3	23.825	24.1	19.1	18.5	94	99
Morse 4	31.267	31.6	25.2	24.5	117.5	124
Morse 5	44.399	44.7	36.5	35.7	149.5	156
Morse 6	63.348	63.8	52.4	51	210	218
Metric 80	80	80.4	69	67	220	228
Metric 100	100	100.5	87	85	260	270
Metric 120	120	120.6	105	102	300	312
Metric 160	160	160.8	141	138	380	396
Metric 200	200	201	177	174	460	480

Designation of taper	a	b $h13$	e Max.	R Max.	R_1	α	Taper on dia	% Taper
Morse 0	3	3.9	10.5	4	1	$1°29'27''$	1:19.212	5.205
Morse 1	3.5	5.2	13.5	5	1.2	$1°25'43''$	1:20.047	4.988
Morse 2	5	6.3	16	6	1.6	$1°25'50''$	1:20.02	4.995
Morse 3	5	7.9	20	7	2	$1°26'16''$	1:19.922	5.02
Morse 4	6.5	11.9	24	8	2.5	$1°29'15''$	1:19.254	5.194
Morse 5	6.5	15.9	29	10	3	$1°30'26''$	1:19.022	5.263
Morse 6	8	19	40	13	4	$1°29'36''$	1:19.18	5.214
Metric 80	8	26	48	24	5	$1°25'56''$	1:20	5
Metric 100	10	32	58	30	5	$1°25'56''$	1:20	5
Metric 120	12	38	68	36	6	$1°25'56''$	1:20	5
Metric 160	16	50	88	48	8	$1°25'56''$	1:20	5
Metric 200	20	62	108	60	10	$1°25'56''$	1:20	5

Note: *   Approximate values of diameters D_1 and d_2 of taper at distances a and l_3 from gauge plane end are given for guidance. Their actual values result from the actual values of D, a, *taper* and l_3

Table 230 Dimensions for external self-holding tapers   All dimensions in *mm* for tapped and plain end shanks (as per IS 1715-1973)

Designation of taper	D	D₁*	d*	d₄ Max.	d₁	d₇	d₈ Max.	d₉ Max.	l₁ Max.	l₂ Max.
Metric 4	4	4.1	2.9	2.5	—	—	—	—	23	25
Metric 6	6	6.2	4.4	4	—	—	—	—	32	35
Morse 0	9.045	9.2	6.4	6	—	—	—	—	50	53
Morse 1	12.065	12.2	9.4	9	M6	6.4	8	8.5	53.5	57
Morse 2	17.78	18	14.6	14	M10	10.5	12.5	13.2	64	69
Morse 3	23.825	24.1	19.8	19	M12	13	15	17	81	86
Morse 4	31.267	31.6	25.9	25	M16	17	20	22	102.5	109
Morse 5	44.399	44.7	37.6	35.7	M20	21	26	30	129.5	136
Morse 6	63.348	63.8	53.9	51	M24	25	31	36	182	190
Metric 80	80	80.4	70.2	67	M30	31	38	45	196	204
Metric 100	100	100.5	88.4	85	M36	37	45	52	232	242
Metric 120	120	120.6	106.6	102	M36	37	45	52	268	280
Metric 160	160	160.8	143	138	M48	50	60	68	340	356
Metric 200	200	201	179.4	174	M48	50	60	68	412	432

Designation of taper	l₉ Min.	t Max.	l₈	a	R	∝	Taper on dia.	% Taper
Metric 4	—	2	—	2	0.2	1°25′56″	1:20	5
Metric 6	—	3	—	3	0.2	1°25′56″	1:20	5
Morse 0	—	4	—	3	0.2	1°29′27′	1:19.212	5.205
Morse 1	16	5	4	3.5	0.2	1°25′43′	1:20.047	4.988
Morse 2	24	5	5	5	0.2	1°25′50′	1:20.02	4.995
Morse 3	28	7	6	5	0.6	1°26′16″	1:19.922	5.02
Morse 4	32	9	8	6.5	1	1°29′15″	1:19.254	5.194
Morse 5	40	10	11	6.5	2.5	1°30′26′	1:19.002	5.263
Morse 6	50	16	12	8	4	1°29′36′	1:19.18	5.214
Metric 80	65	24	14	8	5	1°25′56′	1:20	5
Metric 100	80	30	16	10	5	1°25′56′	1:20	5
Metric 120	80	36	16	12	6	1°25′56′	1:20	5
Metric 160	100	48	20	16	8	1°25′56′	1:20	5
Metric 200	100	60	20	20	10	1°25′56′	1:20	5

Note:

* Approximate diameters D₁ and d of taper at distances a and l₁ from gauge plane end are given for guidance. Their actual values result from the actual values of D, a, taper and l₁

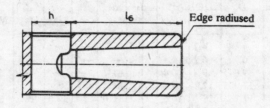

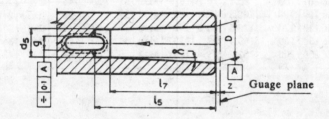

Socket for tanged end shank

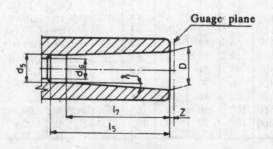

Socket for tapped or plain end shank

Fig. 162

Table 231 Dimensions for internal self-holding tapers (as per IS 1715-1973)

All dimensions in *mm*

Designation of taper	D	d_5 $H11$	d_6	g $A13$	h
Metric 4	4	3	–	2.2	8
Metric 6	6	4.6	–	3.2	12
Morse 0	9.045	6.7	–	3.9	15
Morse 1	12.065	9.7	7	5.2	19
Morse 2	17.78	14.9	11.5	6.3	22
Morse 3	23.825	20.2	14	7.9	27
Morse 4	31.267	26.5	18	11.9	32
Morse 5	44.399	38.2	23	15.9	38
Morse 6	63.348	54.6	27	19	47
Metric 80	80	71.5	33	26	52
Metric 100	100	90	39	32	60
Metric 120	120	108.5	39	38	70
Metric 160	160	145.5	52	50	90
Metric 200	200	182.5	52	62	110

Designation of taper	l_5 Min.	l_6	l_7 Approx.	Z^*	α	Taper on dia.
Metric 4	25	21	20	0.5	$1°25'56''$	1:20
Metric 6	34	29	28	0.5	$1°25'56''$	1:20
Morse 0	52	49	45	1	$1°29'27''$	1:19.212
Morse 1	56	52	47	1	$1°25'43''$	1:20.047
Morse 2	67	62	58	1	$1°25'50''$	1:20.02
Morse 3	84	78	72	1	$1°26'16''$	1:19.922
Morse 4	107	98	92	1.5	$1°29'15''$	1:19.254
Morse 5	135	125	118	1.5	$1°30'26''$	1:19.002
Morse 6	188	177	164	2	$1°29'36''$	1:19.18
Metric 80	202	186	170	2	$1°25'56''$	1:20
Metric 100	240	220	200	2	$1°25'56''$	1:20
Metric 120	276	254	230	2	$1°25'56''$	1:20
Metric 160	350	321	290	3	$1°25'56''$	1:20
Metric 200	424	388	350	3	$1°25'56''$	1:20

Note: * Z is the maximum permissible deviation of diameter D from the end face.

Table 232   Dimensions for Drill chuck tapers (as per IS 2243 - 1971)   All dimensions in *mm*

	Taper Designation	D N9/k9	d	d_1 H13	d_2	d_3	l	l_1	l_2	l_3	z	Taper on dia.
SHORT MORSE	B6	6.35	5.954	6.5	5.85	6.39	7.92	3.5	10	0.8	0.5	0.05
	B10	10.094	9.441	9.8	9.37	10.13	13.11	3.5	14.5	0.8	1	0.04988
	B12	12.065	11.214	11.5	11.143	12.12	17.07	3.5	18.5	0.8	1	(Morse 1)
	B16	15.733	14.602	15	14.534	15.77	22.63	3.5	24	0.8	1.5	0.04995
	B18	17.78	16.253	16.8	16.182	17.83	30.56	3.5	32	0.8	1.5	(Morse 2)
	B22	21.793	19.901	20.5	19.761	21.84	37.69	3.5	40.5	0.8	2	0.0502
	B24	23.825	21.395	22	21.3	23.87	48.41	3.5	50.5	0.8	2	(Morse 3)
JACOBS	0	6.35	5.88	—‡	5.802	6.39	9.52	—‡	11.11	0.8	0.5	0.04928
	1	9.754	8.59	8.92	8.469	9.82	15.08	3.17	16.67	0.8	1	0.07709
	2 Sh	13.94	12.515	12.83	12.386	14	17.46	3.17	19.05	0.8	1	0.08155
	2	14.199	12.515	13.08	12.386	14.26	20.64	3.17	22.22	0.8	1	0.08155
	33	15.85	14.338	14.66	14.237	15.9	23.81	3.17	25.4	0.8	1.5	0.06349
	6	17.17	15.933	16.26	15.852	17.21	23.81	3.17	25.4	0.8	1.5	0.05191
	3	20.599	19.078	19.3	18.951	20.64	28.57	4.76	30.96	0.8	2	0.05325
	4	28.55	26.471	27.1	26.346	28.59	39.67	4.76	42.07	0.8	2.5	0.0524
	5	35.89	33.3	34.49	33.422	35.93	50.01	4.76	47.62	0.8	3	0.05183

Notes:
1. Details of JACOBS 4 and 5 tapers are given for information but these tapers shall be avoided wherever possible.
2. B6 taper is not included in IS.
3. *The gauge plane diameter D for the taper bore is the diameter at the wide end before chamfering.
4. **The diameter of the small end of the taper shank d_2 is that calculated before chamfering.
5. Calculated values of d, d_2 and d_3 are given for information.
6. ‡The JACOBS No. 0 taper bore is not normally recessed.

Guage plane

Spindle end with taper shank

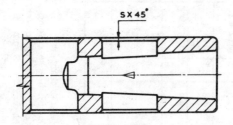

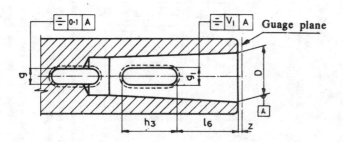

Socket for spindles of boring machines and radial drilling machines

Table 233 Dimensions for sockets for spindles of boring machines and radial
drilling machines   All dimensions in *mm*

Designation of taper	D	g A13	g₁ A13	h₃	l₆	S	Z*	V₁
Morse 3	23.825	7.9	8.2	31	34	1.6	1	0.2
Morse 4	31.267	11.9	8.2	36	30	1.6	1.5	0.2
Morse 5	44.399	15.9	12.2	41	30	2	1.5	0.2
Morse 6	63.348	19	16.2	46	30	2.5	2	0.2
Metric 80	80	26	19.5	44	30	3	2	0.2
Metric 100	100	32	26.5	52	30	4	2	0.3
Metric 120	120	38	32.6	60	30	4	2	0.3
Metric 160	160	50	44.8	76	40	4	3	0.4
Metric 200	200	62	56.8	92	60	4	3	0.4

Note: *Z   is the maximum permissible deviation of diameter *D* from the end face.

Table 234 Dimensions for taper shaft ends of grinding wheel spindles (as per IS 2996 - 1964)

Nominal diameter of grinding spindle $d \pm 0.1$	l	d_1	Type A d_2	l_2	Type B $d_3{}^{**}$
32	40 / 50	M16	13	1.5	M12 x 1.5
40	40 / 50	M24	19.8	1.5	M16 x 1.5
50	50 / 63	M24	19.8	3	M16 x 1.5
63	63 / 80	M36 x 3	31.8	3	M20 x 1.5
80	80 / 100	M36 x 3	31.8	4	—
100	100 / 125	M48 x 3	43.8	4	—

Centre hole B as per IS 2473

1:5

$0.6 \times 45°$

$0.4 \times 45°$

Type A

Protected centre hole as per IS 2540

1:5

Type B

Notes:

1. *Dimensions l_1 and l_3 are according to the design of clamping.
2. ** As a protection against the self-releasing of grinding wheel flanges, the threads should be either left hand or right hand depending upon the direction of rotation. In such cases where the direction of rotation is changing, protection against self-releasing may be done by keying or similar means according to the designer's choice.

All dimensions in *mm*

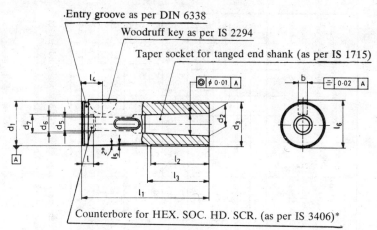

*Form D-Without the counterbore for HEX. SOC. HD. SCR.
Form E-With the counterbore for HEX. SOC. HD. SCR.

Fig. 163   Short type, form D and E

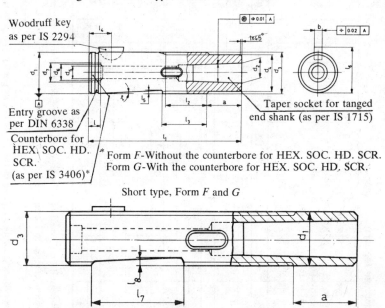

Design of adjustable adaptors of sizes16×1, 25×2 and 32×3 (Other dimensions same
as for the other sizes)

Long type, form F and form G

Fig. 164

Table 235 Dimensions for adjustable adaptors form D and form E (as per DIN 6327 Blatt 1)   All dimensions in *mm*

d_1 $h6$	Taper socket for tanged end shank (as per IS 1715)	d_2 (as per IS 1715)	Keyway width b P9 (as per IS 2294)	Trapezoidal thread d_3	l_1	l_2	l_3	l_4	l_5	l_6 (as per series B of IS 2294)		Woodruff key (as per IS 2294)	Entry groove (as per DIN 6338)	Counterbore for form E only (as per IS 3406)				Range of adjust ability
														d_5	d_6	d_7	l	
10	Metric 6	6	3	Tr 10 × 1.5	62	28	30	10	1	10.9	0	3 × 5	0.8	–	–	–	–	14
12	Metric 6	6	3	Tr 12 × 1.5	62	28	30	10	1	12.9	−0.2	3 × 5	1.2	–	–	–	–	14
16	Morse 0	9.045	5	Tr 16 × 1.5	85	40	42	11	1.3	17.1		5 × 6.5	1.2	–	–	–	–	28
20	Morse 1	12.065	5	Tr 16 × 1.5	85	40	42	11	1.3	17.1	0	5 × 6.5	1.2	6.6	–	11	6.8	28
	Morse 1	12.065	5	Tr 20 × 2	88	40	42	13	1.3	21.1		5 × 7.5	2	6.6	–	11	6.8	28
25[1]	Morse 1	12.065	6	Tr 25 × 2	95	42	44	15	1.5	26.5	−0.25	6 × 9	2	6.6	–	11	6.8	30
	Morse 2	17.78	6	Tr 25 × 2	95	42	44	15	1.5	26.5		6 × 9	2	11	–	17.5	11	30
28	Morse 1	12.065	6	Tr 28 × 2	95	42	44	15	1.5	29.5		6 × 9	3.2	6.6	–	11	6.8	30
	Morse 2	17.78	6	Tr 28 × 2	95	42	44	15	1.5	29.5		6 × 9	3.2	11	–	17.5	11	30
32[1]	Morse 2	17.78	8	Tr 32 × 2	118	50	53	20	1.7	33.5		8 × 11	3.2	11	–	17.5	11	36
	Morse 3	23.825	8	Tr 32 × 2	118	50	53	20	1.7	33.5	0	8 × 11	3.2	14	16	20	13	36
36	Morse 2	17.78	8	Tr 36 × 2	118	50	53	20	1.7	37.5		8 × 11	3.2	11	–	17.5	11	36
	Morse 3	23.825	8	Tr 36 × 2	118	50	53	20	1.7	37.5	−0.35	8 × 11	3.2	14	16	20	13	36
48	Morse 3	23.825	10	Tr 48 × 2	144	65	68	24	2.2	49.9		10 × 13	5	14	16	20	13	47
	Morse 4	31.267	10	Tr 48 × 2	144	65	68	24	2.2	49.9		10 × 13	5	18	20	26	17.5	47

Notes: 1)[1] To be used only when holes with close centre distance must be machined in one station. In case of thin walled adjustable adaptors of sizes 25 × 2 and 32 × 3, care should be taken that these adjustable adaptors are not overloaded through the tool.

2) In case of adjustable adaptors of form E, a key as per IS 2710 can be used in place of woodruff key, if required by design.

Table 236 Dimensions for adjustable adaptors form F and form G (as per DIN 6327 Blatt 2)   All dimensions in *mm*

d_1 h6	Taper socket for tanged end shank (as per IS 1715)	d_4	l_7	l_8	Length l_1 for length a												
					10	20	25	30	40	50	60	75	80	90	100	120	160
10	Metric 6	8	—	—	72	82	—	92	—	—	—	—	—	—	—	—	—
12	Metric 6	10	—	—	72	82	—	92	102	—	—	—	—	—	—	—	—
16	Morse 0	14	38	2.2	—	—	110	—	—	135	—	160	—	—	185	—	—
16	Morse 1 [1]	14	38	2.2	—	—	110	—	—	135	—	160	—	—	185	—	—
20	Morse 1	17	—	—	—	—	113	—	—	138	—	163	—	—	188	—	—
25 [1]	Morse 1	22	44	2.7	—	—	120	—	—	145	—	170	—	—	195	—	—
25	Morse 2 [1]	22	44	2.7	—	—	120	—	—	145	—	170	—	—	195	—	—
28	Morse 1	25	—	—	—	—	120	—	—	145	—	170	—	—	195	—	—
28	Morse 2	25	—	—	—	—	120	—	—	145	—	170	—	—	195	—	—
32 [1]	Morse 2	29	50	2.9	—	—	—	148	—	—	178	—	—	208	—	238	—
32	Morse 3 [1]	29	50	2.9	—	—	—	148	—	—	178	—	—	208	—	238	—
36	Morse 2	33	—	—	—	—	—	148	—	—	178	—	—	208	—	238⁻	—
36	Morse 3	33	—	—	—	—	—	148	—	—	178	—	—	208	—	238	—
48	Morse 3	45	—	—	—	—	—	—	184	—	—	—	224	—	—	264	304
48	Morse 4	45	—	—	—	—	—	—	184	—	—	—	224	—	—	264	304

Notes:

1) [1] To be used only when holes with very close centre distance must be machined in one station. In case of adjustable adaptors of sizes 16 x 1, 25 x 2 and 32 x 3, care should be taken that they are not overloaded through the tool. In case of these sizes, the turning of d_4 is abandoned to increase the stability.

2) In case of adjustable adaptors of form G, a key as per IS 2710 can be used in place of woodruff key if required by design.

3) Other dimensions are same as for forms D and E.

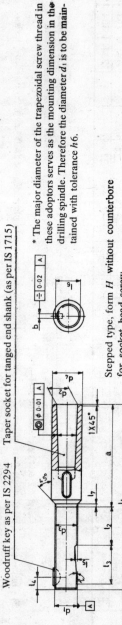

Woodruff key as per IS 2294

Taper socket for tanged end shank (as per IS 1715)

Stepped type, form *H* without counterbore
for socket head screw

Tools which are used on the stepped type adjustable adaptors are allowed to be loaded only to that extent which they can transfer to the driving spindles of the drilling spindle heads.

* The major diameter of the trapezoidal screw thread in these adaptors serves as the mounting dimension in the drilling spindle. Therefore the diameter d_1 is to be maintained with tolerance $h6$.

Table 237 Dimensions for adjustable adaptors, form H (as per DIN 6327 Blatt 3)

All dimensions in *mm*

d_1 $h6$	Taper socket for tanged end shank (as per IS 1715)	a	b $p9$	d_2	Trapezoidal thread d_3 *	d_4	l_1	l_2	l_3	l_4	l_5	l_6		l_7	Woodruff key as per IS 2294	Range of adjustability
8	Metric 6	46	2	6	Tr 8 x 1	12	96	24	22	10	1.5	8.8	0	2	2 x 3.7	14
10	Morse 0	73	3	9.045	Tr 10 x 1.5	18	135	30	28	10	2	10.9	0.2	3	3 x 5	18
12	Morse 0	73	3	9.045	Tr 12 x 1.5	18	135	30	28	10	2	12.9		3	3 x 5	18
16	Morse 1	79	5	12.065	Tr 16 x 1.5	20	164	42	36	11	2.3	17.1		3	5 x 6.5	28
16	Morse 2	94	5	17.78	Tr 16 x 1.5	25	179	42	36	11	2.3	17.1	0	3	5 x 6.5	28
20	Morse 2	94	5	17.78	Tr 20 x 2	28	182	42	38	13	2.5	21.1	−0.25	3	5 x 7.5	28
25	Morse 3	117	6	23.825	Tr 25 x 2	36	212	44	44	15	2.7	26.5		3	6 x 9	30
28	Morse 3	117	6	23.825	Tr 28 x 2	36	212	44	44	15	2.7	29.5		3	6 x 9	30
32	Morse 4	146	8	31.267	Tr 32 x 2	45	264	53	50	20	2.9	33.5	0	3	8 x 11	36
36	Morse 4	146	8	31.267	Tr 36 x 2	48	264	53	50	20	2.9	37.5	0.35	3	8 x 11	36

Adjustment nut for adaptors

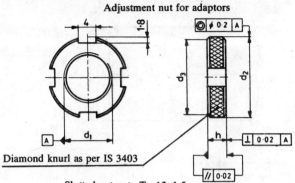

Diamond knurl as per IS 3403

Slotted nut upto Tr. 12x1.5

HEX. SOC. GRUB.
SCR. *E* as per
IS 6094

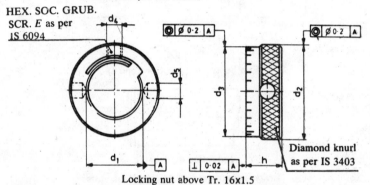

Diamond knurl
as per IS 3403

Locking nut above Tr. 16x1.5

*The distance between graduations should correspond to a setting of 0.1 *mm*. Every fifth graduation is distinguished by a longer length.

Table 238 Dimensions for adjustment nuts (as per DIN 6327 Blatt 4)

Trapezoidal Thread d_1	d_2		d_3 h13	d_4	d_5	h	HEX. SOC. GRUB SCR. as per IS 6094
Tr 8x 1	14.8	0	14	–	–	5	–
Tr 10 x1.5	17.8	– 0.2	17	–	–	6	–
Tr 12 x 1.5	19.7		19	–	–	6	–
Tr 16 x 1.5	24.6		24	M5	5.1	12	E M5 x 5
Tr 20 x 2	31.6		31	M5	6.1	12	E M5 x 5
Tr 25 x 2	36.6		36	M6	6.1	12	E M6 x 6
Tr 28 x 2	39.6	0	39	M6	6.1	12	E M6 x 6
Tr 32 x 2	44.6	–0.4	44	M6	6.1	12	E M6 x 6
Tr 36 x 2	49.6		49	M6	6.1	14	E M6 x 6
Tr 48 x 2	66.6		66	M8	6.1	18	E M8 x 8

Note: 1. The HEX. SOC. GRUB SCR. is shortened to be flush with diameter d_2.

All dimensions in *mm*

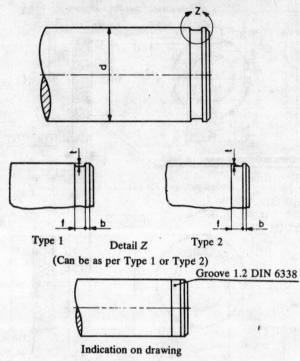

Type 1   Detail Z   Type 2

(Can be as per Type 1 or Type 2)

Groove 1.2 DIN 6338

Indication on drawing

Entry grooves used on adjustable adaptors

Table 239 Dimensions for entry grooves (as per DIN 6338)

f	Range of diameter		b	t
	Above	Upto		
0.5	4	6.3	0.2	0.2
0.8	6.3	10	0.2	0.2
1.2	10	16	0.3	0.2
2	16	25	0.4	0.4
3.2	25	40	0.6	0.4
5	40	63	1	0.6
8	63	100	1.6	1
12.5	100	160	2	2
20	160	250	3	2

All dimensions in *mm*

Table 240 Limiting dimensions for trapezoidal screw threads used on adjustable adaptors and adjustment nuts as per DIN 6327

Nominal size of thread	Major diameter h6*	Limiting dimensions for bolt threads				Limiting dimensions for nut threads				
		Pitch diameter		Minor diameter		Major diameter	Pitch diameter		Minor diameter	
		Max.	Min.	Max.	Min.	Min.	Min.	Max.	Min.	Max.
Tr 8 x 1	8	7.44	7.3	6.8	6.565	8.2	7.5	7.69	7	7.15
Tr 10 x 1.5	10	9.183	9.013	8.2	7.921	10.3	9.25	9.474	8.5	8.69
Tr 12 x 1.5	12	11.183	11.003	10.2	9.908	12.3	11.25	11.486	10.5	10.69
Tr 16 x 1.5	16	15.183	15.003	14.2	13.908	16.3	15.25	15.486	14.5	14.69
Tr 20 x 2	20	18.929	18.729	17.5	17.179	20.5	19	19.265	18	18.236
Tr 25 x 2	25	23.929	23.717	22.5	22.164	25.5	24	24.28	23	23.236
Tr 28 x 2	28	26.929	26.717	25.5	25.164	28.5	27	27.28	26	26.236
Tr 32 x 2	32	30.929	30.717	29.5	29.164	32.5	31	31.28	30	30.236
Tr 36 x 2	36	34.929	34.717	33.5	33.164	36.5	35	35.28	34	34.236
Tr 48 x 2	48	46.929	46.705	45.5	45.149	48.5	47	47.3	46	46.236

All dimensions in *mm*

Notes: 1. The profile of the trapezoidal screw thread corresponds to that as per IS 7008
2. The thread limiting dimensions have been calculated based on the formulae for the dimensions and tolerances for ISO metric trapezoidal screw threads as per IS 7008.
3. *The tolerance h6 for the shank dismeter d_1 of the adjustable adaptor has been retained for the thread major diameter also.

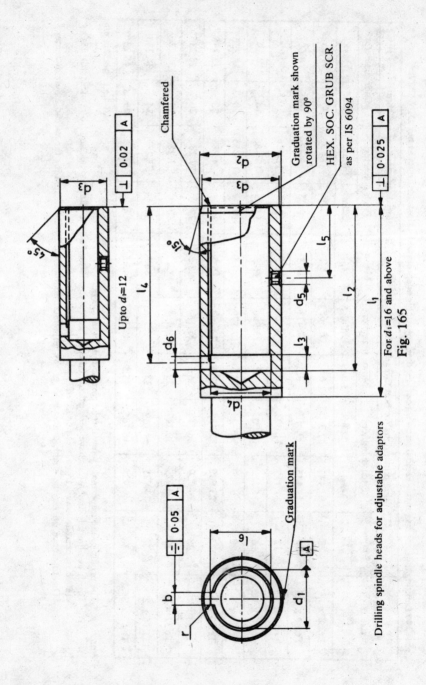

Chamfered

Graduation mark shown
rotated by 90°

HEX. SOC. GRUB SCR.

as per IS 6094

Upto d=12

For d_1=16 and above
Fig. 165

Graduation mark

Drilling spindle heads for adjustable adaptors

Table 241 Dimensions for drilling spindle heads for adjustable adaptors (as per DIN 55058)

All dimensions in mm

d_1 H7	d_2 f7	d_3	d_4	d_s	d_6	l_1 Min.	l_2 Min.	l_3	l_4	l_5 ±0.1	l_6 +0.3 0	b C11	r Max.	HEX. SOC. GRUB. SCR. as per IS 6094[2]
8	15	14.4	8.6	M4	3.5	46	42	8	35	16	9	2	0.2	E M4 x 5
10	18	17.4	10.6	M5	5	60	52	8	48	22	11.1	3	0.2	E M5 x 5
12	20	19.2	12.6	M5	5	60	52	8	48	22	13.1	3	0.2	E M5 x 5
16	25	24	16.6	M6	6	85	74	8	70	34	17.3	5	0.2	E M6 x 6
20	32	31	20.6	M8	6	90	77	8	73	35	21.3	5	0.2	E M8 x 8
25[1]	37	36	25.6	M8	8	100	85	10	80	38	26.7	6	0.4	E M8 x 8
28	40	39	28.6	M8	8	100	85	10	80	38	29.7	6	0.4	E M8 x 8
32[1]	45	44	32.8	M8	10	128	106	10	101	45	33.7	8	0.4	E M8 x 8
36	50	49	36.8	M8	10	128	106	10	101	45	37.7	8	0.4	E M8 x 8
48	67	66	48.8	M10	12	152	129	12	123	57	50.1	10	0.4	E M10 x 10

Notes: 1)[1] Used only when holes with very close centre distance must be finished in one station.

2)[2] The HEX. SOC. GRUB. SCR. is shortened so that it does not project beyond diameter d_2 after an adjustable adapter is clamped in the spindle head.

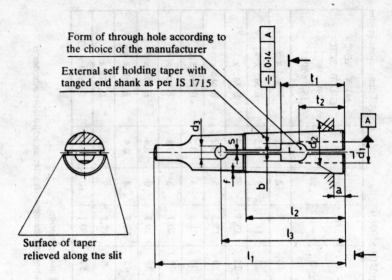

Split taper sockets for tools with parallel shank and square tang

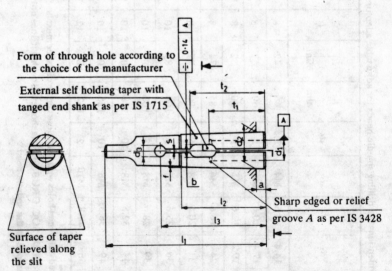

Split taper sockets for tools with parallel shank and flat tang

Table 242 Dimensions for split taper sockets for tools with parallel shank and square tang (as per DIN 6328)

Taper as per IS 1715	Range of diameters d_1 H7		a	d_2	d_3	f	l_1 Max.	l_2* Approx.	l_3 Approx.	s
	Over	Upto								
Morse 0	2.47	4.53	3	9.045	3	0.2	59.5	32	40	0.8
Morse 1	2.47	8.27	3.5	12.065	4	0.2	65.5	36	44	1
Morse 2	4.53	12	5	17.78	5	0.2	80	42	52	1.2
Morse 3	8.27	17.33	5	23.825	6	0.4	99	50	63	1.6
Morse 4	12	21.33	6.5	31.267	8	0.4	124	63	80	2
Morse 5	17.33	29.33	6.5	44.399	10	0.4	156	80	103	2

* 1. The length l_2 of the split taper socket is less than that as per IS1715.
2. These split taper sockets are not suitable for milling work.

Range of diameters d_1 H7		b H11	t_1	t_2
Over	Upto			
2.47	2.83	2.2	19	15
2.83	3.2	2.5		
3.2	3.6	2.8	21	16
3.6	4.01	3.1		
4.01	4.53	3.5		
4.53	5.08	4	24	18
5.08	5.79	4.5		
5.79	6.53	5.1	26	19.5
6.53	7.33	5.7		
7.33	8.27	6.4	27	19.5
8.27	9.46	7.3	30	22
9.46	10.67	8.3	32	23
10.67	12	9.3	34	24
12	13.33	10.3	36	25
13.33	14.67	11.3	38	26
14.67	16	12.3	40	27
16	17.33	13.3	44	30
17.33	19.33	14.9	48	33
19.33	21.33	16.4	52	35
21.33	24	18.4	56	37
24	26.67	20.4	62	42
26.67	29.33	22.4	66	44

Table 243 Dimensions for split taper sockets for tools with parallel shank and flat tang (as per DIN 6329)   All dimensions in *mm*

Range of diameters d_1 H7		b H11	t_1	t_2
Over	Upto			
1.6	2.9	—	16	—
2.9	3.5	1.8	20	25
3.5	4	2.2	20	25
4	4.5	2.4	20	26
4.5	5.5	2.7	20	26
5.5	6.5	3.2	22	29
6.5	8	3.8	22	29
8	9.5	4.8	25	33
9.5	11	5.3	28	37
11	13	6.3	28	39
13	15	7.4	32	44
15	18	8.4	32	46
18	21	10.4	36	53
21	24	11.4	40	58
24	27	13.4	45	67
27	30	14.5	50	73

Taper as per IS 1715	Range of diameters d_1 H7		a	d_2	d_3	f	l_1 Max.	l_2^* $\approx$	l_3 $\approx$	s
	Over	Upto								
Morse 0	1.6	5.5	3	9.045	3	0.2	59.5	32	40	0.8
Morse 1	3	8	3.5	12.065	4	0.2	65.5	36	44	1
Morse 2	5.5	13	5	17.78	5	0.2	80	42	52	1.2
Morse 3	8	18	5	23.825	6	0.4	99	50	63	1.6
Morse 4	13	21	6.5	31.267	8	0.4	124	63	80	2
Morse 5	18	30	6.5	44.399	10	0.4	156	80	103	2

Notes: *1.   The length l_2 of the split taper socket is less than that as per IS 1715.
        2.   These split taper sockets are not suitable for milling work.

LUBRICANTS AND LUBRICATION

Machine tools generally work to a high degree of accuracy and are expected to sustain this accuracy over a long period. This requires proper control of friction and wear of parts which are in relative motion and calls for effective lubrication of the vital elements like bearings, slideways, gears, etc.

Effective lubrication is achieved by using proper lubricants. Lubricants not only reduce the friction and the consequent heat generation, they also aid in transporting the heat generated.

Types of lubricants

The types of lubricants commonly used in machine tools are (i) oil, (ii) grease (iii) oil mist and (iv) solid lubricants.

However, since oil and grease are more commonly used, this chapter is limited only to these lubricants. Mist lubrication is resorted to only for very high speed spindles. Solid lubricants are used in situations where lubrication point is inaccessible or likely to be neglected. Bearing bushes impregnated with PTFE or sintered bushes impregnated with oil or graphite, nylon and other plastics are some of the bearings where solid lubricants are used in machine tools. It should be noted that they are used only for low speed and light load applications. The desired properties of the various machine tool oils and their general application are given in Table 244

Modes of Lubrication

A proper choice of the lubricant depends upon the mode of lubrication. Operating conditions like speed, load, lubricant properties, surface quality, etc. determine the mode of lubrication. There are three modes of lubrication:

(a) Boundary lubrication,
(b) Mixed lubrication and
(c) Fluid film lubrication

In boundary lubrication there is considerable amount of asperity contact and interaction between the sliding surfaces. The friction and wear characteristics of the surfaces are governed more by the chemical properties of the lubricant rather than by the physical properties of the lubricant like viscosity and the role of additives, like polar or extreme pressure (EP) is very important. Hypoid gears and machine tool slideways predominently operate under this mode of lubrication.

In mixed lubrication both boundary and fluid film lubrication exist. Here both physical and chemical properties of the lubricant are important. This mode of lubrication is generally observed in gears, in plain bearings during start and stop, etc.

In fluid film lubrication the sliding surfaces are completely separated by an oil film. This mode of lubrication is observed in plain bearings, gears, rolling bearings, etc. Viscosity of the lubricant is the most important factor to be considered and it is one of the governing factors in deciding film thickness, load capacity, friction torque, flow, etc.

Selection of lubricants

The general choice on the lubricant is between oil and grease. The advantages and disadvantages between oil and grease are as follows.

Lubricant	Advantages	Disadvantages
Oil	Effective cooling, carries away dirt, wear debris, etc. Suitable for wide operating conditions Easy to drain and refill. Coefficient of friction & frictional torque is low.	Invariably requires a pump source. Requires good sealing High maintenance and initial cost.
Grease	Simplicity in system design. Easy sealing arrangements. Easily retained in the housing. Low maintenance and initial cost.	Poor cooling property. Not suitable for very high speed applications. Retains dirt and wear debris.

Bearings

ROLLING BEARINGS

Rolling bearings are generally grease lubricated due to the availability of premium quality grease which permits longer interval of relubrication, often sufficient enough to last between normal overhaul periods. The limiting speed factors for grease lubrication is given in Table 245 . When grease is used, for high speed and horizontal spindle applications grease of consistency grade NLGI No. 2 and for vertical spindle grease of consistency grade NLGI No. 3 may be used. Lithium hydroxystearate grease will meet most of the operating conditions in machine tools though some situations may not warrant such high quality grease. However, with a view to rationalise the grease requirement in machine tools, use of this grease will simplify the lubricant requirement of the machine tools. To meet the requirements of both high and low temperature conditions greases with a synthetic base oil instead of mineral oil are used. In this respect silicone grease is finding wide application. When lubricating with grease the quantity of grease packed inside the bearing should not be more than that required, otherwise excessive temperature would result. Normally bearings are packed to about 25 to 30% full.

When efficient cooling is required (a situation often encountered in taper roller bearings) oil lubrication is resorted to. Higher viscosity oils promote load capacity and larger film thickness but increase the heat generation. For high speed application a thinner oil is preferred. The limiting speed factors for oil lubrication are given in Table 246 . Guidance for the selection of appropriate viscosity grades is given in Table 247 .

SLIDING BEARINGS

For hand operation or slow speed and light loads grease will be adequate. When the operating speed exceeds 2 *m/sec.* or when better cooling is required oil lubrication is necessary. Viscosity is one of the important design parameters which influences the value of film thickness, load capacity, friction torques, etc. For a given speed a thicker oil forms a larger film thickness and improves load capacity but increases the heat generation whereas a thinner oil lowers the film thickness and load capacity but enables cooler running. The oil which is used should have a high Viscosity Index (*VI*) of 90 minimum. Straight mineral oil or any hydraulic oil can be used. For spindle bearing lubrication the viscosity of the oil generally lie in the range of 5 to 10 *cSt* at 40°C.

GEARS

Oil used for gear lubrication must have a high viscosity index and adhesion properties. The important factor in the selection of gear oil is its viscosity. Viscosity is decided depending upon the pitch line velocity. For high speed application a low viscosity oil should be used and vice-versa. The viscosity of an oil for spur and helical gears in relation to the pitch line velocity is as given below.

$$\eta = 500/\sqrt{v}$$ where η is the viscosity in centistokes at 40°C and v is the pitch line velocity in *m/sec.* The value of viscosity should be increased in case of high load coupled with high speed or when any shock load exists or when the ambient

temperature exceeds $30^{\circ}C$. The viscosity could be reduced when the ambient temperature is low ($< 10^{\circ}C$). Another important point to be noted is a thicker oil will give a higher film thickness i.e., it could prevent metal to metal contact thereby reducing the possible wear. On the other hand from the point of view of cooling, an oil of low viscosity is desired. Weighing the advantages the obvious choice of lubricant is based on minimum wear, namely more viscous oil. When a gear box is clustered with a number of gears and plain bearings working under varying operating conditions it is safer to base the selection of lubricant viscosity on the slowest and the most heavily loaded element.

For normally loaded spur and helical gears straight mineral oil without any additives will be sufficient. However for high speed gears mineral oil of turbine quality or hydraulic quality will be better as these oils will have a high oxidation stability at elevated temperatures. Heavy or shock load calls for mild or full EP additive oil. Similarly for worms under normal working load straight mineral oil will suffice if the bulk temperature does not exceed $75^{\circ}C$. To meet the requirements of high temperature applications synthetic oils of polyglycols are found good. Hypoid gears demand the most stringent lubricant requirement among all the gears due to high sliding combined with rolling and heavy load. Full EP oil is required for all hypoid gears.

SLIDEWAYS

The movement of machine tool slides must be smooth and precise without any stick slip in order to maintain fine tolerance on the workpieces. Stick slip is pronounced especially at slow speeds. Slides operate under wide operating conditions and hence they work under boundary as well as fluid film lubrication condition. Therefore both the physical and the chemical properties of the lubricant are important. To effect a favourable friction condition at slow speeds polar additives (oleic acid and stearic acid) are added to the lubricant. These additives enable the oil to adhere to the surface and are not sensitive to changes in oil viscocity. The viscosity of the slideway oils generally range from 30 to 150 cSt (at $40^{\circ}C$). For heavily loaded or slow speed applications oils of high viscosity may be used and for heavily loaded and high speed applications oils of low viscosity may be used. Grease is seldom used in slideways. However they can be used when the slideways are operated intermittently or when they run at slow speeds. Roller guideways are an example where grease is commonly used.

LEAD SCREW AND NUTS

The slideway lubricants can generally be used for leadscrews as well. Plain leadscrew and nuts are often lubricated with grease. For precision screws and for the highly streassed screws oil may be a better choice. Ball nuts are commonly packed with grease for life time. Alternatively thinner oils may be used.

Table 244 Desired properties of machine tool oils and their applications

Type of Oil	Application	Remarks
1. *Straight mineral oil* Oil containing additives to prevent rust and corrosion. Available as inhibited or uninhibited oils.	Used mainly in total loss system for general lubrication of all types of lightly loaded bearings, gears (except hypoid), slideways, lead screws, etc. Can also be used for bath lubrication if the bulk temperature of the oil is kept below 60° C.	In circulating systems, inhibited oils of minimum *VI* 35 should be used. To be preferred wherever possible as they are chemically inert and hence compatible with all bearing metals.
2. *Spindle oil* Low viscosity mineral oil with superior anti corrosion, anti oxidation and anti wear properties.	Used for all spindle bearing lubrication. Suitable for all types of lubrication systems like bath, pressure and mist lubrication.	Can also be used for electromagnetic clutches, hydrostatic bearings, etc.
3. *Hydraulic oil* Mineral oils with superior anti corrosion, anti oxidation, anti wear, anti foam, demulsification characteristics.	Generally meant for hydraulic systems but can be used for spindles, slideways and gear lubrication also.	Oil must possess a minimum *VI* of 90-95. If the oil contain Z.D.D.P as additive, check for its compatibility with bearing materials (for example it can attack white metals and silver lined bearings).
4. *Gear oil* Highly refined oil with mild EP additives to promote load carrying capacity. Contains usual additives to prevent rust and corrosion and to improve oxidation resistance, etc.	Suitable for both pressure and bath lubrication of heavily loaded gears except hypoid gears.	Can be used for leadscrews as well. Operating temperature should be limited to 70° C.
5. *Slideway oil* Highly refined oil with good boundary lubricating performance to prevent stickslip.	For lubrication of all plain slideways. Can be used for leadscrews and lightly loaded wormgears and feed gears.	*If the oil possesses high VI* in addition to slideway oil properties, it can be used for hydraulic system also. In such case bulk oil temperature must be limited to 50° C.

Table 245   Limiting speed factors for grease lubrication

Type of bearing	Speed factor nd_m [1] $n\sqrt{DH}$ [2]
Deep groove ball bearing standard design	500 000
Deep groove ball bearings and angular contact ball bearings having a contact angle of 15°; with plastic cages	750 000
Angular contact ball bearings, series 72B and 73B	400 000
Cylindrical roller bearings, single row	400 000
Double row NN30K [3] and NNU 49	500 000
Taper roller bearings	200 000
Angular contact thrust ball bearings	250 000
Thrust ball bearings	80 000

Notes: 1) n is the speed in *rpm* and d_m is the mean diameter of the bearing in *mm*.

        2)   When calculating DH for angular contact thrust ball bearings, H is the height of a pair of bearings and D is outside diameter of bearing in *mm*.

        3)   When the bearing arrangement incorporates an angular contact ball bearing, the speed factor of this bearing is decisive.

Courtesy SKF

Table 246   Limiting speed factors for different systems of oil lubrication

System of lubrication	Rolling bearings	Gears
	Speed factor	Pitch line velocity
Oil bath lubrication	⩽ 100 000	⩽ 12 *m/sec.*
Free flow lubrication	⩽ 400 000	—
Forced circulation lubrication	⩽ 750 000	—
Drip feed lubrication	⩽ 500 000	⩽ 3 *m/sec.*
Oil mist lubrication	⩽ 1000 000	—
Spray lubrication	⩽ 1300 000	> 12 *m/sec.*

Table 247 Oil viscosity recommendations for machine tool spindle bearings [1]

Type of machine	Speed factor ndm	Viscosity cSt at 40°C
Grinding machines	<300 000 >300 000 [2]	12-23 6-10
Lathes	< 100 000	34-55
Single and multi-spindle automatics.	> 100 000 > 300 000	17-32 12-23
Boring machines etc.	<100 000 > 100 000	34-55 17-32
Drilling and milling machines	<200 000 >200 000	17-32 12-23

Note: 1. If the bearings are lubricated in common with the gears, or are heavily loaded, a thicker oil is usually needed.
2. For oil mist lubrication a slightly thicker oil should be used, for example 15-25 cSt at 40°C.

Lubrication grooves

Lubrication grooves are provided for the retention and supply of lubricant in plain journal and thrust bearings and in slideways. Shape, size and location of lubrication grooves are dependent on type and size of the bearings, method of lubrication, type of lubricant, direction of load, operating speed, etc.

JOURNAL BEARINGS

In journal bearings the grooves are located at 90° from the load line in the direction of rotation. When the load is fixed and the rotation is unidirectional, a single axial groove covering 90% of the bearing length will suffice. If the load is fixed but the journal rotation is in both directions then two axial grooves at 180° to each other will be needed. When the load is variable in direction by more than 180° and also when the journal rotates in both directions, a circumferential groove at the centre of the bearing will be required. For short bearings with $L \leqslant D$ and running at moderate speeds, grooves can be replaced by a single hole located at the middle of the bearing.

THRUST BEARINGS

Radial grooves are provided in thrust bearings in order to provide hydrodynamic lubrication. Grooves approximately occupy about 20% of the thrust area. The lubricant supply point should be located at the inner diameter of the thrust washer so that the flow takes place outward along the grooves.

SLIDEWAYS

Grooves are normally provided at right angles to the direction of movement P and sometimes connected together by longitudinal grooves for the distribution of the lubricant over the entire bearing area. The lubrication groove shown in Fig. 167 is the best and the type shown in Figs. 168 and 169 can be resorted to only if it is not possible to effect the lubricant supply to all the grooves individually.

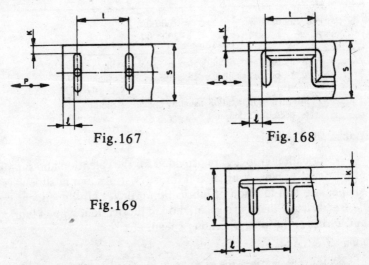

Fig.167   Fig.168

Fig.169

Lubricating grooves are provided only on surfaces which are always covered and only on one surface of a pair. The same type of the groove as on the slideways is provided on the gib.

a) The locating dimensions k and l from the end face on a flat surface are as follows:

$$k = s/6 \qquad l = t/2$$

b) The pitch t of the grooves is fixed depending upon the stroke length of the moving slide. It is recommended that t be less than the most frequently used length of stroke: greater the length of stroke, smaller the number of grooves and vice-versa. However, when it is not possible to fix the pitch on the basis of the stroke length, it can be calculated as follows.

$$t = s \text{ for low sliding speeds}$$
$$t = 1.5s \text{ for high sliding speeds}$$

The shape of the grooves are shown in Figs. 170 , 171 and 172 and their dimensions are given in Tables 248 , 249 , 250 and 251

SURFACE FINISH OF GROOVES

The surface roughness of lubricating grooves is kept in the range of $R_a = 1.6$ to 6.3 μm. Rounded edges (transition to the sliding surface) should be atleast as smooth as the sliding surface.

Fig. 170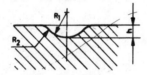

Table 248 Sizes of oil grooves for journal bearings

Internal dia. d of bearing over - upto	16-32	32-45	45-70	70-100	100-125	125-180	180-200
h	1	1.25	1.5	2	2.5	3	3.5
R_1	2	2.5	3	4	5	6	7
R_2	5	7	8	10	13	15	18

Table 249 Sizes of grease grooves for journal bearings

Internal dia. d of bearing over - upto	16-32	32-45	45-70	70-100	100-125	125-180	180-200
h	1.25	1.5	2	2.5	3	3.5	4
R_1	2.5	3	4	5	6	7	8
R_2	7	8	10	13	15	16	20

Fig.171

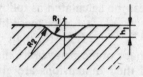

Table 250 Grooves on sliding surfaces of thrust bearings

Internal dia. d of bearing over - upto	16-32	32-45	45-70	70-100
h_1	0.75	1	1.25	1.5
R_1	1.5	2	2.5	3
R_2	4	5	6	8

Fig.172

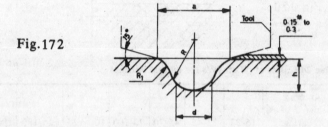

Table 251 Grooves for plane sliding surfaces

Nominal size of groove h	R	R_1	a	d	Recommended for width of guideway S over	upto
0.8	1	0.5	3	1.6	25	40
1.2	1.6	0.8	4	2.5	40	60
2	2.5	1.2	7	4	60	100
3	4	2	10	6	100	160
5	6	3	16	10	160	250

* Finishing allowance on the sliding surface when required.

GEAR BOXES OF MACHINE TOOLS

Nomenclature

$$\text{Transmission ratio} = \frac{rpm \text{ of the driver}}{rpm \text{ of the driven}} = \phi^m$$

$$\text{In the case of Ruppert drives } i = \frac{rpm \text{ of gear on shaft I}}{rpm \text{ of gear on shaft II}}$$

i_E Overall transmission ratio $= \dfrac{\text{Input } rpm}{\text{Max. output } rpm}$

n_o Input speed rpm

$n_1, n_2, \ldots n_n$ Output speeds (maximum to minimum) rpm

R Speed range ratio $= \dfrac{\text{Max. output } rpm}{\text{Min. output } rpm}$

$z_1, z_2, \ldots z_n$ Number of teeth of gears $1, 2, \ldots n$

Z Number of steps in the output speeds

ϕ Step ratio of the output speeds $(\geqslant 1)$

Gear boxes provide for a wide range of cutting speeds and torques from a constant speed power input enabling proper cutting speeds or torques to be obtained at the spindle as required in the case of cutting drives and the desired feed rates in the case of feed drives. Gear boxes are also used to effect inter-related motions between the workpiece and the tool as in the case of screw cutting and gear cutting.

Choice of parameters for gear boxes

The gear boxes of general purpose machine tools should provide a wide range of speeds and feeds consistent with the materials of tools and workpieces, the shape of tool, the type of machining process and the required quality of surface finish. Optimum cutting speeds and feed rates enable the operator to obtain optimum rate of metal removal and operation time.

When the specifications for the working range of the machine requires that a diameter range of $D_{Max.}$ to $D_{Min.}$ and a surface speed range from $V_{Max.}$ to $V_{Min.}$ be covered, the maximum and minimum obtainable speeds can be established as follows:

$$n_{Max.} = \frac{1000 V_{Max.}}{\pi D_{Min.}} \quad \text{and} \quad n_{Min.} = \frac{1000 V_{Min.}}{\pi D_{Max.}}$$

The speed range required is then

$$R = \frac{n_{Max.}}{n_{Min.}} = \frac{V_{Max.}}{V_{Min.}} \times \frac{D_{Max.}}{D_{Min.}}$$

where V in m/min

D in mm

With a constant speed power source there is a need for some method of varying the speed over this range. Stepless mechanical and electrical drive can provide infinitely variable speed variation. However, the torque/speed characteristics of available stepless drives do not meet the requirements of spindle drives which demand an increased driving torque to the spindle at lower output speeds in order to maintain a constant rate of metal removal. The stepless drives which do possess the required torque/speed characteristics are limited by the speed range over which these characteristics can be maintained. In order to provide for a wide range of operating speeds together with adequate torque at lower spindle speeds, it is necessary to use gear boxes which enable the required spindle speed range to be covered in a number of discrete steps.

Gear boxes with a large number of steps within a given range would be bulky and expensive. Hence they should be so designed that while fulfilling the functional requirement they are also economical to manufacture. The cost of a gear box is related to the number of shafts and bearings required and the total number and size of the gears. From different possible arrangements of gears, the layout which promotes compact size and lower cost while still fulfilling the technical requirement of the system should be chosen.

Standard speeds and feeds for machine tools

IS 2218 specifies a standard series of spindle speeds under full load based on geometric progression. The ratio between any two consecutive speeds in a geometric progression is constant. Basic geometric series for these speeds is $R20$ series of preferred numbers, with a step ratio, $\phi = \sqrt[20]{10}$ or 1.12 i.e., ratio between first and twenty first speed is 10. By selecting every second, third, fourth, or sixth speed in $R20$ series additional derived series known as $R20/2$, $R20/3$, $R20/4$, $R20/6$ can be obtained with step ratios of 1.26, 1.41, 1.58, 2 respectively. For calculation of gear transmission the permissible variation in speed is $+3\%$ to -2%. The motor shaft speeds of AC induction motors under full load are contained within the standard speed ranges. Standard steps and spindle speeds are indicated in page 145.

Standard feed rate values are specified in IS 2219 which are also based on geometric progression. The feed rates can be referred either to the spindle speed (*mm/rev.*) as in the case of turning, drilling or boring or independent of spindle speed (*mm/min.*) as in milling and planing. Standard steps and feeds are indicated in page 146.

Choice of step ratio

The relationship between step ratio ϕ, the number of steps Z and the speed range ratio R is given by,

$$\phi^{(Z-1)} = n_{Max}/n_{Min} = R \text{ or } Z = 1 + \log R/\log \phi$$

This is illustrated graphically, in Fig. 174. Hence it is evident that for a given speed range ratio R, the number of steps Z increases rapidly with a smaller step ratio ϕ. Thus in selecting ϕ and Z it is necessary to strike a compromise between the effort to minimise loss in cutting speed by making more steps and thereby complicating the design and the effort to reduce cost of gear box by making it as simple as possible. Step ratios $\phi = 1.26, 1.41$ and 1.58 for speed and feed series are commonly used in machine tools. The most frequently used number of steps Z are 3, 4, 6, 9, 12, 18 and 24.

For cutting threads of various pitches the lathe feed boxes have number of steps 48 to 60.

In order to develop the actual gear box layout, it is essential to establish: (a) transmission ratios, (b) layout of intermediate reduction gears and (c) number of teeth for the gears.

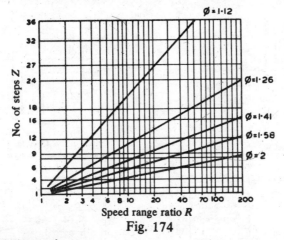

Fig. 174

Transmission ratios

The basic unit of speed change device is the two axes drive. Different transmission ratios can be arranged between output shaft and input shaft so that at constant speed of the input shaft, required speeds of the output shaft are produced. Table 76 gives number of teeth of gear for total number of teeth of gear pair for various step ratios.

Due to space limitations or when the number of teeth on gear or the tangential velocities on the pitch circle have to be limited, it becomes necessary to limit the transmission ratios.

The following can be taken as a guide

$i_{Max.} = 4$ and $i_{Min.} = \frac{1}{2}$ for spur gears, 1/2.5 for helical gears.

The accepted range for feed gear boxes (with slow speed gearing and small diameter gears) is $1/2.8 \leqslant i \leqslant 5$. Thus the limiting maximum range ratio in a two-axes gear box would be

$R_{Max.} = i_{Max.}/i_{Min.} = 8$

Gear layout and the ray diagram

LAYOUT OF INTERMEDIATE REDUCTION GEARS

Gear boxes with larger speed ranges and with greater number of steps can be designed by arranging several two axes gear drives (part drives) in series.

The number of steps in a gear box with more than two axes is equal to the product of the numbers of steps of the part drives (intermediate transmissions).

Two variants for obtaining six speeds are shown in Fig. 175 and Fig. 176.

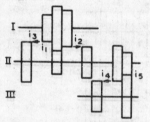

Fig. 175 3x2 Arrangement   Fig. 176 2x3 Arrangement

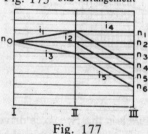 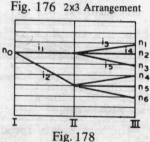

Fig. 177   Fig. 178

Ray diagram of 6 speed gear box

The layout of a gear box with more than two axes can be shown graphically on a ray diagram or speed diagram. Ray diagrams for 6 speed gear box are shown in Figs 177 and 178. Speeds are plotted vertically on a logarithemic scale and the shafts are shown as vertical parallel lines at equal distance from each other. Distance between consecutive speed lines is constant because of geometric progression. The speed values appear on the logarithemic scale at equal distance and the transmission ratios between two axes are indicated by the vertical distances between the corresponding speed values. As the distances between the axes are shown equal, the slope of the lines joining the speed values on different axes are an indication of the corresponding transmission ratios. The ray diagram shows at a glance the kinematic arrangement, shaft speeds at different stages and torque at various speeds.

The use of multi speed induction motors, mechanical stepless speed variators or DC motors, reduces the complexity of the gear box. Variable speed AC motors with frequency control are nowadays coming into prominance for spindle drives.

Methods of speed changing and layout

CHANGE GEARS

Speeds are changed by changing the gears of a group transmission between adjacent shafts with a constant centre distance.

In drives with inverse values of the transmission ratios, the same pair of change gears can be installed in the reverse order. This reduces the total number of gears required for the group transmissions.

Change gears are used in spindle drives where speeds are to be changed infrequently. The drawback of speed changing by change gears is that it is time consuming.

SLIDING GEARS

Gear boxes using sliding gears are very commonly used in machine tools. A cluster of two, three or occasionally four gears is arranged to slide on a spline shaft and any of these gears can be made to engage with the corresponding gear on second shaft.

In a three gear cluster the middle gear should be the largest to keep shaft length a minimum. The two side gears should be sufficiently small so that their tip diameters are smaller than the root diameter of the middle gear. A four gear cluster is rarely used as it results in excessive shaft length. In lathe feed boxes it is sometimes required to get more than three ratios between a shaft pair. In such cases a pair of cluster gears sliding on the same shaft are used. This ofcourse requires two shifting levers and an arrangement to block the movement of one shifter unless the other is in neutral position, is essential. To eliminate the danger of cluster gears coming into accidental engagement from neutral position by vibrations or otherwise, it is advisable to provide positive arrestation.

Normally it is necessary to stop the rotation of shafts before changing speed by sliding gears. However speed changes without stopping can be permitted, if surface speed of gears is below 0.5 *m/sec.*

COMPACT LAYOUT OF GEAR BOXES WITH SLIDING GEARS

The shortest constructional length is obtained if a narrow sliding gear block is axially shifted between a wide fixed gear block. The minimum constructional length is determined by the fact that during axial movement of the sliding gears one set of gears must be completely disengaged before the other set begins to come into mesh. The distance between two gears of the fixed block must, therefore be equal to two gear widths and the minimum length of a two stepped sliding gear drive is therefore, four gear widths (Fig. 179) and that of a three stepped sliding gear drive is seven gear widths. The length of the sliding gear block would increase if a flange with circular groove for gear shifting fork is provided. To eliminate this increased length a sliding fork which acts on the faces of one gear is often used.

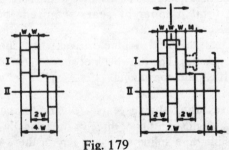

Fig. 179

NON-COMPOSITE LAYOUT

It is possible to cover larger speed ranges with greater numbers of steps by arranging several two shaft gear arrangements in series where each gear is either the driver or the driven member. Though this type of layout permits different modules to be chosen for different gear pairs, the axial space requirement and number of gears required are more. Non-composite layout, properly designed, offers a better solution over the composite or Ruppert drives, when the drive moment of inertia has to be a minimum. Fig. 180 shows a

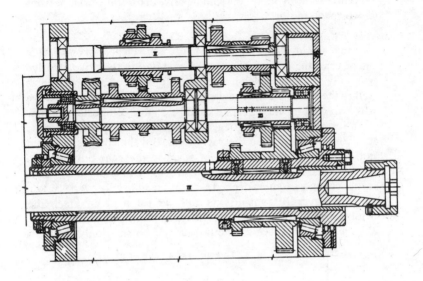

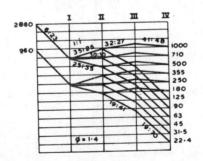

Fig.180 12 Speed gear box (using 2 speed motor) of a horizonal boring machine.

non composite layout for the spindle drive of a boring machine using a two speed motor.

COMPOSITE LAYOUT

In many layouts it is possible to use a driven gear of one part of the drive as the driving gear for the next part of the drive. A drive with only one gear which has the dual function of being driven as well as the driver is called single composite gearing whereas a drive with two dual function gears is called a double composite gearing.

Normally the number of gears in a gear box is equal to twice the total number of steps in the part drives. But the composite layout not only reduces the number of gears in the gear box but also reduces shaft lengths (Figs. 181 and 182). A double composite layout giving a geometric speed series is possible only over a certain range of overall transmission ratio i_E for a given kinematic arrangement as represented by the type of its ray diagram. Even in this range of i_E, space saving layouts are possible only around a

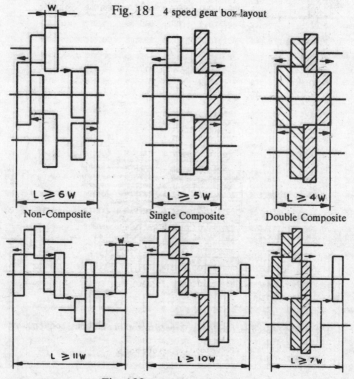

Fig. 181   4 speed gear box layout

Non-Composite   Single Composite   Double Composite

Fig. 182   6 Speed gear box layout

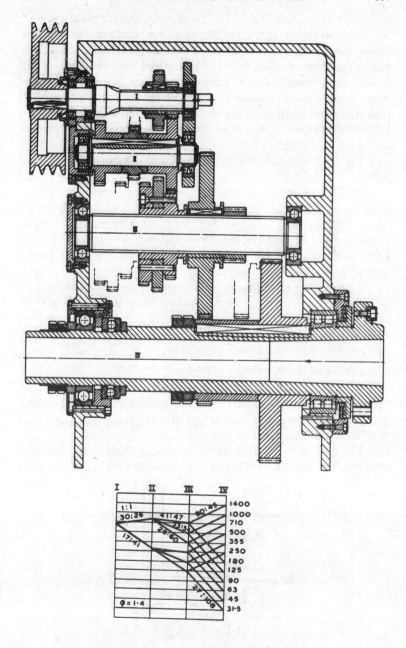

Fig.183 12 Speed gear box of a universal lathe

particular value of i_E. Table 252 gives equations for calculation of transmission ratios. If the equations give a negative value of i, it shows that such a layout is not possible. Table 252 also gives possible and favourable values of overall transmission ratio i_E for double composite drives. If output speeds are decided upon, the input speed can be chosen (or vice versa) to suit a value of i_E favourable for this type of layout wherever conditions permit. Triple composite systems do not produce a clean geometrically stepped speed range, hence are not used in machine tool gear boxes.

Fig. 183 shows an example of a double composite layout for a lathe.

Table 253 gives a survey of non-composite, single and double composite sliding gear drives for various number of steps. It shows that a nine step drive requires the same number of gears as an eight step drive and an eighteen step drive requires the same number of gears as sixteen step drive. This is the reason for preferring drives with 4, 6, 9, 12, 18, 27 and 36 steps.

The number of shafts required depends upon the numbers of gears which can be arranged in one sliding block and on the number of sliding gear blocks which can be arranged in parallel. In addition, the available space must be considered because the axial and radial dimensions of the drive may be limited.

SPEED CHANGING USING TOOTHED CLUTCHES

The positive toothed type clutches require small axial movement for engagement and disengagement. These toothed clutches can be engaged only at rest or when the relative speed of rotation does not exceed 0.7 m/sec.

Fig. 184 shows a two speed branching drive using toothed clutches. The gear block is keyed to the driving shaft I. The meshing

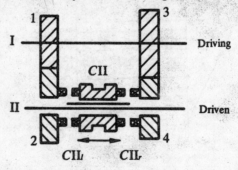

Fig. 184   Clutch drive

Table 252 Values of overall transmission ratio, i_E for double composite drives

No. of steps Z	Gear layout and type of ray diagram	Transmission ratios	Step ratio ϕ	Possible range of overall transmission ratio i_E	Value of i_E to give least centre dist.
4		$\left(1 - \dfrac{1}{\phi}\right)$ $i_1 = \left(\dfrac{1}{\phi^2} - 1\right)\dfrac{1}{i_E} + (\phi - 1)$ $i_2 = \phi^2 i_1 \qquad i_4 = \phi i_3$ $i_3 = \dfrac{i_E}{i_1} \qquad i_E = \dfrac{n_o}{n_1}$	1.12	2 to 5	3.54
			1.26	1·88 to 4.2	2.98
			1.41	1.68 to 3.35	2.51
			1.58	1.58 to 2.8	2.11
			2.00	1.18 to 2	1.58
6		$\left(1 - \dfrac{1}{\phi}\right)$ $i_1 = \left(\dfrac{1}{\phi^3} - 1\right)\dfrac{1}{i_E} + (\phi^2 - 1)$ $i_2 = \phi^3 i_1 \qquad i_3 = \dfrac{i_E}{i_1}$ $i_4 = \phi i_3 \qquad i_E = \dfrac{n_o}{n_1}$ $i_5 = \phi^2 i_3$	1.12	1.33 to 2.8	2.11
			1.26	1 to 2.11	1.67
			1.41	0.75 to 1.33	1.26
			1.58	0.6 to 1	1
			2.00	0.4 to 0.56	0.56

Table 252 Values of overall transmission ratio, i_E for double composite drives (Concld.)

No. of steps Z	Gear layout and type of ray diagram	Transmission ratios	Step ratio ϕ	Possible range of transmission ratio i_E	Value of i_E to give least centre distance
9		$i_1 = \dfrac{\left(1-\frac{1}{\phi}\right)}{\left(\frac{1}{\phi^3}-1\right)\frac{1}{i_E}+(\phi^2-1)}$ $i_2 = \phi^3 i_1 \qquad i_5 = \phi i_4$ $i_3 = \phi^6 i_1 \qquad i_6 = \phi^2 i_4$ $i_4 = \dfrac{i_E}{i_1} \qquad i_E = \dfrac{n_0}{n_1}$	1.12 1.26	1.88 to 2.66 1.26 to 1.88	2.37 1.26
		$i_1 = \dfrac{i_2}{\phi^3}$ $i_2 = \dfrac{\left(1-\frac{1}{\phi}\right)}{\left(\frac{1}{\phi^3}-1\right)\frac{1}{\phi^3}\frac{1}{i_E}+(\phi^2-1)}$ $i_3 = \phi^3 i_2,\ i_4 = \dfrac{i_E}{i_1};\ i_5 = \phi i_4$ $i_6 = \phi^2 i_4 \qquad i_E = \dfrac{n_0}{n_1}$	1.12 1.26	0.94 to 1.67 0.5 to 0.94	1.41 0.63

Note. Dotted Lines show the non-composite pairs.

Table 253 Survey of non-composite, single and double composite sliding gear drives

Number of steps Z	Possible division of Part drives a.b.c.d.e	Number of shafts	Non-composite drive		Single Composite		Double composite		Number of sliding gear blocks
			Number of gears = 2(a+b+c+d+e)	Total width minimum	Number of gears = 2(a+b+c+d+e) −1	Total width minimum	Number of gears = 2(a+b+c+d+e) −2	Total width minimum	
1	1.1	2	2	1	–	–	–	–	0
2	2.1	2	4	4	–	–	–	–	1
3	3.1	2	6	7	–	–	–	–	1
4	2.2	3	8	6	7	5	6	4	2
6	3.2	3	10	11	9	8	8	7	2
8	2.2.2	4	12	10	11	9	10	8	3
9	3.3	3	12	14	11	13	10	8	2
12	3.2.2	4	14	15	13	12	12	11	3
16	2.2.2.2	5	16	10-14	15	9-13	14	8-12	4
18	3.3.2	4	16	14-18	15	13-17	14	8-12	3
24	3.2.2.2	5	18	15-19	17	12-16	16	11-15	4
27	3.3.3	4	18	14-21	17	13-20	16	8-15	3
32	2.2.2.2.2	6	20	10-16	19	9-15	18	8-14	5
36	3.3.2.2	5	20	14-29	19	13-19	18	8-14	4

Note. ⊿ Preferred No. of steps.

gears 2 and 4 are idling on the driven shaft II and can be connected with it by clutch $C\ II_l$ on the left or by clutch $C\ II_r$ on the right so that shaft II is driven either by gears 1-2 and clutch $C\ II_l$ or gears 3-4 and clutch $C\ II_r$. It is advantageous to arrange the clutches on the driven shaft, if possible. Otherwise the idling gears would be driven by the fixed gear block at an excessive relative rotational speed on its shaft.

If gear 4 is made the driving gear and output is obtained at shaft 2, then it forms a back gear system, which is commonly used for speed reduction

SPEED CHANGING USING FRICTION CLUTCHES

Rapid, smooth speed changing without stopping rotation of shafts is possible with friction clutches. Gear pairs are constantly in engagement; one of the gears in each pair runs free on its shaft. The free gear in the gear pair can be coupled to the shaft by engaging its clutch.

Frictional losses and wear during idle rotation of continuously meshed gears and disengaged clutches are the main shortcomings of gear boxes, in which speeds are changed by means of friction clutches. Mechanically operated friction clutches require readjustment of clearance between the plates. The transmission of frictional heat from the clutches to the main spindle might affect the machining accuracy, if the clutches are built into the headstock and are operated frequently

Mechanically operated friction disc clutches are often used for disconnecting gear box from the drive to enable stopping the spindle without stopping the motor or for getting a quick change of transmission ratio.

Electromagnetically and hydraulically operated clutch type drives are particularly suitable for preselection gear boxes because the clutches required to be operative for a particular output speed can be set ready for engagement while the drive is still working at the previous output speed. At the moment of speed change, only the mechanism engaging the clutches is put into operation and the gear sets producing the desired transmission ratio are engaged. Many varients of clutch type gear boxes are possible. The two important ones are Ruppert and winding drives.

RUPPERT DRIVES

These are gear arrangements on two shafts using clutches. The gears are in constant mesh. The Ruppert drives provide simple and space saving layouts. Fig. 185 shows schematically the synthesis of an 8 step Ruppert drive from 3 part drives. This type of gear drive can be conveniently used for 4, 8 and 16 speeds.

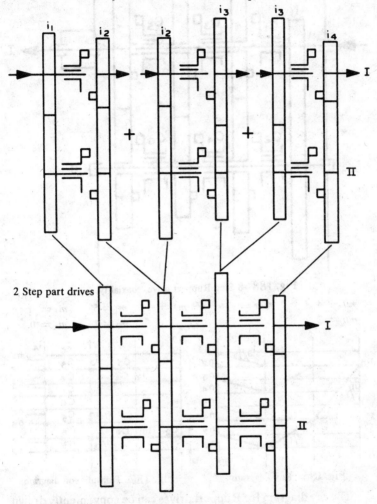

Fig. 185 Synthesis of 8 step Ruppert drive from three 2 step part drives (in back gear form) linked in series.

$$i = \frac{rpm \text{ of gear on shaft } I}{rpm \text{ of gear on shaft } II} \quad \begin{array}{l} i_1 = \phi^{m1}, \quad i_2 = \phi^{m2}, \quad i_3 = \phi^{m3} \\ i_4 = \phi^{m4} \end{array}$$

$$\frac{i_1}{i_2} = \phi^{mI}, \frac{i_2}{i_3} = \phi^{mII}, \frac{i_3}{i_4} = \phi^{mIII} ; \quad \frac{\text{Input } rpm}{\text{Max. output } rpm} = \frac{n_0}{n_1} = \phi^{mE}$$

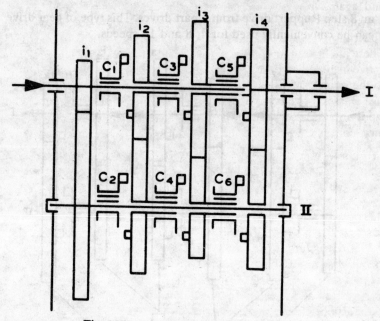

Fig. 188 8 Step Ruppert drive, coaxial type.

$m_I = 4$ $m_{III} = -1$ $m_E = -1$ $m_1 = 5$ $m_3 = -1$

$m_{II} = 2$ $m_2 = 1$ $m_4 = 0$

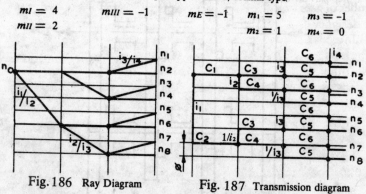

Fig. 186 Ray Diagram   Fig. 187 Transmission diagram

The ray diagram for Ruppert drives can be conveniently drawn using step ratios of the part drives   i.e., ϕ^{mI} , ϕ^{mII} , etc. instead of individual transmission ratios of the gear pairs (Fig. 186). However this has the disadvantage in that, that the speeds of the individual gears and the power flow in the different clutches cannot be easily recognised. Hence the ray diagram is represented as transmission diagram (Fig. 187). In this example (Fig. 188) the first part drive has an output once through clutch C_1

and again through clutch C_2 but reduced by ϕ^5 ($m_1 = 5$). The transmission ratio i_2 has been active twice but each time in opposite direction. Thus at clutches C_3 and C_4 two speeds each are available. These are further multiplied twice at the third part drive where i_3 is active in both directions. The transmission ratio i_4 brings the speeds to the output shaft.

Basically two types of drives are possible where (a) output and input are coaxial, (b) input on one shaft and output on the other (non-coaxial). Both have distinct advantages and disadvantages. Fig. 188 shows an 8 step coaxial Ruppert drive and Fig. 189 an 8 step non-coaxial drive.

Co-axial type

Output speeds must contain the input speed. As each of the part drives has a ratio 1:1, one of the overall transmission ratios is also 1:1.

The structure of the ray diagram and the position of the output speeds with respect to the input speed depends upon the values of the part drive step ratio exponents m_1, m_{II}, m_{III}, etc., and the overall transmission ratio exponent m_E respectively. Thus for an eight speed Ruppert drive there are 48 possibilities of the step ratio ray diagrams. Each of these ray diagrams can be used for innumerable number of gear box layouts. Acutal choice is left to the discretion of the designer.

In a coaxial drive, the shaft I (Fig. 188) has to be in two parts as the input gear has to be connected to one end and the output gear to the other. Though the separation can be anywhere along the shaft it is advantageous to have it near the walls to get a favourable bearing arrangement. Shaft II can be either stationary or coupled to any one of the gears.

Non-coaxial type

The ray diagram based on part drive step ratios for this type of drive is shown in Fig. 190 in which for the last part drive the effective step ratio is used (i.e., $i_3/i_4 \times i_4$). As the output speeds depend upon the transmission ratio i_4 of the last pair, theoretically any desired position of the output speeds is possible, the limitation being the magnitude of i_4. Fig. 191 shows the transmission diagram.

The non-coaxial type of drives offer simple layout possibilities and a convenient position of the output speed range can be chosen with respect to the input speed. However the coaxial type offers many variant layout possibilities.

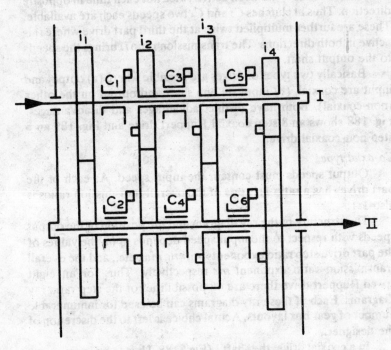

Fig. 189 8 Step Ruppert drive, non-coaxial type

$$m_I = -1 \qquad m_{III} = -4 \qquad m_1 = 1 \qquad m_3 = 0 \qquad m_E = -1$$
$$m_{II} = 2 \qquad \qquad \qquad m_2 = 2 \qquad m_4 = 4$$

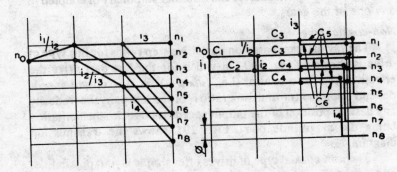

Fig. 190 Ray diagram   Fig. 191 Transmission diagram

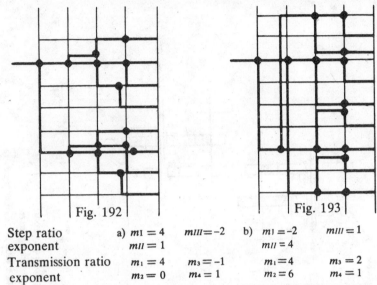

Fig. 192		Fig. 193	

| Step ratio exponent | a) $m_I = 4$ $m_{II} = 1$ | $m_{III} = -2$ | b) $m_I = -2$ $m_{II} = 4$ | $m_{III} = 1$ |
| Transmission ratio exponent | $m_1 = 4$ $m_2 = 0$ | $m_3 = -1$ $m_4 = 1$ | $m_1 = 4$ $m_2 = 6$ | $m_3 = 2$ $m_4 = 1$ |

Favourable (Fig. 192) and unfavourable (Fig. 193) transmission diagram of a 8 speed Ruppert drive

Selection of transmission ratios for Ruppert drives

The choice of step ratio exponents and transmission ratio exponents should be such as to promote favourable power flow between the input and output shafts. The transmission diagram (Fig. 192) shows a favourable power flow and Fig. 193 shows an unfavourable power flow.

In the coaxial drive (Fig. 188) the transmission ratio i_4 transmits power from shaft II to shaft I whereas in the non-coaxial drive (Fig. 189), it drives in the opposite direction. Only in idle running the power flow direction will be reversed. It is advantageous to choose the transmission ratio exponent m_4 negative in the case of coaxial drives and positive in the case of non-coaxial drives (Fig. 191), if the majority of the output speeds are to be maintained on the step down side. If the idle running at high speeds is to be avoided, a smaller value of the exponent has to be chosen.

The gear pairs having ratios i_2 and i_3 will be running under load in both directions (i.e., from shaft I to II and vice versa). It is always advantageous to maintain these ratios close to 1. Only the gear pair having the ratio i_1, drives always in the same direction i.e., from shaft I to II and therefore a larger value of transmission ratio can be

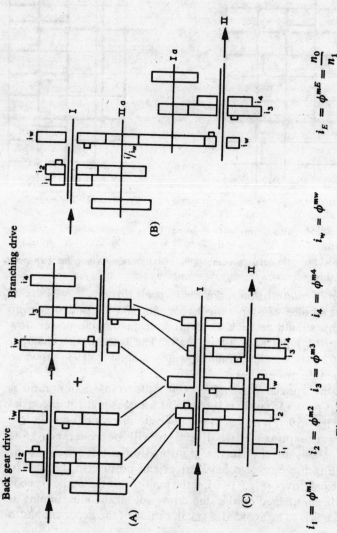

Fig. 194 Synthesis of a 9 step winding drive (C) from

A) a back gear drive and a branching drive

B) two branching drives with a ratio of $1/i_w$ linking them

$i_1 = \phi^{m1}$ $i_2 = \phi^{m2}$ $i_3 = \phi^{m3}$ $i_4 = \phi^{m4}$ $i_w = \phi^{mw}$ $i_E = \phi^{mE} = \dfrac{n_0}{n_1}$

chosen. However the disadvantage is that the drive will require more space due to larger torque coming on the shaft II. For minimum space requirement, the transmission ratio i_1 should be small and the transmission ratio i_4 should be large. But this is at the cost of high speed idle running of gear pair in i_4.

WINDING DRIVES

These are two shaft gear drives. Unlike Ruppert drives, these drives contain sliding gears and the input and output shafts are not coaxial. Toothed clutches are used to couple the gear clusters.

Fig. 194 shows synthesis of a nine step winding drive. The part drives can be either a back gear drive and a branching drive (Fig.194A) having a common ratio i_w or two branching drives (Fig.194B) having one transmission ratio i_w in common and linked by the inverted common ratio (i.e., $1/i_w$). If the part drives are limited to 2 or 3 steps, it is possible to layout 4, 6 and 9 step winding drives. The transmission ratios of the part drives i_1 and i_2 can be interchanged with i_3 and i_4 and if the common ratio i_w is maintained, the output speeds will remain unchanged (Figs. 195 A & B).

Fig. 195A shows the ray diagram for a nine step winding drive as four shaft version. In Fig. 195B the same drive is represented as a two shaft version. Fig. 195C shows the diagram with the part drive ratios interchanged

Though these drives save considerable space, two gears and a shaft, the savings are offset to some extent by the requirement of hollow shafts and toothed clutches. Also there is the need for slightly longer shafts than in the case of conventional drives.

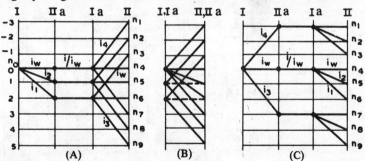

Fig. 195 Ray diagram of a 9 step winding drive

A) Developed form
B) Two shaft version
C) Developed form (i.e., i_1 and i_2 interchanged with i_3 and i_4)

$m_E = -3$; $m_1 = 2$;
$m_2 = 1$; $m_w = 0$;
$m_3 = 3$ and $m_4 = -3$

Out of the many variants possible, a few winding drives which enable compact layouts are given in Table 254.

Fig. 196 shows a 9 step winding drive used in a special purpose machine. Fig. 197 shows a gear box for a co-ordinate table using electromagnetic clutches. Fig. 198 is an example of a planing machine drive using a friction disc clutch, a tooth clutch and a pair of change gears.

Feed gear boxes

Feed gear boxes provide the feed movement which is either dependent on (turning, drilling) or independent of (milling) the cutting movement and it may be either continuous (turning, drilling, milling) or intermittent (planing). The power requirement for the feed drives is usually relatively low. Depending on the purpose of the machine tool, the number of feed steps, range of feeds, frequency of feed change are established.

Feed gear boxes are classified in accordance with the type of geared mechanism used to set up the feeds.

FEED GEAR BOXES WITH CHANGE GEARS

Change gears are commonly used in feed gear boxes, especially for cutting threads of non-metric pitches (inch threads, Dp threads etc.). It is advantageous to use two pairs as shown in Fig. 199. Drive is transmitted from shaft I to shaft II, through gears a, b, c and d. The gears b and c are carried on a cantilever pin in the swinging bracket to provide for various gear ratios between the two shafts.

Table 254   Favourable winding drives

No. of steps	4			6					9				
Exponents of ϕ (m_E)	m_1	m_w	m_3	m_1	m_2	m_w	m_3	m_4	m_1	m_2	m_w	m_3	m_4
2				4	3	2	5		4	3	2	8	5
1	2	1	3	3	2	1	4		3	2	1	7	4
0	1	0	2	2	1	0	3		2	1	0	6	3
-1	0	-1	1	1	0	-1	2		1	0	-1	5	2
-2	0	1	-1		1	0	2	-2	0	-1	-2	4	1
-3	-1	0	-2		-1	0	2	-2	2	1	0	3	-3
-4	-2	-1	-3	0	-1	1	-2		1	-1	0	3	-3
-5				-1	-2	0	-3		-1	-2	0	3	-3
-6				-2	-3	-1	-4		1	0	2	-1	-4

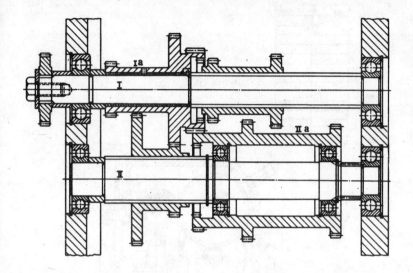

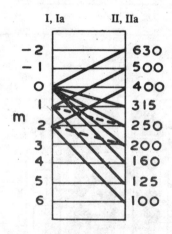

$m_E = -2$

$m_1 = 3$ $m_2 = 2$ $m_w = 1$

$m_3 = -2$ $m_4 = 4$ $\phi = 1 \cdot 25$

$$i = \phi^m$$

Fig. 196   9 Step winding drive of a SPM

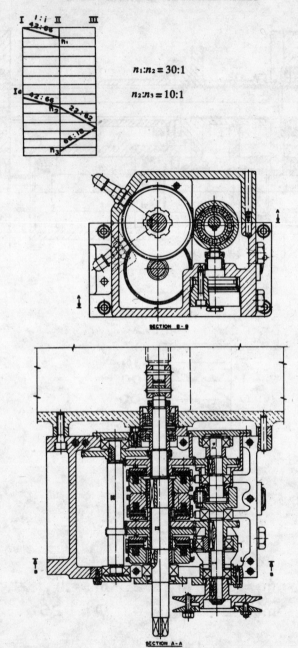

$$n_1 : n_2 = 30 : 1$$

$$n_2 : n_3 = 10 : 1$$

Fig. 197 High reduction gear box using electromagnetic clutches and a planetary drive

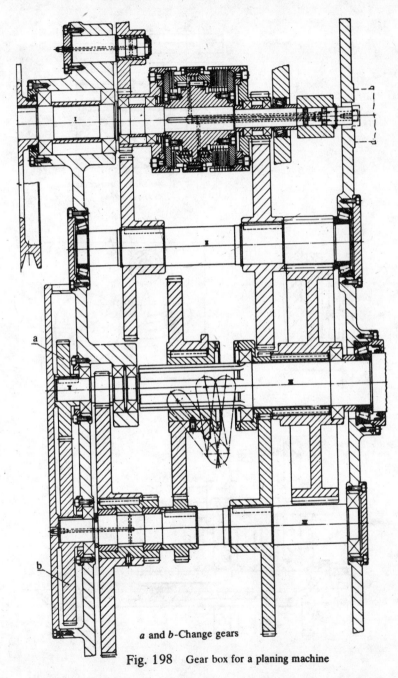

a and *b*-Change gears

Fig. 198   Gear box for a planing machine

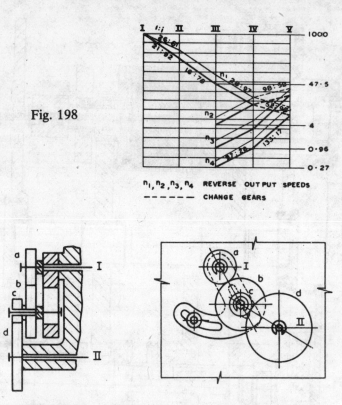

Fig. 198

n_1, n_2, n_3, n_4 REVERSE OUTPUT SPEEDS

– – – – –   CHANGE GEARS

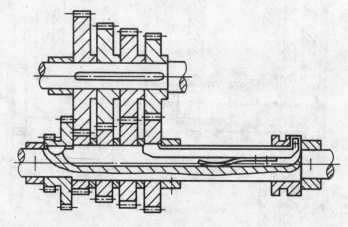

Fig. 199 Swinging bracket for change gears in feed gear boxes

Fig. 200 Sliding key drive

FEED GEAR BOXES WITH SLIDING GEARS

This type of feed gear boxes are more suitable for frequent feed changing than feed boxes with change gears and are therefore widely used in general purpose machine tools

FEED GEAR BOXES WITH CONSTANT MESH GEARS AND SLIDING KEY
Fig. 200

Compact construction and short length of assembly and the control of engagement of a gear pair with a single lever are the main advantages of this type feed gear boxes. In this type a number of gears run free on a shaft. Any one of these can be connected with the shaft by a sliding key which can be axially moved to engage one of the gears. However, the long and deep keyway weakens the shaft. A certain clearance between the key and keyway is unavoidable and local bearing area of the key is also limited. Hence this type is used for transmitting relatively small torques.

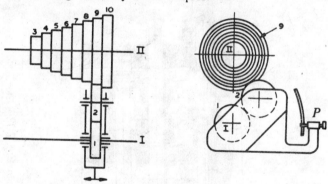

Fig. 201   Norton type gear box

FEED GEAR BOXES WITH TUMBLER GEAR (NORTON TYPE)

The principle of Norton type of arrangement is shown in Fig. 201 . Tumbler gear 1 which can slide on shaft I can engage through the intermediate gear 2 with any of gears 3, 4 to 10 on shaft II. Gear 2 is located on a swinging and sliding lever so that it can engage gears 3 to 10 of different diameters. The lever can be fixed in any desired ratio position with the help of stop pin *P*. All load coming on the intermediate gear is taken up by this pin.

Less number of gears and short length of shaft are the main advantages of this arrangement. But it has low rigidity and requires a long, odd shaped opening in the gear box wall for the operation of the swinging lever and it is difficult to protect the mechanism against dust and to ensure proper lubrication. Closed type of Norton gear

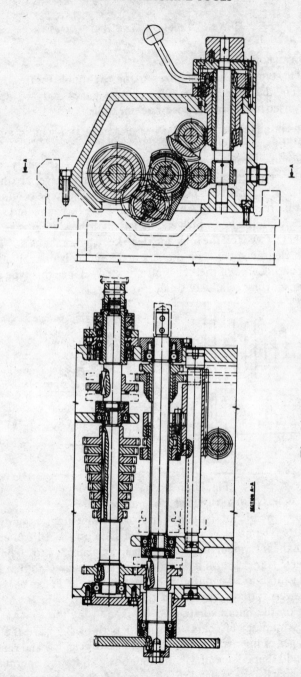

Fig. 202   Closed Norton type gear box for a lathe

box eliminates some of the disadvantages. An example as shown in Fig. 202 is used in a centre lathe.

MEANDER DRIVE

A Meander drive is a three shaft mechanism made up of a series of identical double cluster gears and a sliding carrier with a tumbler gear (Fig. 203). Such features as the single lever controls, small axial overall size and wide range ratio enable this drive to be conveniently used as the first extension group of the feed mechanism.

In Fig. 203 the gear 1 is fixed to the driving shaft I and the remaining gears, always engaged in pairs, are freely mounted on shaft II and I. Gear X can be engaged with any of the gears on shaft II through the intermediate gear Y by shifting the swinging lever on shaft III. All the cluster gears rotate continuously in mesh including those which do not participate in a particular engagement.

Another variant of Meander drive where the tumbler gear is replaced by sliding gear which engages only the larger gears of the clusters is shown in Fig. 204 . The gear X slides on shaft III. Ofcourse this increases the required number of gears and length of feed box.

In Meander drive for speeds to be in geometrical series, the following conditions are to be fulfilled.

$$z_1 = z_3 = z_5 = z_7 \quad \dots \quad z_x$$

$$z_2 = z_4 = z_6 = z_8$$

$$\frac{z_1}{z_2} \cdot \frac{z_3}{z_4} = \frac{z_5}{z_6} \cdot \frac{z_7}{z_8} = \phi^m$$

where $m = 1, 2, 3, 4$, etc.

CONTROL MECHANISMS IN GEAR BOXES

Control mechanisms are used for shifting of gears and operation of clutches and brakes for change of speeds and feeds, starting and stopping of drives.

Shifting of gears and clutches are done by shifters and levers. Shifters, usually single sided are used for small shifting stroke. The swinging movement of the shifter is converted into axial movement of the gear through a freely mounted slider engaging in the groove in the flange of the sliding gear (Fig. 205). The slider is made of bronze, freely rotating on a steel pin permanently fixed to the arm of shifter lever or of steel in highly stressed shifters in heavy machine tools (Fig. 206). Sliders with integral pins (Fig. 207) can also be used. In lightly stressed shifters e.g., in feed boxes, it is possible to use cast iron sliders. The radial clearance between the slider and the flange groove is 0.25 to 0.6 *mm*. The various typified sizes of sliders without and with pin are given in Table 255 and 256 respectively.

Double sided swinging lever (Fig. 208) is used only when it is required to exert considerable shifting force (in some friction clutches, brakes, etc.).

A number of other methods are also used for carrying out shifting operations e.g., rack and pinion arrangements, joystick controls (Fig. 209) drum cams (Fig. 210), hydraulic or pneumatic cylinders, etc. Joystick controls are suitable for small shifting strokes. Joysticks and drum cams also provide centralized controls and interlocks.

Gear shifting can be done by means of hydraulic or pneumatic cylinders. A hydraulic scheme for gear shifting in a 4 step gear box where individual speed step is obtained by engaging two double sided toothed clutches is shown in Fig. 211 . Clutches C_1 and C_2 are controled by hydraulic cylinders H_1 and H_2 through a 4 position rotary valve R. The valve contains an interlock which prevents its operation when the main clutch and brake is still giving drive to the gear box. Hydraulic preselection of speeds can be incorporated in the circuit.

General considerations in the design of gear box

GEAR BOX LOCATION

The design of gear box is intimately linked with the whole structure of the spindle drive. The gear box can be built integral into the spindle head housing. This type of arrangement promotes more compact spindle drive, higher localisation of controls, fewer housings and less assembly work involving the fitting of joining surfaces. Main drawback is the possibility of transmitting vibration from the gear box to the spindle, heating of the spindle head by the heat generated in the gear box.

The gear box can also be arranged in separate housing and linked to the spindle head through belt transmission. This type of arrangement has the advantage that neither the heat generated by friction losses nor vibrations developed in the gear box are transmitted to the spindle head.

GEAR BOX SIZE

If the size reduction is one of the primary criterion then winding or Ruppert drives offer advantageous solutions. Also maintaining

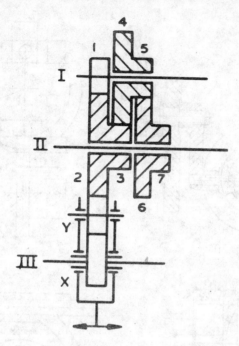

Fig. 203 Meander drive with swinging lever

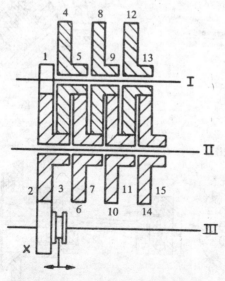

Fig. 204 Meander drive with sliding gear

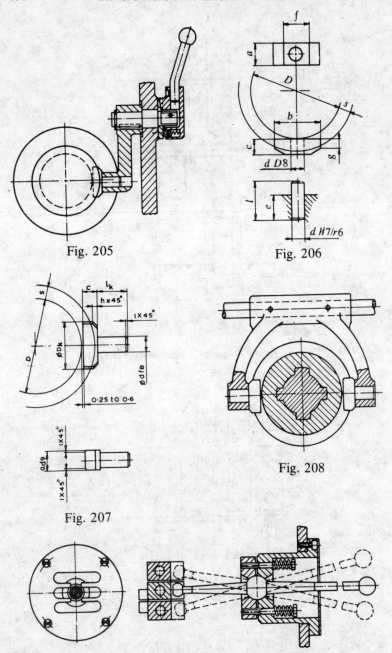

Fig. 205

Fig. 206

Fig. 207

Fig. 208

Fig. 209

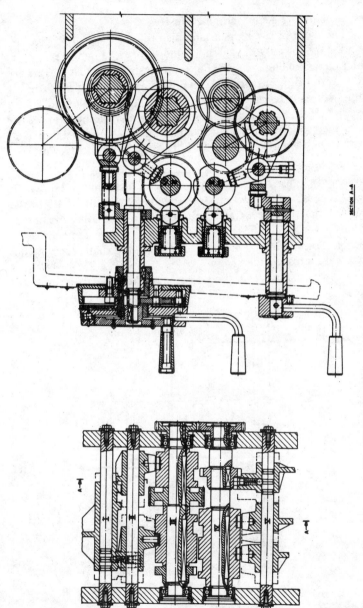

SECTION A-A

Fig. 210 Gear shifting arrangement using drum cams and levers

the transmission ratios in the part drives such that $i_{Max.} \times i_{Min.} = 1$ (i.e. a symmetrical ray diagram), helps in reducing radial dimensions.

Design of sliding gears and gear clusters

The teeth of sliding gear and also that of the mating gear are rounded off for easy engagement and to prevent damage of gears under impact of engagement. Different designs for cluster gears are used depending upon the space avaiable and method of manufacture. A few are shown in Figs. 180, 183, 196 and 198.

The clearance between the fixed gear and sliding gear, in the disengaged condition, in small and medium machine tools should be 1.5 to 3 mm. In heavy machine tools it is upto 5 mm and even more.

GENERAL PRINCIPLES

Gear boxes are not generally designed to transmit full power at lowest speeds. Lower two or three speeds (maximum of 1/3 of the total number of speeds) are omitted while calculating maximum torque. If the calculation is made as per the lowest speeds, gear forces will be very high and the size of gears and shafts carrying the gears will increase excessively.

The kinematic layout should be such as to avoid overlapping of output speeds as far as possible.

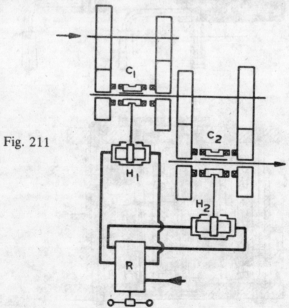

Fig. 211

Table 255

D		a	b	c	d	l	e	f	g	S
Over	Upto									
20	40	10	20	8	6	18	11	12	5	7
30	50	12	25	8	7	20	13	14	5	7
40	70	16	36	10	10	25	16	20	6	9
60	90	20	45	12	12	32	21	25	8	10

Table 256

D		a	D_k	c	d	l_k	L	i	S
Over	Upto								
30	50	10	25	8	8	16	2.5	0.4	7
50	75	12	32	10	10	20	4	0.4	8
75	100	14	40	12	12	25	4	0.4	9
100	140	18	50	14	16	32	4	1	10
140	200	22	63	18	20	40	6	1	14
200	300	28	80	24	25	50	6	1	20

MACHINING DATA

MACHINING DATA

The machining recommendations are merely guidelines as speeds and feeds for any machining operation may vary considerably. The optimum performance or efficiency of any machining operation includes factors other than the proper selection of speeds and feeds. Variables such as part configuration, condition of machine, type of fixturing, dimensional tolerance and surface finish requirements all affect performance.

POWER AND FORCE REQUIREMENTS IN MACHINING

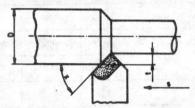

Table 257 Turning

Diameter of workpiece	D	mm	
Revolutions per minute	n	rpm	
Cutting speed	v	m/min.	$v = \pi Dn/1000$
Feed per revolution	s	mm/rev.	
Feed per minute	S_m	mm/min.	$S_m = sn$
Depth of cut	t	mm	
Metal removal rate	Q	cm³/min.	$Q = stv$
Approach angle	x	deg.	
Average chip thickness	a_s	mm	$a_s = s \sin x$
Unit power	U	kW/cm³/min.	Table 269
Correction factor for flank wear	K_h	-	Table 270
Side rake angle	γ	deg.	
Correction factor for rake angle	K_γ	-	Table 271
Power at the spindle	N	kW	$N = UK_hK_\gamma Q$
Efficiency of transmission	E	%	
Power of the motor	N_{el}	kW	$N_{el} = N/E$
Tangential cutting force	P_z	kgf	$P_z = 6120N/v$
Torque at spindle	T_s	kgf.m	$T_s = 975N/n$

POWER AND FORCE REQUIREMENTS IN MACHINING

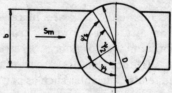

Table 258   Milling — Face milling

Diameter of cutter	D	mm	
Revolutions per minute	n	rpm	
Cutting speed	v	m/min.	$v = \pi Dn/1000$
Feed per tooth	S_z	mm	
Number of teeth	Z		
Feed per minute	S_m	mm/min.	$S_m = S_z.Z.n$
Depth of cut	t	mm	
Width of cut	b	mm	
Metal removal rate	Q	cm³/min.	$Q = b.t.S_m/1000$
Approach angle	x	deg.	
Average chip thickness **mm**	$a_s = 57.3\ S_z.\sin x\ (\cos\psi_1 - \cos\psi_2)/\psi_s^{\circ}$		
Unit power	U	kW/cm³/min.	Table 269
Correction factor for flank wear	K_h		Table 270
Radial rake angle	γ	deg.	
Correction factor for radial rake angle	K_γ	-	Table 271
Power at the spindle	N	kW	$N = UK_hK_\gamma Q$
Efficiency of Transmission	E	%	
Power of the motor	N_{el}	kW	$N_{el} = N/E$
Tangential cutting force	P_z	kgf	$P_z = 6120\ N/v$
Torque at spindle	T_s	kgf.m	$T_s = 975\ N/n$

POWER AND FORCE REQUIREMENTS IN MACHINING

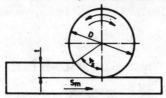

Table 259 Milling–Slab milling, End milling, Slot milling

Diameter of cutter	D	mm	
Revolutions per minute	n	rpm	
Cutting speed	v	m/min.	$v = \pi Dn/1000$
Feed per tooth	S_z	mm	
Number of teeth	Z		
Feed per minute	S_m	mm/min.	$S_m = S_z Zn$
Depth of cut	t	mm	
Width of cut	b	mm	
Metal removal rate	Q	cm³/min.	$Q = btS_m/1000$
Average chip thickness	a_s	mm	$a_s = 114.6\,S_z t/D\,\psi_s^{\circ}$
Unit power	U	kW/cm³/min.	Table 269
Correction factor for flank wear	K_h		Table 270
Radial rake angle	γ	deg.	
Correction factor for radial rake angle	K_γ		Table 271
Power at the Spindle	N	kW	$N = UK_h K_\gamma Q$
Efficiency of transmission	E	%	
Power of the motor	N_{el}	kW	$N_{el} = N/E$
Tangential cutting force	P_z	kgf	$P_z = 6120N/v$
Torque at spindle	T_s	kgf.m	$T_s = 975N/n$

POWER AND FORCE REQUIREMENTS IN MACHINING

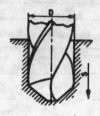

Table 260   Drilling

Diameter of Drill	D	mm	
Revolutions per minute	n	rpm	
Cutting speed	v	$m/min.$	$v = \pi Dn/1000$
Feed per revolution	S	$mm/rev.$	
Material factor	K	-	Table 272
Power at the spindle	N	kW	$N = 1.25D^2Kn(0.056 + 1.5S)/10^5$
Efficiency of transmission	E	$\%$	
Power of the motor	N_{el}	kW	$N_{el} = N/E$
Torque	T_s	$kgf.m$	$T_s = 975\ N/n$
Thrust	T_h	kgf	$T_h = 1.16KD(100S)^{0.85}$

1.   The power and torque values include an allowance of 30% for dull tools.
2.   The thrust values include an allowance of 40% for dull tools.
3.   The above formulae are applicable to drilling with tools conforming to accepted workshop standards.

POWER AND FORCE REQUIREMENTS IN MACHINING

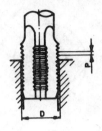

Table 261 Tapping

Thread diameter	D	mm	
Revolutions per minute	n	rpm	
Thread pitch	p	mm	
Meterial factor	K	-	Table 272
Power at the spindle - 60%thread engagement	N N_1	kW	$N_1 = 0.231DP^2nK/10^4$
- 75% thread engagement	N_2	kW	$N_2 = 0.326DP^2nK/10^4$
- 90% thread engagement	N_3	kW	$N_3 = 0.433DP^2nK/10^4$
Efficiency of transmission	E	$\%$	
Power of the motor	N_{el}	kW	$N_{el} = N/E$
Torque	T_s	$kgf.m$	$T_s = 975N/n$

Note: The above formulae are applicable to sharp tools only. For dull tools and chip clogging add 30% more allowance

POWER AND FORCE REQUIREMENTS IN MACHINING

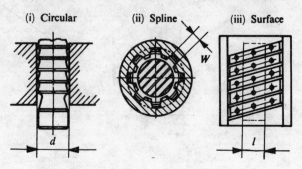

(i) Circular   (ii) Spline   (iii) Surface

Table 262 Broaching

Cutting speed	v	m/min.	
Maximum length of workpiece	$l_{Max.}$	mm	
Pitch of the broach teeth	p	mm	
Number of teeth cutting at a time	Z		$Z = \dfrac{l_{max}}{p} + 1$
Rake angle	γ	deg.	
Tensile strength of workpiece	σ	kgf/mm²	
Rise per tooth	S_z	mm	
Specific cutting force	K_s	kgf/mm²	$K_s = 450 + 3\sigma - 11\gamma - 2500 S_z$
Sum of the lengths of all the teeth engaged at any instant	L^*	mm	
Cutting force	P_z	kgf	$P_z = K_s L S_z$

* $L = \pi d Z$

    $L = W \times Z$

    $L = l \times Z$ (l = width to be broached)

POWER AND FORCE REQUIREMENTS IN MACHINING

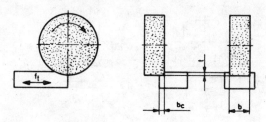

Table 263 Grinding-Surface grinding, horizontal spindle

Diameter of the wheel	D	mm	
Wheel revolutions per minute	n	rpm	
Peripheral wheel speed	v	$m/sec.$	$v = \dfrac{\pi D n}{1000 \times 60}$
Depth of grind	t	$mm/pass$	
Cross feed-traverse grinding	b_c	$mm/pass$	
Width of cut-plunge grinding	b	mm	
Table traverse feed rate	f_t	$mm/min.$	
Metal removal rate - traverse grinding	Q Q_1	$cm^3/min.$	$Q_1 = b_c.t.f_t./1000$
- plunge grinding	Q_2	$cm^3/min.$	$Q_2 = b.t.f_t/1000$
Unit power	U	$kW/cm^3/min.$	Table 273
Power at the spindle	N	kW	$N = U.Q$
Efficiency of transmission	E	$\%$	
Power of the motor	N_{el}	kW	$N_{el} = N/E$
Tangential cutting force	P_z	kgf	$P_z = 102N/v$
Torque at spindle	T_s	$kgf.m$	$T_s = 975\ N/n$

POWER AND FORCE REQUIREMENTS IN MACHINING

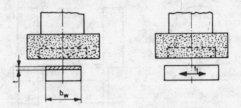

Table 264   Grinding-Surface grinding, vertical spindle, reciprocating table

Diameter of the wheel	D	mm	
Wheel revolutions per minute	n	rpm	
Peripheral wheel speed	v	$m/sec.$	$v = \dfrac{\pi D n}{1000 \times 60}$
Depth of grind	t	$mm/pass$	
Width of workpiece	b_w	mm	
Table traverse feed rate	f_t	$mm/min.$	
Metal removal rate	Q	$cm^3/min.$	$Q = \dfrac{b_w.t.f_t}{1000}$
Unit power	U	$kW/cm^3/min.$	Table 273
Power at the spindle	N	kW	$N = U.Q$
Efficiency of transmission	E	$\%$	
Power of the motor	N_{el}	kW	$N_{el} = N/E$
Tangential cutting force	P_z	kgf	$P_z = 102\, N/v$
Torque at the spindle	T_s	$kgf.m$	$T_s = 975\, N/n$

POWER AND FORCE REQUIREMENTS IN MACHINING

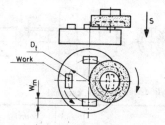

Table 265 Grinding-Surface grinding, vertical spindle, rotary table

Diameter of wheel	D	mm	
Wheel revolutions per minute	n	rpm	
Peripheral wheel speed	v	$m/sec.$	$v = \dfrac{\pi D n}{1000 \times 60}$
Table revolutions per minute	n_t	rpm	
Average diameter of workpiece path on rotary table	D_t	mm	
Peripheral work speed	v_w	$m/min.$	$v_w = \dfrac{\pi D_t n_t}{1000}$
Maximum contact width at wheel on work	W_m	mm	
Plunge infeed per table revolution	S	$mm/rev.$	
Plunge infeed rate per minute	f_p	$mm/min.$	$f_p = S.n_t$
Metal removal rate	Q	$cm^3/min.$	$Q = \pi D_t f_p W_m / 1000$
Unit power	U	$kW/cm^3/min.$	Table 273
Power at the spindle	N	kW	$N = U.Q$
Efficiency of transmission	E	$\%$	
Power of the motor	N_{el}	kW	$N_{el} = N/E$
Tangential cutting force	P_z	kgf	$P_z = 102 \, N/v$
Torque at the spindle	T_s	$kgf.m$	$T_s = 975 \, N/n$

POWER AND FORCE REQUIREMENTS IN MACHINING

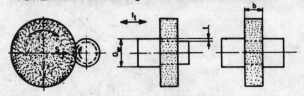

Table 266   Grinding-Cylindrical grinding, external

Diameter of wheel	D	mm	
Wheel revolutions per minute	n	rpm	
Peripheral wheel speed	v	m/sec.	$v = \dfrac{\pi\,Dn}{1000 \times 60}$
Diameter of workpiece	D_w	mm	
Workpiece revolutions per minute	n_w	rpm	
Peripheral work speed	v_w	m/min.	$v_w = \pi\,D_w n_w / 1000$
Depth of grind-traverse grinding	t	mm/pass	
Table traverse feed rate — traverse grinding	f_t	mm/min.	
Width of cut-plunge grinding	b	mm	
Plunge infeed rate per work revolution	S	mm/rev.	
Plunge infeed rate per minute	f_p	mm/min.	$f_p = S.n_w$
Metal removal rate — traverse grinding — plunge grinding	Q Q_1 Q_2	 cm³/min. cm³/min.	 $Q_1 = \pi\,D_w\,tf_t/1000$ $Q_2 = \pi\,D_w b f_p /1000$
Unit power	U	kW/cm³/min.	Table 273
Power at the spindle	N	kW	$N = UQ$
Efficiency of transmission	E	%	
Power of the motor	N_{el}	kW	$N_{el} = N/E$
Tangential cutting force	P_z	kgf	$P_z = 102\,N/v$
Torque at spindle	T_s	kgf.m	$T_s = 975\,N/n$

POWER AND FORCE REQUIREMENTS IN MACHINING

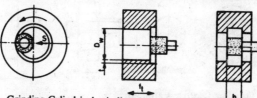

Table 267 Grinding-Cylindrical grinding, internal

Diameter of wheel	D	mm	
Wheel revolutions per minute	n	rpm	
Peripheral wheel speed	v	$m/sec.$	$v = \pi Dn/1000 \times 60$
Diameter of workpiece	D_w	mm	
Workpiece revolutions per minute	n_w	rpm	
Peripheral work speed	v_w	$m/min.$	$v_w = \pi D_w n_w/1000$
Depth of grinding-traverse grinding	t	$mm/pass$	
Table traverse feed rate -traverse grinding	f_t	$mm/min.$	
Width of cut-plunge grinding	b	mm	
Plunge infeed rate per work revolution	S	$mm/rev.$	
Plunge infeed rate per minute	f_p	$mm/min.$	$f_p = Sn_w$
Metal ramoval rate -traverse grinding	Q Q_1	$cm^3/min.$	$Q_1 = \pi D_w t\, f_t/1000$
-plunge grinding	Q_2	$cm^3/min.$	$Q_2 = \pi D_w b\, f_p/1000$
Unit power	U	$kW/cm^3/min.$	Table 273
Power at the spindle	N	kW	$N = UQ$
Efficiency of transmission	E	$\%$	
Power of the motor	N_{el}	kW	$N_{el} = N/E$
Tangential cutting force	P_z	kgf	$P_z = 102\, N/v$
Torque at spindle	T_s	$kgf.m$	$T_s = 975\, N/n$

POWER AND FORCE REQUIREMENTS IN MACHINING

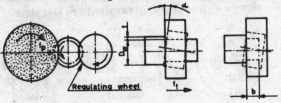

Regulating wheel

Table 268 Grinding-Centreless grinding

Diameter of wheel	D	mm	
Wheel revolutions per minute	n	rpm	
Peripheral wheel speed	v	$m/sec.$	$v = \pi Dn/1000 \times 60$
Diameter of workpiece	D_w	mm	
Depth of grind-thrufeed grinding	t	mm	
Diameter of regulating wheel	D_r	mm	
Regulating wheel revolutions per minute	n_r	rpm	
Regulating wheel inclination angle	α	$deg.$	
Thrufeed rate	f_t	$mm/min.$	$f_t = \pi D_r n_r \sin\alpha$
Width of cut-plunge grinding	b	mm	
Plunge infeed rate	f_p	$mm/min.$	
Metal removal rate -thrufeed grinding	Q Q_1	$cm^3/min.$	$Q_1 = \pi D_w t f_t /1000$
-plunge grinding	Q_2	$cm^3/min.$	$Q_2 = \pi D_w b f_p /1000$
Unit power	U	$kW/cm^3/min.$	Table 273
Power at the spindle	N	kW	$N = UQ$
Efficiency of transmission	E	$\%$	
Power of the motor	Nel	kW	$Nel = N/E$
Tangential cutting force	P_z	kgf	$P_z = 102\, N/v$
Torque at the spindle	T_s	$kgf.m$	$T_s = 975\, N/n$

POWER AND FORCE REQUIREMENTS IN MACHINING

Table 269 Average unit power U, for turning and milling

Work material	Tensile strength kgf/mm^2 Hardness HB	Unit power U. $kW/cm^3/min.$*								
		Average chip thickness, mm.								
		0.025	0.05	0.075	0.1	0.15	0.2	0.3	0.5	0.8
Free machining	40	54	45	41	39	35	33	30	26	23
steels	50	60	50	45	42	39	36	32	29	26
Mild steels	60	66	55	50	47	42	39	35	31	28
Medium carbon	70	69	59	53	50	45	42	37	33	30
steels	80	73	63	56	52	48	44	40	35	32
Alloy steels	90	78	65	59	56	50	47	42	38	34
Tool steels	100	80	69	62	59	53	49	44	39	35
	110	85	72	65	61	56	53	51	44	36
	150	80	71	66	61	57	52	48	44	40
	160	86	76	72	67	62	58	54	50	46
	170	92	82	78	73	68	61	56	52	48
Stainless steels **	180	99	90	84	80	75	69	62	59	52
	190	104	96	91	86	81	78	69	64	58
	200	110	101	96	91	88	85	78	71	60
	160	30	26	24	22	21	19	18	16	14
	170	31	28	25	24	22	20	19	17	15
Cast iron: **	180	35	30	27	25	23	22	21	19	17
Grey,	190	36	31	29	27	24	23	21	20	17
Ductile,	200	38	33	30	28	26	24	22	20	18
Malleable	220	42	36	33	31	29	26	24	22	20
	240	46	40	36	34	31	29	27	24	21
	260	50	43	39	37	33	31	29	26	23
	280	53	46	42	39	36	34	31	28	25
	10	13	11	9	9	8	7	6	6	5
Aluminium	20	19	16	14	13	12	11	10	8	7
alloys	30	24	20	17	16	14	13	12	10	9
	40	28	23	21	19	17	16	14	12	11
	50	32	26	23	22	19	18	16	14	12
Copper alloys	–	25	21	19	17	16	15	13	12	10
	10	9	7	6	6	5	5	4	4	3
Magnesium	15	10	9	8	7	6	6	6	5	4
alloys	20	12	10	9	8	7	7	6	5	4
	25	13	11	10	9	8	7	7	6	5
Titanium alloys — Ti Al Cr	110	59	51	47	45	41	39	36	32	30
Pure Ti	–	61	52	48	45	41	38	35	31	28
Ti Al Mn	–	67	58	53	50	45	43	39	35	31
Ti Al V	–	68	59	54	52	47	45	41	37	34
Ti Al Cr Mo	–	77	66	60	57	52	49	45	40	36

* Multiply the table values by 10^{-3} ** Values in HB

POWER AND FORCE REQUIREMENTS IN MACHINING

Table 270 Correction factor for flank wear

Flank wear mm	Average chip thick- ness mm	Correction coefficient, K_h									
		Hardness of work material									
		HB							HRC		
		125	150	200	250	300	350	400	51	56	61
0.2	0.1	1.16	1.17	1.18	1.19	1.2	1.21	1.22	1.25	1.33	1.38
	0.3	1.06	1.07	1.08	1.08	1.09	1.09	1.09	1.13	1 16	1.18
	0.5	1.04	1.05	1.05	1.05	1.05	1.06	1.07	1.08	1.12	1.13
	1	1.02	1.02	1.03	1.03	1.03	1.03	1.03	1.04	1.06	1.07
0.4	0.1	1.5	1.5	1.5	1.53	1.57	1.67	1.78	1.8	1.92	2.12
	0.3	1.2	1.2	1.2	1.22	1.23	1.27	1.32	1.36	1.41	1.52
	0.5	1.12	1.12	1.14	1.15	1.16	1.19	1.24	1.26	1.3	1.38
	1	1.06	1.06	1.07	1.07	1.08	1.1	1.12	1.14	1.16	1.2
0.6	0.1	1.68	1.71	1.73	1.84	1.94	2.09	2.2	2.43	2.72	2.82
	0.3	1.26	1.25	1.29	1.33	1.37	1.44	1.5	1.61	1.78	1.85
	0.5	1.17	1.19	1.2	1.23	1.26	1.3	1.37	1.47	1.57	1.61
	1	1.09	1.1	1.1	1.12	1.14	1.16	1.19	1.25	1.3	1.33
0.8	0.1	1.91	2.04	2.1	2.34	2.47	2.54	2.65	2.99	3.26	–
	0.3	1.35	1.41	1.42	1.52	1.56	1.62	1.7	1.9	2.02	–
	0.5	1.23	1.28	1.32	1.36	1.38	1.43	1.52	1.66	1.74	–
	1.0	1.12	1.14	1.15	1.17	1.18	1 23	1.27	1.35	1.4	–
1	0.1	2.18	2.32	2.39	2.54	2.65	2.84	3.15	3.46	–	–
	0.3	1.45	1.5	1.56	1.67	1.7	1.74	1.9	2.16	–	–
	0.5	1.3	1.34	1.39	1.47	1.45	1.51	1.67	1.84	–	–
	1	1.15	1.16	1.17	1.2	1.23	1.27	1.35	1.44	–	–

Table 271 Correction factor for rake angle

Rake angle, γ degrees	–15	–10	–5	0	+5	+10	+15	+20
Correction coefficient, K_γ	1.35	1.29	1.21	1.13	1.07	1	0.93	0.87

POWER AND FORCE REQUIREMENTS IN MACHINING

Table 272 Material factors K, for Drilling, Reaming & Tapping

Work Material	Hardress HB	UTS kgf/mm²	Material factor K
Free-machining steels	167	59.9	1.03
	183	63	1.42
Mild steels	121	44.1	1.07
	160	56.7	1.22
Medium carbon steels	152	55.1	1.15
	197	67.7	1.45
Alloy steels Tool steels	163	58.3	1.56
	174	61.4	2.02
	229	78.8	2.1
	241	81.9	2.32
Stainless steels	187	64.6	1.56
	269	92.6	2.41
Cast iron: Grey, Ductile, Malleable	177	21.3	1
	198	28.4	1.5
	224	35.1	2.03
Aluminium alloys	–	–	0.55
Copper alloys	–	–	0.55
Magnesium alloys	–	–	0.45

Table 273 Average unit power U, for grinding

Work material	Unit power U, kW/cm³/min. Depth of grinding, mm per pass Infeed, mm per revolution of work							
	0.0125	0.025	0.05	0.075	0.1	0.25	0.5	0.75
Free-machining steels Mild steels Medium carbon steels	1.4	0.88	0.7	0.6	0.51	0.35	0.23	0.18
Alloy steels	1.3	0.85	0.68	0.58	0.49	0.34	0.25	0.19
Tool steels	1.15	0.82	0.65	0.56	0.46	0.32	0.26	0.21
Stainless steels	1.4	0.84	0.65	0.58	0.51	0.37	0.29	0.26
Cast iron: Grey, Ductile, Malleable	1.15	0.79	0.6	0.51	0.44	0.3	0.23	0.19
Aluminium alloys	0.58	0.45	0.35	0.33	0.29	0.21	0.17	0.15
Titanium alloys	0.93	0.79	0.6	0.56	0.51	0.37	0.3	0.28

POWER AND FORCE REQUIREMENTS IN MACHINING

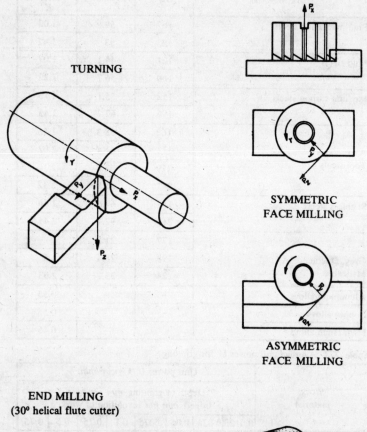

TURNING

SYMMETRIC
FACE MILLING

ASYMMETRIC
FACE MILLING

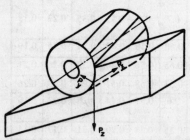

END MILLING
(30° helical flute cutter)

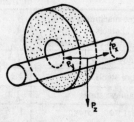

CYLINDRICAL GRINDING

Fig. 212

POWER AND FORCE REQUIREMENTS IN MACHINING

Table 274 Turning

Approach angle degrees	Rake angle degrees	P_x/P_z		
		Nose radius, *mm*		
		0.2 to 0.4	0.4 to 0.8	0.8 to 1.2
	−10	0.7 to 0.8	0.6 to 0.7	0.5 to 0.6
45	0	0.55 to 0.6	0.45 to 0.55	0.4 to 0.45
	+10	0.4 to 0.45	0.35 to 0.4	0.3 to 0.35
	−10	0.75 to 0.9	0.65 to 0.75	0.55 to 0.65
60	0	0.6 to 0.75	0.5 to 0.6	0.4 to 0.5
	+10	0.5 to 0.6	0.4 to 0.5	0.3 to 0.4
	−10	0.8 to 0.95	0.7 to 0.8	0.6 to 0.7
90	0	0.7 to 0.8	0.65 to 0.7	0.55 to 0.65
	+10	0.55 to 0.6	0.5 to 0.55	0.45 to 0.5
		P_y/P_z		
	−10	0.75 to 0.8	0.8 to 0.9	0.9 to 0.95
45	0	0.6 to 0.7	0.7 to 0.75	0.75 to 0.8
	+10	0.4 to 0.5	0.5 to 0.55	0.55 to 0.6
	−10	0.65 to 0.7	0.7 to 0.8	0.8 to 0.9
60	0	0.5 to 0.6	0.6 to 0.65	0.65 to 0.75
	+10	0.35 to 0.4	0.4 to 0.5	0.5 to 0.55
	−10	0.25 to 0.3	0.3 to 0.4	0.4 to 0.45
90	0	0.2 to 0.25	0.25 to 0.3	0.3 to 0.4
	+10	0.15 to 0.2	0.2 to 0.3	0.3 to 0.35

Milling:

SYMMETRICAL FACE MILLING:

$P_x = 0.5$ to $0.55 P_z$ $P_y = 0.25$ to $0.35 P_z$

ASYMMETRICAL FACE MILLING:

$P_x = 0.5$ to $0.55 P_z$ $P_y = 0.3$ to $0.40 P_z$

END MILLING:

(30° helical flute cutters)

$P_x = 0.15$ to $0.25 P_z$ $P_y = 0.45$ to $0.55 P$.

CYLINDRICAL GRINDING:

$P_x = 0.1$ to $0.2 P_z$ $P_y = 1.5$ to $3 P_z$

P_z = Tangential cutting force

P_x = Axial component of tangential force

P_y = Radial component of tangential force

CUTTING CONDITIONS

Table 275 Turning

Work material	Tool material	Cutting speed, *m/min.*			
		Depth, *mm*			
		5 - 10	2 - 5	0.5 - 2	0.1 - 0.5
		Feed *mm/rev.*			
		0.4 - 0.6	0.25 - 0.5	0.2 - 0.3	0.05 - 0.2
Free-machining steels	*HSS*	20 - 40	40 - 70	40 - 110	50 - 120
	Carbide	90 - 150	120 - 180	150 - 250	200 - 500
Mild steels	*HSS*	25 - 35	30 - 50	30 - 60	40 - 80
	Carbide	60 - 120	80 - 150	120 - 200	150 - 450
Medium carbon steels	*HSS*	15 - 25	25 - 45	25 - 50	30 - 70
	Carbide	50 - 110	60 - 120	90 - 150	120 - 300
Alloy steels	*HSS*	10 - 15	15 - 25	15 - 35	20 - 45
	Carbide	30 - 65	40 - 80	60 - 100	80 - 180
Tool steels	*HSS*	15 - 20	20 - 25	20 - 30	30 - 60
	Carbide	50 - 110	60 - 120	90 - 150	120 - 300
Stainless steels	*HSS*	15 - 20	15 - 25	15 - 30	20 - 50
	Carbide	40 - 60	40 - 70	50 - 80	50 - 90
Cast iron: Grey, Ductile Malleable.	*HSS*	20 - 25	25 - 30	35 - 45	40 - 60
	Carbide	60 - 90	70 - 100	80 - 110	80 - 120
Aluminium alloys	*HSS*	40 - 70	70 - 100	90 - 120	100 - 200
	Carbide	60 - 150	80 - 180	90 - 450	150 - 600
Copper alloys	*HSS*	40 - 60	60 - 100	90 - 120	100 - 200
	Carbide	50 - 110	60 - 150	90 - 180	120 - 310
Magnesium alloys	*HSS*	40 - 70	70 - 100	90 - 120	100 - 200
	Carbide	60 - 150	80 - 180	90 - 450	150 - 600
Titanium alloys	*HSS*	10 - 15	15 - 30	30 - 50	50 - 90
	Carbide	15 - 30	30 - 50	50 - 90	60 - 120

Table 276 Milling

Work material	Tool material	Speed m/min.	Feed, mm per tooth				
			Face mills	Slab mills	Slotting and side mills	End mills	Form cutters
Free-machining steels	HSS	30-40	0.3	0. 25	0.175	0.15	0.1
	Carbide	100-200					
Mild steels	HSS	25-40	0.25	0.2	0.15	0.125	0.1
	Carbide	90-130					
Medium carbon steels	HSS	20-30	0.2	0.15	0.125	0.1	0.075
	Carbide	60-90					
Alloy steels	HSS	10-20	0.15	0.1	0.075	0.06	0.05
	Carbide	40-55					
Tool steels	HSS	15-25	0.2	0.15	0.1	0.075	0.05
	Carbide	60-80					
Stainless steels	HSS	15-20	0.15	0:1	0.1	0.075	0.05
	Carbide	30-60					
Cast iron: Grey, Ductile, Malleable	HSS	20-30	0.35	0.3	0.2	0.175	0.1
	Carbide	70-100					
Aluminium alloys	HSS	60-100	0.5	0.4	0.3	0.25	0.175
	Carbide	60-180					
Copper alloys	HSS	40-75	0.3	0.25	0.2	0.175	0.15
	Carbide	60-100					
Magnesium alloys	HSS	60-100	0.5	0.4	0.25	0.2	0.175
	Carbide	60-180					
Titanium alloys	HSS	10-30	0.15	0.125	0.1	0.075	0.05
	Carbide	30-50					

CUTTING CONDITIONS

Table 277 Drilling-Reaming-Tapping

Work material	Cutting speed, m/min.		
	Drilling	Reaming	Tapping
Free machining steels	20-30	11-15	9-12
Mild steels	20-23	11-14	11-12
Medium carbon steels	14-20	9-14	8-11
Alloy steels	18-22	10-14	10-12
Tool steels	5-8	3-5	3-5
Stainless steels	12-15	9-25	8-9
Cast iron:			
Grey, Ductile and Malleable	20-23	12-17	9-12
Aluminium alloys	35-55	25-30	14-18
Copper alloys	30-45	20-40	9-12
Magnesium alloys	60-105	30-40	15-25
Titanium alloys	12-15	9-25	8-9

Table 278

Hole dia. mm	Feed, mm/rev.	
	Drilling	Reaming
1.5-2.5	0.04-0.06	0.08-0.13
3-4	0.05-0.1	0.1 -0.2
4.5-5.5	0.05-0.13	0.15-0.3
6-8.5	0.1 -0.18	0.2 -0.4
9-11.5	0.12-0.2	0.3 -0.51
12-14.5	0.15-0.25	0.41-0.61
15-18	0.18-0.28	0.46-0.66
18.5-20.5	0.2 -0.3	0.5 -0.71
21-24	0.23-0.33	0.56-0.76
25-29	0.25-0.36	0.61-0.81
30-38	0.28- 0.41	0.71-0.91
over 38	0.3 -0.41	0.81-1.00

Table 279 Broaching

Work material	Rise per tooth mm	Cutting speed m/min.
Free machining steels	0.1	9
Mild steels	0.1	6-9
Medium carbon steels	0.075	3-8
Alloy steels	0.085	5-9
Tool steels	0.05	3-6
Stainless steels	0.075	4-6
Cast iron:		
Grey, Ductile & Malleable	0.1	7-9
Aluminium alloys	0.15	9-15
Copper alloys	0.125	8-9
Magnesium alloys	0.15	9-15
Titanium alloys	0.025	1.5-3

CUTTING CONDITIONS

Table 280 Surface grinding, hortizontal spindle, reciprocating table

Work material	Hardness	Wheel speed m/sec.	Table speed m/min.	Down feed mm/pass	Cross feed mm/pass	Wheel
Free-machining steels Mild steels Medium carbon steels	upto 48HRC	28-32	15-30	Rough 0.075 Finish 0.025 Max.	1.25 to 12.5 (¼ W)*	A46JV
	48-65 HRC	28-32	15-30	Rough 0.05 Finish 0.0125 Max.	0.65 to 12.5 (1/10W)*	A46IV
Alloy steels	Upto 48HRC	28-32	15-30	Rough 0.075 Finish 0.0125 Max.	1.25 to 12.5 (1/ 4 W)*	A46IV
Tool steels	100-250 HB	28-32	15-30	Rough 0.05 Finish 0.0125 Max.	0.625 to 6.25 (1/10W)*	A46IV
	56-65 HRC	28-32	15-30	Rough 0.05 Finish 0.0125 Max.	0.625 to 12.5 (1/10W)*	A46IV
Stainless steels	135-275 HB	28-32	15-30	Rough 0.05 Finish 0.0125 Max.	1.25 to 12.5 (1/ 4 W)*	A46JV
Cast iron: Grey, Ductile, Malleable	upto 45HRC	28-32	15-30	Rough 0.05 Finish 0.0125 Max.	1.25 to 12.5 (1/5W)*	A46IV
Alluminium alloys	30-150 HB	20-25	15-30	Rough 0.075 Finish 0.025 Max.	1.25 to 12.5 (1/3W)*	C46KV
Copper alloys	10-100 HRB	28-32	15-30	Rough 0.075 Finish 0.0125	1.25 to 12.5 (⅓W)	C46KV
Magnesium alloys	40-90 HB	20-25	15-30	Rough 0.075 Finish 0.025 Max.	1.25 to 12.5 (1/3W)*	C46KV
Titanium alloys	300-380 HB	7.5-30	12	Rough 0.025 Finish 0.0125 Max.	0.625 to 12.5 (1/10W)*	C46KV

* Values in brackets refer to maximum cross feed in terms of wheel width. W being the maximum wheel width.

CUTTING CONDITIONS

Table 281   Surface grinding, vertical spindle and rotary table

Work material	Hardness	Wheel speed *m/sec.*	Table speed *m/min.*	Down feed *mm*/revolution of table	Wheel
Free machining steels, Mild steels, Medium carbon steels	100-275 *HB*	16.5-25	25-60	0.025-0.125	A80I or A80H
Alloy steels	175-350 *HB*	16.5-25	25-60	0.025-0.075	A80I or A80H
Tool steels	48-62*HRC*	16.5-25	30-75	0.0125-0.025	A100H or A100G
Stainless steels	135-275 *HB*	16.5-25	30-100	0.025-0.075	A80G or A80I
Cast iron: Grey, Ductile Malleable	Upto45 *HRC*	16.5-25	30-100	0.075-0.2	A80I or C80I
Aluminium alloys	30-150 *HB*	16.5-25	30-75	0.025-0.1	C60I or C60H

CUTTING CONDITIONS

Table 282 Cylindrical grinding

Work material	Hardness	Wheel speed m/sec.	Work speed m/min.	Infeed mm/pass on dia.		Traverse per work revolut- ion	Wheel
Free machining steels	Upto 48HRC	27-32	20-30	Rough 0.05	0.5 W*	A60KV
				Finish 0.0125 Max.	0.16 W	
Mild steels Medium carbon steels	48-65 HRC	27-32	20-30	Rough 0.05	0.25 W	A60KV
				Finish 0.0125 Max.	0.125 W	
Alloy steels	Upto 48HRC	27-32	20-30	Rough 0.05	0.5 W	A60KV
				Finish 0.0125 Max.	0.16 W	
Tool steels	100-250 HB	27-32	18-30	Rough 0.05	0.5 W	A60KV
				Finish 0.0125 Max.	0.16 W	
	56-65 HRC	27-32	18-30	Rough 0.05	0.25 W	A60JV
				Finish 0.0125 Max.	0.125 W	
Stainless steels	135-275 HB	27-32	15-30	Rough 0.05	0.5 W	A60KV
				Finish 0.0125 Max.	0.16 W	
Cast iron: Grey, Ductile, Malleable	Upto 45HRC	27-32	20-30	Rough 0.05	0.5 W	A60LV
				Finish 0.025 Max.	0.16 W	
Aluminium alloys	30-150 HB	27-32	15-45	Rough 0.05	0.5 W	C46KV
				Finish 0.0125 Max.	0.16 W	
Copper alloys	10-100 HRB	27-32	20-30	Rough 0.05	0.3 W	C60KV
				Finish 0.0125 Max.	0.16 W	
Magnesium alloys	40-90 HB	27-32	20-40	Rough 0.05	0.3 W	C46JV
				Finish 0.0125 Max.	0.16 W	
Titanium alloys	300-380 HB	7.5-30	15-30	Rough 0.025	0.2 W	C60JV
				Finish 0.0125 Max.	0.1 W	

* W = Wheel width

CUTTING CONDITIONS

Table 283 Internal grinding

Work material	Hardness	Wheel speed m/sec.	Work speed m/min.	Infeed mm/pass on dia.	Traverse per work revolution	Wheel
Free machining steels Mild steels Medium carbon steels	Upto 48HRC	25-32	23-60	Rough 0.0125	0.3 W*	A60MV
				Finish 0.005 Max.	0.16 W	
	48-65 HRC	25-32	23-60	Rough 0.0125	0.3 W	A60KV
				Finish 0.005 Max.	0.16 W	
Alloy steels	Upto 18HRC	25-32	23-60	Rough 0.0125	0.3 W	A60KV
				Finish 0 005 Max.	0.16 W	
Tool steels	100-250 HB	25-32	23-60	Rough 0.0125	0.3 W	A80KV
				Finish 0.005 Max.	0.16 W	
	56-65 HRC	25-32	23-60	Rough 0.0125	0.3 W	A80JV
				Finish 0.005 Max.	0.16 W	
Stainless steels	135-275 HB	25-32)	23-60	Rough 0.0125	0.3 W	A60JV
				Finish 0.005 Max.	0.16 W	
Cast iron: Grey, Ductile, Malleable	Upto 45HRC	25-32	23-60	Rough 0.05	0.3 W	A60JV
				Finish 0.005 Max.	0.16 W	
Aluminium alloys	30-150 HB	25-32	23-60	Rough 0.075	0.3 W	A60JV
				Finish 0.005 Max.	0.16 W	
Copper alloys	10-100 HRB	25-32	23-60	Rough 0.05	0.3 W	C60KV
				Finish 0.005 Max.	0.16 W	
Magnesium alloys	40-90 HB	25-32	23-60	Rough 0.075	0.3 W	A60KV
				Finish 0.005 Max.	0.16 W	
Titanium alloys	300-380 HB	7.5-30	15-45	Rough 0.0125	0.3 W	A60JV
				Finish 0.005 Max.	0.16 W	

*W = Wheel width
    Maximum hole length = 2½ x Hole diameter
    Maximum wheel width = 1½ x Wheel diameter

CUTTING CONDITIONS

Table 284 Centreless grinding

Work material	Hardness	Grinding wheel speed m/sec.	Thru feed of work m/min.	Infeed mm/pass on dia.	Regulating wheel		Grinding wheel
					Angle deg.	rpm	
Free machining Steels Mild steels	Upto 48HRC	27-32	1.25-3.75	Rough 0.125 Finish 0.04 Max.	3°	50-70	A 60MV
Medium carbon Steels	48-66 HRC	27-32	1.25-3.75	Rough 0.125 Finish 0.04 Max.	3°	50-70	A 60MV
Alloy steels	Upto 48HRC	27-32	1.25-3.75	Rough 0.125 Finish 0.04 Max.	3°	50-70	A 60MV
Tool steels	100-250 HB	27-32	1.25-3.75	Rough 0.125 Finish 0.04 Max.	3°	50-70	A 60MV
	56-65 HRC	27-32	1.25-3.75	Rough 0.125 Finish 0.04 Max.	3°	50-70	A 80LV
Stainless steels	135-275 HB	27-32	1.25-3.75	Rough 0.125 Finish 0.04 Max.	3°	50-70	C 60KV
Cast iron: Grey, Ductile, Malleable	Upto 45HRC	27-32	1.25-3.75	Rough 0.125 Finish 0.05 Max.	3°	50-70	C 60JV or A 60JV
Aluminium alloys	30-150 HB	27-32	1.25-3.75	Rough 0.125 Finish 0.05 Max.	3°	50-70	C 60JV
Copper alloys	10-100 HRB	27-32	1.25-3.75	Rough 0.125 Finish 0.04 Max.	3°	50-70	A 60LV
Magnesium alloys	40-90 HB	27-32	1.25-3.75	Rough 0.125 Finish 0.04 Max.	3°	50-70	C 60KV or A 60JV
Titanium alloys	300-380 HB	7-30	1.25-3.75	Rough 0.025 Finish 0.0125 Max	3°	50-70	A 60JV

TOOL GEOMETRY

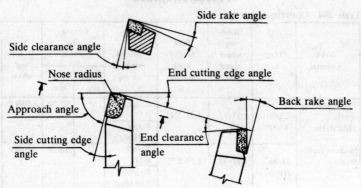

Table 285 Turning

Work Material	Hardness HB	Tool material	Back rake angle degrees	Side rake angle degrees	End clearance angle degrees	Side clearance angle degrees	Side & end cutting edge angle degrees
Free machining steels,	85 to 225	HSS	10	12	5	5	15
		Brazed carbides	0	6	5	5	15
		Throwaway carbides	−5	−5	5	5	15
Mild steels,	225 to 325	HSS	8	10	5	5	15
		Brazed carbides	0	6	5	5	15
		Throwaway carbides	−5	−5	5	5	15
Medium carbon steels, Alloy steels,	325 to 425	HSS	0	10	5	5	15
		Brazed carbides	0	6	5	5	15
		Throwaway carbides	−5	−5	5	5	15
Tool steels.	45HRC to 58HRC	HSS	0	10	5	5	15
		Brazed carbides	−5	−5	5	5	15
		Throwaway carbides	−5	−5	5	5	15
Stainless steel,	135 to 275	HSS	0	10	5	5	15
		Brazed carbides	0	6	5	5	15
		Throwaway carbides	−5	−5	5	5	15

TOOL GEOMETRY

Table 285 (Concld.)

Work material	Hardness HB	Tool material	Back rake angle degrees	Side rake angle degrees	End clearance angle degrees	Side clearance angle degrees	Side & end cutting edge angle degrees
Cast iron: Grey, Ductile, Malleable	100 to 200	HSS	5	10	5	5	15
		Brazed carbides	0	6	5	5	15
		Throwaway carbides	-5	-5	5	5	15
	200 to 300	HSS	5	8	5	5	15
		Brazed carbides	0	6	5	5	15
		Throwaway carbides	-5	-5	5	5	15
	300 to 400	HSS	5	5	5	5	15
		Brazed carbides	-5	-5	5	5	15
		Throwaway carbides	-5	-5	5	5	15
Aluminium alloys	30 to 150	HSS	20	15	12	10	5
		Brazed carbides	3	15	5	5	15
		Throwaway carbides	0	5	5	5	15
Copper alloys	40 to 200	HSS	5	10	8	8	5
		Brazed carbides	0	8	5	5	15
		Throwaway carbides	0	5	5	5	15
Magnesium alloys	40 to 90	HSS	20	15	12	10	5
		Brazed carbides	3	15	5	5	15
		Throwaway carbides	0	5	-5	5	15
Titanium alloys	110 to 440	HSS	0	5	5	5	15
		Brazed carbides	0	6	5	5	5
		Throwaway carbides	-5	-5	5	5	5

Notes: 1. Nose radius will generally be dictated by type of operation being performed. When not specified use 1.2 *mm.*

2. Approach angle = 90°—side cutting edge angle.

TOOL GEOMETRY

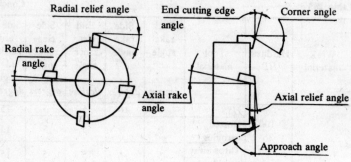

Approach angle = 90° − Corner angle

Table 286   Milling—face mills

Work material	Hard-ness HB	Tool material	Axial rake angle deg.	Radial rake angle deg.	Corner angle deg.	End cutting edge angle deg.	Axial relief angle deg.	Radial relief angle deg.
Free machining steels	85 to 270	*HSS*	10-15	10-15	30	5-10	5-7	3-7
		Brazed carbides	5-7	−5 to −14	30	5-10	5-7	3-7
		Throwaway carbides	0 to −7	0 to −7	30	5-10	5-7	3-7
Mild steels	270 to 325	*HSS*	10-15	10-15	30	5-10	5-7	3-7
		Brazed carbides	−4 to −8	−3 to −11	30	5-10	5-7	3-7
		Throwaway carbides	0 to −7	0 to −7	30	5-10	5-7	3-7
Medium carbon steels Alloy steels	325 to 425	*HSS*	10-12	10-12	30	5-10	5-7	3-7
		Brazed carbides	−4 to −8	−3 to −11	30	5-10	5-7	3-7
		Throwaway carbides	0 to −10	0 to −10	30	5-10	5-7	3-7
Tool steels	43 *HRC* to 50 *HRC*	*HSS*	5-10	5-10	45	4-7	5-7	3-7
		Brazed carbides	−4 to 8	−3 to −11	45	4-7	5-7	3-7
		Throwaway carbides	−5 to −15	−5 to −15	45	4-7	5-7	3-7
Stainless steels	135 to 275	*HSS*	10-15	10-12	45	5	8-10	8-10
		Brazed carbides	5-11	−5 to −11	45	5	8-10	8-10
		Throwaway carbides	0-5	0 to −5	45	5	8-10	8-10

Table 286 (Concld.)

Work material	Hardness HB	Tool material	Axial rake angle deg.	Radial rake angle deg.	Corner angle deg.	End cutting edge angle deg.	Axial relief angle deg.	Radial relief angle deg.
Cast iron: Grey, Ductile, Malleable	100 to 400	HSS	20-30	−5 to −10	45	5-10	4-7	4-7
		Brazed carbides	5-11	−5 to −11	45	5-10	4-7	4-7
		Throwaway carbides	5-10	5 to -10	45	5-10	4-7	4-7
Aluminium alloys	30 to 150	HSS	20-35	20-35	45	7-12	3-5	10-12
		Brazed carbides	5-7	0-5	45	7-12	3-5	10-12
		Throwaway carbides	10-20	10-20	45	7-12	3-5	10-12
Copper alloys	40 to 200	HSS	12-25	10-12	45	7-12	3-5	5-10
		Brazed carbides	5-7	0-5	45	7-12	3-5	5-10
		Throwaway carbides	3-10	3-10	45	7-12	3-5	5-10
Magnesium alloys	40 to 90	HSS	20-35	20-35	45	7-12	3-5	10-12
		Brazed carbides	5-7	0-5	45	7-12	3-5	10-12
		Throwaway carbides	10-20	10-20	45	7-12	3-5	10-12
Titanium alloys	110 to 440	HSS	5	5	45	6-12	10-12	10-12
		Brazed carbides	0 to -5	0 to -5	45	6-12	10-12	10-12
		Throwaway carbides	0	-10	45	6-12	10-12	10-12

TOOL GEOMETRY

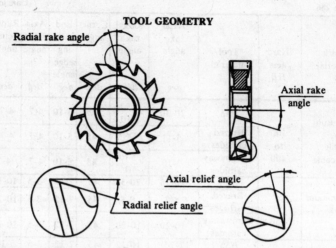

Radial rake angle

Axial rake angle

Axial relief angle

Radial relief angle

Table 287 Milling—side and slot mills

Work material	Hardness HB	Tool material	Axial rake angle *degrees*	Radial rake angle *degrees*	Axial relief angle *degrees*	Radial relief angle *degrees*
Free machining steels	85-325	HSS	10 to 15	10 to 15	3 to 5	4 to 8
		Carbide	0 to −5	−5 to 5	2 to 4	5 to 8
Mild steels	325-425	HSS	10 to 12	5 to 12	3 to 5	4 to 8
		Carbide	0 to −5	−5 to 5	2 to 4	5 to 8
Medium carbon steels	45HRC, to 52 HRC	HSS	10 to 12	5 to 12	2 to 4	3 to 7
		Carbide	−5 to −10	0 to −10	2 to 4	5 to 8
Alloy steels	125-425	HSS	10 to 12	5 to 12	3 to 5	4 to 8
		Carbide	−5 to −10	0 to −10	2 to 5	5 to 8
Tool steels	45 HRC to 52 HRC	HSS	10 to 12	5 to 12	3 to 5	4 to 8
		Carbide	−5 to −10	0 to −10	2 to 5	5 to 8
Stainless steels	135-425	HSS	10 to 12	5 to 12	3 to 5	4 to 8
		Carbide	0 to 5	−5 to 5	2 to 4	5 to 8
Cast iron Grey, Ductile, Malleable	100-400	HSS	10 to 12	10 to 12	2 to 4	3 to 7
		Carbide	0 to −10	5 to −10	3 to 5	5 to 8
Aluminium alloys	30-150	HSS	12 to 25	10 to 20	5 to 7	5 to 11
		Carbide	10 to 20	5 to 15	5 to 7	7 to 10
Copper alloys	40-200	HSS	12 to 25	10 to 20	5 to 7	5 to 11
		Carbide	10 to 20	5 to 10	4 to 7	5 to 8
Magnesium alloys	40-90	HSS	12 to 25	10 to 20	5 to 7	5 to 11
		Carbide	10 to 20	5 to 15	5 to 7	7 to 10
Titanium alloys	110-440	HSS	10 to 15	5 to 10	5 to 7	5 to 11
		Carbide	0 to −10	0 to −10	5 to 7	5 to 8

TOOL GEOMETRY

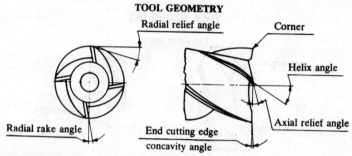

Table 288 Milling—end mills (*HSS*)

Material	Hardness HB	Helix angle degrees	Radial rake angle degrees	End cutting edge concavity angle degrees	Axial relief angle degrees	Radial* relief angle degrees
Free-machining steels, Mild steels, Medium carbon steels	85-325	30	10-20	3	3-7	A
Alloy steels	125-425	30	15	3	3-7	A
Tool steels	52-100*HRC*	30	10-12	3	3-7	A
Stainless steels	135-425	30	15	3	3-7	A
Cast iron: Grey, Ductile, Malleable	100-400	30	12	3	3-7	A
Alluminium alloys	30-150	30-45	15-20	5	8-12	B
Copper alloys	40-200	30	10-20	5	8-12	B
Magnesium alloys	40-90	30-45	15-20	5	8-12	B
Titanium alloys	110-440	30	10	3	8-12	B

*Radial relief angles for various diameters of end mills

Diameter *mm*	6	10	12	16	25
A	12°	11°	10°	9°	8°
B	15°	13°	13°	12°	10°

TOOL GEOMETRY

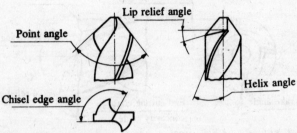

Table 289 Drilling—*HSS* drills

Work material	Hardness *HB*	Point angle *degrees*	Lip relief angle *degrees*	Chisel edge angle *degrees*	Helix angle *degrees*
Free-machining steels	85-225	118	10-15	125-135	24-32
Mild steels	225-325	118	10-12	125-135	24-32
Medium carbon steels	325-425	118-135	8-10	125-135	24-32
Alloy steels Tool steels	45HRC-52HRC	118-135	7-9	125-135	24-32
Stainless steels	135-325	118	7-10	125-135	24-32
Cast iron: Grey, Ductile, Malleable	110-440	118	8-12	125-135	24-32
Aluminium alloys	30-150	90-118	12-15	125-135	24-48
Copper alloys	40-200	118	12-15	125-135	10-30
Magnesium alloys	40-90	70-118	12-15	120-135	30-45
Titanium alloys	110-440	118-135	7-10	125-135	15-32

Note: The angles recommended are valid for depth of holes = 3 x dia. of holes For deep holes high helix, wider fluted drills should be used.

TOOL GEOMETRY

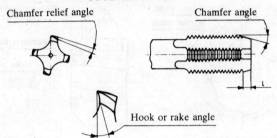

Chamfer relief angle   Chamfer angle

Hook or rake angle

Table 290 Tapping

Work material	Hardness HB	Number* of flutes	Hook or rake angle *degrees*	Chamfer relief angle *degrees*	Chamfer length *l*
Free-machining steels, Mild steels, Medium carbon steels	100-425	2-3	9-12	8	
Alloy steels	125-425	2-3	10-12	8	
Tool steels	100-375	2-3	5-8	4-8	2 ½ to 3 threads for blind holes 3 to 4 threads for through holes
Stainless steels	135-275	2-3	15-20	10	
Cast iron: Grey, Ductile, Malleable	100-320	4	0-3	6	
Aluminium alloys	30-150	2-3	15-18	12	
Copper alloys	40-200	4	3-6	10	
Magnesium alloys	40-90	2-3	15-18	12	
Titanium alloys	110-440	2-3	6-10	12	

* For taps 12*mm* or smaller

TOOL GEOMETRY

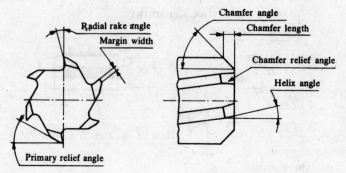

Table 291 Reaming-*HSS*

Work material	Hardness HB	Margin width mm	Chamfer Angle deg.	Chamfer length mm	Chamfer Relief angle deg.	Radial rake angle deg.	Helix angle deg.	Primary relief angle deg.
Free machining steels Mild steels	85-225	0.1-0.25	45	1.6	7	2-3	0-10	4-5
Medium carbon steels	225-325	0.17-0.21	45	1.6	7	2-3	0-10	4-5
Alloy steels Tool steels	325-425	0.17-0.21	45	1.6	7	2-3	0-10	2-4
Stainless steels	135-425	0.125-0.37	30-40	1.6	4-5	3-8	0-10	4-8
Cast iron: Grey, Ductile, Malleable	110-330	0.1-0.625	45	1.6	7-23	0-10	0-10	5-6
Aluminium alloys	30-150	0.5-i.5	45	1.6	10-15	5-10	0-10	7-8
Copper alloys	40-200	0.125-0.37	40	2.4	10-15	0-5	0-12	5-7
Magnesium alloys	40-90	0.15-0.30	45	1.6	10-15	7	0 to -10	5-6
Titanium alloys	110-440	0.05-0.125	45	1.6	10-15	2-3	0-10	10-15

TOOL GEOMETRY

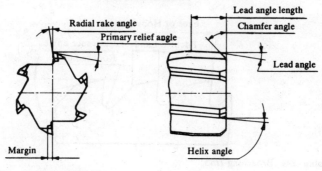

Table 292   Reaming-Carbide

Work material	Hardness HB	Margin width mm	Lead angle		Radial rake angle deg.	Helix angle deg.	Primary relief angle deg.
			Angle deg.	Length mm			
Free machining steels	85-225	0.125-0.25	2	4.8	7-10	5-8	7-15
Mild steels	225-325	0.05-0.125	2	4.8	5-7	5-8	7-15
Medium carbon steels	325-425	0.05-0.125	0	0	5-7	0	7-15
Alloy steels Tool steels	45HRC-52HRC	0.05-0.125	0	0	0-5	0	7-15
Stainless steels	135-275	0.125-0.25	0	0	7-10	5-8	7-15
Cast iron:	110-225	0.375-0.5	2	4.8	5-7	0	7-15
Grey, Ductile, Malleable	225-330	0.05-0.125	2	4.8	0-5	0	7-15
Aluminium alloys	30-150	0.375-0.5	0	0	7-10	5-8	7-15
Copper alloys	40-200	0.375-0.5	0	0	5-7	5-8	7-15
Magnesium alloys	40-90	0.375-0.5	0	0	7-10	5-8	7-15
Titanium alloys	110-440	0.125-0.25	0	0	5-7	5-8	7-15

TOOL GEOMETRY

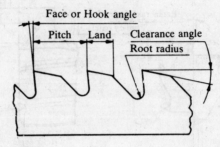

Table 293 Broaching-*HSS*

Work material	Hardness *HB*	Hook angle *deg.*	Clearance angle *deg.*
Free-machining steels Mild steels Medium carbon steels	100-375	15-20	2-3
Alloy steels	120-375	8-15	1-3
Tool steels	100-375	8-12	1-2
Stainless steels	135-275	12-18	2-3
Cast iron: Grey, Ductile, Malleable	100-320	6-8	2-3
Aluminium alloys	30-150	10-15	1-3
Copper alloys	40-200	−5-15	1-3
Magnesium alloys	40-90	10-15	1-3
Titanium alloys	110-440	8-20	2-8

MATERIALS AND HEAT TREATMENT

MATERIALS AND HEAT TREATMENT

Material selection is a matter of quality and cost. The properties of the material must be adequate to meet design requirements and service conditions.

Machine tool construction materials can be mainly classified into three types: (1) Ferrous (2) Non-ferrous (3) Non-metals.

Ferrous metals and alloys consist of various grades and types of cast-iron and different types of steels. However, from the point of view of reducing the inventory, the varieties of cast iron and steels used in machine tools are limited to a few selected items.

Non-ferrous metals and alloys are used for their special properties such as friction and wear characteristics, low specific gravity, etc.

Non metallic materials such as rubber, PTFE, perspex, etc., are also used in the construction of machine tools. Some machine tool manufacturers use concrete for machine tool bases.

The procedure for material selection for machine tool components should include consideration of the following points:

Mechanical properties

Tensile strength
Bending strength
Shear strength
Impact strength

Compressive strength
Bearing strength
Torsion strength
Damping properties, etc.

Physical properties

Wear Resistance
Hardenability
Thermal conductivity
Specific gravity

Coefficient of friction
Coefficient of thermal expansion
Rust and corrosion inhibition, etc.

Technological properties

Machinability
Weldability
Castability

Formability
Malleability, etc.

Tables 294 and 296 indicate the above properties of materials generally used in machine tools.

Certain salient features of the materials used in the construction of machine tools are as follows:

Cast iron

Cast iron is primarily an alloy of iron, carbon and silicon.

Carbon content is between 1.7 and 4.5%. This carbon in most cases is in either free form as graphite or combined form as cementite. The physical properties of the castings mainly depend upon the relative amounts of graphite or cementite that is present in the cast iron. Also, manganese, chromium, nickel, phosphorus, sulphur, etc., are alloyed with cast iron to give special properties.

The general term *Cast iron* includes Grey cast iron, Meehanite, Malleable iron, Spheroidal graphite iron, etc.

GREY CAST IRON

Grey cast iron is the least expensive of all the metals that could be used for castings and hence is considered first when a cast metal is being selected. Other metals are selected only when the mechanical and physical properties of grey cast iron are inadequate. Mechanical properties and chemical composition of grey cast iron of grades 15, 20 and 25 are given in Tables 297, 298 and 299. These grades are based on the minimum tensile strength in kgf/mm^2 of 30 mm diameter cast test bar as per IS 210. Elastic modulus of cast iron is only 0.9 to 1 x 10^4 kgf/mm^2 as compared to 2.1 x $10^4 kgf/mm^2$ of steel. Even then major structural components of machine tools such as bed, saddle, headstock, tailstock body, base, column, slides, gearbox body as also the smaller components such as brackets, housings, covers, pulleys, etc., are made of grey cast iron in preference to steel due to its excellent castability and better damping properties against vibration. Grey iron castings have a tendency to warp and distort. This is mainly due to internal stresses generated by uneven cooling of the casting in the foundry, as well as due to redistribution of stresses in the machine shop during machining. Uneven cooling can be reduced by avoiding abrupt changes in sections and also by having the thickness of walls as uniform as possible. Grey iron castings are often stress relieved by annealing. Grey cast iron *Grade* 15 of IS 210 is used for brackets, covers, etc.; *Grade* 20 is used for bed, column, saddle, headstock, etc. and *Grade* 25 is used for components subjected to severe loading. The slideways on the column, bed, etc., are generally cast integral with the structures and are normally hardened either by flame or induction hardening to *HB* 450-500.

MEEHANITE

Meehanite is basically a cast iron produced by a controlled process with the addition of certain alloying elements. There are various grades of meehanite suitable for individual application. Table 301 indicates some of the important properties of general engineering grades of meehanite, and Table 302 indicates the properties of wear resistant grades of meehanite. Tables 303 and 304 refer to the heat treatment and application of both the types of meehanite.

MALLEABLE IRON

Malleable iron is an iron-carbon alloy which solidifies in as-cast condition in a graphite free structure; that is, total carbon content is present in its combined form as cementite (Fe_3C). Malleable iron castings may either be whiteheart, blackheart or pearlitic, according to the chemical composition, temperature and time cycle of annealing process and properties resulting therefrom. It is quite ductile with good castability. Blackheart malleable iron is used in machine tools for handwheels, hand cranks, shifting levers, etc., also it is used widely for automotive components. Mechanical properties of Blackheart malleable iron Grade B of IS 2108 are given in Table 305.

Table 305 Properties of grade B blackheart malleable iron castings

Diameter of bar	Tensile strength kgf/mm² Min.	0.5% proof stress kgf/mm²	Elongation Min.%	Hardness HB Max.
15	32	19	10	149

SPHEROIDAL GRAPHITE IRON (SG IRON OR NODULAR IRON)

In this iron, the graphite is precipitated in the form of spheroids by addition of magnesium or cerium to the grey cast iron. SG iron can be cast into intricate shapes like grey cast iron and still obtain better strength and toughness which are nearly equal to those of medium carbon steel. It exhibits good pressure tightness, and can be welded and brazed. It can be softened by annealing or hardened by heat treatment. There are six grades of SG iron as per IS 1865. Table 306 lists the detailed properties of these six grades of SG iron together with the heat treatment data.

SG iron has the following advantages and disadvantages over grey cast iron:
1) SG iron at present costs nearly 50% more than grey cast iron.
2) Ultimate tensile strength of SG iron is higher.
3) Impact properties are better.
4) Section sensitivity of SG iron is much less as compared to grey cast iron.
5) Component weight of SG iron and hence cost may be less as compared to grey cast iron for the same external forces on the component.

6) Damping properties of SG iron are similar to those of grey cast iron.

Mechanical properties of SG iron are comparable to those of low and medium carbon steels. SG iron has better damping capacity, wear resistance, machinability, lower specific gravity as compared to low and medium carbon steels. Use of SG iron is considered for parts such as components of 3 jaw and 4 jaw chucks, gears, vices, cutter body, brackets, handles, levers, pulleys, shifting forks, slides, tool post body, etc., and even spindles in some cases.

Steels

Steel is very widely used for machine components as it can be manufactured and processed into a number of different specifications each of which has a definite use.

Two properties of steel apart from strength are of special significance in selection of proper grades of steel. First, the maximum hardness that can be achieved and second, the hardenability which determines the depth of hardened zone of a cross-section which can be through hardened. In the case of direct hardening steels the hardness that can be achieved is exclusively dependent upon its carbon content. The hardenability is essentially a function of the total alloy content and also to a certain extent, of grain size. Fig. 213 shows the maximum hardness as a function of carbon content.

Steel is an alloy of iron and carbon and depending upon the carbon content it is classified as low carbon steel (carbon content less than 0.3%), medium carbon steel (carbon 0.3 to 0.6%) and high carbon steel (more than 0.6%). Steels commonly used in machine building are (a) free cutting steels (b) carbon steels and (c) alloy steels. Indian steels used in machine tools and their nearest foreign equivalents together with the rationalised steels are given in Table 307.

FREE CUTTING STEELS

These steels have good machinability and a good surface finish can be achieved on the components. Slightly higher sulphur content in the composition imparts this property of good machinability. 14Mn1S14 and 40Mn2S12 are the preferred steels of this category and are used for handles, levers, spacers, hydraulic fittings, screws, etc. Tables 308 and 309 give the detailed properties of these steels.

MILD STEEL

This is a low carbon steel with no precise control over the composition or mechanical properties. The cost is low in comparison with other steels and it is used for covers, sheet metal work, tanks, fabricated items, etc.

STRUCTURAL STEEL

This corresponds to the *St* grades of steel where the main criterion in the selection and inspection of the steel is the tensile strength which is used as the basis of design. Generally load carrying welded structures such as frames, beds, columns, etc., are fabricated using this type of steel. Table 310 gives the properties of *St* 42W as per IS 2062, a weldable quality structural steel with a tensile strength in the range of 42 to 54 kgf/mm^2, which is frequently used in machine tools for welded structures.

CARBON STEELS

(1) *C*40, *C*45, *C*55, etc.

These are medium carbon steels with a carbon percentage varying between 0.35% and 0.6%. *C*45 is the preferred steel of this category and is suitable for applications such as shafts, gears, keys, toothed clutch, threaded fasteners requiring high strength, pins,etc. *C*45 can be induction hardened for wear resistance. Table 311 gives the properties of the above steel.

(2) *C*75, *C*80, *C*85

These are high carbon steels with carbon content from 0.7 to 0.9%, and are mainly used for springs, clutch plates, etc. *C*75 is the preferred steel and can be heat treated to a high hardness, generally in the range of *HRC* 60-64. Table 312 gives the properties of this steel.

ALLOY STEELS

Apart from iron and carbon, certain alloying elements such as nickel, manganese, chromium, tungsten, molybdenum, etc., are added to enhance certain desirable properties of steel. Alloy steels could be mainly grouped as direct hardening steels and case hardening steels.

103*Cr*1

This is popularly known as bearing steel which is a high carbon low chromium alloy steel generally used for ball and roller bearing races, lathe guideways, tools and dies, etc. Table 313 gives the properties of this steel.

40Ni2Cr1Mo28

This is the preferred direct hardening alloy steel. This is a medium carbon alloy steel with carbon content 0.35 to 0.45% and can be hardened to HRC 56-60 with very little distortion. Highly stressed components and also components with high wear resistance such as shafts, gears, etc., can be made from this steel. Control of the thickness of the hardened layer is more difficult in case of direct hardening steels as compared to case carburizing steels. Table 314 gives the properties of this steel.

15Ni2Cr1Mo15

This is the preferred case hardneing (carburizing) alloy steel, with carbon content 0.12 to 0.18% which could be, with very little distortion, case carburized and hardened to HRC 58-63. Portions of a component that are to be hardened could be case carburized and hardened leaving the core soft. This steel is mainly used for machine tool spindles, gears, cams, precision lead screws and other critical components where the case should be hard and wear resistant leaving the core tough. Table 315 gives the properties of this steel.

17Mn1Cr95

This is another case hardening (carburizing) alloy steel used in the machine tool industry. However, this steel is not generally recommended to be used for parts of diameter or width more than 100 mm. Table 316 gives the properties of this steel.

40Cr2Al1Mo18

This is the preferred case hardening (nitriding) alloy steel, which is used for components requiring high resistance to abrasion, higher surface hardness combined with high fatigue strength and freedom from distortion. Table 317 gives the properties of this steel.

STEEL CASTINGS

Intricate castings which are subjected to high working stresses are manufactured using steel. IS 1030 specifies five grades of steel for castings. Table 318 gives the important properties.

Non ferrous metals

Following are the main non ferrous metals used in machine tools owing to their unique characteristics such as low specific gravity, high wear resistance, low coefficient of friction, anticorrosive properties, etc. Some of them find particular application as bearing materials. Detailed analysis and requirement for bearing materials are separately covered in the chapter on *Hydrodynamic bearings*.

ALUMINIUM

Aluminium and its alloys have been used in machine tools owing to their light weight, corrosion resistance, castability into intricate shapes and thin wall thickness, forgeability and ability to be extruded and rolled. Some of these alloys respond to solution treatment and precipitation hardening by means of which the mechanical properties are improved. Aluminium and its alloy castings are standardised as per IS 617. The grades of Aluminium which are frequently used in machine tools are 4223M & WP (A4), 4600M (A6) and 2280W & WP (A11) for general castings and 5230M(A5) where decorative anodising is necessary. Table 319 gives some of the important properties of these grades of aluminium castings and wrought aluminium.

COPPER AND ITS ALLOYS

Copper: Copper in its pure form is used in electrical industry since it has the best electrical conductivity of any commercially priced metal. It also possesses good resistance to corrosion and can be formed into any shape easily.

Owing to their good corrosion resistance, good thermal conductivity, ductility and ease with which they can be brazed, copper tubes find an application for oil circulation in machine tools.

Brass: It is an alloy of copper and zinc. It is resistant to corrosion by water and atmosphere. Brass tubes are used for hydraulic cylinders. Brass strips are used for guideway wipers. Brass also finds application where free cutting steels are not permitted as in the case of fittings and other threaded fasteners in corrosive atmospheres.

Bronze: It is used as bearing material for worm wheels and lead screw nuts. It is an alloy of copper and tin. It is heavier, stronger and more resistant to corrosion than brass.

Phosphor bronze: A small percentage of phosphorous is added to bronze to act as a cleanser to the metal so that sound castings can be produced. Cast phosphor bronze to grade 2 of IS 28 is used for bearing bushes, lead screw nuts and worm wheels. Table 319 gives important properties as per IS 28 and IS 7811.

Aluminium bronze: It is harder, stronger and more wear resistant than phosphor bronze. It is also costlier than phosphor bronze. This is used for bearing bushes and for lead screw nuts where strength and wear are important. Table 319 gives important properties as per IS 305.

Leaded tin bronze: Because of the lead content this material possesses good lubricating property compared to phosphor bronze and aluminium bronze. This material is also more wear reisistant than phosphor bronze. Leaded tin bronze to grade 4 of IS 318 is used as a bearing material. Table 319 gives important properties as per IS 318.

Sintered bronzes: Sintered bronzes manufactured by powder metallurgy techniques offer advantages of simplicity of design and ease of manufacture. Owing to their inherent property of high porosity they can be very conveniently used as filtering media. Sintered bronzes impregnated with oil, graphite, molybdenum disulphide or PTFE are used as bearing materials.

Non metallic materials

Non metallic materials find their applications in machine tools in specific applications such as synthetic rubbers for seals, packings and belts; bakelite for control levers and knobs: nylon for lightly loaded gears and tubes for low pressure pneumatic and hydraulic systems and coolants; etc. PTFE can also be used as a guideway surfacing material. Concrete in some cases is used for machine tool beds.

Some examples of material selection for certain components are given as general guidance. The designer should always make the proper choice keeping in view the technical requirements, availability and cost.

Hardness conversion

Table 320 pertains to hardness conversion numbers together with the tensile strength for steels. For non ferrous materials the approximate relation between the tensile strength and the Brinell hardness is given is Table 321.

Conversion of hardness values should be used only when it is impossible to test the material under the conditions specified and when conversion is made it should be done with discretion. Each type of hardness test is subjected to certain errors, but if precautions are carefully observed the reliability of hardness readings will be found comparable.

The conversion values specified in the tables are only approximate. It is emphasized that there are a number of factors which may influence the accuracy of a hardness test. Moreover, deviations from the test conditions, namely, load, size of indentor, testing procedure used in deriving these tables may affect the accuracy of the hardness conversions.

Test data for different hardness tests are as follows:

1. *Brinell:* Ball 10*mm* diameter, load 3000 *kgf* for steel.
2. *Vickers:*
   - a) Diamond pyramid with dihedral angle of 136°, load 50 *kgf* for steel in normalised or annealed condition.
   - b) Diamond pyramid with dihedral angle of 136°, load 30 *kgf* for steel in hardened and tempered condition.
3. *Rockwell B:* Ball 1/16″ diameter, load 100 *kgf* for steel
4. *Rockwell C:* Diamond cone with angle of point of 120°, load 150 *kgf* for steel.

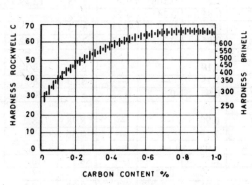

Fig. 213

Table 294 Physical constants of materials

Material	Modulus of elasticity $E \times 10^{-4}$ kgf/mm²	Modulus of rigidity $G \times 10^{-4}$ kgf/mm²	Poisson's ratio	Specific weight $\gamma \times 10^6$ kgf/mm³	Coefficient of linear expansion $\alpha \times 10^6$ mm/mm/°C	Thermal conductivity k cal/sec.cm.°C
Steel	2.1	0.83	0.3	7.85	10.98	0.12
Meehanite:						
GM	1.55	0.665	0.33	7.48	12.43	0.122
GA	1.4	0.613	0.32	7.43	12.37	0.121
GB	1.25	—	—	7.37	12.31	0.118
GC	1.2	0.508	0.3	7.25	12.19	0.112
GE	0.85	0.385	0.27	7.02	—	—
GF	0.63	0.28	0.24	6.73	—	—
Grey cast iron	0.85	0.385	0.27	7.5	10.62	0.13
Brass :						
Wrought	1.19	—	—	8.84	18.1	0.31
Red	1.19	—	—	8.73	18.7	0.31
Free cutting	0.98	0.35	0.33	8.47	20.5	0.31
Bronze :						
Commercial	0.98	—	—	8.78	18.2	
Aluminium	1.19	—	—	7.36	17.1	
Manganese	1.05	—	—	8.5	21.2	
Leaded Tin	0.91	—	—	8.7	18.5	0.16
Phosphor	1.12	0.42	0.349	8.78	18.2	
Gun metal	0.91	0.35	0.3	8.29	18.1	
Babbit :						
Lead base	0.294	—	—	9.66	19.6	
Tin base	0.532	—	—	7.36	—	0.16
Monel	1.295	0.49	0.32	8.61	12.9	0.06
Inconel	2.17	0.77	0.29	8.29	14	—
Copper	0.98	0.368	0.33	8.95	16.5	0.94
Beryllium copper	1.288	0.49	0.285	8.2	16.6	—
Aluminium	0.721	0.27	0.33	2.7	22	0.53
Nickel	2.1	—	—	8.85	13.3	—
Duralumin	0.68	—	—	2.8	—	—
Perspex	0.0316	0.0117	0.35	1.19	—	—

Table 295 Coefficient of friction μ for various material combinations

Sl. No.	Materials	Condition	μ
1	Steel on steel	dry	0.1 to 0.15
2	Steel on steel	in oil	0.05
3	Steel on cast iron	in oil	0.06
4	Cast iron on cast iron	dry	0.15 to 0.2
5	Cast iron on cast iron	in oil	0.05
6	Cast bronze on cast iron or steel	in oil	0.05
7	Hardened steel on hardened steel	in oil	0.05
8	Disc clutches	in oil	0.1

GREY CAST IRON: GRADE 15

Table 297a Mechanical properties

Sectional thickness of castings	mm	Over 4 Upto 8	Over 8 Upto 15	Over 15 Upto 30	Over 30 Upto 50
Dia. of test bars, as cast	mm	13	20	30	45
Tensile strength, Min.[1]	kgf/mm²	19	17	15	13
Compressive strength	kgf/mm²	63 approx.			
Impact strength *, Min.	kgf.m/cm²	0.83 to 1.38			
Breaking load, Min.	kgf	180	400	800	1700
Transverse rupture stress	kgf/mm²	41.7	38.2	34	28.5
Deflection, Min.	mm	2	2.5	4	6
Elastic modulus in tension** E	kgf/mm²	$(0.77 \text{ to } 1.05) \times 10^4$			
Modulus of rigidity	kgf/mm²	approx. 0.35-0.38 of modulus of elasticity in tension			
Torsion/Tensile ratio (solid bar)	—	1.15 approx.			
Shear strength/Tensile ratio	—	1.46 to 1.5 approx. varies inversely with strength of iron			
Fatigue/Tensile ratio	—	0.46 to 0.475 approx. Lower with higher strength			
Brinell hardness	HB	130 to 180			
Machinability rating ***	—	125 (100)			
Structure	—	Coarse open grained with large flakes of graphite.			

*   Un-notched $\phi 20mm$ test piece in Izod machine.
**   Values are approximate. E falls as graphite content increases and stress rises.
***   Machinability ratings given are for turning operation. Ratings for milling and drilling are given in brackets (approx. value)

Characteristics:

A soft grade of grey iron having very good fluidity and castability, very easily machinable. Generally advisable where only moderate strength is required. Particularly used for medium and thin sectioned castings. Preferred casting wall thickness 5 to 30 *mm*. For note 1 refer page 692

Applications

Parts subjected to average bending stresses upto 1 *kgf/mm²* such as columns and pedestals of machines; beds and other parts of complicated shape which should not suffer any distortion or deformation and which cannot be seasoned artificially. Parts working at pressures upto 0.05*kgf/mm²*, covers, arms, housings, transmission boxes, but not for direct bearing surfaces. For friction surfaces, slides, tables, only if subjected to wear and work with mating parts fabricated from grade 20.
Not desirable for guideways.

Wear resistance

Generally grey iron attains favourable wear resistance properties above *HB* 200.

For abrasive wear, the wear resistance can be further increased by heat treatment and alloying.

Table 297b

Equivalent grey cast irons	CSN	BS	DIN	ASTM	JIS	GOST	AFNOR
	42 2415	1452 Grade 10	1691 Grade GG 14	A 48 Class 20	G 5501 Grade 2 (FC 15)	1412 Grade SCh12-28	A 32-101 Grade Ft. 15

Table 297c Typical chemical composition *²

Section	Total carbon	Graphitic carbon	Combined carbon	Si	Mn	P	S	CE	Sc
Light**	3.4-3.6	—	—	1.75 to 2.5	0.4 to 0.6	0.3 to 0.9	0.07 to 0.14		
Medium**	3.5	2.9	0.6	2.25	0.5	0.4	0.1	4.3	0.98 to 1.04
Heavy**	3.3-3.6	—	—	2.2 to 2.4	0.4 to 0.6	0.3 to 0.6	0.07 to 0.14		

Carbon equivalent $(CE) = \% \text{ Total } C + 0.3\,(\%Si + \%P)$

Degree of eutectivity $(Sc) = \dfrac{\%C}{4.23 - 0.312\% \, Si - 0.275\% P}$

*All values expressed as a percentage.

**Light   upto 12 *mm*
  Medium   12 to 25*mm*
  Heavy   25 to 50 *mm*

For note 2 refer page 692.

GREY CAST IRON: GRADE 20

Table 298a Mechanical properties

Sectional thickness of castings	mm	Over 4 Upto 8	Over 8 Upto 15	Over 15 Upto 30	Over 30 Upto 50
Dia. of test bars, as cast	mm	13	20	30	45
Tensile strength, Min[1].	kgf mm²	24	22	20	17
Compressive strength	kgf/mm²	63 approx.			
Impact strength *, Min	kgf.m/cm²	0.83 to 1.38			
Breaking load, Min.	kgf	200	450	900	2000
Transverse rupture stress	kgf/mm²	46.4	43	38.2	33.5
Deflection, Min.	mm	2	3	4.5	6.5
Elastic modulus in tension** E	kgf/mm²	$(0.84 \text{ to } 1.12) \times 10^4$			
Modulus of rigidity	kgf/mm²	approx. 0.35-0.38 of modulus of elasticity in tension			
Torsion/Tensile ratio(solid bar)		1.2 approx.			
Shear strength/Tensile ratio		1.34 to 1.4 approx. Varies inversely with strength of iron			
Fatigue/Tensile ratio		0.42 to 0.455 approx. Lower with higher strength			
Brinell hardness	HB	160 to 220			
Machinability rating ***		100 (80)			
Structure		Medium open grained with large flakes of graphite			

*   Un-notched ϕ20 mm test piece in Izod machine.
**   Values are approximate. E falls as graphite content increases and stress rises.
***   Machinability ratings given are for turning operation. Ratings for milling and drilling are given in brackets (approx. value)

Characteristics
     A soft grade of grey iron with similar characteristics as those of grade 15, but generally advisable where higher strength than grade 15 is required. Preferred casting wall thickness 8 to 30 mm.
For note 1 refer page 692.

Applications

Parts subjected to bending stresses upto 3 *kgf/mm²* such as beds of slotting machines, planers and lathes; columns of milling, planing and boring machines; gear wheels working at low speeds and loads; parts working at pressures above 0.05 *kgf/mm²* On friction surfaces, beds with guideways (unhardened and hardened), cross beams, spindle stock housings, tailstocks and some parts of hydraulic systems upto 0.8 *kgf/mm²* pressures.

Wear resistance

Generally grey iron attains favourable wear resistance properties above *HB* 200.

For abrasive wear. the wear resistance can be further increased by heat treatment and alloying.

Table 298b

	CSN	BS	DIN	ASTM	JIS	GOST	AFNOR
Equivalent grey cast irons	42 2420	1452 Grade 12	1691 Grade GG-18	A 48 Class 25	G 5501 Grade 3 (FC 20)	1412 Grade SCh18-36	A 32-101 Grade Ft 20

Table 298c Typical chemical composition*²

Section	Total carbon	Graphitic carbon	Combined carbon	Si	Mn	P	S	CE	Sc
Light**	3.25-3.55	—	—	2-2.4	0.4-0.75	0.15-0.5	0.07-0.14		
Medium**	3.4	—	—	2.2	0 5	0.3	0.1	4.15	0.92 to 0.98
Heavy**	3-3.55	—	—	2-2.3	0.5	0.15-0.5	0.07-0.14		

Carbon equivalent $(CE) = \%$ Total $C + 0.3(\%Si + \%P)$

Degree of eutectivity $(Sc) = \dfrac{\%C}{4.23 - 0.312\% \, Si - 0.275\%P}$

* All values expressed as a percentage.

**Light   Upto 12 *mm*

  Medium   12 to 25 *mm*

  Heavy   25 to 50 *mm*

For note 2 refer page 692.

GREY CAST IRON: GRADE 25

Table 299a Mechanical properties

Sectional thickness of casting	mm	Over 4 Upto 8	Over 8 Upto 15	Over 15 Upto 30	Over 30 Upto 50
Dia. of test bar, as cast	mm	13	20	30	45
Tensile strength, Min.	kgf/mm²	28	26	25	22
Compression strength	kgf/mm²	80 approx.			
Impact strength*, Min.	kgf.m/cm²	0.83 to 1.66			
Breaking load, Min.	kgf	220	500	1000	2300
Transverse rupture stress	kgf/mm²	51	47.8	42.4	38.6
Deflection, Min.	mm	2	3	5	7
Elastic modulus in tension** E	kgf/mm²	$(0.984 \text{ to } 1.195) \times 10^4$			
Modulus of rigidity	kgf/mm²	0.35 to 0.38 of modulus of elasticity in tension			
Torsion/tensile ratio (solid bar)		1.25 approx.			
Shear strength/Tensile ratio		1.29 to 1.36 approx. Varies inversely with strength of iron.			
Fatigue/Tensile ratio		0.38 to 0.42 approx. Lower with higher strength			
Brinell hardness	HB	180 to 230			
Machinability rating***		80 (63)			
Structure		Close grained with relatively fine flakes of graphite.			

*   Un-notched ϕ 20 mm test piece in Izod machine
**   Values are approximate. E falls as graphite content increases and stress rises.
***   Machinability ratings given are for turning operation. Ratings for milling and drilling are given in brackets (approx. value).

Characteristics

     Readily castable, high machinability characteristics, close grained structure and sound in relatively heavy sections. Perferred casting wall thickness 15 to 45 mm. For note 1 refer page 692.

Applications

Parts subjected to high bending stresses upto 5 *kgf/mm²* such as columns of presses, chucks for lathes, gear wheels, etc.; parts working at pressures upto 0.2 *kgf/mm²* on friction surfaces, parts which are surface hardened, guideways, beds with guiding surfaces for turrets, semi-automatic and automatic lathes, etc.; parts which should have a high degree of nonporosity such as hydraulic cylinders, bodies of hydraulic pumps, high pressure pumps, etc.

Wear resistance:

Generally grey iron attains favourable wear resistance properties above *HB* 200.

For abrasive wear, the wear resistance can be further increased by heat treatment and alloying.

Table 299b

Equivalent grey cast irons	CSN	BS	DIN	ASTM	JIS	GOST	AFNOR
	42 2425	1452 Grade 14	1691 Grade GG 22	A 48 Class 35	G 5501 Grade 4 (F C 25)	1412 Grade SCh 24-44	A 32-101 Grade Ft 25

Table 299c Typical chemical composition*²

Section	Total carbon	Graphitic carbon	Combined carbon	Si	Mn	P	S	CE	Sc
Light**	3.2-3.4	2.55	0.7	2	0.7	0.2	0.1		
Medium**	3.1-3.4	2.35-2.7	0.6-0.8	1.5-2.5	0.55-0.75	0.1-0.25	0.07-0.14	3.9	0.86-0.92
Heavy**	3.15-3.35	—	—	1.5-2.2	0.5-0.75	0.15-0.25			

Carbon equivalent $(CE) = \%$ total $C + 0.3\,(\%Si + \%P)$

Degree of eutectivity $(Sc) = \dfrac{\%C}{4.23-0.312\%Si-0.275\%P}$

*All values expressed as a percentage.

**Light   Upto 12 *mm*

  Medium 12 to 25 *mm*

  Heavy   25 to 50 *mm*

For note 2 refer page 692.

Notes:

1. Generally the damping capacity decreases with decreasing amount of graphite and increasing tensile strength.
2. Typical chemical composition is given for information only. Acceptance tests should be based only on mechanical properties. Actual compositions should be chosen according to foundry practice, raw materials and production conditions, maintaining the degree of eutectivity in the given range.

Table 300   Heat treatment for grey cast irons

Operation	Temperature°C	Remarks
Stress relieving	510-565	Hold for 1 *hr.* per 25 *mm* of max. wall thickness. Furnace cool to 290°C, then air cool.
Low temperature anneal	700-760	Hold for 45 minutes to 1 *hr.* per 25 *mm* of max. wall thickness. Furance cool at 55°C/hr. between 540°C and 290°C.
Medium anneal*	790-900	Hold for about 45 minutes per 25 *mm* of max. wall thickness. Furnace cool from annealing temperature to 290°C.
Flame hardening	—	Surface hardness of between 400 and 450 Brinell may be obtained with the depth of hardness layer between 1.25 and 3.8 *mm*. Grey iron with 0.5 to 0.7 % combined carbon is recommended. Irons with more than 0.8% combined carbon tend to crack on surface hardening.
Induction hardening	—	Hardness of 59 to 60 Rockwell *C* may be obtained Grey irons with 0.5 to 0.7% combined carbon may be induction hardened.
Hardening by oil quenching	760-900	Hold for about 20 minutes per 25*mm* of max. wall thickness at temperature. Quench in oil bath at 230°C-290°C.
Tempering**	150-550	Hold at temperature for about 1 *hr.* per 25 *mm* of max. wall thickness and cool.
	200-250	For max. wear resistance.
	350-450	For strength and wear resistance. Corresponding average hardness 321-418 *HB*.

*   Adopted for grey irons which do not respond to low temperature anneal.
**   Tempering temperature should be chosen depending on the composition of the iron and final properties required. However, the chosen temperature should be higher than the actual service temperature. Two specific examples are indicated.

Table 301 Meehanite for general engineering purposes

Properties		Grade Unit	GM	GA	GB	GC	GD	GE
Normal minimum casting section		mm	18	13	10	7	5	3
Tensile strength Min.		kgf/mm²	38	35	32	28	25	21
Modulus of elasticity		kgf/mm² E×10⁻⁴	1.5	1.45	1.3	1.2	1.05	0.85
Transverse strength (test bar 30 mm dia.) distance between supports 50 mm	Load P	kgf	1500	1350	1100	1100	1000	900
	Deflection δ	mm	6.6	6.1	5.6	6	6.3	7
Compression strength		kgf/mm²	140	125	115	105	95	80
Shear strength		kgf/mm²	38	34	30	28	25	21
Fatigue strength (10⁷ reversals)		kgf/mm²	18	16	14	12	10	9
Impact strength (Izod) (20mm dia. unnotched bar)		kgf.m	4	3.4	2.7	2	1.3	0.9
Hardness Min. (depending on section thickness)		HB	230	220	210	195	185	170
Damping capacity 14 kgf/mm². Torsional stress energy dissipated in first cycle		%	21	24	25	28	30	32
Pattern marker's shrinkage allowance		%	1.3 to 1.5	1.3 to 1.5	1 to 1.3	1 to 1.3	1 to 1.3	0.8 to 1.0
Nearest equivalent CI grade as per IS 210		—	40	35	30/35	30	25	20

Table 302 Wear resisting meehanite metal

Type	Condition	Tensile strength kgf/mm²	Brinell hardness HB
WA	Sand cast	35	200 - 300
WAH	Heat treated	upto 50	upto 550
WH	Sand cast	upto 32	upto 650
WB	Sand cast	upto 27	350 - 550
WBC	Chill cast	upto 32	upto 550 on chill face
WEC	Chill cast	22	upto 550 on Chill face

Table 303 Heat treatment of meehanite

Description	Treatment	Type of Meehanite
Anneal for improved machinability without serious loss of strength	Anneal for 1 hour/25mm section at 660°-700°C.	GE,GD and GC.
Full anneal for maximum machinability	Heat slowly to 850°-870°C hold for 1 hour/25 mm section. Cool slowly in furnace.	GE
Heat treatment for maximum strength	Heat to 870°C. Quench in oil. Temper at 500°C.	GA,GM and WA.
Heat treatment for maximum hardness	Heat to 870°C. Quench in oil. Temper at 250°C.	GA, GM and WA.
Stress relief	Heat Slowly to 550°C.	GE,GD, GC.
	Heat Slowly to 650°C. Hold for 4 hours. Cool Slowly in furnace.	GA and GM

Table 304 Guide for selection of Meehanite for machine tool components

Service condition	Recommended type	Examples of application
High fluid pressure	GC, GB, GA or GM according to stress condition and section thickness	Pump bodies, cylinders, valves, pistons, pipes
Vibratory stresses	GE, GC, GA or GM according to stress condition and section thickness	Machine tool parts, gears, crankshafts, bed plates, machine frames
Dimensional stability	GD, GC, GB, GA or GM according to stress condition and section thickness. Stress relief is necessary	Gauges, jigs, machine tool parts, gears
Lubricated wear	GC, GB, GA for moderate wear condition and machinability required; WA for heavier wear, harder but machinable; WB for heavy wear and machinability not required	Machine tool slides, cylinder liners, gears, pistons, cams
Wear with generation of heat	GD, GC, GA or WA according to severity of wear and strength required. Stress relief is necessary	Heavy brake drums, clutch plates, Hot forming dies, cylinder liners.

Table 306 Mechanical properties of spheroidal graphite iron. (IS 1865)

Characteristics / Grade	SG 800/2	SG 700/2	SG 600/3	SG 500/7	SG 400/12	SG 370/17
Tensile strength kgf/mm^2	80	70	60	50	40	37
Proof stress 0.2% kgf/mm^2	48	44	37	32	25	23
Elongation % Min.	2	2	3	7	12	17
Typical Brinell hardness HB	248-352	229-302	192-269	170-241	201 Max.	179 Max.
Predominant structural constituent	Pearlite or tempered structure	Pearlite	Pearlite and ferrite	Ferrite and pearlite	Ferrite	Ferrite

Heat treatment:

Operation	Temperature °C	Remarks
Stress relieving	510 -650	Hold for 1 hour +1 hour per 25 mm of max. wall thickness. Furnance cool to 290°C, then air cool.
Annealing	900 -950	Hold for 1 hour +1 hour per 25 mm of max. wall thickness. For light sections 1 to 3 hours is sufficient but for heavy sections 3 to 8 hours are needed. Uniformly cool at a rate not more than 20°C/hour.
Sub. critical annealing	700 -710	Hold for 5 hours +1 hour per 25mm of max. wall thickness. Furnance cool to 600°C.
Normalising	870 -940	Hold for 1 hour min. or 1 hour +1 hour per 25 mm of max. wall thickness. Actual holding time and temperature depend upon the composition of the S G iron.
Hardening	850 -930	Actual holding time and temperature depends upon composition. Quench in oil. Quenching in water or brine is also possible for simple shapes.
Tempering	350 -700	Temper for 1 hour per 25 mm of max. wall thickness or min. 2 hours. Actual tempering temperature depends upon the properties desired.
Flame or Induction hardening	—	Hardness of HRC 53-60 may be obtained depending on the structure prior to hardening.

Table 307 Indian steels and their equivalents

IS Indian	BS British	AISI American	DIN German	JIS Japanese	GOST Russian	CSN Czech	UNI Italian	AFNOR French
14Mn1S14	En202	C1118	15S20	SUM2	A12	11110	–	15F2
40Mn2S12	En15AM	C1139	45S20	SUM5	A40G	11140	–	45MF4
C40	–	C1040	40Mn4	S40C	40	12040	–	–
C45	En8D	C1045	Ck45	S45C	MSt6	12050	C45	XC45F
C75	S513	C1074	MK75	–	75	12081	–	–
17Mn1Cr95	En207	5115	16MnCr5	5Cr21	20ChGA	14220	16MC5	16MC5
T50Cr1V23	–	6150	50CrV4	SUP10	50ChFA	15260	50CrV4	50CV4
15Ni2Cr1Mo15	En354	4317	15CrNi6	SNCM22	12ChN2	16220	–	18NCD6
40Ni2Cr1Mo28	En24	4340	36CrNiMo4	SNCM8	40ChNMA	16341	–	–
13Ni3Cr80	En36A	E9310	14NiCr14	SNC22	12ChN3A	16420	18NiCr13	14NC12
30Ni4Cr1	En30A	–	35NiCr18	–	–	16640	–	35NC15
T215Cr12	–	–	X210Cr12	SKD1	Ch12	19436	UX200C13	Z200C12
T110W2Cr1	–	–	105WCr6	SKS2	ChVG	19710	U100WC	–
T83MoW6Cr4V2	–	M2	DM05	SKH9	R9	19800	UX82WD65	6-5-2
T123W14Co5CrV4	–	–	EV4Co	–	R14F4	19810	–	–
T75W18Co6Cr4V1Mo75	–	T4	E18Co5	SKH3	R18K5F2	19855	UX80WK185	18-0-1-5
103Cr1	En31	E52100	100Cr6	SUJ1	ShCh15	14109	–	–

Note: Steels indicated in the thick frames are rationalized steels for machine tool industry.

Table 308 Free cutting steel 14Mn1S14

Steel designation			14Mn1S14			
Chemical composition, %		C	Mn	Si	S	P Max.
		0.1-0.18	1.2-1.5	0.05-0.3	0.1-0.18	0.06
Forms of material		Bars, billets and forgings				
Supply condition		Hot rolled or cold drawn				
Condition of material		Hot rolled or normalised	Cold drawn			Refined and quenched
Limiting ruling section, mm			Upto 20	Over 20 Upto 40	Over 40 Upto 63	30
Tensile strength kgf/mm^2		44-54	Min. 55	Min. 52	Min. 48	Min. 60
Yield strength kgf/mm^2 Min.		—	38	38	36	—
Elongation in % Min.	Gauge length 5.65 $\sqrt{So}$	22	10	11	12	17
	Gauge length 4 $\sqrt{So}$	26	12	14	15	20
Izod impact strength, $kgf.m$ Min.		—	—	—	—	4.1
Hardness	Brinell HB	121	137	137	137	—
	Vickers HV	—	—	—	—	—
	Rockwell HRC	—	—	—	—	—
Fatigue limit kgf/mm^2. 10^7 reversals		—	—	—	—	—
Machinability and machinability rating		Very good	200* 160**	—	—	—	@
Application		Parts requiring good machinability and finish. Threaded fasteners				
Remarks		@ Good in refined and quenched condition. *For turning operation only. **For milling, drilling, reaming & tapping operations				
Equivalent steels	CSN	BS	AISI	GOST	JIS	DIN	AFNOR
	11110	En 202	C1118	A12	SUM2	15S20	15F2

Table 309 Free cutting steel 40Mn2S12

Steel designation			40Mn2S12				
Chemical composition, %		C	Mn	Si	S	P	
		0.35-0.45	1.3-1.7	0.25 Max.	0.08-0.15	0.06 Max.	
Form of material		Bars, billets and forgings					
Supply condition		Hot rolled or cold drawn					
Condition of material		Hot rolled or normalised	Cold drawn			Hardened and tempered	
Limiting ruling section, mm (dia.)		—	Upto 20	Over 20 Upto 40	Over 40 Upto 63	30	60
Tensile strength in kgf/mm^2		60-70	Min. 68	Min. 64	Min. 62	80-95	70-85
Yield strength in kgf/mm^2 Min		—	—	—	—	56	50
Elongation in %,Min.	Gauge length 5.65 $\sqrt{S_o}$	15	7	8	10	16	18
	Gauge length 4 $\sqrt{S_o}$	18	9	10	12	20	22
Izod impact strength in $kgf.m$,Min.		—	—	—	—	4.1	4.8
Hardness	Brinell HB	187	212	212	212	—	—
	Vickers HV	—	—	—	—	Ref. hardenability curve	
	Rockwell HRC	—	—	—	—	Ref. tempering curve	
Fatigue limit,kgf/mm^2, 10^7 reversals		33-35	—	—	—	38	35
Machinability and Machinability rating		Very good	* 150	Good		Good in toughened condition	
Application	Parts requiring good machinability and finish. Threaded fasteners						
Remarks	Not suitable for forgings where transverse properties are important * For turning, shaping and planing operations only						
Equivalent steels	CSN	BS	AISI	DIN	JIS	GOST	
	Approx 11140	En15AM	C1139	45S20	Approx. SUM5	A40G	

Table 309 Free cutting steel 40Mn2S12   (Contd.)

Recommended temperatures for forging and heat treatment

Operation		Temperature °C	Remarks
Forging		1200 to 850	Start forging at 1150°C & finish above 850°C; cool forged parts slowly in air
Normalizing		840 to 870	Slowly cool in air after heating
Sub-critical annealing		680 to 720	About 4 hours, slowly cool in furnace
Hardening	Water	840 to 870	Water quench for thicker workpieces
	Oil	850 to 880	Oil quench for thinner workpieces
Flame hardening	—	—	With oil or water
Tempering	Air	550 to 660	Ref: Tempering curve
	—	—	—

Note: Mechanical properties under hardened and tempered condition correspond to a tempering temperature in the range of 550 to 660°C.

Tempering curve

Toughness in kgf/mm^2

Hardness HRC

Temperature in °C

Hardenability curve (End quench)

Hardness HV

Distance in mm

Table 310 Mechanical properties of structural steel
(Fusion welding quality St 42W) (IS 2062)

Class of steel product	Nominal thickness/ diameter, mm	Tensile strength @ kgf/mm^2	Yield stress* Min. kgf/mm^2	Percentage** elongation Min.
Plates, sections (e g., tees, beams, angles, channels, etc.) & flats	$\leqslant$ 6	Bend test only shall be required *		
	> 6 $\leqslant$ 20	42 to 54	26	23
	> 20 $\leqslant$ 40	42 to 54	24	23
	> 40	42 to 54	23	23
Bars (Round, square and hexagonal)	$\leqslant$ 10	Bend test only shall be required @@		
	> 10 $\leqslant$ 20	42 to 54	26	23
	> 20	42 to 54	24	23

Notes:

@ Provided the yield stress and elongation requirements are complied with, the upper limit may be raised by 3 kgf/mm^2

* In case of plates, section, flats below 6 mm, yield stress shall be assumed to be at least the same as that for thickness between 6 mm & 20 mm

** Values given for a guage length of 5.65 $\sqrt{S_0}$

@@ In case of bars below 10 mm dia. the yield stress shall be assumed to be at least the same as that for bars of dia. between 10 mm & 20 mm

Table 311 Medium carbon steel for machine tools *C45*

Steel designation		*C45*					
Chemical composition,%		*C*	*Mn*	*Si*	*·P*	*S*	
		0.4-0.5	0.6-0.9	0.1-0.35	0.035 Max.	0.035 Max.	
Form of material		Bars, plates, billets, sections and forgings					
Supply condition		Rolled or forged					
Condition of material		Hot rolled, normalised or annealed		Hardened and tempered*			
Limiting ruling section,*mm*		—		100		30	
Tensile strength in *kgf/mm²*		63-71		60-75		70-85	
Yield strength in *kgf/mm²*Min.		34		38		48	
Elongation in %,Min.	Gauge length 5.65$\sqrt{S_o}$	15		17		15	
	Gauge length 4$\sqrt{S_o}$	18		20		18	
Izod impact strength in *kgf.m*, Min.		—		4.1		3.5	
Hardness	Brinell *HB*	207 Max.		—		—	
	Vickers *HV*	—		Ref. hardenability curve			
	Rockwell *HRC*	—		Ref. tempering curve			
Fatigue limit *kgf/mm²* 10⁷ reversals		32-35		30-37		35-42	
Machinability & machinability rating		Good	100	Good in toughened condition upto 30*HRC* (95*kgf/mm²*)			
Application		Spindles, gears, threaded fasteners requiring high strength, shafts, keys etc.					
Equivalent steels	BS	AISI	DIN	JIS	AFNOR	GOST	CSN
	Approx En 8D	C1045	Ck45	S45C	XC45F	MSt6	12050

Table 311 Medium carbon steel for machine tools C45 (Concld.)

Recommended temperatures for forging and heat treatment

Operation		Temperature °C	Remarks
Forging (hot working)		1200 - 850	Start forging at 1200°C, after forging cool forged part slowly in air
Normalizing		830 - 860	After heating, slowly cool in air
Sub-critical annealing		680 - 720	About 4 hours, slowly cool in furnace
Hardening	Water	830 - 860	Water quench for thicker workpieces
	Oil	830 - 860	Oil quench for thinner workpieces
Flame hardening	Water quenched	—	For ϕ upto 30: 2-3mm depth. 55-58HRC For ϕ 30 to 60: 2-3mm depth. 50-55HRC For ϕ 60 to 100: 2-3mm depth. 45-50HRC
	Oil	—	
Tempering	Air	530 - 670	Ref. tempering curve

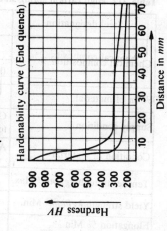

Notes:

1) Mechanical properties under hardened and tempered condition corresponds to a tempering temperature in the range of 530 to 670°C. Avoid tempering in the range of 230-400°C.

2) This steel can also be induction hardened. 2 to 3 mm depth and a case hardness of HRC 57-59 can be normally obtained on all sizes when quenched in water.

3) Ideal critical diameter range is 22 to 32 mm

Table 312 High carbon spring steel C 75

Steel designation			C 75			
Chemical composition %		C	Mn	Si	S	P
		0.7-0.8	0.5-0.8	0.1-0.35	0.05Max.	0.05Max.
Form of material		Strip				
Supply condition		Cold rolled, annealed or hardened and tempered				
Condition of material		Annealed or rolled	Hardened and tempered			
Tensile strength kgf/mm^2 Max.		65	120-160			
Yield strength kgf/mm^2 Min.		30	110			
Elongation % Min. (Gauge length 5.65 $\sqrt{So}$)		25	6			
Hardness	Brinell HB	—	—			
	Vickers HV	220 Max.	350 - 475			
	Rockwell HRC	—	—			
Fatigue limit kgf/mm^2 10^7 reversals		—				
Machinability and Machinability rating		—				
Application		For light flat spring from annealed stock				
Remarks		Shall be free from defects such as scales, rust, blisters, laminations, cracked edges, etc.The surface condition of cold rolled, annealed or hardened and tempered steel strip may be either dull bright, polished or polished and tempered. Strips shall be adequately coated with rust preventive oil.				
Equivalent steels		BS	CSN	AISI	DIN	GOST
		S513	12081	C1074	MK75	75

Recommended temperatures for forging and heat treatment

Operation	Temperature ^0C	Remarks
Hot working	1100 - 850	—
Soft annealing	600 - 650	Cool in furnace
Normalizing	—	—
Hardening	780 - 810	Quench in oil
Tempering	420 - 500	According to the required properties

Table 313 Bearing steel 103Cr1

Steel designation		103Cr1					
Chemical composition,%		C	Mn	Si	S Max.	P Max.	Cr
		0.95-1.1	0.25-0.45	0.15-0.35	0.025	0.025	0.9-1.2
Allowable tolerance,%		±0.03	±0.03	±0.02	+0.005	+0.005	±0.05
Form of material		Bars, billets & forgings		Bars		Bars & wires	
Condition of material		Annealed or forged		Annealed or forged		Drawn	
Tensile strength,*kgf/mm²,* Min.		62 -74		62 - 74		64 - 78	
Elongation,%, Min.		(18)		(18)		(18)	
Hardness	Brinell *HB* (annealed condition)	207 Max.		207 Max.		225 Max.	
	Rockwell *HRC* (after quenching in oil)	62 Min.		62 Min.		62 Min.	
Ductility	Hot	Good					
	Cold	—					
Microstructure		The annealed material shall show a completely spheroidized structure of uniformly distributed globular carbides. Shall be as per grades 2, 3 and 4 of IS 4398. The structure shall be free from segregation and shall not reveal presence of carbides in cellular form. The acceptable limits of carbide bonding shall be as per grades 1 and 2 of IS 4398.					
Equivalent steels		BS	AISI	DIN	CSN	JIS	GOST
		En31	E52100	100Cr6	14109	SUJ 1	ShCh15

24(45-83/1972) MTDH

Table 313 Bearing steel 103*Cr*1　　　　　　　　　　　　　Concld.)

Recommended temperatures for forging and heat treatment		
Operation	Temperature °C	Procedure
Forging	1100-850	After forging cool forged part in air.
Soft annealed	750-800	For about 4 hrs, slowly cool in furnace
Normalizing	870-900	Cool in air.
Quenching — Oil	830-870	
Quenching — Water	800-830	
Tempering	150-170	After heating in oil for 3 hrs.

Note:　1. Values given in brackets are for information only.

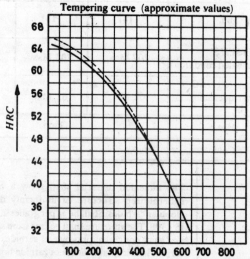

Tempering curve (approximate values)

HRC

Tempering temperature °C (time 1 hour air cooled)

Table 314a Direct hardening steel for machine tools 40*Ni2Cr1Mo*28

Steel designation		40*Ni2Cr1Mo*28					
Chemical composition* %		C	Si	Mn	Ni	Cr	Mo
		0.35- 0.45	0.1- 0.35	0.4- 0.7	1.25- 1.75	0.9- 1.3	0.2- 0.35
Form of material		Bars, billets and forgings					
Supply condition		Rolled or forged					
Condition of material		As supplied	Hardened and tempered				
Limiting ruling section *mm*			30	63	100	150	
Tensile strength *kgf/mm²*		80	120-135 (**155 Min.**)	110-125	100-115	80-95 (90-105)	
Yield strength *kgf/mm²* ** Min.			100 (130)	88	80	60 (70)	
Elongation in % Min.	Gauge length 5.65 $\sqrt{So}$ **	—	10(6)	11	13	16(15)	
	(Gauge length 4 $\sqrt{So}$)**	—	13 (8)	14	16	20 (18)	
Izod impact strength *kgf.m* **Min..		—	3(1.1)	4.1	4.8	5.5	
Hardness	Brinell *HB* **	230 Max.	341-401 (**440 Min.**)	311-363	285-341	229-277 (255-311)	
	Vickers *HV*		Ref. hardenability curve				
	Rockwell *HRC*		Ref. tempering curve				
Fatigue limit *kgf/mm²*. 10⁷ reversals			50-60	45-55	40-50	35-45	
Machinability and machinability rating		Fair	80	Fair in toughened condition up to a hardness of *HRC* 30			
Application		Highly stressed machine parts, shafts, gears, etc.					
Remarks		* Phosphorus and sulphur each 0.035 Max. ** **For note refer page 708.**					

Equivalent steels	BS	CSN	AISI	JIS	DIN	GOST
	En 24	16341	4340	SNCM8	36CrNiMo4	40ChNMA

Table 314b Direct hardening steel for machine tools 40Ni2Cr1Mo28
Recommended temperatures for forging and heat treatment

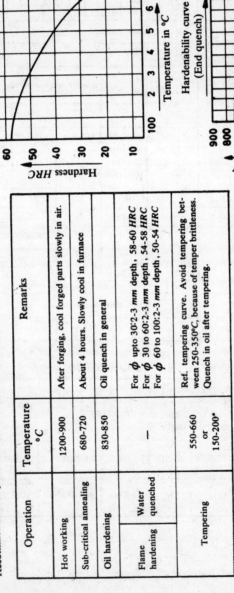

Operation		Temperature °C	Remarks
Hot working		1200-900	After forging, cool forged parts slowly in air.
Sub-critical annealing		680-720	About 4 hours. Slowly cool in furnace
Oil hardening		830-850	Oil quench in general
Flame hardening	Water quenched	–	For φ upto 30:2-3 mm depth, 58-60 HRC For φ 30 to 60:2-3 mm depth, 54-58 HRC For φ 60 to 100:2-3 mm depth, 50-54 HRC
Tempering		550-660 or 150-200*	Ref. tempering curve. Avoid tempering between 250-350°C, because of temper brittleness. Quench in oil after tempering.

Notes: Mechanical properties under hardened and tempered condition corresponds to a tempering temperature in the range of 550 to 660°C

*Depending on hardness required

**   Bracketed values correspond to oil hardening and others to air hardening. It is preferable to oil quench for sections above dia. 50 mm

Table 315 Carburizing steel for machine tools 15 Ni2Cr1Mo15

Steel designation	15Ni2Cr1Mo15							
	C	Si	Mn	Ni	Cr	Mo	S	P
Chemical composition %	0.12-0.18	0.1-0.35	0.6-1	1.5-2	0.75-1.25	0.1-0.2	0.05 Max.	0.035 Max.
Form of material	Bars, billets and forgings							
Supply condition	Rolled or forged							
Condition of material	As supplied, annealed		Refined and quenched					
Limiting ruling section *mm*	—		90		60		30	
Tensile strength *kgf/mm²* Min.	63		95		100		110	
Yield strength *kgf/mm²* Min.	52		80		90		95	
Elongation in % Min. — Gauge length $5.65\sqrt{S_o}$	17		—		—		9	
Elongation in % Min. — Gauge length $4\sqrt{S_o}$	—		—		—		12	
Izod impact strength *kgf.m* Min.	—		—		—		3.5	
Hardness — Brinell *HB*	185*		230		250		300	
Hardness — Vickers *HV*	Ref. hardenability curve							
Hardness — Rockwell *HRC*	59-63 case							
Fatigue limit *kgf/mm²*. 10⁷ reversals	25-31		40-47		40-50		44-55	
Machinability and Machinability rating	Fair	80						
Application	Heavy duty components, gears, etc.							
Remarks	*Hardness value should not exceed 217							

	BS	CSN	DIN	AISI	JIS	GOST	AFNOR
Equivalent steels	En 354	16220	15CrNi6	4317	SNCM22	12ChN2	18NCD6

Table 315 Carburizing steel for machine tools 15Ni2Cr1Mo15   (Concld.)

Recommended temperatures for forging and heat treatment

Operation		Temperature	Procedure
Forging		1150-900°C	Forging to start at 1150°C. Forged parts to cool slowly in air.
Normalizing		860-880°C	Slowly cool in air
Sub-critical annealing		680-720°C	4 hours, cool in furnance
	Carburize	900-930°C	Duration of carburizing as per penetration curve. After carburizing cool in furnace. — Any machining operation on the carburized parts to be completed before further heating.
Hardening	Oil (Double quench)	Core toughen 850-880°C Case harden 780-820°C	After core toughening, the components to be quenched in oil and reheated. After reheating components to be again quenched in oil.
	Oil (Single quench)	820-850°C	Alternatively, the parts are heated to the single quenching temperature range and quenched in oil. However, this method can be adopted only in exceptional cases, as the properties cannot be guaranteed.
Tempering	Air	180-200°C	Cool in air

Notes: After carburizing the parts should be cooled only in the furnace. Air cooling will cause decarburization.

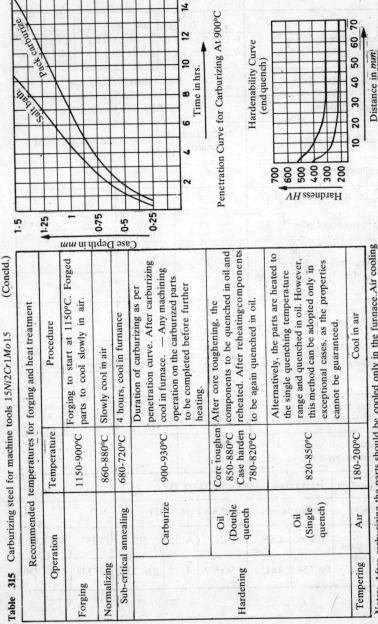

Penetration Curve for Carburizing At 900°C

Hardenability Curve (end quench)

Table 316 Carburizing steel for machine tools $17Mn1Cr95*$

Steel designation		17Mn1Cr95					
Chemical composition		C	Si	Mn	Cr	S	P
		0.14-0.19	0.1-0.35	1-1.3	0.8-1.1	0.035 Max.	0.035 Max.
Form of material		Bars, billets and forgings					
Supply condition		Rolled or forged					
Condition of material		As supplied annealed	Refined and quenched				
Limiting ruling section mm		—	100	50	30		
Tensile strength kgf/mm² Min		60	65	75	80		
Yield strength kgf/mm² Min		45	50	55	60		
Elongation % Min.	Gauge length $5.65\sqrt{S_0}$	22	18	13	10		
	Gauge length $4\sqrt{S_0}$	—	—	—	13		
Izod impact strength kgf. m Min.		—	—	4	3.5		
Hardness	Brinell HB	170	180	200	240		
	Vickers HV	—	Ref. Hardenability curve				
	Rockwell HRC	—	59-63 case				
Fatigue limit kgf/mm² 10^7 reversals		24-30	25-30	28-32	32-40		
Machinability and Machinability rating		Fair	63	—			
Application		Used for shafts, gears and spindles. Generally applicable where wear properties coupled with core strength are required.					
Remarks		In general not recommended to be used for parts of dia. or width more than 100mm * Preferred Carburizing Steel is $15Ni2Cr1Mo15$					

Equivalent steels	CSN	SAE or AISI	DIN	GOST	BS		
	14220	5115	16MnCr5	20XGA	En 207		

Table 316 Carburizing steel for machine tools 17Mn1Cr95   (Concld.).

Recommended temperatures for forging and heat treatment

Operation	Temperature	Procedure
Forging	1100–850°C	Forging to start at 1150°C. Forged parts to cool slowly in air.
Normalizing	850–880°C	Slowly cool in air
Sub-critical annealing	680–720°C	4 hours. cool in furnace
Carburize	890–910°C	Duration of carburizing as per penetration curve. After carburizing cool in furnace. Any machining operations on the carburized parts to be completed before further heating
Hardening — Oil (Double quench)	Pre-heat 890°–910°C Re-heat 760°–780°C	After pre-heating, the components to be quenched in oil and reheated. After reheating components to be again quenched in oil
Hardening — Oil (Single quench)	810–830°C	Alternatively, the parts are heated to the single quenching temperature range and quenched in oil. However, this method can be adopted only in exceptional cases as the properties cannot be guaranteed.
Tempering — Air	150–175°C	Cool in air.

Note: After carburizing the parts should be cooled only in the furnace. Air cooling will cause decarburization.

Case depth in mm vs Time in hrs (Salt bath, Pack carburize); Hardenability curve (End quench) — Hardness HV vs Distance in mm.

Table 317 Nitriding steel for machine tools 40*Cr*2*Al*1*Mo*18

Steel designation	40*Cr*2*Al*1*Mo*18						
Chemical composition,* %	*C*	*Si*	*Mn*	*Ni*	*Cr*	*Mo*	*Al*
	0.35 0.45	0.1 0.45	0.4 0.7	0.3 Max.	1.5 1.8	0.1 0.25	0.9 1.3
Form of material	Bars, billets and forgings						
Supply condition	Rolled or forged						

Condition of material		As supplied (Annealed)		Hardened and tempered		
Limiting ruling section, *mm*				63	100	150
Tensile strength *kgf/mm*² Min.		—		90 to 105	80 to 95	70 to 85
Yield strength *kgf/mm*² Min.		—		70	60	54
Elongation in %, Min.	Gauge length $5.65\sqrt{S_o}$	—		15	16	18
	Gauge length $4\sqrt{S_o}$	—		18	20	22
Izod impact strength, *kgf.m*, Min.		—		4.8	5.5	5.5
Hardness	Brinell *HB*	230 Max.		255-311	229-277	201-248
	Vickers *HV*	—		Ref: Hardenability curve case 900-930		
	Rockwell *HRC*	—		case 66-67		
Fatigue limit, *kgf/mm*². 10^7 reversals		—		60-65	50-55	45-50
Machinability and machinability rating		Fair	63**	—		
Application	Applications requiring high resistance to abrasion, maximum surface hardness, high fatigue strength and freedom from distortion; such as boring bars, etc. Cannot be used for highly concentrated loads and shock loads.					
Remarks	* Phosphorus maximum 0.035 and sulphur maximum 0.03. ** <u>For turning, shaping and planing operations only.</u>					

Equivalent steels	BS	CSN	DIN	JIS	ASTM
	EN41B	15340	Approx. 34*CrAlMo*5	G4202 Class 1 SACM1	A355 Class A

Table 317 Nitriding steel for machine tools 40Cr2Al1Mo18   (Concld.)

Recommended temperatures for forging and heat treatment		
Operation	Temperature °C	Remarks
Forging (hot working)	1200-850	After forging, cool forged parts slowly in air.
Normalizing	—	Not normally normalized since the material is highly susceptible to air hardening.
Sub-critical annealing	680-720	About 4 hours–slowly cool in air.
Hardening	850-900	Oil quench in general.
Nitriding Oil	550-700	The hardened component is finish-machined to the required size and then nitrided for about 8 to 10 hours to get a layer of 0.4 to 0.5 mm. After nitriding, the material is oil quenched.

Notes:

1)   If the material is not hardened before nitriding, the nitrided layer may peel off.

2)   Machinability will be improved by sub-critical annealing.

3)   No further heat treatment after nitriding.

4)   No further machining after nitriding.

5)   During nitriding the temperature should be rigidly controlled; any deviation from the specified range will result in a decrease in efficiency of nitriding.

6)   The dissociation rate of about 15-30% with a constant gas flow should be maintained from the beginning of the cycle for about 4 to 8 hours and then gradually increased to 80-85% depending on the case depth required.

7)   If a good dimensional stabilization is required, then the hardened component is tempered at a temperature of about 20°C above the nitriding temperature.

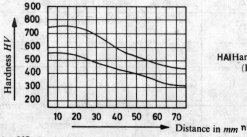

HA1Hardenability curve /E (End quench))

Table 318 Steel castings (IS 1030-1974)

Grade	Yield stress, Min. kgf/mm²	Tensile strength Min. kgf/mm²	Elongation % on guage length 5.65 √S₀, Min.	Reduction of area, % Min.	Impact strength, Min. kgf.m.cm²
20-40	20	40	25	40	3
23-45	23	45	22	31	2.5
26-52	26	52	18	25	2.2
27-54	27	54	16	23	2
30-57	30	57	15	21	1.8

Table 319 Composition and mechanical properties of typical non ferrous metals and alloys for machine tools (Contd.)

a) Wrought aluminium and aluminium alloys

Material section	Specification	Condition	Chemical composition %						Elongation on gauge length $5.65\sqrt{S_o}$ % Min.	Tensile strength kgf/mm²	Brinell hardness HB
			Al	Si	Cu	Fe	Mg	Mn			
Bars, rods and sections	IS 733 19000 (E1C) 19500 (E1B)	M	99 Min	0.5	0.1	0.7	0.2	0.1	18	6.5	70-85
			99.5 Min	0.3	0.05	0.4	—	0.05	23	6.5	
Plates	IS 736 19000 (P1C)	M	99 Min.	0.5	0.1	0.7	0.2	0.1	7 Min.	7 Min.	
		O							>6.3 ≤ 25 is 28 >3.15 ≤ 6.3 is 30 >6.3 ≤ 25 is 28	7-11	
		H2							>3.15 ≤ 6.3 is 5 >6.3 ≤ 12.5 is 7	11-14	
	19500 (P1B)	M	99.5 Min.	0.3	0.05	0.4	—	0.05	>6.3 ≤ 25 is 28	6.5 Min.	
		O							>3.15 ≤ 25 is 30	10 Max.	
		H2							>3.15 ≤ 6.3 is 6 >6.3 ≤ 12.5 is 7	10 to 13.5	
Sheets and strips	IS 737 19000 (S1C)	O							0.5 – 0.8 is 20 >0.8 ≤ 1.3 is 25 >1.3 ≤ 2.6 is 29 >2.6 ≤ 6.3 is 30	7-11	
		H1	99 Min.	0.5	0.1	0.7	0.2	0.1	0.5-0.8 is 5 >0.8 ≤ 1.3 is 6 >1.3 ≤ 6.3 is 8	9-13	70-85
		H2							0.5-0.8 is 3 >0.8 ≤ 1.3 is 4 >1.3 ≤ 6.3 is 5	10.5-14	
		H3							0.5-0.8 is 2 >0.8 ≤ 1.3 is 3 >1.3 ≤ 6.3 is 4	12.5-15	

Table 319 Composition and mechanical properties of typical non ferrous metals and alloys for machine tools

a) Wrought aluminium and aluminium alloys (contd.)

Material section	Specification	Condition	Chemical composition % Al	Si	Cu	Fe	Mg	Mn	Elongation on gauge length $5.65\sqrt{So}$ %		Tensile strength kgf/mm²		Brinell hardness HB
Sheets and strips	IS 737 19000 (S1C)	H4	99 Min.	0.5	0.1	0.7	0.2	0.1	0.5–1.3 is 2 >1.3 ≤6.3 is 3		14 Min.		70-85
	IS 737 19500 (S1B)	O							0.5–0.8 is 22 >0.8 ≤1.3 is 25 >1.3 ≤2.6 is 29 >2.6 ≤6.3 is 30		5.5-9.5		70-85
		H2	99.5 Min.	0.3	0.05	0.4	—	0.05	0.5–0.8 is 4 >0.8 ≤1.3 is 5 >1.3 ≤6.3 is 6		10-13.5		
		H4							0.5–1.3 is 3 >1.3 ≤6.3 is 4		13 Min.		
									Sand cast	Chill cast	Sand cast	Chill cast	
Ingots and castings	4223 (A-4)	M	Remainder	4-6	2-4	0.8	0.15	0.2-0.6	—	—	14	16	55-70
IS 617		WP							2	2	23	28.5	
	4600 (A-6)	M	Remainder	10-13	0.6	0.1	0.1	0.5	5	7	16.5	19	
	2280 (A-11)	W	Remainder	0.25	4.5	0.25	0.1	0.1	7	13	22	27	
		WP							4	9	28	31.5	
	5230 (A-5)	M	Remainder	0.3	0.1	0.6	3-6	0.3-0.7	3	5	14	17	

Notes: O   represents annealed.

H   represents strain hardened, digits 1, 2, 3 or 4 in ascending order of tensile strength.

M   represents as manufactured.

W   and/ or P represents thermally treated to produce tempers other than M, O or H.

WP represents solution heat treated and precipitation treated.

Table 319 Composition and mechanical properties of typical non ferrous metals and alloys for machine tools   (Contd.)

b) Phosphor-bronze

Material section	Specification	Sn	P	Pb Max.	Ni Max.	Total impurities Max.	Cu	Zn	Size (Diameter or minor sectional dimension) Over	Upto	Elongation % on gauge length 5.65√So, Min. sand cast	chill cast	continuously cast	Tensile strength, kgf/mm² Min. sand cast	chill cast	continuously cast
Phosphor-bronze rods and bars as manufactured	IS 7811	4.6-5.5	0.02-0.4	0.02	—	0.2	Remainder	—	10	18	10			50.5		
									18	38	10			47		
									38	75	15			39.5		
									75	110	20			32		
									110	—	25			28		
Phosphor-bronze ingots and castings	IS 28										sand cast	chill cast	continuously cast	sand cast	chill cast	continuously cast
	Grade 1	6-8	0.3-0.5	0.25	0.7	1.2	Remainder	0.5 Max.	—	—	3	5	8	19	21	28
	Grade 2	10	0.5 Min.	0.25	0.7	1.2	Remainder	0.05	—	—	3	2	7	22.5	31.5	36.5
	Grade 3	6.5-8.5	0.3 Min.	2.5	1	0.5	Remainder	2 Max.	—	—	3	2	5	19	22.5	27.5
	Grade 4	9-11	0.15 Max.	0.25	0.25	0.8	Remainder	0.25	—	—	6	5	9	23.5	27.5	31.5

c) Aluminium-bronze:

Material	Specification	Chemical composition % Al	Fe	Mn Max.	Ni	Zn Max.	Cu	Elongation % Min.** Sand cast	Chill cast	Tensile strength kgf/mm² Min. Sand cast	Chill cast
Aluminium-bronze ingots and castings	Grade 1 (IS 305)	8.5-10.5	3.5-5.5	1.5	4.5-6.5	0.5	Remainder	15	12	66	66
	Grade 2	8.5-10.5	1.5-3.5	1	1 Max.	0.5	Remainder	20	20	50	55

d) Leaded – tin – bronze:

Material	Specification	Chemical composition % Sn	Zn Max.	Pb	Sb Max	Fe Max.	Fe+Sb Max	Cu + incidental Ni	Elongation % Min.** Sand cast	Chill cast	Tensile strength, kgf/mm² Min. Sand cast	Chill cast
Leaded-tin-bronze ingots and castings	Grade 3 (IS 318)	6-8	0.5	9-11	0.5	0.35	0.7	Remainder	2	5	16	20.5
	Grade 4	6-8	0.5	14-16	0.5	0.35	0.7	Remainder			14	19

** on gauge length 5.65√So

Table 319 Composition and mechanical properties of typical non ferrous metals and alloys for machine tools   (Concld.)

e) Brass:

| Material section | Specification | Condition | Chemical composition % | | | | | Size (Diameter or minor sectional dimension.) | | Elongation on gauge length $5.65\sqrt{S_o}$ Min. % | Tensile strength kgf/mm^2 Min. | Hardness HB Max. |
			Cu	Pb	Fe Max.	Impurities Max.	Zn	Over	Upto			
Free cutting brass bars, rods and sections	IS 319	Annealed	56-59	2-3.5	0.35	0.7*	Remainder	10	25	12	35	88
								25	50	17	32	80
								50	—	22	29	72
	Type I	Half hard						10	12	4	41	103
								12	25	6	40	100
								25	50	12	36	90
								50	—	17	33	82
		Hard						10	12	—	56	140
								12	21	4	50	125
	Type II	Annealed	60-63	2.5-3.7	0.35	0.5*	Remainder	10	25	15	34	85
								25	50	20	31	77
								50	—	25	28	70
		Half hard						10	12	7	40	100
								12	25	10	39	98
								25	50	15	35	88
								50	—	20	32	80
		Hard						10	12	—	56	140
								12	21	4	49	137
Brass rods for general engineering purposes	Cu Zn20	As manufactured	79-81	0.1	0.05	0.3**	Remainder	5 and above.		25	32	—
		Annealed								50	25	90
	IS 4170 Cu Zn30	As manufactured	68-72	0.03	0.05	0.3**	Remainder			25	35	100
		Annealed								50	28	90
	Cu Zn40	As manufactured	59-62	0.75	0.1	0.3**	Remainder			25	35	—
		Annealed								30	28	90

* excluding Fe   ** excluding Pb

TENSILE STRENGTH & HARDNESS
FOR STEELS AND NONFERROUS METALS

Table 320 Tensile strength and hardness in various scales for steel

Tensile strength kgf/mm^2	Dia. of ball impression mm.	Brinell hardness	Vickers hardness	Rockwell hardness scale B	Rockwell hardness scale C	Tensile strength kgf/mm^2	Dia. of ball impression mm.	Brinell hardness	Vickers hardness	Rockwell hardness scale B	Rockwell hardness scale C
σ_t	d	HB	HV	HRB	HRC	σ_t	d	HB	HV	HRB	HRC
34	6	95.5		53		71	4.27	199	203	93.5	
35	5.92	98.4		56		72	4.24	202	206	94	
36	5.85	101		58		73	4.21	206	208	95	15
37	5.77	104		61		74	4.19	208	210	95	16
38	5.7	107		63		75	4.16	211	213	96	16
39	5.64	110		64.5		76	4.13	214	216		17
40	5.57	113		66		77	4.1	217	220	97	18
41	5.51	115		67.5		78	4.08	219	222		
42	5.45	118		69		79	4.05	223	225	98	19
43	5.39	121		70.5		80	4.03	225	227		
44	5.33	124		71.5		81	4.01	228	230	99	20
45	5.28	127		73		82	3.98	231	233		21
46	5.23	129		73.5		83	3.96	234	236	100	
47	5.18	132		74.5		84	3.94	236	238		22
48	5.13	135		75.5		85	3.92	239	240	101	
49	5.08	138		76.5		86	3.9	241	243		23
50	5.03	141		77.5		87	3.87	245	247	102	
51	4.98	144		78.5		88	3.85	248	249		24
52	4.94	146		79.5		89	3.83	250	252		
53	4.9	149		80		90	3.81	253	254	103	25
54	4.85	152		81		91	3.79	256	257		
55	4.81	155		82		92	3.77	259	260	104	26
56	4.77	158		83		93	3.75	262	263	104	
57	4.73	161		84		94	3.73	265	265		27
58	4.7	163		84.5		95	3.71	268	268	105	27
59	4.66	166		85		96	3.69	271	271		
60	4.62	169	173	86		97	3.67	274	274		28
61	4.58	172	176	87		98	3.66	275	276	106	
62	4.55	174	178	87.5		99	3.64	278	279		29
63	4.52	177	181	88		100	3.62	282	282		
64	4.48	180	184	89		101	3.6	285	285	107	
65	4.45	183	187	89.5		102	3.59	287	287		30
66	4.42	185	189	90.5		103	3.57	290	290		
67	4.39	188	192	91		104	3.55	293	293		31
68	4.36	191	194	91.5		105	3.54	295	295	108	
69	4.33	194	197	92		106	3.52	298	298		
70	4.3	197	200	93		107	3.51	300	300		32

Table 320 Tensile strength and hardness in various scales for steel   (Contd.)

Condition: Normalised or annealed

Tensile strength kgf/mm²	Dia. of ball impression mm.	Brinell hardness	Vickers hardness	Rockwell hardness scale B	Rockwell hardness scale C
σ$_t$	d	HB	HV	HRB	HRC
108	3.49	304	303		32
109	3.47	307	307	109	
110	3.46	309	309		33
111	3.44	313	312		
112	3.43	315	314		
113	3.41	319	318		34
114	3.4	321	320	110	
115	3.38	325	323		
116	3.37	327	325		35
117	3.36	329	327		
118	3.34	333	331		
119	3.33	335	333		36
120	3.31	339	337	111	
121	3.3	341	339		
122	3.29	343	341		
123	3.28	345	343		37
124	3.26	350	348		
125	3.25	352	350		
126	3.24	354	352	112	38
127	3.23	356	354		
128	3.21	361	359		
129	3.2	363	361		
130	3.19	366	363	112.5	39

Condition: Hardened & tempered or only hardened

σ$_t$	d	HB	HV	HRB	HRC
60	4.53	176	190	87.5	
61	4.49	179	194	89	
62	4.46	182	197	90	
63	4.42	185	200	90	
64	4.38	189	204	91	
65	4.35	192	207	92	
66	4.32	195	209	92	
67	4.29	198	212	93	
68	4.25	201	216	94	
69	4.22	204	219	95	
70	4.19	208	223	95	15
71	4.16	211	226	96	16
72	4.13	214	229	96	17

Condition: Hardened & tempered or only hardened

Tensile strength kgf/mm²	Dia. of ball impression mm.	Brinell hardness	Vickers hardness	Rockwell hardness scale B	Rockwell hardness scale C
σ$_t$	d	HB	HV	HRB	HRC
73	4.11	216	231	97	17
74	4.08	219	235	97	18
75	4.05	223	238	98	
76	4.02	226	242		19
77	4	229	244	99	20
78	3.97	232	248		21
79	3.95	235	251	100	
80	3.92	239	254		22
81	3.9	241	257	101	23
82	3.87	245	261		23
83	3.85	248	264	102	24
84	3.83	250	266		
85	3.81	253	269	103	25
86	3.78	257	273		
87	3.76	260	276	104	26
88	3.74	263	279		
89	3.72	266	282		27
90	3.7	269	285	105	28
91	3.68	272	289		28
92	3.66	275	292	106	29
93	3.64	278	295	106	29
94	3.62	282	298		30
95	3.6	285	302	107	30
96	3.58	288	305		
97	3.56	292	308		31
98	3.54	295	312	108	31
99	3.53	297	314	108	
100	3.51	300	317		32
101	3.49	304	321		
102	3.47	307	325	109	
103	3.46	309	327		33
104	3.44	313	330		
105	3.42	317	334		34
106	3.41	319	336	110	
107	3.39	323	340	110	
108	3.38	325	342		35
109	3.36	329	346		35
110	3.35	331	348		

Table 320 Tensile strength and hardness in various scales for steel (Contd.)

Tensile strength kgf/mm^2	Dia. of ball impression $mm.$	Brinell hardness	Vickers hardness	Rockwell hardness scale B	Rockwell hardness scale C	Tensile strength kgf/mm^2	Dia. of ball impression $mm.$	Brinell hardness	Vickers hardness	Rockwell hardness scale B	Rockwell hardness scale C
σ_t	d	HB	HV	HRB	HRC	σ_t	d	HB	HV	HRB	HRC
111	3.34	333	350			150	2.92	438	463		
112	3.32	337	354	111	36	151	2.91	441	467		
113	3.31	339	357			152	2.9	444	470		46
114	3.3	341	359			153					
115	3.28	345	363		37	154	2.89	448	474		
116	3.27	347	365			155	2.88	451	478		
117	3.26	350	368			156	2.87	454	482		
118	3.25	352	370	112		157					47
119	3.23	356	374		38	158	2.86	457	486		
120	3.22	359	377			159	2.85	461	490		
121	3.21	361	379			160					
122	3.2	363	382			161	2.84	464	494		
123	3.19	366	384		39	162	2.83	467	498		48
124	3.17	370	389			163					
125	3.16	373	391	113		164	2.82	471	502		
126	3.15	375	394		40	165	2.81	474	506		
127	3.14	378	396			166	2.8	477	511		
128	3.13	380	399			167	2.8	477	511		
129	3.12	383	401			168	2.79	481	515		49
130	3.11	385	404		41	169	2.78	485	520		
131	3.1	388	407		41	170	2.78	485	520		
132	3.09	390	410			171	2.77	488	524		
133	3.07	395	415	114		172					
134	3.06	398	418		42	173	2.76	492	529		
135	3.05	401	421			174	2.75	495	533		50
136	3.04	404	424			175					
137	3.03	406	427			176	2.74	499	538		
138						177					
139	3.02	409	431		43	178	2.73	503	543		
140	3.01	412	434			179	2.72	507	548		51
141	3	415	437	115		180	2.72	507	548		51
142	2.99	417	440			181	2.71	510	553		
143	2.98	420	443		44	182					
144	2.97	423	446			183	2.7	514	558		
145	2.96	426	450			184					
146	2.95	429	453			185	2.69	518	564		
147						186					52
148	2.94	432	457		45	187	2.68	522	570		
149	2.93	435	460			188					

Table 320 Tensile strength and hardness in various scales for steel (Concld.)

Condition: Hardened & tempered or only hardened									
Tensile strength kgf/mm^2	Dia. of ball impression $mm.$	Brinell hardness	Vickers hardness	Rockwell hardness scale C	Tensile strength kgf/mm^2	Dia. of ball impression $mm.$	Brinell hardness	Vickers hardness	Rockwell hardness scale C
σ_t	d	HB	HV	HRC	σ_t	d	HB	HV	HRC
189	2.67	526	575		228	2.51	597	688	58
190					231	2.5	602	697	
191	2.66	530	581		235	2.49	606	707	59
192					238	2.48	611	717	
193	2.65	534	587	53	242	2.47	616	727	60
194					245	2.46	622	737	
195	2.64	538	593		249	2.45	627	748	61
196					253	2.44	632	759	
197	2.63	543	599		257	2.43	637	771	62
198					261	2.42	643	784	
200	2.62	547	606	54	266	2.41	648	798	63
202	2.61	551	612		271	2.4	654	813	63
204	2.6	555	619		277	2.39	659	828	64
206	2.59	560	626	55	283	2.38	665	845	64
209	2.58	564	633		289	2.37	670	863	65
211	2.57	569	640		296	2.36	676	882	66
214	2.56	573	647	56	303	2.35	682	904	66
216	2.55	578	654		312	2.34	683	931	67
219	2.54	582	662	57	324	2.33	694	965	68
222	2.53	587	671	57	340	2.32	700	1008	68
225	2.52	592	679						

Table 321 Tensile strength and Brinell hardness for non ferrous metals

Sl. No.	Material	Relationship
1	Cu, Ni, Bronze (cold worked condition)	$\sigma_t = 0.40\,HB$
2	Cu, Ni, Bronze (annealed condition)	$\sigma_t = 0.55\,HB$
3	Cast Bronze	$\sigma_t = 0.23\,HB$
4	Al-Cu-Mg Alloy	$\sigma_t = 0.35\,HB$
5	Al-Mg Alloy	$\sigma_t = 0.44\,HB$
6	Cast Aluminium	$\sigma_t = 0.26\,HB$

HEAT TREATMENT OF STEELS

Heat treatment is generally applied to steels to impart specific mechanical properties such as increased strength or toughness or wear resistance. Heat treatment is also resorted to relieve internal stresses and to soften hard metals to improve machinability. Heat treatment is essentially a process of heating the steels to a predetermined temperature followed by a controlled cooling at a predetermined rate to obtain desired end results.

The heat treatment process can be classified into four important groups:

i) *Recrystallization annealing* which is employed to relieve internal stresses, reduce the hardness and to increase the ductility of strain hardened metal. At first, upon an increase in the heating temperature the elastic distortions of the crystal lattices are eliminated. At higher temperature new grains form and begin to grow (recrystallization).

ii) *Full annealing* which involves phase recrystallization and is achieved by heating alloys above the temperature required for phase transformation. This is followed by slow cooling. Full annealing substantially changes the physical and mechanical properties and may refine a coarse grained structure.

iii) *Quenching* wherein hardening alloys are heated above the phase transformation temperature and are then rapidly cooled (quenched).

iv) *Tempering* involves the reheating of hardened steel to a temperature below that required for phase transformation so as to bring it nearer to an equilibrium state.

Annealing

Annealing is the process necessary to obtain softness, improve machinability, increase or restore ductility and toughness, relieve internal stresses, reduce structural non-homogeneity and to prepare for subsequent heat treatment operations.

The process consists of heating the metal to the required temperature depending upon the carbon content and other alloying elements of the steel and then cooling in the furnace at a slow rate. Most of the cast iron components are annealed at a low temperature before final machining.

Normalizing

This is the process necessary to eliminate coarse-grained structure obtained in previous working, to increase the strength of medium carbon steels to a certain extent (in comparison with annealed steel), to improve the machinability of low carbon steels, to reduce internal stresses; etc.

More rapid cooling in air used in normalizing causes the austenite to decompose at lower temperatures. This increases the disparity of the ferrite-cementite mixture (pearlite) and increases the amount of eutectoid constituent. Therefore normalized steel has a higher strength and is harder than annealed steel.

Hardening and tempering

In this process steel is heated to a predetermined temperature and then quenched in water, oil or molten salt baths.

Hardenability

Hardenability is defined as the capacity to develop a desired degree of hardness usually measured in terms of depth of penetration.

The depth of hardness depends upon the critical rate; since this is not the same for the whole cross-section, full hardening may be achieved if the actual cooling rates, even at the core, exceed the critical values. The higher the carbon content, the harder a steel will be after hardening owing to a martensitic structure.

Hardening followed by tempering is done to improve the mechanical properties of steel. The aim in structural steels is to obtain a good combination of strength, ductility and toughness.

Hardened steel is in a stressed condition and is very brittle so that it cannot be employed for practical purposes. After hardening, steel must be tempered to reduce the brittleness, relieve the internal stresses due to hardening and to obtain predetermined mechanical properties.

In spite of the high hardness, hardened steel has a low cohesive strength, a lower tensile strength and particularly a low elastic limit, due to the stress conditions after hardening. The impact strength, relative elongation and reduction of area are also considerably reduced by hardening.

Tempering the steel at a suitable temperature will enable the steel to attain the desired mechanical properties. Tempering consists of reheating the hardened steels to a temperature below lower critical values followed by cooling at a desired rate.

At low tempering temperatures, the hardness changes only by a small extent but the true tensile strength and bending strength are attained. As the tempering temperature is gradually increased, the steel regains its true strength and resistance to shock with a gradual decrease in hardness value. However where toughness is the criterion, tempering in the range of 230-400°C is avoided to overcome the condition of temper brittleness.

Surface hardening

This is a selective heat treatment in which the surface layer of metal is hardened to a certain depth whilst a relatively soft core is maintained. The principal purpose of surface hardening is to increase the hardness and wear resistance of the surface. Surface hardening may be accomplished with or without changing the chemical composition of the surface. While carburizing and nitriding correspond to the first type, flame and induction hardening correspond to the second type, i.e. without changing the chemical composition. Steels with carbon content less than 0.25% can be generally carburized while steels with a minimum carbon content of 0.4% only can be flame or induction hardened.

Induction hardening

Induction hardening has the advantage that it reduces the time required for heat treatment. Parts may be hardened with practically no scaling, so that allowance for further machining can be reduced. Deformation due to heat treatment is only marginal.

In comparison with other processes for a given tensile strength, induction hardened steels have higher hardness, wear resistance, impact strength and fatigue limit.

The increase in the fatigue limit after induction hardening is associated with the appearance of residual compressive stresses in the hardened layer. These stresses reduce the effect of tensile stresses arising from the application of external forces.

Flame hardening

In this process, the surface of the part to be hardened is heated by an oxy-acetylene flame at temperature of 3000°-3200°C. The large amount of heat transferred to the surface rapidly heats it to a hardening temperature before the core is appreciably heated. Subsequent quenching hardens the layer.

Carburizing

This is a process for saturating the surface layer of low carbon steels with carbon. Several methods are employed for this purpose such as pack carburizing, gas carburizing and liquid carburizing.

The advantages of gas carburizing over pack carburizing are:
a) possibility of better regulation of the process and of obtaining more accurate case depth
b) less time is required for the process
c) the operation is clean and simpler and
d) process can be mechanised

Liquid carburizing has the advantage that it provides a uniform heating combined with least deformation of the part.

After carburizing, regardless of the process employed, the material is heat treated to produce a hard surface resistant to wear. The heat treatment process for carburized parts consists of the following:
a) normalizing after carburizing at temperatures of 880^0-900^0C to improve the core structure of the work which is overheated by carburizing
b) hardening at 750-780^0C to eliminate the effects of overheating and to impart a high hardness to the carburized layer and
c) tempering at 150^0 to 180^0C.

Nitriding

This is a process of saturating the surface of steel with nitrogen by holding it for a prolonged period at a temperature from 480^0 to 650^0C in an atmosphere of ammonia. Nitriding increases the hardness of the surface to a high degree. It also increases the wear resistance and fatigue limit. When high hardness and wear resistance are the chief requirements the part is made of steel containing aluminium. Aluminium, chromium and molybdenum in steel impart an exceptionally high hardness and wear resistance to the nitrided case.

Sub-zero treatment of steel

A certain amount of retained austenite may always be found in hardened steel. Retained austenite reduces the hardness, wear resistance and thermal conductivity of steel and makes its dimensions unstable.

A sub-zero treatment has been devised to reduce the retained austenite in hardened steel. It consists in cooling the metals being treated to sub-zero temperatures. Sub-zero treatment is usually conducted in the temperature range from-70^0 to-120^0C. The holding time at this temperature is from 1 to 1.5 hours.

Prolonged holding at room temperature after hardening will stabilize the austenite of many grades of steel and reduce the effect of sub-zero treatment. It is therefore, advisable to perform sub-zero treatment directly following the hardening operation.

Sub-zero treatment is most frequently used for high speed steel tools, measuring tools, carburized gears, machine tool spindles, etc.

ELECTRICAL EQUIPMENT FOR MACHINE TOOLS

DESIGN AND INSTALLATION OF ELECTRICAL EQUIPMENT OF GENERAL PURPOSE MACHINE TOOLS

Design and installation of electrical equipment of general purpose machine tools should ensure uniform design and installation procedure, ease of maintenance and safety of personnel and equipment.

Certain basic information necessary for the designers are dealt with in this chapter. However for a detailed code of practice reference should be made to any of the following standards: IS 1356(1); BS 2771; CSN 34 1630, VDE 0113 or JIC EGP1.

Connections to supply network

The electrical equipment is as far as possible connected to a single source of power supply by an ON-OFF switch. Other voltages if required are obtained from apparatus such as transformers, rectifiers, etc., forming an integral part of the electrical equipment of the machine tools. In the OFF position of the switch all electrical equipment are disconnected from the supply, except those which may cause danger if disconnected, such as magnetic chuck, brake system, etc. Sometimes the switch is locked in the OFF position to prevent unauthorised persons operating the machine. Some switches have the provision of being interlocked with the door of the electrical cabinet such that the door cannot be opened in the ON position of the switch. The current and voltage ratings of the switch are decided by the connected load.

Emergency stopping device

In addition to the mains isolating switch it may also be required to provide an emergency stopping device which is clearly visible and easily accessible to stop the machine tool as quickly as possible in order to avoid danger to the operator or equipment. This is a red-coloured mushroom push button and provided in as many places as required in machine tools with multi control stations.

However circuits for magnetic chucks, braking systems, etc. are not to be interrupted by the emergency stopping device. Withdrawal motions required for safety may also be initiated by actuating the emergency stopping device.

Protective measures

The machine tool motors require short-circuit protection, no-voltage protection and overload protection.

Due to either insulation failure or faulty connection short-circuit may occur. To prevent such short-circuits, HRC type fuses are provided. The rating of the fuse should be such as to take care of the overcurrent due to motor starting. Refer under *Guide for selection of fuses and wires* for correct fuse selection.

No voltage protection prevents automatic restarting of the motor after a supply interruption. Electromagnetically operated devices such as contactors, starters and circuit breakers provide this safety.

Motors for machine tools are not permitted to be overloaded unless they are specifically designed for overload. To prevent the motor from being overloaded, overload protection is used. Generally the bi-metallic type overload relay is used for this purpose which senses the execessive current and trips off the motor.

CONTROL CIRCUITS

Control circuits control the switching-on pattern for the contactors to switch ON or OFF the electric motors as well as control the actuating elements such as solenoid valves, electro-magnets, electro-magnetic clutches, etc.

It is suggested to use a control transformer having separate windings and connected to the load side of the supply disconnecting switch. For *ac* control circuits supplied by a transformer the recommended secondary voltages are 110, 220 and 240 volts at 50 Hz. For *dc* control circuits, the recommended voltages are 24, 48, 110 and 220 volts. To prevent earth faults, it is recommended that one side of the control circuit be earthed.

Design of control circuit

The design of the electrical control circuit requires a thorough understanding of the working of the complete machine. Safety of the operator and the machine should always be remembered and it should be ensured that damage in case of failure of any signalling device is a minimum.

A few typical cases of circuit design and conditions for interlocking are given below; the list is not comprehensive, but is given only for guidance. An example of a typical circuit diagram is given in Fig. 229. The symbols used in the circuit are covered in Table 341.

Opposing motions interlocked

Starters, relays, contactors and solenoids which are mechanically interlocked are also electrically interlocked to prevent simultaneous energization.

Movement initiation by limit switches

Control circuits are so designed that, when the machine tool is not in its working cycle, the actuation of any limit-switch does not initiate the movement of any part of the machine tool.

Covers and doors interlocked

Hinged covers may be interlocked with the machine tool control to prevent operation of the machine tool while the cover is left open.

Spindle drive interlocked with feed

Interlocking is provided to ensure that the spindle drive motor is switched on before the tool is driven into the workpiece while in the automatic cycle.

Non-repetition of the cycle

On all equipment where automatic repetition of the cycle is dangerous, the circuit is so designed that this repetition does not occur and the machine is brought to a total stop at the end of the cycle without requiring the action of the operator.

Reverse current braking

When reverse current braking is used on a motor, all measures are taken to avoid the motor re-starting in the opposite direction at the end of braking, when this inversion may endanger the personnel or damage the workpiece; in such cases the use of a device operating exclusively as a function of time is not allowed.

Furthermore, all measures are taken to prevent false starting of the motor caused by the rotation of the motor shaft.

STOP functions are generally initiated through de-energization rather than energization of control devices wherever possible.

One station for motor starting

Only one station is made effective at any time for starting all motors concurrently. However multiple STOP stations may be used.

Control enclosures and compartments

Control enclosures and compartments are to be so enclosed as to give adequate protection against ingress of dust, oil, coolant or chips and against machanical damage. All control devices like contactors, fuses, etc., are front mounted on a rigid metal panel and it should be possible to remove the entire panel through the opening in the enclosure. The control devices in the enclosure are so installed that they are readily accessible when the doors or covers are opened.

All connection terminals should be easily accessible and located at least 200 *mm* from the floor or servicing level.

Apparatus requiring easy access for maintenance and adjustment are not to be situated below 400 *mm* and above 2000 *mm* from the servicing level.

There shall be no opening between compartments containing electrical apparatus and reservoirs holding coolant, lubricating or hydraulic oils.

Mechanical parts to which access is necessary during the normal operation of the machine tool and moving parts (rotating shafts) are not to be housed in control enclosures and compartments.

Heat generating components like resistors, valves, etc., are so located, that the temperature rise of components inside the control enclosure is kept within permissible limits, otherwise the available space can be divided into a ventilating section containing the heat generating components and a protected section containing the other components.

Control and Operating Devices

CONTROL DEVICES

All control devices external to the control enclosure such as limit switches, magnetically operated valves, pressure switches, etc., are so mounted that they are readily accessible and located in reasonably dry and clean locations and free from accidental operation by normal movement of machine components or operator.

Limit switches or position sensors are so mounted that they will not be damaged in the event of accidental overtravel.

Pipe lines, tubing or devices for handling air, gases or liquids must not be located in electrical control enclosures or compartments.

OPERATING DEVICES

The operating devices should be mounted in a dust and oil free location and within easy reach of the machine tool operator in his normal working position.

PUSH-BUTTONS

General

Push-buttons are mounted so that the movement of the buttons is either in a horizontal plane or does not exceed 45° from the horizontal. All push-buttons other than STOP buttons should be shrouded to avoid the danger of unintentional operation.

Colour of push-button

The recommended colours for push-buttons used on machine tools are indicated in Table 322.

Illuminated push buttons

The recommended colours and functions of illuminated push buttons are indicated in Table 323.

SIGNALLING LAMPS

The recommended colours of signalling lamps used on machine tools are indicated in Table 324.

Table 322 Colour and functions of push-buttons

Colour	Function	Example of application
Red	Stop	— Stop of one or several motors — Stop of machine elements — De-energizing of magnetic chucks — Stop of the cycle (if the operator pushes the button during a cycle, the machine stops after the relevant cycle is completed)
	Emergency stop	General stop.
Yellow	Start of a return motion not in the usual operating sequence or Start of an operation intended to avoid dangerous conditions	Return of machine elements to the starting point of the cycle, if the cycle has not been completed. Pressing the yellow push-button may override other functions which have been selected previously.
Green	Start (preparation)	— Energizing of the control circuits — Start of one or several motors for auxiliary functions — Start of machine elements — Energizing of magnetic chucks
Black	Start (execution)	— Start of a cycle or a partial sequence — Inching, jogging
White or Light Blue	Any function not covered by the above colours	— Control of auxiliary functions which are not directly related to the working cycle — Reset of protective relays (if the same button is used for STOP, it shall be RED).

Table 323 Colour and functions of illuminated push-buttons

Colour and mode of use	Significance of the lighted button	Function of the button	Examples of application and remarks
Red	See note 1	STOP (see note 2) and in some instances RESET (only if this same button is also used for STOP)	
Yellow (Amber)	Attention or caution	Start of an operation intended to avoid dangerous conditions	Some value (current, temperature) is approaching its permissible limit. Pressing the yellow push-button may override other functions, which have been selected previously.
Green	Machine or unit ready for operation	Start after authorisation by the lighted button	− Start of one or several motors for auxiliary functions − Start of machine elements − Energizing of magnetic chucks or plates − Start of a cycle or a partial sequence (see also note 3)
Blue	Any signification not covered by the above colours and by white	Any function not covered by the above colours and by white	Indication or order to the operator to perform a certain task, for example to make an adjustment (after having fulfilled this requirement, he presses the button as an acknowledgement).
White (clear)	Permanent confirmation that a circuit has been energized or that a function or a movement has been started or preselected	Closing of a circuit or Start or preselection	Engergizing of an auxiliary circuit not related to the working cycle Start or preselection of direction of feed motion of speeds, etc.

Notes: 1. The use of RED illuminated push-buttons is not recommended.
2. Emergency stop buttons are never illuminated push-buttons.
3. For inching and jogging, non-illuminated black push-buttons should be used.

Table 324 Colour and significance of signalling lamps

Colour	Significance	Example of application
Red	Abnormal conditions requiring immediate action by the operator (see notes 1 and 2)	Order to stop the machine immediately (for example because of an overload) or To indicate that a protective device has stopped the machine (for example because of an overload, overtravel or another failure)
Yellow (Amber)	Attention or caution (see note 1)	Some value (current, temperature) is approaching its permissible limit or Automatic cycle running
Green	Machine ready	Machine ready for operation; all necessary auxiliaries functioning, units in starting position and hydraulic pressure or output voltage of a motor-generator in the specified range, etc. Cycle completed and machine ready to be restarted.
White (clear)	Circuit energized Normal conditions	Main switch in ON position (see note 2) — Choice of the speed or the direction of rotation — Auxiliaries not related to the working cycle are functioning.
Blue	Any signification not covered by the above colours	— Selector switch in SET UP position — A unit in forward position — Microfeed of a carriage or unit

Notes: 1. For the significations *Abnormal conditions requiring immediate action* or *Attention*, a flashing signal of the appropriate colour may be used accompanied by an audible signal, if desired.

2. For *Main switch in ON position*, RED may be used if the signalling lamp is not on the operator's control station.

WIRING

General

The cables and wires used must be suitable for the loads they supply taking into account current, voltage drop, etc.

Grade of cables

When the voltage to earth exceeds 250 volts, the cable must be of 650/1100 volts grade and when the voltage to earth does not exceed 250 volts, either 250/440 volts grade or 650 volts grade may be used.

For mechanical reasons, the cross-sections used must not be less than those indicated in Table 325.

Wiring methods and practices:

Conductors of different colours are selected as indicated in Table 326.

All wiring, other than those of suitably protected cables, outside control enclosures and compartments, are generally laid in conduits or raceways.

Flexible cables are generally used for connections to moving or adjustable machine components in which electrical equipment is incorporated.

Conductors are identified at each termination by marking with a number to correspond with the circuit diagram.

Table 325 Minimum cross section of copper conductors

Sl. No.	Type of connection	Min. area mm^2
1	Conductors outside control enclosures and compartments: i) for stranded conductors ii) for solid conductors (used only in exceptional cases)	 1 1.5

2	Multi-core cables outisde control enclosures and compartments
	 i) for two and more cores in general
	 ii) for connecting very low current circuits
	 (electronic logic and similar circuits)
	   a) cables with two or more cores
	   b) cables with three or more cores
	   c) screened two-core cables
	   terminals of such cables shall be fixed in such a
	   manner as to prevent strain on the ends of conductors.
	 iii) for flexible cables and connecting parts subject
	   to frequent movement
		
		0.75
		
		0.5
		0.3
		0.3
		
		1

3	Conductors inside control enclosures and compartments: i) in general ii) for very low current circuits (electronic logic and similar circuits)	 0.75 0.2

Note: Conductors with smaller cross-section may be used only where sizes indicated above affect the proper functioning of the equipment.

Table 326 Colour coding of conductors

Colour	Type of circuit
Black	*ac* or *dc* power circuits
Red	*ac* control circuit
Blue	*dc* control circuit
Green	Equipment earthing conductor Earthed circuit conductor

Terminals on terminal blocks are permanently marked to correspond with the identification shown on the circuit diagram and are conveniently numbered.

It is recommended that on complex machine tools, when several switching devices (such as limit switches, push-buttons, etc.) are connected in series and/or parallel, the conductors between them shall be returned to terminals forming intermediate test points conveniently placed and adequately protected. These test points are shown on the relevant diagrams.

Earthing

It is necessary to earth all electrical elements such as motors, switches, etc., which may become dangerously live in case of faults. In order to earth these, a main earthing terminal must be provided close to the main input terminals. It should be of such a size as to enable the connection of an earth continuity conductor of the following cross-section.

Cross-section of the main conductors supplying the equipment	Cross-section for which the main earthing terminal has to be dimensioned
Upto and including 16 mm^2	Equal to that of the main conductor
Larger than 16 mm^2	At least 50% of the main conductor with a min. of 16 mm^2

The internal earth connection wires within the machine tool must all be connected to the main earthing terminal.

Local lighting of the machine

For local lighting in machine tools, adjustable lamps preferably of low voltage are used.

Tests

Eventhough the individual electrical elements are tested by the manufacturers to satisfy the specifications, the complete electrical system of the machine tool is tested for proving its suitability. The following tests are generally conducted for this purpose.

INSULATION RESISTANCE TEST

The insulation resistance measured with *dc* voltage of 500 volts between each conductor of the main circuits, the individual conductors of the control circuits and earthed frame should not be less than one *megaohm*.

When the control circuits are not directly connected to the main circuits, separate tests are conducted between the main circuits and the earthed frame, between the main circuits and the control circuits, and between the control circuits and the earthed frame.

VOLTAGE TEST

All the equipment are subjected to a voltage test of one minute duration by applying a test voltage as defined below between the short-circuited conductors of the main circuits, including any control circuits directly connected to the main circuits and the earthed frame.

The test voltage shall be atleast equal to 85% of the lowest test voltage to which all components are already tested before assembly on the machine, with a minimum of 1500 V.

This voltage is supplied from a transformer with a rating of atleast 500 VA

Components which are not designed to withstand such high test voltages (rectifiers, capacitors, electronic apparatus, etc.) are disconnected during the test.

Any radio interference capacitors fitted between parts normally alive and accessible metal parts are not disconnected and should withstand the above test.

RESISTANCE TO EARTH:

The resistance between the main earth terminal and any metallic part of the machine containing electrical equipment should not exceed 0.1 Ohm.

OPERATING TEST:

No-load operating test:

With the electrical equipment normally energized within the prescribed condition (for example, maximum variation of supply voltage) it should be proved that the sequence of operations is normal. In particular, the correct operation of the emergency stopping device should be checked. This test must be performed on each machine tool.

On-load operating test

When the machine tool operates under normal load, continuously or at the duty agreed to between the purchaser and the manufacturer, the temperature rise of all the equipment above the ambient temperature in which the machine is intended to operate should not exceed that permitted by the Standards appropriate for the apparatus concerned.

The correct operation of all the equipment must be proved and in particular, that interruption and restoration of power supply does not endanger personnel or adversely affect the equipment.

It is essential to prove that the emergency stopping of motors under load by means of the stopping devices occurs safely.

Safety practices

The safety of machine tool operation is a thing which should be assured by each and every one connected with the machine tool, the designer, the maintenance engineer, the operator and the electrician. Some of the safety practices concerning the electrical equipment are as follows.

Machine tools be suitably designed to prevent undesirable hazardous movements of parts due to a drop in the supply voltage or an interruption to the electric supply.

When using electrically operated clamping unit, suitable equipment be provided to stop the machine if there is a drop in the clamping force due to an undervoltage or interruption of the electric supply.

The control elements such as push-buttons, switches, etc., be properly located to ensure that the operator is not endangered by the hazardous moving parts while operating the controls.

Control elements of the machine must be guarded against unauthorised operation by providing suitable locking units necessitating the use of special tools.

Electric motors for machine tools

CHOICE OF MOTORS

The characteristics of the motors are determined in accordance with the service conditions under which they are required to operate. In this respect, distinction is made between the following three classes of motors.

a) Continuously operating motors;

b) Motors for frequent starting and reverse-current braking; and

c) Motors for driving machine tools with large inertia which shall have a suitable slip (for example, motors for presses).

In general the following refer to three phase squirrel cage induction motors, the most used motors in machine tools.

MOUNTING OF MOTORS

Motors are so mounted as to be easily accessible for inspection, maintenance, lubrication and to allow for easy wiring of the conductors and for mounting of the motor. It should also be possible to tension or replace belts or chains and to align couplings easily.

Unless sealed-for-life or similar bearings are incorporated, motors are provided with readily accessible lubricating points.

If a motor is mounted inside the machine tool, it is essential that the cooling air to the motor is not restricted and the motor is sufficiently ventilated to keep the cooling air temperature inside the compartment within corresponding limits specified in the relevant Indian Standards.

DATA REQUIRED BY THE MANUFACTURERS

The details of the data required by the manufacturer are given in Table 328. The designer should completely fill-in this and submit along with the enquiry to enable the motor manufacturers to offer the most appropriate motor for the machine tool to be designed.

PREFERRED OUTPUTS

The preferred output ratings for induction motors are (in kW): 0.06, 0.09, 0.12, 0.18, 0.25, 0.37, 0.55, 0.75, 1.1, 1.5, 2.2, 3.7, 5.5, 7.5, 11, 15, 18.5, 22, 30, 37, 45, 55, 75, 90 and 110.

SYNCHRONOUS SPEEDS AND RATED SPEEDS

The rated speeds in relation to the synchronous speed is specified between two limits as given in Table 327. The minimum rated speed is the nominal speed of the motor and is to be taken as the input speed for calculating the output speeds in a drive. The maximum limit is 3% above the minimum. The values in the Table are given for guidance only, while the actual values may vary from manufacturer to manufacturer.

Table 327 Limits for rated speeds

No. of poles	output, kW		Declared speed, rpm	
	over	upto	Min. (Nominal)	Max.
2 pole (3000 rpm)	0.37 3.7	3.7 37	2785 2870	2870 2955
4 pole (1500 rpm)	0.37 3.7	3.7 37	1370 1430	1430 1483
6 pole (1000 rpm)	0.37 2.2	2.2 37	910 950	940 980
8 pole (750 rpm)	0.37 1.5	1.5 37	680 700	705 720

CLASS OF INSULATION

Class E insulated motors are preferred for general purpose applications. For more severe duty applications Class B or Class F motors may be used.

PROTECTION
The degree of protection provided generally corresponds to either IP 23, IP 44 or IP 55 of IS 4691.

COOLING
The degree of cooling provided generally corresponds to IC 0141 (totally enclosed fan cooled) of IS 6362.

DIMENSIONS
Standard dimensions of foot, flange and face mounted motors are indicated in Tables 329 330 and 331. Maximum overall dimensions are indicated in Tables 332 and 333. This is intended to enable the designer to provide a suitable enclosure for the motor.

Accuracy requirements for face and flange mounted motors are covered in Tables 334 and 335.

THREADED CENTRE HOLES IN SHAFTS OF THE MOTORS
The dimensions of the threaded centre holes are given in Table 336. Generally no centre holes are provided and if they are required for a particular design, this should be specifically indicated in the enquiry form. Shaft extension and maximum torque details are given in Table 337.

SIZES OF TERMINALS
The sizes of the terminal studs shall be suitable for fixing the lugs as given below:

Frame Size	63, 71	80 to 100L	112M to 160M & L	180M & L
Size of Phase terminal	M 4	M 4	M 6	M 8
Lug for Earth terminal	M 4	M 5	M 6	M 8

PERFORMANCE OF MOTOR Vs AMBIENT TEMPERATURE
Generally the motors are rated for 100% output at 40°C ambient temperature. At temperatures other than 40°C, a correction factor is to be applied to the rated output as given below:

Cooling medium temperature °C	Output correction factor %
40	100
45	95
50	88
55	83
60	75

DUTY CYCLES TO BE PERFORMED BY MOTORS FOR DIFFERENT MACHINE TOOLS
The application of different duty cycles as per IS 325 encountered in different machine tools is given in Table 338 for general guidance. However, assigning the exact duty is a matter of agreement between the manufacturer and the user.

Table 328 Data for supplying electrical motors

Rated output Single speed .. *kW* at ... *rpm* Multi speed* .... *Supplier to indicate type of winding	*Load*
Motor characteristic Constant power/Constant torque Variable torque/Variable power	
	Insulation Class E/...
Surface temp. Max. ...°C	*Protection* IS 4691 Motor IP 23/44/55/... Terminal box IP 55/...
Site condition Max. ambient temp.°C: 40/45/50/55/60 Humidity: Max....%; Min....% Altitude, *m*: 1000/2000/3000/4000 Atmosphere: Acid/Alkali Area of operation: Hazardous/Non hazardous (after consulting supplier)	

	Cooling
	      IS 6362
	      IC 0141/...
	      (TEFC)

	Mounting IS 2253 *H* .../*V* .../*U* .../ ..

Motor location
Open/Enclosed within the machine tool
            with/without ventilation
	If flange mounted
	      Type of flange: B/C/D

	Terminal box Positioned Right/Left/Any other ... (when viewed from driving end)

Duty
a) Starting: On load/No load
b) *GD²* w.r.t. motor shaft: .. *kgf.m²*
c) Cycle: Enclose a graph of load vs
       time
d) Accelerating time
   rest to rated speed: ... *sec.*
e) Starting current
   not to exceed: ...*Amps.*
f) Method of starting: Direct on/
g) Braking:Electromechanical/Mech./
      dc/Capacitor/Reverse
      current/...
h) Reversal: Reverse current/Starting
      from rest
	Vibration limits IS 4729
	     Normal ... *μm*
	     Precision *A/B/C/*... *μm*
	     Other ...
	Supplier to indicate method of balanc-
	ing: with full/half key.

	Shaft extension and accuracies a) Normal/Precision as per IS 2223 b) Non standard (In both the cases indicate the values in the diagram)
Drive Flat belt/V belt/Chain Gear—Spur/Helical*/Bevel* *Shaft axial play .. *mm*	

Mains supply
  ...*V* ...*ph* ...*Hz*
  Tol. on *V*±6%/...
  Tol. on *Hz* ±3%/...
	Accuracies in *μm*

Score out whatever is not required and mark ✓ for the required statement or value.	*Inspection* At our end/At supplier's end by our staff
	Any other data

Dimensions of foot mounted motors for machine tools (TEFC Squirrel cage rotor with class *E* insulation) (As per IS 1231 - 1974)

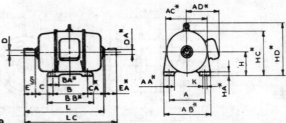

Table 329

Frame designation	kW Rating for ** Synchronous speed (*rpm*)			H Nom.		D		Fixing dimensions			
	3000 & 1500	1000	750	Size	Tol.	For 3000 rpm	Upto 1500 rpm	A	B	C	K
56	0.06 0.09	—	—	56		9j6	9j6	90	71	36	6
63	0.12 0.18	—	—	63		11j6	11j6	100	80	40	7
71	0.25 0.37	—	—	71		14j6	14j6	112	90	45	
80	0.55 0.75	0.37 0.55	—	80		19j6	19j6	125	100	50	9
90 S	1.1	0.75	0.37	90		24j6	24j6	140	100	56	
90 L	1.5	1.1	0.55						125		
100 L	2.2	1.5	0.75 1.1	100		28j6	28j6	160	140	63	12
112M	3.7	2.2	1.5	112	0			190		70	
132 S	5.5	—	2.2	132	-0.5	38k6	38k6	216	178	89	
132 M	7.5	3.7 5.5	—								
160 M	11	7.5	3.7 5.5	160		42k6	42k6	254	210	108	14
160 L	15	11	7.5						254		
180 M	18.5	—	—	180		48k6	48k6	279	241	121	
180 L	22	15	11						279		
200 L	30	18.5 22	15	200			55m6	318	305	133	
225 S	37	—	18.5	225		55m6	60m6	356	286	149	18
225 M	45	30	22						311		
250 M	55	37	30	250		60m6	65m6	406	349	168	22

* Dimensions not standardized. Manufacturers catalogues to be referred.
§ For shaft extension details, refer Table 337
** Wherever two values are indicated for *kW* ratings, manufacturers catalogues to be referred for exact values.
Performance requirements as per IS 325 - 1970.

Dimensions of flange mounted motors for machine tools (TEFC squirrel cage rotor with class *E* insulation) (As per IS 2223 - 1971)

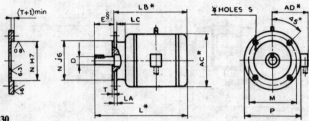

Table 330

Flange Designation**	Frame Size	kW Rating ††			D	Dimensions								
						M		N	P	S		T	LA	LC
		3000 & 1500	1000	750		Nom.	Tol.	H7/j6	Max.	Size	No.	Max.		
F 115B	56	0.06 0.09	–	–	9j6	115		95	140			3		
	63	0.12 0.18	–	–	11j6					10			9	16
F 130B	71	0.25 0.37	–	–	14j6	130		110	160					
F 165B	80	0.55 0.75	0.37 0.55	–	19j6	165	± 0.3	130	200	12		3.5	10	20
	90S	1.1	0.75	0.37	24j6									
	90L	1.5	1.1	0.55							4			
F 215B	100L	2.2	1.5	0.75 1.1	28j6	215		180	250				11	
	112M	3.7	2.2	1.5						15		4		24
F 265B	132S	5.5	–	2.2	38k6	265		230	300				12	
	132M	7.5	3.7 5.5	–										
F 300B	160M	11	7.5	3.7 5.5	42k6	300		250	350				13	
	160L	15	11	7.5										32
	180M	18.5	–	–	48k6					19				
	180L	22	15	11								5		
F 350B	200L	30	18.5 22	15	55 m6 † 55 m6 ‡	350	± 0.5	300	400				15	
F 400B	225S	37	–	18.5	60 m6 ‡	400		350	450		8		16	
	225M	45	30	22										

For notes refer page 747.

Notes:

* Dimensions not standardized. Manufacturers' catelogues to be referred.
** These motors are designated as B type flange motors in IS 2223 - 1971
§ For shaft extension details, refer Table 337
†† Wherever two values are indicated for *kW* ratings, manufacturers' catalogues to be referred for exact values. output as per IS 1231 - 1967.
† Values for 3000 *rpm*   ‡ Values for 1500 *rpm*.
For accuracy requirements, refer Tables 334 and 335.

Dimensions of face mounted motors for machine tools
(TEFC with class *E* insulation) (As per IS 2223 - 1971)

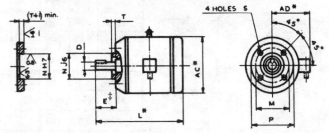

Table 331

Flange ** designation	Frame Size	kW Rating for † synchronous speed (rpm)			D j6	M		N H7/j6	P	S		T Max.
		3000 & 1500	1000	750		Nom.	Tol.			Size	Depth	
F 65 C	56	0.06 0.09	—	—	9	65		50	80			
F 75 C	63	0.12 0.18	—	—	11	75		60	90	M5	6	2.5
F 85 C	71	0.25 0.37	—	—	14	85	± 0.3	70	105			
F 100C	80	0.55 0.75	0.37 0.55	—	19	100		80	120	M6	8	3
F 115 C	90 S	1.1	0.75	0.37	24	115		95	140			
	90 L	1.5	1.1	0.55								
F 130 C	100L	2.2	1.5	0.75 1.1	28	130		110	160	M8	10	3.5
	112 M	3.7	2.2	1.5								

** Designated as C type flange motors in IS 2223 - 1971. Performance requirements as per IS 325 - 1970
* Dimensions not standardized. Manufacturers' catalogues to be referred.
‡ For shaft extension details, refer Table 337
† Wherever two values are indicated for *kW* rating, manufacturers' catalogues to be referred to for exact values. Output as per IS 1231 - 1967.
For accuracy requirements, refer Tables 334 and 335.

OVERALL SIZES OF ELECTRIC MOTORS

Table 332 Foot mounted motors

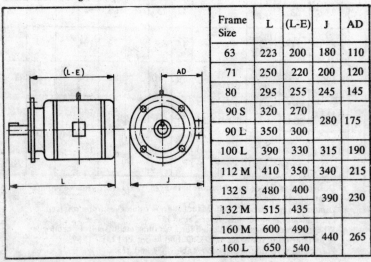

Frame Size	L	(L-E)	J	AD
63	223	200	165	115
71	250	220	185	125
80	295	255	220	145
90 S	290	260	280	175
90 L	345	295		
100 L	400	340	310	190
112 M	410	350	340	205
132 S	480	400	400	235
132 M	520	440		
160 M	610	540	440	260
160 L	650	550		
180 M	680	590	485	285
180 L	710	615		
200 L	765	655	590	355

Table 333 Flange and face mounted motors

Frame Size	L	(L-E)	J	AD
63	223	200	180	110
71	250	220	200	120
80	295	255	245	145
90 S	320	270	280	175
90 L	350	300		
100 L	390	330	315	190
112 M	410	350	340	215
132 S	480	400	390	230
132 M	515	435		
160 M	600	490	440	265
160 L	650	540		

Accuracy requirements for flange and face mounted motors
for machine tools (As per IS 2223 - 1971)

SHAFT RUNOUT

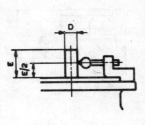

Table 334 Shaft runout

D		Permissible runout in μm	
Over	Upto	Normal*	Precision
—	10	30	15
10	18	35	18
18	30	40	20
30	50	50	25
50	80	60	30
80	120	70	35

CONCENTRICITY AND PERPENDICULARITY

Concentricity test
indicator

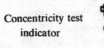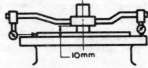

Perpendicularity test
indicator

Note: Tests shall be carried out with the motor set vertical to eliminate the axial clearance in the bearings.

Table 335 Concentricity and Perpendicularity:

Flange Designation	Concentricity, μm		Perpendicularity **	
	Normal *	Precision	Normal*	Precision
F 65 C F 75 C F 85 C F 100 C F 115 B & C	40	20	80	40
F 130 B & C F 165 B F 215 B F 265 B	50	25	100	50
F 300 B F 350 B F 400 B & D F 500 B & D	63	32	125	63

*   Unless otherwise specified, motors are available with normal accuracy.
**   For this test, the pointer to be positioned between the pitch circle dia. of bolt holes and outside dia. of flange.

Table 336 Threaded centre holes for motor shafts

d		D	t_1	t_2
over	Upto			
7	10	M 3	9	13
10	13	M 4	10	14
13	16	M 5	12.5	17
16	21	M 6	16	21
21	24	M 8	19	25
24	30	M 10	22	30
30	38	M 12	28	37.5
38	50	M 16	36	45
50	85	M 20	42	53
85	130	M 24	50	63

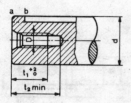

Notes:

1. The dimensions specified for centre holes M 3 and M 5 are applicable only for blind keyways *b* . For open keyways *a* , the diameter of the shaft must be atleast 10 mm for hole M 3, atleast 12 mm for M 4 and atleast 16 mm for M 5.

2. Threaded centre holes for shaft diameters upto 30 *mm* are provided only if specifically requested for. Otherwise shafts will be without threaded holes.

Table 337 Shaft extension and maximum torque details of electric motors for machine tools. (TEFC with class E insulation) (As per IS 1231-1974)

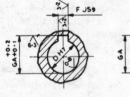

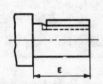

COUPLING OR PULLEY   SHAFT EXTENSION

D		E	F	GA		Max. torque for continuous duty, Nm
Nom.	Tol.		h9	Nom.	Tol.	
9		20	3	10.2	+ 0.0045 − 0.1295	0.63
11		23	4	12.5	+ 0.0055 − 0.1325	1.25
14	*j*6	30	5	16		2.8
19		40	6	21.5	+ 0.0065 − 0.2365	9
24		50	8	27	+ 0.0065 − 0.2965	18
28		60		31		31.5
38		80	10	41	+ 0.018 − 0.288	90
42	*k*6		12	45		125
48		110	14	51.5		200
55			16	59	+ 0.03 − 0.279	355
60	*m*6	140	18	64	+ 0.03 − 0.299	450
65				69		630

Table 338 Duty type and examples of application

Duty type	Applications
S₁ (Continuous duty)	a) Hydraulic pump motor b) Lubrication pump motor c) Chip conveyor motor d) Grinding machine wheel head motor e) Coolant pump motor
S₂ (Short time intermittent rating)	Rapid traverse motor
S₃ (Intermittent periodic duty)	a) Main drive motor for gear shapers b) Main drive motor for drilling machines
S₄ (Intermittent periodic duty with starting)	a) Main drive motor for lathes without clutch in the drive b) Work head motor in grinding machines c) Main drive motor for gear hobbers d) Coolant pump motor - when frequent starting and stopping is resorted to
S₅ (Intermittent periodic duty with starting & electrical braking)	Work head motor in grinding machines with electrical braking
S₆ (Continuous duty with intermittent periodic loading)	Main and feed drive motor with clutch in the drive

Table 339 Recommended vibration class of motors for different applications

Type of machine	Vibration limits on spindle in the direction of cut, in microns (double amplitude)	Recommended* vibration class of motor
Lathe	15 - 25	A
Drilling machine		
φ1 mm-φ5 mm, high speed	4 - 8	B or C
φ5 mm and above, low speed	12 - 24	A
Milling machine	5 - 25	A
Grinding machine		
Thread	1.5	Special
Profile	1 - 2.5	Special
Cylindrical	1 - 2.5	Special
Surface	1 - 5	C
Centreless	1 - 2.5	Special
Boring machine		
Jig boring machine	1	Special
Normal	1.5 - 2.5	Special

* Refer IS 4729-1968 - Measurement and evaluation of vibration of rotating electrical machines.

Table 339 gives the recommended vibration class of motors for
different applications with the vibration limits on the spindle in the
direction of cut. The vibration limits of the special class of motors is
a matter of agreement between the manufacturer and the user.

Electrical elements used in machine tools

Increasing use is made of electrical elements in the present day
machine tools. The majority of the demands of automatic machine
tools are met by electrical controls. The electrical elements can be
remotely controlled and can be used as indicators of pressure,
cutting force, dimensional control, position of table, etc., with
inherent advantage of being used for further functions. Many of the
electrical elements permit themselves to be placed outside the
machine tool thereby making the size of the machine tool smaller.

Electrical elements generally used in machine tools are dealt
with here.

ELECTROMAGNETIC CLUTCHES AND BRAKES

Electromagnetic clutches and brakes are devices, especially
suited for programming and pre-selection. Very short engaging and
disengaging timings can be had and they can be varied too.
Operation (Fig. 214)

When the coil is energized, the current sets up an
electro-magnetic field which can close its circuit only via the outer
plates, the armature plate and the inner plates. This magnetic field
of force compresses the outer and inner plates which are alternately
arranged between the body and the armature plate so that a positive
connection between the parts to be coupled is established.

When the clutch is de-energized pressure is removed from the
plates enabling the driving portion to revolve while the driven
portion comes to rest.

It is essential that the clutch plates be lubricated. Oil splash or
oil mist is usually sufficient for this purpose. Care must be taken not
to flood oil on to the slip rings.

Brake clutches for quick stopping of gear wheels or shafts are
generally arranged on a casing wall in such a manner that the
magnet body is stationary and the inner plates brake the shaft or
gear (Fig. 215). Braking clutches with a stationary magnet body are
not fitted with slip rings, they are provided with terminals only.

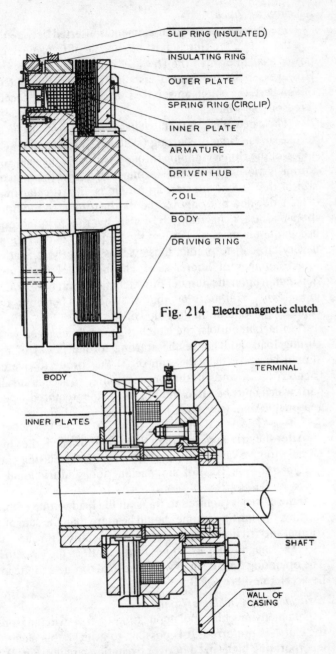

SLIP RING (INSULATED)

INSULATING RING

OUTER PLATE

SPRING RING (CIRCLIP)

INNER PLATE

ARMATURE

DRIVEN HUB

COIL

BODY

DRIVING RING

Fig. 214 Electromagnetic clutch

TERMINAL

BODY

INNER PLATES

SHAFT

WALL OF CASING

Fig. 215 Brake clutch

Selection of clutch

If a flywheel or a similar element is inserted between the motor and the clutch, the clutch must be capable of transmitting the extra power available when the flywheel decelerates during an overload However, if the flywheel is installed between the clutch and the driven part, the clutch must have sufficient power to accelerate it in a reasonable time.

The electromagnetic clutch has two values of torque namely (i) torque at the time of starting (switching), (ii) torque at the rated speed of the clutch (running torque). Normally the value of running torque is more than that of starting torque. For the selection of the size of the clutch, the starting torque is of particular importance. This is decided not only by the motor torque as modified by the intermediate gear train, but also by the torque required to accelerate the masses as well as the frequency of operation. Since it is very difficult to predict these values precisely, it is advisable to select clutches of liberal size. The power of the clutch varies depending upon the number of switchings in unit time. This is shown in Fig. 216. With more number of switchings per hour, the power capacity per operation goes down.

In machine tools, the clutch is generally engaged without any cutting load, and hence the starting torque (T_{start}) of the clutch should be more than the sum of (i) the torque to overcome the friction ($T_{friction}$) and (ii) the torque required to accelerate all driven parts which must be set in motion upto the required speed in a reasonable time (say 1 *sec.*) ($T_{acceleration}$).

$$T_{start} = T_{friction} + T_{acceleration}.$$

After the driven parts attain the required speed, the clutch must be able to give out the torque required for the cutting conditions. (i.e., load torque T_{load}) of the machine tool and frictional torque.

$$T_{running} = T_{load} + T_{friction}.$$

If the clutch is required to transmit the load torque at the time of starting, the starting torque should be more than the sum of the load torque, the friction torque and the acceleration torque.

The magnitudes of above torques affect the operating time. The operating time is small when the load torque and the masses to be accelerated are small.

Switching on-switching off time

In many machine tool applications, a fast switching-on time of the clutch is preferred. It is possible to achieve this electrically by incorporating high speed or over excitation circuits. Fig. 217 shows

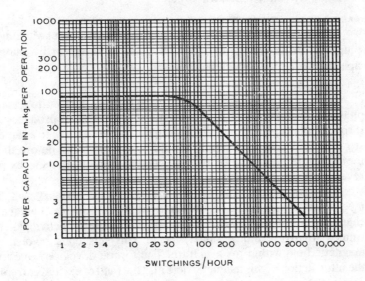

Fig. 216

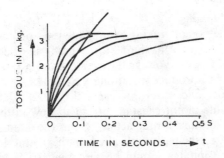

Fig. 217

the switching-on time curves of the same clutch with variation of electrical parameters. Similarly the switching-off time can be shortened by suitable quench circuits.

BRAKING OF INDUCTION MOTORS

Induction motors can be braked by reverse current switching (plugging) or *dc* voltage injection (dynamic braking) or by using capacitors or electromagnetic drums.

Reverse current braking

In this method (Fig. 218), on pressing the stop push button, two phases of the three phase supply are interchanged which produces a torque in the reverse direction causing the motor to come to a rapid stop. The interchanged supply should be disconnected before the motor comes to rest. The reverse current braking is done by a reverse current braking switch.

Reverse current braking switch (Fig. 219) is a rotary switch mechanically connected to the shaft of the motor. Its contacts change over at a preset speed. The rotor made of permanent magnet rotates when the motor shaft rotates. This sets up a revolving magnetic field within the armature. The torque developed due to the inter-action of this magnetic field and the field of excited current rotates the armature as in a motor. The rotary movement of the cam causes deflection of the spring contact, restricted by adjustable stops. This deflection causes breaking of the contact at C 1 and making of contact at C 2. The stiffness of the contact springs and the distance between the contacts can be adjusted for operation at a particular speed.

dc Injection braking

In this method of braking by pressing the stop push button, the three phase *ac* supply to the stator winding is disconnected and simultaneously a *dc* voltage (variable through resistances) is supplied so as to produce a stationary *dc* field.

The rotor conductors interesect the *dc* field and voltage is generated in the motor causing current to flow in the rotor. This rotor current sets up a torque acting against the direction of roation of the motor i.e., the motor is braked. In this method the *dc* supply to the stator windings should be disconnected before the rotor comes to rest by a suitable time circuit.

The maximum number of brakings per hour by these methods is limited by the heating of the motor.

ELECTROMAGNETS

Electromagnets are used for functions like tool lifting, braking,

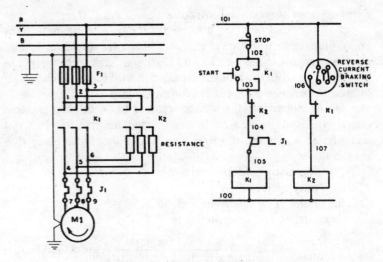

Fig. 218 Circuits for reverse current braking

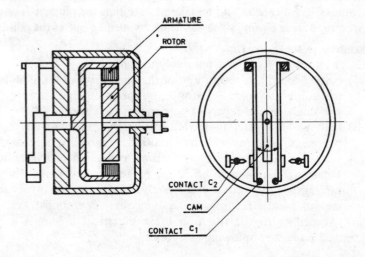

Fig. 219 Reverse current braking clutch

indexing and operating hydraulic valves. An electromagnet essentially consists of a coil, a frame for the flow of magnetic lines and a plunger to which mechanical linkages are made. *ac* and *dc* electromagnents are shown in Figs. 220 and 221 respectively. On energizing the coil, the plunger is pulled in with force which can be utilised to perform different functions.

On de-energizing, the plunger returns to its original position due to the spring force of the connected mechanism. For the selection of a magnet, the following points should be given due consideration.

    i)   force-stroke characteristic;
    ii) duty cycle;
    iii) number of switchings per hour;
    iv) closing and releasing time.

$$\text{Duty cycle as percentage} = \frac{\text{switching on time}}{\text{switching on time} + \text{rest period}} \times 100$$

With a smaller percentage of duty cycle, the magnet can be used for a higher force. With higher number of switchings per hour, the heating in the coil of the *ac* magnet is more. In case of *ac* magnets, great care should be taken to see that the plunger is not prevented from closing since this leads to burning out of the coil.

ELECTRICAL EQUIPMENT FOR CONTROL

    In machine tools, it is necessary to control:

 i) cutting speed i.e., revolutions of the main drive of the spindle including its starting, stopping, speed-change, braking, pre-selection and reversing (programme of speeds).

 ii) feed and traverse, their rate, direction, pre-selection and precise stopping in the desired position (programme of feeds).

 iii) further auxiliary operations such as movement of hydraulic tailstock, clamping, lubrication of guideways, tool-lifting, movement of cross rail, coolant system, on-process measurement etc. (programme of auxiliary operations).

According to the function these electrical elements are grouped under the following categories:

Switching devices

These carry out the desired connections in the circuits e.g., to close or open the connections of the electric motors to the mains or change over the speed or to set the braking in action. Switching devices are operated either manually, or more frequently, by coils

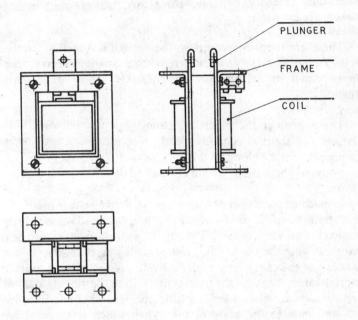

Fig. 220   ac electromagnet

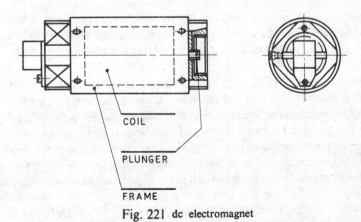

Fig. 221   dc electromagnet

remotely controlled by hand operated switches or push buttons. Under this, category switches, contactors, starters and circuit breakers are available.

Control devices

These are used to accomplish the desired connections in the electric circuit. Limit switches, micro switches, auxiliary relays, time relays, push buttons, pressure switches, etc., come under this category.

Protective devices

These protect the electrical circuit against the effects of overload and short circuit. Overload protector, fuses and single phasing preventor belong to this category.

Some of these elements are covered in detail.

Switches

In machine tools generally cam operated switches are used. On the shaft of the switch are fixed bakelite or tough nylon cams which act upon the moving contacts. Fig. 222 shows the aspects of a switch clearly showing the positioning and switching mechanisms. When the recess of the cam comes under the roller of the moving contact its spring forces it against the fixed contact. Both the moving and the fixed contacts are placed inside a little open chamber and fixed by bolts and nuts to the body of the switch which is made of an insulating material like bakelite. Cam operated switches can be used for different functions such as on-off, reversing, star-delta starting and pole-changing of motors, distribution, etc.

Other types of switches are: Toggle switches, Knife switches, Lever (or Key) switches, Slider switches, Rotary switches, Limit switches, Push-button switches, Micro-switches, Pressure switches and Mercury switches. Out of these limit switches, micro-switches and pressure switches find considerable applications in machine tools.

Limit Switches (Fig. 223a)

Limit Switches (Fig. 223a): These are usually operated by dogs fixed on the moving parts of the machine. When a moving part of the machine reaches the limit position of travel, the switch is pressed by the dog to open its normally closed contact and close its normally open contact. For the correct design of cam, the closing and opening of the contacts should be precisely known with respect to the travel of the plunger. Different types of limit switches such as normal, snap action, over-lapping and extended stroke are available to suit all possible design requirements (Figs. 223b, c, d, e). Definitions relating to limit switches are indicated in Fig. 224.

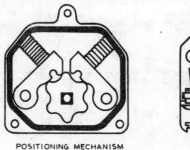
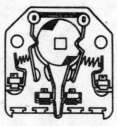

POSITIONING MECHANISM   SWITCHING MECHANISM

Fig. 222 Cam operated switch

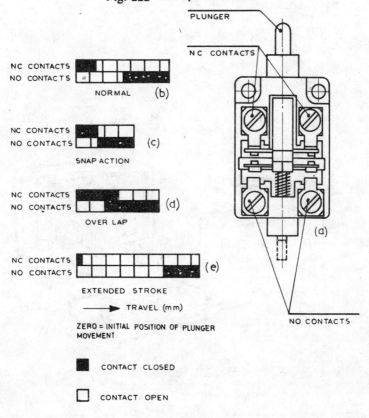

PLUNGER

NC CONTACTS

NC CONTACTS
NO CONTACTS
NORMAL   (b)

NC CONTACTS
NO CONTACTS   (c)
SNAP ACTION

NC CONTACTS
NO CONTACTS   (d)
OVER LAP

NC CONTACTS
NO CONTACTS   (e)
EXTENDED STROKE

⟶ TRAVEL (mm)

ZERO = INITIAL POSITION OF PLUNGER MOVEMENT

■ CONTACT CLOSED

☐ CONTACT OPEN

(a)

NO CONTACTS

Fig. 223

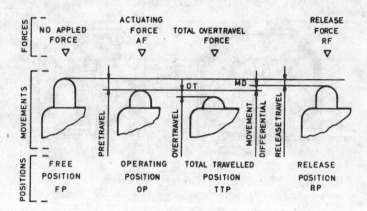

Fig. 224

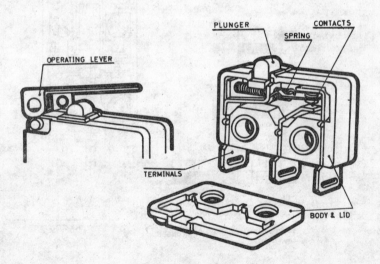

Fig. 225 Micro switch

Micro switches (Fig. 225): The principal function of the micro-switch is the same as that of a limit switch. They are used where small dimensions, high accuracy of switching and very short stroke are the deciding criteria of the design. However their current carrying capacity is low.

Push buttons: These are one of the most widely used control elements. It consists of a knob which is pressed by the finger and a set of contacts. They are available in different colours to distinguish different functions of the machine tool.

Pressure switches: These are mainly microswitches operated by a diaphragm dependent on pressure to give a signal to the control circuitry when the pressure of a system reaches or crosses a certain adjusted value.

Contactors: These are switching devices, generally operated by a coil, for repeatedly making and breaking an electric power circuit; Fig. 226 illustrates two views of a three phase *ac* contactor. A contactor operated by a coil in conjunction with push buttons provides an inherent no-voltage protection i.e., the closed contacts open when the electric supply goes off and do not close when the supply is restored. A good contactor operates reliably from 85% to 110% of the normal coil voltage. For very low voltages, the contactor does not operate, safeguarding the motor against overload. The life of the contactor depends on the type of electrical loading.

Overload protectors: These are normally bi-metalic relays. The current in the power circuit passes through the bi-metallic strips. When the current exceeds the set value, they expand and open the auxiliary contact (used in the control circuit). Two versions of overload protectors are available:

i) automatic resetting type (in which the protector resets to its original condition after the cooling of the bi-metallic elements)

ii) hand resetting type (in which manual pressing is required to reset).

Relays (Fig. 227)

An electrically operated relay is a device, which on energization of its coil switches on the control circuits of other electrical devices such as contactors, electro-magnetic clutches, etc., but it does not switch on directly any heavy current equipment like motors. There are different types of relays. Auxiliary relays serve for amplifying or multiplying of contacts. Control relays control the working process in dependence on a certain electrical value. Reed-

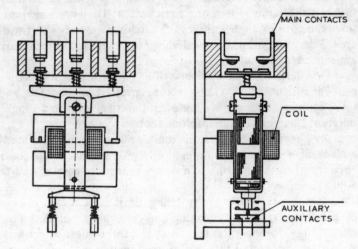

Fig. 226 Contactor

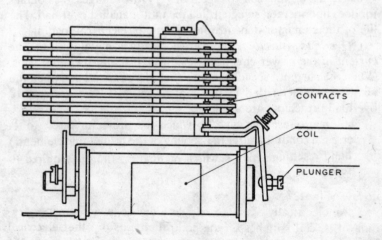

Fig. 227 Relay

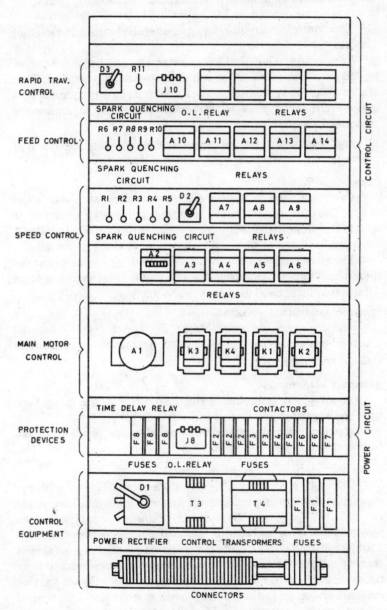

Fig. 228 Control panel

relay is essentially a small ferromagnetic cantilevered beam that bends to make contact under the force of a magnetic field.

The control and reed relays are extensively used as switching devices.

Time relays: They find use in automatic and programme controlled machine tools and for star delta switching of motors. They can be used for switching ON or OFF an electrical circuit after a lapse of time. The types available include pneumatic, electronic and motorized time relays.

Signal lamps: Signal lamps can be used as visual indication of functions like supply on, motor running, clamping, etc.

FUSES

Fuses are used to protect electrical equipment against short circuit. The principle of its function is fusing of easily fusible wire or strip at a current higher than the nominal current of the fuse itself. They are provided at the origin of the circuit to be protected.

The characteristics of the fuse and the device to be protected should be suitably matched electrically. Guide for selection of fuses is indicated in Table 340.

STARTERS AND CIRCUIT BREAKERS

These are units incorporating within themselves contactors and overload protector. They provide safety against no-voltage, under voltage and overload.

Assembly of elements

Assembly and mounting of electrical elements are carried out with due regard to the factors indicated earlier.

A typical installation scheme of the electrical panel inside the control console is shown in Fig. 228.

Guide for selection of fuses and wires

Table 340 gives the guidelines for selection of fuses and copper conductors for induction motors for machine tools. The motors are rated for 415 V, 3 phase, 50 Hz and are of squirrel cage type suitable for operation at an ambient of 40°C. HRC fuses correspond to IS 2208, BS 88 or VDE 0660 and this selection is based on a maximum starting current of 6 times full load current and a maximum starting time of 5 *secs*. for direct on line starting and a maximum starting current of twice the full load current and a maximum starting time of 15 *secs*. for star delta starting. The full load currents indicated in the table are applicable for motors of 1500 *rpm*. For other values reference should be made to the motor name plate. Motors with higher full load current and/or higher starting current and/or longer starting times as well as motors with higher frequency of operations may require fuses of a higher rating.

Table 340 Guide for selection of fuses and wires

Motor rating at full load current		Motor full load current Amps.	Fuse Rating (HRC) Amps		Recommended size of copper conductors, sq. mm
kW	HP		Direct-On-Line starting.	Star-delta starting.	
0.06	0.08	0.2	2	—	
0.09	0.12	0.3	2	—	
0.12	0.16	0.4	2	—	
0.18	0.25	0.7	2	—	1.5
0.25	0.34	0.8	2	—	
0.37	0.5	1.2	4	—	
0.55	0.75	1.6	4	—	
0.75	1	1.8	6	—	
1.1	1.5	2.6	6	—	
1.5	2	3.5	10	—	
2.2	3	5	15	—	
3.7	5	7.5	15	—	2.5
5.5	7.5	11	25	20	
7.5	10	14	30	20	
11	15	21	35	25	4
15	20	28	50	35	6
18.5	25	35	60	50	
22	30	40	80	50	10
30	40	55	100	60	
37	50	66	125	80	25
45	60	80	160	100	
55	75	100	200	100	
75	100	135	200	160	35

Item	Element	Type	No.off
S1	Mains on-off switch	16 *Amps*, 500 *V*	1
S2	Three position selector switch	(1 NO + 1 NC) contacts	1
F1	Triple pole fuse unit	6 *Amps* (Delayed action)	1
K1	Air break contactor, 220 *V*, ac coil, 50 *Hz*	2 NO Aux. contacts	1
J1	Overload relay	3 to 6 *Amps*. range	1
P1	Push button switch, mushroom actuator, Red colour	(1 NO + 1 NC) contacts	1
P2, P5 P6	Push button switch, Shrouded actuator, Green colour	(1 NO + 1 NC) contacts	3
P3	Push button switch, Shrouded actuator, Green colour	(2 NO + 2 NC) contacts	1
P4	Push button switch, Un-shrouded actuator, Red colour	(1 NO + 1 NC) contacts	1
A1 A2 A3 A4	Auxiliary contactor, open execution, 220*V*, ac Coil, 50 *Hz*	(4 NO + 4 NC) contacts	4
A5	Time delay relay, 220 *V*, ac coil, 50 *Hz*	On-delay, 0-60 *secs*.	1
H1	Indicator lamp, yellow colour with 220 *V*, 7 *W bulb*.	For rear connection	1
F2, F3, F4, F5, F6	Fuse	Glass cartridge 1 *Amp*	5
F7	Fuse	2 *Amps*.	2
F8	Fuse	4 *Amps*.	2

CIRCUIT FUNCTIONS

The circuit is for an electrical coil pressing machine.

The functioning of the circuit is as follows:

1) Switch on S1 to connect power to all equipment.

2) Start motor M1 by pressing push button P2 which energizes contactor K1 and self holds.

3) Select the direction (horizontal, vertical or both) by switch S2

4) Clamp by pressing push-button P6 which energizes solenoid valve 9A. The coil is pressed.

   If a high pressure is desired, press push-button P3. Indicator H1 indicates high pressure.

   (Note that after high pressure is selected, the changing of switch position S2 does not have any effect. This is specifically provided to avoid damage to hydraulic elements.)

5) Press P5 to declamp which is accomplished by the solenoid valve 9B being kept energized for a specified time selected in pneumatic time delay relay A5.

Note: Necessary no-voltage protection and short circuit protection are provided for all elements. P1 serves as master stop.

Table 341 Graphical symbols for electrical circuits on machine tools

Graphical symbols	Meaning	Graphical symbols	Meaning	Graphical symbols	Meaning
—	Direct Current	(earth symbol)	Earth	Z	Impedance
∿	Alternating Current	(hatched symbol)	Frame or chassis connection	(symbol)	Resistor with moving contact (General symbol)
3N ∿ 50 Hz 415 V	Alternating current 3 phase with neutral 50 Hz, 415 V.	(arrow symbol)	Variability (General symbol)	(symbol)	Voltage divider with moving contact
+	Positive polarity	(arrow symbol)	Variability by steps	(symbol)	Temperature dependent resistor with negative resistance coefficient (Thermistor)
−	Negative polarity	(symbol)	Preset adjustment (General symbol)	(symbol)	Variable resistor (General symbol)
△	3 phase winding - delta	(symbol)	Preset adjustment by steps	(symbol)	Resistor with fixed tappings
Y	3 phase winding - star	(terminal strip)	Terminal strip	(symbol)	Voltage divider with fixed tapping
Y (symbol)	3 phase winding - star with neutral broughtout	⊗	Luminous push button	(coil symbol)	Inductance or winding
(symbol)	3 phase winding - zig-zag or interconnected star	(box symbol)	Resistance, resistor (when not necessary to specify whether it is reactive or not)	(symbol)	Capacitance, capacitor
∿	Flexible conductor	R	Non-reactive resistor	(symbol)	Electrolytic capacitor
O	Terminal				
⊥	Junction of conductors				

(Contd.)

Table 341 Graphical symbols for electrical circuits on machine tools

Graphical symbols	Meaning	Graphical symbols	Meaning	Graphical symbols	Meaning
	Signal lamp		Relay coil with two windings		Brush on commutator
	Local lighting		OFF-Delay relay coil		ac motor (M) ac generator (G) (General symbol)
	Gearing		ON-Delay relay coil		dc motor (M) dc generator (G) (General symbol)
	Manually operated control		Relay coil of combined ON/OFF Delay relay		dc two wire generator (G) or motor (M) Separately excited
	Cam operated control		Relay coil of a polarized relay		
	Control operated by electric motor		Electromagnetic short circuit release		dc two wire shunt generator (G) or motor (M)
	Single acting pneumatic or hydraulic control		Thermal overload release (bimetal elements)		Induction motor, 3 phase, squirrel cage.
	Double acting pneumatic or hydraulic control		Fuse		Induction motor, 3 phase with wound rotor
	Operating coil of electromagnetic actuator.		Brush on slip ring		
	Coil of relay or contactor (General symbol)				

(Contd.)

Table 341 Graphical symbols for electrical circuits on machine tools

Graphical symbols	Meaning
(symbol)	Induction motor, 3 phase, squirrel cage, both leads of each phase broughtout.
(symbol)	Induction motor, single phase, squirrel cage
(symbol)	Auto-transformer (General symbol)
(symbol)	Single phase transformer with two separate windings
(symbol)	Transformer with three separate windings
(symbol)	Primary cell or accumulator The long line represents the positive pole and the short line the negative pole

Graphical symbols	Meaning
(symbol)	Battery of accumulators or primary cells
(symbol)	Rectifier, Diode
(symbol)	Zener Diode
(symbol)	Transistor type PNP
(symbol)	Transistor type NPN
(symbol)	Thyristor (General symbol)
(symbol)	Contactor - make contact
(symbol)	Contactor - break contact

Graphical symbols	Meaning
(symbol)	Make contact (relay)
(symbol)	Break contact (relay)
(symbol)	Changeover contact (relay) break before make
(symbol)	Break contact of thermal overload relay
(symbol)	Switch (General symbol)
(symbol)	Three pole switch

(Concld.)

Table 341 Graphical symbols for electrical circuits on machine tools

Graphical symbols	Meaning	Graphical symbols	Meaning	Graphical symbols	Meaning
	Twoway contact with neutral position		Pressure switch make contact		OFF-Delay, normally open contact
	Make before break contact		Pressure switch break contact.		OFF-Delay, Normally closed contact
	Change over switch with many contacts		Circuit breaker with protective devices (General symbol)		Socket (female)
	Make contact push button		Circuit breaker 3 pole		Plug (male)
	Break contact push button		Mushroom head, push button-break contact (Master stop)		Plug and socket
	Limit switch make contact		ON-Delay, Normally open contact		Volt meter - Insert V Ammeter-Insert A Watt meter - Insert W Tachometer - Insert n
	Limit switch breakcontact		ON-Delay, Normally closed contact		Rectifier Bridge

Coolant Pumps for machine tools (As per IS 2161 - 1962)

Position of conduit entry adjustable through every 90° by rotating terminal box.

Relative position of terminal box to coolant outlet adjustable through 90° by turning the terminal box.

Conduit entry

Table 342

Parameter	1/120	1/170	1/220	2/220	2/270
Size	1	1	1	2	2
Type	1/120	1/170	1/220	2/220	2/270
Immersion depth H	120	170	220	220	270
H_1	30	80	130	130	180
H_2 Max.	160			200	
$A\ {}^{0}_{-1}$	100			140	
B Max.	80			80	
D Max.	130			180	
$d \pm 0.2$	115			160	
d_1	FP3/4"			FP1"	
$E \mp 1$	30			30	
F	130			180	
L_1 Max.	100			120	
L_2 Max.*	125			150	
L_3	70			100	
Min. pumping Capacity in l/Min. For water head in meters — 1	36			70	
2	30			63	
3	24			55	
4	12			45	
5				30	

*Dimension L_2 is for coolant pumps with switch.

Outlet
Max. fluid level
Min. fluid level
4 Holes, $\phi\,7$

HYDRAULIC SYSTEMS FOR MACHINE TOOLS

OIL HYDRAULICS IN MACHINE TOOLS

Nomenclature

a Width of piston ring, cm

A Area of cross-section, cm^2

A_e Effective area of piston, cm^2

A_o Area of orifice, cm^2

A_p Projected area of seal, cm^2

b Face width of gear, cm

b_r Width of rotor, cm

B_{ss} Isentropic secant bulk modulus, kgf/cm^2

B_{st} Isentropic tangent bulk modulus, kgf/cm^2

C_1 Width of sealing, cm

C_d Coefficient of discharge

d Pitch circle diameter of gear, cm

D_h Hydraulic diameter, cm ($=4A/s$)

D_p Diameter of piston, cm

D_r Diameter of rotor, cm

e Eccentricity between rotor and stator centres, cm

e' ¼ (major diameter – minor diameter), cm

f Friction factor

f_c Friction due to O-ring compression , kgf/cm length

f_h Friction due to fluid pressure, kgf/cm^2

f_l Friction factor between lip seal and cylinder wall

f_r Friction between ring and cylinder wall

F_p Friction force, kgf

ΣF Sum of friction force and force due to back pressure, kgf

F_a Actual output force, kgf

F_b Breakout friction, kgf

F_t Theoretical output force, kgf

i Number of piston rings

K Loss coefficient

K_1 A constant, $cm^2/(kgf)^{\frac{1}{2}}/sec$

L Length of passage or pipe, cm

L_p Length of seal rubbing surface, cm

m Module, cm

n Speed of pump/motor, rpm

N Number of pistons

p	Pressure on piston, kgf/cm^2
p_k	Preload due to the ring, kgf/cm^2
Δp	Pressure difference across unit, kgf/cm^2
P_h	Hydraulic power, kW
P_i	Power input to the cylinder, kW
P_m	Mechanical power, kW
P_o	Power output from cylinder, kW
q	Derived capacity, cm^3/rev
Q	Actual discharge from pump /into motor, l/min
q	Flow rate through passage, cm^3/sec
Q_c	Flow into cylinder, l/min
Q_l	Leakage flow from motor, l/min
Q_o	Discharge at no load, l/min
Q_r	Flow through orifice, l/min
Q_t	Theoretical discharge from pump/into motor, l/min
R	Pitch circle radius of barrel, cm
R_e	Reynolds number
s	Perimeter of cross-section, cm
T	Temperature at the start of the test, $^\circ K$
T_p	Input torque to pump, $kgf.m$
T_m	Output torque from motor, $kgf.m$
v	Velocity of flow, cm/sec
v_a	Actual velocity of piston, m/sec
v_e	Velocity at the pipe outlet, cm/sec
v_t	Theoretical velocity of piston, m/sec
V	Volume at pressure p and temperature T, cm^3
dv	Change in volume, cm^3
α	Inclination of swash plate to the drive shaft axis, $degrees$
ϵ	Eccentricity ratio
η	Dynamic viscosity of oil, $kgf\, sec./cm^2$
η_o	Overall efficiency, %
η_v	Volumetric efficiency, %
θ	Inclination of cylinder barrel axis to the drive shaft axis, $degrees$
ρ	Mass density, $kgf\, sec^2/cm^4$

Introduction

Oil hydraulics–the science of transmitting and controlling energy through the medium of pressurised oil has several advantages over other methods of energy control. Oil hydraulic systems:

- pack high power in small light components,
- have flat load (torque)–speed characteristics,
- can operate continuously under stall conditions safely,
- provide stepless variation of speeds and
- have longer life due to the lubricating properties of the working medium and can withstand heavy duty cycles without undue heating of the medium.

More important, oil hydraulic systems can be easily built, using readily available standard elements together with electrical or/and pneumatic interface to perform any complicated sequence of operations. These merits, far outweighing the disadvantages of contaminant sensitivity and noisy operation, have influenced the machine tool designer to employ oil hydraulics for not only the basic tool and work piece movements but also for auxiliary functions such as tool indexing and clamping, loading/unloading of workpieces, etc. They have found applications to varying degrees in grinding, honing, broaching, shaping and planing machines. The use of copying attachments on lathes and milling machines is well established. The innovation of electrohydraulic servovalve which could conveniently interface with electrical measuring and signalling devices has led to the popular use of electrohydraulic servodrives in tracer controlled machines, numerically controlled machines, electrodischarge and electrochemical machines. The latest is the application of electrohydraulic stepping motors–electric stepping motors with hydraulic torque amplifiers for feed drives in an open loop configuration.

HYDRAULIC FLUIDS AND FLUID FLOW

Hydraulic fluids meant primarily to effect energy transmission also help in:

- lubricating the elements,
- carrying away the heat generated in the system,
- sealing of moving parts and
- inhibiting rust and corrosion.

Paraffin based mineral oils are generally used in machine tool hydraulics. These are more stable and have a higher viscosity index compared to aromatic or naphthanic based oils. However, where high temperature and fire hazardous conditions prevail these oils are not suitable. Under such conditions, invert emulsions (water-in-oil), water glycols, phosphate esters and silicones are used.

Properties

VISCOSITY

Viscosity is the most important property of the oil. Viscosity of the oil selected should be such as to yield optimum pump performance, due consideration being given to the type of pump, its operating speed and pressure. Use of an oil with too low a viscosity increases leakage and may affect the life of the components because of inadequate lubrication, while that with too high a viscosity can cause inefficient operation due to large pressure drops and viscous drag with subsequent overheating. Pressure losses may be so high as to result in cavitation in pumps.

Viscosity changes with temperature and pressure. The effect of pressure on viscosity can be neglected in machine tool hydraulics. The effect of temperature on the oil is indicated by its viscosity index. An oil with higher viscosity index exhibits less change in viscosity for unit temperature change and is hence preferred. The viscosity index of mineral oils is improved by certain additives.

BULK MODULUS/COMPRESSIBILITY

Bulk modulus which is the reciprocal of compressibility is defined as change in pressure required to cause unit volumetric strain. It is an important parameter in the system design, figuring in calculations of pump output, decompression volumes, pressure surges due to sudden valve operations and drive stiffness resonance.

Two values of bulk moduli–the isothermal and the isentropic–are admissible. Isothermal bulk modulus refers to the value at constant temperature. The isentropic bulk modulus is applicable when pressure changes are rapid allowing no time for entropy change and is also referred to as the dynamic bulk modulus.

These values are further defined as tangent and secant bulk moduli. The isentropic tangent bulk modulus (B_{st}) at pressure p and temperature T is given by the relation:

$$B_{st} = -V \, (dp/dV)_s$$

The isentropic secant bulk modulus (B_{ss}) at pressure p and temperature T is given by

$$B_{ss} = \left(\frac{-dp}{d\,V/V_o} \right)_s$$

where dV change in volume,
dp change in pressure,
V_o initial volume at atmospheric pressure,
and subscript s refers to constant entropy.

At low pressures, the difference between the two values can be ignored. The isentropic bulk modulus of oil under ideal conditions may be as high as 20,000 kgf/cm^2. However, presence of even small quantities of free air and flexing of hoses, tubes and containers, etc., will considerably reduce the value. For general engineering calculations, a value of 7000 kgf/cm^2 is reckoned practical.

RESISTANCE TO FOAMING

Hydraulic oil contains about 8% of dissolved air by volume. The dissolved air by itself has no harmful effect on the system. However, presence of air in free state does considerably reduce the bulk modulus. The effect is less pronounced at high working pressures, since the free air tends to dissolve in oil at high pressures. On reduction of pressure, the dissolved air is released, promoting the formation of bubbles which may result in loss of drive control as well possible breakdown of pump due to cavitation. Antifoam agents are added to oils for increasing the rate of collapse of the bubbles.

RESISTANCE TO OXIDATION

Hydraulic fluids being composed of hydrocarbons tend to oxidise. The rate of oxidation increases with high operating temperature, ingress of water and metallic particles which act as catalysts. The products of oxidation which are acidic in nature can be either soluble or insoluble in oil. Soluble oxidation products tend to thicken the oil while the insoluble ones, generally known as sludge, may clog lines, orifices and filters. The oil rapidly degrades with oxidation leading to total breakdown. The extent of oxidation in a fluid sample is assessed by measuring its *neutralization number* which is the number of milligrams of potassium hydroxide needed to neutralise one gram of oil sample. The rate of increase of neutralisation number is a good measure of the progress of oxidation. A neutralisation number of 1 (a value of 1 *mg* кон/*gm*) is considered as the point for changing/reconditioning of oil. Certain inhibitors added to the oil improve the oil resistance to oxidation.

Some of the properties of fluids used in machine tools are listed in Table 343

Table 343 Properties of hydraulic oils used in machine tools

Properties	Unit	Type/values	Remarks
Base oil		Paraffinic	
Specific gravity at 15°C		0.865-0.88	
Kinematic viscosity at 40°C	cSt	30-65	
Viscosity index		90-95 Min.	
Flash point (closed)	°C	200 (Min.)	
Pour point	°C	-6	
Thermal conductivity at 40°C	W/m°C	0.13	
Specific heat at 40°C	J/kgf °C	1966	
Neutralization number	mg.KOH/gm	0.1	
Aniline point	°C	85	
Average isentropic tangent bulk modulus at presure 0-35 kgf/cm²			
(i) at 25°C	kgf/cm²	20,000	
(ii) at 40°C	kgf/cm²	18,700	
Average isentropic secant bulk modulus at pressure 0-35 kgf/cm²			
(i) at 25°C	kgf/cm²	19,200	
(ii) at 40°C	kgf/cm²	17,800	
Operating temperature			Usually limited to 60°C to have a long life.
Compatibility: Bearing materials			Usually compatible with all bearing materials. If Zinc Dialkyl Dithiophosphate is present as antiwear additive it reacts with white metal and silver lining. Most hydraulic oils react with zinc, magnesium alloys and copper.
Sealing materials			Compatible with many synthetic rubbers like nitrile, neoprene, polyacrylic, fluorocarbon rubbers etc. Not compatible with natural rubbers.
Paints and insulation			Oil resistant paint is required.

FLUID FLOW

Fluid flow is basically governed by a set of equations called the *Navier stokes equations* and *Equation of continuity*. These are non linear partial differential equations having complex boundary conditions and hence have no general solutions. But for practical applications certain approximations have been made to reduce the complexities and make the solutions accurate enough for most of the purposes. Calculations in hydraulics are generally based on (1) Reynolds equations, (2) Hagen–Poiseuille equation for flow through capillaries and (3) Bernoulli's equation for steady state flow.

The flow of fluid in hydraulic systems may be either laminar, being physically characterised by orderly, smooth, parallel line motion, or turbulent being irregular, erratic and eddy-like motion. The internal fluid friction (viscous) forces dominate in laminar flow, whereas the inertia forces are predominant in turbulent flow. The nature of flow is dependant on the velocity of flow (v), density of fluid (ρ) viscosity (μ) and the characteristic dimension of the particular flow passage (D_n), and is expressed as a non-dimensional number called *Reynolds number* (R_e) being equal to $\rho v D_n / \mu$

Laminar or turbulent flow can be either steady or unsteady depending on whether the velocities of fluid particles at a section is independent or dependent on time. Generally the flow is assumed to be steady, incompressible and one dimensional. Also the cubical expansion coefficients for liquids are small and hence the effect of temperature on fluid density and flow is negligible. Based on these assumptions, equations for flow and pressure losses have been derived for different flow passages. Table 344 gives the resistance values (R) of different flow passages and the flow can be obtained from the relation– $Q' = \Delta p/R$. Table 345 gives the pressure loss due to various pipe configurations for steady turbulent flow. For steady laminar flow, coefficient of loss (K) is multiplied by a correction factor b which increases with the decrease in R_e. Fig 230 gives the range of Reynolds number over which the flow is laminar and turbulent and also the value of b as a function of R_e.

Laminar flow, though desirable in systems to minimise the pressure losses, renders the system bulky. It prevails in leakage paths as well in capillaries used for stabilization of valves, hydro-static bearings and drive systems. A capillary is characterised by a large length to diameter ratio ($l/d \geq 400$) of the flow passage. The capillaries are temperature sensitive and hence are unsuitable for control of flow rates in hydraulic systems. Orifices associated with turbulent flow are commonly used for this purpose. An orifice is defined as an opening of short length causing sudden restriction in a flow passage.

Table 344 Equations for laminar flow through different configurations

Sl. No.	Configuration of passage cross section	Cross sectional view	Resistance for flow R $kgf\,sec\,/\,cm^5$
1	Capillary		$\dfrac{128\eta\,l}{\pi d_R^4}$
2	Elliptical		$\dfrac{4\eta\,L(a^2+b^2)}{\pi\,a_1{}^3 b_1{}^3}$
3	Rectangular $w \geqslant h$		$\dfrac{12\eta\,L}{wh^3\left[1-\dfrac{192h}{\pi^5 w}\,tanh\left(\dfrac{\pi w}{2h}\right)\right]}$
4	Square $w=h$		$\dfrac{28{\cdot}4\eta\,L}{w^4}$
5	Between two parallel plates $(w>>h)$		$\dfrac{12\eta\,L}{wh^3}$
6	Equilateral triangle		$\dfrac{185\eta\,L}{S^4}$
7	Right angled isosceles triangle		$\dfrac{155{\cdot}5\eta\,L}{S^4}$
8	Concentric annular ring $(c<<d_R/2)$		$\dfrac{12\eta\,L}{\pi\,d_R\,C^3}$
9	Eccentric annular ring $(c<<d_R/2)$		$\dfrac{12\eta\,L}{\pi\,d_R\,C^3(1+1.5\varepsilon^2)}$

Table 345 Equations for pressure losses in pipe configurations

Sl. No.	Configuration	Cross Sectional view	Pressure loss kgf/cm^2
1	Straight Pipe		$$\frac{\rho f L v^2}{2 D_h}$$
2	Entry into pipe		$$\frac{\rho K v^2}{2}$$ $K = 1$ for right angle entry edge. $= 0.1$ for rounded inlet edge.
3	Reservoir inlet with a change of velocity from v to v_o	—	$$\frac{\rho K(v - v_o)^2}{2}$$ $K = 1$ for $v_o = 0$
4	Sharp bend in pipe		$$\frac{\rho K v^2}{2}$$ For values of K refer Table 346
5	Elbows		$$\frac{\rho K \phi v^2}{180^0}$$ For values of K refer Table 347
6a	Branches	$K = 0.15$	$$\frac{\rho K v^2}{2}$$

Table 345 Equations for pressure losses in pipe configurations   (Concld.)

Sl. No.	Configuration	Cross Sectional view	Pressure loss kgf/cm^2
6 b		 K = 0.5	
6 c	Branches	 K = 1.3	$$\dfrac{\rho K v^2}{2}$$
6 d		 K = 3	

Table 346

θ°	10°	20°	30°	40°	50°	60°	70°	80°	90°
K	0.04	0.1	0.17	0.27	0.4	0.55	0.7	0.9	1.12

Table 347

$d/2r_e$	0.1	0.2	0.3	0.4	0.5
K	0.13	0.14	0.15	0.21	0.29

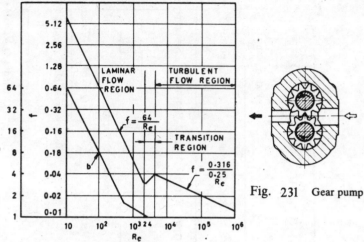

Fig. 230   Relation between factors f and b
and Reynold's number

Fig. 231   Gear pump

HYDRAULIC ELEMENTS

A hydraulic system consists of a number of elements arranged in a sequence of logic to perform the desired function. Based upon the function these elements are broadly classified as:

   i) Power generating elements;

   ii) Power utilizing elements;

   iiit Power controlling elements and

   iv) Accessories.

Power generating elements

Pumps inject energy into the hydraulic system by converting the kinetic energy at the input shaft into hydraulic energy. This energy is a combination of potential energy (pressure) and kinetic energy (flow). Pump creates the flow and the pressure is due to the external resistance. These pumps are called positive displacement pumps, because they trap fluid in discrete segments and force it out at the delivery, thus displacing a fixed volume of fluid per revolution. Different types of pumps used in fluid power systems are given in Table 348 . The most commonly used types are discussed here.

GEAR PUMPS

The external gear pump is the simplest and the most commonly used variety of pumps. It consists of two spur gears of equal

Table 348   Classification of pumps

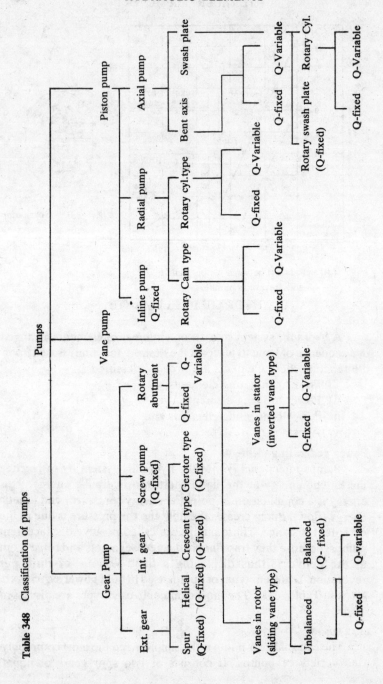

diameter meshing with each other and enclosed in a body with two intersecting bores, with suitable bearings in the end covers (Fig. 231). One of the gears is driven by an external source such as an electric motor. The rotation of the gears creates a partial vacuum in the chamber formed by the gear teeth coming gradually out of mesh. Oil from the pump is sucked into this chamber and carried around between the teeth to be forced out through the delivery port by the pair of teeth coming into engagement.

The capability of a pump is based on its ability to operate at a reasonably high efficiency (of the order of about 80%) at the rated continuous working pressure. The performance is dependent both on the volumetric and mechanical efficiencies. The delivery from the pump reduces with pressure owing to the internal leakage, flow back to suction through the different gaps viz:

– the axial clearance between the gear and bearing faces,
– the radial clearance between the body bore and the outer diameter of the gear,
– the sealing between the mating teeth and
– the clearance between the bearing and the gear shaft.

The first two are the major sources of leakage. The order of clearances built into the pump and permissible geometrical inaccuracies on the components are indicated in Table 349.

The clearances will, nevertheless, tend to grow with the wear of the rubbing surfaces. Present day pumps for working pressures in the range of 100-300 kgf/cm^2 are therefore provided with built in features to compensate for wear and thereby continuously eliminate the undersirable clearances. Fig. 232 shows the constructional details of pumps with these features.

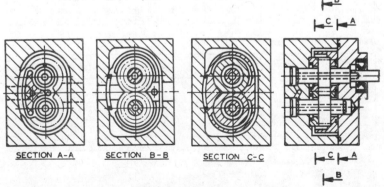

SECTION A-A   SECTION B-B   SECTION C-C

Fig. 232   Axial and radial play elimination in gear pumps.

Table 349a Materials of construction (Gear Pump)

Component	Housing and end covers	Bearing bush and sliding plate	Gears and shaft	Shaft Seal
Materials used	i) *Aluminium alloy* For low pressure application; 4223(A4) of IS 617 For high pressure application; 5500 (A10) or 2280(A11) of IS 617 (ii)*Grey Cast Iron* Gr. 25 of IS 210	(i) 70/20 Copper/Lead alloy and aluminium alloy 24345 (HF15) of IS 734. (ii) Liner bushes of steel strip with a layer of sintered bronze impregnated with PTFE.	Alloy Steel 13*Ni*3*Cr*80 or 15*Ni*2*Cr*1*Mo*15 of IS 1570, hardened and ground to *HRC* 60-65	High grade nitrile rubber supported in mild steel case. Acceptable leakage past the seal is 0.15 cc/hr.

Table 349b Geometrical accuracies

Error in	Permissible values in μm					
	Shaft axis and gear face	Bearing bore and face	Body bore and face	Gear faces	Bearing faces	Body faces
Perpendicularity	5	5	5	—	—	—
Parallelity	—	—	—	2	2	2
Coaxiality	Between shaft axis and gear OD: 5 Between bearing bush bore and OD: 5					

Table 349c Surface finish values on the components

Component	Body bore	Bearing bush bore and faces	Gear OD and faces	Tooth flanks	Journal diameter
Ra in μm	0.8 - 1.6	0.4	0.4	0.4 - 0.8	0.2 - 0.4

Table 349 d Values of working clearance in μm

Type of construction	Axial clearance	Radial clearance
Uncompensated type	20 to 30	15 to 20
Axial clearance compensated type	0.5 to 5 (effective)	15 to 20
Axial and radial clearance compensated type	0.5 to 5 (effective)	0.5 to 5 (effective)

The bearings for the gear shafts are of both the bush and the antifriction types. The sleeve bearing is popular, for it can be used as a sliding member for the purpose of axial clearance compensation. Use of needle bearings or bushes offers the advantage of low starting and running friction-a feature desirable in units used as motors. These bearings are housed either as sliding or fixed bushes. If the bush is fixed, sliding plates are used between the gear and bush faces for the purpose of axial clearance compensation.

In the early designs stub tooth gears were used since the wide crests of these gears assisted in reducing the internal leakage. However, stub tooth gears suffer from a major disadvantage in that the discharge of pumps using these gears is approximately 20% less than that of a pump with a standard full depth involute gear of equivalent diameter. Hence modern gear pumps of compact designs use standard full depth involute gears of 20° pressure angle. Eventhough the minimum number of teeth to avoid undercutting is limited to eight, use of nine number of teeth is found to be satisfactory from the point of view of reducing flow ripple.

A good sealing between the meshing teeth can be obtained by finishing the gear tooth form to the highest standard of accuracy including concentricity and alignment with the shaft axis.

Yet another factor to be considered during the construction of gear pumps is the compression and expansion of the volume of oil trapped in between the pair of meshing teeth which results in heavy cyclic loading on the bearings and noisy operation. To overcome this effect, pressure relieving grooves are provided on the faces of the end bushes as shown in Fig. 233 .

The details on the materials of construction of different components of a gear pump, the tolerances on component dimensions, geometrical accuracies and clearances are given in Table 349 . Equations for calculation of flow, power and

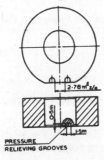

Fig. 233   Pressure relieving grooves

efficiency are given in Table 352

Applications

Gear pumps find common application in constant volume hydraulic systems, for drive, control and auxiliary applications, supercharging of pumps and for lubrication in gear boxes.

Advantages

1) Simplicity and compactness,
2) Low cost,
3) Less sensitive to contamination and
4) High operating speeds.

VANE PUMPS

Sliding vane type pump is the most commonly used types in machine tool hydraulic systems. The working principle of the pump is shown in Fig. 234 . The pump consists of a circular rotor mounted eccentrically inside a circular stator ring, thus providing the suction and delivery chambers. The rotor has suitable slots for accomodating radially moving vanes. The vanes of similar width as the rotor, press against the stator ring to provide the radial sealing between the adjacent chambers and hence between the inlet and outlet ports. The pump output is dependent on the eccentricity of the rotor with respect to the stator. The pump construction lends the feasibility of providing variable delivery feature built into these pumps. In this type of construction, the suction and delivery ports are diagonally opposite and hence the rotor will be subjected to an unbalanced load, thus limiting the working pressure to about 70 kgf/cm^2. In the balanced vane pump design, the load on the rotor is balanced by providing a pair of diametrically opposite suction/delivery ports as shown in Fig. 235 . A disadvantage of this construction is that it cannot be built with the variable delivery feature. The various paths of leakage in the pump are:

i)   the clearance between the tip of the vane and the stator ring,

ii)   the axial clearance between the rotor/the vane/and faces of the side plates,

iii)   the bearing clearance and

iv)   the clearance between the vanes and their respective slots.

The first two cited above are the major sources of leakage contributing towards the marked reduction in volumetric efficiency.

In a simple vane pump, the centrifugal forces of the vanes provide contact between the vane tips and the cam track. This, being dependent on the speed, does not provide for uniform sealing conditions. This drawback is overcome by pressurising the undersides of the vanes (Fig. 236) taking sufficient care to balance the forces on the vane to avoid excessive wear of the cam track. Different design practices of vanes are shown in Fig. 237 .

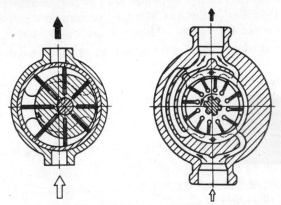

Fig. 234   Vane pump, unbalanced type   Fig. 235   Vane pump, balanced type

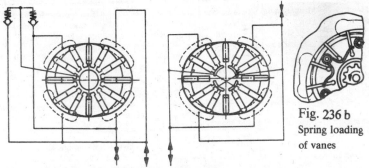

Fig. 236 b
Spring loading
of vanes

Fig. 236 a   Pressurising the undersides of vanes

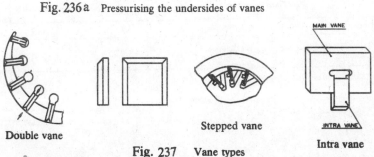

Double vane

Stepped vane

Intra vane

Fig. 237   Vane types

The method of axial play elimination is on similar lines as in the case of gear pumps.

The materials of construction for the different parts of the pump and the desired component accuracies are listed in Table 350. The equations for calculation of flow, power and efficiency are given in Table 352.

Applications

The unbalanced type of pumps are generally used for pressures upto 100 kgf/cm^2, while the balanced construction extends their usage to about 170 kgf/cm^2. These pumps are also built as single, double and triple pumps and find similar applications as the gear pumps.

Advantages

1) Low cost with respect to power output,
2) Less noisy,
3) Long service life,
4) Variable delivery and pressure compensation features are possible.

Table 350 a Material of Construction (Vane pump)

Component	Housing and End Cover	Bearings	Rotor and Shaft	Vanes	Side plate
Material used	Meehanite or high strength cast iron	Sleeve bearing of bronze or sintered bronze impregnated with PTFE liner and roller bearings	Carburizing grade of alloy steel; hardened and ground. If the rotor is split type, the shaft can be of medium carbon alloy steel hardened and ground	High speed tool steel, properly hardened and ground	Manganese or Silicon bronze

Table 350 b Working clearances and geometrical accuracies

Type of pump	Clearance in μm			Geometrical accuracies in μm		
	Axial	Radial	Vane and slot widths	Facial runout	Coaxiality	Parallelity
Uncompensated	5 to 15	0.5 to 5	5	5	5	3
Axial clearance Compensated	0.5 to 5	0.5 to 5	5	5	5	3

Table 351 Equation for derived Capacity

Sl. No.	Type of pump		Derived capacity (q) (cm^3/revolution)
1	Gear pump		$2\pi dmb$
2	Vane pump	Unbalanced	$2\pi D_r eb_r$
		Balanced	$4\pi D_r e'b_r$
3	Piston pump	Radial	$\dfrac{\pi}{2} Ne D_p^2$
		Axial (Swash plate type)	$\dfrac{\pi}{2} \cdot N D_p^2 R \tan\alpha$
		Axial (Bent axis type)	$\dfrac{\pi}{2} \cdot N D_p^2 R \sin\theta$

Table 352 Equations for flow, power and efficiency

(Pump)	(Motor)
$Q_t = qn/1000$	$T_m = \Delta pq/2\pi$
$P_h = \Delta pQ/612$	$P_h = \Delta pQ/612$
$P_m = 2\pi nT_p/6120$	$P_m = 2\pi n\, T_m/6120$
$\eta_v = \dfrac{Q}{Q_o} 100$	$\eta_v = \dfrac{qn}{(qn+Q_l)} 100$
$\eta_o = \dfrac{P_h}{P_m} 100$	$\eta_o = \dfrac{P_m}{P_h} 100$

Table 353 Typical characteristics of pumps

Type of pump		Flow range l/min	System pressure kgf/cm²	Speed Range rpm	Weight/Power kgf/kW	% Efficiencies (Approximate)	
						Vol.	Overall
Gear pumps	Spur or helical	1.8 to 2800	2 to 200	500 to 3500	0.27 to 1.59	80 to 95	70 to 80
	Screw	1.8 to 4950	20 to 170	1170 to 3400	—	60 to 75	60 to 70
	Rotary abutment	32 to 80	50 to 100	1000	4.25 to 6.75	—	—
Vane pumps	Single-stage	4.5 to 225	70	1000 to 1800	1.25 to 5	—	—
	Two-stage	4.5 to 200	140 to 200	1200	2.6 to 10	—	—
	High pressure pumps	4.5 to 225	80 to 140	1200 to 1800	0.75 to 1.83	—	—
	High performance pumps	90 to 675	140 to 175	2000 to 2700	0.25 to 0.50	85 to 95	75 to 85
	Inverted vane	1 to 250	70 to 140	1500	—	—	—
Piston pumps	In-line	2 to 340	70 to 600	60 to 1500	2.6 to 120	—	—
	Radial—Rotating cam	1.5 to 450	70 to 400	1200 to 4000	0.6 to 34	—	—
	Radial—Rotating cyl. block	1 to 4500	40 to 140	450 to 1500	2.4 to 24	—	—
	Axial—Bent axis	4.5 to 13500	70 to 170	200 to 3000	3 to 14	—	—
	Axial—Rotary swash plate	1 to 700	1400 to 670	1000 to 4000	0.21 to 6.1	—	—
	Axial—Rotating cylinder	15 to 450	70 to 300	1500 to 4000	0.20 to 2.4	90 to 99	85 to 95

PISTON PUMPS

Table 348 shows the classification of piston type pumps. The simplest of these is a reciprocating pump Fig. 238 driven by an eccentric and with one-way valves both at the inlet and outlet. These pumps are generally restricted to lubricating systems because of high flow fluctuations associated with them. Multipiston pumps each with individual valves, are frequently devised for high pressure applications. The swash plate design of axial piston pump with port plate valving is however a commonly used design, because of simplicity of valving and compactness. The principle of operation of this pump is shown in Fig. 239 . It consists of a cylinder block with several pistons placed in bores parallel to the axis of rotation. As the cylinder block rotates, the pistons held pressed against the inclined plate (known as swash plate) reciprocate in their respective bores, thereby drawing in oil from the inlet port and forcing it out through the outlet port. The inlet and outlet ports in the stationary port plate are shaped as shown in Fig. 239 . The widths of the lands in between the two kidney ports are so maintained as to avoid any interport leakage during the crossover of the flow passages on the cylinder block from low to high pressure side and vice versa.

The physical construction of an axial piston swash plate type pump is shown in Fig. 240 . The cylinder block/rotor with a splined bore is held on the drive shaft, supported at the swash plate end by a needle bush bearing. The pistons are provided with slipper pads to slide over the swash plate. The slipper pads are held in position by a retainer ring located centrally on the spherical portion of the drive shaft. The rotor is kept pressed against the stationary port plate by means of a spring load, so as to maintain zero gap between the two faces at starting. The spring loaded pistons render the unit self sucking. The rotor carries odd number of pistons, generally 7, 9 or 11, to provide low flow and torque ripples.

The performance of these pumps is mainly dependent on the frictional and leakage lossess at the portplate-rotor block and slipper pads–swash plate interfaces. The interport leakage, the leakage past the pistons, the leakage at the ball joints of slipper pads (piston shoes) are other sources of loss of performance. Hydrostatic bearings are provided for the valve plate and the slipper pads to achieve-high static and dynamic stiffness of bearings, reduced starting and stopping friction and hence low wear rate of sliding surfaces and minimum noise. The pressurised fluid from the unit is supplied to these bearing pockets through suitable inlet restrictors. Low

radial clearances between the pistons and bores and good surface
finish minimise the leakage past the pistons. Ball joints on pistons
are required to be with minimum clearance. The land between the
kidney ports in valve plate is generally wider by about 1° of arc
relative to the ports on the rotor so as to minimise the interport
leakages. This overlap condition may however give rise to
compression/expansion of trapped fluid during transition, resulting
in noise and loss of energy. Many of the present day designs provide
relief grooves at the ends of the kidney ports (Fig. 241). An
effective underlap between the ports helps in gradual changeover
between the ports. The ends of the notches restrict the leakage flow
between the ports.

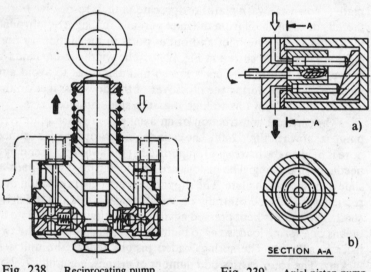

Fig. 238   Reciprocating pump   Fig. 239   Axial piston pump,
                                                    (swash plate type)

SECTION A-A

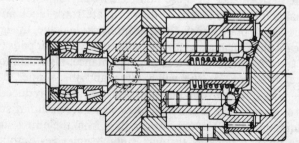

Fig. 240   Constructional details of swash plate type pump

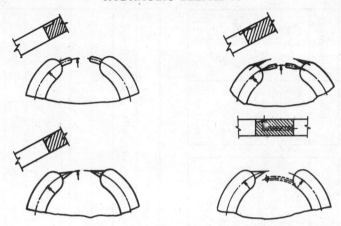

Fig. 241   Relief grooves on valve plates

The inclination of swash plate is limited to 20° since higher angles tend to provide a locking effect on the piston. The angle of inclination determines the stroke of the piston and hence the flow rate. The pump can therefore be rendered a variable delivery one by having an adjustable swash plate which is set to different angles. The materials of construction of the different components and the desired working clearances of the unit are given in Table 354 . Equations for calculation of flow, power and efficiency are given in Table 352 .

Applications

These pumps are compact and can work with high operating efficiencies (volumetric efficiency of 90% to 95%, overall efficiency 80% to 85%) at operating pressures ranging upto 400 kgf/cm^2. They are therefore commonly used in all high pressure applications such as presses. The variable delivery feature of the pumps renders their use as pressure compensated pumps for econominal power utilisation in different types of systems.

Power utilising elements

Power utilising elements or drives as generally referred to, convert hydraulic energy into mechanical energy. They may be motors, partial rotation/oscillating motors or cylinders.

MOTORS

All the types of pumps discussed earlier can principally function as motors. The speed and torque ranges of these motors are given in Table 355. Hydraulic motors are in principle constant

Materials, geometrical accuracies and working clearances; (Piston pumps)

Table 354a Materials of Construction

Component	Housing and end covers	Rotor barrel	Valve plate	Piston	Piston shoe	Swash plate
Materials used	Meehanite or Grey Cast Iron.	Forged steel, carburizing grade of alloy steel, hardened and ground and the bores are provided with bronze liner	i) Chill Cast phosphor bronze ii) Steel blank coated with a thin layer of phosphor bronze	Carburizing grade alloy steel hardened and ground. ii) Bronze piston or steel piston coated with bronze	Cold drawn bronze tensile strength 35 kgf/cm^2 hardness 130 to 160 HB	Carburizing grade of alloys steel, hardened to 60 to 65 HRc and the surface is lapped

Table 354c Working clearances

Item	Value in μm
Valve plate bearing	0 to 15
Piston shoe bearing	10 to 15
Cylinder bore and piston (radial clearance)	5 to 10 μm (depends on the diameter)

Table 354b Geometrical accuracies and surface finish

Component	Values in microns			
	Flatness	Parallelism	Circularity & Cylindricity	Surface finish R_a
Surface of the valve plate.	0.4	2	—	0.1 to 0.2
Bearing face of the piston shoe	0.4	—	—	-do-
Bearing face of the cylinder barrel	0.4	—	—	-do-
Bearing face of the swash plate	0.4	—	—	-do-
Cylinder bore	—	—	Depends on the bore size	0.2 to 0.4
Piston	—	—	Depends on the diameter	0.4

torque devices and have flat torque speed characteristics (Fig. 242) for a given specific output flow and a set system pressure. However, all types of motors do not display good low speed characteristics. It may be noted here that in applications such as feed drives in machine tools with copy control and numerical control, drive speeds of the order of even less than 1 *rpm* are needed. Some special contructions of motors such as the roll vane motors, multilobe piston and ball motors are used to meet these requirements.

Fig. 243 shows the construction and working principle of a roll vane motor. The rotor carries two fixed blades as well as the inlet and outlet connections. The pressurised fluid acts always on the trailing faces of both the vanes while the leading faces are under return line pressure. The four rolling vanes are geared mechanically to the rotor so as to provide the necessary synchronisation to effect the sealing between return and pressure chambers and also to accommodate the fixed vanes in the axial recesses cut on them. The continuous presence of a constant pressure behind the fixed vanes ensures ripple free torque and speed characteristics. The radial forces on the roll vane are hydraulically balanced and the vanes are mounted on needle bearings to reduce stiction. These motors are suitable for a wide speed range–from under 1 *rpm* to about 2500 *rpm*.

The principle of operation of multilobe piston motors is shown in Fig. 244 . The motor has an output shaft attached to a cylinder barrel which usually contains odd number of pistons. These pistons at the outer end bear against the multilobed cam of a stationary outer ring. The inner ends of the cylinder bores communicate to a non-rotating pintle valve which conveys the fluid to and from the external ports. The object of having a multilobed cam is to minimise the fluctuation of torque (Fig. 245) and radial force on the shaft. The multilobe feature results in compact units.

BALL MOTORS

Multilobed ball motors offer similar advantages as the multilobed piston motors. Both radial and axial types are commercially available. Fig. 246 shows the construction of a multilobe radial ball motor wherein the balls bearing against the cam track do also act as pistons.

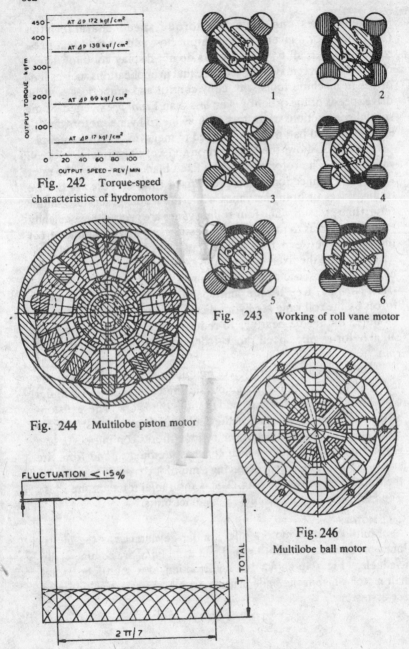

Fig. 242 Torque-speed
characteristics of hydromotors

Fig. 243 Working of roll vane motor

Fig. 244 Multilobe piston motor

Fig. 246
Multilobe ball motor

Fig. 245 Torque ripple in multilobe piston motor

Table 355 Typical performance of motors

Type of Motor	Brake horse power or output power in kW	Maximum speed rpm	Maximum pressure kgf/cm²	Torque range kgf.m	Weight/power Kg/kW	Running torque % of theoretical	Stalled torque %theoretical	% Vol. efficiency
Gear motor	1.1 to 37.5	1500 to 3000	70 to 170	0.42 to 22	0.61 to 6.1	60 to 88	50 to 85	70 to 90
Vane motor	2.5 to 25	1000 to 2200	50 to 100	1.4 to 18	1.8 to 4	70 to 90	75 to 90	85 to 95
Inverted vane motor	5 to 80	3000	140	4 to 84	—	—	—	—
Radial piston motor (i) Low speed	7.5 to 75	50 to 750	170	10 to 1400	1.5 to 3.5	85 to 95	85 to 95	90 to 98
(ii) Medium speed	0.75 to 100	50 to 1600	20 to 100	1.5 to 1400	6.0 to 12	85 to 90	80 to 90	90 to 98
Axial piston motor	0.75 to 180	240 to 4000	70 to 275	0.3 to 16000	0.04 to 3 and 6 to 7.2	85 to 95	90 to 95	95 to 99

Applications and advantages

Ease of regulation of speed, reversibility and constant torque-speed characteristics lend application of motors to a variety of functions in machine tools such as feed drives of slides in contouring and die-sinking machines (copying, numerically controlled and spark erosion machines), indexing of tool turrets, and plasticiser screw drives in pressure die casting machines.

Motor drives are convenient for ready retrofitting of copying and numerically controlled systems onto the manually operated machines. The hydromotor drives offer the added advantage of high drive stiffness because of low trapped volume of oil and are applicable even for long strokes. Hydraulic spindle drives are also practised.

OSCILLATING MOTORS

These are so termed because they transmit movement over an arc only. Depending upon the construction these are classified as vane type and piston type.

Fig. 247 shows a vane type construction. Fluid pressure on one side of the vane produces an unbalanced force on the output shaft and hence the torque. The partitioning vane provides sealing between the two chambers. These units have been built for arcs upto 270° and torque capacities upto 2500 *kgf.m.* The piston type oscillating motors derive rotary movement by rack and pinion arrangement, with the rack cut on the rod or piston of a cylinder. Oscillating motors find applications such as indexing, clamping, feeding, pick-feeding, transfering, etc.

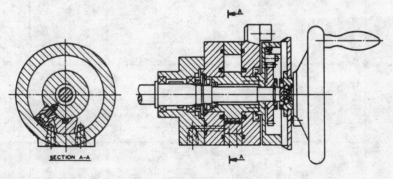

SECTION A-A

Fig. 247   Oscillating motor vane type

CYLINDERS

Cylinders provide a linear drive and are the most commonly used of hydraulic drives. A cylinder essentially consists of a piston located in a tubular housing and a piston rod passing through one of the end covers. The ports provided in the end covers permit entry and return of hydraulic oil. Classification of cylinders based on material, function, body and mounting styles is shown in Tables 356 to 358.

The detailed constructional features of the commonly used piston type cylinder is shown in Fig. 248 . Standard cylinders are generally made of cold drawn seamless steel tubes. The tubes are bored/ground and finish honed to the required size. A surface finish between 0.2 to 0.4 μm is generally desired, especially when using *rubber* seals for the piston. The pistons are of high grade cast iron, meehanite or bronze. Pistons for low pressure applications may depend upon the fineness of clearance between the piston and the bore for sealing. However, generally, sealing elements such as O rings (dynamic), piston rings, cupseals, etc., are used for the pistons. The allowable tolerances on the dimensions of the finished bore and the piston using seals can be between $H7$ to $H11$ and $f7$ to $e8$ respectively depending upon the pressure rating and the type of seal used. Some configurations are shown in Fig. 249.

The piston rods are made of medium carbon steel or case hardening steels depending upon the application and service conditions. The rods are hardened and ground and occassionally chrome plated and polished. The rods should be strong enough to prevent buckling. Tubular or hollow rods are preferred to solid ones in case of long strokes. The free end of the rod is usually threaded to facilitate connecting the load. The rod is supported and guided by a bearing in the end cover. The bearing bush is either of bronze or cast iron depending upon the load conditions. O-rings or multilip seals are generally provided for rod sealing. The multilip packings,when used, are stacked so as not to cause heavy preloads on the rod. It must be noted that the material of the seals for the piston and rod are to be compatible with the fluid medium.

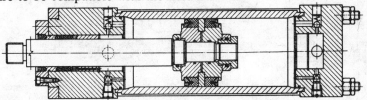

Fig. 248   Construction of a cylinder

The end covers are made of steel or high grade cast iron. They are generally either screwed to the tube ends or held together by tie rods. Supply ports are provided on the end covers, locating them at the top most points with respect to the cylinder mounting to enable automatic scavenging of trapped air. Otherwise separate bleeding points are to be provided.

Cushioning of piston movement at the two ends of its stroke is generally built in to avoid damaging of components due to impact. Cushioning is effected by reducing the velocity of the piston towards the end of the stroke. The flow passage of the oil is blocked by a suitable collar on the piston and the trapped oil is then led through a restrictor (adjustable or fixed). Different methods of providing end cushioning are shown in Fig. 250 .

Applications

Application of cylinders in machine tools are numerous; table drives of grinding machines, ram drives for broaching machines, slides for tracer controlled operations and clamping devices being the common examples. Simplicity and ease of manufacture, ease of speed control over a wide range and dispensing with lead screws have been the reasons of the popularity of cylinder drives. They can be conveniently used as feed drives for position control systems with strokes upto a metre. Longer strokes generally suffer from inadequate stiffness due to the larger trapped volume of oil.

Table 359 gives details on calculation of force, velocity, power and frictional losses due to the seals.

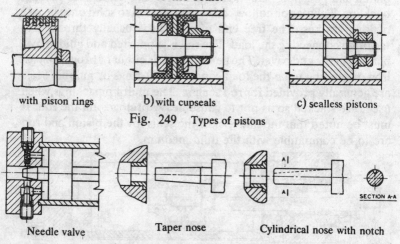

with piston rings   b) with cupseals   c) sealless pistons

Fig. 249   Types of pistons

Needle valve   Taper nose   Cylindrical nose with notch

SECTION A-A

Fig. 250   End cushioning in cylinders

Table 356 Classification based on type of piston

Configuration	Description
	Single acting
	Spring return cylinder
	Double acting cylinder
	Double end rod cylinder
	Ram type cylinder
	Telescopic cylinder
	Rolling diaphragm cylinder

Table 357 Classification based on body style

Configuration	Description
	Tie-rod construction
	Threaded head
	One piece cylinder

Table 358 Classification based on mounting style

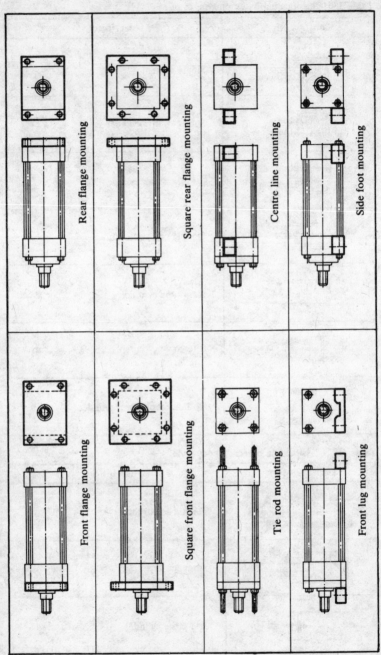

Rear flange mounting

Square rear flange mounting

Centre line mounting

Side foot mounting

Front flange mounting

Square front flange mounting

Tie rod mounting

Front lug mounting

(Concld.)

Table 358 Classification based on mounting style

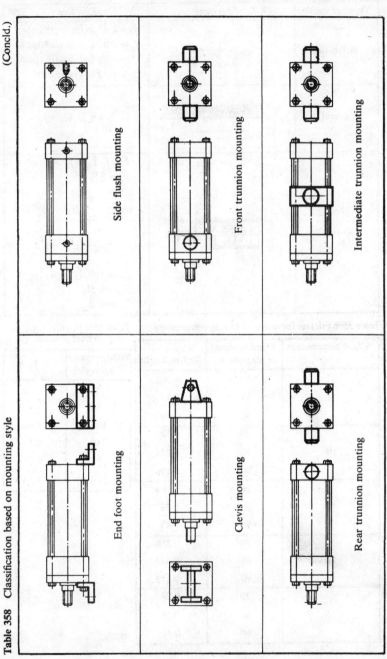

Side flush mounting

Front trunnion mounting

Intermediate trunnion mounting

End foot mounting

Clevis mounting

Rear trunnion mounting

Table 359a Friction forces with different types of seals

Type of seal	Configuration	Equation
Piston ring		$F_f = \pi D_p a(i p_k + p) f_r$ $f_r = 0{\cdot}07$ for high speeds $= 0{\cdot}15$ for small speeds $p_k = (0{\cdot}8 \text{ to } 1) kgf/cm^2$
Lip seal		$F_f = \pi D_p C_1\, p f_1$
O-ring		$F_f = f_c L_p + f_h A_p$ $F_b = 3 F_f$ $A_p = \dfrac{\pi}{4}(D_1{}^2 - D_2{}^2)$ $= \dfrac{\pi}{4}(d_1{}^2 - d_2{}^2)$ $= \pi/4(d_1{}^2 - d_2{}^2)$ $L_p = \pi D_1$ $= \pi d_2$

Table 359b Friction force due to O ring compression

% Seal compression	Shore hardness degrees A	f_c Kgf/cm length
5	70	0.063
	80	0.12
	90	0.142
10	70	0.12
	80	0.224
	90	0.296
15	70	0.185
	80	0.33
	90	0.43
20	70	0.256
	80	0.445
	90	0.575
25	70	0.28
	80	0.55
	90	0.715

Table 359c Friction force due to fluid pressure

Fluid pressure kgf/cm^2	f_h
35	1.62
70	2.5
105	3.38
140	4.1
175	4.7
210	5.15

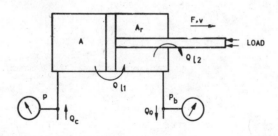

Table 359 d

Force: $F_t = p A - P_b A_r$ kgf

              (A_r - annular piston area)

         $F_a = p A - \Sigma F$ kgf

         $\Sigma F = F_f + p_b A_r$ kgf

Velocity: $V_t = Q_c / 6A$ m/sec

         $V_a = (Q_c - Q_{l1}) / 6A$ m/sec

         (Output flow from cylinder $Q_o = v_a A_r + Q_{l1} - Q_{l2}$)

              Q_{l1} = Interport leakage in lit/min

              Q_{l2} = External leakage in lit/min

     $P_{input} = p\, Q_c / 612$ kW

     $P_{output} = F_a\, v_a / 102$ kW

Control of pumps and motors

Control means the variation of displacement of the unit (pump or motor) by change of eccentricity or inclination of swash plate, etc., depending upon the construction. There are several methods of effecting the control, based on either an external control signal or by the system pressure.

Control by external signal is generally suitable for varying the discharge/speed of the unit. The simplest form of this type is the handwheel actuator to vary the displacement manually and is suitable for slow operations. A simple extension of this to automatic operation is provided by using an electric motor to drive the handwheel through a reduction gear. This method is not suitable for accurate control and rapid changes of displacement. To meet these stringent requirements, a force amplifier normally consisting of a spool valve and a follow-up piston is used. Fig. 251 shows a servo assisted manual control.

The control based on the system pressure is adopted in case of pumps to obtain variable flow-pressure (torque) characteristics to provide pressure/power and flow compensation. Fig. 252 shows the working of a pressure compensated pump together with input torque and flow characteristics. The delivery pressure acts on the spring loaded piston which is linked to the displacement mechanism of the pump. When the pressure exceeds the level set by the spring, the delivery from the pump reduces as shown in the figure. At any pressure setting the pump output matches the load flow demand, thereby resulting in economy of power.

The power compensating (torque limiting) pumps are similar to the pressure compensated type, excepting that the control piston is loaded with several springs of different lengths and stiffness in parallel. The characteristic curves obtainable from such a system are shown in Fig. 253. These pumps, in addition to economising the motor size and power requirements, enable gradual pressurisation of chambers.

In the case of the differential pressure control, the pump displacement is regulated by sensing the pressure drop across an orifice placed in series with the delivery line from the pump. These controls are further classified as the constant flow and load sensing type. The former employs a constant area orifice and the latter a variable one.

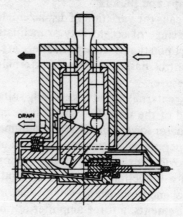

Fig. 251 Manual servocontrol of swash plate type pump/motor

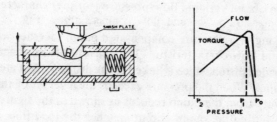

Fig. 252 Pressure compensation

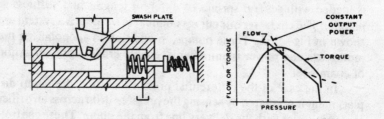

Fig. 253 Power compensation

Control valves

Valves are devices to control pumps / actuators in hydraulic systems. They are basically assemblies of one or more flow restricting elements which fall into three main classes–sliding (spool and plate type), seating (poppet, ball and flapper) and flow dividing elements. The spool valves find application in almost all types of control valves. The poppet and ball types are popular for application in check valves and the first stage of pressure control valves. The flapper valves are successfully employed in two stage electrohydraylic servovalves. Control valves are broadly classified according to their function as pressure, flow and direction control valves.

PRESSURE CONTROL VALVES

Relief valve

Relief valves protect the other elements in the system from excessive pressure by diverting the excess fluid to the tank when the system pressure tends to exceed the set level. The direct acting type of relief valves have a ball, poppet or a sliding spool working against a spring. The preload on the spring determines the system pressure and can be adjusted. The pressure at which a relief valve cracks open is termed as the *cracking pressure* and the pressure when the valve is fully open to by-pass the full rated flow is *full flow pressure*. The difference between the two– *the pressure differential* or *the pressure override* should necessarily be small for close pressure control. The pressure override is due to the extra compression of the spring at higher valve openings to accommodate increased flow rates and is therefore dependent on the stiffness of the spring used. Long, weak springs are therefore preferred for the application.

Ball and poppet valves suffer from high pressure override and tendency to chatter. Spool type relief valves of the direct acting type (Fig. 254) provide smooth and stable operation with superior pressure-flow characteristics. Suitable restrictor in the pressure sensing line helps achieving the stability of operation. For low pressure override at higher pressures and flow rates, the two stage balanced piston (compound) relief valves are used(Fig. 255). This valve consists of two stages–a small poppet type relief valve for determining the pressure and the main spool valve for handling the flow. At a pressure lower than the set value, the main piston is held on to its seat by a light spring, with the system pressure acting on

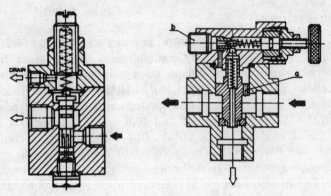

Fig. 254 Relief valve, direct acting type Fig. 255 Compound relief valve

either side of the piston because of the capillary *a*. When the pressure reaches the set value, the poppet is forced off its seat venting the oil in the upper chamber of the main valve. The restricted flow through the orifice into the upper chamber results in a pressure differential on the main piston, thereby lifting it off to bypass the flow to reservoir. The light spring acting on the piston restores it to original position when the system pressure falls below the set value. The light return spring results in low pressure override for dumping the flow. The vent connection *b* provided in the valve allows the valve to be manipulated remotely by an auxiliary valve, operated manually or otherwise.

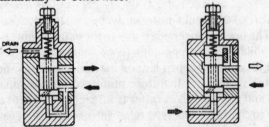

Fig. 256 Sequence valve   Fig. 257 Unloading valve

Sequence valves

These valves are used to ensure action in a particular order of priority in the system. They maintain a preset minimum pressure level in the primary line (upstream of the valve) prior to effecting any function in the secondary line (downstream of the valve). These are basically relief valves with minor modifications and may be either of the direct acting or compound type. In the direct acting sequence valve (Fig. 256) fluid flows freely through the primary passage to operate the first phase until the pressure setting of the valve is reached. On reaching the set pressure, the spool is lifted to divert flow to the secondary circuit. Unlike in a relief valve, the sequence valve necessitates an external drain connection from the bias spring chamber. Sequence valves are sometimes built with check valves to permit flow from secondary to primary circuit.

Unloading valves

These valves are used to off load the pump flow back to tank at a low pressure when not required by the system. Unloading helps in reducing considerably the generation of heat due to unused energy in the system. Fig. 257 shows the construction of an unloading valve. The primary port pressure is independent of the spring setting in the valve. On receiving the pilot signal, the spool lifts against the spring to unload the pump flow into the reservoir.

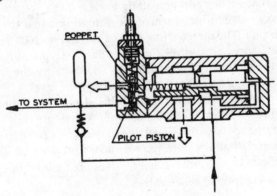

Fig. 258   Unloading relief valve

A compound relief with pilot control of the vent port can also be used as an unloading valve. The balance piston type relief valve together with a differential area mechanism at the vent port of the valve forms an unloading relief valve or a differential type of unloading valve which can be used in accumulator circuits to provide automatic cut-out (unloading) and cut-in (recharging) of pump flow. Fig. 258 illustrates the application. When the system pressure rises to the value set by the poppet valve, the poppet lifts to bypass the fluid thereby permitting the auxiliary (pilot) piston to push the poppet further up to dump the pump flow at a very low pressure. The reseating of the poppet is then dependent on the area of the pilot piston and the pressure behind it. The pilot piston is generally 15% larger by area in relation to the poppet. Accordingly, the poppet can be reseated at a proportionally lower pressure level behind the piston as compared to the relief setting by the poppet valve. The check valve shown in the figure holds the pressure in the accumulator, when the pump is unloaded.

Counter balance valves

These valves provide sufficient back pressure in a hydraulic system to balance a weight or load to prevent its descent due to gravity. A direct acting type of valve consisting of a relief valve and a free flow check valve is shown in Fig. 259 . The check valve permits flow in the reverse direction.

Pressure reducing valves

These valves are used to maintain a secondary pressure which is lower than the relief valve setting, in any part of a hydraulic system. Unlike relief, sequence and unloading valves, these are normally open two way valves which receive the actuating signal from the downstream side. The direct acting type of valve (Fig. 260) uses a spring loaded spool to control the secondary line pressure. The pressure in the secondary line acts on the spool against the spring. With the increase of secondary pressure above the level set by the spring, the spool lifts to control the metering at a so as to reduce the pressure to the set level. When the secondary pressure falls below the set value, the spool moves down due to the spring force to allow more flow through and restore the secondary pressure to the set level.Pilot operated pressure reducing valves are built on principles similar to the compound relief valves.

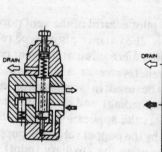

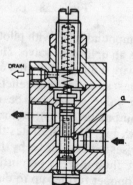

Fig. 259 **Counter balance valve**

Fig. 260 **Pressure reducing valve, direct acting type**

FLOW CONTROL VALVES

The control of load speed in most hydraulic systems is done by flow control valves. Having sized the pump and actuator for maximum load speed, the lower speeds of load are derived by throttling, bypassing the excess flow through the relief valve. Variable displacement pumps could perform the same task, but are not an economic proposition in many applications. Throttling is hence commonly resorted to.

The flow through an orifice is given by the equation:

$$Q_r = K_1 A_o \sqrt{\Delta p}$$
$$K_1 = C_d \sqrt{2/\rho}$$
$$= 955 \; cm^2(kgf)-\tfrac{1}{2} \; sec^{-1}$$

$C_d = (0.61$ for sharp edged orifice$)$

$\rho = 0.834 \times 10^{-6} \; kgf \, sec^2 \, cm^{-4}$ for petroleum based fluids

For a given pressure differential across the orifice, the flow rate can be varied by changing the flow cross section A. A few of the basic designs for varying the orifice area are shown in Fig. 261 Needle valves are the simplest of these valves finding applications for fine metering such as in cushioning of cylinders, gauge shut off valves and for very fine feed control. Cross section of a rotary type spool valve is shown in Fig. 262. The orifice cut along the length of the sleeve is gradually opened or cut out by the rotation of the spool. The helical groove on the spool and the shape of the orifice are selected so as to provide a gradual regulation of flow. Also the length of the orifice is kept as small as possible to minimise the effect of temperature (oil viscosity) on the set flow rates.

A major disadvantage of these simple devices is that any variation in load pressure affects the flow through the valve because of the change of pressure difference across the orifice. Pressure compensated flow control valves (Fig. 263) provide constant flow for any particular valve setting by maintaining a constant pressure drop across the throttle valve. A common approach of achieving this is by having a pressure reducing valve in series with the throttle valve. The pressures at the inlet and outlet of the orifice are fed on to either sides of the reducing valve spool so as to effect a constant pressure differential, equal to the spring value, across the flow orifice.

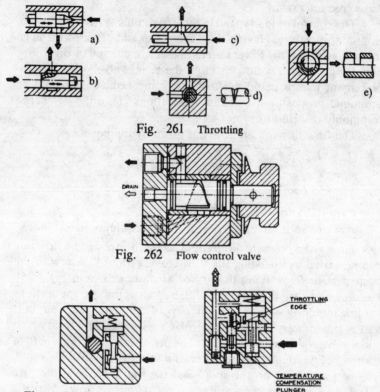

Fig. 261   Throttling

Fig. 262   Flow control valve

Fig. 263   Flow control valve,   Fig. 264   Flow control valve, pressure
pressure compensated type   and temperature compensated type

Changes in temperature and the consequent changes in the viscosity of the oil do affect the flow rates from the orifice type valves. To overcome this problem, temperature compensated valves are used. A flow control valve with both pressure and temperature compensation is shown in Fig. 264 . The length of the temperature compensating element made of an aluminium alloy varies with the oil temperature and the throttle is accordingly adjusted automatically to achieve a constant flow rate. The compensation is brought about by the dissimilar thermal expansion rates of the compensating plunger and the matching components. The built-in check valve allows for free flow in the reverse direction.

The flow controls mentioned above are known to offer regulation of flow from about 0.05 *l/min.* to about 100 *l/min.* Very low flow rate using small orifice is impractical because of *plugging* of orifice due to contamination of fluids. However, with a fine filter at the inlet to the valve, control of flow of the order of 5 to 20 $cm^3/min.$ are claimed to have been achieved.

Deceleration valves

These valves are devised for external deceleration of slides moving at high speeds with heavy loads to achieve shockless braking and reversal at midstrokes of cylinders. A deceleration valve (Fig. 265) is essentially a linear type spool valve operated by a cam on the sliding member. As the cam presses the roller, the tapered land on the spool moves along the orifice to cut the flow to the drive. The rate of cut off is dependent both on the shape of the orifice and the cam profile. When the roller on the spool is off the cam, the valve permits free flow.

Flow dividing valves

These are incorporated in circuits which require approximately equal flows in two of its branches. A floating spool controlling two variable orifices is the heart of the flow divider (Fig. 266). Any increase in load pressure in one branch of the circuit causes the spool to shift so as to reduce the restriction to the particular branch while at the same time increasing the resistance in the other branch The spool takes up such a position as to divide the flow equally

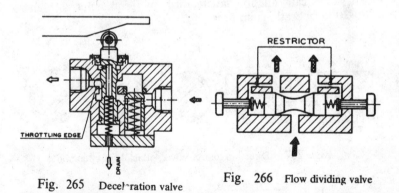

Fig. 265   Deceleration valve

Fig. 266   Flow dividing valve

DIRECTION CONTROL VALVES

These valves are deployed to steer the flow to selected flow paths in any part of a hydraulic circuit. The spool type valves both of the linear as well as the rotary movement are devised for the purpose. Principle of operation of these valves is illustrated in Fig. 267 . Rotary type direction control valves (Fig. 268) are commonly seen as applied to machine tool table reversals such as in cylindrical and surface grinding machines. Owing to the feasibility of application of different modes of control for their operation, the linear spool valves are the most commonly used type. These valves based on their functions and mode of control are classified as in Table 360 & 361.

The physical construction of a solenoid valve is shown in Fig. 269 These valves are operated either on *AC* or *DC*. The *AC* operated valves have a drawback in that they tend to burn due to heavy currents drawn, in the event of improper closure of the plunger (as in the case of spool sticking). They are however popular from the point of ease of deriving the control voltage. At high pressures and flow rates,large operating forces are required making solenoid control untidy. Two stage electrohydraulic valves are therefore devised, whereby low pilot pressure supply from the first stage solenoid valve is used to shift the spool of the main valve (Fig. 270). The solenoid pilot operated valve shown in Fig. 271 is built with restrictions in the pilot lines to regulate the speed of spool shift, controlling either braking or reversing of the load.

Minimum interport leakage, low pressure drop due to flow through the valve and fast response of operation of solenoid valves are the important requirements of direction control valves. Solenoid valves can function satisfactorily at frequencies as high as 1500-2000 operations an hour.

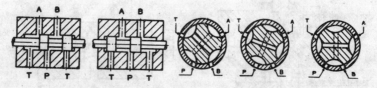

Fig. 267   Direction control valves, principle of operation

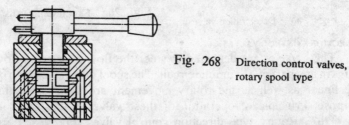

Fig. 268   Direction control valves, rotary spool type

Table 360   Classification of direction control valves

Path of flow					
	Two functional ports		Three functional ports		Four functional ports
Two way		Three way		Four way	

Position	
Two position	Three position

Multiway, multiposition

Control					
	Manual	General		Electrical	Solenoid actuated
		Pushbotton actuated			
		Lever actuated			Motor actuated
		Foot pedal actuated			
	Mechanical	Plunger actuated		Pilot	Pilot pressure actuated
		Roller trip actuated		Combined	Two stage valve; solenoid shifts first stage spool for pilot pressure actuation of main spool.
		Roller actuated			

Spring		
Spring offset	Spring centred	No spring

Table 361   Valve classification

	Classification	Description
Spool	Open centre	These are five of the more common spool types; refers to the pattern of flow permitted when spool is in center position (three position valves) or during cross-over (two position valves) (Table 360) The open center valves are suited for single actuator application so that the pump can be off-loaded under very low pressure during the idle periods. Also, the actuator can be freely moved as per table movement by hand wheel. The closed center valve holds the pressure during *null* so as not to disturb the other functions drawing the flow from the same source. The tandem arrangement has the advantage of both-holding the pump pressure as well the free actuator movement at *null*. The other types of valves can be built to meet specific system requirements.
	Closed center	
	Tandem center	
	Partially closed center	
	Semi-open center	

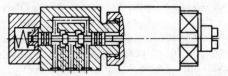

Fig. 269   4-way, 2 position direction control valve, solenoid operated

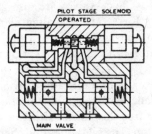

Fig. 270   4-way, 3 position direction control valve, solenoid operated

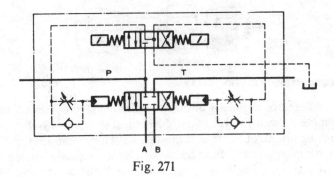

Fig. 271

Check valves

Check valves (non-return valves or one-way valves) are used to control the direction of flow in a circuit to the extent that they permit flow in only one direction. Provision of reversed free flow feature in flow control valves, counterbalance valves and pressure reducing valves and interlock between different pumps of a multipump system are some of the examples of the wide range of application of these valves. These valves consist of a poppet/ball held on to its seat by a spring. A poppet type in-line check valve is shown in Fig. 272 . Check valves are also available with O-ring sealing. The spring force on the sealing element determines the *cracking* pressure or the back pressure prior to allowing the flow in the intended direction. Valves without spring are suitable for vertical mounting only.

Two commonly used modifications from the standard design are the restriction check valve and the pilot operated check valve. The former permits free flow in one direction and restricted flow in the opposite direction. The pilot operated check valve operates as a standard valve but can be controlled by a pilot pressure to permit free flow in the reverse direction.

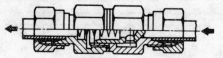

Fig. 272 Check valve, inline type

ELECTROHYDRAULIC SERVOVALVES

These are the heart of electrohydraulic servosystems which combine the control flexibility of electrical systems with the power handling ability of the hydraulic systems. These perform functions of both direction and flow control valves in response to electrical input signals. The most commonly used valve is the two stage flapper type flow control servovalve (Fig. 273a). Current input to torque motor coils causes either clockwise or anticlockwise torque on the armature which displaces the flapper between the nozzles. This movement away from the equilibrium position causes unequal leakage flows past the flapper valves giving rise to a differential pressure on the spool. The hydraulic amplifier multiplies the force output of the torque motor to meet the flow, friction and acceleration forces on the spool. The spool moves until the feedback torque provided by the spring wire counteracts the electromagnetic torque. The

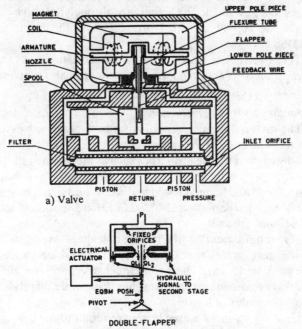

a) Valve

b) Hydraulic amplifier

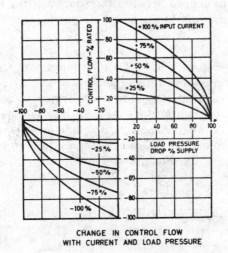

CHANGE IN CONTROL FLOW
WITH CURRENT AND LOAD PRESSURE

c) Output flow characteristics

Fig. 273   Electrohydraulic servovalve

spool is held in this particular position until the electrical signal changes to a new level. The actual flow through the valve is dependent on the supply and load pressures and the electrical input (Fig. 273b).

These valves are rated by their flow capacities at a valve pressure drop of 70 *kgf/cm²* and are available upto about 60*l/min*. and for working pressures upto 210 *kgf/cm²*.

Valve materials and construction

The material requirement of different types of valves is similar in nature. The body of the valve is generally of non porous cast iron of good wear and tensile properties such as meehanite or high grade cast iron of grade 25 as per IS 210. The valve bores are bored/ground and honed/lapped to the required size. The spools are made of case hardening steel 15*Ni*2*Cr*1*Mo*15 hardened to around 60 *HRC* and ground.

The requirement of low interport leakage over a reasonably long life necessitates maintenance of diametral clearances in the range of 5 to 10 μm. The geometrical tolerances on circularity, cylindricity and concentricity are therefore to be maintained very fine, of the order of 2 μm.

Grooves on spool and in bores are required to be square as to provide sharp corners (unless otherwise intended) as to prevent dirt particles getting lodged between the spool and the bore. Sharp edges are also essential functionally for obtaining the desired orifice control.

The lands subjected to pressure are generally provided with grooves on the spool (Fig. 274) to minimise the spool shifting force. The form errors viz, taper and ovality of the spool and the bore cause unbalanced lateral forces on the spool, following the non uniform pressure gradient along the landwidth. Grooves cut on the sealing land help in overcoming the hydraulic lock by equalising the pressure around the spool. Generally, three equally spaced grooves per land are considered to give optimum results.

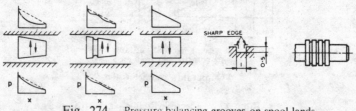

Fig. 274   Pressure balancing grooves on spool lands

Accessories

ACCUMULATORS

Accumulators are employed in hydraulic circuits to store the excess flow (hydraulic power) from a pump to meet any of the following requirements:

- to meet the demand for increased flow rates during a portion of the working cycle, thus allowing a smaller pump to be used for the system,
- to provide an emergency source of energy to render a fail-safe system in the event of failure of the pump,
- to maintain the system pressure within tolerable limits in locked circuits by compensating for leakage flow or increase in pressure due to thermal expansion and
- to absorb pump flow ripples and pressure surges in a system.

Accumulators can be of weight, spring or hydropneumatic type. The hydropneumatic type of accumulators using precharged nitrogen to act as a spring is commonly used. Nitrogen is separated from the oil side by a piston, bag or diaphragm. Hydraulic fluid pressure, when admitted into the oil side of the accumulator, moves the separator compressing the nitrogen till a balance of pressure is reached on the two sides. Nitrogen is used in preference to air since the latter has the tendency to cause *dieselling* of air-oil vapour. The constructional details of piston and bag type accumulators are shown in Fig. 275. The charging unit shown alongside enables regulated precharging. Accumulators are specified by the maximum swept volume of oil and the maximum operating pressure. The diaphragm and bag type accumulators are generally supplied with a small precharge of gas to protect the separator element.

Presuming the availability of sufficient flow from the pump, the flow into and from the accumulator is governed by the law $pv^n = $ constant, the index n varying between 1 to 1.3 depending upon the construction and operation. The accumulator is chosen based on the requirements of maximum flow and operating pressure. The precharge pressure of the gas is determined based on the flow and operating pressure range, aided by the p-v diagram supplied by the manufacturer of the accumulator. Operation at higher precharge pressures provides for larger flow for a given operating pressure range (Fig. 276) and is therefore advantageous particularly in accumulator applications for leakage compensation.

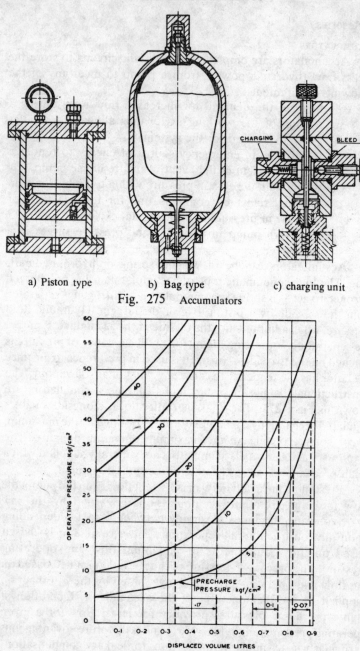

a) Piston type   b) Bag type   c) charging unit

Fig. 275   Accumulators

Fig. 276   Flow characteristics of accumulators

PRESSURE SWITCHES

These are auxiliary control elements required in a hydraulic system to sense the pressure level in any branch of the system to electrically trigger any other function in the system. It consists essentially of a microswitch whose contacts are opened or closed by a pressure sensing mechanism. Based on the pressure sensing mechanism the switches are classified as–diaphragm,plunger and bellows type. A diaphragm type construction is shown in Fig. 277 The deflection of the diaphragm under pressue is utilised to operate the microswitch. The initial gap between the diaphragm and the plunger of the switch is adjusted for the desired pressure setting.

Pressure switches can be built for providing signals both at low and high limits of system pressure and also to operate on a pressure differential.

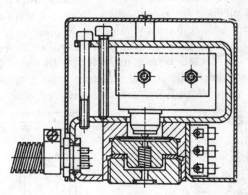

Fig. 277

FILTRATION AND FILTERS

Contamination of oil in the hydraulic system is the single biggest source of failure of hydraulic components and systems. The close fits in most hydraulic components make it essential that the level of contamination of fluid in the system be adequately low. Contamination of system fluid is traced to sources both internal and external. The internal sources include–the dirt particles such as the lapping compounds, burrs, scales from hardened components entering into the system due to improper cleaning and seal chippings, rags from *cleaning* cloth and the wear particles generated during the use of the system. The external sources are–the dirt contained in the *fresh* oil being filled in and the dust, chips, etc., blown in from the surrounding area apart from the unclean handling

by the personnel. The effects of contamination are:
- – plugging of orifices,
- – scoring of polished surfaces,
- – damage to seals and
- – silting and wear of critical edges of valves leading to loss of control in servo systems.

To render a system reliable and long lasting, adequate degree of filtration of system fluid is essential.

Filters

Filtering media: These offer resistance to flow whereby the contaminants in the fluid are retained back from flowing through with the fluid because of the fineness of flow passages and the tortuous flow paths. A wide range of materials such as wires, wiremesh, paper, felt, metal discs and sintered powder materials are used for the media. Wire meshes made from stainless steel, copper and phosphor bronze form the simplest of filtering media. Wire cloth, paper and felt are commonly used in pleated form for filter elements. The elements are provided with suitable backup to withstand the differential pressure across them due to fluid flow. Wire cloth filters are manufactured with a filtration

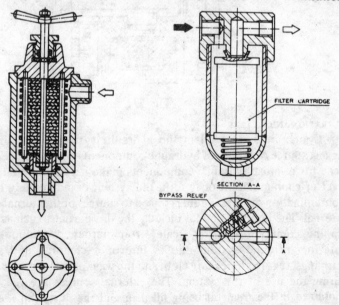

Fig. 278 Disc type filter   Fig. 279 Construction of a filter

capacity of 15 *um* absolute and for pressure differential as high as 300 *kgf/cm²*. They can be cleaned and reused and are suitable for a very wide temperature range. Backwash by reversed flow, manual cleaning and ultrasonic cleaning are the methods used for cleaning these elements. Paper elements are available upto 3´ μm absolute filtration capacity and are suitable for a temperature range of 0 to 125°C. Generally paper filter elements are of throwaway type.

Metal and paper discs are used to provide edge type filtration. Discs are stacked one above the other with discrete gaps in between (Fig. 278) for the fluid flow. The dirt retained at the edges can be cleaned periodically by a suitable brushing device or by reversed flow.

Filters: Filters (Fig. 279) are integral units housing the filter elements to enable routing of fluid flow through the filtering medium in the intended direction. Filter housings are designed to withstand the system pressure they are subjected to. The filter cartridges are generally held in position by a spring, with suitable sealing rings on the end covers of the element to prevent any bypass of unfiltered oil to the downstream.

A fully clogged element may mean either total blockage of flow into the system (in case of elements made to full system pressure differential) or bursting/collapsing of the element and eventual releasing of the entire bulk of dirt retained earlier, leading to total damage of the system. Often, the filter housings include a valve to bypass the unfiltered fluid to the downstream at a predetermined pressure differential. The bypass valve is set well below the burst collapse rating of the filter elements. Indicators and switches sensing the pressure differential serve better since they indicate as to when exactly the filter cartridge needs to be cleaned/replaced, rather than resorting to *adhoc* practices on periodic checks.

SEALS AND PACKINGS

Seals are devices for closing gaps to prevent leakage or make pressure tight joints and also to prevent the entry of air and dirt from outside into the system. A wide variety of seals of different shapes and materials are used. The material of the seal must be compatible with the fluid medium. Table 364 indicates the classification of seals. They are broadly classified as static and dynamic seals.

Static seals

These are employed to provide a seal between two relatively static or nearly static components. Compression gaskets sandwiched between two surfaces provide sealing at the interface. Nonmetallic gaskets of cork and paper can be used for pressures in the range of 1 to 2 kgf/cm^2. Metallic gaskets in the form of flat sheets, corrugations, round cross-sections, etc., are used for pressures as high as 1000 kgf/cm^2. Gaskets and beadings are commonly used for applications such as door flanges and covers of reservoirs. Elastomeric O-rings are commonly used for static sealing applications such as–flanges, flange fittings, end covers of valves and cylinders and in manifold mountings of valves. Different methods of sealing using O-rings are shown in Fig. 280 . These rings are generally circular and mounted with a certain preload. Static O-rings are known to have been used for sealing pressure upto 2000 kgf/cm^2. The static preload of about 10% to 15% of O-ring section provides the initial seal with the fluid pressure providing further sealing action. The groove volume provided should be about 15% to 20% more than the volume of the O-ring so as to permit free flow of the material of O-ring at right angles to the squeeze.

Dynamic seals

Dynamic seals are required to seal the annular gap between two components with relative movement. The movement may be rotary or reciprocating.

Reciprocating seals

The dynamic seals for sealing between two relatively reciprocating parts can be O-rings, lipped seals (single or multilipped) and piston rings. The O-rings and lipped seals provide intimate contact of the seal with the surface it is rubbing against, filling even the undulations due to surface roughness and ovality of the bore to provide thorough sealing. Elastomeric O-rings are very popular for dynamic applications because of low cost and small size. Generally a hardness value of *HS* 70 is used for the material of these rings. These rings as in the case of static applications are mounted with a preload to provide the initial sealing. The use of these rings for working pressures beyond 100 kgf/cm^2 is not recommended since they tend to get extruded through the clearance. For use at higher pressures, O-rings are to be provided with antiextrusion devices by way of backup rings-generally made of PTFE (Fig. 282). Also, O-rings in short stroke applications are found to wear rapidly. Reduced friction and wear can be effected by use of coaxial rings of PTFE (Fig. 281). Rings of other cross sections such as delta, square, multilobed/quad-rings (Fig. 283) are used in place of O-rings, for minimum twisting and squeezing through.

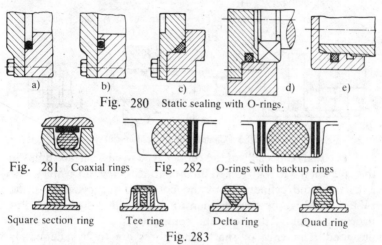

Fig. 280 Static sealing with O-rings.

Fig. 281 Coaxial rings   Fig. 282 O-rings with backup rings

Square section ring   Tee ring   Delta ring   Quad ring

Fig. 283

Lip types are by far the most commonly used seals for sealing between two relatively sliding parts. The seals are moulded from synthetic rubber (with or without fabric reinforcement) or leather (plain or impregnated). Y packings (also referred to as V–rings), the most popular shape of the lip seals, are used in stacks with suitable support rings (Tab. 362). The lip seals are assembled with light preload and proper arrangements for pressuring and lubricating the seals. The general survey on seals of sealing reciprocating parts and their guidelines for application are presented in Tab. 386 to Tab. 401.

Use of piston rings made of cast iron or PTFE is popular for piston seals in providing low load friction and long life and in applications permitting a certain amount of interport leakage. The PTFE rings are either solid or split. The solid rings are mounted with an interference of about 0.2% on diameter for small pistons and about 0.1% for large pistons. The ends of the split rings of cast iron or PTFE are cut as shown in Fig. 284 . One ring for every 70 kgf/cm^2, with a minimum of two rings is recommended. The ends of the adjacent rings are staggered to minimise leakage. These rings are mounted in grooves with minimum play, of about 0.01 to 0.02 mm. The rings with butt ends are assembled to have a gap of not more than 0.05 mm between the two ends.

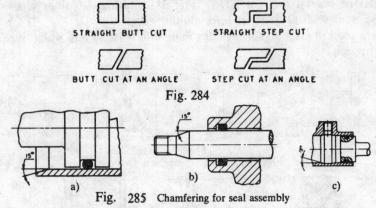

STRAIGHT BUTT CUT   STRAIGHT STEP CUT

BUTT CUT AT AN ANGLE   STEP CUT AT AN ANGLE

Fig. 284

a)   b)   c)

Fig.   285   Chamfering for seal assembly

The seals especially of the rubber type need a surface finish in the range of 0.3 to 0.4 $\mu m\ R_a$ on the surfaces they rub against. The circularity and cylindricity of the bores for the cylinders using rubber seals can be coarser than for units with piston rings. For proper assembly, the relevant components are to be suitably designed. The ends of shafts and bores are to be adequately chamfered (Fig. 285) to avoid chipping of the seal during assembly. Units are to be designed eliminating any possibility of the seals passing over grooves and ports (either during assembly or working). Otherwise, the ports and grooves are also to be chamferred to enable the seals move over them freely. Seals are required to be well lubricated prior to assembly.

Rotary seals

The seals for rotary applications can be either lip seals or mechanical seals. The rotary shaft seals of nitrile rubber are commonly used for sealing leakage of oil from gear boxes through bearings, in pumps for preventing the external leakage of oil through the bearings and further for preventing air from being sucked into the pump. They are recommended for peripheral speeds upto 18 *m/sec.* Good surface finish on the shaft and adequate lubrication for the seals are essential for proper functioning

Mechanical seals are employed in applications where rotary shaft seals are inadequate. These are applicable for pressure ranges upto 200 *kgf/cm²* and shaft peripheral speeds as high as 250 *m/sec.* Mechanical seals can be either face or bushing seals depending upon whether the leakage flow is radial between plain surfaces or axial between cylindrical surfaces. While axial force controls the leakage in face seals, the radial clearance limits the same in case of bushing seals. Fig. 286 shows the construction of a simple face seal. An elastomeric diaphragm supported by a fixed metal casing provides the sealing, by pressing against the shoulder of the rotary shaft. Large area of sealing surface as compared with lip seals backed up by a spring force results in positive sealing. Being spring loaded, the seal face is automatically compensated for wear. This type of seal is suitable for low pressure applications (about $1.7 \, kgf/cm^2$). For more arduous duties, seals employing two rigid rings–one stationary and the other rotating–held in close contact by spring pressure are used. In addition these seals carry secondary seals which are relatively static (Fig. 287). In some instances one of the sealing members is press fitted on to the shaft or into the bore of the housing thus eliminating the need for a secondary seal. It is also possible to make one of the sealing faces an integral part of the shaft or the housing. Application of these seals with springs is limited to low pressures (about $7 \, kgf/cm^2$). For high pressure applications the fluid pressure itself may be conveniently used to provide the sealing with judicious design to keep down the friction power loss and wear rate. Materials of construction for sealing faces can be carbon, plastic or ceramics against metals and metallic oxides or carbides. For oils, the combination of hardened steel with graphite and sintered or cast bronze may be used. These seals can withstand high pressures and speeds and can operate even under conditions of poor surface finish and lubrication.

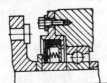 

Fig. 286 Face seal   Fig. 287 Face seal with secondary seals

Table 362 Recommended no. of packings per set*

Pressure kgf/cm^2	Leather (Natural and synthetic)	Homogeneous (No. of rings/sets)	Fabricated
Upto 33	3	3	3
33 to 100	4	4	4
100 to 200	4	5	4
200 to 333	4	5	5
333 to 660	5	—	6
660 and above	6	—	—

*Based on solid rings

CONNECTING ELEMENTS

Interconnection of various elements in a hydraulic system is obtained through several methods–tubing, hoses and panel mounting. The panel mounting of elements connecting each other in a desired pattern by passages within a block (Fig 288) is a popular method. It has the advantage of making the assembly neat and compact since it dispenses with the pipes, hoses and fittings. Static O-rings and gaskets are used in the interface to provide the sealing. The panel may be a single block with drilled interconnecting passages or an assemblage of grooved plates held together by a suitable bonding material. In general, panel mounting needs to be supplemented by tube or/and hose connections to provide flexibility of making connections and also to meet the remoteness of disposition of control and drive elements.

Tubings and tube fittings

Connecting two elements by G I pipes with threads at either ends has its limitations. Tube connections with end fittings are commonly employed for this purpose in fluid power applications. Various practices of effecting these joints are known. The ferrule type compression fittings are the current standard practice. Fig 289 shows a male stud coupling for connecting a body with a tube end. The body end of the coupling is screwed into the body with a sealing washer in between to provide a leak proof joint at the body-coupling interface. Body end sealing without a sealing washer is also at times resorted to by using taper threads on the body end of the coupling. The sealing at the coupling-tube interface is due to the biting action of the ferrule on the tube. The ferrule is forced into the conical portion of the coupling by the nut, thereby causing its sealing edge to bite into the tube in order to take the hydraulic load and to prevent the leakage along the tube. It is standard practice to classify ferrule type steel compression couplings based on the pressure rating as light and heavy (Refer Page 898). BS 4368 and DIN 2353 may be referred to for details.

Tubes for the above couplings are specified by their outer diameter. Cold drawn precision seamless steel tubes (as per BS 3601, 3602, DIN 2391, ISO R 560) of adequate wall thickness to withstand the system pressure are chosen for the purpose. The tubes can generally be mounted with a bend radius equal to three times their outside diameter.

Table 366 gives different types of couplings to meet the general industrial requirements. Banjo bolt connections are made suitable for hose ends and brazed or ferrule type tube connection. These connectors find application in surroundings with space limitation.

Hoses

Hoses provide for flexible connection demanded of linking two relatively moving members. They are made of nylon, PVC, flexible metallic tubes and elastomeric or rubber tubes of reinforced construction. The elastomeric hoses are the most commonly used type in the field of oil hydraulics. The basic construction of these hoses consists of an oil resistant seamless tube of synthetic rubber (neoprene) with textile, rayon or steel wire braiding for reinforcement and an outer cover of oil and weather resistant material. Hoses used in oil hydraulics are classified by SAE into 11 series (100 R 1 to 100 R 11). The eleven series cover the wide range of hoses manufactured for various applications (low, medium and high pressure) based on the material and type of braiding. Each of the series lists the size range (bores), the recommended working pressures, test pressures, burst rating and other properties of hoses.

The end fittings attached to the hose ends (Fig 290) may be of the permanent type or reusable type. The latter facilitates fitting over and over again, permitting assembling required lengths of hoses in situ. Hoses need adequate care to be exercised in their application and installation as regards operation within permissible limits of pressure, temperature and bend radius. They should in no case be used as load carrying members. They are to be mounted without any twist and in configurations shown in Fig 291

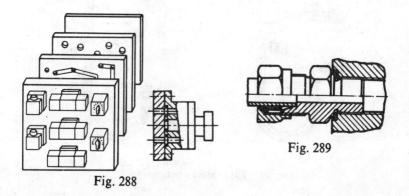

Fig. 289

Fig. 288

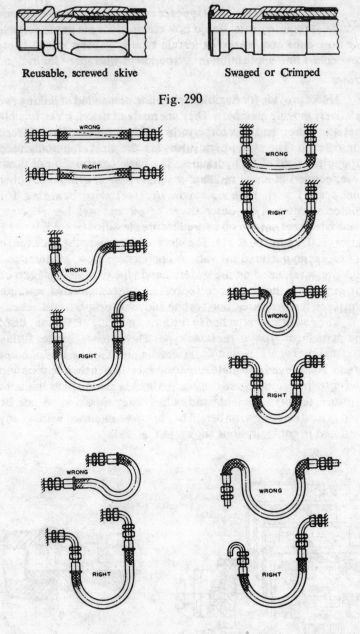

Reusable, screwed skive   Swaged or Crimped

Fig. 290

Fig. 291 Hose arrangements

HEATERS AND HEAT EXCHANGERS

The temperature of fluid in the system is required to be maintained within certain limits. Industrial hydraulic systems are required to operate satisfactorily at ambient temperatures ranging from 5°C onwards. The maximum temperature of the system fluid is limited to about 60°C. Too low a temperature of fluid means high viscosity and eventual loss of system pressure. Pumps may not be able to suck at very low temperatures of oil. To overcome the problem, suitable heaters with controls to limit the temperature may have to be provided. A working temperature range of 40°C to 50°C is generally preferred. It may be noted that heaters can be employed for achieving the stable thermal condition within a very short warm up period.

Too high a temperature of fluid means loss of viscosity and hence increased leakage, lowered efficiency, oxidation of the fluid and deterioration of the additives in the fluid resulting in formation of gums and sludges. Even more critical could be the thermal distortion of machine members causing machining inaccuracies. The heating up of oil in a system is due to the combined effect of ambient conditions and the loss of energy in the system. The loss of energy in the system may be due to the pump inefficiency, losses in pipelines and other connecting passages, valves and the bypassing of unused fluid from the pump under pressure through the relief valves. Often coolers may therefore be needed to keep the system fluid temperature sufficiently low. Three types of heat exchanges are commonly used—the liquid to air, liquid to liquid and refrigeratory types.

The liquid to air or the air cooled type is similar to the automobile radiators. Oil is circulated through the finned passages to be cooled by the forced circulation of air from a suitable fan over the radiator fins. Air cooled radiators are especially suitable for applications in areas with scarcity of water supply. Also they are free from the possibility of any water coming into direct contact with oil.

The liquid to liquid or the water cooled type use water for carrying the heat away from the fluid. They are more effective even at a lower differential between ambient and fluid temperature. These coolers are either of the shell type or tube type. These are made as single pass parallel flow, single pass reverse flow and multipass counterflow types.

Refrigeration units are used in cases where the temperature of oil has to be controlled in a very limited range, and especially where ambient temperatures are high. These units are compact and highly efficient. They may be either separate units with inlet and outlet connections for circulation of oil or the refrigerant may be circulated through cooling coils immersed inside the reservoir.

HYDRAULIC SYSTEMS

The hydraulic elements discussed earlier are deployed in building systems to achieve a variety of functions, effective at a time or otherwise and of varying degree of complexity. A system design is primarily based on the desired functions in a logical sequence. Besides it is an essential requirement that the approach be so as to provide an economical proposition from the point of view of utilization of minimum number of components and conservation of energy, without however affecting the reliability of operation.

Hydraulic power supply

The fundamental requirement of a system is a suitable pressure supply or a power pack as is generally referred to. A powerpack may be a simple pump-motor unit with a reservoir or a package consisting of the entire system. Power pack generally implies a source of supply of pressurized fluid in a condition acceptable to the drive and control circuitry. It consists essentially of a reservoir, a pump and its drive, a strainer at the pump inlet, a relief valve, a pressure gauge with shut-off and filter/filters for the system fluid and a manifold block with pressure outlet/outlets, return and drain-line connections. The pump is either mounted within the reservoir or on the top cover. Both the methods are widely practised. The former method provides better suction conditions for the pump and suppression of pump noise to some extent. The latter arrangement as recommended by the JIC and other standards is however considered a better practice from the point of ease of servicing.

RESERVOIRS

These are essentially storage tanks for the system fluid, although they may often facilitate mounting of atleast a part of the hydraulic system. Reservoirs are to be generally kept separated from the machine actual for reasons of isolation of thermal conditions and ease of servicing. Points to be reckoned in designing a reservoir are:

  i)   sealing of fluid chamber from external source of contamination;

 ii) sizing of the reservoir

     a) to hold adequate volume of fluid reckoning the amount of oil that may drain back from the system either during a portion of the cycle or at the time of servicing.

     b) to have sufficient radiating area for dissipating the heat generated in the system so that the fluid temperature in the tank does not exceed $60^{\circ}C$. A separate cooler may have to be included if the tank tends to be too large.

iii) the bottom of the reservoir is to be kept 120 to 200 *mm* above the ground level to facilitate draining, cleaning, transportation and improved heat dissipation,

iv) drain plug or some other means for draining almost the entire content of oil,

 v) convenient access for cleaning the inside of reservoirs–cleanout openings are to be provided in case of reservoirs with permanently fitted top covers,

vi) provision of a breather hole with a filter,

vii) provision of a filler cup with wiremesh screen,

viii) fluid level indicator showing the minimum and maximum permissible levels of the fluid in the reservoirs,

ix) inside of the reservoir to be painted with an oil resistant paint and

 x) baffle plates to separate pump suction and return lines.

Reservoirs are generally sized to hold 3 to 5 times the pump discharge per minute. Tanks are generally rectangular in shape. Too shallow a reservoir does not provide enough surface area for heat dissipation and too deep and narrow a tank does not provide enough surface for proper separation of entrained air bubbles or foam. Pumps are sometimes mounted at the base of the reservoir on a suitable extension to provide positive suction head due to the column of oil in the reservoir.

Heaters and coolers to maintain the fluid temperature within the maximum permissible limits are required to be provided based on the heat generation in the system, the ambient temperature range (at the work-site) and dissipation through the tank walls. Thermostat controllers to this effect, if essential, and temperature indicators are to be provided.

Pressure gauges with snubbers and gauge shut off valves are to be provided at a suitable location with respect to the relief valve and operator's position.

The fluid in the reservoir is required to be cleaned prior to entry into the system to a level of contamination acceptable for satisfactory functioning of the circuit. Location of filters, filtration capacity, flow and pressure ratings and dirt holding capacity of filter elements are to be reckoned while providing filtration. There are various possible locations for a filter in a hydraulic circuit (Table 363). The general practice in most of the hydraulic circuits for machine tools is to provide a strainer at the pump inlet. A 125 μm (equivalent of 120 mesh) rating for strainers is considered a good practice. The strainers need to be of adequate capacity, not creating more than 0.15 kgf/cm^2 pressure drop at the prevailing flow rate. Apart from a strainer, either a pressure line filter or return line filter of 30-40 μm absolute rating is adequate for most of the common applications. However, when critical components like servovalves are used, a pressure line filter of suitable micron rating immediately behind the valve is essential. Servo drive manufacturers recommend the use of a filter rated at 25 μm absolute for the purpose. These filters protecting the servo equipment may not necessarily clean up the entire system. A suitable pressure line prefilter is generally included in these systems (Fig 292). Use of prefilter is also an economical proposition since the filters protecting servovalves are recommended to be without by pass valves and preferably being capable of withstanding a pressure differential equal to the full system pressure. These costly filter elements do therefore need to be protected from dirt.

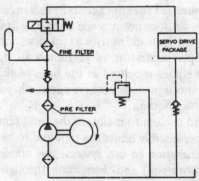

FINE FILTER

PRE FILTER

SERVO DRIVE PACKAGE

Fig. 292

Table 363 Location of filter

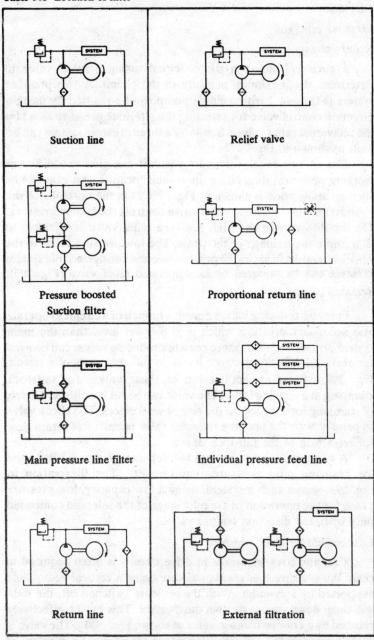

Suction line	Relief valve
Pressure boosted Suction filter	Proportional return line
Main pressure line filter	Individual pressure feed line
Return line	External filtration

Circuits

PRESSURE CONTROL

Relief valves

Figure 293 shows a simple circuit using a relief valve to determine the maximum pressure in the circuit. A two pressure system (Fig. 294) with a single pump can be realised by using a direction control valve for selecting the alternate pressure as set by the individual relief valves. Similarly a three pressure system can be built as/shown in Fig. 295 .

Use of two relief valves in a circuit can give two different working pressures depending upon their location. An example of such an application is shown in Fig. 296 . On the upstroke of the cylinder, the low pressure relief valve controls the system pressure. On the downstroke the high pressure relief valve is effective to determine the tonnage of the press. The low pressure relief in the upstroke results in saving of power. A continuously variable system pressure can be achieved by cam operated relief valve (Fig. 297).

Reduced pressure

Pressure reducing valves permit a branch of a circuit to operate at a secondary pressure which is at a lower level than the main system pressure. One or more pressure reducing valves can be used for realising different pressure levels, in different parts of a circuit. Fig. 298 shows an application of these valves for tailstock clamping in a copying lathe. The valve can be set for different levels of clamping force based on the size of workpieces. The check valve in parallel with the pressure reducing valve permits free return flow for retraction of the tailstock sleeve.

A two pressure system with two separate pumps is often used for ensuring pilot pressure requirements. The illustration in Fig. 299 shows such a system, using a low capacity, low pressure pump for the operation of the pilot stage of the solenoid controlled pilot operated direction control valve.

Counterbalance/Back pressure

Creating back pressures in drive circuits is often required to resist the negative loads, a typical case being a vertical heavy slide supported by a cylinder. With the pressure switched off, the slide will drop down due to its own deadweight. This can be effectively resisted by a counterbalance valve as shown Fig. 300 . The valve is set to a pressure level slightly higher than the pressure equivalent of the deadweight to be supported. The slide can move down only if the supply pressure on the cover end is adequate. A preloaded check valve can also be used for this purpose.

Back pressure valves can be used for ensuring a minimum system pressure that is required to operate pilot operated direction control valves (Fig. 301).

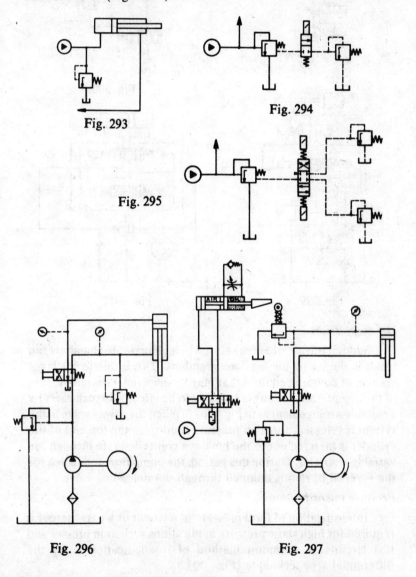

Fig. 293

Fig. 294

Fig. 295

Fig. 296   Fig. 297

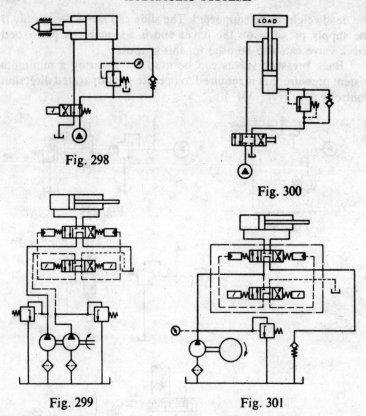

Fig. 298

Fig. 300

Fig. 299

Fig. 301

Decompression

Sudden release of energy stored in high pressure chambers can result in damaging the machine members. This is due to the large volume of compressed fluid that may be involved. Gradual release of the compressed volume is required to be effected in such cases by a suitable arrangement as in Fig. 302 . When the 4-way valve in the system is released, the high pressure fluid from the top end of the cylinder is metered out to the tank at a controlled rate through the variable restrictor. During this period, the pump flow connected for the reversal of ram is dumped through the unloading valve.

Pressure intensification

Intensification of fluid pressure in a circuit or a part thereof is required for high static pressure applications such as in presses and test circuits. A common method of intensification is by the differential area technique (Fig. 303).

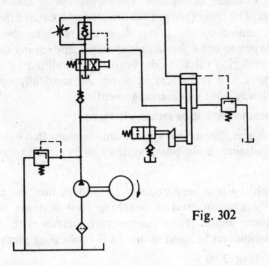

Fig. 302

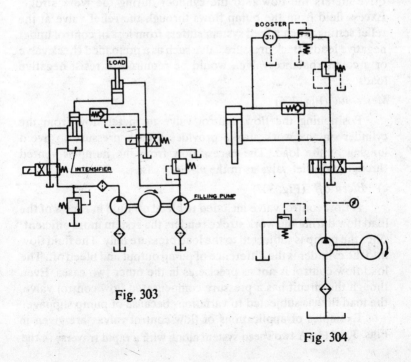

Fig. 303

Fig. 304

The example in Fig. 304 shows the use of another type of intensifier. The ram of the press is lowered to approach the work at a rate determined by the pump flow. On reaching the load, the pressure rises to open the sequence valve. The supply through the pressure reducing valve to the booster is amplified by a set ratio. The booster is a reciprocating pump automatically cycled by a built-in 4-way pilot valve arrangement.

FLOW CONTROL/CONTROL OF SPEEDS AND FEEDS

Feeds and speeds in a system can be controlled by–throttling, pump regulation, multi-pump systems or by circuit nuances.

Throttling

Throttling is a very popular method of flow control and is commonly encountered in machine tool systems with fixed displacement pumps. Flow control valves either with or without compensation can be used in any of the following methods.

Meter-in (Fig 305)

The flow control valve connected between the pump and the drive meters the flow into the cylinder during the work stroke. Excess fluid from the pump flows through the relief valve at the relief setting. Such a feed system suffers from loss of control under negative loads. Back pressure valve such as a preloaded check valve or a counterbalance valve., would be required to resist negative loads.

Meter-out (Fig 306)

Positioning the flow control valve to meter flow from the cylinder on the work stroke provides a back pressure to avoid lunging of the load. The excess flow from the pump is vented through the relief valve as in the *meter-in* case

c) Bleed off (Fig. 307)

A flow control valve included to bleed off fluid in excess of the load flow during the work stroke renders the system more efficient, since the pump is subjected to the load pressure only. The fluid flow into the cylinder is the difference of pump output and bleed off. The load flow control is not as precise as in the other two cases. Even though the circuit has a pressure compensated flow control valve, the load flow is subjected to variations because of pump slippage.

Examples of applications of flow control valves are given in Figs. 308-311. A two speed system along with a rapid traverse in the forward direction, obtained by suitable switching of the two solenoid operated valves is shown in Fig. 308. It is to be noted that the control is effective, only if the setting of F_2 is finer than that of F_1. An alternative arrangement to this is shown in Fig. 309 . Figs. 310 & 311 show equivalent circuits built with flow control valves connected in parallel.

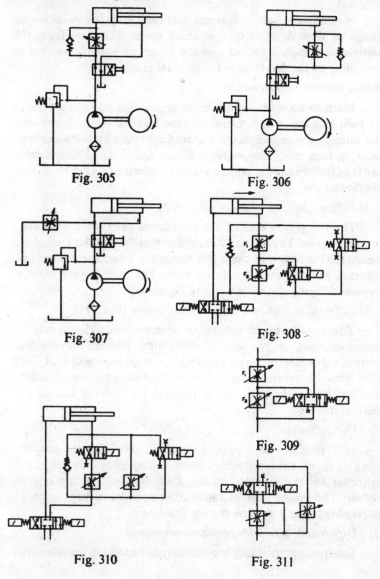

Fig. 305

Fig. 306

Fig. 307

Fig. 308

Fig. 309

Fig. 310

Fig. 311

Pump control

Speed control in a system can be achieved by using a variable delivery pump which can be regulated to provide the desired range of flow rate. The system is preferable to *throttling*, as it delivers approximately the same amount of flow as needed and hence more efficient.

A multipump system is commonly used in systems with a wide range of flow demand in a working cycle. The flow from the individual pumps is routed into the circuit by suitable valving or switching of the drives to the individual pumps.

Rapid traverse requirements

Machine tools often need rapid approaches and retractions to be built into the system. Generally, the systems are designed with the minimum pump capacity, dictated normally by the maximum working feed rate. Ingenuity of circuit design as explained later can lead to fulfilling the rapid traverse requirements under low load conditions.

a) *Differential area cylinder* (Fig. 312)

The full piston area of the cylinder is used for the forward working stroke. The rod size is chosen so that the annular area of the piston for the return stroke is low enough to effect the desired fast traverse with the available pump flow. The flow control valve is bypassed during this stroke by the check valve.

b) *Differential area with regenerative action* (Fig. 313)

The rapid return stroke is obtained in this circuit by interconnecting the two ends of the cylinder through the direction control valve thereby supplementing the pump flow with the return flow from the rod end of the cylinder. The mixing of flow is possible due to the higher pressure at the rod end arising out of the differential area.

c) *Two cylinders*

In a system with a pair of cylinders providing the forward working speed, the rapid return is effected by using only one of the cylinders for the reversed stroke, thereby doubling the rate of return. The *return* port of the second cylinder is kept open to prevent creating a vacuum during this stroke.

d) *Differential area with prefill arrangements*

Separate/external kicker cylinders can be used to provide rapid movement, along with an arrangement to filling the large piston area during the movement. Where space limitations make them impractical, a main cylinder with an axial bore accomplishes the same (Fig. 314). This feature provides a reduced area and a corresponding higher speed for the available pump flow. During the rapid advance, the area behind the main piston is kept filled by sucking oil through the prefill valve. The overhead tank would assist this function. The small return area on the main ram provides for the rapid return.

e) *Accumulator aided circuits*

Use of an accumulator for meeting the high flow demand during a portion of the working cycle is shown in Fig. 315 . The

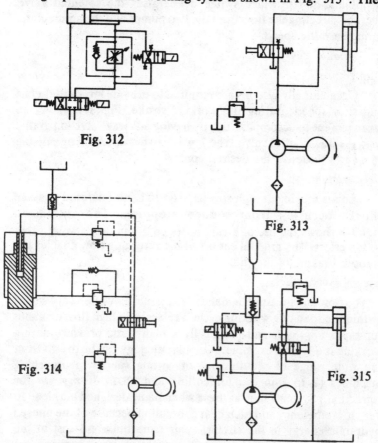

Fig. 312

Fig. 313

Fig. 314

Fig. 315

pilot check valve operated by the pressure in the return line permits drawing the stored fluid in the accmulator into the system during the return stroke, thereby augmenting the speed. In the other position of the direction control valve, the accumulator is charged to the pressure setting of the unloading valve.

Another example of using an accumulator for obtaining rapid traverse is shown in Fig. 316 . The system enables rapid traverse in both the directions of movement. With the piston at one end of the cylinder, pressure is built up in the system for the pressure switch to energise the solenoid. This directs the flow from the pump II to charge the accumulator. On shifting the manually operated direction control valve to the other position, the pressure in the line drops and the pressure switch deenergises the solenoid valve, thereby directing the flow from the two pumps and the accumulator to increase the speed.

f) *Variable feed*

Cam control

Occassionally systems are required to provide intermittent fast and slow speeds during the working stroke. Fig. 317 shows an arrangement to accomplish this by having a 2-way valve in parallel with the flow control valve. The 2-way valve is operated by suitable cam/cams to derive the desired control.

Deceleration

Loads moving at high speeds need to be decelerated to avoid shocks occuring from sudden stopping. The system in Fig. 318 shows the use of a cam operated 2-way decelerating valve to the effect. The gradual cut off of the returning flow enables the smooth braking of the slide.

ENERGY SAVING CIRCUITS

The flow and pressure demands of a system may often vary widely within the span of a working cycle. Presses with high flow demands for rapid approach and practically a high static pressure during working is a typical example. A similar analogy can be drawn from the wide range of requirement of cutting speeds in broaching machines (2 to 8m/min., for cutting and 25 to 30 m/min for retraction.) Systems such as these when badly designed may lead to very low efficiency and high heat generation because of the unused hydraulic energy in the circuit, apart from bulk and cost of the equipment. Circuits are designed with due care to minimise the

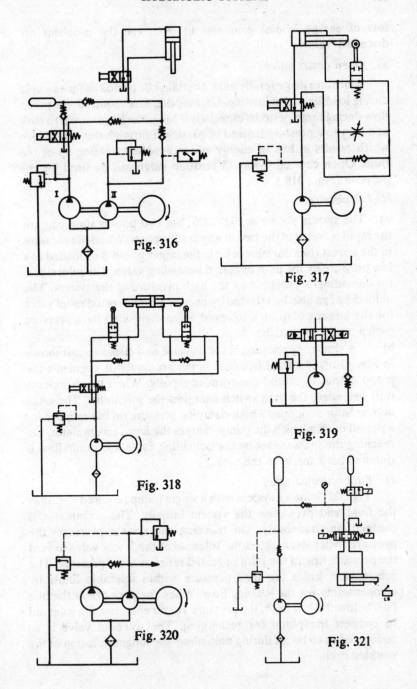

Fig. 316

Fig. 317

Fig. 318

Fig. 319

Fig. 320

Fig. 321

loss of energy. Some examples to overcome the problems are discussed below.

a) *Open centre valve*

Machines do generally have a certain idle period as for example during loading and unloading. Unless otherwise required, the pump flow during these periods, especially if long, can be returned to tank at a very low pressure instead of passing it through the relief valve which results in loss of energy and subsequent heating up of the fluid. Open centred 4-way, 3 position valve can be used for the purpose (Fig. 319).

HI-LO circuits

a) The system shown in Fig. 320 has two pumps supplying for the rapid traverse of the ram in a hydraulic press. When the pressure in the circuit rises during working, the larger pump is unloaded at a low pressure by the pilot operated unloading valve. The pilot signal for unloading is provided by the high pressure of the system. The unloading can also be effected by using a pilot operated relief valve for the larger pump or a solenoid valve controlled by a pressure switch or a limit switch.

b) A similar application is the traverse and clamp circuit shown in Fig. 321 . The fluid from the larger accumulator augments the pump output to extend the cylinder rapidly. When the cylinder is fully extended, the limit switch energises the solenoid c. The small accumulator maintains a high clamping pressure on the cylinder for a period during which the pump charges the large accumulator. On reaching the pressure set by the unloading valve, the pump flow is dumped back freely to the tank.

c) *Pilot operated relief*

Fig 322 shows a system with a single pump required to supply the fluid and pressurise the system initially. The accmulator is charged simultaneously. On reaching the system pressure, the pressure switch deenergises the solenoid of the 2-way valve to vent the pressure line of the pilot operated relief at a low pressure. The accumulator holds the load pressure within tolerable limits by compensating for the leakage flow. When the pressure in the line falls below the low limit, the pressure switch energises the solenoid to connect the pump for recharging. The solenoid valve is so connected as to be off during unloading–the longer duration of the working cycle.

A similar application shown in Fig. 323 uses a sequence valve to control the 2-way pilot valve for the dumping.

d) *Unloading relief valve* (Fig. 324)

An unloading relief valve used in an accumulator circuit provides for automatic dumping of the pump flow on reaching a set pressure. The construction of the valve gives a unique advantage in that the cut-off pressure of the relief valve corresponding to the full system pressure is higher than the cut-in pressure for recharging. This difference allows for longer idle periods for the pump.

e) *Pressure compensated pump*

Use of pressure compensated pumps, with internal feedback to match the pump output with load flow demand, is the ideal solution to applications such as the above.

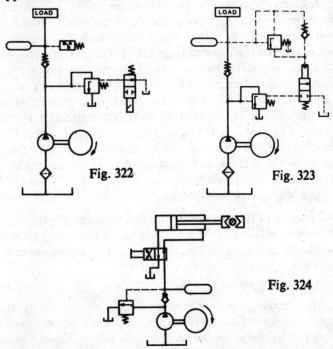

Fig. 322

Fig. 323

Fig. 324

SYNCHRONISING CIRCUITS

Often applications involve two drive units to be controlled to have the same feed rate. With a common supply into the two lines, the flow takes the path of least resistance and the drive with lower friction tends to move ahead of the other. Use of a flow divider valve provides an ideal solution to the problem. The following are several other methods to achieve the synchronization to different degrees of accuracy.

a) *Metering valves* (Fig. 325)

This is the simplest of the methods. Two meter-out flow control valves with pressure compensation are independently tuned to achieve synchronization of the two pistons. Unequal leakages across the pistons cause inaccuracy of control.

b) *Cylinders in series*

Better synchronization than that with metering valves is obtainable in a system with two equal double rod cylinders by using the displacement flow from one cylinder to drive the second cylinder. The leakage conditions may however cause one of the cylinders to bottom first, resulting in loss of stroke of the other cylinder. The cumulative errors may build up large differences. This is overcome by using a replenishing circuit (Fig. 326   ). The cylinders connected in series are controlled by the 4-way manually operated valve. On the return stroke, if cylinder 1 bottoms first, the valve A is actuated by the limit switch LS1 to open the check valve permitting extra fluid from cylinder 2 to flow to tank thereby completing its stroke. If the piston 2 returns first, the valve B is actuated by the switch LS2 to direct fluid to complete the stroke of the cylinder. The circuit diagram in Fig. 327 is an arrangement to effect replenishing at either ends of the cylinders.

c) *Rack and pinion arrangements*

Tying up of the two drive cylinders mechanically by having a rack on each piston rod driving a common pinion shaft gives very good results, provided the linkage is rigid. This arrangement however restricts the location of the cylinders.

d) *Coupled motors* (Fig. 328)

An effective flow divider can be made up of two hydraulic motors of matched flow characteristics coupled together. Since the motors run at the same speed, they will deliver equal volumes. Difference in internal leakage of the two motors will affect the synchronization. Variations in load do not greatly affect the performance.

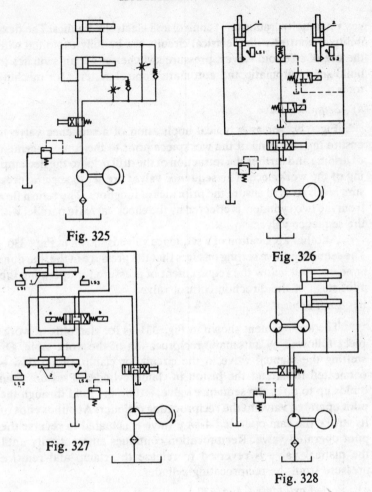

Fig. 325

Fig. 326

Fig. 327

Fig. 328

e) *Two pump control*

Two variable displacement pumps can be used to move cylinders in unison. Drive speed and leakage differences can be compensated for by adjusting the pump settings. The circuit readily permits changing drive speeds as well as synchronizing their movements

SEQUENCING CIRCUIT

Functioning of different elements and units in a system need to be invariably in a desired order of priority. Depending upon the complexity of the circuit, the sequence may be brought about by various methods–special valve configurations, cam controlled valves, pneumohydraulic pilot control and electrohydraulics. The flexibility of control with electrical circuits can be fully exploited with the use of solenoid valves, pressure switches and limit switches to build semi-automatic and automatic control systems for machine tools.

a) *Sequence valve*

Fig. 329 shows a typical application of a sequence valve to ensure the clamping of the workpiece prior to the commencement of drilling and further the retraction of the drill prior to the declamping of the workpiece. The sequence valves are set to specific pressure levels so as to ensure the priorities of functions. The return flow from the two cylinders is effected by the check valves in parallel with the sequence valves.

Another application of a sequence valve is shown in Fig. 330 . The sequence valve setting ensures that the pressure in the line does in no case fall below the requirement of pressure for operating the pilot stage of the direction control valve.

b) *Cam control*

The arrangement shown in Fig. 331 is for clamping of workpiece followed by automatic reciprocation of the work table. On shifting the manual valve in the circuit for clamp, pump flow is connected to extend the piston in clamp cylinder. Pressure then builds up to open the sequence valve, letting the fluid through the pilot operated valve to the reciprocating cylinder. At either ends of its stroke the cam operated 4-way valve is actuated to reverse the pilot operated valve. Reciprocation continues automatically until the manual valve is reversed to release the clamp and remove pressure from the reciprocating cylinder.

c) *Ported cylinders* (Fig. 332)

Operation of the lock cylinder in the circuit sequences the feed cylinder for indexing a rotary table. When indexing, the solenoid a is energised to direct flow into the left side of the lock cylinder. After the lock pin is out of location, the port 1 is uncovered to let the flow into the feed cylinder for indexing. At the end of the operation, the solenoid b is energised to move the piston in lock cylinder to the left. After the lock pin has entered into the location, the port 2 is uncovered to let the flow into the feed cylinder for retraction via a unidirectional clutch.

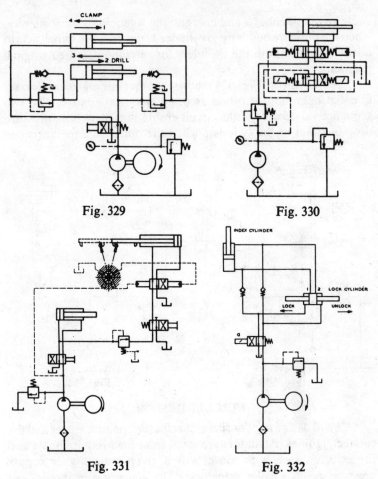

Fig. 329

Fig. 330

Fig. 331

Fig. 332

c) *Electrical interface*

Use of electrical control along with solenoid valves and pressure switches enable achieving any degree of complexity in sequencing. A simple example of such an application is shown in Fig. 333 . The piston in cylinder 1 is required to advance and dwell while piston in cylinder 2 rises to position 4. Both the pistons then retract together. Pressing the start switch energises the solenoid *b* advancing the piston in cylinder 1. Limit Switch LS1 is then actuated to energise the solenoid *d* to extend the piston in cylinder 2 and de-energise solenoid *b* to stop the piston in cylinder 1. On extension of piston 2, the limit switch LS2 is pressed to simultaneously de-energise the solenoid d and energise the solenoid b. Subsequently, when the pressure builds up in cylinder 1, the pressure switch acts to admit flow into both the cylinders for retraction to their original positions.

The circuit in Fig. 334 shows another arrangement for sequential operation of cylinders using electrical control. The two sequence valves used in the circuit ensure that there is no pressure drop in either of the cylinders when the other is in operation.

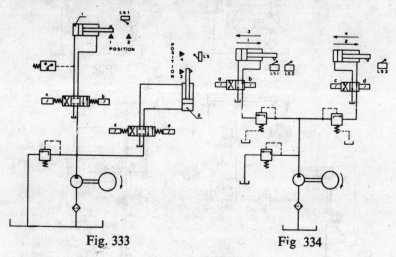

Fig. 333   Fig 334

CIRCUIT DESIGN

Circuit design implies sizing of actuators, pumps, valves and the connecting lines. Actuators are sized from load requirements and the selected system pressure. For a given load, higher supply pressure means smaller actuators. The actuator size should also meet the low speed requirement based on the minimum flow rate that can be effectively controlled. Supply pressures ranging from 10 to 100 kgf/cm^2 are generally used in machine tools. The load requirement should include the cutting loads, friction forces, pressure losses in valves, filters and pipelines as well as any intended back pressure in the circuit. The maximum operating speed determines the maximum flow demand of the actuator. The pressure and flow requirements of the individual actuators in the system over a full cycle of operation are estimated.

Fig. 335 shows an imaginary circuit with two cylinders and their requirements of pressure and flow over a cycle. The maximum

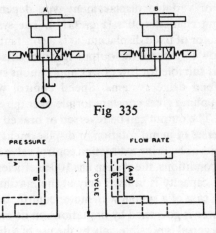

Fig 335

Fig. 336

simultaneous demand of the system is then arrived at by graphically summing up the two. The power supply to the system is sized to meet these demands. The power supply may be from–1) a constant delivery pump 2) a constant delivery pump with an accumulator 3) multi pumps or 4) a pressure compensated pump, depending upon the economy of power and minimisation of wasted energy. Demand graphs typical for these power supplies are shown in Fig. 336 .

The control valves and filters included in the circuit are to be consistent with the maximum pressure and flow rates in the system. The pressure drop across the valves and filters while passing the maximum system flow should be as low as possible. Flow *vs* pressure drop characteristics of these elements are generally supplied by the manufactures.

The connecting lines and passages are sized on the basis of the permissible flow rates. Flow velocities of $0.8 m/sec.$ in suction lines, 5 to 6 $m/sec.$ in other lines and velocities as high as 25 to 20 $m/sec.$ in short passages of manifold blocks are admissible. The number of bends in a system is to be kept to a minimum. Pressure losses in pipelines and bends can be evaluated as detailed in Table 345.

Hydrostatic transmission

A hydrostatic transmission is essentially a pump-motor combination to enable torque transmission utilizing the advantages of hydraulic drive and control. The units may be integral or remotely connected to each other. Pumps and motors in these can be of fixed or variable displacement type depending upon the desired output characteristics (Fig. 337). The system with both pump and motor of fixed displacement is a constant torque system. Bypass of flow to control the output speed results in inefficiency. The system is suitable for low power applications such as the valve controlled feed drive systems. Speed control with a variable displacement pump gives constant torque over the entire operating speed range. The output can be reversed or braked either by pump control (reversal of pump rotation or displacement setting) in case of *close* circuits or by using a direction control valve in *open* circuits. Under ideal conditions, the system has 100% efficiency, but the full prime mover capacity is utilised only at the maximum pump flow condition. In case of a system with motor control, the displacement setting of motor is governed by limitations on its maximum speed. The output reversal is possible only by the use of a direction control valve both in *open* and *close* circuits. The constant horsepower characteristics forthcoming from the arrangement is ideal for spindle drives of machine tools. The maximum torque capacity of the motor does limit the minimum output speed (Fig. 337b). Lower speeds can be obtained by resorting to pump control in the lower range, with the torque value remaining constant at the maximum rated value. The constant power characteristics at very low speeds has not much of a practical significance, since the maximum torque requirements are limited by the spindle design. If however the feature is essential, it can be accomplished by suitable gearing. Gearing would incidentally obviate any limitations on the low speed performance of the hydromotor.

The pumps and motors in a hydrostatic transmission system can be linked in *open* or *close* circuit. In case of the open circuit (Fig. 338) the pump sucks oil from the sump or draws from a booster pump and delivers at the motor. The fluid returns to the tank after passing through the system. The open circuit has the following disadvantages:

i)   the booster pump, when used, needs to be of a discharge capacity matching the system pump,

i)   self sucking pumps limit the maximum speed of the pump and hence of the transmission and

iii)   torque reversal and braking are possible only by use of a direction control valve between the pump and motor

In the closed circuit (Fig. 339), the return flow from the system is connected back to the pump inlet, the leakage losses only being made up by a booster pump operating at 2 to 3 kgf/cm^2. The system has the disadvantage of locking up the fluid within the closed circuit, without venting, filtration and cooling. These systems are therefore required to have a suitable bleed-off into the reservoir. The reversal and braking of output shaft can be effected by pump control. Under high intertia conditions, reversal and braking of motors lead to high pressure surges and cavitation in the lines connecting the motor. It would be appropriate to mention here that the inertia effects are far more critical in motor drives than in cylinder applications because of the low trapped volume in case of the former. It is therefore normal to include cross port relief in hydrostatic transmissions.

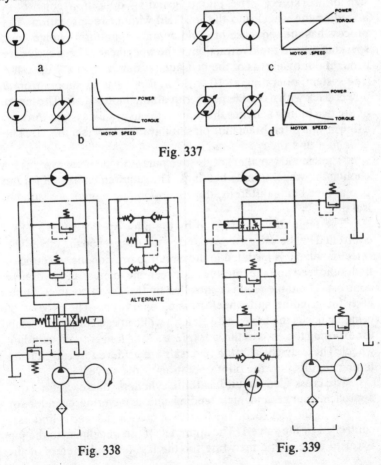

Fig. 337

Fig. 338   Fig. 339

APPLICATIONS IN MACHINE TOOLS

Some of the hydraulic systems adopted in machine tools are briefly explained.

Cylindrical grinding machines

A cylindrical grinding machine requires two types of operational cycles to be built-in–the traverse grinding and the plunge grinding. The traverse grinding has the sequence of–wheel head approach to the job, table stroking, pick feeding of the wheelhead at the table reversals dwell of table at reversals, sparking off and rapid retraction of the wheelhead after attaining the desired size of workpiece. The plunge grinding operation consists in infeeding the wheel into the job held without reciprocation. The process has the sequence of rapid approach, uniform plunge feed, sparking off and rapid retraction of the wheelhead. These cycles are brought out by means of the hydraulic system shown in Fig. 340 . The system, operating at 10 to 12 kgf/cm^2, has its supply from a power pack with a double pump run off the motor shaft. The larger of the two pumps supplies the fluid to the hydraulic system through a fine in-line filter, branching off suitably for the two sectors *viz*. The *table feed* and the *cross feed*.

The second (smaller) of the two pumps in the powerpack is for cooling the wheel spindle bearings. The requirements of oil for the two pumps being different, the corresponding pumps are suitably isolated.

The table drive is effected by the pilot operated valve 11 as controlled by the valve 10 operated by the dogs set on the table. The selector valve 9 is included in the circuit to provide the selection of hydraulic feed and the interlock between the manual and hydraulic controls. The table feed is controlled by two throttle valves 7–one each for grinding and wheel dressing operations. The process of grinding needs the table to dwell for a while at each of the reversals for exposing the ends of the workpiece a little longer to the grinding wheel. These dwells at table reversals are achieved by introducing timing circuits into the pilot operation of the valve 11.

The cross feed circuit has three cylinders, one each for rapid approach, plunge and pick feeding and a metering cylinder for controlling the pick feed. The flow to approach and feed cylinders is controlled by the valve 1. The approach of the wheelhead to the job is set to have a fine feed rate for the last few millimeters of the

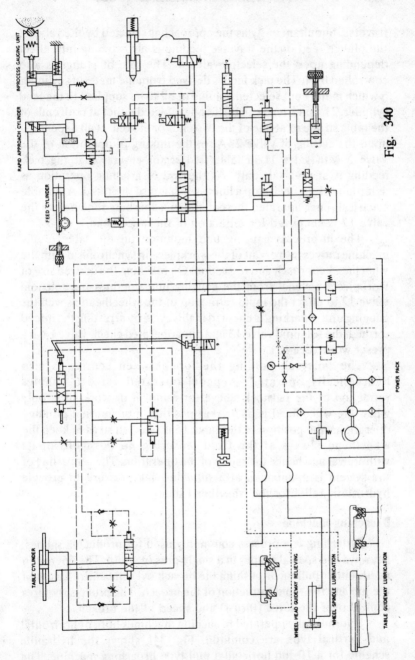

Fig. 340

traverse. Simultaneously as the approach is selected by the valve 1, the plunge feed or the traverse feeding is also coming into effect depending upon the selector valve 4. The rate of plunge is also controlled by 4. The pick feed is derived from the metering cylinder 5 which meters out selected volume of oil to be supplied to the feed cylinder. The pick feed can further be effected either at both ends of the table stroke or at one of the two ends or at none at all depending upon the setting of valve 2. Also, the linking of operation of the valve 3 with valve 11 in table feed sector ensures that the pick feeding is at reversals only. At the end of grinding operation as determined by a dead stop, a limit switch is also operated. A suitable electrical timer operated thereof allows a sparking time before the valve 12 is energized for retraction of the wheelhead.

The in-process gauging unit mounted on the table of the machine moves into or out of the workpiece in synchronism with the wheelhead approach and retraction. On reaching the desired size of workpiece, a signal from the gauging unit operates the solenoid valve 12 to effect the rapid retraction of the wheelhead as well the gauging unit. Sparking is effected in this cycle by arresting the infeed through the solenoid valve 13 and pilot operated check valve 14 at a preset workpiece size.

The tailstock holding the job between centres is also hydraulically operated. A pedal operated valve 6 enables retraction of the tailstock sleeve for removal of workpiece, only when the wheelhead is at its rear position. For internal grinding operations, the position of the valve 15 is selected so as to keep the wheelhead always at the front, ensuring no possibility of its withdrawal and hence the safety of the operation. The return line of the system is maintained at a suitable back pressure to provide hydrostatic relief for the wheelhead slide.

Broaching machine

Broaching is a process commonly used for producing splines, keyways and special shapes in a single pass of a tool. The operation needs either pulling or pushing the broach over the entire length of the job and the rapid retraction of the broach. The process requires control over the force (thrust) and speed of the broach.

Hydraulically operated broaching machines both of horizontal and vertical type are common. Fig. 341 shows the hydraulic scheme for a 10-ton horizontal pull type broaching machine. The machine is provided with a hydraulic cylinder which holds the broaching tool. The power pack for the system has two pumps (P1 and P2) which are run either independently or together to get three different ranges of broaching speeds. The pumps with their individual motors can be so selected for the required range of speeds so as to avoid wastage of power. The flow control valve 1 provides the bleed off of the pump output to get the required broaching speed. In the return stroke the circuit accommodates for a regenerative cycle by combining the oil from rod side to the main pump flow on the piston side through the interconnecting valve 6 to effect rapid return of the ram. The cylinder can be reciprocated for broaching and return through limit switches to operate the solenoids of the pilot operated direction control valve 2. Needle valve 3 provides for bypassing a portion of the flow during return so as to control the rapid rate. Needle valve 4 in the circuit ensures smooth changeover of the direction control valve to result in smooth reversals.

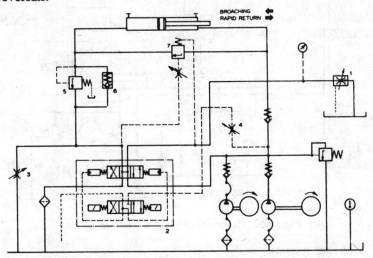

Fig. 341   Hydraulic drive and control of a broaching machine

Flow forming lathe

An adoption of hydraulic drive and control for obtaining flow forming facility built on to a centre lathe is shown in Fig. 342. The process consists in holding the workpiece with a definite clamping force between the mandrel on spindle and the tailstock and forming it over the mandrel by feeding a forming roller mounted on the carriage. The formed container is ejected out after the completion of the forming cycle. This complete cycle of operation is effected by the hydraulic system shown in Fig 342 together with suitable electrical controls. The hydraulic system has a power pack with all the control elements mounted on a panel fixed onto the top plate of the tank. Standard cylinders are used for providing the workpiece clamping (tailstock cylinder), the carriage feed and workpiece ejection. The sequence of operation–workpiece clamping, spindle rotation, carriage feeding, stopping and braking of spindle and of the feed stroke, withdrawal of tailstock cylinder, ejection of workpiece and carriage withdrawal are controlled in that order by the combination of limit switches, electrical control circuitry and the solenoid valves. The rapid stroke of the carriage is effected by the large rod area of the drive cylinders and further due to the supply for the return stroke being only to one of the saddle cylinders.

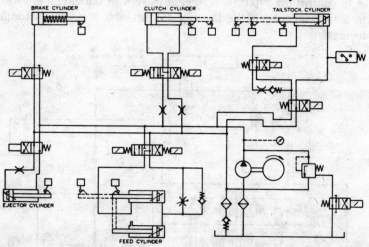

Fig. 342 Hydraulic circuit of a flow forming lathe

Copying systems

Hydromechanical positional servo systems are used in machine tools for tracer controlled mechanisms to enable reproduction of profiled parts. These are valve controlled drive systems, with a tracer attached to the valve for sensing the master. Linear type spool valves with single edge, 2-edge or 4-edge metering are commonly used. The principles of operation of systems with these valves are shown in Fig. 343 .

Fig. 343a shows an asymmetrical system with a single edge valve and a differential area cylinder with 2:1 area ratio. The rod end is connected directly to the supply pressure. The head end is connected through the valve to the tank. The two ends of the cylinder are interconnected through an orifice type restrictor. The pressure on the head end of the chamber is dependent on the flow across the orifice and the opening of the valve to the cylinder. At *null* flow through the valve is such that it creates a pressure of $p_s/2$ on the headside to balance the loads on either side of the piston. A shift in the position of the spool, controlling the valve opening, increases or decreases the head chamber pressure either to advance or retract the slide. A similar system is shown in Fig. 343b wherein the constant load equivalent of the pressure at the rod end is provided by the deadweight of the moving member. The 2-edge system (Fig. 343c) does also generally operate on the differential area principle with an area ratio of 2:1. The two edges of the central land on the spool operate to determine the load flow into or out of the head end of the cylinder. At *null* the leakage flows into and out of the cylinder match to give the necessary balancing pressure on the full area of the piston. The system can also be built with a deadweight (Fig. 343d) instead of the hydraulic force from the rod end of the cylinder. The 4-edge system is a symmetrical system and has four metering edges operating to determine the load flow.

Performance of the system is dependent on the precision of metering of the load flow. Single edge systems are therefore the least expensive of the three types. The 2-edge valve has a pair of dimensions–*viz* the width of central land and corresponding port–to be controlled and matched. The 4-edge valve is the most expensive as it has six dimensions to be accurately controlled. The differential area systems are less sensitive and have only half the maximum load capacity as that of the symmertical system, for a given piston size and system pressure. Further the 4-edge systems are applicable to motor drives also. All the three types are in use, although the 2-edge and 4-edge systems are more common.

The flow and pressure characteristics of the copying valves and the drive size are based on load and load speed requirements, with due consideration to response and stability of the servosystem. The valves are built so that a small shift in the spool position can produce high forces and flow rates resulting in increased copying accuracies. Valves are generally mounted close to the drive with rigid flow paths to minimise the compressibility effects on the system response.

Copying systems find application in practically every type of machine tool–lathes, milling, shaping, planing, boring and grinding machines. Depending upon the number of axes controlled, they may be single axis, 2-axis and 3-axis systems. Single axis systems work with a basic machine feed along an axis called the *feed axis* inclined to the copying or the tracer axis to derive the required angle of the profile.

The resultant cutting feed rate varies with the angle of copying. A two axis system provides for maintaining the resultant feed rate constant as well for copying every possible combination of angles formed by the two axes. The 3-axis copying systems enable copying in all the 3 axes at a time.

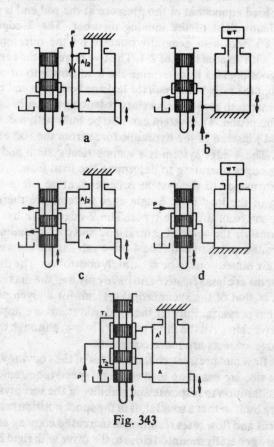

Fig. 343

COPY TURNING

Use of a single axis copying unit for production of stepped shafts and profiles on a centre lathe is shown in Fig. 344 . The unit is mounted perpendicular to the feed axis–the longitudinal traverse–or at any other angle depending upon the desired limits on angle of copying (Fig. 345). The unit shown in Fig. 344 has a valve mounted directly on the slide housing the cylinder. For a fixed setting of the distance between the tool tip and the stylus finger and the fixed position of the template/master, the position of the tool tip is unique. Hence by feeding the carriage along, the relative contour/step dimensions on the template can be accurately transferred on to the workpiece. The infeed drum moves the servovalve along enabling advancing or receding the tool tip with respect to the stylus finger thereby increasing or decreasing the depth of cut. The turret drum mounted on top of the valve actuates the stylus when any of the stoppers presses against the dog post. This arrangement provides for biasing the stylus finger control depending upon the settings of the individual stoppers on the index drum. The stoppers can be set to provide the desirable number of rough cuts and finally a finish copying cut as per the template. Automatic indexing of the stop drum linked up with longitudinal carriage traverse provides automatic programmed cycles (Fig. 346).

Copying units are usually provided with features for mounting boring bars to facilitate internal copying. A desirable if not an essential feature of the unit in such cases would be an arrangement for limiting the back position of the slide to avoid damaging of boring bars and work pieces during withdrawal. Copying units are also made with the valve mounted at the end of an arm fixed to the tool post. The arm can be swivelled to vary the disposition for setting to different sizes of workpieces. The tool slides are often provided with separate tool holders with tool infeed machanism for adjustment of depth of cut.

These units can also be used for facing operations by mounting them suitably and using the transverse feed for the feed axis. Their application can be further extended to rotary copying as well (for cams) by rotating the master in synchronisation with the workpiece by a suitable drive mechanism.

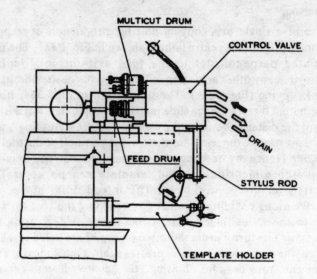

Fig. 344 Copying attachment for turning

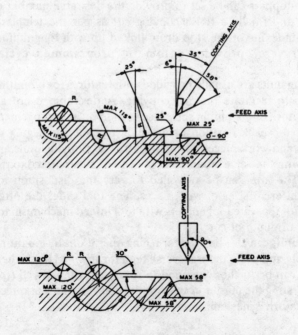

Fig 345 Limitation on angle of copying

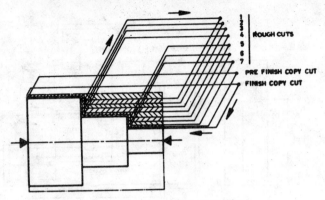

Fig. 346   Programmed cycle of copy turning

COPY MILLING

Copy milling operations find a wide range of applications as for example in making punches and dies and cams of simple to complicated profiles, either of two or three dimensional nature. These systems are popular with the vertical spindle milling machines. Fig. 347 shows a vertical milling machine with a single axis copying system, generally referred to as a 180°-1D system. A differential area cylinder is provided for the actuation of the knee. Actuation is controlled by a 2-edge valve mounted at the end of a bracket fixed to the headstock. The basic feature of manual control of the knee is retained by suitably designing the piston rod. The system enables copying undulations in the vertical axis on the master. The profile on the master at the scanned cross-section is thereby reproduced 1:1 by the cutter on the workpiece. Using this arrangement a three dimensional object can be reproduced by scanning the entire width of the profile in suitable rectangular cycles as shown in Fig. 348 . The pickfeed motor mounted for the transverse axis can be energised to this effect at either ends of the longitudinal strokes through limit switches or *fence control*. The force required to operate the spool being low (0.25 to 1.5 *kgf*), the masters can be made out of wood, plaster of paris etc., which can be easily worked to required intricate shapes. Machines with hydraulic table feed can be provided with a mechanism for modulation of the table feed in relation to the angle of copying so as to effect a fairly constant resultant feed rate. Such a system does also permit copying even 90° walls as against the normal limitation to about 85°.

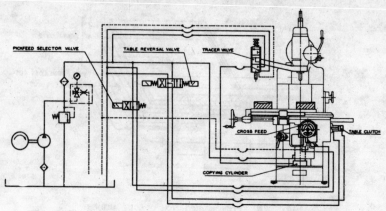

Fig. 347   Single axis copying attachment on a vertical milling machine

A 2-axis (360°-2D) system has two axes under simultaneous copy control to provide for 360° profiling (Fig. 349). An application of the same is shown in Fig. 350a . The table X and cross slide Y are hydraulically driven with one cylinder for the former and two for the latter. The flow into and out of the cylinders and hence the slide movements are controlled by a pair of servovalves arranged horizontally and parallel to the respective axes. The two systems are matched for similarity of flow and pressure characteristics. Deflection of the stylus in any direction effects shifting of the spools of the two valves and thereby the actuation of the two slides at velocities proportional to the respective valve openings. The resultant velocity is in a direction opposite to that of stylus deflection, whereby the approaching profile does always tend to push the stylus to *null*. The stylus is held in contact with the profile and steered around it to produce a similar workpiece. The steering around can be manual or by a bias force which can be kept changing continuously depending upon the profile by an automatic steering machanism.

In a 360°-3D system, all the 3-axes (X, Y & Z) are under copy control to enable contouring as shown in Fig. 350b . The valves for the longitudinal and transverse axes are placed horizontally at 90° to each other as in a 2-axis system, while the valve for the vertical axis is mounted either vertically or horizontally. A stylus deflection in the horizontal plane actuates the X and Y valves and the vertical movement of the stylus actuates the Z-axis valve. The angle of copying with the Z-axis in the 3D contouring is limited to about 30° to the horizontal. The copying in the Z-axis can be extended for die sinking operations as in 180°-1D with a separate stylus attachment. The two horizontal axes are then selected to function as feed and pick feed axes.

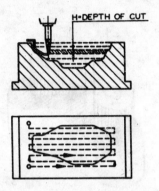

Fig. 348 Component scanning and copying

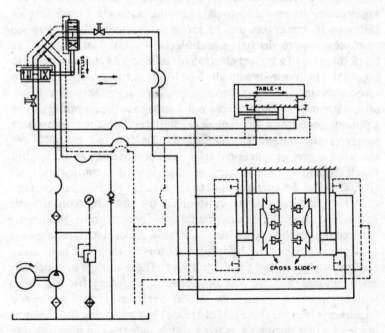

Fig. 349 A 2 axis copying system

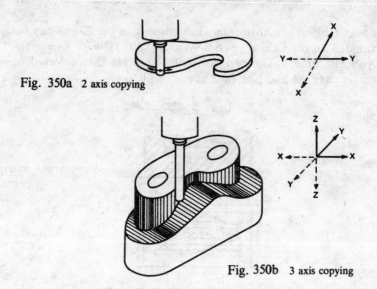

Fig. 350a 2 axis copying

Fig. 350b 3 axis copying

Electrohydraulic servomechanisms

Electrohydraulic servomechanisms are popular for applications in machine tools requiring accurate positioning of slides as in tracer controlled machines, numerically controlled machines, electro discharge and electro-chemical machines. The block diagram of a typical electrohydraulic servo system is shown in Fig. 351 . In common with all close loop control systems these are error actuated i.e. they have a negative feedback. The servo programmer providing the command voltage to the amplifier can be a potentiometer or a NC tape or a computer. Feedback transducer measures the output position of the system and generates an electrical signal proportional to it. The comparator and amplifier together form the servocontroller which is used to control and limit to a safe level the current input to the electrohydraulic servovalve.

The drive element controlled by the electrohydraulic servovalve can either be a hydraulic cylinder or a motor. Motors are preferred from the point of drive stiffness. Axial piston, rolling vane and gerotor types of hydraulic motors with good low speed performance are used for the purpose. These motors are specially manufactured to have low breakaway to running friction ratios which ensure a good low speed performance. Low speed requirements of machine tables are met by connecting the motor to the lead screw through a gearing which also helps in motor seeing reduced load inertia. The reduction of the rapid traverse speeds due to gearing can be countered by having dual gain servovalves or switchable displacement motors. A servo drive package–hydraulic motor close coupled to an electrohydraulic servovalve–is shown in Fig. 352 . Accessories to these packages include crossport relief valves and adjustable interport leak orifices. Crossport relief valves protect the package from high transient pressures which result when slides moving at high speeds are brought to rest. Adjustable metering orifices serve to improve the damping of the load-motor hydraulic resonances. A schematic arrangement of a feed drive system is shown in Fig. 353 . The inner loop providing the velocity feedback from the tachogenerator serves to improve the system resolution.

Stepping motors transforming an electrical input pulse to a mechanical displacement are used to control the position and speed of movement of machine tables. Stepper motor drives avoid the use of position measuring devices, simplifiying the system considerably. An Electrohydraulic stepping motor (Fig. 354) consists essentially of an electric stepping motor and a hydraulic valve-motor combination to multiply the torque output of the electric motor. The electric stepping motor is mechanically connected to the torque amplifier through a gear train. The driven gear is mounted on a threaded extension of the valve spool and is restrained against axial movement. Any rotation of the gear train causes an axial movement of the spool thereby directing the flow to the motor. The motor shaft is connected by splines to the valve spool. Consequently, the rotation of motor shaft turns the valve spool to restore equilibrium. The position and speed of movement of the table are determined by the number of pulses and pulse rate given to the electric stepping motor.

A linear electro hydraulic stepper motor drive (Fig. 355) has the added benefit of not requiring motion converting mechanism, as it can be directly connected to the slide. Stepper motor (1) rotates the servo screw (2) relative to nut (3) in the piston which shifts the spool (4) of the hydraulic servovalve (5). This shifting of the spool opens the valve port to admit pressure oil to one side of the piston and moves it in such a way as to bring the spool back to its null position.

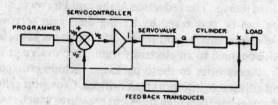

Fig. 351 Block diagram of a electrohydraulic servo system

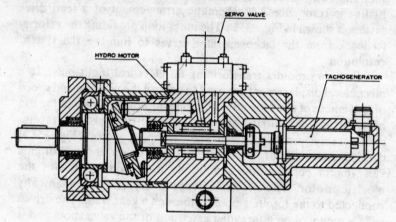

Fig. 352 Servo drive package

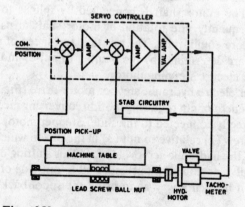

Fig. 353 Schematic diagram of a feed drive system

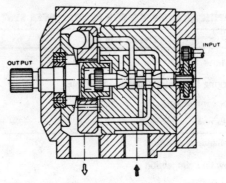

Fig. 354 Electrohydraulic stepping motor

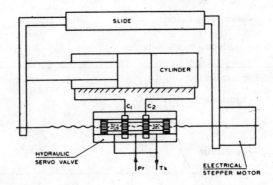

Fig. 355 Linear electro hydraulic stepping motor

Table 364 Classification of seals

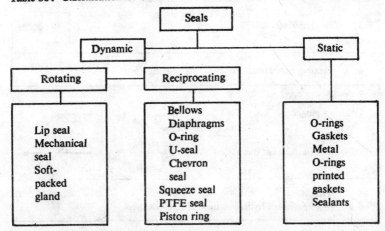

Table 365 GRAPHICAL SYMBOLS FOR FLUID POWER SYSTEMS

No.	Description	Symbol
1	Flow lines	
1.1	—working line, return line and feed line	
1.2	—pilot control line	
1.3	—drain line	
2	Miscellaneous symbols	
2.1	flow line connection	
2.2	— spring	
2.3	— restriction	
2.3.1	— affected by viscosity	
2.3.2	— unaffected by viscosity	
3	Pressure source:	
3.1	— hydraulic	
3.2	— pneumatic	
4	Electric motor	
5	Heat engine	
6	Flexible hose, usually connecting moving parts	
7	Power take-off	
7.1	— plugged	
7.2	— with take off line	
8 8.1	Triangle — direction of hydraulic flow in one direction	

Table 365 GRAPHICAL SYMBOLS FOR FLUID POWER SYSTEMS (Contd.)

No.	Description	Symbol
8.2	— direction of hydraulic flow in both directions	
8.3	— direction of pneumatic flow or exhaust to atmosphere	
9	Arrows	
9.1	— indication of direction	
9.2	— indication of direction of rotation	
9.3	— indication of the possibility of a regulation or of a progressive variability	
10	Pumps and compressors	
10.1	— fixed capacity hydraulic pump with one direction of flow (a)	
10.2	— fixed capacity hydraulic pump with two directions of flow (b)	
10.3	— variable capacity hydraulic pump with one direction of flow (a) — variable capacity hydraulic pump with two directions of flow (b)	
10.4	— fixed capacity compressor (always one direction of flow)	
10.5	— vacuum pump	

Table 365 GRAPHICAL SYMBOLS FOR FLUID POWER SYSTEMS (Contd.)

No.	Description	Symbol
11	Motors	
11.1	— fixed capacity hydraulic motor with one direction of flow (a)	
11.2	— fixed capacity hydraulic motor with two directions of flow (b)	
11.3	— variable capacity hydraulic motor with one direction of flow (a)	
11.4	— variable capacity hydraulic motor with two directions of flow (b)	
11.5	— fixed capacity pneumatic motor with one direction of flow	
11.6	— fixed capacity pneumatic motor with two directions of flow	
11.7	— variable capacity pneumatic motor with one direction of flow	
11.8	— variable capacity pneumatic motor with two directions of flow	
11.9	— hydraulic oscillating motor	
11.10	— pneumatic oscillating motor	

Table 365 GRAPHICAL SYMBOLS FOR FLUID POWER SYSTEMS (Contd.)

No.	Description	Symbol
12	Cylinders	
12.1	— single acting cylinder returned by an unspecified force	
12.2	— single acting cylinder returned by spring	
12.3	— double acting cylinder with single piston rod	
12.3	— double acting cylinder with double ended piston rod.	
12.4	— cylinder with single fixed cushion	
12.5	— cylinder with double fixed cushions	
12.6	— cylinder with double adjustable cushions	
13	Hydraulic pressure intensifier	
14	Variable speed drive unit (torque converter)	
14.1	— with variable control of pump	
14.2	— with variable control of both pump and motor	

Table 365 GRAPHICAL SYMBOLS FOR FLUID POWER SYSTEMS (Contd.)

No.	Description	Symbol
15	Control valves	
15.1	— two position valve (No. of squares is equal to the no. of positions)	
15.2	— two position valve with port positions shown	
15.3	— direction control valve, four way, two positions with electromagnetic control and return by spring	
15.4	— direction control valve, four way, three positions with manual operation and spring centered	
15.5	— direction control valve, four way, two positions, combined with solenoid operated pilot valve against a return spring	
15.6	— tracer valve with two ports (one throttling orifice) plunger operated against a return spring	
15.7	— tracer valve with three ports (two throttling orifices) plunger operated against a return spring	
15.8	— two stage electrohydraulic servo valve, with hydraulic feed back and indirect pilot operation	
16	Non-return valve	
16.1	— free (opens if the inlet pressure is higher than the outlet pressure)	
16.2	— spring loaded (opens if the inlet pressure is greater than the outlet pressure plus the spring pressure)	

Table 365 GRAPHICAL SYMBOLS FOR FLUID POWER SYSTEMS (Contd.)

No.	Description	Symbol
16.3 (a)	— pilot controlled to prevent closing of the valve	
16.4 (b)	— pilot controlled to prevent opening of the valve	
17	Shuttle valve	
18	Rapid exhaust valve	
19	Pressure relief valve (safety valve)	
19.1	— with remote pilot control	
20	Sequence valve	
21	Pressure regulator or reducing valve	
22	Flow control valves	
22.1	— throttle valve	
22.1.1	— with mechanical control against a return spring (braking valve)	

Table 365 GRAPHICAL SYMBOLS FOR FLUID POWER SYSTEMS (Contd.)

No.	Description	Symbol
22.2	— flow control valve with fixed output	
22.3	— flow control valve with variable output	
23	Flow dividing valve	
24	Shut-off valve	
25	Silencer	
26	Reservoirs	
26.1	— with inlet pipe above fluid level	
26.2	— with inlet pipe below fluid level	
26.3	— with a header line	
27	Accumulators	
28	Filter or strainer	
29	Water-trap	
29.1	— with manual control	
29.2	— Automatically drained	

Table 365 GRAPHICAL SYMBOLS FOR FLUID POWER SYSTEMS (Contd.)

No.	Description	Symbol
30	Air dryer	
31	Lubricator	
32	Heater	
33	Cooler	
34	Temperature controller	
35	Manual control	
35.1	— general (without indication of control type)	
35.2	— by push button	
35.3	— by lever	
35.4	— by pedal	
36	Mechanical control	
36.1	— by plunger or tracer	
36.2	— by spring	
36.3	— by roller	

Table 365 GRAPHICAL SYMBOLS FOR FLUID POWER SYSTEMS (Concld.)

No.	Description	Symbol
36.4	— by roller, opening in one direction only	
37	Electrical control	
37.1	— by solenoid	
37.2	— by electric motor	
38	Control by application or release of pressure	
38.1	— by application of pressure (direct acting)	
38.2	— indirect control, pilot actuated by application of pressure	
38.3	— indirect control, pilot actuated by release of pressure	
39	Pressure gauge	
40	Thermometer	
41	Flow meter	
42	Integrating flow meter	
43	Pressure electric switch	

Table 366   OIL HYDRAULIC COUPLINGS – SURVEY

	Male stud coupling-assembly
	Male stud elbow-assembly
	Male stud tee (stud branch)-assembly
	Banjo stud elbow-assembly
	Banjo stud tee-assembly
	Straight coupling-assembly

Table 366   OIL HYDRAULIC COUPLINGS—SURVEY

(Contd.)

	Equal elbow-assembly
	Equal tee-assembly
	Unequal tee-assembly
	Bulkhead assembly (tube to tube)
	Elbow bulkhead assembly (tube to tube)

Table 366   OIL HYDRAULIC COUPLINGS—SURVEY   (Contd.)

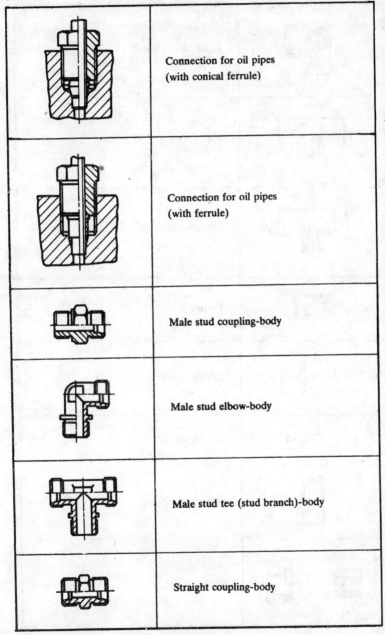

	Connection for oil pipes (with conical ferrule)
	Connection for oil pipes (with ferrule)
	Male stud coupling-body
	Male stud elbow-body
	Male stud tee (stud branch)-body
	Straight coupling-body

Table 366   OIL HYDRAULIC COUPLINGS – SURVEY   (Contd.)

	Equal elbow-body
	Equal tee-body
	Unequal tee-body
	Ferrule
	Conical ferrule
	Sealing washer
	Hexagonal locknut
	Coupling nut

Table 366   OIL HYDRAULIC COUPLINGS – SURVEY   (Contd.)

	Bulkhead body (tube to tube)
	Elbow bulkhead body (tube to tube)
	Eyes for banjo stud elbow assembly
	Eye for banjo stud tee assembly
	Bolt for banjo assembly
	Cap screw for oil pipe connection with conical ferrule
	Cap screw for oil pipe connection with ferrule
	Socket for body to hose connection

Table 366   OIL HYDRAULIC COUPLINGS SURVEY   (Contd.)

	Socket for tube to hose connections
	Bulkhead assembly (hose to tube)
	Bulkhead body (hose to tube)
	Screwed plug
	Cylindrical screwed plug
	Cylindrical plug

Table 366   **OIL HYDRAULIC COUPLINGS – SURVEY**   (Concld.)

	End connection details -tube end
	End connection details -stud end
	End connection details for hoses
	Male hose end
	Banjo assembly (stud elbow & stud tee) -brazed type

MALE STUD COUPLING ASSEMBLY TYPE - A

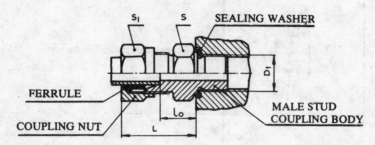

Table 367

Series	Outside dia. of tube	D	l_o $\approx$	L	Washer size
Light L	6	G 1/8	8.5	23	10 x 13.5
	8	G 1/4	10	25	14 x 18
	10	G 1/4	11	26	14 x 18
	12	G 3/8	12.5	27	17 x 21
	15	G 1/2	14	29	21 x 26
	18	G 1/2	14.5	31	21 x 26
	22	G 3/4	16.5	33	27 x 32
	28	G 1	17.5	34	33 x 39
	35	G 1 1/4	17.5	39	42 x 49
	42	G 1 1/2	19	42	48 x 55
Heavy H	6	G 1/4	13	28	14 x 18
	8	G 1/4	15	30	14 x 18
	10	G 3/8	15	31	17 x 21
	12	G 3/8	17	33	17 x 21
	14	G 1/2	19	37	21 x 26
	16	G 1/2	18.5	37	21 x 26
	20	G 3/4	20.5	42	27 x 32
	25	G 1	23	47	33 x 39
	30	G 1 1/4	23.5	50	42 x 49
	38	G 1 1/2	26	57	48 x 55

Notes: 1. For details for Ferrule refer **Page 902**

2. For details for coupling nut and port details refer **Pages 901 & 900**

3. For details of male stud coupling body refer **Page 897**

4. For details of sealing washer refer **Pages 903 & 904**

MALE STUD COUPLING BODY

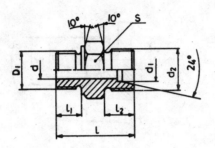

Table 368

Series	Outside dia. of tube	D_1	l_1 ±0.2	D_2	l_2 ±0.2	d_1 B11	d	S	Tol. on S	l ± 0.3
Light L	6	G 1/8	8	M 12 x 1.5	10	6	4	14	0 0.11	23.5
	8	G 1/4		M 14 x 1.5		8	6	19		29
	10		12	M 16 x 1.5	11	10	7			30
	12	G 3/8		M 18 x 1.5		12	9	22		31.5
	15	G 1/2	14	M 22 x 1.5	12	15	11	27	0 0.13	35
	18			M 26 x 1.5		18	14			36
	22	G 3/4	16	M 30 x 2	14	22	18	32		40
	28	G 1	18	M 36 x 2		28	23	41		43
	35	G 1¼	20	M 45 x 2	16	35	30	50		48
	42	G 1½	22	M 52 x 2		42	36	55	0 −0.16	52
Heavy H	6	G 1/4		M 14 x 1.5		6	4	19		32
	8		12	M 16 x 1.5	12	8	5			34
	10	G 3/8		M 18 x 1.5		10	7	22		34.5
	12			M 20 x 1.5		12	8			36.5
	14	G 1/2	14	M 22 x 1.5	14	14	10	27	0 −0.13	41
	16			M 24 x 1.5		16	12			41
	20	G 3/4	16	M 30 x 2	16	20	16	32		47
	25	G 1	18	M 36 x 2	18	25	20	41		53
	30	G 1¼	20	M 42 x 2	20	30	25	50		57
	38	G 1½	22	M 52 x 2	22	38	32	55	0 −0.16	64

Material: 14Mn1S14

TUBE END DETAILS

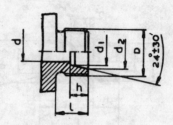

Table 369

Series	Outside dia. of tube	Nom. pressure kgf/cm²	D	d	d_1 B11	d_2 +0.1 0	l ±0.2	h +0.3 0
Light L	4	40	M 8 x 1	3	4	5	8	4
	6	250	M 12 x 1.5	4	6	8.1	10	7
	8		M 14 x 1.5	6	8	10.1	10	7
	10		M 16 x 1.5	8	10	12.3	11	7
	12		M 18 x 1.5	10	12	14.3	11	7
	15		M 22 x 1.5	12	15	17.3	12	7
	18	160	M 26 x 1.5	15	18	20.3	12	7.5
	22		M 30 x 2	19	22	24.3	14	7.5
	28	100	M 36 x 2	24	28	30.3	14	7.5
	35		M 45 x 2	30	35	38	16	10.5
	42		M 52 x 2	36	42	45	16	11
Heavy H	6	400	M 14 x 1.5	4	6	8.1	12	7
	8		M 16 x 1.5	5	8	10.1	12	7
	10		M 18 x 1.5	7	10	12.3	12	7.5
	12		M 20 x 1.5	8	12	14.3	12	7.5
	14		M 22 x 1.5	10	14	16.3	14	8
	16		M 24 x 1.5	12	16	18.3	14	8.5
	20		M 30 x 2	16	20	22.9	16	10.5
	25	250	M 36 x 2	20	25	27.9	18	12
	30		M 42 x 2	25	30	33	20	13.5
	38		M 52 x 2	32	38	41	22	16

Note:

1) Undercut details Refer Page 126

STUD END DETAILS

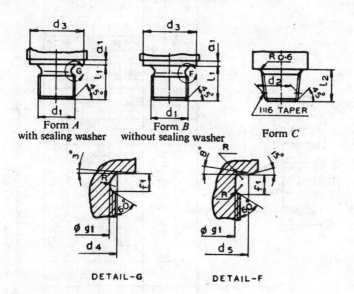

Form *A*
with sealing washer

Form *B*
without sealing washer

Form *C*

DETAIL-G

DETAIL-F

Table 370

d_1 Forms A & B	d_2 Form C	d_3 h14	d_4 0 −0.3	d_5 0 −0.5	a_1 Min.	f_1 +0.3 0	g_1 0 −0.2	l_1 ±0.2	l_2 Min.	R
G 1/8	R 1/8	14	10	13	1.5	2	8.3	8	8	1
G 1/4	R 1/4	18	13.4	17	2	3	11.2	12	12	1.2
G 3/8	R 3/8	22	17	21	2.5	3	14.7	12	12	1.2
G 1/2	R 1/2	26	21.3	25	3	4	18.4	14	14	1.6
G 3/4	R 3/4	32	26.7	30	3	4	23.9	16	16	1.6
G 1	R 1	39	33.5	37	3	5	29.9	18	—	2.5
G 1¼	R 1¼	49	42.2	47	3	5	38.6	20	—	2.5
G 1½	R 1½	55	48.1	53	3	5	44.5	22	—	2.5

PORT DETAILS FOR STUD ENDS OF COUPLINGS

FOR FORMS A, B & C

FOR FORM C ONLY

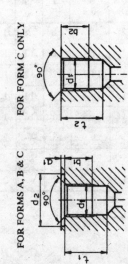

Table 371

d_1	$d_2 +0.4 \atop 0$	a_1	b_1	b_2	t_1 Min.	t_2 Min.
G⅛	15	1	8	5.5	13	9.5
G¼	19	1.5	12	8.5	18.5	13.5
G⅜	23	2	12	8.5	18.5	13.5
G½	27	2.5	14	10.5	22	16.5
G¾	33	2.5	16	13	24	19
G1	40	2.5	18	—	27	—
G1¼	50	2.5	20	—	29	—
G1½	56	2.5	22	—	31	—

Note: Parallel pipe threads as per IS 2643 untruncated

END CONNECTION DETAILS FOR HOSES

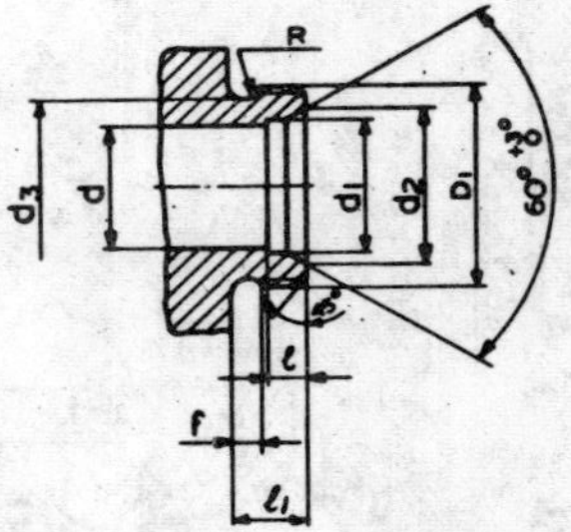

Table 372

Nom. size of hose	d	D_1 Metric	D_1 Pipe	d_1 H11	d_2 H11	l	l_1	Details of undercut d_3	f	R
4	3	M12x1.5	G¼	6	9	5	10	11.4	3	1
6	4	M14x1.5	G¼	8	11	5	10	11.4		
8	7	M16x1.5	G⅜	10	13	5	10	14.9		
10	8	M18x1.5	G½	12	15	5	10	18.6	4	1.2
12	12	M22x1.5	G⅝	15	19	6	12	20.5		
16	16	M26x1.5	G¾	18	22	6	12	24.1		
20	20	M30x1.5	G1	22	26	6	12	29.2	5	1.6
25	25	M38x1.5	G1¼	28	33	7	14	37.9		
32	32	M45x1.5	G1½	35	40	7	14	43.8		

Notes:

1. Details of undercut are given for pipe threads only. For undercut details of metric threads refer Page 126

COUPLING NUTS FOR OIL HYDRAULIC COUPLINGS

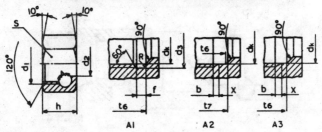

A1   A2   A3

Table 373

Series	Outside dia. of tube	d_1	d_2 B11	b Min.	h	S	Tol. on S	t_6 + 0.2 0	t_7 +0.2 0
	4	M 8 x 1	4	5	11	10	0 −0.09	7.5	8
	6	M 12 x 1.5	6	7	14.5	14	0	10	10.5
	8	M 14 x 1.5	8	7	14.5	17	−0.11		
	10	M 16 x 1.5	10	8	15.5	19		11	11.5
Light	12	M 18 x 1.5	12	8	15.5	22			
L	15	M 22 x 1.5	15	8.5	17	27	0	11.5	12.5
	18	M 26 x 1.5	18	8.5	18	32	−0.13	11.5	13
	22	M 30 x 2	22	9.5	20	36		13.5	14.5
	28	M 36 x 2	28	10	21	41		14	15
	35	M 45 x 2	35	12	24	50		16	17
	42	M 52 x 2	42	12	24	60	0 −0.16	16	17
	6	M 14 x 1.5	6		16.5	17	0 −0.11		
	8	M 16 x 1.5	8	8.5	16.5	19		11	12.5
	10	M 18 x 1.5	10		17.5	22			
	12	M 20 x 1.5	12		17.5	24	0		
Heavy	14	M 22 x 1.5	14	10.5	20.5	27	−0.13		
H	16	M 24 x 1.5	16	10.5	20.5	30		13	14.5
	20	M 30 x 2	20	12	24	36		15.5	17
	25	M 36 x 2	25	14	27	46		17	19
	30	M 42 x 2	30	15	29	50		18	20
	38	M 52 x 2	38	17	32.5	60	0 −0.16	19.5	22.5

Notes: 1. For dimensions d_3, f, X and R Refer IS 1369
2. d_k Minor diameter of thread

Coupling nuts (Contd.)

Form	Series	Application for OD
A1	L	18 *mm* & above
	H	16 *mm* & above
A2	H	6-14 *mm*
	L	All sizes
A3	L	Upto 15 *mm*
	H	Not applicable

Notes:
1. Material: 14*Mn* 1S14
2. Workmanship: Blackened or phosphated.

FERRULES FOR OIL HYDRAULIC COUPLINGS

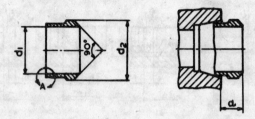

Table 374

Series	Outside dia. of tube	Nom. pressure kgf/cm^2	a		d_1 B11	d_2
			Min.	Max.		
Light L	4	40	3.5	4.5	4	5.5
	6			6	6	9
	8		5	6	8	11
	10	250		6.5	10	13
	12			6.5	12	15
	15			6.5	15	18
	18	160	5.5	7	18	21
	22		6	8	22	25
	28			8	28	31
	35	100	7	9	35	40
	42			9	42	47
Heavy H	6			6.5	6	9
	8			6.5	8	11
	10			6.5	10	13
	12	400	5	6.5	12	15
	14			6.5	14	17
	16			6.5	16	19
	20		6.5	8.5	20	24
	25		6.5	8.5	25	29
	30	250	7	9	30	35
	38		7.5	9.5	38	43

Note: Sealing edge detail *A* is left to the manufacturer

SEALING WASHERS

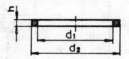

Table 375

Nominal size $d_1 \times d_2$	d_1	Tol on d_1	d_2	Tol. on d_2	h	Applicable for thread D		
						Metric	Pipe	Metric
3.5x6	3.7		5.9					M8x1
5x7.5	5.2		7.4			—		M10x1
5.5x8	5.7		7.9				—	
6.5x9.5	6.7		9.4					M12x1.5
8x11.5	8.2		11.4		1±0.2	M8x1		M14x1.5
8x12	8.2		11.9					—
10x13.5	10.2		13.4			M10x1	$G\frac{1}{8}$	M16x1.5
12x15.5	12.2		15.4			M12x1.5		M18x1.5
14x18	14.2		17.9			M14x1.5	$G\frac{1}{4}$	—
15x19	15.2		18.9			—		M22x1.5
16x20	16.2		19.9			M16x1.5	—	
17x21	17.2	+0.3	20.9	0	1.5±0.2	—	$G\frac{3}{8}$	M24x1.5
18x22	18.2	0	21.9	−0.2		M18x1.5	—	M26x1.5
20x24	20.2		23.9			M20x1.5	—	M27x2
21x26	21.2		25.9			—	$G\frac{1}{2}$	M30x2
22x27	22.2		26.9			M22x1.5	—	M30x1.5
								M30x2
23x28	23.3		27.9			—	$G\frac{5}{8}$	—
24x29	24.3		28.9			M24x1.5		M33x2
25x30			29.9					M33x1.5
25x33	25.3		32.9		2±0.2	—	—	M36x1.5
								M36x2
26x31	26.3		30.9			M26x1.5		—
27x32			31.9					M36x2
27x35	27.3		34.9			M27x1.5 M27x2	$G\frac{3}{4}$	M38x1.5 M38x2

Table 375 (Concld.)

Nominal size $d_1 \times d_2$	d_1	Tol. on d_1	d_2	Tol. on d_2	h	Applicable for Thread D		
						Metric	Pipe	Metric
28x33	28.3		32.9			—	—	M36x2
30x36	30.3		35.9			M30x1.5 M30x2	G⅞	M39x2 M42x2
32x38	32.3		37.9			—	—	M42x2
33x39	33.3		38.9			M33x2	G1	M42x1.5 M42x2 M45x2
35x41	35.3	+0.3 0	40.9	0 −0.2	2±0.2	—		M45x2
36x42	36.3		41.9			M36x1.5 M36x2	—	M45x1.5 M45x2 M48x2
38x44	38.3		43.9			M38x1.5	G1⅛	M48x2
39x46	39.3		45.9			M39x2	—	M52x2
39x48			47.9					
40x47	40.3		46.9			—	—	
42x49	42.3		48.9			M42x1.5 M42x2	G1¼	M52x1.5 M52x2
45x52	45.3		51.9			M45x1.5 M45x2	—	—
48x55	48.3		54.9			M48x1.5 M48x2	G1½	

Material: a) Copper of *HB* Max. 45 or aluminium of *HB* 32-45
b) Rubber or Alkathene for coolant connections

BANJO STUD ELBOW-ASSEMBLY

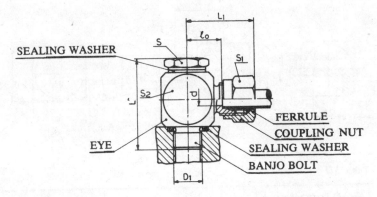

Table 376

N_b	O.D. of tube	d	D_1	l_0	L	L_1 $\approx$	S $h\,11$	S_1 $h\,11$	S_2 $h\,14$
3	4 ± 0.1	3	$G\,\frac{1}{8}$	12	32	24.5	14	12	18
4	6 ± 0.1	4	$G\,\frac{1}{4}$	15	38	31.5	17	14	20
6	10 ± 0.1	7	$G\,\frac{3}{8}$	19.5	50	37	22	22	30
8	12 ± 0.1	8				37.5		24	
12	16 ± 0.1	12	$G\,\frac{1}{2}$	21.5	54	40	27	30	32
16	20 ± 0.1	16	$G\,\frac{3}{4}$	28	74	53	32	36	45
20	25 ± 0.1	20	$G\,1$	34	87	59	41	46	56
25	30 ± 0.1	25	$G\,1\frac{1}{4}$	36.5	98	64	50	50	60
32	38 ± 0.15	32	$G\,1\frac{1}{2}$	43.5	115	75	55	60	75
40	42 ± 0.2	36		42.5		68			

Notes:
1) For Banjo bolt details refer **Page 907**
2) For Eye details refer **Page 906**
3) For Sealing washer details refer **Pages 903 & 904**
4) For Ferrule details refer **Page 902**
5) For coupling nut refer **Page 901**

EYES FOR BANJO STUD ELBOW-ASSEMBLY

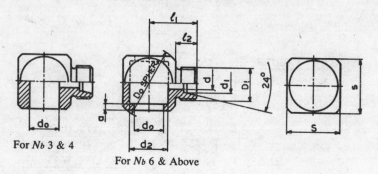

For N_b 3 & 4

For N_b 6 & Above

Table 377

N_b	O D of tube	d	d_o $H11$	d_1 $B11$	d_2	D_o	D_1	l_1	l_2	a	S $h14$
3	4 ± 0.1	3	10	4	—	22	M10x1 *	17	8	—	18
4	6 ± 0.1	4	14	6	—	26	M12x1.5	22	12	—	20
6	10 ± 0.1	7	17	10	22	39	M18x1.5	27	12	3.5	30
8	12 ± 0.1	8	17	12	22	39	M20x1.5	27	12	3.5	30
12	16 ± 0.1	12	21	16	27	42	M24x1.5	32	14	4.5	36
16	20 ± 0.1	16	27	20	35	60	M30x2	38.5	16	5	45
20	25 ± 0.1	20	34	25	45	73	M36x2	46	18	7	56
25	30 ± 0.1	25	42	30	50	80	M42x2	50	20	7	60
32	38 ± 0.15	32	48	38	60	100	M52x2	59.5	22	9	75
40	42 ± 0.2	36	48	42	60	100	M52x2	53.5	16	9	75

Notes:
1) For details of tube end refer **Page 898**
2) Material: 14Mn1S14
3) Workmanship: Blackened or Phosphated
   * Non preferred

BOLT FOR BANJO ASSEMBLY

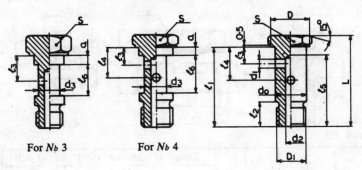

For N_b 3   For N_b 4

For N_b 6 and above

Table 378

N_b	d_o	D_1	d_1	d_2	d_3	D	l_1	l_2	l_3	l_4	l_5	l_6	a	L	S $h11$
3	10	G ⅛	2.5	3	7	14	27	8	9	—	22	13	3	32	14
4	14	G ¼	2.8	4	9	17	34	12	7	12	30	15	3.5	38	17
6	17	G ⅜	4.5	7	—	22	44	12	9	18	41	—	—	50	22
8	17	G ⅜	5.5	8	—	22	44	12	9	18	41	—	—	50	22
12	21	G ½	7	12	—	26	48	14	10	22	46	—	—	54	27
16	27	G ¾	10	16	—	32	65	16	14	27	61	—	—	74	32
20	34	G 1	13	20	—	39	77	18	17	32	73	—	—	87	41
25	42	G 1¼	16	25	—	49	88	25	20	46	84	—	—	98	50
32	48	G 1½	20	32	—	55	103	25	22	53	103	—	—	115	55
40	48	G 1½	20	36	—	55	103	25	22	53	103	—	—	115	55

Notes:

1) Material: $14Mn1S14$
2) Workmanship: Blackened or Phosphated
3) Surface corresponding to diameter d_o and d_3 should have a surface roughness Ra$=1.6\mu m$ and other surfaces may have surface roughness Ra $=3.2$ to $6.3\mu m$.

SCREWED PLUGS

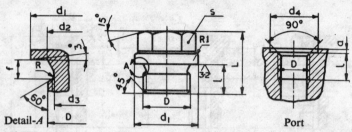

Detail-*A*

Port

Table 379

D	l	d₁	L	S h11	Details of undercut				Details of port		
					d_2 0 −0.3	d_3 0 −0.2	f +0.3 0	R	d_4 +0.4 0	a	l Min.
G ⅛	8	14	17	11	10	8.3	2	1	15	1	8
G ¼	12	18	21	14	13.4	11.2	3	1.2	19	1.5	12
G ⅜	12	22	21	17	17	14.7	3	1.2	23	2	12
G ½	14	26	26	19	21.3	18.4	4	1.6	27		14
G ¾	16	32	30	24	26.7	23.9	4	1.6	33		16
G 1	18	39	34	27	33.5	29.9			40	2.5	18
G 1¼	20	49	37	30	42.2	38.6	5	2.5	50		20
G 1½	22	55	39	30	48.1	44.5			56		22

CYLINDRICAL PLUGS

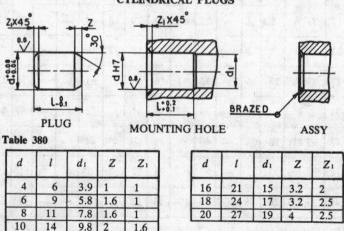

PLUG   MOUNTING HOLE   ASSY

Table 380

d	l	d₁	Z	Z₁
4	6	3.9	1	1
6	9	5.8	1.6	1
8	11	7.8	1.6	1
10	14	9.8	2	1.6
12	16	11.2	2.5	1.6
14	19	13.2	2.5	2

d	l	d₁	Z	Z₁
16	21	15	3.2	2
18	24	17	3.2	2.5
20	27	19	4	2.5

Notes: 1) These plugs are applicable for oil pressures upto 70 *kgf/cm²*
2) For pressures above 30 *kgf/cm²*, plugs are brazed.

CONNECTION FOR OIL PIPES WITH CONICAL FERRULES

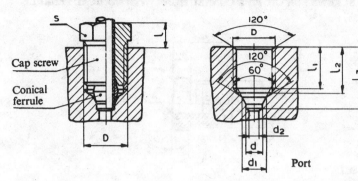

Table 381

Nb	For O.D. of tube	D	S	d H11	d₁	d₂	l₁	l₂	l₃	l ≈
3	4±0.1	M 10 x 1	10	4.1	6	3	8	10	13	7
4	6±0.1	M 12 x 1.5	12	6.1	7	5	10	13	16	8
6	10±0.1	M 18 x 1.5	19	10.1	11	8	9.5	12.5		10

Note: For details of cap screw and tubes Refer **Page 910**

CONICAL FERRULES FOR OIL PIPE CONNECTIONS

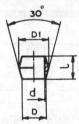

Table 382

Nb	For O.D. of tube	d H11	D	D₁	l
3	4±0.1	4.1	4.5	5.7	5.5
4	6+0.1	6.1	6.5	8.1	6
6	10±0.1	10.1	10.5	12.5	6.5

CAP SCREWS FOR OIL PIPE CONNECTIONS WITH CONICAL FERRULE

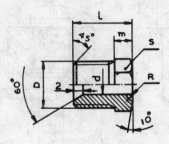

Table 383

Nb	For O D of tube	d H11	D	l ±0.2	m	R	s
3	4 ±0.1	4.2	M 10 x 1	15	4		10
4	6 ±0.1	6.2	M 12 x 1.5	18	4.5	1.6	12
6	10 ±0.1	10.2	M 18 x 1.5	19	6	2	19

PRECISION COLD DRAWN SEAMLESS STEEL TUBES FOR HYDRAULIC PRESSURE LINES

Table 384

D Nominal	4	6	10	12	16	20	25	30	38	42
Tolerance	± 0.1								±0.15	±0.2
Steady max. operating pressure in *kgf/cm²*	Recommended wall thickness *t* (tol. ± 10%)*									
63	1	1	1	1	1.5	1.5	2	2	3	3
100	1	1	1	1	1.5	1.5	2	2	3	3
160	1	1	1	1.5	1.5	2	3	3	4	6
250	1	1	1.5	2	2	2.5	3	4	5	6
400	1	1.5	2	2.5	3	4	5	6	7	9

Notes: * Tolerance on *t* is ±20% for diameter $D = 4$ *mm* and ±15% for diameter $D = 6$ *mm* tubes.

PRESSURE GAUGE CONNECTOR

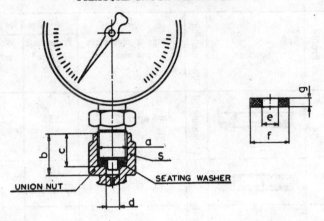

Table 385

Nominal size a	b	c	d	S	f	e	Thickness g
G ⅛	14.45	10	4.47	12.79	8.3	4.5	
G ¼	16.45	12	5.47	18.79	11.1	5.5	Varies between 1.6 to 3.2 for all sizes depending on application
G ⅜	23.05	17.5	6.47	21.58	14.3	6.5	
G ½	26.55	21	6.97	26.58	18.3	6.5	

Notes:
Dimension *c* is based on the dimension of the threaded length of shank of pressure gauge as per IS 3624-1966.

Material for seating washer:
Leather, red or grey vulcanized fibre, rubber-bonded asbestos and copper asbestos are suitable upto working pressure of 60 kgf/cm^2. For pressures beyond 60 kgf/cm^2 one of the following metal washers is used.
  a) Annealed copper
  b) Annealed aluminium
  c) Metal-bonded seals

Table 386   **SURVEY OF PACKING SEALS**

Designation of Profile	Section	Workmanship	Range of diameters in *mm*	Used for	
				Pressure p kgf/cm^2 Max.	Working temperature upto $^{\circ}C$
U		Unlaminated	$d=6$ upto 220	150	100
		Laminated	$d=6$ upto 1000	100 upto 400	230
Y		Laminated	$d=6$ upto 1000	600	
M		Unlaminated	$D=8$ upto 70	50	100
		Laminated	$D=80$ upto 630	100	230
B		Unlaminated	$D=25$ upto 43.6		
A		Unlaminated	$D=17$ upto 36	200	100
O			Dynamic 4 to 165		
			Static 6 to 500		

ASSEMBLING LENGTHS FOR U TYPE RUBBER CUPS

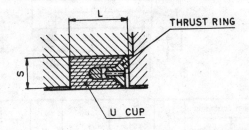

Table 387

S	Nominal length L	Manufacturing tolerance for length of sealing area
4	6.5	+ 0.3 + 0.23
5.5		
6	10	
6.5		+ 0.31 +0.26
7.5		
8	13.5	
8.5		
9	16	+ 0.33 +0.26
10		
12	20	
12.5		
15	24	+ 0.45 + 0.35
20	29.5	
25	37	

Notes: 1. For thrust rings refer **Pages 917 & 920**
2. For U-Cups refer **Pages 914 & 915**

UNLAMINATED U-TYPE CUPS

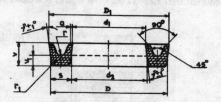

Table 388

Nominal dimension		S	D_1	d_1	r	r_1	a	V	V_1
d	D		js 13	Js 13					
6	14		14.6	5.4					
7	15		15.6	6.4					
8	16	4	16.6	7.4	0.8		0.3	5	2
9	17		17.6	8.4					
10	18		18.6	9.4					
11	22	5.5	23	10					
12	25	6.5	26	11					
14	25	5.5	26	13					
16	28		29	15					
18	30	6	31	17	1		0.5	7	3
20	32		35	19					
22	34		35	21		0.2			
25	36	5.5	37	24					
25	38	6.5	39	24					
25	40	7.5	41.1	23.6					
28	45	8.5	46.4	26.6					
32	48		49.4	30.6			0.7	10.5	4.5
36	52	8	53.4	34.6					
40	56		57.4	38.6	1.5				
45	63	9	64.8	43.2					
45	65		66.8	43.2				13	5
50	70	10	71.8	48.2					
55	75		76.8	53.2					
60	80		81.8	58.2			0.9		
70	95		96.8	68.2					
80	105	12.5	106.8	78.2				16	6.5
90	115		116.8	88.2	2				
100	125		126.8	98.2					
110	140		142	108		0.5			
125	155		157	123					
140	170		172	138			1	19	7.5
160	190	15	192	158	2.5				
180	210		212	178					
200	230		232	198					
220	250		252	218					

LAMINATED U-TYPE CUPS

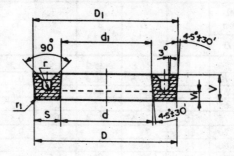

Table 389

. Nominal dimension		S	D_1 js 13	d_1 Js 13	r	r_1	V	V
d	D							
6	14		14.4	5.6				
7	15		15.4	6.6				
8	16	4	16.4	7.6	0.8		5	2
9	17		17.4	8.6				
10	18		18.4	9.6				
10	25	7.5	25.8	9.2	1.5		10.5	4.5
11	22	5.5	22.6	10.4				
12	25	6.5	25.6	11.4	1		7	3
12	28	8	28.8	11.2	1.5		10.5	4.5
14	25	5.5	25.6	13.4	1		7	3
14	30	8	30.6	13.4	1.5		10.5	4.5
16	28	6	28.6	15.4	1	0.2	7	3
16	32	8	32.8	15.2	1.5		10.5	4.5
18	30	6	30.6	17.4	1		7	3
18	34	8	34.8	17.2	1.5		10.5	4.5
20	32	6	32.6	19.4	1		7	3
20	36	8	36.8	19.2	1.5		10.5	4.5
20	40	10	41	19	1.5		13	5
22	34	6	34.6	21.4	1		7	3
22	40	9	41	21	1.5		13	5
25	36	5.5	36.6	24.4	1		7	3
25	38	6.5	38.6	24.4	1		7	3
25	40	7.5	40.8	24.2	1.5		10.5	4.5
25	45	10	46	24	1.5		13	5
28	45	8.5	45.8	27.2	1.5		10.5	4.5
28	48	10	49	27	1.5		13	5
32	48	8	48.8	31.2	1.5	0.2	10.5	4.5
32	56	12	57.3	30.7	2	0.5	16	6.5
36	52	8	52.8	35.2	1.5	0.2	10.5	4.5

Table 389 (Concld.)

Nominal dimension		S	D_1 js 13	d_1 Js 13	r	r_1	V	V_1
d	D							
40	56	8	56.8	39.2	1.5	0.2	10.5	4.5
40	70	15	71.6	38.4	2.5	0.5	19	7.5
45	63	9	64	44	1.5	0.2	13	5
45	65	10	66	44	1.5	0.2	13	5
45	75	15	76.6	43.4	2.5	0.5	19	7.5
50	70	10	71	49	1.5	0.2	13	5
50	80	15	81.6	48.8	2.5	0.5	19	7.5
55	75	10	76	54	1.5	0.2	13	5
55	85	15	86	54	2.5	0.5	19	7.5
60	80	10	81	59	1.5	0.2	13	5
60	90	15	91.6	58.4	2.5		19	7.5
70	95	12.5	96.3	68.7	2		16	6.5
70	100	15	101.6	68.4	2.5		19	7.5
80	105	12.5	106.3	78.7	2		16	6.5
80	110	15	111.6	78.4	2.5		19	7.5
90	115	12.5	116.3	88.7	2		16	6.5
90	120	15	121.6	88.4	2.5		19	7.5
100	125	12.5	126.3	98.7	2		16	6.5
100	130	15	131.6	98.4	2.5		19	7.5
110	140	15	141.6	108.4	2.5		19	7.5
110	150	20	152	108	3		24	9.5
125	155	15	156.6	123.4	2.5		19	7.5
140	170	15	171.6	138.4	2.5		19	7.5
140	180	20	182	138	3		24	9.5
160	190	15	191.6	158.4	2.5		19	7.5
160	200	20	202	158	3	0.5	24	9.5
180	210	15	211.6	178.4	2.5		19	7.5
200	230	15	231.6	198.4	2.5		19	7.5
220	250	15	251.6	218.4	2.5		19	7.5
250	280	15	281.6	248.4	2.5		19	7.5
280	310	15	311.6	278.4	2.5		19	7.5
320	350	15	351.6	318.4	2.5		19	7.5
360	390	15	391.6	358.4	2.5		19	7.5
400	430	15	331.6	398.4	2.5		19	7.5
450	490	20	492	448	3		24	9.5
500	540	20	542	498	3		24	9.5
560	600	20	602	558	3		24	9.5
630	680	25	682.4	627.6	4		30	12.
710	760	25	762.4	707.6	4		30	12
800	850	25	852.4	797.6	4		30	12
900	950	25	952.4	897.6	4		30	12
1000	1050	25	1052.4	997.6	4		30	12

METAL THRUST RINGS FOR U TYPE CUPS WITH INSIDE CENTRING

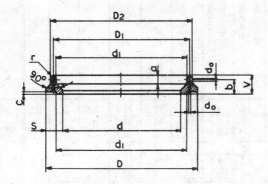

Table 390

d H8	Nominal dimension D h11	Nominal dimension S	d_1	D_1	D_2	a	b	c	r	V js12	d_o Diameter	d_o No. of holes
6	14		8.8	10	11.2							
7	15		9.8	11	12.2							
8	16	4	10.8	12	13.2	2.3	3	0.8	0.6	4.5		
9	17		11.8	13	14.2							
10	18		12.8	14	15.2							
10	25	7.5	15.1	17.5	19.9	4.4	6.5	1.2	1.2	9		
11	22	5.5	14.9	16.5	18.1	2.8	5	1	0.8	7		
12	25	6.5	16.9	18.5	20.1							
12	28	8	17.6	20	22.4	4.4	6.5	1.2	1.2	9		
14	25	5.5	17.9	19.5	21.1	2.8	5	1	0.8	7	1	3
14	30	8	19.6	22	24.4	4.4	6.5	1.2	1.2	9		
16	28	6	20.4	22	23.6	2.8	5	1	0.8	7		
16	32	8	21.6	24	26.4	4.4	6.5	1.2	1.2	9		
18	30	6	22.4	24	25.6	2.8	5	1	0.8	7		
18	34	8	23.6	26	28.4	4.4	6.5	1.2	1.2	9		
20	32	6	24.4	26	27.6	2.8	5	1	0.8	7		
20	36	8	25.6	28	30.4	4.4	6.5	1.2	1.2	9		
20	40	10	27.6	30	32.4	5.4	8	1.2	1.2	11		
22	34	6	26.4	28	29.6	2.8	5	1	0.8	7		
22	40	9	28.6	31	33.4	5.4	8	1.2	1.2	11		

Table 390   **(Contd.)**

d H8	Nominal dimension D h11	Nominal dimension S	d_1	D_1	D_2	a	b	c	r	V js12	d_0 Diameter	No. of holes
25	36	5.5	28.9	30.5	32.1	2.8	5	1	0.8	7		
25	38	6.5	29.9	31.5	33.1							
25	40	7.5	30.1	32.5	34.9	4.4	6.5			9		
25	45	10	32.6	35	37.4	5.4	8			11		
28	45	8.5	34.1	36.5	38.9	4.4	6.5	1.2	1.2	9		
28	48	10	35.6	38	40.4	5.4	8			11	1	
32	48	8	37.6	40	42.4	4.4	6.5			9		
32	56	12	41	44	47	6.5	10	1.5	1.5	13.5		
36	52	8	41.6	44	46.4	4.4	6.5	1.2	1.2	9		
40	56	8	45.6	48	50.4	4.4						
40	70	15	51	55	59	7.7	12.5	2	2	16.5		3
45	65	10	52.6	55	57.4	5.4	8	1.2	1.2	11		
45	63	9	51.6	54	56.4	5.4						
45	75	15	56	60	64	7.7	12.5	2	2	16.5		
50	70	10	57.6	60	62.4	5.4	8	1.2	1.2	11		
50	80	15	61	65	69	7.7	12.5	2	2	16.5		
55	75	10	62.6	65	67.4	5.4	8	1.2	1.2	11		
55	85	15	66	70	74	7.7	12.5	2	2	16.5		
60	80	10	67.6	70	72.4	5.4	8	1.2	1.2	11		
60	90	15	71	75	79	7.7	12.5	2	2	16.5		
70	95	12.5	79.5	82.5	85.5	6.5	10	1.5	1.5	13.5	2	
70	100	15	81	85	89	7.7	12.5	2	2	16.5		
80	105	12.5	89.5	92.5	95.5	6.5	10	1.5	1.5	13.5		
80	110	15	91	95	99	7.7	12.5	2	2	16.5		
90	115	12.5	99.5	102.5	105.5	6.5	10	1.5	1.5	13.5		
90	120	15	101	105	109	7.7	12.5	2	2	16.5		
100	125	12.5	109.5	112.5	115.5	6.5	10	1.5	1.5	13.5		
100	130	15	111	115	119	7.7	12.5			2	16.5	
110	140	15	121	125	129							
110	150	20	125	130	135	9.2	15			2.5	20	
125	155	15	136	140	144	7.7	12.5	2	2	16.5		4
140	170	15	151	155	159							
140	180	20	155	160	165	9.2	15			2.5	20	
160	190	15	171	175	179	7.7	12.5			2	16.5	
160	200	20	175	180	185	9.2	15			2.5	20	

Table 390   (Concld.)

d $H8$	Nominal dimension D $h11$	Nominal dimension S	d_1	D_1	D_2	a	b	c	r	V $js12$	d_o Diameter	No. of holes
180	210		191	195	199							
200	230		211	215	219							
220	250		231	235	239							
250	280	15	261	265	269	7.7	12.5		2	16.5	2	4
280	310		291	295	299							
320	350		331	335	339			2				
360	390		371	375	379							
400	430		411	415	419							
450	490		465	470	475							
500	540	20	515	520	525	9.2	15		2.5	20		
560	600		575	580	585							
630	680		648	655	662							
710	760		728	735	742							
800	850	25	818	825	832	11.5	19	3	3.5	25	2.5	6
900	950		918	925	932							
1000	1050		1018	1025	1032							

Notes:

1) Actual dimension for D $h11$ is 1 *mm* less than the nominal dimension tabulated.

2) Actual dimension for S is 0.5 *mm* less than the nominal dimension tabulated.

3) For material, workmanship, etc, refer *Metal thrust rings for Y-type cups with outside centring*

METAL THRUST RINGS FOR U TYPE CUPS WITH OUTSIDE CENTRING

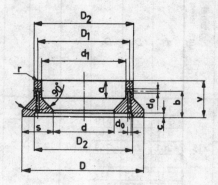

Table 391

Nominal dimension d $H11$	D $f8$	Nominal dimension S	d_1	D_1	D_2	a	b	c	r	V $js12$	d_o Dimension	No. of holes
6	14		8.8	10	11.2							
7	15		9.8	11	12.2							
8	16	4	10.8	12	13.2	2.3	3	0.8	0.6	4.5		
9	17		11.8	13	14.2							
10	18		12.8	14	15.2							
10	25	7.5	15.1	17.5	19.9	4.4	6.5	1.2	1.2	9		
11	22	5.5	14.9	16.5	18.1	2.8	5	1	0.8	7		
12	25	6.5	16.9	18.5	20.1							
12	28	8	17.6	20	22.4	4.4	6.5	1.2	1.2	9		
14	25	5.5	17.9	19.5	21.1	2.8	5	1	0.8	7		
14	30	8	19.6	22	24.4	4.4	6.5	1.2	1.2	9		
16	28	6	20.4	22	23.6	2.8	5	1	0.8	7	1	3
16	32	8	21.6	24	26.4	4.4	6.5	1.2	1.2	9		
18	30	6	22.4	24	25.6	2.8	5	1	0.8	7		
18	34	8	23.6	26	28.4	4.4	6.5	1.2	1.2	9		
20	32	6	24.4	26	27.6	2.8	5	1	0.8	7		
20	36	8	25.6	28	30.4	4.4	6.5	1.2	1.2	9		
20	40	10	27.6	30	32.4	5.4	8	1.2	1.2	11		
22	34	6	26.4	28	29.6	2.8	5	1	0.8	7		
22	40	9	28.6	31	33.4	5.4	8	1.2	1.2	11		

Table 391 (Contd.)

Nominal dimension d H11	D f8	Nominal dimension S	d_1	D	D	a	b	c	r	V js12	d_0	
											Dimension	No. of holes
25	36	5.5	28.9	30.5	32.1	2.8	5		0.8	7	1	
25	38	6.5	29.9	31.5	33.1							
25	45	10	32.6	35	37.4	5.4	8			11		
25	40	7.5	30.1	32.5	34.9	4.4	6.5	1.2	1.2	9		
28	45	8.5	34.1	36.5	38.9							
26	48	10	35.6	38	40.4	5.4	8			11		
32	48	8	37.6	40	42.4	4.4	6.5			9		
32	56	12	41	44	47	6.5	10	1.5	1.5	13.5		
36	52	8	41.6	44	46.4	4.4	6.5	1.2	1.2	9		
40	56	8	45.6	48	50.4							
40	70	15	51	55	59	7.7	12.5	2	2	16.5		
45	63	9	51.6	54	56.4	5.4	8	1.2	1.2	11		
45	65	10	52.6	55	57.4							
45	75	15	56	60	64	7.7	12.5	2	2	16.5		3
50	70	10	57.6	60	62.4	5.4	8	1.2	1.2	11		
50	80	15	61	65	69	7.7	12.5	2	2	16.5		
55	75	10	62.6	65	67.4	5.4	8	1.2	1.2	11		
55	85	15	66	70	74	7.7	12.5	2	2	16.5		
60	80	10	67.6	70	72.4	5.4	8	1.2	1.2	11		
60	90	15	71	75	79	7.7	12.5	2	2	16.5	2	
70	95	12.5	79.5	82.5	85	6.5	10	1.5	1.5	13.5		
70	100	15	81	85	89	7.7	12.5	2	2	16.5		
80	105	12.5	89.5	92.5	95.5	6.5	10	1.5	1.5	13.5		
80	110	15	91	95	99	7.7	12.5	2	2	16.5		
90	115	12.5	99.5	102.5	105.5	6.5	10	1.5	1.5	13.5		
90	120	15	101	105	109	7.7	12.5	2	2	16.5		
100	125	12.5	109.5	112.5	115	6.5	10	1.5	1.5	13.5		
100	130	15	111	115	119	7.7	12.5		2	16.5		4
110	140		121	125	129					16.5		
110	150	20	125	130	135	9.2	15	2	2.5	20		
125	155	15	136	140	144	7.7	12.5		2	16.5		
140	170		151	155	159	7.7	12.5			16.5		

Table 391   (Concld.)

Nominal dimension d $H11$	D $f8$	Nominal dimension S	d_1	D_1	D_2	a	b	c	r	V $js\,12$	d_o Dimension	d_o No. of holes
140	180	20	155	160	165	9.2	15		2.5	20		
160	190		171	175	179							
180	210		191	195	199							
200	230		211	215	219							
220	250		231	235	239							
250	280	15	261	265	269	7.7	12.5		2	16.5		
280	310		291	295	299						2	4
320	350		331	335	339			2				
360	390		371	375	379							
400	430		411	415	419							
450	490		465	470	475							
500	540	20	515	520	525	9.2	15		2.5	20		
560	600		575	580	582							
630	680		648	655	665							
710	760		728	735	742							
800	850	25	818	825	832	11.5	19	3	3.5	25	2.5	6
900	950		918	925	932							
1000	1050		1018	1025	1032							

Note:
1) Actual dimensions for $dH11$ is 1 *mm* more than the nominal dimension tabulated.
2) Actual dimension for S is 0.5 *mm* less than the nominal dimension tabulated. For material, workmanship etc refer Metal thrust rings for Y-type cups with outside centring.

ASSEMBLING LENGTHS FOR Y CUPS

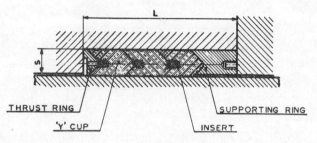

THRUST RING 'Y' CUP   SUPPORTING RING INSERT

Table 392

S	Number of cups									
	2		3		4		5		6	
	Nominal length of sealing area L									
	Nominal length	Tolerance	Nominal length	Tolerance	Nominal length	Tolerance	Nominal length	Tolerance	Nominal length	Tolerance
4	13.7	+0.55 +0.4	17.5	+0.8 +0.6	21.2	+1 +0.8	25	+1.25 +1	28.7	+1.45 +1.15
5.5	19.9		25.2		30.5		35.8		41.1	
6	19.3		24.3		29.3		34.3		39.3	
6.5	19.1		24		28.9		33.8		38.7	
7.5	29.2	+0.8 +0.6	37.4	+1.1 +0.85	45.6	+1.45 +1.15	53.8	+1.8 +1.4	62	+2.1 +1.7
8	28.8		36.8		44.8		52.8		60.8	
8.5	28.2		35.9		43.6		51.3		59	
9	35.4		45.5		55.6		65.7		75.8	
10	34.5		44.1		53.7		63.3		72.9	
12.5	42.8		54.6		66.4		78.2		90	
15	50.9	+1.1 +0.85	64.9	+1.55 +1.2	78.9	+2 +1.6	92.9	+2.5 +2	106.9	+3 +2.3
20	63.1		80		96.9		113.8		130.7	
25	77.4		98.7		120		141.3		162.6	

Number of Cups		2	3	4	5	6
Pressure	Over	—	20	40	200	400
kgf/cm²	upto	20	40	200	400	600

Notes:

    For thrust ring details refer Page 928
    For supporting ring details refer Page 930
    For Y cup details refer Page 924

LAMINATED Y TYPE CUPS

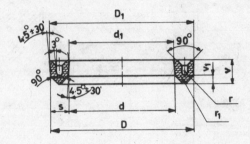

Table 393

Nominal dimension		S	D_1	d_1	r	r_1	V	V_1
d	D		$js\,13$	$Js\,13$				
6	14		14.4	5.6				
7	15		15.4	6.6				
8	16	4	16.4	7.6	0.8	1	5	2
9	17		17.4	8.6				
10	18		18.4	9.6				
11	22	5.5	22.6	10.4				
12	25	6.5	25.6	11.4				
14	25	5.5	25.6	13.4				
16	28		28.6	15.4				
18	30	6	30.6	17.4	1	1.5	7	3
20	32		32.6	19.4				
22	34		34.6	21.4				
25	36	5.5	36.6	24.4				
25	38	6.5	38.6	24.4				
25	40	7.5	40.8	24.2				
28	45	8.5	45.8	27.2				
32	48		48.8	31.2	1.5	2	10.5	4.5
36	52	8	52.8	35.2				
40	56		56.8	39.2				
45	63	9	64	44		2.5	13	5

Table 393 (Concld.)

Nominal dimension		S	D_1	d_1	r	r_1	V	V_1
d	D		js 13	Js13				
45	65		66	44				
50	70	10	71	49	1.5	2.5	13	5
55	75		76	54				
60	80		81	59				
70	95		96.3	68.7				
80	105		106.3	78.7	2	3	16	6.5
90	115	12.5	116.3	88.7				
100	125		126.3	98.7				
110	140		141.6	108.4				
125	155		156.6	123.4				
140	170		171.6	138.4				
160	190		191.6	158.4				
180	210		211.6	178.4				
200	230		231.6	198.4				
220	250	15	251.6	218 4	2.5	3.5	19	7.5
250	280		281.6	248.4				
280	310		311.6	278.4				
320	350		351.6	318.4				
360	390		391.6	358.4				
400	430		431.6	398.4				
450	490		492	448				
500	540	20	542	498	3	4	24	9.5
560	600		602	558				
630	680		682.4	627.6				
710	760		762.4	707.6				
800	850	25	852.4	797.6	4	5	30	12
900	950		952.4	897.6				
1000	1050		1052.4	997.6				

METAL THRUST RINGS FOR Y-TYPE CUPS
WITH INSIDE CENTRING

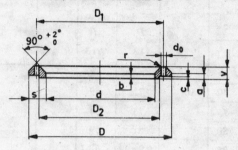

Table 394

d H8	Nominal dimension D	Nominal dimension S	D	D_2	a	b	c	r	V js 12	d_o Diameter	d_o No. of holes
6	14		10	8.5							
7	15		11	9.5							
8	16	4	12	10.5	1.5	1	0.8	1.5	2.4		
9	17		13	11.5							
10	18		14	12.5							
11	22	5.5	16.5	15					3.9		
12	25	6.5	18.5	17					4.4		
14	25	5.5	19.5	18					3.9		
16	28		22	20.5							
18	30		24	22.5							
20	32	6	26	24.5			1	2	4.2	1	3
22	34		28	26.5							
25	36	5.5	30.5	29					3.9		
25	38	6.5	31.5	30					4.4		
25	40	7.5	32.5	30.5	2.5	2			4.7		
28	45	8.5	36.5	38.5					5.2		
32	48		40	38				2.5			
36	52	8	44	42					5		
40	56		48	46							
45	63	9	54	52			1.2		5.3		
45	65		55	52.5							
50	70	10	60	57.5				3	5.8		
55	75		65	62.5							
60	80		70	67.5							

Table 394 (Concld.)

d H8	Nominal dimension D	Nominal dimension S	D_1	D_2	a	b	c	r	V js 12	d_o	
										Diameter	No. of holes
70	95		82.5	79.5							
80	105	12.5	92.5	89.5	3.5	3	1.5	3.5	7.8	1	3
90	115		102.5	99.5							
100	125		112.5	109.5							
110	140		125	121.5							
125	155		140	136.5							
140	170		155	151.5							
160	190		175	171.5							
180	210	15	195	191							
200	230		215	211	4	3.5		4	9.4		4
220	250		235	231							
250	280		265	261							
280	310		295	291			2			1.5	
320	350		335	331							
360	390		375	371							
400	430		415	411							
450	490		470	465							
500	540	20	520	515	5.5	5		5	12.9		
560	600		580	575							
630	680		655	648							6
710	760		735	728							
800	850	25	825	818	5	4.5	3	6	14.5	2.5	
900	950		925	918							
1000	1050		1025	1018							

Notes:
1) Actual dimension for D is 1 mm less than the nominal dimension tabulated.
2) Actual dimension for S is 0.5 mm less than the nominal dimension tabulated
3) For material, workmanship, etc., refer *Metal thrust rings for Y-type cups with outside centring.*

METAL THRUST RINGS FOR Y - TYPE CUPS
WITH OUTSIDE CENTRING

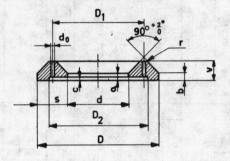

Table 395

Nominal dimension d H11	D f8	Nominal dimension S	D_1	D_2	a	b	c	r	V js12	d_o Diameter	d_o No. of holes
6	14		10	11.5							
7	15		11	12.5							
8	16	4	12	13.5	1.5	1	0.8	1.5	2.4		
9	17		13	14.5							
10	18		14	15.5							
11	22	5.5	16.5	18					3.9		
12	25	6.5	18.5	20					4.4		
14	25	5.5	19.5	21					3.9		
16	28		22	23.5					4.2		
18	30		24	25.5				2			
20	32	6	26	27.5			1		4.2		
22	34		28	29.5							
25	36	5.5	30.5	32					3.9		
25	38	6.5	31.5	33	2.5	2			4.4	1	3
25	40	7.5	32.5	34.5					4.7		
28	45	8.5	36.5	38.5					5.2		
32	48	8	50	42				2.5			
36	52	8	44	46					5		
40	56	8	48	50							
45	63	9	54	56			1.2		5.3		
45	65		55	57.5							
50	70		60	62.5				3	5.8		
55	75	10	65	67.5							
60	80		70	72.5							

Table 395 (Concld.)

Nominal dimension d $H11$	D $f8$	Nominal dimension S	D_1	D_2	a	b	c	r	V $js12$	d_o Diameter	d_o No. of holes
70	95		82.5	85.5							
80	105	12.5	92.5	95.5	3.5	3	1.5	3.5	7.8	1	3
90	115		102.5	105.5							
100	125		112.5	115.5							
110	140		125	128.5							
125	155		140	143.5							
140	170		155	158.5							
160	190		175	178.5							
180	210		195	199							
200	230	15	215	219							
220	250		235	239	4	3.5	2	4	9.4	1.5	4
250	280		265	269							
280	310		295	299							
320	350		355	339							
360	390		375	379							
400	430		415	419							
450	490		470	475							
500	540	20	520	525	5.5	5	2	5	12.9	1.5	
560	600		580	585							
630	680		655	662							6
710	760		735	742							
800	850	25	825	832	5	4.5	3	6	14.5	2.5	
900	950		925	932							
1000	1050		1025	1032							

Notes:
1) Actual dimension for d $H11$ is 1 *mm* more than the nominal dimension tabulated.
2) Actual dimension for S is 0.5 *mm* less than the nominal dimension tabulated.
3) Material Mild steel
4) Workmanship: Corrosion resistant, cylindrical and tapered surfaces should be phosphated, all sharp edges rounded off.
5) Surface roughness of rings shall be Ra=1.6 *μm* and functional cylindrical surfaces shall have Ra= 0.8 *μm*.

METAL SUPPORTING RINGS FOR Y TYPE CUPS

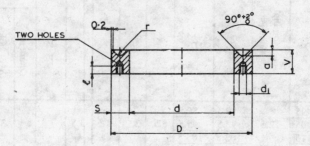

Table 396

d H8	D f8	S	a	r	V js 12	d₁	l
6	14						
7	15						
8	16	4	1.4	1	5	M1.7	2
9	17						
10	18						
11	22	5.5	1.9				
12	25	6.5	2.4				
14	25	5.5	1.9				
16	28						
18	30						
20	32	6	2.2	1.5	7	M2.6	3
22	34						
25	36	5.5	1.9				
25	38	6.5	2.4				
25	40	7.5	2.7				
28	45	8.5	3.2				
32	48			2	10.5		
36	52	8	3				
40	56					M4	4
45	63	9	3.3				
45	65						
50	70						
55	75	10	3.7	2.5	13		
60	80						

Table 396 (Concld.)

d $H8$	D $f8$	S	a	r	V $js\,12$	d_1	l
70	95						
80	105	12.5	4.8	3	16		
90	115						
100	125						
110	140						
125	155						
140	170						
160	190						
180	210						
200	230	15	5.8	3.5	19	M6	6
220	250						
250	280						
280	310						
320	350						
360	390						
400	430						
450	490						
500	540	20	8.1	4	24		
560	600						
630	680						M8	10
710	760						
800	850	25	10.2	5	30		
900	950						
1000	1050						

Notes:

1. Material: Mild steel
2. Workmanship: Cylindrical surface shall be lined with bronze.
3. Surface roughness of cylindrical surface of ring shall be Ra $= 0.4\ \mu m$ and other surfaces shall have Ra $= 1.6\ \mu m$

O-RINGS FOR SEALING NON-MOVING PARTS

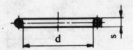

Table 397

Nominal size	Tolerance	Thickness s			Nominal size	Tolerance	Thickness s		
		2	3	5			2	3	5
		Tolerance					Tolerance		
		+0.2 −0.1		±0.2			+0.2 −0.1		±0.2
6					90				
8					95	+0.4			
10					100	−0.5			
12					105				
14					110				
16	+0.2				120				
18	−0.3				125	+0.5			
20					130	−0.8			
22					140				
25					150				
26					160				
28					170	+0.8			
30					180	−1.2			
32					190				
34					200				
36					210				
38					220				
40	+0.3				240	+1.2			
42	−0.4				250	−1.6			
45					260				
48					280				
50					300				
52					320				
55					340				
60					360	+1.6			
63	+0.4				380	−2			
65	−0.5				400				
70					420				
75					450				
80					480				
85					500				

(Thickness s columns for both halves are marked: "Recommended diameter and thickness combination.")

O - RINGS FOR SEALING NON - MOVING PARTS (Concld.)

Groove details

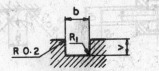

Table 398

s	V ± 0.05	b ± 0.1	h ± 0.1	R_1
2	1.5	2.7	2.8-	0.2
3	2.3	4	4	0.3
5	3.9	6.7	7	0.5

Material: Synthetic rubber

Remarks: 1) O-Rings are used for sealing flanges covers, etc., for a maximum pressure upto 200 kgf/cm^2.

2) Surface roughness of sealing surfaces of O-rings is recommended as $Ra = 6.3\,\mu m$.

O-RINGS FOR SEALING MOVING PARTS

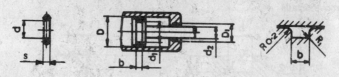

Table 399

Dimensions of O-Ring				Dimensions for sealing pistons		b		R_1	δ
d		s		D $H8/f\,8$	d_1				
				Dimensions for sealing piston rod					
Nominal diameter	Permissible deviation	Nominal diameter	Permissible deviation	D_1	d_2 $H8/f\,8$	Nominal width	Permissible deviation		Max. permissible error of coaxiality between holes D and d_2
3.6				8	4				
4.6				9	5				0.007
5.6	±0.1	2.3	±0.1	10	6	3	±0.1	0.5	
6.6				11	7				
7.6				12	8				0.008
9.6				14	10				
10.6				15	11				
11.6				16	12				
13.6				18	14				
14.6				19	15				
15.6	±0.15	2.3	±0.1	20	16	3	±0.1	0.5	0.011
17.6				22	18				
19.6				24	20				
20.6				25	21				
19.3				28	20				
21.3				30	22				
23.3				32	24				
24.3				33	25				
26.3				35	27				0.014
27.3	±0.2	4.6	±0.1	36	28	6	±0.1	0.8	
29.3				38	30				
31.3				40	32				
34.3				43	35				
35.3				44	36				0.017
36.3				45	37				
37.3				46	38				

Table 399

Dimensions of O-Ring				Dimensions for sealing pistons		b		R_1	δ
d		S		D H8/f8	d_1				
Nominal diameter	Permissible deviation	Nominal diameter	Permissible deviation	Dimensions for sealing piston rod		Nominal width	Permissible deviation		Max. permissible error of coaxiality between holes D and d_2
				D_1	d_2 H8/f 8				
39.1				50	40				
41.1				52	42				
44.1				55	45				0.017
45.1				56	46				
49.1				60	50				
52.1				63	53				
54.1				65	55				
55.1				66	56				
59.1	±0.3	5.8	±0.12	70	60	7.5	±0.1		0.021
62.1				73	63				
64.1				75	65				
69.1				80	70				
74.1				85	75				
79.1				90	80				
84.1				95	85			1	
89.1				100	90				
94.1				105	95				
99.1				110	100				0.025
104.1				115	105				
109.1				120	110				
108.7				125	110				
113.7				130	115				
118.7				135	120				
123.7				140	125				
128.7				145	130				
133.7	±0.5	8.6	±0.15	150	135	11	±0.2		
138.7				155	140				
143.7				160	145				0.031
148.7				165	150				
153.7				170	155				
158.7				175	160				
163.7				180	165				

O - RINGS FOR SEALING MOVING PARTS　　(Concld.)

Grooves in piston

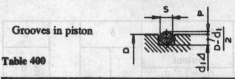

Table 400

s	Diameter d_1		Min. grip, p and tolerances on dia. d_1 for pressures					
			Over 40 upto 200		Over 8 upto 40		Upto 8 kgf/cm^2	
	Over	Upto	p	Δ	p	Δ	p	Δ
2.3	4	21	0.22	+0.08 0	0.18	0 -0.08	0.12	-0.12 -0.2
4.6	20	38	0.5	+0.1 0	0.35	-0.2 -0.3	0.25	-0.4 -0.5
5.8	40	110	0.68	+0.15 0	0.43	-0.35 -0.5	0.33	-0.55 -0.7
8.6	110	165	0.95	+0.2 0	0.65	-0.4 -0.6	0.45	-0.8 -1

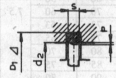

Table 401　　Grooves for piston rod

s	Diameter D_1		Min. grip p and tolerances on dia. D_1 for pressures					
			Over 40 Upto 200		Over 8 Upto 40		Upto 8 kgf/cm^2	
	Over	Upto	p	Δ	p	Δ	p	Δ
2.3	8	25	0.22	0 -0.08	0.18	+0.08 0	0.12	+0.2 +0.12
4.6	28	46	0.5	0 -0.1	0.35	+0.3 +0.2	0.25	+0.5 +0.4
5.8	50	120	0.68	0 -0.15	0.43	+0.5 +0.35	0.33	+0.7 +0.55
8.6	125	180	0.95	0 -0.2	0.65	+0.6 +0.4	0.45	+1 +0.8

Material: Synthetic rubber.

Remarks:　1)　O-rings are used for a maximum pressure upto 200 kgf/cm^2 and
a maximum temperature of 100°C.

2)　The surface roughness of grooves for O-rings is Ra = 1.6 to 3.2 μm.

List of Indian Standards that have been revised since the first publication of this Handbook.

1	IS	210-1978	Grey iron castings
2	IS	325-1978	Three Phase induction motors
3	IS	1365-1978	Slotted countersunk head and slotted raised countersunk head screws (dia. range 1.6 to 20 mm)
4	IS	1368-1980	Dimensions of ends of bolts and screws
5	IS	2014-1977	T-bolts
6	IS	2015-1977	T-nuts
7	IS	2062-1980	Structural steel (fusion welding quality)
8	IS	2403-1975	Transmission steel roller chains and chain wheels
9	IS	2535-1978	Basic rack and modules of cylindrical gears for general engineering and heavy engineering
10	IS	3428-1980	Dimensions for relief grooves
11	IS	5129-1979	Rotary shaft oil seal units (related dimensions)
12	IS	5519-1979	Deviations for untoleranced dimensions of grey iron castings.

INDEX

INDEX